Medicinal Natural Products

A Biosynthetic Approach

Second Edition

Medicinal Natural Products

A Biosynthetic Approach

Second Edition

Paul M Dewick
School of Pharmaceutical Sciences
University of Nottingham, UK

JOHN WILEY & SONS, LTD

Other Wiley Editorial Offices

John Wiley & Sons, Inc., 605 Third Avenue,
New York, NY 10158-0012, USA

Wiley-VCH Verlag GmbH, Pappelallee 3,
D-69469 Weinheim, Germany

John Wiley, Australia, Ltd, 33 Park Road, Milton,
Queensland 4064, Australia

John Wiley & Sons (Canada) Ltd, 22 Worcester Road,
Rexdale, Ontario, M9W 1L1, Canada

John Wiley & Sons (Asia) Pte Ltd, 2 Clementi Loop #02-01,
Jin Xing Distripark, Singapore 129809

Library of Congress Cataloguing in Publication Data

Dewick, Paul M.
    Medicinal natural products : a biosynthetic approach / Paul M. Dewick. – 2nd ed.
       p. cm.
    Includes bibliographical references and index.
    ISBN 0 471 49640 5 (cased) – ISBN 0 471 49641 3 (pbk.)
       1. Natural products.   2. Biosynthesis.   3. Pharmacognosy.   I. Title.

RS160 .D48 2001
615′.321 – dc21

                                        2001026766

British Library Cataloguing in Publication Data

A catalogue record for this book is available from the British Library

ISBN 0 471 49640 5 (cased)   ISBN 0 471 49641 3 (pbk.)

Typeset by Laserwords Private Ltd, Chennai, India
Printed and bound in Great Britain by Antony Rowe Ltd, Chippenham, Wilts
This book is printed on acid-free paper responsibly manufactured from sustainable forestry,
in which at least two trees are planted for each one used for paper production.

*Dedicated to the many hundreds of
pharmacy students I have taught over
the years, who, unknowingly, have
encouraged me to write this text*

CONTENTS

1

ABOUT THIS BOOK, AND HOW TO USE IT

An introductory chapter briefly describing the subject, the aim, the approach and the topics. Students are offered advice on how to select material for study, and encouraged to understand the information rather than learn the factual material. General information on ionization in biochemicals and on nomenclature is given, together with a list of common abbreviations.

THE SUBJECT

This book has been written primarily for pharmacy undergraduates to provide a modern text to complement lecture courses dealing with pharmacognosy and the use of natural products in medicine. Nevertheless, it should be of value in most other courses where the study of natural products is included, although the examples chosen are predominantly those possessing pharmacological activity.

For centuries, drugs were entirely of natural origin and composed of herbs, animal products, and inorganic materials. Early remedies may have combined these ingredients with witchcraft, mysticism, astrology, or religion, but it is certain that those treatments that were effective were subsequently recorded and documented, leading to the early Herbals. The science of pharmacognosy – the knowledge of drugs – grew from these records to provide a disciplined, scientific description of natural materials used in medicine. Herbs formed the bulk of these remedies. As chemical techniques improved, the active constituents were isolated from plants, were structurally characterized, and, in due course, many were synthesized in the laboratory. Sometimes, more active, better tolerated drugs were produced by chemical modifications (semi-synthesis), or by total synthesis of analogues of the active principles.

Gradually synthetic compounds superseded many of the old plant drugs, though certain plant-derived agents were never surpassed and remain as valued medicines to this day. Natural drugs derived from microorganisms have a much shorter history, and their major impact on medicine goes back only about 60 years to the introduction of the antibiotic penicillin. Microbially produced antibiotics now account for a very high proportion of the drugs commonly prescribed. There is currently a renewed interest in pharmacologically active natural products, be they from plants, microorganisms, or animals, in the continued search for new drugs, particularly for disease states where our present range of drugs is less effective than we would wish. Herbal remedies are also enjoying a revival as many sufferers turn away from modern drugs and embrace 'complementary medicine'.

THE AIM

Many modern university pharmacy courses include a pharmacognosy component covering a study of plant-derived drugs, and traditionally this area of natural products has been taught separately from the microbially derived antibiotics, or the animal-related steroidal and prostanoid drugs. These topics usually form part of a pharmaceutical chemistry course. The traditional boundaries may still remain, despite a general change in pharmacognosy teaching from a descriptive study to a phytochemical-based approach, a trend towards integrating pharmacognosy within pharmaceutical chemistry, and the general adoption of modular course structures. A chemistry-based teaching programme encompassing all types of natural product of medicinal importance, semi-synthetic derivatives, and synthetic analogues based on natural product templates, is a logical development, and one we have practised at Nottingham for several years. This coursebook provides a suitable text for such a programme, and attempts to break down the artificial divisions.

THE APPROACH

This book establishes a groundwork in natural product chemistry/phytochemistry by considering biosynthesis – the metabolic sequences leading to various selected classes of natural products. This allows application of fundamental chemical principles and shows the relationships between the diverse structures encountered in nature, thus giving a rationale for natural products and replacing the traditional descriptive approach with one based more on deductive reasoning. Subdivision of the topics is predominantly via biosynthesis, not class or activity, and this provides a logical sequence of structural types and avoids a catalogue effect. There is extensive use of chemical schemes and mechanism, with detailed mechanistic explanations being annotated to the schemes, as well as outline discussions in the text. Extensive cross-referencing is used to emphasize links and similarities. As important classes of compounds or drugs (indicated by an asterisk) are reached, more detailed information is then provided in the form of short separate monographs in boxes, which can be studied or omitted as required, in the latter case allowing the main theme to continue. The monograph information covers sources, production methods, principal components, drug use, mode of action, semi-synthetic derivatives, synthetic analogues, etc, as appropriate. Those materials currently employed as drugs are emphasized in the monographs by the use of bold type.

THE TOPICS

A preliminary chapter is used to outline the main building blocks and the basic construction mechanisms employed in the biosynthesis of natural products. Many of these fundamental principles

should be familiar, having been met previously in courses dealing with the fundamentals of organic chemistry and biochemistry. These principles are then seen in action as representative natural product structures are described in the following chapters. These are subdivided initially into areas of metabolism fed by the acetate, shikimate, mevalonate and deoxyxylulose phosphate pathways. Remaining chapters then cover alkaloids, peptides and proteins, and carbohydrates. The book tries to include a high proportion of those natural products currently used in medicine, the major drugs that are derived from natural materials by semi-synthesis, and those drugs which are structural analogues. Some of the compounds mentioned may have a significant biological activity which is of interest, but not medicinally useful. The book is also designed to be forward looking and gives information on possible leads to new drugs. A selection of supplementary reading references is provided at the end of each chapter; these are limited as far as possible to recent review articles in easily accessible journals rather than books, and have been chosen as student friendly.

BE SELECTIVE

Coverage is extensive to allow maximum flexibility for courses in different institutions, but not all the material will be required for any one course. However, because of the many subdivisions and the highlighted keywords, it should be relatively easy to find and select the material appropriate to a particular course. On the other hand, the detail given in monographs is purposely limited to ensure students are provided with enough factual information, but are not faced with the need to assess whether or not the material is relevant. Even so, these monographs will undoubtedly contain data which exceed the scope of any individual course. It is thus necessary to apply selectivity, and portions of the book will be surplus to immediate requirements. The book is designed to be user friendly, suitable for modular courses and student-centred learning exercises, and a starting point for later project and dissertation work. The information presented is as up to date as possible, but undoubtedly new research will be published that modifies or even contradicts some of the statements made. The reader is asked always to be critical and to maintain a degree of flexibility when reading the scientific literature, and to appreciate that science is always changing.

TO LEARN, OR TO UNDERSTAND?

The primary aim of the book is not to rely just on factual information, but to impart an understanding of natural product structures and the way they are put together by living organisms. Rationalization based on mechanistic reasoning is paramount. The sequences themselves are not important; the mechanistic explanations for the processes used are the essence. Students should concentrate on understanding the broad features of the sequences, and absorb sufficient information to be able to predict how and why intermediates might be elaborated and transformed. The mechanistic explanations appended to the schemes should reinforce this approach. Anyone who commits to memory a sequence of reactions for examination purposes has missed the point. Of course, passing exams is probably the main reason why students are prompted to read this book, and the retention of some factual information will be essential. There is no alternative to memory for some of the material covered in the monographs, but wherever possible, information should be reduced to a concept that can be deduced, rather than remembered. The approach used here should help students to develop such deductive skills.

CONVENTIONS REGARDING ACIDS, BASES, AND IONS

In many structures, the abbreviation **P** is used to represent the phosphate group and **PP** the diphosphate (or pyrophosphate) group:

P (phosphate)   **PP** (diphosphate)

At physiological pHs, these groups will be ionized as shown, but in schemes where structures are given in full, the unionized acids are usually depicted. This is done primarily to simplify structures, to eliminate the need for counter-ions, and to avoid mechanistic confusion. Likewise, amino acids are shown in unionized form, although they will typically exist as zwitterions:

Ionized and unionized forms of many compounds are regarded as synonymous in the text, thus acetate/acetic acid, shikimate/shikimic acid, and mevalonate/mevalonic acid may be used according to the author's whim and context, and have no especial relevance.

NOMENCLATURE

Natural product structures are usually quite complex, some exceedingly so, and fully systematic nomenclature becomes impracticable. Names are thus typically based on so-called trivial nomenclature, in which the discoverer of the natural product exerts his or her right to name the compound. The organism in which the compound is found is frequently chosen to supply the root name, e.g. podophyllotoxin and peltatins from *Podophyllum peltatum*. Name suffixes might be -in to indicate 'a constituent of', -oside to show the compound is a sugar derivative, -genin for the aglycone released by hydrolysis of the sugar derivative, -toxin for a poisonous constituent, or may reflect chemical functionality, such as -one or -ol. Traditionally -ine is always used for alkaloids (amines). Structurally related compounds are then named as derivatives of the original, using standard prefixes, such as hydroxy-, methoxy-, methyl-, dihydro-, homo-, etc for added substituents, or deoxy-, demethyl-, demethoxy-, dehydro-, nor-, etc for removed substituents, positioning being indicated from systematic numbering of the carbon chains or rings. Some groups of compounds, such as steroids, fatty acids, and prostaglandins, are named semi-systematically from an accepted root name. In this book, almost all structures depicted in the figures carry a name; this is primarily to help identification and, for the student, structural features are more pertinent than the names used. It will soon become apparent that drug names chosen by pharmaceutical manufacturers are quite random, and have no particular relationship to the chemical structure.

We are also currently experiencing a transitional period during which many established drug names are being changed to recommended international non-proprietary names (rINN); both names are included here, the rINN preceding the older name.

SOME COMMON ABBREVIATIONS

ACP	acyl carrier protein
ADP	adenosine diphosphate
Api	apiose
Ara	arabinose
ATP	adenosine triphosphate
B:	general base
CoA	coenzyme A as part of a thioester, e.g. acetyl-CoA ($CH_3COSCoA$)
CDP	cytidine diphosphate
CTP	cytidine triphosphate
Dig	digitoxose
DMAPP	dimethylallyl diphosphate (dimethylallyl pyrophosphate)
Enz	enzyme
FAD	flavin adenine dinucleotide
$FADH_2$	flavin adenine dinucleotide (reduced)
FMN	flavin mononucleotide
$FMNH_2$	flavin mononucleotide (reduced)
FPP	farnesyl diphosphate (farnesyl pyrophosphate)
Fru	fructose
Gal	galactose
GFPP	geranylfarnesyl diphosphate (geranylfarnesyl pyrophosphate)
GGPP	geranylgeranyl diphosphate (geranylgeranyl pyrophosphate)
Glc	glucose
GPP	geranyl diphosphate (geranyl pyrophosphate)
HA	general acid
HMG-CoA	β-hydroxy-β-methylglutaryl coenzyme A
HSCoA	coenzyme A
IPP	isopentenyl diphosphate (isopentenyl pyrophosphate)
LT	leukotriene
Mann	mannose
NAD^+	nicotinamide adenine dinucleotide
NADH	nicotinamide adenine dinucleotide (reduced)
$NADP^+$	nicotinamide adenine dinucleotide phosphate
NADPH	nicotinamide adenine dinucleotide phosphate (reduced)
O	oxidation – in schemes
P	phosphate – in text
P	phosphate – in structures
PEP	phosphoenolpyruvate
PG	prostaglandin
PLP	pyridoxal 5′-phosphate
PP	diphosphate (pyrophosphate) – in text
PP	diphosphate (pyrophosphate) – in structures
Rha	rhamnose
Rib	ribose
SAM	*S*-adenosyl methionine
TDPGlc	thymidine diphosphoglucose
TPP	thiamine diphosphate (thiamine pyrophosphate)
TX	thromboxane
UDP	uridine diphosphate
UDPGlc	uridine diphosphoglucose
UTP	uridine triphosphate
W–M	Wagner–Meerwein rearrangement
Xyl	xylose
hν	electromagnetic radiation; usually UV or visible
Δ	heat

FURTHER READING

Pharmacognosy, Phytochemistry, Natural Drugs

Books

Bruneton J (1999) *Pharmacognosy*. Lavoisier, Andover.

Evans WC (1996) *Trease and Evans' Pharmacognosy*. Saunders, London.

Newall CA, Anderson LA and Phillipson JD (1996) *Herbal Medicines. A Guide for Health-care Professionals*. Pharmaceutical, London.

Robbers JE, Speedie MK and Tyler VE (1996) *Pharmacognosy and Pharmacobiotechnology*. Williams and Wilkins, Baltimore, MD.

Robbers JE and Tyler VE (1999) *Tyler's Herbs of Choice. The Therapeutic Use of Phytomedicinals*. Haworth Herbal, New York.

Samuelsson G (1999) *Drugs of Natural Origin*. Swedish Pharmaceutical, Stockholm.

Schultz V, Hansel R and Tyler VE (1998) *Rational Phytotherapy. A Physician's Guide to Herbal Medicine*. Springer, Berlin.

Reviews

Briskin DP (2000) Medicinal plants and phytomedicines. Linking plant biochemistry and physiology to human health. *Plant Physiol* **124**, 507–514.

Cordell GA (2000) Biodiversity and drug discovery – a symbiotic relationship. *Phytochemistry* **55**, 463–480.

Cragg G and Newman D (2001) Nature's bounty. *Chem Brit* **37** (1), 22–26.

Cragg GM, Newman DJ and Snader KM (1997) Natural products in drug discovery and development. *J Nat Prod* **60**, 52–60.

Newman DJ, Cragg GM and Snader KM (2000) The influence of natural products upon drug discovery. *Nat Prod Rep* **17**, 215–234.

Shu Y-Z (1998) Recent natural product based drug development: a pharmaceutical industry perspective. *J Nat Prod* **61**, 1053–1071.

Sneader W (1990) Chronology of drug introductions. *Comprehensive Medicinal Chemistry*. Vol 1. Pergamon, Oxford, 7–80.

Tyler VE (1999) Phytomedicines: back to the future. *J Nat Prod* **62**, 1589–1592.

2

SECONDARY METABOLISM: THE BUILDING BLOCKS AND CONSTRUCTION MECHANISMS

Distinctions between primary and secondary are defined, and the basic building blocks used in the biosynthesis of secondary natural products are introduced. The chemistry underlying how these building blocks are assembled in nature is described, subdivided according to chemical mechanism, including alkylation reactions, Wagner–Meerwein rearrangements, aldol and Claisen reactions, Schiff base formation and Mannich reactions, transaminations, decarboxylations, oxidation and reduction reactions, phenolic oxidative coupling, and glycosylations.

PRIMARY AND SECONDARY METABOLISM

All organisms need to transform and interconvert a vast number of organic compounds to enable them to live, grow, and reproduce. They need to provide themselves with energy in the form of ATP, and a supply of building blocks to construct their own tissues. An integrated network of enzyme-mediated and carefully regulated chemical reactions is used for this purpose, collectively referred to as **intermediary metabolism**, and the pathways involved are termed **metabolic pathways**. Some of the crucially important molecules of life are carbohydrates, proteins, fats, and nucleic acids. Apart from fats, these are polymeric materials. Carbohydrates are composed of sugar units, whilst proteins are made up from amino acids, and nucleic acids are based on nucleotides. Organisms vary widely in their capacity to synthesize and transform chemicals. For instance, plants are very efficient at synthesizing organic compounds via photosynthesis from inorganic materials found in the environment, whilst other organisms such as animals and microorganisms rely on obtaining their raw materials in their diet, e.g. by consuming plants. Thus, many of the metabolic pathways are concerned with degrading materials taken in as food, whilst

others are then required to synthesize specialized molecules from the basic compounds so obtained.

Despite the extremely varied characteristics of living organisms, the pathways for generally modifying and synthesizing carbohydrates, proteins, fats, and nucleic acids are found to be essentially the same in all organisms, apart from minor variations. These processes demonstrate the fundamental unity of all living matter, and are collectively described as **primary metabolism**, with the compounds involved in the pathways being termed **primary metabolites**. Thus degradation of carbohydrates and sugars generally proceeds via the well characterized pathways known as glycolysis and the Krebs/citric acid/tricarboxylic acid cycle, which release energy from the organic compounds by oxidative reactions. Oxidation of fatty acids from fats by the sequence called β-oxidation also provides energy. Aerobic organisms are able to optimize these processes by adding on a further process, oxidative phosphorylation. This improves the efficiency of oxidation by incorporating a more general process applicable to the oxidation of a wide variety of substrates rather than having to provide specific processes for each individual substrate. Proteins taken in via the diet provide amino acids, but the proportions of each will almost certainly vary from the organism's requirements. Metabolic pathways are thus available to interconvert amino acids, or degrade those not required and thus provide a further source of energy. Most organisms can synthesize only a proportion of the amino acids they actually require for protein synthesis. Those structures not synthesized, so-called essential amino acids, must be obtained from external sources.

In contrast to these primary metabolic pathways, which synthesize, degrade, and generally interconvert compounds commonly encountered in all organisms, there also exists an area of metabolism concerned with compounds which have a much more limited distribution in nature. Such compounds, called **secondary metabolites**, are found in only specific organisms, or groups of organisms, and are an expression of the individuality of species. Secondary metabolites are not necessarily produced under all conditions, and in the vast majority of cases the function of these compounds and their benefit to the organism is not yet known. Some are undoubtedly produced for easily appreciated reasons, e.g. as toxic materials providing defence against predators, as volatile attractants towards the same or other species, or as colouring agents to attract or warn other species, but it is logical to assume that all do play some vital role for the well-being of the producer. It is this area of **secondary metabolism** that provides most of the pharmacologically active natural products. It is thus fairly obvious that the human diet could be both unpalatable and remarkably dangerous if all plants, animals, and fungi produced the same range of compounds.

The above generalizations distinguishing primary and secondary metabolites unfortunately leave a 'grey area' at the boundary, so that some groups of natural products could be assigned to either division. Fatty acids and sugars provide good examples, in that most are best described as primary metabolites, whilst some representatives are extremely rare and found only in a handful of species. Likewise, steroid biosynthesis produces a range of widely distributed fundamental structures, yet some steroids, many of them with pronounced pharmacological activity, are restricted to certain organisms. Hopefully, the blurring of the boundaries will not cause confusion; the subdivision into primary metabolism ($\equiv$ biochemistry) or secondary metabolism ($\equiv$ natural products chemistry)

is merely a convenience and there is considerable overlap.

THE BUILDING BLOCKS

The building blocks for secondary metabolites are derived from primary metabolism as indicated in Figure 2.1. This scheme outlines how metabolites from the fundamental processes of photosynthesis, glycolysis, and the Krebs cycle are tapped off from energy-generating processes to provide biosynthetic intermediates. The number of building blocks needed is surprisingly few, and as with any child's construction set a vast array of objects can be built up from a limited number of basic building blocks. By far the most important building blocks employed in the biosynthesis of secondary metabolites are derived from the intermediates acetyl coenzyme A (acetyl-CoA), shikimic acid, mevalonic acid, and 1-deoxyxylulose 5-phosphate. These are utilized respectively in the **acetate**, **shikimate**, **mevalonate**, and **deoxyxylulose phosphate** pathways, which form the basis of succeeding chapters. **Acetyl-CoA** is formed by oxidative decarboxylation of the glycolytic pathway product pyruvic acid. It is also produced by the β-oxidation of fatty acids, effectively reversing the process by which fatty acids are themselves synthesized from acetyl-CoA. Important secondary metabolites formed from the acetate pathway include phenols, prostaglandins, and macrolide antibiotics, together with various fatty acids and derivatives at the primary/secondary metabolism interface. **Shikimic acid** is produced from a combination of phosphoenolpyruvate, a glycolytic pathway intermediate, and erythrose 4-phosphate from the pentose phosphate pathway. The reactions of the pentose phosphate cycle may be employed for the degradation of glucose, but they also feature in the synthesis of sugars by photosynthesis. The shikimate pathway leads to a variety of phenols, cinnamic acid derivatives, lignans, and alkaloids. **Mevalonic acid** is itself formed from three molecules of acetyl-CoA, but the mevalonate pathway channels acetate into a different series of compounds than does the acetate pathway. **Deoxyxylulose phosphate** arises from a combination of two glycolytic pathway intermediates, namely pyruvic acid and glyceraldehyde 3-phosphate. The mevalonate and deoxyxylulose phosphate pathways are together

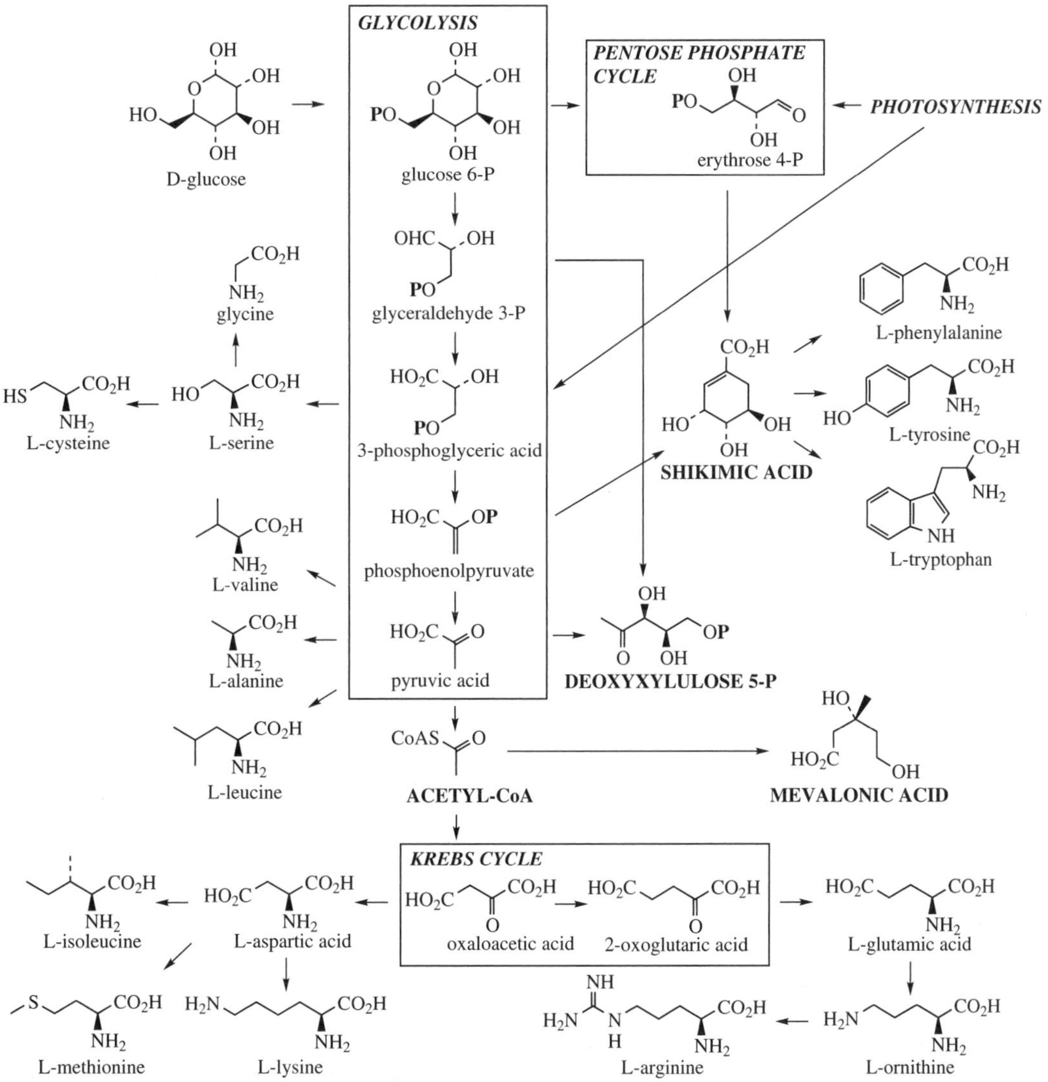

Figure 2.1

responsible for the biosynthesis of a vast array of terpenoid and steroid metabolites.

In addition to acetyl-CoA, shikimic acid, mevalonic acid, and deoxyxylulose phosphate, other building blocks based on amino acids are frequently employed in natural product synthesis. Peptides, proteins, alkaloids, and many antibiotics are derived from amino acids, and the origins of the most important amino acid components of these are briefly indicated in Figure 2.1. Intermediates from the glycolytic pathway and the Krebs cycle are used in constructing many of them, but the aromatic amino acids **phenylalanine**, **tyrosine**,

and **tryptophan** are themselves products from the shikimate pathway. **Ornithine**, a non-protein amino acid, along with its homologue **lysine**, are important alkaloid precursors having their origins in Krebs cycle intermediates.

Of special significance is the appreciation that secondary metabolites can be synthesized by combining several building blocks of the same type, or by using a mixture of different building blocks. This expands structural diversity, and consequently makes subdivisions based entirely on biosynthetic pathways rather more difficult. A typical natural product might be produced by combining elements from the acetate, shikimate, and deoxyxylulose phosphate pathways. Many secondary metabolites also contain one or more sugar units in their structure, either simple primary metabolites such as glucose or ribose, or alternatively substantially modified and unusual sugars. To appreciate how a natural product is elaborated, it is of value to be able to dissect its structure into the basic building blocks from which it is made up, and to propose how these are mechanistically joined together. With a little experience and practice, this becomes a relatively simple process, and it allows the molecule to be rationalized, thus exposing logical relationships between apparently quite different structures. In this way, similarities become much more meaningful than differences, and an understanding of biosynthetic pathways allows rational connecting links to be established. This forms the basic approach in this book.

Relatively few building blocks are routinely employed, and the following list, though not comprehensive, includes those most frequently encountered in producing the carbon and nitrogen skeleton of a natural product. As we shall see, oxygen atoms can be introduced and removed by a variety of processes, and so are not considered in the initial analysis, except as a pointer to an acetate (see page 62) or shikimate (see page 123) origin. The structural features of these building blocks are shown in Figure 2.2.

- **C_1:** The simplest of the building blocks is composed of a single carbon atom, usually in the form of a methyl group, and most frequently it is attached to oxygen or nitrogen, but occasionally to carbon. It is derived from the S-methyl of L-**methionine**. The methylenedioxy group (OCH_2O) is also an example of a C_1 unit.

- **C_2:** A two-carbon unit may be supplied by **acetyl-CoA**. This could be a simple acetyl group, as in an ester, but more frequently it forms part of a long alkyl chain (as in a fatty acid) or may be part of an aromatic system (e.g. phenols). Of particular relevance is that in the latter examples, acetyl-CoA is first converted into the more reactive **malonyl-CoA** before its incorporation.

- **C_5:** The branched-chain C_5 'isoprene' unit is a feature of compounds formed from **mevalonate** or **deoxyxylulose phosphate**. Mevalonate itself

is the product from three acetyl-CoA molecules, but only five of mevalonate's six carbons are used, the carboxyl group being lost. The alternative precursor deoxyxylulose phosphate, a straight-chain sugar derivative, undergoes a skeletal rearrangement to form the branched-chain isoprene unit.

- **C_6C_3:** This refers to a phenylpropyl unit and is obtained from the carbon skeleton of either L-**phenylalanine** or L-**tyrosine**, two of the shikimate-derived aromatic amino acids. This, of course, requires loss of the amino group. The C_3 side-chain may be saturated or unsaturated, and may be oxygenated. Sometimes the side-chain is cleaved, removing one or two carbons. Thus, C_6C_2 and C_6C_1 units represent modified shortened forms of the C_6C_3 system.

- **C_6C_2N:** Again, this building block is formed from either L-**phenylalanine** or L-**tyrosine**, L-tyrosine being by far the more common. In the elaboration of this unit, the carboxyl carbon of the amino acid is removed.

- **indole.C_2N:** The third of the aromatic amino acids is L-**tryptophan**. This indole-containing system can undergo decarboxylation in a similar way to L-phenylalanine and L-tyrosine so providing the remainder of the skeleton as an indole.C_2N unit.

- **C_4N:** The C_4N unit is usually found as a heterocyclic pyrrolidine system and is produced from the non-protein amino acid L-**ornithine**. In marked contrast to the C_6C_2N and indole.C_2N units described above, ornithine supplies not its α-amino nitrogen, but the δ-amino nitrogen. The carboxylic acid function and the α-amino nitrogen are both lost.

- **C_5N:** This is produced in exactly the same way as the C_4N unit, but using L-**lysine** as precursor. The ϵ-amino nitrogen is retained, and the unit tends to be found as a piperidine ring system.

These eight building blocks will form the basis of many of the natural product structures discussed in the following chapters. Simple examples of how compounds can be visualized as a combination of building blocks are shown in Figure 2.3. At this stage, it is inappropriate to justify why a particular combination of units is used, but this aspect should become clear as the pathways are described.

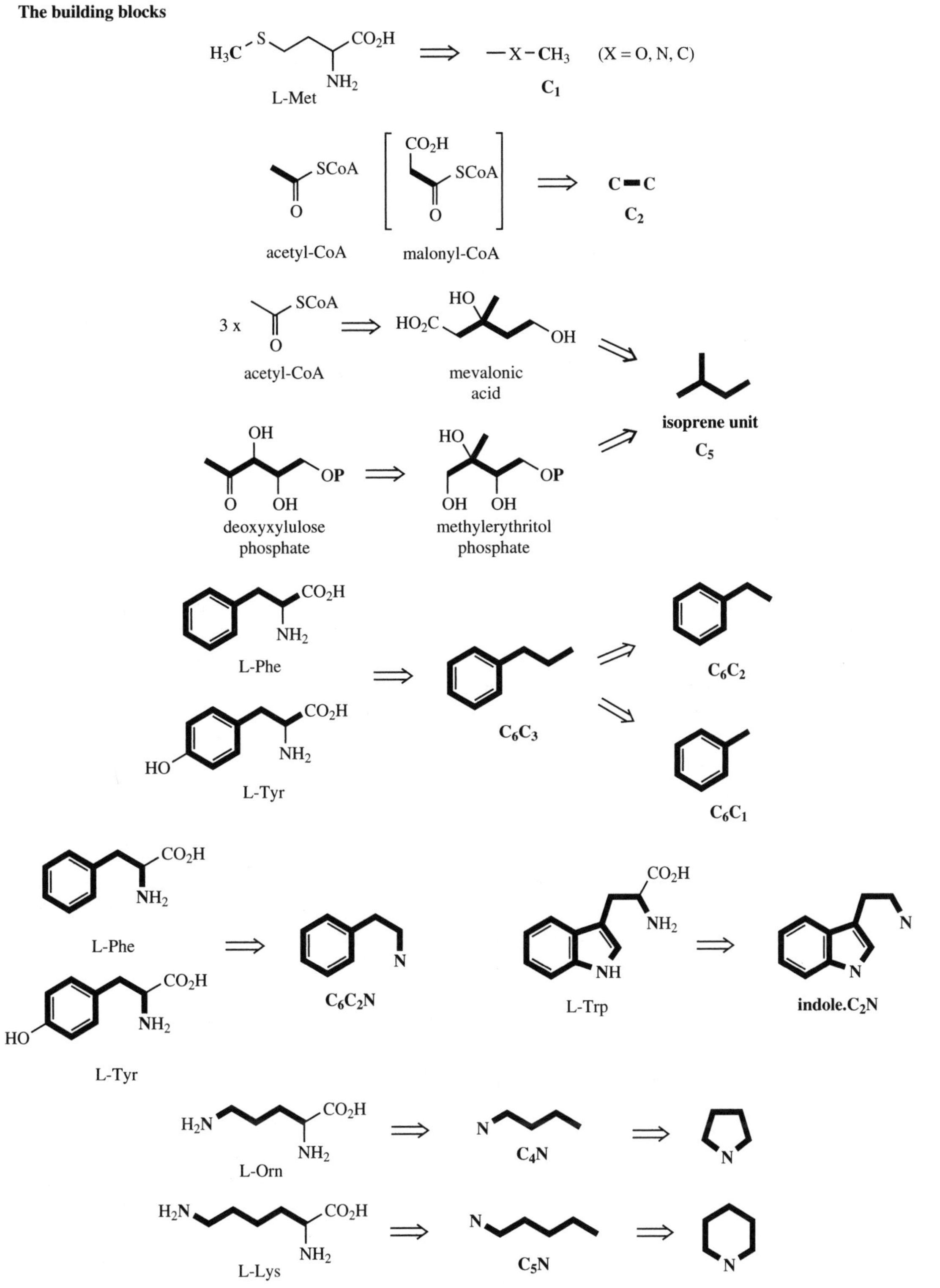
The building blocks
H3C S CO2H
NH2
L-Met
—X—CH3 (X = O, N, C)
C1
CO2H
SCoA
SCoA
O
O
acetyl-CoA
malonyl-CoA
C—C
C2
3 x SCoA
O
acetyl-CoA
HO
HO2C
OH
mevalonic acid
isoprene unit
C5
OH
OP
O OH
deoxyxylulose phosphate
HO
OP
OH OH
methylerythritol phosphate
CO2H
NH2
L-Phe
C6C3
C6C2
C6C1
HO
L-Tyr
N
C6C2N
NH
L-Trp
indole.C2N
H2N
L-Orn
C4N
L-Lys
C5N

Figure 2.2

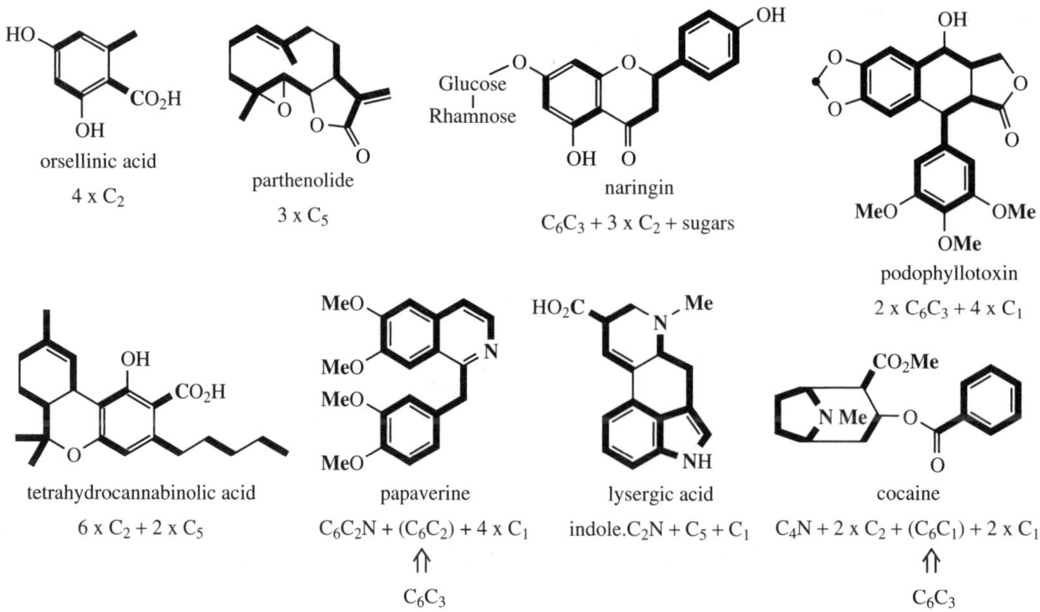

orsellinic acid

$4 \times C_2$

parthenolide

$3 \times C_5$

naringin

$C_6C_3 + 3 \times C_2 + $ sugars

podophyllotoxin

$2 \times C_6C_3 + 4 \times C_1$

tetrahydrocannabinolic acid

$6 \times C_2 + 2 \times C_5$

papaverine

$C_6C_2N + (C_6C_2) + 4 \times C_1$
⇑
C_6C_3

lysergic acid

indole.$C_2N + C_5 + C_1$

cocaine

$C_4N + 2 \times C_2 + (C_6C_1) + 2 \times C_1$
⇑
C_6C_3

Figure 2.3

A word of warning is also necessary. Some natural products have been produced by processes in which a fundamental rearrangement of the carbon skeleton has occurred. This is especially common with structures derived from isoprene units, and it obviously disguises some of the original building blocks from immediate recognition. The same is true if one or more carbon atoms are removed by oxidation reactions.

THE CONSTRUCTION MECHANISMS

Natural product molecules are biosynthesized by a sequence of reactions which, with very few exceptions, are catalysed by enzymes. Enzymes are protein molecules which facilitate chemical modification of substrates by virtue of their specific binding properties conferred by the particular combination of functional groups in the constituent amino acids. In many cases, a suitable cofactor, e.g. NAD^+, PLP, HSCoA (see below), as well as the substrate, may also be bound to participate in the transformation. Although enzymes catalyse some fairly elaborate and sometimes unexpected changes, it is generally possible to account for the reactions using sound chemical principles and mechanisms. As we explore the pathways to a wide variety of natural products, the reactions will generally be discussed in terms of chemical analogies. Enzymes have the power to effect these transformations more efficiently and more rapidly than the chemical analogy, and also under very much milder conditions. Where relevant, they also carry out reactions in a stereospecific manner. Some of the important reactions frequently encountered are now described.

Alkylation Reactions: Nucleophilic Substitution

The C_1 methyl building unit is supplied from L-methionine and is introduced by a nucleophilic substitution reaction. In nature, the leaving group is enhanced by converting L-methionine into **S-adenosylmethionine** (**SAM**) [Figure 2.4(a)]. This gives a positively charged sulphur and facilitates the nucleophilic substitution (S_N2) type mechanism [Figure 2.4(b)]. Thus, O-methyl and N-methyl linkages may be obtained using hydroxyl and amino functions as nucleophiles. The generation of C-methyl linkages requires the participation of nucleophilic carbon. Positions *ortho* or *para* to a phenol group, or positions adjacent to one or more carbonyl groups, are thus candidates for C-methylation [Figure 2.4(c)].

A C_5 isoprene unit in the form of **dimethylallyl diphosphate** (**DMAPP**) may also act as an

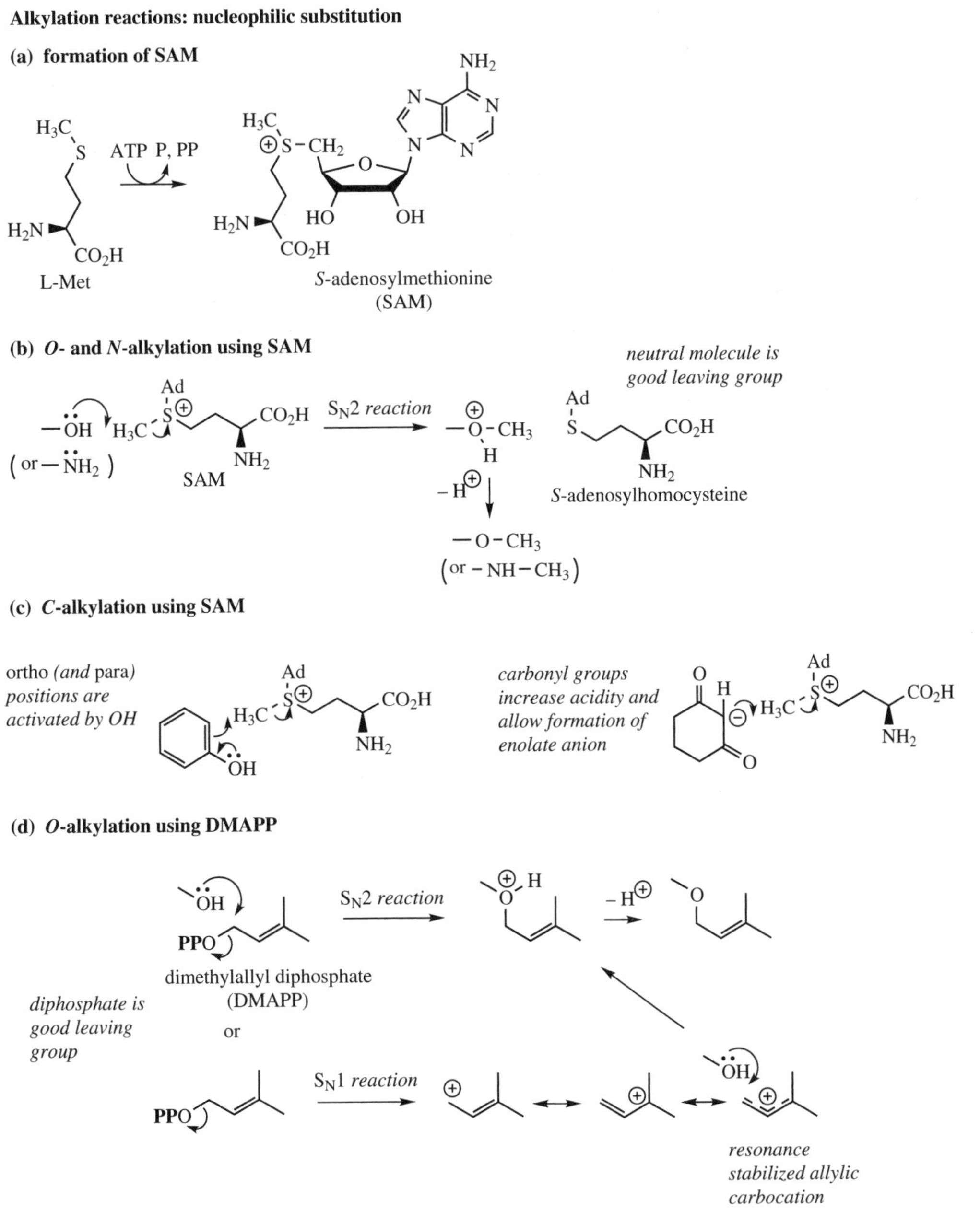

Alkylation reactions: nucleophilic substitution

(a) formation of SAM

L-Met   *S*-adenosylmethionine
(SAM)

(b) *O*- and *N*-alkylation using SAM

*neutral molecule is
good leaving group*

S$_N$2 reaction

S-adenosylhomocysteine

$$-O-CH_3$$
$$\left(or -NH-CH_3 \right)$$

(c) *C*-alkylation using SAM

ortho *(and para)*
*positions are
activated by OH*

*carbonyl groups
increase acidity and
allow formation of
enolate anion*

(d) *O*-alkylation using DMAPP

S$_N$2 reaction $-H^{\oplus}$

dimethylallyl diphosphate
(DMAPP)

*diphosphate is
good leaving
group*

or

S$_N$1 reaction

*resonance
stabilized allylic
carbocation*

Figure 2.4

alkylating agent, and a similar S$_N$2 nucleophilic displacement can be proposed, the diphosphate making a good leaving group [Figure 2.4(d)]. In some cases, there is evidence that DMAPP may ionize first to the resonance-stabilized allylic carbocation and thus an S$_N$1 process operates instead. *C*-Alkylation at activated positions using DMAPP is analogous to the *C*-methylation process above.

Alkylation Reactions: Electrophilic Addition

As indicated above, the C$_5$ isoprene unit in the form of **dimethylallyl diphosphate (DMAPP)** can be used to alkylate a nucleophile. In the elaboration of terpenoids and steroids, two or more C$_5$ units are joined together, and the reactions are rationalized in terms of carbocation chemistry, including electrophilic addition of carbocations on to alkenes. DMAPP may ionize to generate a resonance-stabilized allylic carbocation as shown in Figure 2.4(d), and this can then react with an alkene [e.g. **isopentenyl diphosphate (IPP)**] as depicted in Figure 2.5(a). The resultant carbocation may then lose a proton to give the uncharged product geranyl diphosphate (GPP). Where the alkene and carbocation functions reside in the same molecule, this type of mechanism can be responsible for cyclization reactions [Figure 2.5(a)].

The initial carbocation may be generated by a number of mechanisms, important examples being loss of a leaving group, especially diphosphate (i.e. S_N1 type ionization), protonation of an alkene, and protonation/ring opening of epoxides [Figure 2.5(b)]. *S*-**Adenosylmethionine** may also alkylate alkenes by an electrophilic addition mechanism, adding a C_1 unit, and generating an intermediate carbocation.

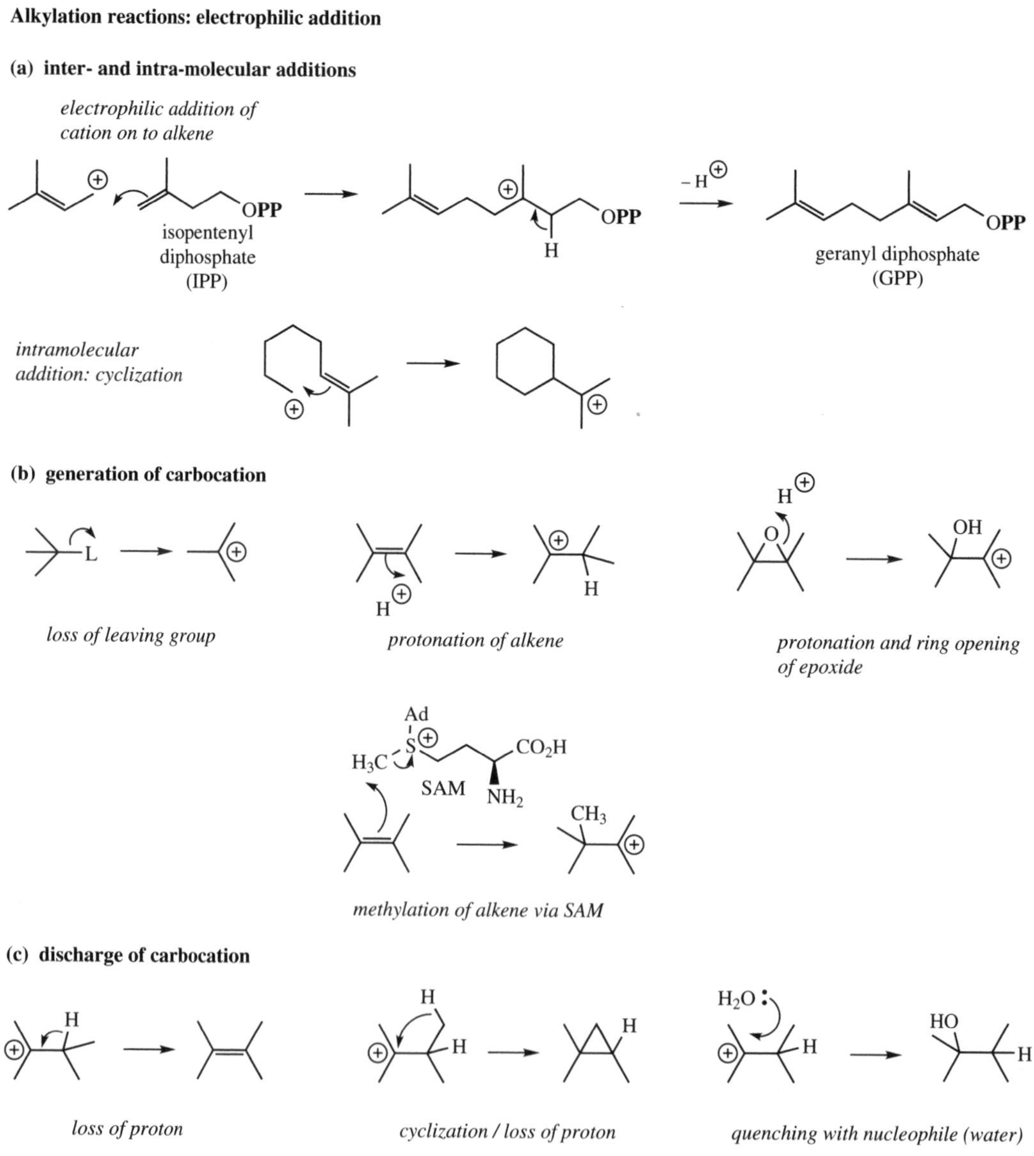

Alkylation reactions: electrophilic addition

(a) inter- and intra-molecular additions

electrophilic addition of cation on to alkene

isopentenyl
diphosphate
(IPP)

geranyl diphosphate
(GPP)

*intramolecular
addition: cyclization*

(b) generation of carbocation

loss of leaving group

protonation of alkene

*protonation and ring opening
of epoxide*

methylation of alkene via SAM

(c) discharge of carbocation

loss of proton

cyclization / loss of proton

quenching with nucleophile (water)

Figure 2.5

The final carbocation may be discharged by loss of a proton (giving an alkene or sometimes a cyclopropane ring) or by quenching with a suitable nucleophile, especially water [Figure 2.5(c)].

Wagner–Meerwein Rearrangements

A wide range of structures encountered in natural terpenoid and steroid derivatives can only be rationalized as originating from C_5 isoprene units if some fundamental rearrangement process has occurred during biosynthesis. These rearrangements have, in many cases, been confirmed experimentally, and are almost always consistent with the participation of carbocation intermediates. Rearrangements in chemical reactions involving carbocation intermediates, e.g. S_N1 and E1 reactions, are not uncommon, and typically consist of 1,2-shifts of hydride, methyl, or alkyl groups. Occasionally, 1,3- or longer shifts are encountered. These shifts, termed **Wagner–Meerwein rearrangements**, are readily rationalized in terms of generating a more stable carbocation, or relaxing ring strain (Figure 2.6). Thus, tertiary carbocations are favoured over secondary carbocations,

and the usual objective in these rearrangements is to achieve tertiary status at the positive centre. However, a tertiary to secondary transition might be favoured if the rearrangement allows a significant release of ring strain. These general concepts are occasionally ignored by nature, but it must be remembered that the reactions are enzyme-mediated and carbocations may not exist as discrete species in the transformations. An interesting feature of some biosynthetic pathways, e.g. that leading to steroids, is a remarkable series of concerted 1,2-migrations rationalized via carbocation chemistry, but entirely a consequence of the enzyme's participation (Figure 2.6).

Aldol and Claisen Reactions

The **aldol** and **Claisen** reactions both achieve carbon–carbon bond formation and in typical base-catalysed chemical reactions depend on the generation of a resonance-stabilized enolate anion from a suitable carbonyl system (Figure 2.7). Whether an aldol-type or Claisen-type product is formed depends on the nature of X and its potential as a leaving group. Thus, chemically, two molecules

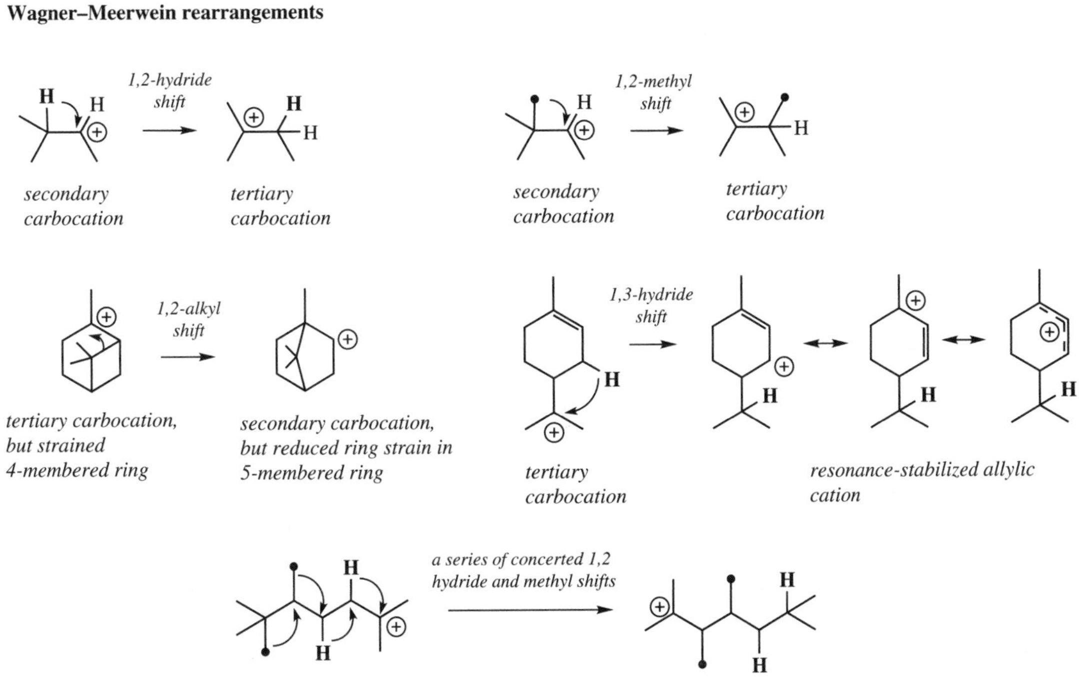

Wagner–Meerwein rearrangements

Figure 2.6

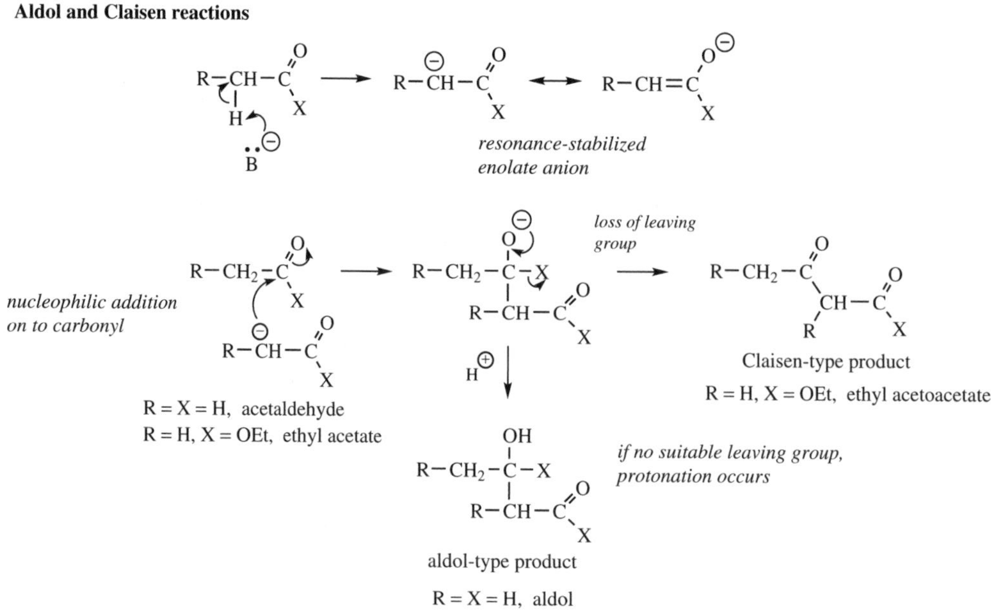

Aldol and Claisen reactions

Figure 2.7

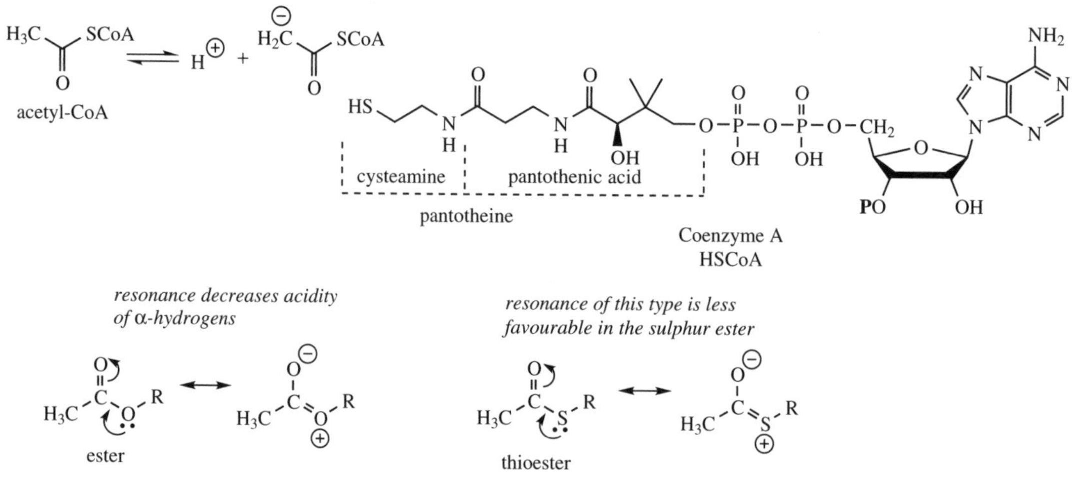

Figure 2.8

of acetaldehyde yield aldol, whilst two molecules of ethyl acetate can give ethyl acetoacetate. These processes are vitally important in biochemistry for the elaboration of both secondary and primary metabolites, but the enzyme catalysis obviates the need for strong bases, and probably means the enolate anion has little more than transitory existence. Nevertheless, the reactions do appear to parallel enolate anion chemistry, and are frequently responsible for joining together of C_2 acetate groups.

In most cases, the biological reactions involve coenzyme A esters, e.g. **acetyl-CoA** (Figure 2.8). This is a thioester of acetic acid, and it has significant advantages over oxygen esters, e.g. ethyl acetate, in that the α-methylene hydrogens are now more acidic, comparable in fact to those in the equivalent ketone, thus increasing the likelihood of generating the enolate anion. This is explained in terms of electron delocalization in the ester function (Figure 2.8). This type of delocalization is more prominent in the oxygen ester than in the sulphur ester, due to oxygen's smaller size and thus closer proximity of the lone pair for overlap with carbon's orbitals. Furthermore, the thioester has a much more favourable leaving group than the oxygen ester, and the combined effect is to increase the reactivity for both the aldol and Claisen-type reactions.

Claisen reactions involving acetyl-CoA are made even more favourable by first converting acetyl-CoA into **malonyl-CoA** by a carboxylation reaction with CO_2 using ATP and the coenzyme biotin (Figure 2.9). ATP and CO_2 (as bicarbonate, HCO_3^-) form the mixed anhydride, which carboxylates the coenzyme in a biotin–enzyme complex. Fixation of carbon dioxide by biotin–enzyme complexes is not unique to acetyl-CoA, and another important example occurs in the generation of oxaloacetate from pyruvate in the synthesis of glucose from non-carbohydrate sources

(gluconeogenesis). The conversion of acetyl-CoA into malonyl-CoA means the α-hydrogens are now flanked by two carbonyl groups, and have increased acidity. Thus, a more favourable nucleophile is provided for the Claisen reaction. No acylated malonic acid derivatives are produced, and the carboxyl group introduced into malonyl-CoA is simultaneously lost by a decarboxylation reaction during the Claisen condensation (Figure 2.9). An alternative rationalization is that decarboxylation of the malonyl ester is used to generate the acetyl enolate anion without any requirement for a strong base. Thus, the product formed from acetyl-CoA as electrophile and malonyl-CoA as nucleophile is acetoacetyl-CoA, which is actually the same as in the condensation of two molecules of acetyl-CoA. Accordingly, the role of the carboxylation step is clear cut: the carboxyl activates the α-carbon to facilitate the Claisen condensation, and it is immediately removed on completion of this task.

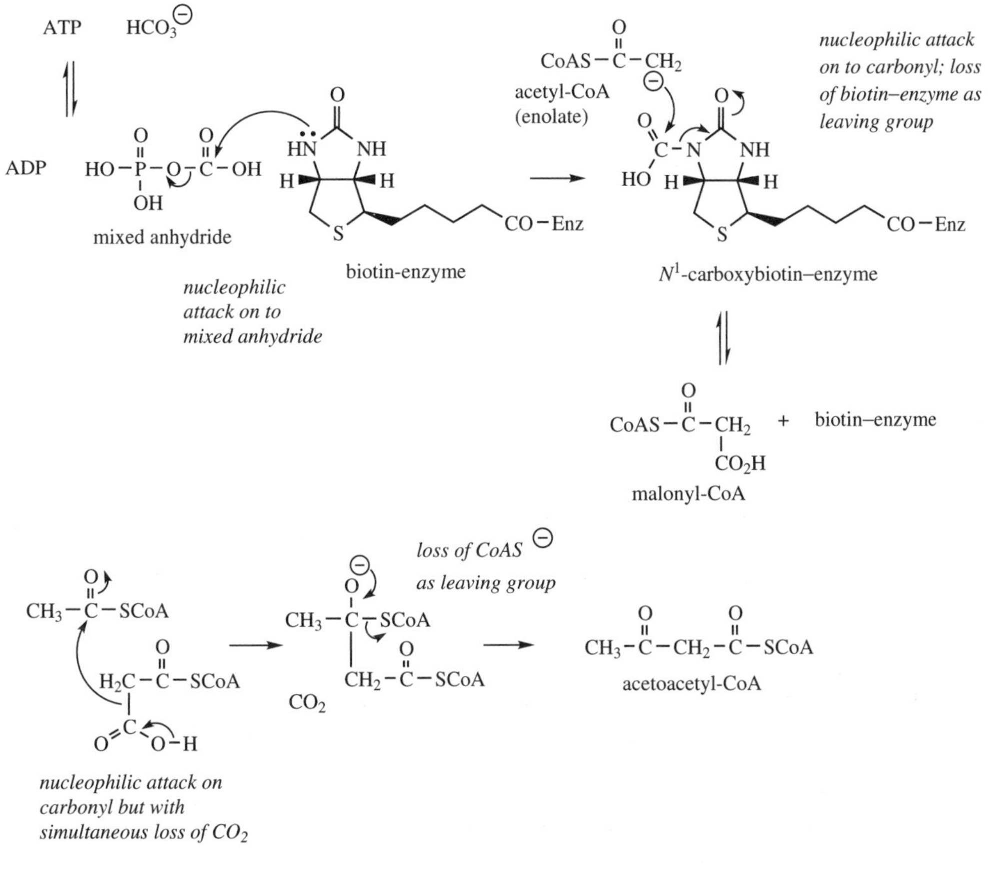

Figure 2.9

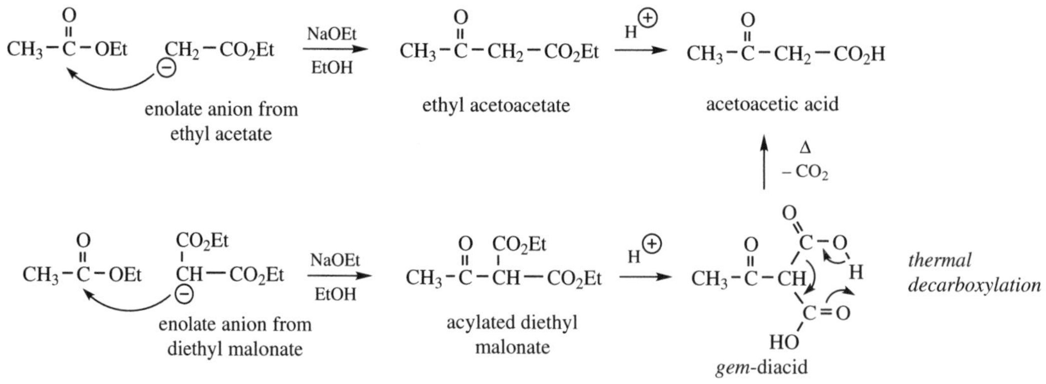

Figure 2.10

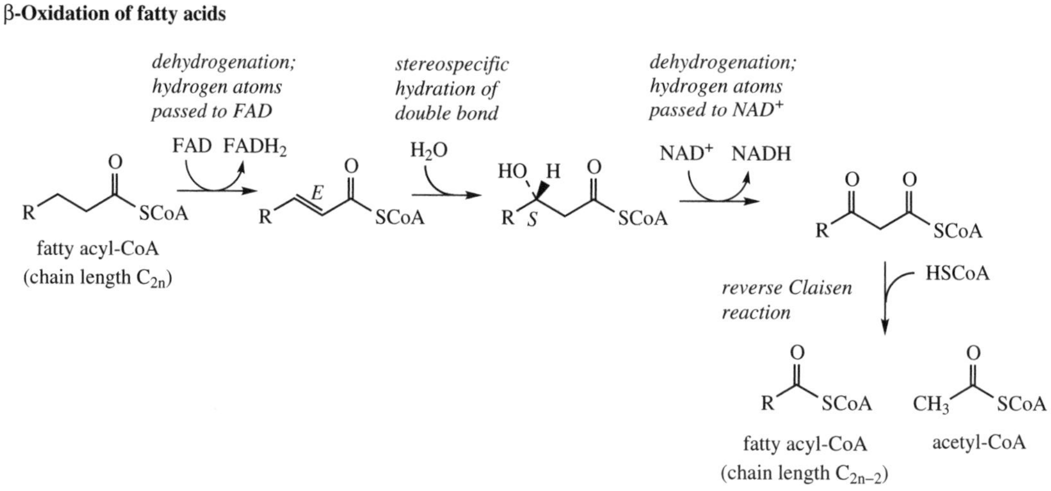

β-Oxidation of fatty acids

Figure 2.11

By analogy, the chemical Claisen condensation using the enolate anion from diethyl malonate in Figure 2.10 proceeds much more favourably than that using the enolate from ethyl acetate. The same acetoacetic acid product can be formed in the malonate condensation by hydrolysis of the acylated malonate intermediate and decarboxylation of the *gem*-diacid.

Both the **reverse aldol** and **reverse Claisen** reactions may be encountered in the modification of natural product molecules. Such reactions remove fragments from the basic skeleton already generated, but may extend the diversity of structures. The reverse Claisen reaction is a prominent feature of the **β-oxidation** sequence for the catabolic degradation of fatty acids (Figure 2.11),

in which a C_2 unit as acetyl-CoA is cleaved off from a fatty acid chain, leaving it two carbons shorter in length.

Schiff Base Formation and the Mannich Reaction

Formation of $C-N$ bonds is frequently achieved by condensation reactions between amines and aldehydes or ketones. A typical nucleophilic addition is followed by elimination of water to give an imine or **Schiff base** [Figure 2.12(a)]. Of almost equal importance is the reversal of this process, i.e. the hydrolysis of imines to amines and aldehydes/ketones [Figure 2.12(b)]. The imine so produced, or more likely its protonated form the

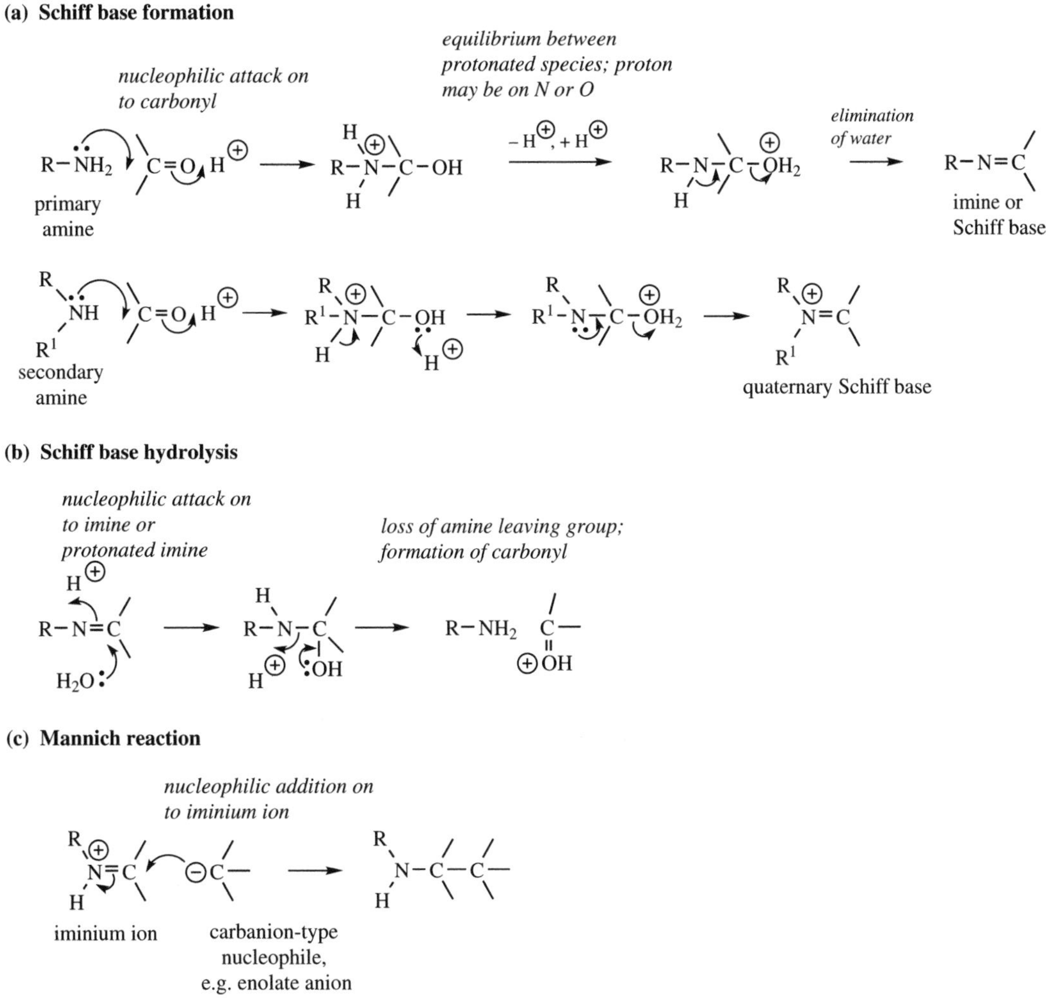

(a) Schiff base formation

(b) Schiff base hydrolysis

(c) Mannich reaction

Figure 2.12

iminium ion, can then act as an electrophile in a **Mannich reaction** [Figure 2.12(c)]. The nucleophile might be provided by an enolate anion, or in many examples by a suitably activated centre in an aromatic ring system. The Mannich reaction is encountered throughout alkaloid biosynthesis, and in its most general form involves combination of an amine (primary or secondary), an aldehyde or ketone, and a nucleophilic carbon. Secondary amines will react with the carbonyl compound to give an iminium ion (quaternary Schiff base) directly, and the additional protonation step is thus not necessary.

It should be appreciated that the Mannich-like addition reaction in Figure 2.12(c) is little different from nucleophilic addition to a carbonyl group. Indeed, the imine/iminium ion is merely acting as the nitrogen analogue of a carbonyl/protonated carbonyl. To take this analogy further, protons on carbon adjacent to an imine group will be acidic, as are those α to a carbonyl group, and the isomerization to the enamine shown in Figure 2.13 is analogous to keto–enol tautomerism. Just as two carbonyl compounds can react via an aldol reaction, so can two imine systems, and this is indicated in Figure 2.13. Often aldehyde/ketone substrates in enzymic reactions become covalently bonded to the enzyme through imine linkages; in so doing they lose none of the carbonyl activation as a consequence of the new form of bonding.

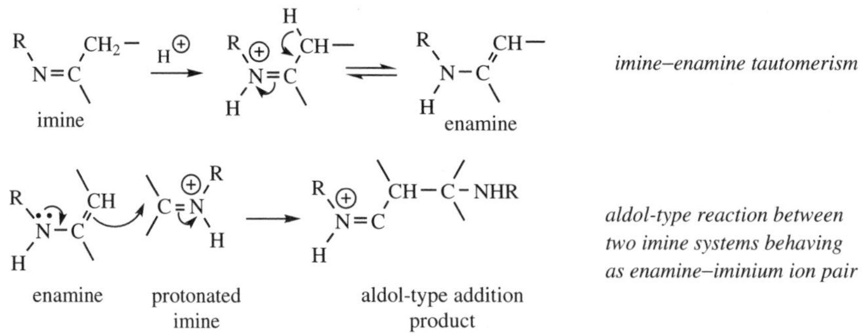

imine–enamine tautomerism

*aldol-type reaction between
two imine systems behaving
as enamine–iminium ion pair*

Figure 2.13

Transamination

Transamination is the exchange of the amino group from an amino acid to a keto acid, and provides the most common process for the introduction of nitrogen into amino acids, and for the removal of nitrogen from them. The couple **glutamic acid/2-oxoglutaric acid** are the usual donor/acceptor molecules for the amino group. Reductive amination of the Krebs cycle intermediate 2-oxoglutaric acid to glutamic acid (Figure 2.14) is responsible for the initial incorporation of nitrogen, a reaction which involves imine formation and subsequent reduction. Transamination then allows the amino group to be transferred from glutamic acid to a suitable keto acid, or in the reverse mode from an amino acid to 2-oxoglutaric acid. This reaction is dependent on the coenzyme **pyridoxal phosphate** (**PLP**) and features a Schiff base/imine intermediate (aldimine) with the aldehyde group of PLP (Figure 2.14). The α-hydrogen of the original amino acid is now made considerably more acidic and is removed, leading to the ketimine by a reprotonation process which also restores the aromaticity in the pyridine ring. The keto acid is then liberated by hydrolysis of the Schiff base function, which generates pyridoxamine phosphate. The remainder of the sequence is now a reversal of this process, and transfers the amine function from pyridoxamine phosphate to another keto acid.

Decarboxylation Reactions

Many pathways to natural products involve steps which remove portions of the carbon skeleton. Although two or more carbon atoms may be

cleaved off via the reverse aldol or reverse Claisen reactions mentioned above, by far the most common degradative modification is loss of one carbon atom by a **decarboxylation** reaction. Decarboxylation is a particular feature of the biosynthetic utilization of amino acids, and it has already been indicated that several of the basic building blocks, e.g. C_6C_2N, indole.C_2N, are derived from an amino acid via loss of the carboxyl group. This decarboxylation of amino acids is also a **pyridoxal phosphate**-dependent reaction (compare transamination) and is represented as in Figure 2.15(a). This similarly depends on Schiff base formation and shares features of the transamination sequence of Figure 2.14. Decarboxylation is facilitated in the same way as loss of the α-hydrogen was facilitated for the transamination sequence. After protonation of the original α-carbon, the amine is released from the coenzyme by hydrolysis of the Schiff base function.

β-Keto acids are thermally labile and rapidly decarboxylated *in vitro* via a cyclic mechanism which proceeds through the enol form of the final ketone [Figure 2.15(b)]. Similar reactions are found in nature, though whether cyclic processes are necessary is not clear. *ortho*-Phenolic acids also decarboxylate readily *in vitro* and *in vivo*, and it is again possible to invoke a cyclic β-keto acid tautomer of the substrate. The corresponding decarboxylation of *para*-phenolic acids cannot have a cyclic transition state, but the carbonyl group in the proposed keto tautomer activates the system for decarboxylation. The acetate pathway frequently yields structures containing phenol and carboxylic acid functions, and decarboxylation reactions may thus feature as further modifications. Although the carboxyl group may originate by hydrolysis of the

Transamination

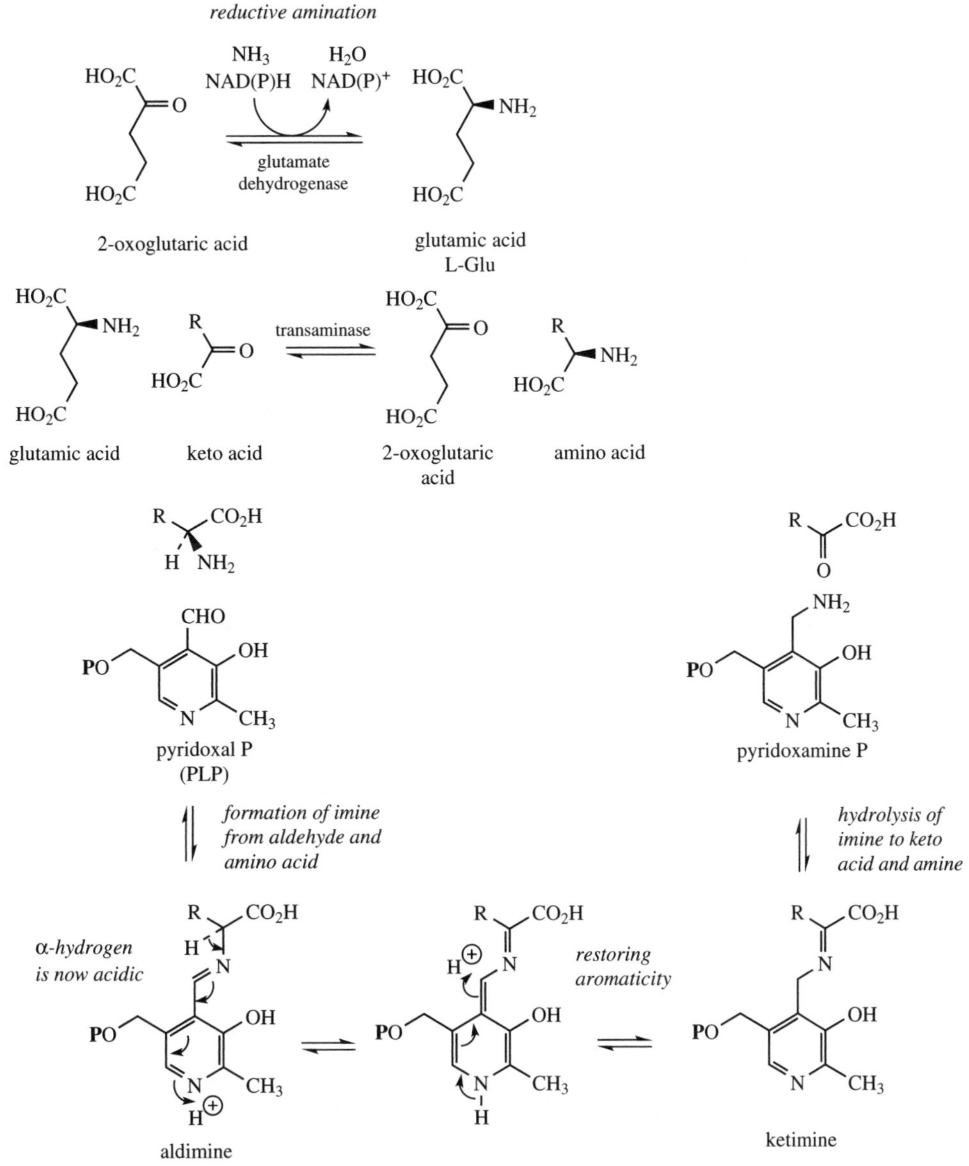

Figure 2.14

thioester portion of the acetyl-CoA precursor, there are also occasions when a methyl group can be sequentially oxidized to a carboxyl, which then subsequently suffers decarboxylation.

Decarboxylation of **α-keto acids** is a feature of primary metabolism, e.g. pyruvic acid → acetaldehyde in glycolysis, and pyruvic acid → acetyl-CoA, an example of overall oxidative decarboxylation prior to entry of acetyl-CoA

into the Krebs cycle. Both types of reaction depend upon **thiamine diphosphate (TPP)**. TPP is a coenzyme containing a thiazole ring, which has an acidic hydrogen and is thus capable of yielding the carbanion. This acts as a nucleophile towards carbonyl groups. Decarboxylation of pyruvic acid to acetaldehyde is depicted as in Figure 2.15(c), which process also regenerates the carbanion. In the oxidation step of oxidative

Decarboxylation reactions

(a) amino acids

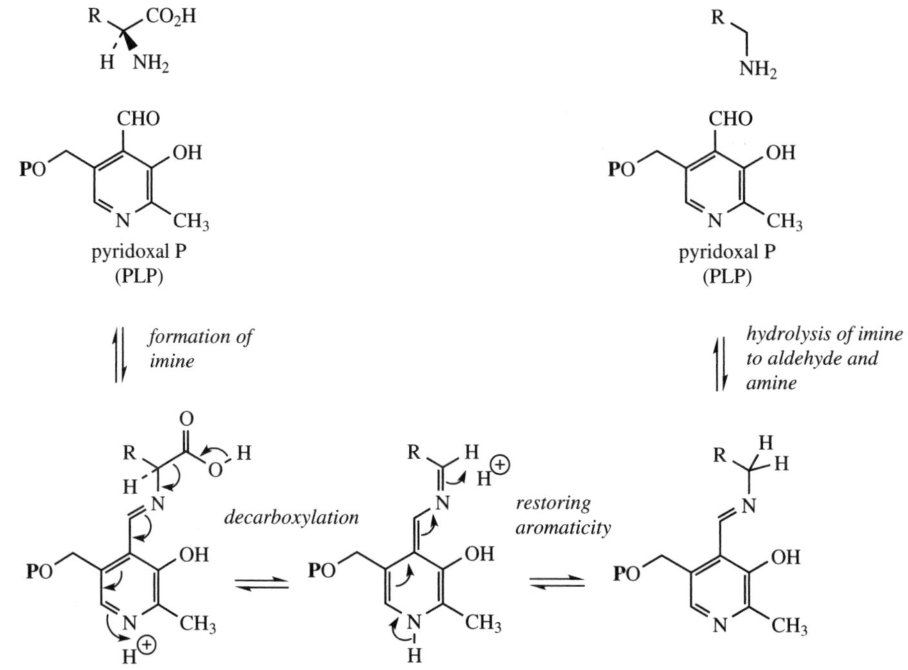

pyridoxal P
(PLP)

pyridoxal P
(PLP)

formation of imine

hydrolysis of imine to aldehyde and amine

decarboxylation

restoring aromaticity

(b) β-keto acids

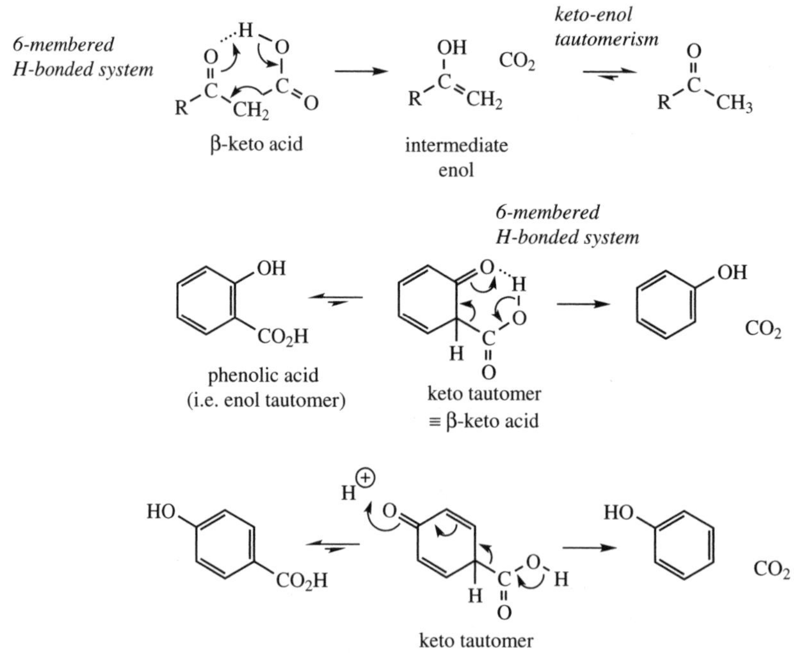

6-membered H-bonded system

β-keto acid

intermediate enol

keto-enol tautomerism

phenolic acid
(i.e. enol tautomer)

6-membered H-bonded system

keto tautomer
≡ β-keto acid

keto tautomer

Figure 2.15

(c) α-keto acids

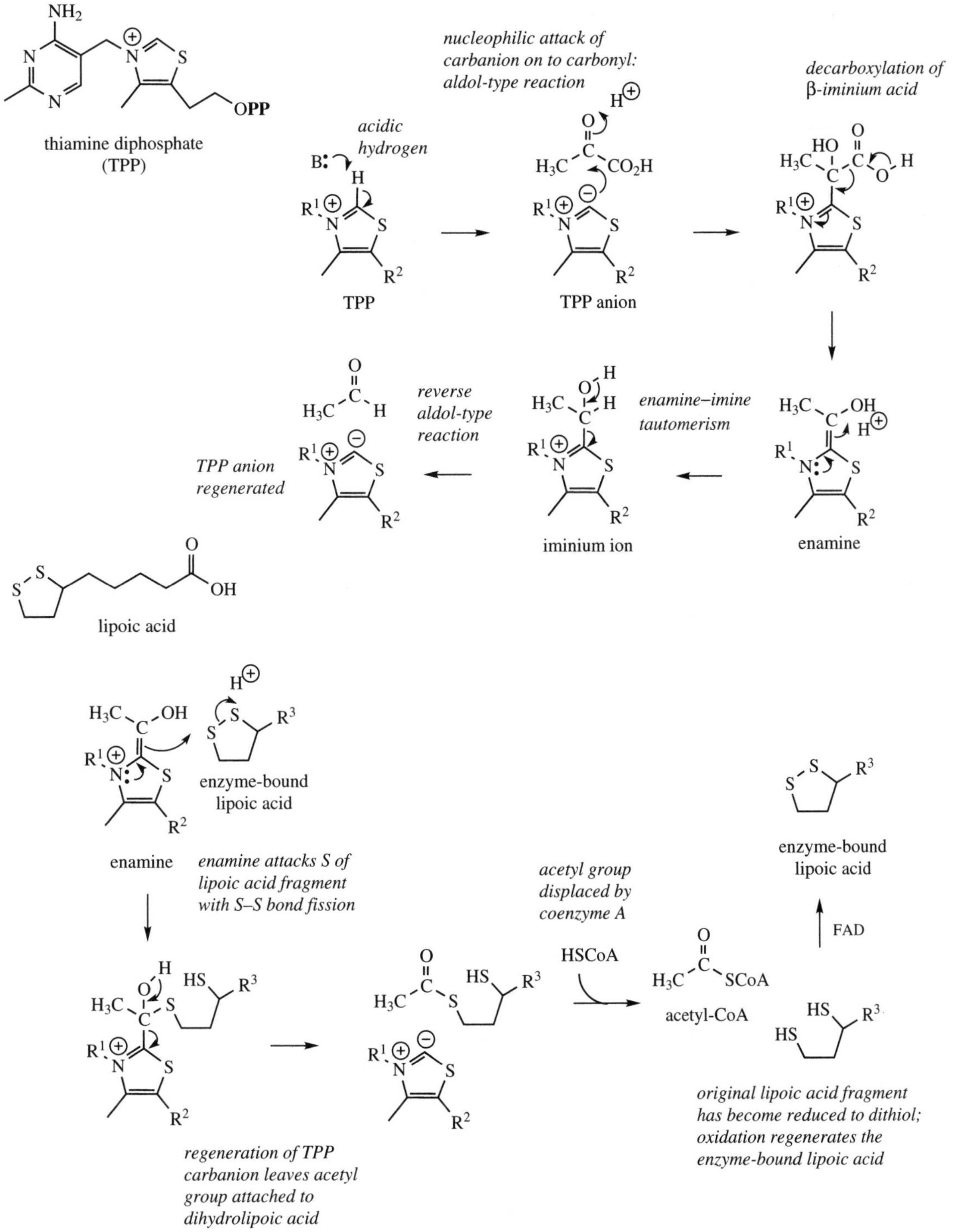

Figure 2.15 (continued)

decarboxylation, the enzyme-bound disulphide-containing coenzyme **lipoic acid** is also involved. The intermediate enamine in Figure 2.15(c), instead of accepting a proton, is used to attack a sulphur in the lipoic acid moiety with subsequent S—S bond fission, thereby effectively reducing the lipoic acid fragment. This allows regeneration of the TPP carbanion, and the acetyl group is bound to the dihydrolipoic acid. This acetyl group is then released as acetyl-CoA by displacement with the thiol coenzyme A. The bound dihydrolipoic acid fragment is then reoxidized to restore its function. An exactly equivalent reaction is encountered in the Krebs cycle in the conversion of 2-oxoglutaric acid into succinyl-CoA.

Oxidation and Reduction Reactions

Changes to the oxidation state of a molecule are frequently carried out as a secondary metabolite is synthesized or modified. The processes are not always completely understood, but the following general features are recognized. The processes may be classified according to the type of enzyme involved and their mechanism of action.

Dehydrogenases

Dehydrogenases remove two hydrogen atoms from the substrate, passing them to a suitable coenzyme acceptor. The coenzyme system involved can generally be related to the functional group being oxidized in the substrate. Thus if the oxidation process is

$$\overset{\backslash}{\underset{/}{C}}H-OH \longrightarrow \overset{\backslash}{\underset{/}{C}}=O$$

then a pyridine nucleotide, **nicotinamide adenine dinucleotide (NAD⁺)** or **nicotinamide adenine dinucleotide phosphate (NADP⁺)**, tends to be utilized as hydrogen acceptor. One hydrogen from the substrate (that bonded to carbon) is transferred as hydride to the coenzyme, and the other, as a proton, is passed to the medium (Figure 2.16). NAD(P)⁺ may also be used in the oxidations

$$\overset{H}{\underset{/}{\backslash}}C=O \longrightarrow -CO_2H$$

$$\overset{\backslash}{\underset{/}{C}}H-NH_2 \longrightarrow \overset{\backslash}{\underset{/}{C}}=NH$$

The reverse reaction, i.e. reduction, is also indicated in Figure 2.16, and may be compared with the chemical reduction process using complex metal hydrides, e.g. LiAlH₄ or NaBH₄, namely nucleophilic addition of hydride and subsequent protonation. The reduced forms NADH and NADPH are conveniently regarded as hydride-donating

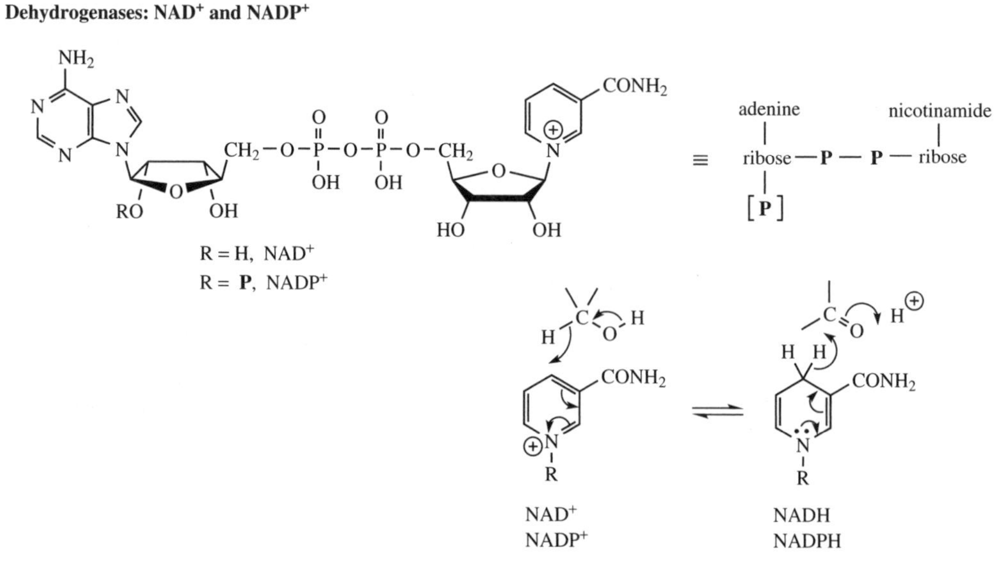

Figure 2.16

Dehydrogenases: FAD and FMN

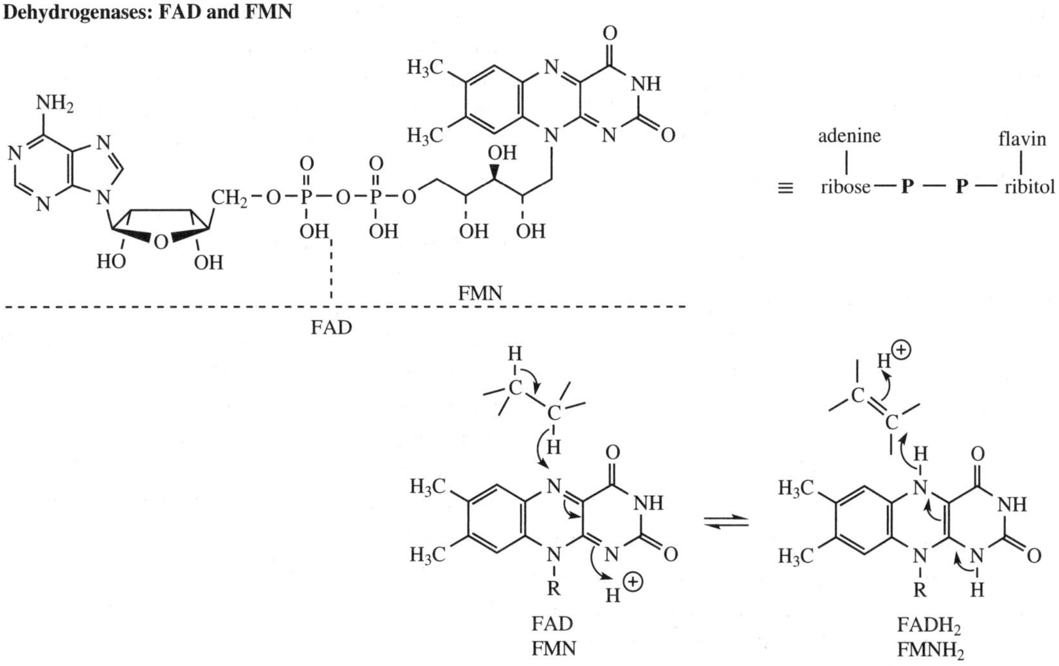

Figure 2.17

reducing agents. In practice, NADPH is generally employed in reductive processes, whilst NAD$^+$ is used in oxidations.

Should the oxidative process be the conversion

$$-CH_2-CH_2- \longrightarrow -CH=CH-$$

the coenzyme used as acceptor is usually a flavin nucleotide, **flavin adenine dinucleotide (FAD)** or **flavin mononucleotide (FMN)**. These entities are bound to the enzyme in the form of a flavoprotein, and take up two hydrogen atoms, represented in Figure 2.17 as being derived by addition of hydride from the substrate and a proton from the medium. Alternative mechanisms have also been proposed, however. Reductive sequences involving flavoproteins may be represented as the reverse reaction in Figure 2.17. NADPH may also be employed as a coenzyme in the reduction of a carbon–carbon double bond.

These oxidation reactions employing pyridine nucleotides and flavoproteins are especially important in primary metabolism in liberating energy from fuel molecules in the form of ATP. The reduced coenzymes formed in the process are normally reoxidized via the electron transport chain

of oxidative phosphorylation, so that the hydrogen atoms eventually pass to oxygen giving water.

Oxidases

Oxidases also remove hydrogen from a substrate, but pass these atoms to molecular oxygen or to hydrogen peroxide, in both cases forming water. Oxidases using hydrogen peroxide are termed **peroxidases**. Mechanisms of action vary and need not be considered here. Important transformations in secondary metabolism include the oxidation of *ortho*- and *para*-quinols to quinones (Figure 2.18), and the peroxidase-induced phenolic oxidative coupling processes (see page 28).

Mono-oxygenases

Oxygenases catalyse the direct addition of oxygen from molecular oxygen to the substrate. They are subdivided into mono- and di-oxygenases according to whether just one or both of the oxygen atoms are introduced into the substrate. With **mono-oxygenases**, the second oxygen atom from O_2 is reduced to water by an appropriate hydrogen

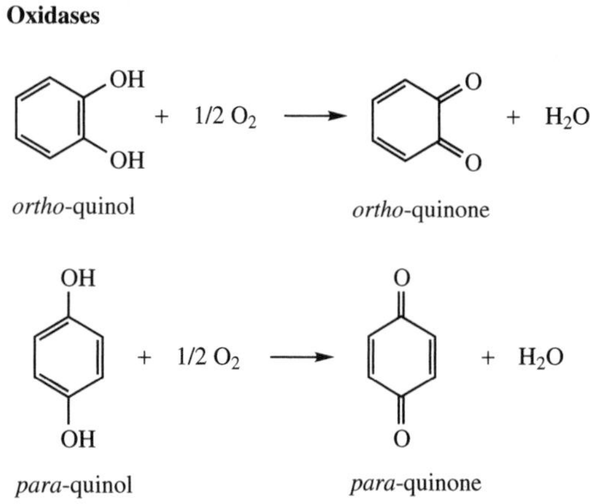

Oxidases

ortho-quinol   ortho-quinone

para-quinol   para-quinone

Figure 2.18

donor, e.g. NADH, NADPH, or **ascorbic acid** (vitamin C). In this respect they may also be considered to behave as oxidases, and the term 'mixed-function oxidase' is also used for these enzymes. Especially important examples of these enzymes are the **cytochrome P-450-dependent mono-oxygenases**. These are frequently involved in biological hydroxylations, either in biosynthesis, or in the mammalian detoxification and metabolism of foreign compounds such as drugs, and such enzymes are thus termed '**hydroxylases**'. **Cytochrome P-450** is named after its intense absorption band at 450 nm when exposed to CO, which is a powerful inhibitor of these enzymes. It contains an iron–porphyrin complex (haem), which is bound to the enzyme, and a redox change involving the Fe atom allows binding and the

cleavage of an oxygen atom. Many such systems have been identified, capable of hydroxylating aliphatic or aromatic systems, as well as producing epoxides from alkenes (Figure 2.19). In most cases, NADPH features as hydrogen donor.

Aromatic hydroxylation catalysed by mono-oxygenases (including cytochrome P-450 systems) probably involves arene oxide (epoxide) intermediates (Figure 2.20). An interesting consequence of this mechanism is that when the epoxide opens up, the hydrogen atom originally attached to the position which becomes hydroxylated can migrate to the adjacent carbon on the ring. A high proportion of these hydrogen atoms is subsequently retained in the product, even though enolization allows some loss of this hydrogen. This migration is known as the **NIH shift**, having been originally observed at the National Institute of Health, Bethesda, MD, USA.

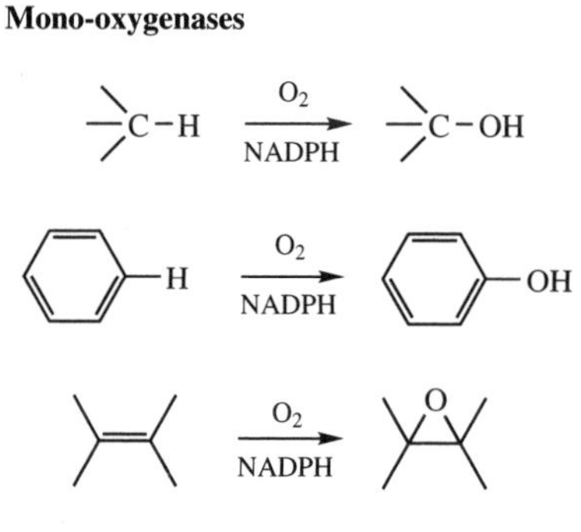

Mono-oxygenases

Figure 2.19

NIH shift

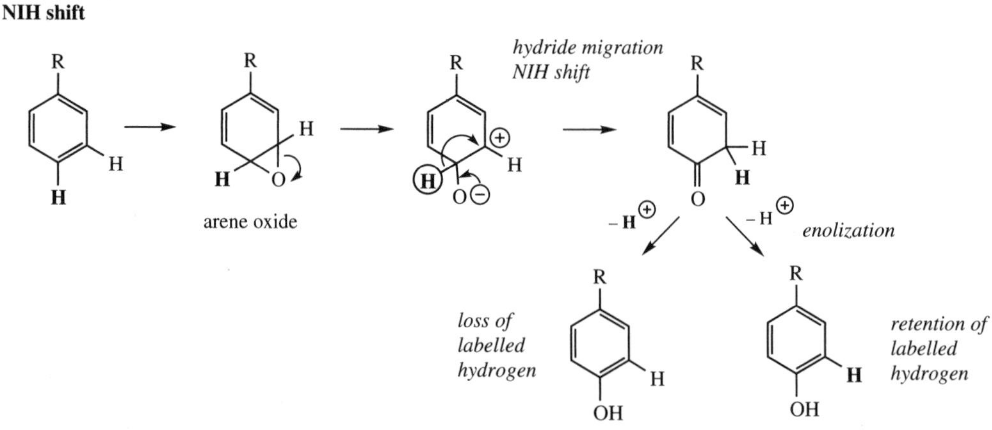

Figure 2.20

The oxidative cyclization of an *ortho*-hydroxy-methoxy-substituted aromatic system giving a **methylenedioxy group** is also known to involve a cytochrome P-450-dependent mono-oxygenase. This enzyme hydroxylates the methyl to yield a formaldehyde hemiacetal intermediate, which can cyclize to the methylenedioxy bridge (the acetal of formaldehyde) by an ionic mechanism (Figure 2.21).

Dioxygenases

Dioxygenases introduce both atoms from molecular oxygen into the substrate, and are frequently involved in the cleavage of bonds, including aromatic rings. Cyclic peroxides (dioxetanes) are likely to be intermediates (Figure 2.22). Oxidative cleavage of aromatic rings typically employs catechol (1,2-dihydroxy) or quinol (1,4-dihydroxy) substrates, and in the case of catechols, cleavage may be between or adjacent to the two hydroxyls, giving products containing aldehyde and/or carboxylic acid functionalities (Figure 2.22).

Some dioxygenases utilize two acceptor substrates and incorporate one oxygen atom into each. Thus, **2-oxoglutarate-dependent dioxygenases** hydroxylate one substrate, whilst also transforming 2-oxoglutarate into succinate with the release of CO_2 (Figure 2.23). 2-Oxoglutarate-dependent dioxygenases also require as cofactors

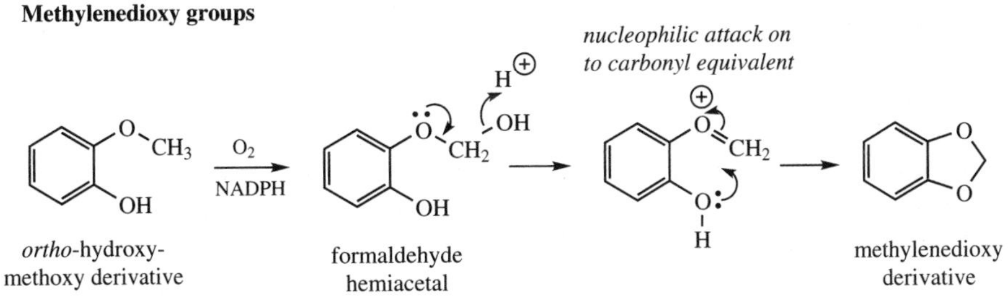

Methylenedioxy groups

Figure 2.21

Dioxygenases

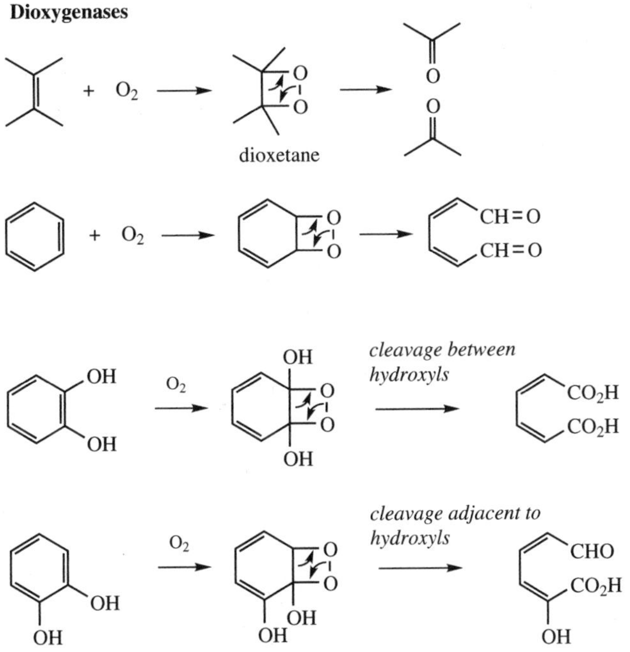

Figure 2.22

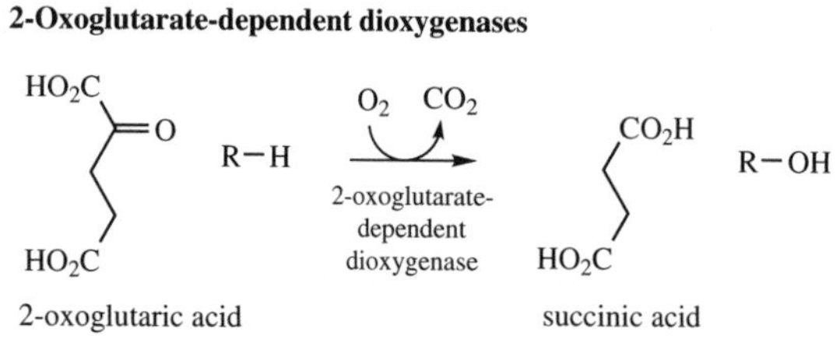

2-Oxoglutarate-dependent dioxygenases

Figure 2.23

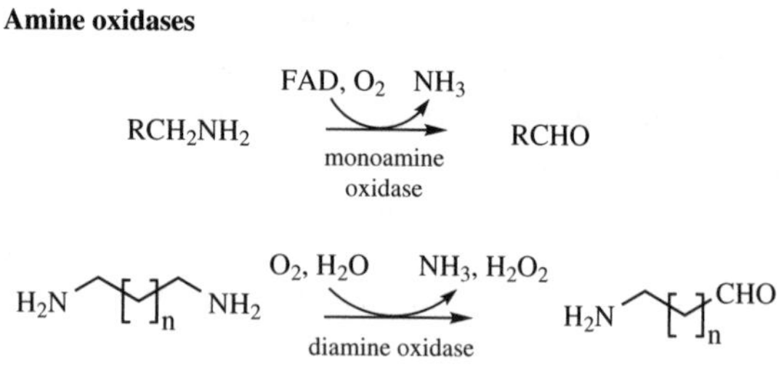

Fe^{2+} to generate an enzyme-bound iron–oxygen complex, and **ascorbic acid** (vitamin C) to subsequently reduce this complex.

Amine Oxidases

In addition to the oxidizing enzymes outlined above, those which transform an amine into an aldehyde, the **amine oxidases**, are frequently involved in metabolic pathways. These include **monoamine oxidases** and **diamine oxidases**. Monoamine oxidases utilize a flavin nucleotide, typically FAD, and molecular oxygen, and involve initial dehydrogenation to an imine, followed by hydrolysis to the aldehyde and ammonia (Figure 2.24). Diamine oxidases require a diamine substrate, and oxidize at one amino group using molecular oxygen to give the corresponding aldehyde. Hydrogen peroxide and ammonia are the other products formed. The aminoaldehyde so formed then has the potential to be transformed into a cyclic imine via Schiff base formation.

Amine oxidases

Figure 2.24

Baeyer–Villiger Oxidations

The chemical oxidation of ketones by peracids, the **Baeyer–Villiger oxidation**, yields an ester, and the process is known to involve migration of an alkyl group from the ketone (Figure 2.25). For comparable ketone → ester conversions known to occur in biochemistry, cytochrome-P-450- or FAD-dependent enzymes requiring NADPH and O$_2$ appear to be involved. This leads to formation of a peroxy–enzyme complex and a mechanism similar to that for the chemical Baeyer–Villiger oxidation may thus operate. The oxygen atom introduced thus originates from O$_2$.

Phenolic Oxidative Coupling

Many natural products are produced by the coupling of two or more phenolic systems, in a process readily rationalized by means of free radical reactions. The reactions can be brought

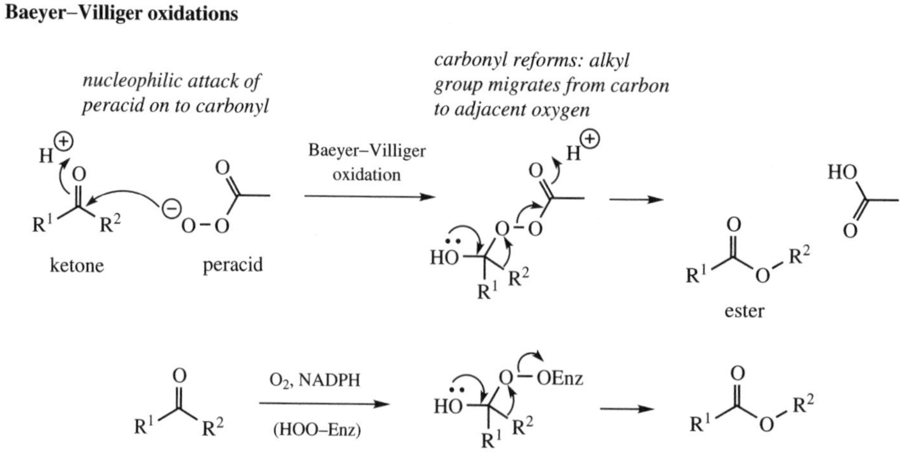

Baeyer–Villiger oxidations

Figure 2.25

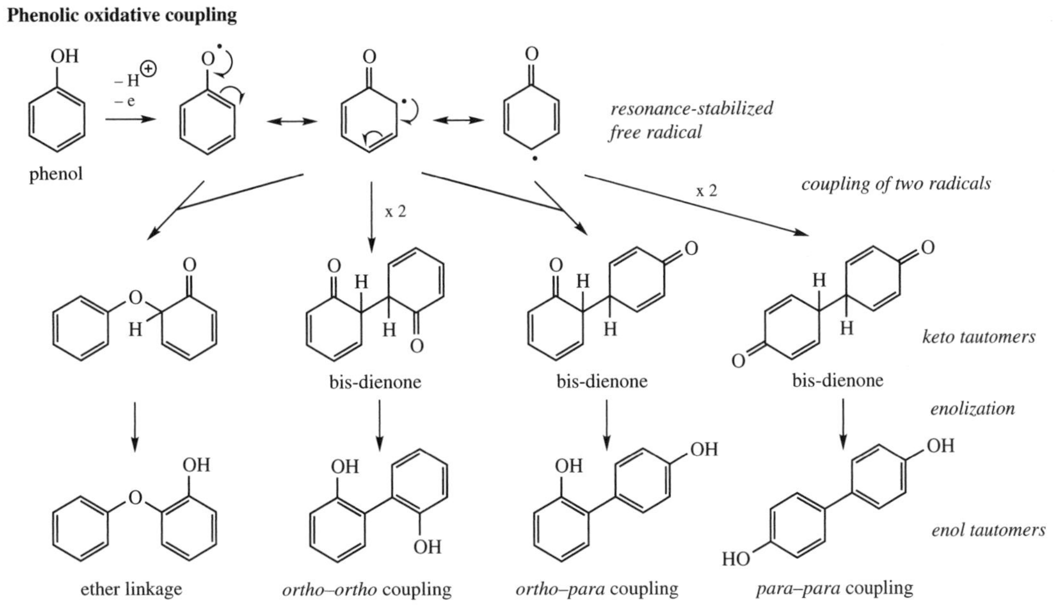

Phenolic oxidative coupling

Figure 2.26

about by oxidase enzymes, including peroxidase and laccase systems, known to be radical generators. Other enzymes catalysing **phenolic oxidative coupling** have been characterized as cytochrome P-450-dependent proteins, requiring NADPH and O_2 cofactors, but no oxygen is incorporated into the substrate. A one-electron oxidation of a phenol gives the free radical, and the unpaired electron can then be delocalized via resonance forms in which the free electron is dispersed to positions *ortho* and *para* to the original oxygen function (Figure 2.26). Coupling of two of these mesomeric structures gives a range of dimeric systems as exemplified in Figure 2.26. The final products indicated are then derived by enolization, which restores aromaticity to the rings. Thus, carbon–carbon bonds involving positions *ortho* or *para* to the original phenols, or ether linkages, may be formed. The reactive dienone systems formed as intermediates may in some cases be attacked by other nucleophilic groupings, extending the range of structures ultimately derived from this basic reaction sequence.

Glycosylation Reactions

The widespread occurrence of **glycosides** and **polysaccharides** requires processes for attaching sugar units to a suitable atom of an aglycone to give a glycoside, or to another sugar giving a polysaccharide. Linkages tend to be through oxygen, although they are not restricted to oxygen, since S-, N-, and C-glycosides are well known. The agent for glycosylation is a **uridine diphosphosugar**, e.g. **UDPglucose**. This is synthesized from glucose 1-phosphate and UTP, and then the glucosylation process can be envisaged as a simple S_N2 nucleophilic displacement reaction [Figure 2.27(a)]. Since UDPglucose has its leaving group in the α-configuration, the product has the β-configuration, as is most commonly found in natural glucosides. Note, however, that many important carbohydrates, e.g. sucrose and starch, possess α-linkages, and these appear to originate via double S_N2 processes (see page 470). Other UDPsugars, e.g. UDPgalactose or UDPxylose, are utilized in the synthesis of glycosides containing different sugar units.

The hydrolysis of glycosides is achieved by specific hydrolytic enzymes, e.g. β-glucosidase for β-glucosides and β-galactosidase for β-galactosides. These enzymes mimic the readily achieved acid-catalysed processes [Figure 2.27(b)]. Under acidic conditions, the α- and β-anomeric hemiacetal forms can also equilibrate via the open chain

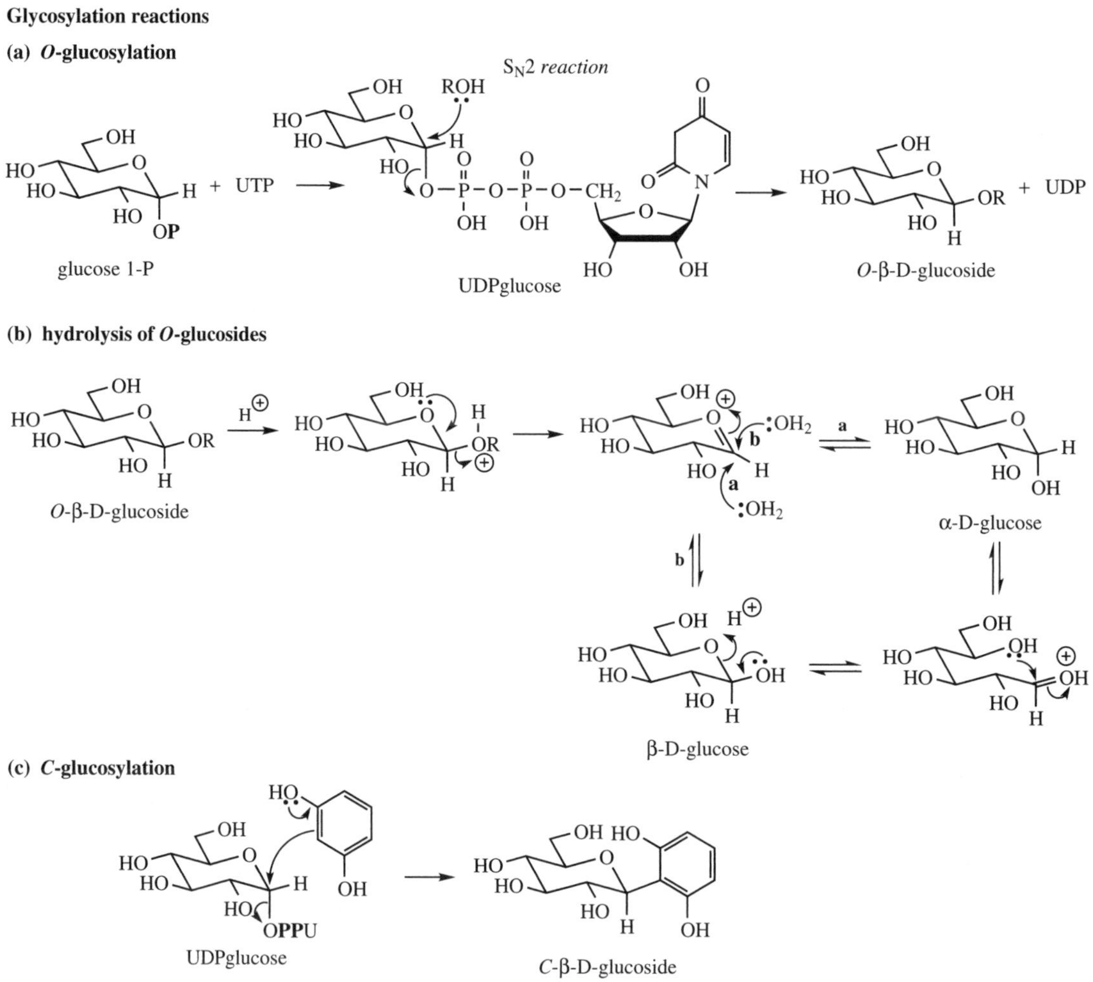
Glycosylation reactions
(a) O-glucosylation
SN2 reaction
glucose 1-P
UDPglucose
O-β-D-glucoside
(b) hydrolysis of O-glucosides
O-β-D-glucoside
α-D-glucose
β-D-glucose
(c) C-glucosylation
UDPglucose
C-β-D-glucoside

Figure 2.27

sugar. Of particular importance is that although *O*-, *N*-, and *S*-glycosides may be hydrolysed by acid, *C*-glycosides are stable to acid. *C*-Glycosides are produced in a similar manner to the *C*-alkylation process described above, where a suitable nucleophilic carbon is available, e.g. aromatic systems activated by phenol groups [Figure 2.27(c)]. The resultant *C*-glycoside thus contains a new carbon–carbon linkage, and cleavage would require oxidation, not hydrolysis.

SOME VITAMINS ASSOCIATED WITH THE CONSTRUCTION MECHANISMS

Vitamin B_1

Vitamin B_1 (thiamine) (Figure 2.28) is a water-soluble vitamin with a pyrimidinylmethylthiazolium structure. It is widely available in the diet, with cereals, beans, nuts, eggs, yeast, and vegetables providing sources. Wheat germ and yeast have very high levels. Dietary deficiency leads to beriberi, characterized by neurological disorders, loss of appetite, fatigue,

(Continues)

(*Continued*)

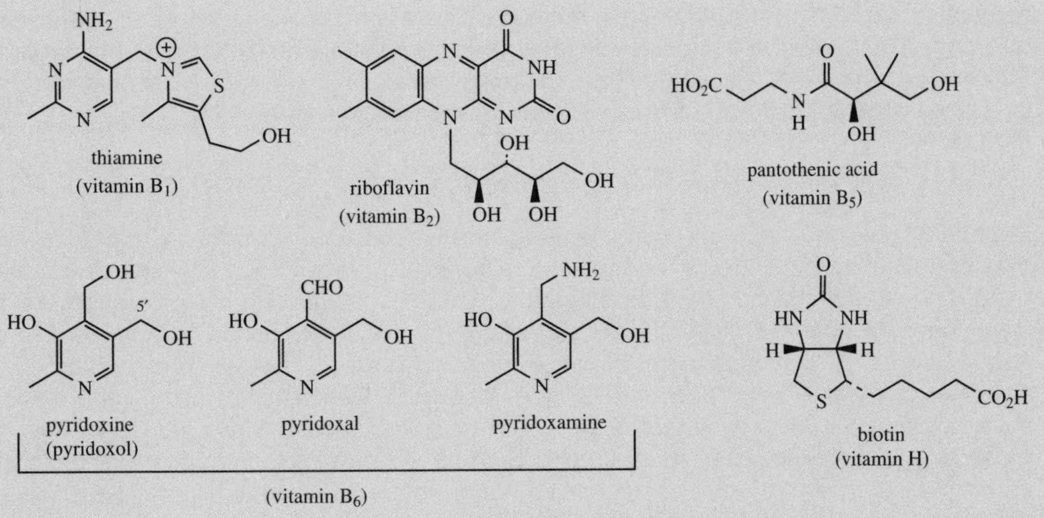

Figure 2.28

and muscular weakness. **Thiamine** is produced synthetically, and foods such as cereals are often enriched. The vitamin is stable in acid solution, but decomposes above pH 5, and is also partially decomposed during normal cooking. As thiamine diphosphate, vitamin B$_1$ is a coenzyme for pyruvate dehydrogenase which catalyses the oxidative decarboxylation of pyruvate to acetyl-CoA (see page 21), and also for transketolase which transfers a two-carbon fragment between carbohydrates in the pentose phosphate pathway (see page 446). Accordingly, this is a very important component in carbohydrate metabolism.

Vitamin B$_2$

Vitamin B$_2$ (riboflavin) (Figure 2.28) is a water-soluble vitamin having an isoalloxazine ring linked to D-ribitol. It is widely available in foods, including liver, kidney, dairy products, eggs, meat, and fresh vegetables. Yeast is a particularly rich source. It is stable in acid solution, not decomposed during cooking, but is sensitive to light. Riboflavin may be produced synthetically, or by fermentation using the yeastlike fungi *Eremothecium ashbyii* and *Ashbya gossypii*. Dietary deficiency is uncommon, but manifests itself by skin problems and eye disturbances. Riboflavin is a component of FMN (flavin mononucleotide) and FAD (flavin adenine dinucleotide), coenzymes which play a major role in oxidation–reduction reactions (see page 25). Many key enzymes containing riboflavin (flavoproteins) are involved in metabolic pathways. Since riboflavin contains ribitol and not ribose in its structure, FAD and FMN are not strictly nucleotides, though this nomenclature is commonly accepted and used.

Vitamin B$_5$

Vitamin B$_5$ (pantothenic acid) (Figure 2.28) is a very widely distributed water-soluble vitamin, though yeast, liver, and cereals provide rich sources. Even though animals must obtain the vitamin through the diet, pantothenic acid deficiency is rare, since most foods provide

(*Continues*)

(Continued)

adequate quantities. Its importance in metabolism is as part of the structure of coenzyme A (see page 16), the carrier molecule essential for carbohydrate, fat, and protein metabolism. Pantothenic acid is specifically implicated in enzymes responsible for the biosynthesis of fatty acids (see page 36), polyketides (page 62) and some peptides (page 421).

Vitamin B_6

Vitamin B_6 covers the three pyridine derivatives pyridoxine (pyridoxol), pyridoxal, and pyridoxamine, and also their 5′-phosphates (Figure 2.28). These are water-soluble vitamins, pyridoxine predominating in plant materials, whilst pyridoxal and pyridoxamine are the main forms in animal tissues. Meat, salmon, nuts, potatoes, bananas, and cereals are good sources. A high proportion of the vitamin activity can be lost during cooking, but a normal diet provides an adequate supply. Vitamin B_6 deficiency is usually the result of malabsorption, or may be induced by some drug treatments where the drug may act as an antagonist or increase its renal excretion as a side-effect. Symptoms of deficiency are similar to those of niacin (vitamin B_3) and riboflavin (vitamin B_2) deficiencies, and include eye, mouth, and nose lesions, and neurological changes. Synthetic **pyridoxine** is used for supplementation. Pyridoxal 5′-phosphate is a coenzyme for a large number of enzymes, particularly those involved in amino acid metabolism, e.g. in transamination (see page 20) and decarboxylation (see page 20). The production of the neurotransmitter γ-aminobutyric acid (GABA) from glutamic acid is an important pyridoxal-dependent reaction.

Vitamin B_{12}

Vitamin B_{12} (cobalamins) (Figure 2.29) are extremely complex structures based on a corrin ring, which, although similar to the porphyrin ring found in haem, chlorophyll, and cytochromes,

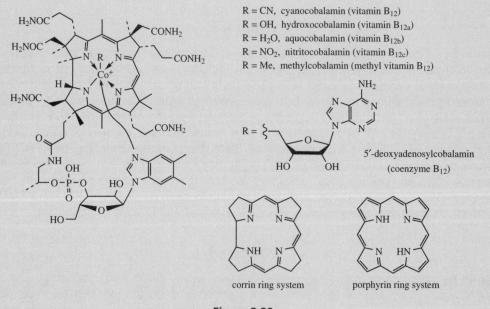

R = CN, cyanocobalamin (vitamin B_{12})
R = OH, hydroxocobalamin (vitamin B_{12a})
R = H_2O, aquocobalamin (vitamin B_{12b})
R = NO_2, nitritocobalamin (vitamin B_{12c})
R = Me, methylcobalamin (methyl vitamin B_{12})

5′-deoxyadenosylcobalamin
(coenzyme B_{12})

corrin ring system   porphyrin ring system

Figure 2.29

(Continues)

(Continued)

has two of the pyrrole rings directly bonded. The central metal atom is cobalt; haem and cytochromes have iron, whilst chlorophyll has magnesium. Four of the six coordinations are provided by the corrin ring nitrogens, and a fifth by a dimethylbenzimidazole moiety. The sixth is variable, being cyano in **cyanocobalamin** (vitamin B_{12}), hydroxyl in **hydroxocobalamin** (vitamin B_{12a}), or other anions may feature. Cyanocobalamin is actually an artefact formed as a result of the use of cyanide in the purification procedures. The physiologically active coenzyme form of the vitamin is 5′-deoxyadenosylcobalamin (coenzyme B_{12}). Vitamin B_{12} appears to be entirely of microbial origin, with intestinal flora contributing towards human dietary needs. The vitamin is then stored in the liver, and animal liver extract has been a traditional source. Commercial supplies are currently obtained by semi-synthesis from the total cobalamin extract of *Streptomyces griseus*, *Propionibacterium* species, or other bacterial cultures. This material can be converted into cyanocobalamin or hydroxocobalamin. The cobalamins are stable when protected against light. Foods with a high vitamin B_{12} content include liver, kidney, meat, and seafood. Vegetables are a poor dietary source, and strict vegetarians may therefore risk deficiencies. Insufficient vitamin B_{12} leads to pernicious anaemia, a disease that results in nervous disturbances and low production of red blood cells, though this is mostly due to lack of the gastric glycoprotein (intrinsic factor) which complexes with the vitamin to facilitate its absorption. Traditionally, daily consumption of raw liver was used to counteract the problem. Cyanocobalamin, or preferably hydroxocobalamin which has a longer lifetime in the body, may be administered orally or by injection to counteract deficiencies. Both agents are converted into coenzyme B_{12} in the body. Coenzyme B_{12} is a cofactor for a number of metabolic rearrangements, such as the conversion of methylmalonyl-CoA into succinyl-CoA in the oxidation of fatty acids with an odd number of carbon atoms, and for methylations, such as in the biosynthesis of methionine.

Vitamin H

Vitamin H (biotin) (Figure 2.28) is a water-soluble vitamin found in eggs, liver, kidney, yeast, cereals, and milk, and is also produced by intestinal microflora so that dietary deficiency is rare. Deficiency can be triggered by a diet rich in raw egg white, in which a protein, avidin, binds biotin so tightly so that it is effectively unavailable for metabolic use. This affinity disappears by cooking and hence denaturing the avidin. Biotin deficiency leads to dermatitis and hair loss. The vitamin functions as a carboxyl carrier, binding CO_2 via a carbamate link, then donating this in carboxylase reactions, e.g. carboxylation of acetyl-CoA to malonyl-CoA (see page 17), of propionyl-CoA to methylmalonyl-CoA (see page 92), and of pyruvate to oxaloacetate during gluconeogenesis.

FURTHER READING

Natural Products, Biosynthesis

Mann J (1994) *Chemical Aspects of Biosynthesis*. Oxford Chemistry Primers, Oxford.

Mann J, Davidson RS, Hobbs JB, Banthorpe DV and Harborne JB (1994) *Natural Products: Their Chemistry and Biological Significance*. Longman, Harlow.

Torssell KBG (1997) *Natural Product Chemistry. A Mechanistic, Biosynthetic and Ecological Approach*. Apotekarsocieteten, Stockholm.

Vitamins

Battersby AR (2000) Tetrapyrroles: The pigments of life. *Nat Prod Rep* **17**, 507–526.

Burdick D (1998) Vitamins [pyridoxine (B_6)]. *Kirk–Othmer Encyclopedia of Chemical Technology*, 4th edn, Vol **25**. Wiley, New York, pp 116–132.

Burdick D (1998) Vitamins [thiamine (B_1)]. *Kirk–Othmer Encyclopedia of Chemical Technology*, 4th edn, Vol **25**. Wiley, New York, pp 152–171.

Kingston R (1999) Supplementary benefits? *Chem Brit* **35** (7), 29–32.

Outten RA (1998) Vitamins (biotin). *Kirk–Othmer Encyclopedia of Chemical Technology*, 4th edn, Vol **25**. Wiley, New York, pp 48–64.

Rawalpally TR (1998) Vitamins (pantothenic acid). *Kirk–Othmer Encyclopedia of Chemical Technology*, 4th edn, Vol **25**. Wiley, New York, pp 99–116.

Scott JW (1998) Vitamins (vitamin B_{12}). *Kirk–Othmer Encyclopedia of Chemical Technology*, 4th edn, Vol **25**. Wiley, New York, pp 192–217.

Yoneda F (1998) Vitamins [riboflavin (B_2)]. *Kirk–Othmer Encyclopedia of Chemical Technology*, 4th edn, Vol **25**. Wiley, New York, pp 132–152.

3

THE ACETATE PATHWAY: FATTY ACIDS AND POLYKETIDES

Polyketides, metabolites built primarily from combinations of acetate units, are described. The biosynthesis of saturated and unsaturated fatty acids is covered, together with prostaglandins, thromboxanes, and leukotrienes. Cyclization of polyketides to give aromatic structures is then rationalized in terms of aldol and Claisen reactions. More complex structures formed via pathways involving alkylation reactions, phenolic oxidative coupling, oxidative cleavage of aromatic rings, and employing starter groups other than acetate are developed. The use of extender groups other than malonate gives rise to macrolides and polyethers, whilst further cyclization of polyketide structures may be achieved through Diels−Alder reactions. The application of genetic engineering to modify products from the acetate pathway is discussed. Monograph topics giving more detailed information on medicinal agents include fixed oils and fats, evening primrose oil, echinacea, prostaglandins and isoprostanes, thromboxanes, leukotrienes, senna, cascara, frangula and allied drugs, St John's wort, mycophenolic acid, khellin and cromoglicate, griseofulvin, poison ivy and poison oak, aflatoxins, cannabis, tetracyclines, anthracycline antibiotics, macrolide antibiotics, avermectins, polyene antifungals, tacrolimus and sirolimus, ansa macrolides, mevastatin and other statins.

Polyketides constitute a large class of natural products grouped together on purely biosynthetic grounds. Their diverse structures can be explained as being derived from poly-β-keto chains, formed by coupling of acetic acid (C_2) units via condensation reactions,

$$\text{i.e.} \quad n\text{CH}_3\text{CO}_2\text{H} \longrightarrow -[\text{CH}_2\text{CO}]_n-$$

Included in such compounds are the fatty acids, polyacetylenes, prostaglandins, macrolide antibiotics and many aromatic compounds, e.g. anthraquinones and tetracyclines.

The formation of the poly-β-keto chain could be envisaged as a series of Claisen reactions, the reverse of which are involved in the β-oxidation sequence for the metabolism of fatty acids (see page 18). Thus, two molecules of acetyl-CoA could participate in a Claisen condensation giving acetoacetyl-CoA, and this reaction could be repeated to generate a poly-β-keto ester of appropriate chain length (Figure 3.1). However, a study of the enzymes involved in fatty acid biosynthesis showed this simple rationalization could not be correct, and a more complex series of reactions was operating. It is now known that fatty acid biosynthesis involves initial carboxylation of acetyl-CoA to malonyl-CoA, a reaction involving ATP, CO_2 (as bicarbonate, HCO_3^-), and the coenzyme biotin as the carrier of CO_2 (see page 17).

The conversion of acetyl-CoA into malonyl-CoA increases the acidity of the α-hydrogens, and thus provides a better nucleophile for the Claisen condensation. In the biosynthetic sequence, no acylated malonic acid derivatives are produced, and no label from [^{14}C]bicarbonate is incorporated, so the carboxyl group introduced into malonyl-CoA is simultaneously lost by a decarboxylation reaction during the Claisen condensation (Figure 3.1). Accordingly, the carboxylation step helps to activate the α-carbon and facilitate Claisen condensation, and the carboxyl is immediately removed on completion of this task. An alternative rationalization is that decarboxylation of the malonyl ester is used to generate the acetyl enolate anion without any requirement for a strong base.

The pathways to fatty acids and aromatic polyketides branch early. The chain extension process of Figure 3.1 continues for aromatics,

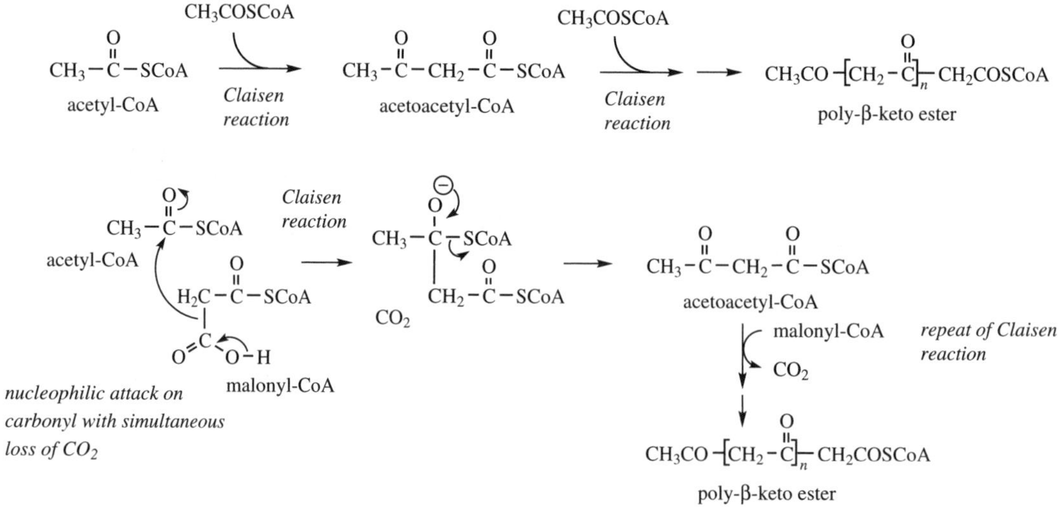

Figure 3.1

generating a highly reactive poly-β-keto chain, which has to be stabilized by association with groups on the enzyme surface until chain assembly is complete and cyclization reactions occur. However, for fatty acids, the carbonyl groups are reduced before attachment of the next malonate group. Partial reduction processes, leading to a mixture of methylenes, hydroxyls, and carbonyls, are characteristic of macrolides (see page 92).

SATURATED FATTY ACIDS

The processes of fatty acid biosynthesis are well studied and are known to be catalysed by the enzyme **fatty acid synthase**. In animals, this is a multifunctional protein containing all of the catalytic activities required, whilst bacteria and plants utilize an assembly of separable enzymes. Acetyl-CoA and malonyl-CoA themselves are not involved in the condensation step: they are converted into enzyme-bound thioesters, the malonyl ester by means of an acyl carrier protein (ACP) (Figure 3.2). The Claisen reaction follows giving acetoacetyl-ACP (β-keto acyl-ACP; R=H), which is reduced stereospecifically to the corresponding β-hydroxy ester, consuming NADPH in the reaction. Then follows elimination of water giving the E (*trans*) α,β-unsaturated ester. Reduction of the double bond again utilizes NADPH and generates a saturated acyl-ACP (fatty

acyl-ACP; R=H) which is two carbons longer than the starting material. This can feed back into the system, condensing again with malonyl-ACP, and going through successive reduction, dehydration, and reduction steps, gradually increasing the chain length by two carbons for each cycle, until the required chain length is obtained. At that point, the fatty acyl chain can be released as a fatty acyl-CoA or as the free acid. The chain length actually elaborated is probably controlled by the specificity of the thioesterase enzymes that subsequently catalyse release from the enzyme.

The fatty acid synthase protein is known to contain an acyl carrier protein binding site, and also an active site cysteine residue in the β-ketoacyl synthase domain. Acetyl and malonyl groups are successively transferred from coenzyme A esters and attached to the thiol groups of Cys and ACP (Figure 3.3). The Claisen condensation occurs, and the processes of reduction, dehydration, and reduction then occur whilst the growing chain is attached to ACP. The ACP carries a phosphopantatheine group exactly analogous to that in coenzyme A, and this provides a long flexible arm, enabling the growing fatty acid chain to reach the active site of each enzyme in the complex, allowing the different chemical reactions to be performed without releasing intermediates from the enzyme (compare polyketide synthesis page 62 and peptide synthesis, page 421). Then the chain is transferred to the thiol of Cys, and the process can

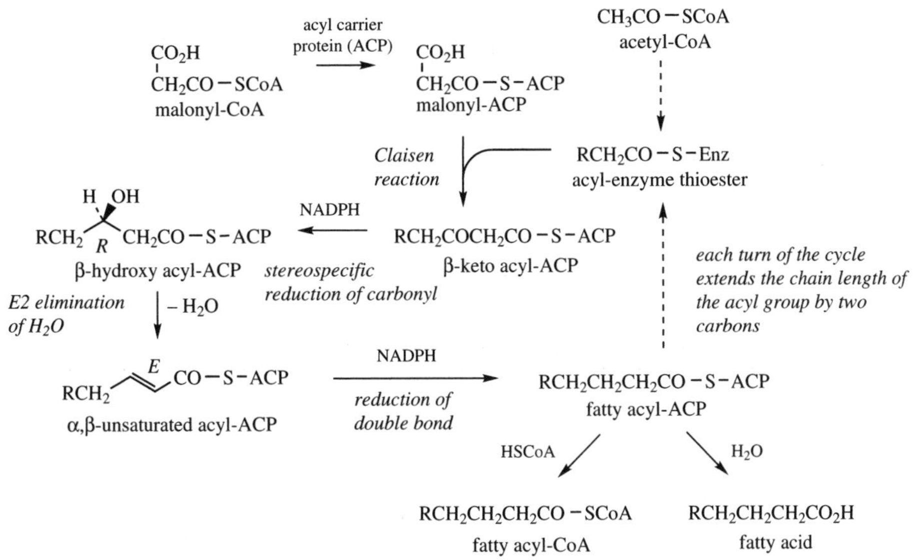

Figure 3.2

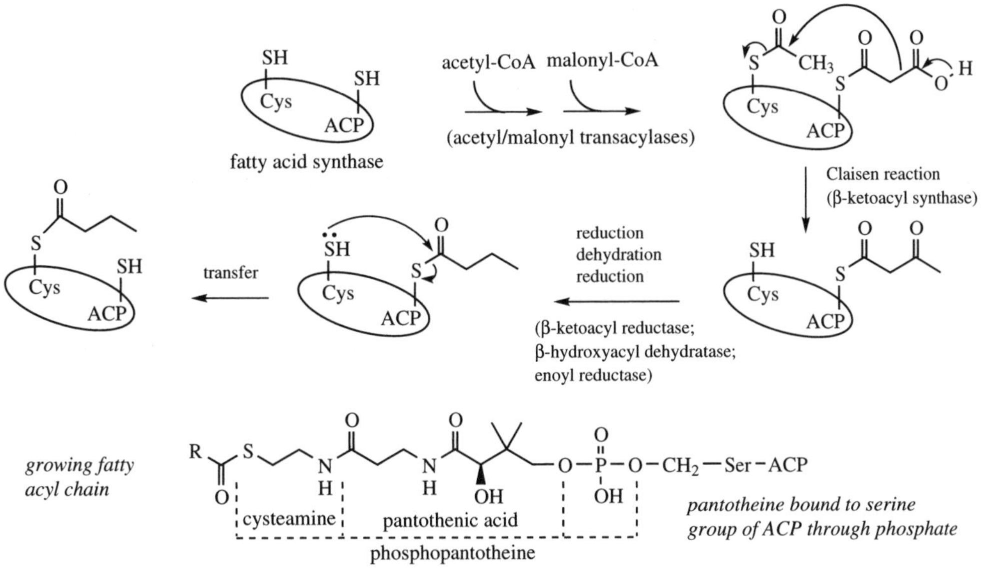

Figure 3.3

continue. Making the process even more efficient, animal fatty acid synthase is a dimeric protein containing two catalytic centres and is able to generate two growing chains. The monomeric subunits are also arranged head to tail so that the acyl group of one unit actually picks up a malonyl extender from the other unit (Figure 3.4). Note that the sequence of enzyme activities along the protein chain of the enzyme complex does

not correspond with the order in which they are employed.

Thus, combination of one acetate starter unit with seven malonates would give the C_{16} fatty acid, palmitic acid, and with eight malonates the C_{18} fatty acid, stearic acid. Note that the two carbons at the head of the chain (methyl end) are provided by acetate, not malonate, whilst the remainder are derived from malonate, which itself

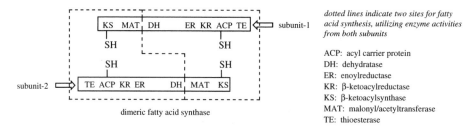

dotted lines indicate two sites for fatty acid synthesis, utilizing enzyme activities from both subunits

ACP: acyl carrier protein
DH: dehydratase
ER: enoylreductase
KR: β-ketoacylreductase
KS: β-ketoacylsynthase
MAT: malonyl/acetyltransferase
TE: thioesterase

dimeric fatty acid synthase

Figure 3.4

Table 3.1 Common naturally occurring fatty acids

Saturated

butyric	$CH_3(CH_2)_2CO_2H$	4:0	stearic	$CH_3(CH_2)_{16}CO_2H$	18:0
caproic*	$CH_3(CH_2)_4CO_2H$	6:0	arachidic	$CH_3(CH_2)_{18}CO_2H$	20:0
caprylic*	$CH_3(CH_2)_6CO_2H$	8:0	behenic	$CH_3(CH_2)_{20}CO_2H$	22:0
capric*	$CH_3(CH_2)_8CO_2H$	10:0	lignoceric	$CH_3(CH_2)_{22}CO_2H$	24:0
lauric	$CH_3(CH_2)_{10}CO_2H$	12:0	cerotic	$CH_3(CH_2)_{24}CO_2H$	26:0
myristic	$CH_3(CH_2)_{12}CO_2H$	14:0	montanic	$CH_3(CH_2)_{26}CO_2H$	28:0
palmitic	$CH_3(CH_2)_{14}CO_2H$	16:0	melissic	$CH_3(CH_2)_{28}CO_2H$	30:0

*To avoid confusion, systematic nomenclature (hexanoic, octanoic, decanoic) is recommended

Unsaturated

palmitoleic	$CH_3(CH_2)_5CH=CH(CH_2)_7CO_2H$	16:1 (9c)
oleic	$CH_3(CH_2)_7CH=CH(CH_2)_7CO_2H$	18:1 (9c)
cis-vaccenic	$CH_3(CH_2)_5CH=CH(CH_2)_9CO_2H$	18:1 (11c)
linoleic	$CH_3(CH_2)_4CH=CHCH_2CH=CH(CH_2)_7CO_2H$	18:2 (9c,12c)
α-linolenic	$CH_3CH_2CH=CHCH_2CH=CHCH_2CH=CH(CH_2)_7CO_2H$	18:3 (9c,12c,15c)
γ-linolenic	$CH_3(CH_2)_4CH=CHCH_2CH=CHCH_2CH=CH(CH_2)_4CO_2H$	18:3 (6c,9c,12c)
gadoleic	$CH_3(CH_2)_9CH=CH(CH_2)_7CO_2H$	20:1 (9c)
arachidonic	$CH_3(CH_2)_4CH=CHCH_2CH=CHCH_2CH=CHCH_2CH=CH(CH_2)_3CO_2H$	
		20:4 (5c,8c,11c,14c)
eicosapentaenoic	$CH_3CH_2CH=CHCH_2CH=CHCH_2CH=CHCH_2CH=CHCH_2CH=CH(CH_2)_3CO_2H$	
(EPA)		20:5 (5c,8c,11c,14c,17c)
erucic	$CH_3(CH_2)_7CH=CH(CH_2)_{11}CO_2H$	22:1 (13c)
docosapentaenoic	$CH_3CH_2CH=CHCH_2CH=CHCH_2CH=CHCH_2CH=CHCH_2CH=CH(CH_2)_5CO_2H$	
(DPA)		22:5 (7c,10c,13c,16c,19c)
docosahexaenoic	$CH_3CH_2CH=CHCH_2CH=CHCH_2CH=CHCH_2CH=CHCH_2CH=CHCH_2CH=CH(CH_2)_2CO_2H$	
(DHA)		22:6 (4c,7c,10c,13c,16c,19c)
nervonic	$CH_3(CH_2)_7CH=CH(CH_2)_{13}CO_2H$	24:1 (15c)

all double bonds are Z (*cis*)

Number of carbon atoms

Abbreviations: 18:2 (9c,12c)

Position of double bonds

Stereochemistry of double bonds
(c = *cis/Z*; t = *trans/E*)

Number of double bonds

is produced by carboxylation of acetate. This means that all carbons in the fatty acid originate from acetate, but malonate will only provide the C_2 chain extension units and not the C_2 starter group. The linear combination of acetate C_2 units as in Figure 3.2 explains why the common fatty acids are straight chained and possess an even number of carbon atoms. Natural fatty acids may contain from four to 30, or even more, carbon atoms, the most abundant being those with 16 or 18 carbons. Some naturally occurring fatty acids are shown in Table 3.1. The rarer fatty acids containing an odd number of carbon atoms typically originate from incorporation of a different starter unit, e.g.

propionic acid, or can arise by loss of one carbon from an even-numbered acid.

Fatty acids are mainly found in ester combination with glycerol in the form of triglycerides (Figure 3.5). These materials are called **fats** or **oils**, depending on whether they are solid or liquid at room temperature. If all three esterifying acids are the same, the triglyceride is termed simple, whereas a mixed triglyceride is produced if two or more of the fatty acids are different. Most natural fats and oils are composed largely of mixed triglycerides. In this case, isomers can exist, including potential optical isomers, since if the primary alcohols are esterified with different fatty acids the central carbon of glycerol will become chiral. In practice, only one of each pair of enantiomers is formed in nature. Triglycerides are produced from glycerol 3-phosphate by esterification with fatty

acyl-CoA residues, the phosphate being removed prior to the last esterification (Figure 3.5). The di-acyl ester of glycerol 3-phosphate is also known as a **phosphatidic acid**, and is the basis of phospholipid structures. In these structures, the phosphate is also esterified with an alcohol, which is usually choline, ethanolamine, serine, or *myo*-inositol, e.g. **phosphatidyl choline** (Figure 3.6). **Phospholipids** are important structural components of cell membranes, and because of the polar and non-polar regions in their structure, they have detergent-like properties. They are also able to form liposomes, which have considerable potential as drug delivery systems. A particularly important natural phospholipid is **platelet-activating factor** (**PAF**) (Figure 3.6), which resembles a phosphatidylcholine, though this compound possesses an ether linkage to a long chain fatty alcohol,

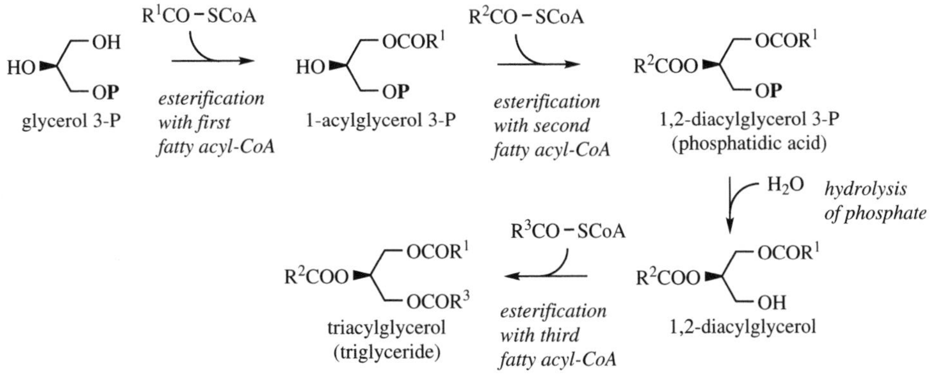

Figure 3.5

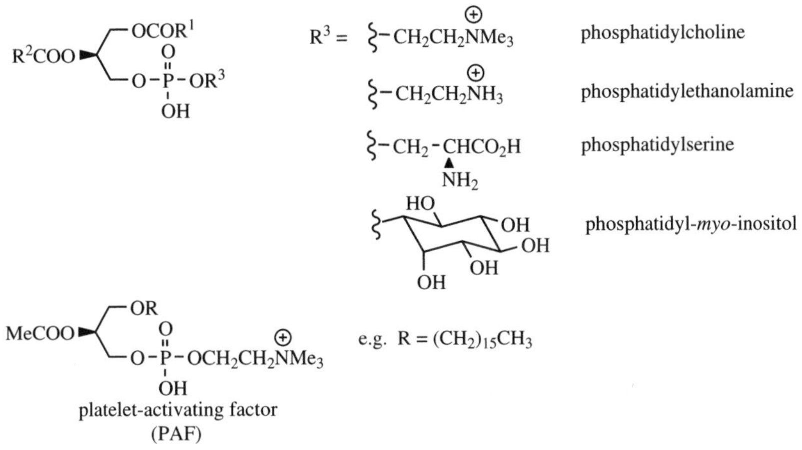

Figure 3.6

usually hexadecanol, rather than an ester linkage. The central hydroxyl of glycerol is esterified, but to acetic acid rather than to a long chain fatty acid. PAF functions at nanomolar concentrations, activates blood platelets and contributes to diverse biological effects, including thrombosis, inflammatory reactions, allergies, and tissue rejection. Long chain alcohols are reduction products from fatty acids and also feature in natural **waxes**. These are complex mixtures of esters of long chain fatty acids, usually $C_{20}-C_{24}$, with long chain monohydric alcohols or sterols.

UNSATURATED FATTY ACIDS

Animal fats contain a high proportion of glycerides of saturated fatty acids and tend to be solids, whilst those from plants and fish contain predominantly unsaturated fatty acid esters and tend to be liquids. Some of the common naturally occurring unsaturated fatty acids are also included in Table 3.1. A convenient shorthand representation for fatty acids indicating chain length with number, position and stereochemistry of double bonds is also presented in Table 3.1. A less systematic numbering starting from the methyl (the ω end) may also be encountered. Major groups of fatty acids are designated ω-3 (omega-3), ω-6 (omega-6), ω-9 (omega-9), etc (or sometimes n-3, n-6, n-9), if there is a double bond that number of carbons from the methyl terminus. This has some value in relating structures when an unsaturated fatty acid is biosynthetically elongated from the carboxyl end as during prostaglandin biosynthesis (see page 45). Double bonds at position 9 are common, but unsaturation can occur at other positions in the chain. Polyunsaturated fatty acids tend to have their double bonds in a non-conjugated array as a repeating unit $-(CH=CHCH_2)_n-$. In virtually all cases, the stereochemistry of the double bond is Z (*cis*), introducing a 'bend' into the alkyl chain. This interferes with the close association and aggregation of molecules that is possible in saturated structures, and helps to maintain the fluidity in oils and cellular membranes.

Fats and oils represent long term stores of energy for most organisms, being subjected to oxidative metabolism as required. Major oils which are produced commercially for use as foods, toiletries, medicinals, or pharmaceutical formulation aids are listed in Table 3.2. Typical fatty acid analyses are shown, though it must be appreciated that these figures can vary quite widely. For instance, plant oils show significant variation according to the climatic conditions under which the plant was grown. In colder climates, a higher proportion of polyunsaturated fatty acids is produced, so that the plant can maintain the fluidity of its storage fats and membranes. The melting points of these materials depend on the relative proportions of the various fatty acids, reflecting primarily the chain length and the amount of unsaturation in the chain. Saturation, and increasing chain length in the fatty acids gives a more solid fat at room temperature. Thus, butterfat and cocoa butter (theobroma oil) contain a relatively high proportion of saturated fatty acids and are solids. Palm kernel and coconut oils are both semi-solids having a high concentration of the saturated C_{12} acid **lauric acid**. A characteristic feature of olive oil is its very high **oleic acid** (18:1) content, whilst rapeseed oil possesses high concentrations of long chain C_{20} and C_{22} fatty acids, e.g. **erucic acid** (22:1). Typical fatty acids in fish oils have high unsaturation and also long chain lengths, e.g. **eicosapentaenoic acid (EPA)** (20:5) and **docosahexaenoic acid (DHA)** (22:6) in cod liver oil.

Unsaturated fatty acids can arise by more than one biosynthetic route, but in most organisms the common mechanism is by desaturation of the corresponding alkanoic acid, with further desaturation in subsequent steps. Most eukaryotic organisms possess a Δ^9-desaturase that introduces a *cis* double bond into a saturated fatty acid, requiring O_2 and NADPH or NADH cofactors. The mechanism of desaturation does not involve any intermediates hydroxylated at C-9 or C-10, and the requirement for O_2 is as an acceptor at the end of an electron transport chain. A stearoyl (C_{18}) thioester is the usual substrate giving an oleoyl derivative (Figure 3.7), coenzyme A esters being utilized by animal and fungal enzymes, and ACP esters by plant systems. The position of further desaturation then depends very much on the organism. Non-mammalian enzymes tend to introduce additional double bonds between the existing double bond and the methyl terminus, e.g. oleate → linoleate → α-linolenate. Animals always introduce new double bonds towards the carboxyl group. Thus **oleate** is desaturated to

Table 3.2 Fixed oils and fats

The term fat or oil has no precise significance, and merely describes whether the material is a solid (fat) or liquid (oil) at room temperature. Most commercial oils are obtained from plant sources, particularly seeds and fruits, and the oil is extracted by cold or hot expression, or less commonly by solvent extraction with hexane. The crude oil is then refined by filtration, steaming, neutralization to remove free acids, washing, and bleaching as appropriate. Many food oils are partially hydrogenated to produce semi-solid fats. Animal fats and fish oils are usually extracted by steaming, the higher temperature deactivating enzymes that would otherwise begin to hydrolyse the glycerides.

Oils and fats feature as important food components and cooking oils, some 80% of commercial production being used as human food, whilst animal feeds account for another 6%. Most of the remaining production is used as the basis of soaps, detergents, and pharmaceutical creams and ointments. A number of oils are used as diluents (carrier or base oils) for the volatile oils employed in aromatherapy.

Oil	Source	Part used	Oil content[†] (%)	Typical fatty acid composition[†] (%)	Uses, notes
Almond	*Prunus amygdalus* var. *dulcis*, or var. *amara* (Rosaceae)	seed	40–55	oleic (62–86), linoleic (7–30), palmitic (4–9), stearic (1–2)	emollient base, toiletries, carrier oil (aromatherapy)
Arachis (groundnut, peanut)	*Arachis hypogaea* (Leguminosae/ Fabaceae)	seed	45–55	oleic (35–72), linoleic (13–43), palmitic (7–16), stearic (1–7), behenic (1–5), arachidic (1–3)	food oil, emollient base
Borage	*Borago officinalis* (Boraginaceae)	seed	28–35	linoleic (38), γ-linolenic (23–26), oleic (16), palmitic (11)	dietary supplement for γ-linolenic acid content (see page 46)
Butterfat	cow *Bos taurus* (Bovidae)	milk	2–5	palmitic (29), oleic (28), stearic (13), myristic (12), butyric (4), lauric (3), caproic (2), capric (2), palmitoleic (2)	food
Castor	*Ricinus communis* (Euphorbiaceae)	seed	35–55	ricinoleic (80–90), oleic (4–9), linoleic (2–7), palmitic (2–3), stearic (2–3)	emollient base, purgative, soap manufacture Castor seeds contain the highly toxic, but heat-labile protein ricin (see page 434)

(Continued overleaf)

Table 3.2 (*Continued*)

Oil	Source	Part used	Oil content† (%)	Typical fatty acid composition† (%)	Uses, notes
Coconut	*Cocos nucifera* (Palmae/ Arecaceae)	seed kernel	65–68	lauric (43–53), myristic (15–21), palmitic (7–11), caprylic (5–10), capric (5–10), oleic (6–8), stearic (2–4)	soaps, shampoos Fractionated coconut oil containing only short to medium length fatty acids (mainly caprylic and capric) is a dietary supplement
Cod-liver	cod *Gadus morrhua* (Gadidae)	fresh liver	50	oleic (24), DHA (14), palmitic (11), EPA (6), palmitoleic (7), stearic (4), myristic (3)	dietary supplement due to presence of EPA and DHA, plus vitamins A (see page 230) and D (see page 259); halibut-liver oil from halibut *Hippoglossus vulgaris* (Pleurnectideae) has similar properties and is used in the same way
Cottonseed	*Gossypium hirsutum* (Malvaceae)	seed	15–36	linoleic (33–58), palmitic (17–29), oleic (13–44), stearic (1–4)	solvent for injections, soaps Cotton seeds also contain 1.1–1.3% gossypol (see page 200) and small amounts of cyclopropenoid fatty acids, e.g. sterculic and malvalic acids (see page 50)
Evening primrose	*Oenothera biennis* (Onagraceae)	seed	24	linoleic (65–80), γ-linolenic (7–14), oleic (9), palmitic (7)	dietary supplement for γ-linolenic acid content (see page 46)
Honesty	*Lunaria annua* (Cruciferae/ Brassicaceae)	seed	30–40	erucic (43), nervonic (25), oleic (24)	nervonic acid is being investigated for the treatment of multiple sclerosis; the disease is characterized by a deficiency in nervonic acid

Lard	pig *Sus scrofa* (Suidae)	abdominal fat		oleic (45), palmitic (25), stearic (12), linoleic (10), palmitoleic (3)	foods
Linseed (flaxseed)	*Linum usitatissimum* (Linaceae)	seed	35–44	α-linolenic (30–60), oleic (39), linoleic (15), palmitic (7), stearic (4)	liniments, dietary supplement for α-linolenic acid content Formerly the basis of paints, reacting with oxygen, polymerizing, and drying to a hard film
Maize (corn)	*Zea mays* (Graminae/ Poaceae)	embryo	33–39	linoleic (34–62), oleic (19–50), palmitic (8–19), stearic (0–4)	food oil, dietary supplement, solvent for injections
Olive	*Olea europaea* (Oleaceae)	fruits	15–40	oleic (56–85), palmitic (8–20), linoleic (4–20), stearic (1–4)	food oil, emollient base
Palm kernel	*Elaeis guineensis* (Palmae/ Arecaceae)	kernel	45–50	lauric (40–52), myristic (14–18), oleic (9–16), palmitoleic (6–10), caprylic (3–6), capric (3–5), stearic (1–4), linoleic (1–3)	soaps Fractionated palm oil is a solid obtained by fractionation and hydrogenation and is used as a suppository base
Rapeseed	*Brassica napus* (Cruciferae/ Brassicaceae)	seed	40–50	erucic (30–60), oleic (9–25), linoleic (11–25), gadoleic (5–15), α-linolenic (5–12), palmitic (0–5)	food oil, using varieties producing lower levels of erucic acid where the main components are now oleic (48–60%), linoleic (18–30%), α-linolenic (6–14%), and palmitic (3–6%) acids Erucic acid is used as a plasticizer in PVC clingfilm

(*Continued overleaf*)

Table 3.2 (*Continued*)

Oil	Source	Part used	Oil content[†] (%)	Typical fatty acid composition[†] (%)	Uses, notes
Sesame	*Sesamum indicum* (Pedaliaceae)	seed	44–54	oleic (35–50), linoleic (35–50), palmitic (7–12), stearic (4–6)	food oil, soaps, solvent for injections, carrier oil (aromatherapy)
Soya (soybean)	*Glycine max* (Leguminosae/ Fabaceae)	seed	18–20	linoleic (44–62%), oleic (19–30), palmitic (7–14), α-linolenic (4–11), stearic (1–5)	food oil, dietary supplement, carrier oil (aromatherapy) Soya oil contains substantial amounts of the sterols sitosterol and stigmasterol (see page 256)
Suet (mutton tallow)	sheep *Ovis aries* (Bovidae)	abdominal fat		stearic (32), oleic (31), palmitic (27), myristic (6)	foods
Suet (beef tallow)	cow *Bos taurus* (Bovidae)	abdominal fat		oleic (48), palmitic (27), palmitoleic (11), stearic (7), myristic (3)	foods
Sunflower	*Helianthus annuus* (Compositae/ Asteraceae)	seed	22–36	linoleic (50–70), oleic (20–40), palmitic (3–10), stearic (1–10)	food oil, carrier oil (aromatherapy)
Theobroma	*Theobroma cacao* (Sterculiaceae)	kernel	35–50	oleic (35), stearic (35), palmitic (26), linoleic (3)	suppository base, chocolate manufacture Theobroma oil (cocoa butter) is a solid

[†]The oil yields and fatty acid compositions given in the table are typical values, and can vary widely. The quality of an oil is determined principally by its fatty acid analysis. Structures of the fatty acids are shown in Table 3.1 (see page 38).

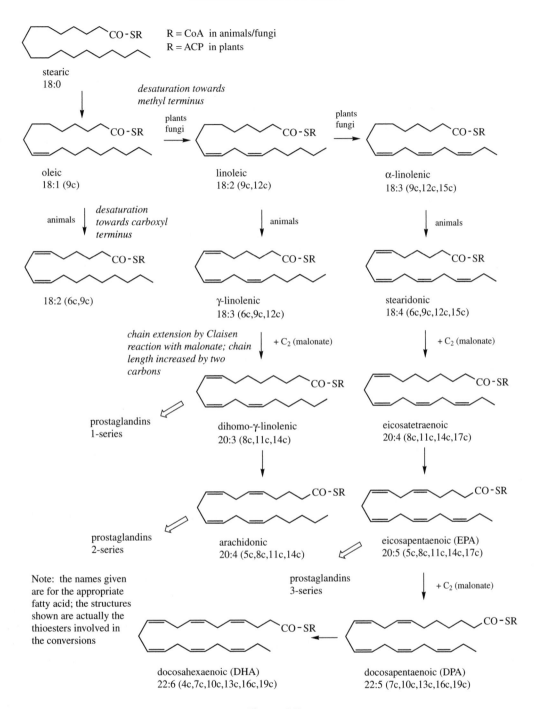

Figure 3.7

$\Delta^{6,9}$-octadecadienoate rather than linoleate. However, animals need **linoleate** for the biosynthesis of **dihomo-γ-linolenate** ($\Delta^{8,11,14}$-eicosatrienoate) and **arachidonate** ($\Delta^{5,8,11,14}$-eicosatetraenoate), C_{20} polyunsaturated fatty acid precursors of prostaglandins in the 'one' and 'two' series respectively (see page 52). Accordingly, linoleic acid must be obtained from plant material in the diet, and it is desaturated towards the carboxyl to yield **γ-linolenate**, which is then used as the substrate for further chain extension, adding a C_2 unit from malonate, and producing dihomo-γ-linolenate. Arachidonate derives from this by additional desaturation, again towards the carboxyl end of the chain (Figure 3.7). **α-Linolenate** is similarly a precursor on the way to $\Delta^{5,8,11,14,17}$-**eicosapentaenoate (EPA)**, required for the synthesis of prostaglandins of the 'three' series, and it is also obtained from the diet. A similar chain extension process using further molecules of malonate is encountered in the sequence from α-linolenate in animal systems (Figure 3.7). Chain extension/dehydrogenations lead to formation of eicosapentaenoate (EPA) with further elaborations producing **docosapentaenoate (DPA)** and then **docosahexaenoate (DHA)**. DHA is a component of lipids in sperm, the retina, and the brain. It is thought to be important for brain development, and deficiency is associated with abnormalities in brain function. Linoleate and α-linolenate are referred to as '**essential fatty acids**' (**EFAs**) since they and their metabolites are required in the diet for normal good health. Some food sources such as the oils present in fish are rich in the later metabolites derived from α-linolenic acid, e.g. EPA and DHA, and are also beneficial to health. Since these fatty acids all have a double bond three carbons from the methyl end of the chain, they are grouped together under the term ω-3 fatty acids (**omega-3 fatty acids**). Regular consumption of fish oils is claimed to reduce the risk of heart attacks and atherosclerosis.

Although most plant-derived oils contain high amounts of unsaturated fatty acid glycerides, including those of linoleic and α-linolenic acids, the conversion of linoleate into **γ-linolenate** can be blocked or inhibited in certain conditions in humans. This restricts synthesis of prostaglandins. In such cases, the use of food supplements, e.g. evening primrose oil* from *Oenothera biennis* (Onagraceae), which are rich in γ-linolenic esters, can be valuable and help in the disorder. Many plants in the Boraginaceae, e.g. borage (*Borago afficinalis*), also accumulate significant amounts of γ-linolenic acid glycerides, as does evening primrose, indicating their unusual ability

---

Evening Primrose Oil

Evening primrose oil is extracted from the seeds of selected strains of the evening primrose (*Oenothera biennis*; Onagraceae), a biennial plant native to North America, which is now widely cultivated in temperate countries. The seeds contain about 24% fixed oil, which has a high content of glycerides of the unsaturated fatty acids linoleic acid (65–80%) and γ-linolenic acid (**gamolenic acid**) (7–14%). Because of this high γ-linolenic acid content, evening primrose oil is widely used as a dietary supplement, providing additional quantities of this essential fatty acid, which is a precursor in the biosynthesis of prostaglandins, which regulate many bodily functions (see page 54). Genetic and a number of other factors may inhibit the desaturation of linoleic acid into γ-linolenic acid. Ageing, diabetes, excessive alcohol intake, catecholamines, and zinc deficiency have all been linked to inhibition of the desaturase enzyme. The conversion may also be inhibited if there is a high proportion of fatty acids in the diet, which compete for the desaturase enzyme, including saturated and *trans*-unsaturated fatty acids. The latter group may be formed during the partial hydrogenation of polyunsaturated fatty acids which is commonly practised during food oil processing to produce semi-solid fats. Evening primrose oil appears to be valuable in the treatment of premenstrual tension, multiple sclerosis, breast pain (mastalgia), and perhaps also in eczema. There is potential for further applications, e.g. in diabetes, alcoholism, and cardiovascular disease. In evening primrose, γ-linolenic acid is usually present in the form of a dilinoleoylmono-γ-linolenylglycerol. This triglyceride is also being explored as a drug material for the treatment of diabetes-related neuropathy and retinopathy. γ-Linolenic acid is also found in the fixed oil of other plants, e.g. blackcurrant, comfrey, and borage, and in human milk. **Borage oil (starflower oil)** from the seeds of *Borago officinalis* (Boraginaceae) is used in the same way as evening primrose oil. It contains higher concentrations of γ-linolenic acid (23–26%), but rather less linoleic acid.

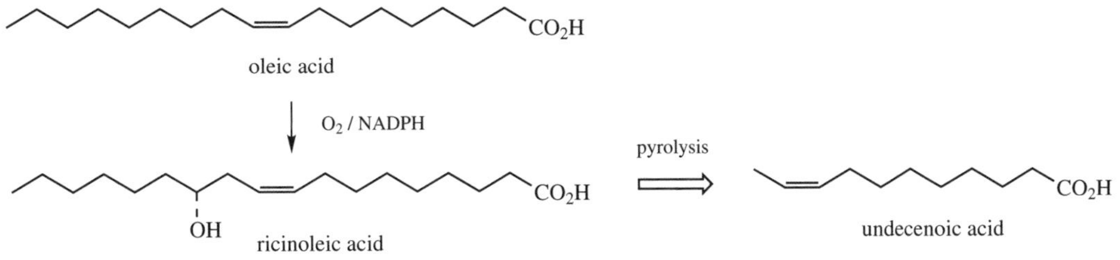

Figure 3.8

to desaturate linoleic esters towards the carboxyl terminus, rather than towards the methyl terminus as is more common in plants. Arachidonic acid itself has not been found in higher plants, but does occur in some algae, mosses, and ferns.

Ricinoleic acid (Figure 3.8) is the major fatty acid found in castor oil from seeds of the castor oil plant (*Ricinus communis*; Euphorbiaceae), and is the 12-hydroxy derivative of oleic acid. It is formed by direct hydroxylation of oleic acid (usually esterified as part of a phospholipid) by the action of an O_2- and NADPH-dependent mixed function oxidase, but this is not of the cytochrome P-450 type. Castor oil has a long history of use as a domestic purgative, but it is now mainly employed as a cream base. **Undecenoic acid** (Δ^9-undecenoic acid) can be obtained from ricinoleic acid by thermal degradation, and as the zinc salt or in ester form is used in fungistatic preparations.

Primary amides of unsaturated fatty acids have been characterized in humans and other mammals, and although their biological role is not fully understood, they may represent a group of important signalling molecules. **Oleamide**, the simple amide of oleic acid, has been shown to be a sleep-inducing lipid, and the amide of erucic acid, **erucamide**, stimulates the growth of blood vessels.

ACETYLENIC FATTY ACIDS

Many unsaturated compounds found in nature contain one or more acetylenic bonds, and these are predominantly produced by further desaturation of olefinic systems in fatty acid-derived molecules. They are surprisingly widespread in nature, and are found in many organisms, but are especially common in plants of the Compositae/Asteraceae, the Umbelliferae/Apiaceae, and fungi of the group

Basidiomycetes. These compounds tend to be highly unstable and some are even explosive if sufficient amounts are accumulated. Since only very small amounts are present in plants, this does not present any widespread hazard. Whilst fatty acids containing several double bonds usually have these in a non-conjugated array, molecules containing triple bonds tend to possess conjugated unsaturation. This gives the compounds intense and highly characteristic UV spectra which aids their detection and isolation.

The processes of desaturation are exemplified in Figure 3.9, in which oleic acid (probably as a thiol ester) features as a precursor of **crepenynic acid** and **dehydrocrepenynic acid**. The acetylenic bond is now indicated by *a* in the semi-systematic shorthand nomenclature. Chain shortening by β-oxidation (see page 18) is often a feature of these pathways, and formation of the C_{10} acetylenic acid **dehydromatricaria acid** proceeds through C_{18} intermediates, losing eight carbons, presumably via four β-oxidations. In the latter part of the pathway, the *Z*-double bond from oleic acid moves into conjugation with the polyacetylene chain via an allylic isomerization, giving the more favoured *E*-configuration. Some noteworthy acetylenic structures (though they are no longer acids and components of fats) are given in Figure 3.10. **Cicutoxin** from the water hemlock (*Cicuta virosa*; Umbelliferae/Apiaceae) and **oenanthotoxin** from the hemlock water dropwort (*Oenanthe crocata*; Umbelliferae/Apiaceae) are extremely toxic to mammals, causing persistent vomiting and convulsions, leading to respiratory paralysis. Ingestion of the roots of these plants may frequently lead to fatal poisoning. **Falcarinol** is a constituent of *Falcaria vulgaris* (Umbelliferae/Apiaceae), *Oenanthe crocata*, *Hedera helix* (Araliaceae), and several other plants, and is known to cause contact dermatitis in certain

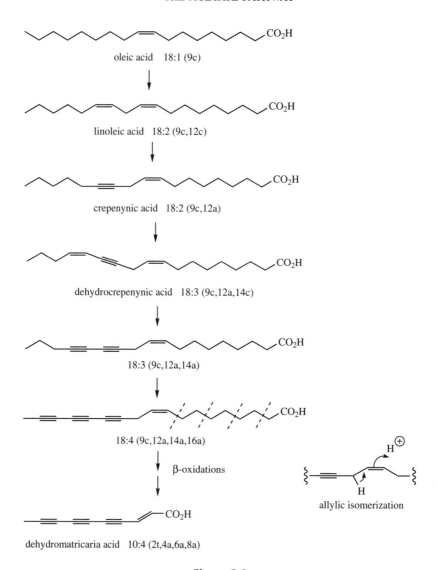

oleic acid   18:1 (9c)

linoleic acid   18:2 (9c,12c)

crepenynic acid   18:2 (9c,12a)

dehydrocrepenynic acid   18:3 (9c,12a,14c)

18:3 (9c,12a,14a)

18:4 (9c,12a,14a,16a)

β-oxidations

allylic isomerization

dehydromatricaria acid   10:4 (2t,4a,6a,8a)

Figure 3.9

individuals when the plants are handled. Falcarinol (sometimes called panaxynol) and the structurally related **panaxytriol** are also characteristic poly-acetylene components of ginseng (*Panax ginseng*; Araliaceae) (see page 222). **Wyerone** from the broad bean (*Vicia faba*; Leguminosae/Fabaceae) has antifungal properties, and its structure exemplifies how the original straight chain may be cross-linked to produce a ring system. The furan ring may originate from a conjugated diyne.

The herbal preparation echinacea* is derived from the roots of *Echinacea purpurea* (Compositae/Asteraceae) and is used for its immunostimulant properties, particularly as a prophylactic and

treatment for the common cold. At least some of its activity arises from a series of alkylamides, amides of polyunsaturated acids with isobutylamine. These acids are predominantly C_{11} and C_{12} diene-diynes (Figure 3.11).

BRANCHED-CHAIN FATTY ACIDS

Whilst straight-chain fatty acids are the most common, branched-chain acids have been found to occur in mammalian systems, e.g. in wool fat and butter fat. They are also characteristic fatty acid constituents of the lipid part of cell walls in some

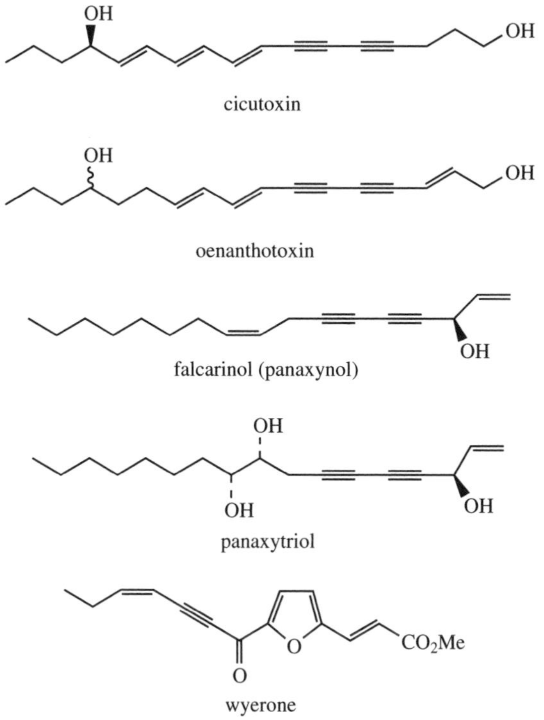

cicutoxin

oenanthotoxin

falcarinol (panaxynol)

panaxytriol

wyerone

Figure 3.10

pathogenic bacteria. Several mechan[...] operate in their formation. Thus, t[...] **corynomycolic acid** from *Coryneb[...] theriae* can be rationalized from a [...] two palmitic acid units (Figure 3.12[...] chains can be introduced by using methylmalonyl-CoA instead of malonyl-CoA as the chain extending agent (Figure 3.13). Methylmalonyl-CoA arises by biotin-dependent carboxylation of propionyl-CoA in exactly the same way as malonyl-CoA was formed (see page 17). **2,4,6,8-Tetramethyldecanoic acid** found in the preen gland wax of the goose (*Anser anser*) is produced from an acetyl-CoA starter, and four methylmalonyl-CoA chain extender units. The incorporation of propionate as well as acetate is also a feature of many microbial antibiotic structures (see page 17). However, in other examples, methyl side-chains can be produced by a *C*-alkylation mechanism using *S*-adenosylmethionine (SAM). **Tuberculostearic acid** (Figure 3.14) found in *Mycobacterium tuberculosis*, the bacterium causing tuberculosis, is derived from oleic acid by alkylation on C-10,

Echinacea

Echinacea consists of the dried roots of *Echinacea purpurea*, *E. angustifolia*, or *E. pallida* (Compositae/Asteraceae), herbaceous perennial plants indigenous to North America, and widely cultivated for their large daisy-like flowers, which are usually purple or pink. Herbal preparations containing the dried root, or extracts derived from it, are promoted as immunostimulants, particularly as prophylactics and treatments for bacterial and viral infections, e.g. the common cold. Tests have validated stimulation of the immune response, though the origins of this activity cannot be ascribed to any specific substance. Activity has variously been assigned to lipophilic alkylamides, polar caffeic acid derivatives, high molecular weight polysaccharide material, or to a combination of these. Compounds in each group have been demonstrated to possess some pertinent activity, e.g. immunostimulatory, anti-inflammatory, antibacterial or antiviral effects. The alkylamides comprise a complex mixture of unsaturated fatty acids as amides with 2-methylpropanamine (isobutylamine) or 2-methylbutanamine, amines which are probably decarboxylation products from valine and isoleucine respectively. The acid portions are predominantly C_{11} and C_{12} diene-diynes or tetraenes (Figure 3.11). These compounds are found throughout the plant though relative proportions of individual components vary considerably. The root of *E. purpurea* contains at least 12 alkylamides (about 0.6%), of which C_{12} diene-diynes predominate; levels of these compounds fall significantly during drying and storage. Caffeic acid derivatives present include caffeic acid (see page 132), chlorogenic acid (5-*O*-caffeoylquinic acid, see page 132), 2-*O*-caffeoyltartaric acid, and cichoric acid (2,3-di-*O*-caffeoyltartaric acid) (Figure 3.11). Cichoric acid is a major component (0.6–2.1%) in *E. purpurea*, but only minor in the other species.

(Continues)

(Continued)

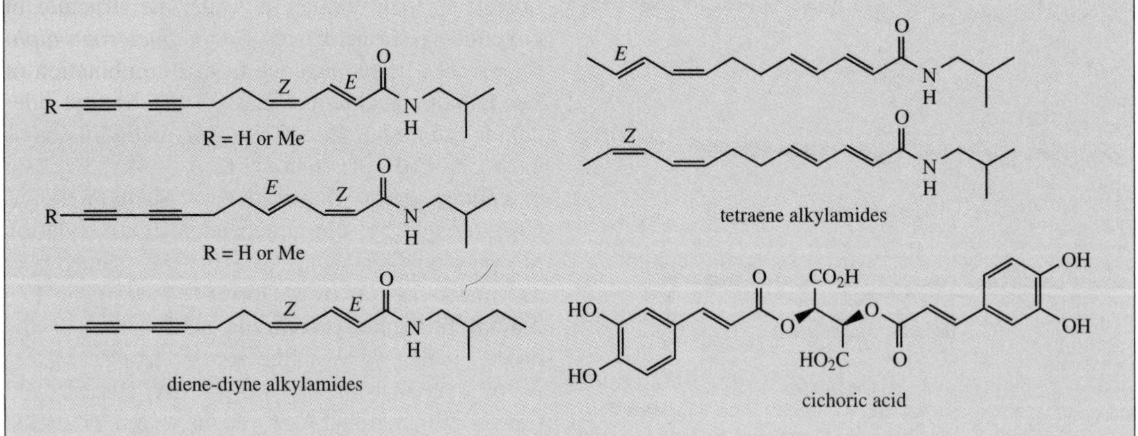
R = H or Me
R = H or Me
diene-diyne alkylamides
tetraene alkylamides
cichoric acid

Figure 3.11

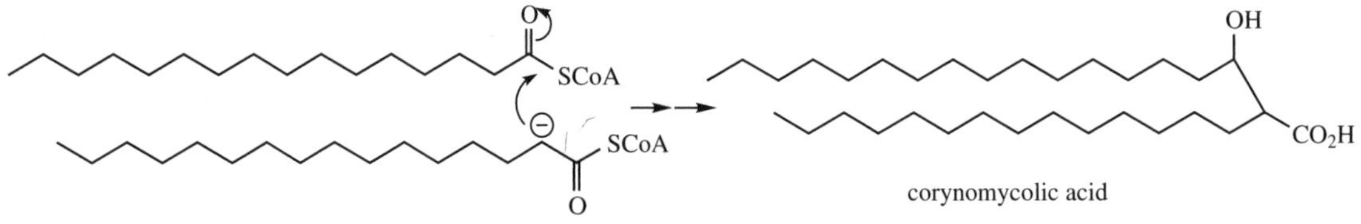
SCoA
SCoA
OH
CO_2H
corynomycolic acid

Figure 3.12

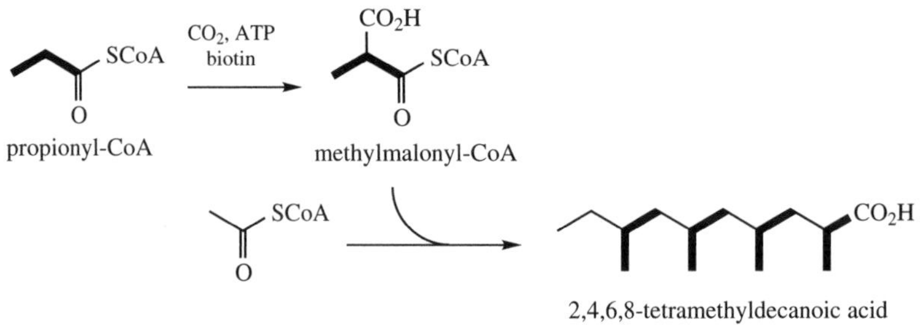
SCoA
CO_2, ATP
biotin
CO_2H
SCoA
propionyl-CoA
methylmalonyl-CoA
SCoA
CO_2H
2,4,6,8-tetramethyldecanoic acid

Figure 3.13

initiated by the double bond electrons. A postulated carbocation intermediate could then be discharged by accepting hydride from NADPH giving tuberculostearic acid. Alternatively, loss of a proton via cyclopropane ring formation could occur giving dihydrosterculic acid. This is known to be dehydrogenated to **sterculic acid**, an unusual fatty acid containing a highly strained cyclopropene ring. Sterculic acid is present in the seed oil from *Sterculia foetida* (Sterculiaceae) and with similar cyclopropene acids, e.g. malvalic acid, is present in edible cottonseed oil from *Gossypium* species (Malvaceae). **Malvalic acid** is produced from sterculic acid by chain shortening from the carboxyl end (Figure 3.14). Sterculic acid is an inhibitor of the Δ^9-desaturase which converts stearic acid into oleic acid and is potentially harmful to humans in that it can alter membrane permeability and inhibit reproduction.

Chaulmoogric and **hydnocarpic acids** (Figure 3.15) are cyclopentenyl fatty acids found in chaulmoogra oil expressed from seeds of *Hydnocarpus wightiana* (Flacourtiaceae). These acids are known to arise by malonate chain extension of the

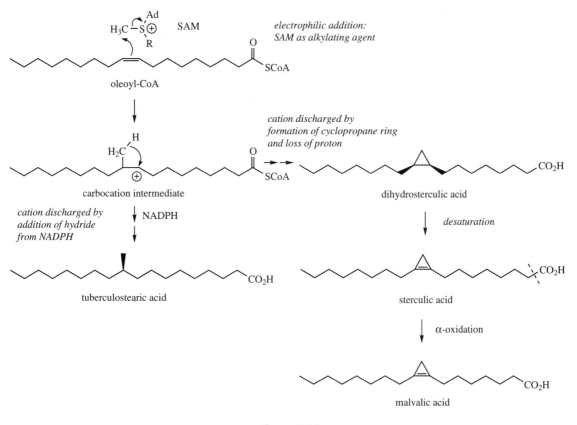

Figure 3.14

Figure 3.15

coenzyme A ester of 2-cyclopentenyl carboxylic acid as an alternative starter unit to acetate, demonstrating a further approach to unusual fatty acids. Chaulmoogra oil provided for many years the only treatment for the relief of leprosy, these two acids being strongly bactericidal towards the leprosy infective agent *Mycobacterium leprae*. Purified salts and esters of hydnocarpic and chaulmoogric acids were subsequently employed, until they were then themselves replaced by more effective synthetic agents.

PROSTAGLANDINS

The prostaglandins* are a group of modified C_{20} fatty acids first isolated from human semen and initially assumed to be secreted by the prostate gland. They are now known to occur widely in animal tissues, but only in tiny amounts, and they have been found to exert a wide variety of pharmacological effects on humans and animals. They are active at very low, hormone-like concentrations and can regulate blood pressure, contractions of smooth muscle, gastric secretion, and platelet aggregation. Their potential for drug use is extremely high, but it has proved difficult to separate the various biological activities into individual agents.

The basic prostaglandin skeleton is that of a cyclized C_{20} fatty acid containing a cyclopentane ring, a C_7 side-chain with the carboxyl function, and a C_8 side-chain with the methyl terminus. Prostaglandins are biosynthesized from three essential fatty acids, $\Delta^{8,11,14}$-**eicosatrienoic acid (dihomo-γ-linolenic acid)**, $\Delta^{5,8,11,14}$-**eicosatetraenoic acid (arachidonic acid)**, and $\Delta^{5,8,11,14,17}$-**eicosapentaenoic acid**, which yield prostaglandins of the 1-, 2-, and 3-series, respectively (Figure 3.16) (see below for principles of nomenclature). The three precursors lead to products of similar structure, but with varying levels of unsaturation in the two side-chains. Some of the structures elaborated from arachidonic acid are shown in Figure 3.17. In the first reaction, arachidonic acid is converted into **prostaglandin G_2 (PGG$_2$)** by an oxygenase (**cyclooxygenase; COX**) enzyme, which incorporates two molecules of oxygen, liberating a compound with both cyclic and acyclic peroxide functions. In arachidonic acid the methylene group flanked by two double bonds is susceptible to oxidation, probably via a free radical process. This may lead to incorporation of oxygen giving the proposed free radical intermediate. Formation of PGG$_2$ is then depicted as a concerted cyclization reaction, initiated by the peroxide radical, in which a second oxygen molecule is incorporated. The

acyclic peroxide group in PGG$_2$ is then cleaved by a peroxidase to yield **prostaglandin H_2 (PGH$_2$)**, which occupies a central role and can be modified in several different ways. These modifications can be rationally accommodated by initial cleavage of the cyclic peroxide to the diradical; alternative ionic mechanisms may also be proposed. Quenching of the free radicals by abstraction of hydrogen atoms gives rise to **prostaglandin $F_{2\alpha}$ (PGF$_{2\alpha}$)**, whilst capture and loss of hydrogen atoms would provide either **prostaglandin E_2 (PGE$_2$)** or **prostaglandin D_2 (PGD$_2$)**. The bicyclic system in **prostaglandin I_2 (PGI$_2$; prostacyclin)** is envisaged as arising by involvement of a side-chain double bond, then loss of a hydrogen atom. Prostaglandin structures representative of the 1-series, e.g. **PGE$_1$**, or of the 3-series, e.g. **PGE$_3$**, can be formed in a similar way from the appropriate fatty acid precursor (Figure 3.16).

The basic skeleton of the prostaglandins is termed **prostanoic acid**, and derivatives of this system are collectively known as prostanoids. The term **eicosanoids** is also used to encompass prostaglandins, thromboxanes, and leukotrienes, which are all derived from C_{20} fatty acids (eicosanoic acids). Semi-systematic nomenclature of prostaglandins is based on the substitution pattern in the five-membered ring, denoted by a letter suffix (Figure 3.18), and the number of double bonds in the side-chains is given by a numerical subscript. Greek letters α and β are used to indicate the configuration at C-9, α indicating the substituent is below the plane (as

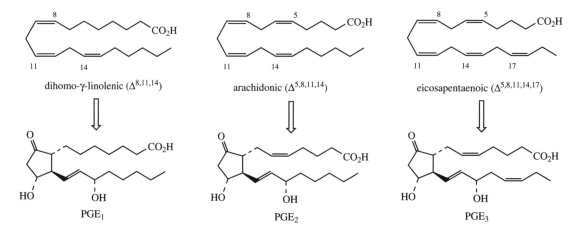

dihomo-γ-linolenic ($\Delta^{8,11,14}$)   arachidonic ($\Delta^{5,8,11,14}$)   eicosapentaenoic ($\Delta^{5,8,11,14,17}$)

PGE$_1$ PGE$_2$ PGE$_3$

Figure 3.16

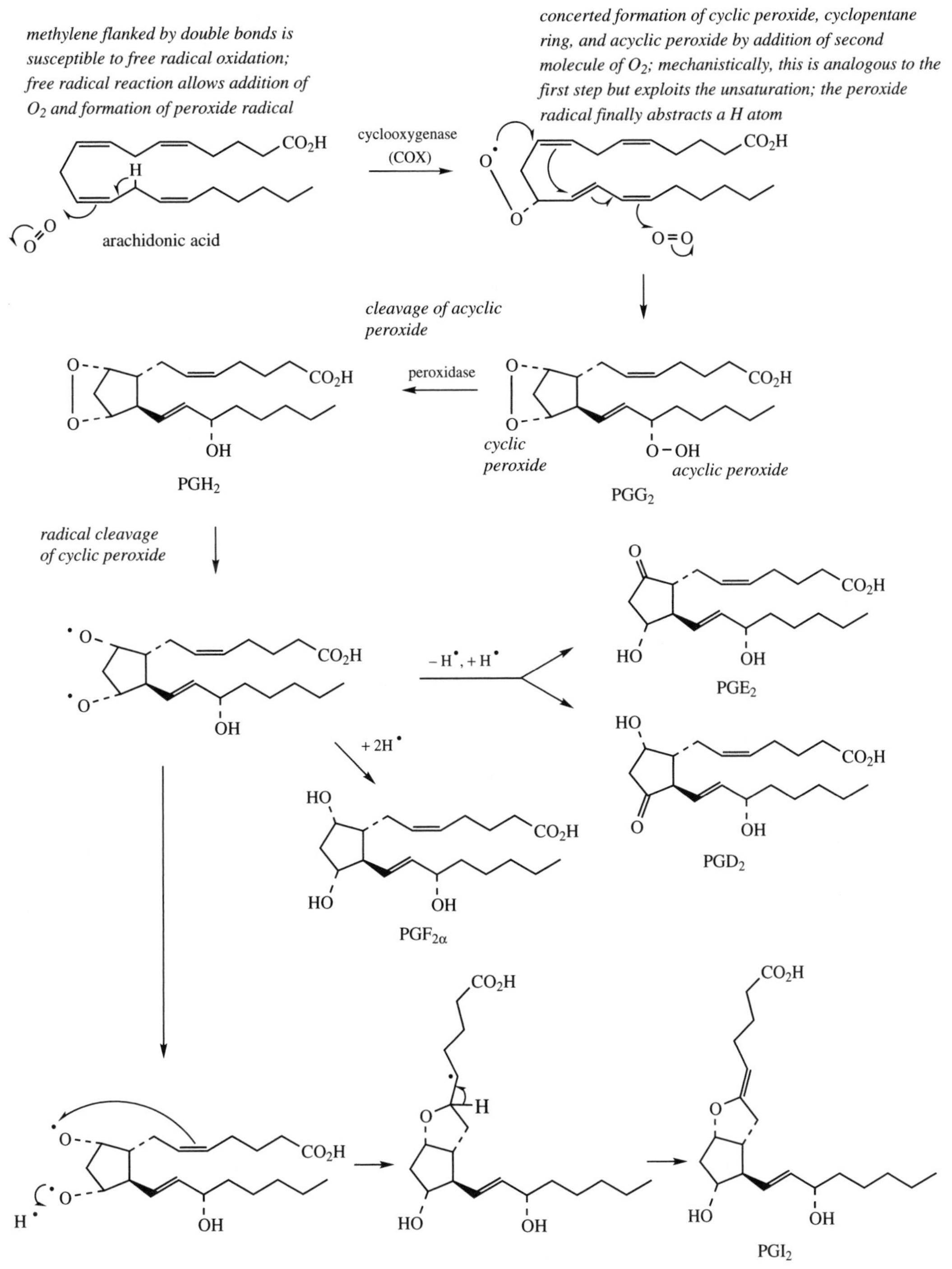

methylene flanked by double bonds is susceptible to free radical oxidation; free radical reaction allows addition of O_2 and formation of peroxide radical

concerted formation of cyclic peroxide, cyclopentane ring, and acyclic peroxide by addition of second molecule of O_2; mechanistically, this is analogous to the first step but exploits the unsaturation; the peroxide radical finally abstracts a H atom

cyclooxygenase (COX)

arachidonic acid

cleavage of acyclic peroxide

peroxidase

PGH$_2$

cyclic peroxide

acyclic peroxide

PGG$_2$

radical cleavage of cyclic peroxide

$- H^{\bullet}, + H^{\bullet}$

PGE$_2$

$+ 2H^{\bullet}$

PGD$_2$

PGF$_{2\alpha}$

PGI$_2$

Figure 3.17

found in natural prostaglandins), and β indicating the substituent is above the plane (as in some synthetic analogues). 'Prostaglandin' is usually abbreviated to PG. Prostaglandins A, B, and C are inactive degradation products from the natural prostaglandins.

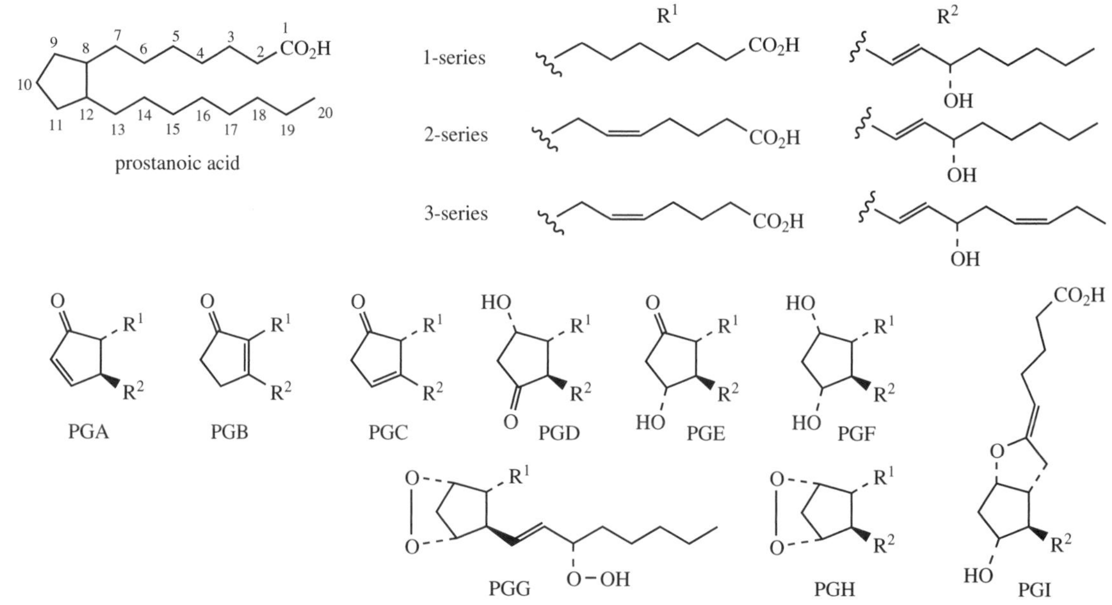

Figure 3.18

Prostaglandins

Prostaglandins occur in nearly all mammalian tissues, but only at very low concentrations. PGE_1 and $PGF_{1\alpha}$ were initially isolated from sheep seminal plasma, but these compounds and PGD_2, PGE_2, and $PGF_{2\alpha}$ are widely distributed. Animal sources cannot supply sufficient amounts for drug usage. The soft coral *Plexaura homomalla* (sea whip) from the Caribbean has been identified as having very high (2–3%) levels of prostaglandin esters, predominantly the C-15 epimer of PGA_2 (1–2%) with related structures. Prostaglandins of the A-, E-, and F-types are widely distributed in soft corals, especially *Plexaura*, but these are unlikely to provide a satisfactory and renewable natural source. Considerable effort has been exerted on the total synthesis of prostaglandins and their interconversions, and the high level of success achieved has opened up the availability of compounds for pharmacological testing and subsequent drug use. Synthetic analogues have also been developed to modify or optimize biological activity. The studies have demonstrated that biological activity is effectively confined to the natural enantiomers; the unnatural enantiomer of PGE_1 had only 0.1% of the activity of the natural isomer.

The prostaglandins display a wide range of pharmacological activities, including contraction and relaxation of smooth muscle of the uterus, the cardiovascular system, the intestinal tract, and of bronchial tissue. They may also inhibit gastric acid secretion, control blood pressure and suppress blood platelet aggregation. Some of these effects are consistent with the prostaglandins acting as second messengers, modulating transmission of hormone stimulation and thus metabolic response. Some prostaglandins in the A and J series have demonstrated potent antitumour properties. Since the prostaglandins control many important physiological processes in animal tissues, their drug potential is high, but the chances of precipitating unwanted side-effects are also high, and this has so far limited their therapeutic use.

(Continues)

(Continued)

There is, however, much additional scope for controlling the production of natural prostaglandins in body tissues by means of specific inhibitors. Indeed it has been found that some established non-steroidal anti-inflammatory drugs (NSAIDs), e.g. aspirin, indometacin, and ibuprofen, inhibit early steps in the prostaglandin biosynthetic pathway that transform the unsaturated fatty acids into cyclic peroxides. Thus aspirin is known to irreversibly inactivate the cyclooxygenase activity (arachidonic acid $\rightarrow$ PGG$_2$), though not the peroxidase activity (PGG$_2$ $\rightarrow$ PGH$_2$), by selective acetylation of a serine residue of the enzyme; ibuprofen and indometacin compete with arachidonic acid at the active site and are reversible inhibitors of the cyclooxygenase. A recent discovery is that two forms of the cyclooxygenase enzyme exist, designated COX-1 and COX-2. COX-1 is expressed constitutively in most tissues and cells and is thought to control synthesis of those prostaglandins important for normal cellular functions such as gastrointestinal integrity and vascular homeostasis. COX-2 is not normally present, but is inducible in certain cells in response to inflammatory stimuli, resulting in enhanced prostaglandin release in the CNS and inflammatory cells with the characteristic inflammatory response. Current NSAIDs do not discriminate between the two COX enzymes, and so this leads to both therapeutic effects via inhibition of COX-2, and adverse effects such as gastrointestinal problems, ulcers, and bleeding via inhibition of COX-1. Because of differences in the nature of the active sites of the two enzymes, it has now been possible to develop agents that can inhibit COX-2 rather than COX-1 as potential new anti-inflammatory drugs. The first of these, meloxicam and rofecoxib, have recently been introduced for relief of pain and inflammation in osteoarthritis. The anti-inflammatory activity of corticosteroids correlates with their preventing the release of arachidonic acid from storage phospholipids, but expression of COX-2 is also inhibited by glucocorticoids.

The role of essential fatty acids (see page 46) such as linoleic and γ-linolenic acids, obtained from plant ingredients in the diet, can now be readily appreciated. Without a source of arachidonic acid, or compounds which can be converted into arachidonic acid, synthesis of prostaglandins would be compromised, and this would seriously affect many normal metabolic processes. A steady supply of prostaglandin precursors is required since prostaglandins are continuously being synthesized and then deactivated. Prostaglandins are rapidly degraded by processes which include oxidation of the 15-hydroxyl to a ketone, reduction of the 13,14-double bond, and oxidative degradation of both side-chains.

A major area of application of prostaglandins as drugs is in obstetrics, where they are used to induce abortions during the early to middle stages of pregnancy, or to induce labour at term. **PGE$_2$ (dinoprostone)** (Figure 3.19) is used in both capacities, whilst **PGF$_{2\alpha}$ (dinoprost)** is now less commonly prescribed and restricted to abortions. PGF$_{2\alpha}$ is rapidly metabolized in body tissues (half-life less than 10 minutes), and the modified version **15-methyl PGF$_{2\alpha}$ (carboprost)** has been developed to reduce deactivation by blocking oxidation at position 15. Carboprost is produced by oxidizing the 15-hydroxyl in a suitably-protected PGF$_{2\alpha}$, then alkylating the 15-carbonyl with a Grignard reagent. Carboprost is effective at much reduced dosage compared with dinoprost, and is of value in augmenting labour at term, especially in cases where ergometrine (see page 375) or oxytocin (see page 415) are ineffective. **Gemeprost** is another unnatural structure and is used to soften and dilate the cervix in early abortions. These agents are usually administered vaginally.

PGE$_1$ (alprostadil) differs from PGE$_2$ by having unsaturation in only one side-chain. Though having effects on uterine muscle, it also has vasodilator properties, and these are exploited for maintaining new-born infants with congenital heart defects, facilitating blood oxygenation prior to corrective surgery. The very rapid metabolism of PGE$_1$ means this drug must be

(Continues)

(Continued)

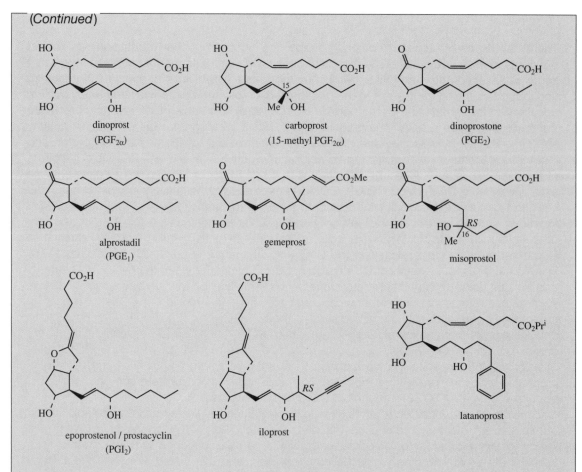

Figure 3.19

delivered by continuous intravenous infusion. Alprostadil is also of value in male impotence, self-injectable preparations being used to achieve erection of the penis. An interesting modification to the structure of PGE$_1$ is found in the analogue **misoprostol**. This compound has had the oxygenation removed from position 15, transferred to position 16, plus alkylation at position 16 to reduce metabolism (compare 15-methyl PGF$_{2\alpha}$ above). These modifications result in an orally active drug which inhibits gastric secretion effectively and can be used to promote healing of gastric and duodenal ulcers. In combination with non-specific NSAIDs, it can significantly lower the incidence of gastrointestinal side-effects such as ulceration and bleeding.

PGI$_2$ (epoprostenol, prostacyclin) reduces blood pressure and also inhibits platelet aggregation by reducing calcium concentrations. It is employed to inhibit blood clotting during renal dialysis, but its very low half-life (about 3 minutes) again necessitates continuous intravenous administration. The tetrahydrofuran ring is part of an enol ether and is readily opened by hydration, leading to 6-ketoprostaglandin F$_{1\alpha}$ (Figure 3.20). **Iloprost** (Figure 3.19) is a stable carbocyclic analogue of potential use in the treatment of thrombotic diseases.

Latanoprost (Figure 3.19) is a recently introduced prostaglandin analogue which increases the outflow of aqueous humour from the eye. It is thus used to reduce intraocular pressure in the treatment of the eye disease glaucoma.

(Continues)

(Continued)

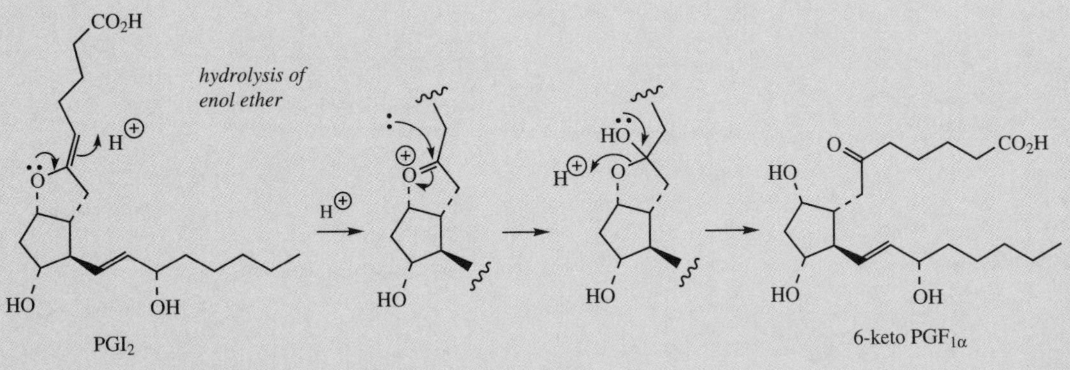

Figure 3.20

Isoprostanes

Isoprostanes represent a new class of prostaglandin-like compounds produced *in vivo* in humans and animals by non-enzymic free-radical-mediated oxidation of membrane-bound polyunsaturated fatty acids. An isomer of $PGF_{2\alpha}$ in which the two alkyl substituents on the five-membered ring were arranged *cis* rather than *trans* was detected in human urine and was the first of these compounds to be characterized. This compound was initially termed 8-*iso*-$PGF_{2\alpha}$, or 8-*epi*-$PGF_{2\alpha}$, though as many more variants in the isoprostane series were discovered it is now termed $iPF_{2\alpha}$-III (Figure 3.21). The last figure refers to the compound being of type III, with eight types being differentiated by the nature of the non-carboxylic chain. Compounds may be formed from linolenic acid and γ-linolenic acid, as well as from arachidonic, eicosapentaenoic, and dihomo-γ-linolenic acids. Structural characteristics of the four classes of isoprostanes derived from arachidonic acid are shown in Figure 3.22; the letter code as in prostaglandin nomenclature is used to define the ring substitution pattern. The four types of isoprostane shown in Figure 3.22 can be viewed as arising by a free radical mechanism which resembles the enzyme-mediated formation of prostaglandins shown in Figure 3.17. The varying side-chain substituents arise by utilizing different double bonds from the several available in the cyclization mechanism, and incorporating an oxygen atom from molecular oxygen at different positions. Many variants are produced because chemical processes rather than enzyme-controlled processes are employed. Free-radical-derived isomers of leukotrienes and thromboxanes have also been reported.

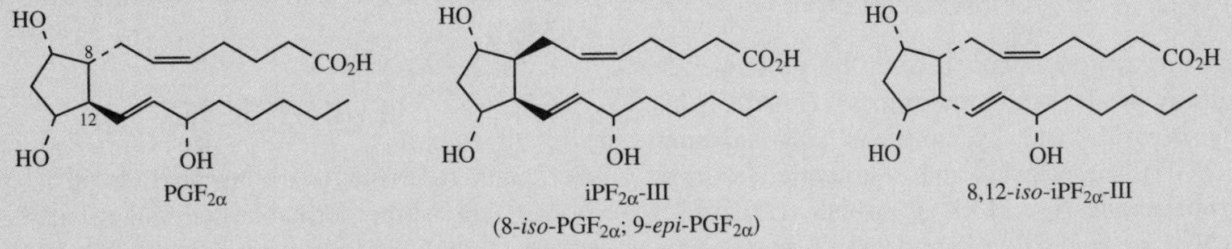

PGF$_{2\alpha}$

iPF$_{2\alpha}$-III
(8-*iso*-PGF$_{2\alpha}$; 9-*epi*-PGF$_{2\alpha}$)

8,12-*iso*-iPF$_{2\alpha}$-III

Figure 3.21

(Continues)

(*Continued*)

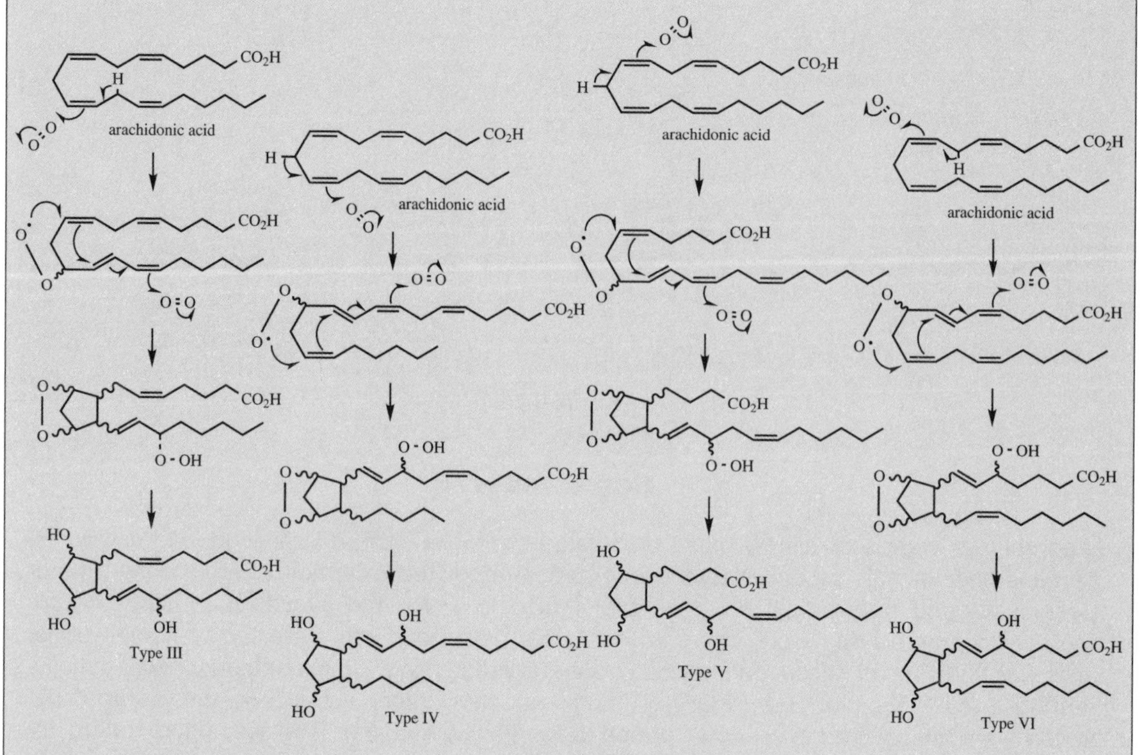

Figure 3.22

Interest in these isoprostanoid derivatives stems partly from the finding that certain compounds possess biological activity, probably via interaction with receptors for prostaglandins. For example, iPF$_{2\alpha}$-III is a potent vasoconstrictor and also aggregates platelets, whilst 8,12-*iso*-iPF$_{2\alpha}$-III (Figure 3.21) possesses activity similar to PGF$_{2\alpha}$. Another potential application relates to their origin via free radical peroxidation of unsaturated fatty acids. Free radicals are implicated in inflammatory and degenerative diseases such as atherosclerosis, cancer, and Alzheimer's disease. Isoprostane analysis of urine or serum may thus allow non-invasive monitoring of oxidative damage as a insight into these disease states.

THROMBOXANES

An intriguing side-branch from the prostaglandin pathway leads to thromboxanes* (Figure 3.23). The peroxide and cyclopentane ring functions of PGH$_2$ are cleaved and restructured to form **thromboxane A$_2$** (**TXA$_2$**), which contains a highly strained four-membered oxetane ring. TXA$_2$ is highly unstable, and reacts readily with nucleophiles. In an aqueous environment, it

reacts to yield the hemiacetal **thromboxane B$_2$** (**TXB$_2$**).

LEUKOTRIENES

Yet another variant for the metabolism of arachidonic acid is the formation of leukotrienes*, a series of fatty acid derivatives with a conjugated triene functionality, and first isolated from leukocytes. In a representative pathway (others have

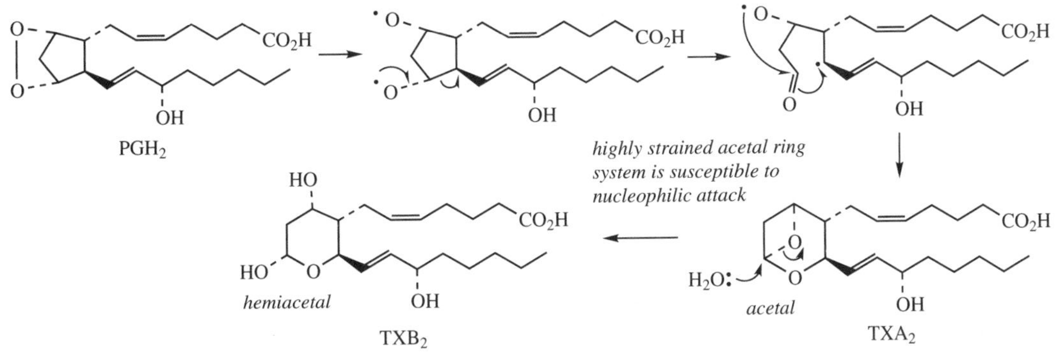

Figure 3.23

Thromboxanes

The thromboxanes were isolated from blood platelets, and whilst TXA_2 showed high biological activity TXB_2 was only weakly active. TXA_2 causes blood platelets to aggregate to form a clot or thrombus, by increasing cytoplasmic calcium concentrations and thus deforming the platelets which then fuse together. It has the opposite effect to PGI_2, and presumably the development of thrombosis reflects an imbalance in the two activities. Both compounds are produced from the same precursor, PGH_2, which is converted in the blood platelets to TXA_2, and in the blood vessel wall to PGI_2. Thromboxanes A_3 and B_3 have also been isolated from blood platelets, are structurally analogous to prostaglandins in the 3-series, and are derived from $\Delta^{5,8,11,14,17}$-eicosapentaenoic acid. TXA_3 is not strongly aggregatory towards blood platelets. The highly unstable nature of the biologically active thromboxanes has made their synthesis difficult, and drug use of natural structures will probably be impracticable. It is likely that most efforts will be directed towards thromboxane antagonists to help reduce blood platelet aggregation in thrombosis patients. The value of aspirin in preventing cardiovascular disease is now known to be related to inhibition of thromboxane A_2 biosynthesis in platelets.

Leukotrienes

The leukotrienes are involved in allergic responses and inflammatory processes. An antigen–antibody reaction can result in the release of compounds such as histamine (see page 379) or materials termed slow reacting substance of anaphylaxis (SRSA). These substances are then mediators of hypersensitive reactions such as hay fever and asthma. Structural studies have identified SRSA as a mixture of LTC_4, LTD_4 and LTE_4. These cysteine-containing leukotrienes are powerful bronchoconstrictors and vasoconstrictors, and induce mucus secretion, the typical symptoms of asthma. LTE_4 is some 10–100-fold less active than LTD_4, so that degradation of the peptide side-chain represents a means of eliminating leukotriene function. LTB_4 appears to facilitate migration of leukocytes in inflammation, and is implicated in the pathology of psoriasis, inflammatory bowel disease, and arthritis. The biological effects of leukotrienes are being actively researched to define the cellular processes involved. This may lead to the development of agents to control allergic and inflammatory reactions. Drugs inhibiting the formation of LTC_4 and LTB_4 are in clinical trials, whilst montelukast and zafirlukast have been introduced as orally active leukotriene (LTD_4) receptor antagonists for the prophylaxis of asthma.

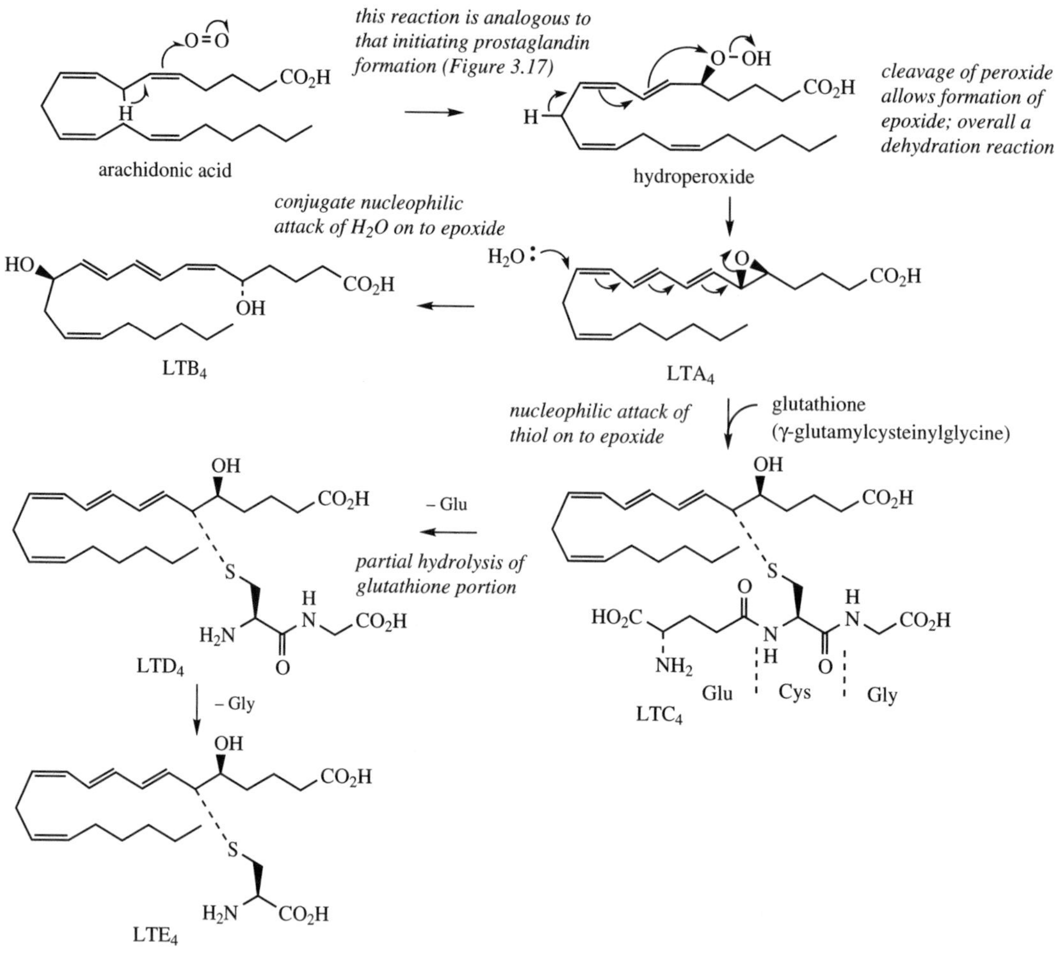

Figure 3.24

been characterized) (Figure 3.24), **arachidonic acid** is converted into a hydroperoxide, the point of oxygenation being C-5, rather than C-11 as in the prostaglandin pathway (Figure 3.17). This compound loses water via formation of an epoxide ring, giving **leukotriene A$_4$ (LTA$_4$)**. This unstable allylic epoxide may hydrolyse by conjugate addition giving **leukotriene B$_4$ (LTB$_4$)**, or alternatively the epoxide may be attacked directly by a nucleophile, in this case the sulphur atom of the tripeptide glutathione (γ-glutamylcysteinylglycine) (Figure 3.24). The adduct produced in the latter reaction is termed **leukotriene C$_4$ (LTC$_4$)**. Partial hydrolysis in the tripeptide fragment then leads to **leukotriene D$_4$ (LTD$_4$)** and **leukotriene E$_4$ (LTE$_4$)**. Analogues, e.g. LTA$_3$ and LTA$_5$, are

also known, and these are derived from $\Delta^{5,8,11}$-**eicosatrienoic acid** and $\Delta^{5,8,11,14,17}$-**eicosapentaenoic acid** respectively. The subscript numeral indicates the total number of double bonds in the leukotriene chain.

AROMATIC POLYKETIDES

For fatty acid biosynthesis, reduction after each condensation step affords a growing hydrocarbon chain. In the absence of this reduction process, the growing poly-β-keto chain needs to be stabilized on the enzyme surface until the chain length is complete, at which point cyclization or other reactions can occur. The poly-β-keto ester is very reactive, and there are various possibilities for undergoing intramolecular Claisen or aldol reactions, dictated of course by the nature of the enzyme and how the substrate is folded. Methylenes flanked by two carbonyls are activated, allowing formation of carbanions/enolates and subsequent reaction with ketone or ester carbonyl groups, with a natural tendency to form strain-free six-membered rings.

Cyclization: Simple Phenols

The polyketo ester (Figure 3.25), formed from four acetate units (one acetate starter group and three

malonate chain extension units) is capable of being folded in at least two ways, A and B (Figure 3.25). For A, ionization of the α-methylene allows aldol addition on to the carbonyl six carbons distant along the chain, giving the tertiary alcohol. Dehydration occurs as in most chemical aldol reactions, giving the alkene, and enolization follows to attain the stability conferred by the aromatic ring. The thioester bond (to coenzyme A or ACP) is then hydrolysed to produce **orsellinic acid**. Alternatively, folding of the polyketo ester as in B allows a Claisen reaction to occur, which, although mechanistically analogous to the aldol reaction, is

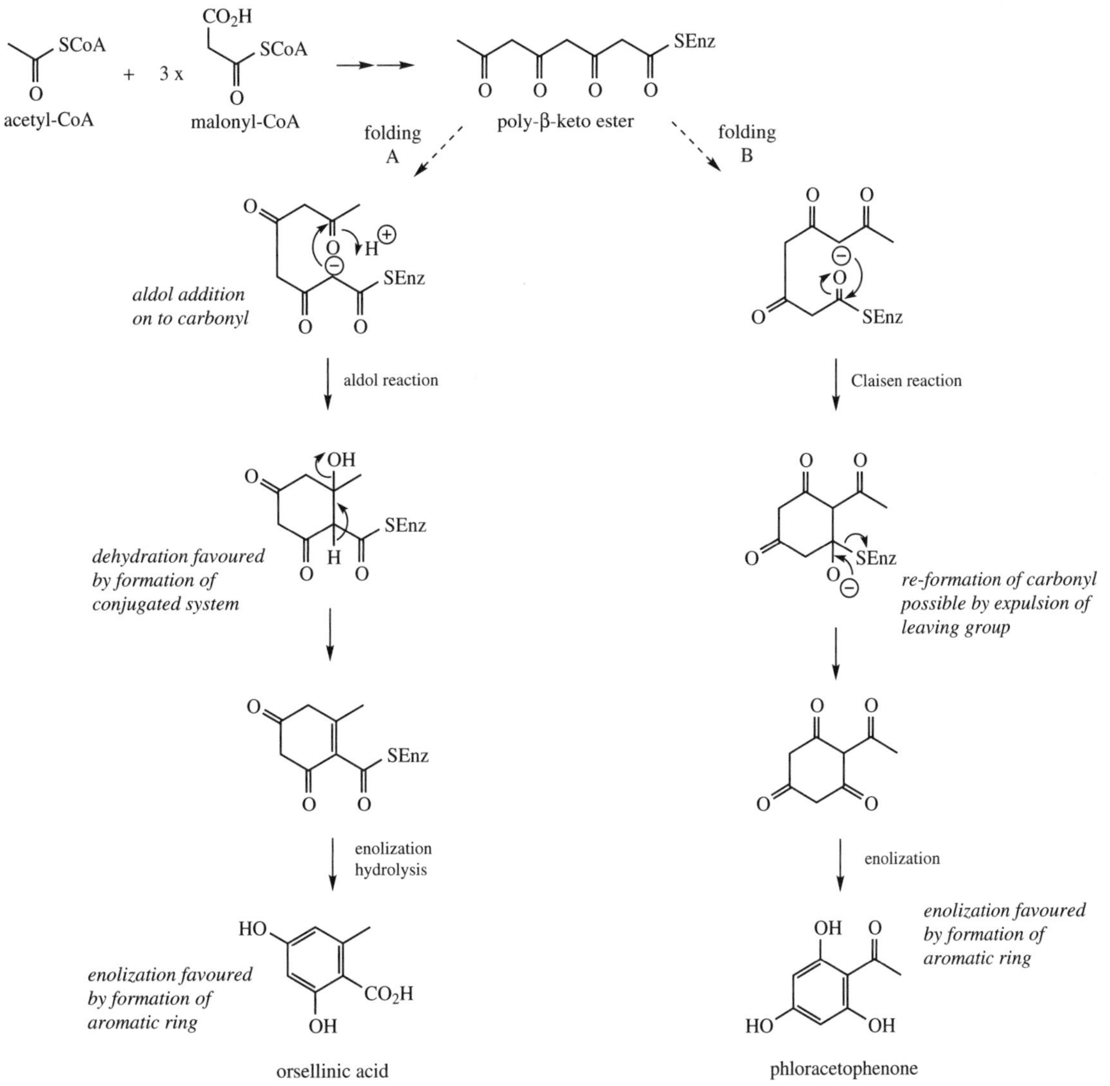

Figure 3.25

terminated by expulsion of the thiol leaving group, and direct release from the enzyme. Enolization of the cyclohexatrione produces **phloracetophenone**. As with fatty acid synthases, the whole sequence of reactions is carried out by an enzyme complex which converts acetyl-CoA and malonyl-CoA into the final product without giving any detectable free intermediates. These enzyme complexes combine **polyketide synthase** and **polyketide cyclase** activities and share many structural similarities with fatty acid synthases, including an acyl carrier protein with a phosphopantatheine group, a reactive cysteine residue, and an analogous β-ketoacyl synthase activity.

A distinctive feature of an aromatic ring system derived through the acetate pathway is that several of the carbonyl oxygens of the poly-β-keto system are retained in the final product. These end up on alternate carbons around the ring system. Of course, one or more might be used in forming a carbon–carbon bond, as in orsellinic acid. Nevertheless, this oxygenation on alternate carbon atoms, a *meta* oxygenation pattern, is usually easily recognizable, and points to the biosynthetic origin of the molecule. This *meta* oxygenation pattern contrasts to that seen on aromatic rings formed via the shikimate pathway (see Chapter 4).

6-methylsalicylic acid (Figure 3.26) is a metabolite of *Penicillium patulum*, and differs from orsellinic acid by the absence of a phenol group at position 4. It is also derived from acetyl-CoA

and three molecules of malonyl-CoA, and the 'missing' oxygen function is removed during the biosynthesis. Orsellinic acid is not itself deoxygenated to 6-methylsalicylic acid. The enzyme 6-methylsalicylic acid synthase requires NADPH as cofactor, and removes the oxygen function by reduction of a ketone to an alcohol, followed by a dehydration step (Figure 3.26). Whilst on paper this could be carried out on an eight-carbon intermediate involved in orsellinic acid biosynthesis (Figure 3.25), there is evidence that the reduction/dehydration actually occurs on a six-carbon intermediate as the chain is growing (compare fatty acid biosynthesis, page 36), prior to the final chain extension (Figure 3.26). Aldol condensation, enolization, and release from the enzyme then generate 6-methylsalicylic acid. Important evidence for reduction occurring at the C_6 stage as shown in Figure 3.26 comes from the formation of triacetic acid lactone if NADPH is omitted from the enzymic incubation.

The folding of a polyketide chain can be established by labelling studies, feeding carbon-labelled sodium acetate to the appropriate organism and establishing the position of labelling in the final product by chemical degradation and counting (for the radioactive isotope ^{14}C), or by NMR spectrometry (for the stable isotope ^{13}C). ^{13}C NMR spectrometry is also valuable in establishing the location of intact C_2 units derived from feeding $^{13}C_2$-labelled acetate. This is exemplified in

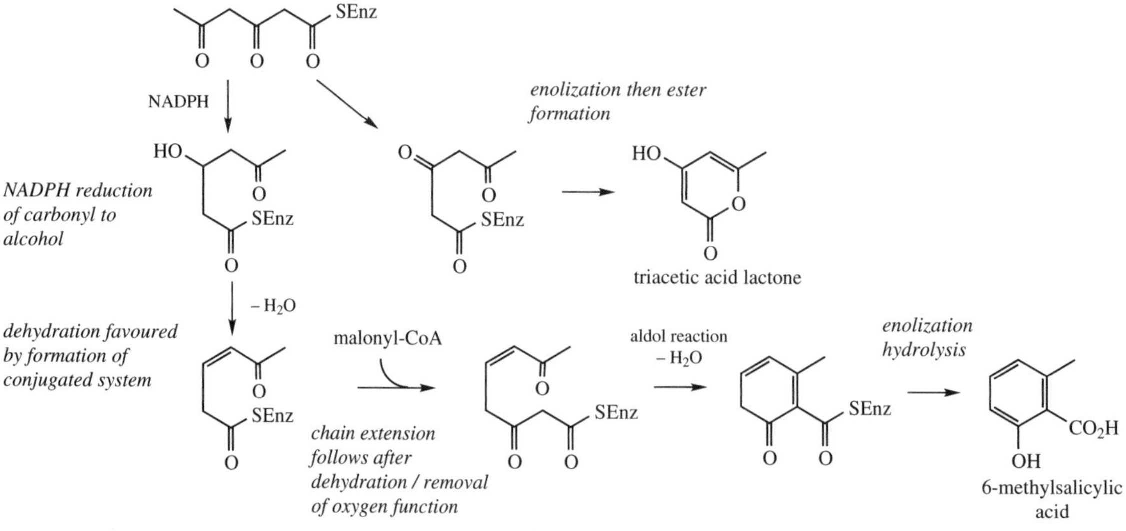

Figure 3.26

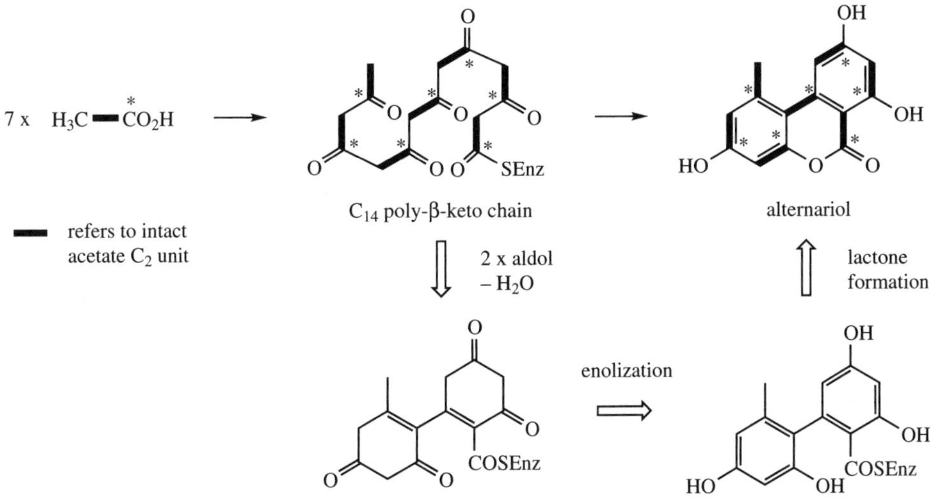

Figure 3.27

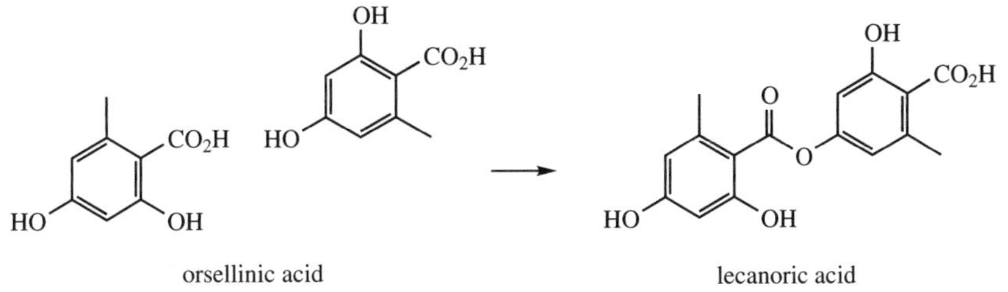

Figure 3.28

Figure 3.27, where **alternariol**, a metabolite from the mould *Alternaria tenuis*, can be established to be derived from a single C_{14} polyketide chain, folded as shown, and then cyclized. Whilst the precise sequence of reactions involved is not known, paper chemistry allows us to formulate the essential features. Two aldol condensations followed by enolization in both rings would give a biphenyl, and lactonization would then lead to alternariol. The oxygenation pattern in alternariol shows alternate oxygens on both aromatic rings, and an acetate origin is readily surmised, even though some oxygens have been used in ring formation processes. The lone methyl 'start-of-chain' is also usually very obvious in acetate-derived compounds, though the carboxyl 'end-of-chain' can often react with convenient hydroxyl functions, which may have arisen through enolization, and lactone or ester functions are thus reasonably common. For example, **lecanoric acid** is a **depside** (an ester

formed from two phenolic acids) found in lichens and produced from two orsellinic acid molecules (Figure 3.28).

Structural Modifications: Anthraquinones

A number of natural anthraquinone derivatives are also excellent examples of acetate-derived structures. **Endocrocin** (Figure 3.29) found in species of *Penicillium* and *Aspergillus* fungi is formed by folding a polyketide containing eight C_2 units to form the periphery of the carbon skeleton. Three aldol-type condensations would give a hypothetical intermediate 1, and, except for a crucial carbonyl oxygen in the centre ring, endocrocin results by enolization reactions, one of which involves the vinylogous enolization $-CH_2-CH=CH-CO- \rightarrow -CH=CH-CH=C(OH)-$. The additional carbonyl oxygen must be introduced at some stage

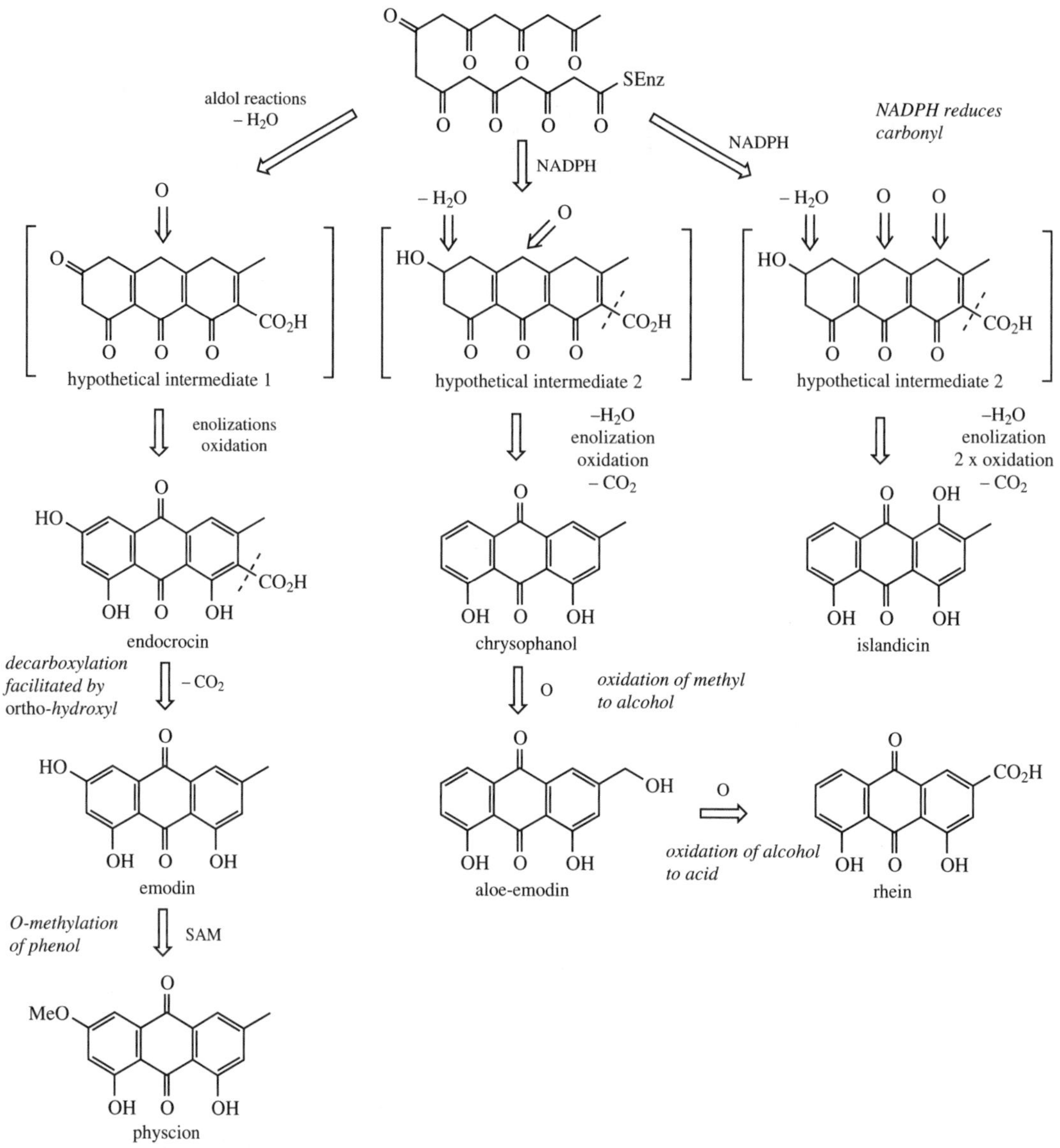

Figure 3.29

during the biosynthesis by an oxidative process, for which we have little information. **Emodin**, a metabolite of some *Penicillium* species, but also found in higher plants, e.g. *Rhamnus* and *Rumex* species, would appear to be formed from endocrocin by a simple decarboxylation reaction. This is facilitated by the adjacent phenol function (see page 20). *O*-Methylation of emodin would then lead to **physcion**. **Islandicin** is another anthraquinone pigment produced by *Penicillium islandicum*, and differs from emodin in two ways.

One hydroxyl is missing, and a new hydroxyl has been incorporated adjacent to the methyl. Without any evidence for the sequence of such reactions, the structure of intermediate 2 shows the result of three aldol condensations and reduction of a carbonyl. A dehydration reaction, two oxidations, and a decarboxylation are necessary to attain the islandicin structure. In **chrysophanol**, **aloe-emodin**, and **rhein**, the same oxygen function is lost by reduction as in islandicin, and decarboxylation also occurs. The three compounds are interrelated by a sequential oxidation of the methyl in chrysophanol to a hydroxymethyl in aloe-emodin, and a carboxyl in rhein.

These structural modifications undergone by the basic polyketide are conveniently considered under two main headings, according to the timing of the steps in the synthetic sequence. Thus, 'missing' oxygen functions appear to be reduced out well before the folded and cyclized polyketide is detached from the enzyme, and are mediated by a reductase component of the enzyme complex during chain elongation *before the cyclization reaction*. On the other hand, reactions like the decarboxylation, *O*-methylation, and sequential oxidation of a methyl to a carboxyl are representative of transformations occurring *after the cyclization reaction*. It is often possible to demonstrate these later conversions by the isolation of enzymes catalysing the individual steps. Most of the secondary transformations are easily rationalized by careful consideration of the reactivity conferred on the molecule by the alternating and usually phenolic oxygenation pattern. These oxygens activate adjacent sites creating nucleophilic centres. Introduction of additional hydroxyl groups *ortho* or *para* to an existing phenol will be facilitated (see page 26), allowing the extra hydroxyl of islandicin to be inserted, for example. *Ortho-* or *para*-diphenols are themselves susceptible to further oxidation in certain circumstances, and may give rise to *o*- and *p*-quinones (see page 25). The quinone system in anthraquinones is built up by an oxidation of the central cyclohexadienone ring, again at a nucleophilic centre activated by the enone system. Methyls on an aromatic ring are also activated towards oxidation, facilitating the chrysophanol → aloe-emodin oxidation, for example. Decarboxylation, e.g. endocrocin → emodin, is readily achieved in the presence of an *ortho* phenol function, though a *para* phenol can also facilitate this (see page 20).

It is now appreciated that the assembly of the anthraquinone skeleton (and related polycyclic structures) is achieved in a step-wise sequence. After the polyketide chain is folded, the ring at the centre of the fold is formed first, followed in turn by the next two rings. The pathway outlined for the biosynthesis of endocrocin and emodin is shown in Figure 3.30. Mechanistically, there is little difference between this and

the speculative pathway of Figure 3.29, but the sequence of reactions is altered. Decarboxylation appears to take place before aromatization of the last-formed ring system, and tetrahydroanthracene intermediates such as atrochrysone carboxylic acid and atrochrysone are involved. These dehydrate to the anthrones **endocrocin anthrone** and **emodin anthrone**, respectively, prior to introduction of the extra carbonyl oxygen as a last transformation in the production of anthraquinones. This oxygen is derived from O_2.

Note that many other natural anthraquinone structures are not formed via the acetate pathway, but by a more elaborate sequence involving shikimate and an isoprene unit (see page 158). Such structures do not contain the characteristic *meta* oxygenation pattern, and often have oxygenation in only one aromatic ring (see page 164).

Emodin, physcion, chrysophanol, aloe-emodin, and rhein form the basis of a range of purgative anthraquinone derivatives found in long-established laxatives such as Senna*, Cascara*, Frangula*, Rhubarb*, and Aloes*. The free anthraquinones themselves have little therapeutic activity and need to be in the form of water-soluble glycosides to exert their action. Although simple anthraquinone *O*-glycosides are present in the drugs, the major purgative action arises from compounds such as **cascarosides**, e.g. cascaroside A (Figure 3.33), which are both *O*- and *C*-glycosides, and **sennosides**, e.g. sennoside A (Figure 3.33), which are dianthrone *O*-glycosides. These types of derivative are likely to be produced from intermediate anthrone structures. This could act as substrate for both *O*- and *C*-glucosylation, employing the glucose donor UDPglucose (see page 29), and would generate a cascaroside structure (Figure 3.31). Alternatively, a one-electron oxidation allows oxidative coupling (see page 28) of two anthrone systems to give a dianthrone (Figure 3.32). This can be formulated as direct oxidation at the benzylic $-CH_2-$, or via the anthranol, which is the phenolic tautomer of the anthrone (Figure 3.32). Glycosylation of the dianthrone system would then give a sennoside-like product. However, further oxidative steps can create a dehydrodianthrone, and then allow coupling of the aromatic rings through **protohypericin** to give a naphthodianthrone, e.g. **hypericin** (Figure 3.32). The reactions of Figure 3.32 can be achieved

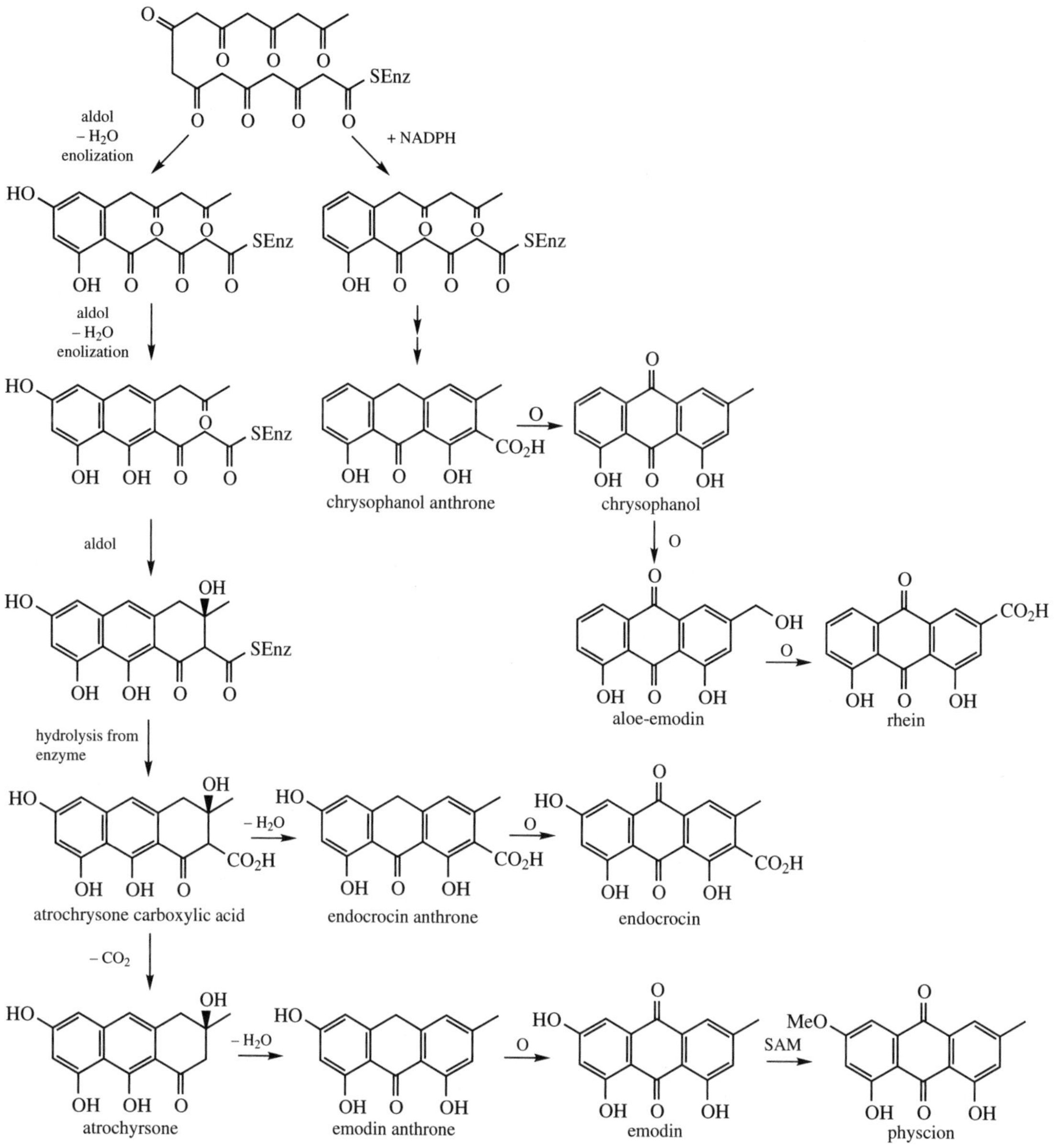

Figure 3.30

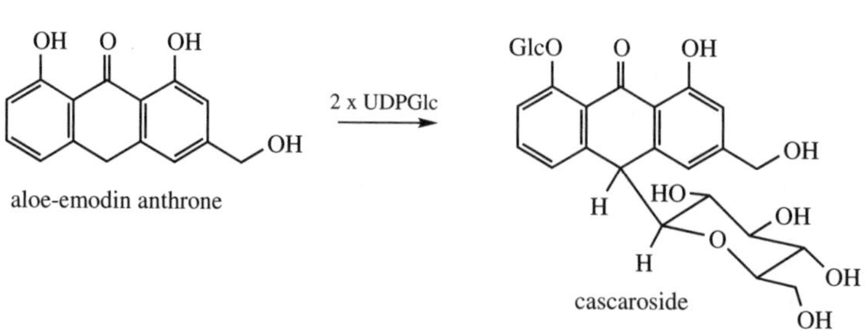

Figure 3.31

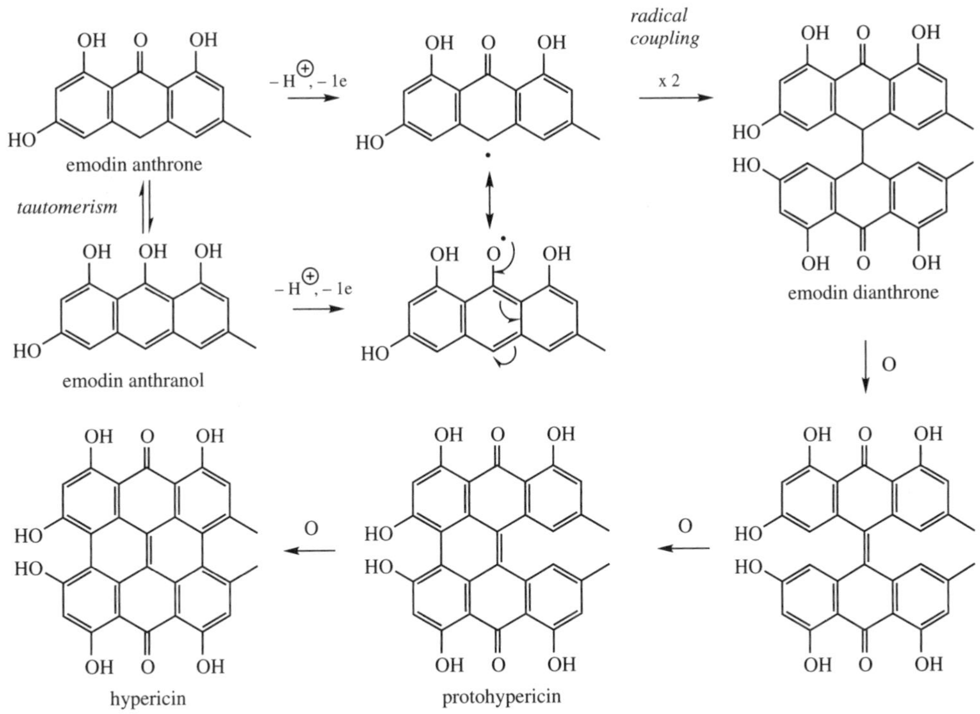

Figure 3.32

chemically by passing air into an alkaline solution of **emodin anthrone**. **Hypericin** is found in cultures of *Dermocybe* fungi, and is also a constituent of St John's Wort, *Hypericum perforatum* (Guttiferae/Hypericaceae), which is a popular herbal medicine in the treatment of depression. The naphthodianthrones have no purgative action, but hypericin can act as a photosensitizing agent in a similar manner to furocoumarins (see page 146). Thus ingestion of hypericin results in an increased absorption of UV light and can lead to dermatitis and burning. Hypericin is also being investigated for its antiviral activities, in particular for its potential activity against HIV.

Senna

Senna leaf and fruit are obtained from *Cassia angustifolia* (Leguminosae/Fabaceae), known as Tinnevelly senna, or less commonly from *Cassia senna* (*syn C. acutifolia*), which is described as Alexandrian senna. The plants are low, branching shrubs, *C. angustifolia* being cultivated in India and Pakistan, and *C. senna* being produced in the Sudan, much of it from wild plants. Tinnevelly senna is cultivated in wetter conditions than Alexandrian senna, which gives more luxuriant growth. Early harvests provide leaf material whilst later on, both leaf and fruit (senna pods) are obtained, a mixture which is separated by sieving (Alexandrian) or hand picking after drying (Tinnevelly). There are no significant differences in the chemical constituents of the two sennas, or between leaf and fruit drug. However, amounts of the active constituents do vary, and appear to be a consequence of cultivation conditions and the time of harvesting of the plant material.

The active constituents in both senna leaf and fruit are dianthrone glycosides, principally sennosides A and B (Figure 3.33). These compounds are both di-*O*-glucosides of rhein

(Continues)

(*Continued*)

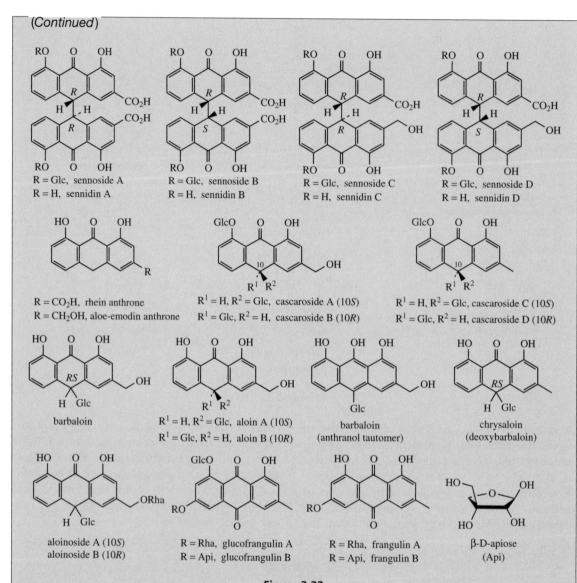

Figure 3.33

dianthrone (sennidins A and B), and liberate these aglycones on acid hydrolysis, or the anthraquinone rhein (Figure 3.30) on oxidative hydrolysis (e.g. aq HNO_3 or H_2O_2/HCl). Sennidins A and B are optical isomers: sennidin A is dextrorotatory (+) whilst sennidin B is the optically inactive *meso* form. Minor constituents include sennosides C and D (Figure 3.33), which are also a pair of optical isomers, di-*O*-glucosides of heterodianthrones sennidins C and D. Sennidin C is dextrorotatory, whilst sennidin D is optically inactive, approximating to a *meso* form in that the modest change in substituent does not noticeably affect the optical rotation. Oxidative hydrolysis of sennosides C and D would produce the anthraquinones rhein and aloe-emodin (Figure 3.30). Traces of other anthraquinone glycoside derivatives are also present in the plant material. Much of the sennoside content of the dried leaf appears to be formed by enzymic oxidation of anthrone glycosides during the drying process. Fresh leaves

(*Continues*)

and fruits also seem to contain primary glycosides which are more potent than sennosides A and B, and which appear to be partially hydrolysed to sennosides A and B (the secondary glycosides) by enzymic activity during collection and drying. The primary glycosides contain additional glucose residues.

Senna leaf suitable for medicinal use should contain not less than 2.5% dianthrone glycosides calculated in terms of sennoside B. The sennoside content of Tinnevelly fruits is between 1.2 and 2.5%, that of Alexandrian fruits being 2.5–4.5%. Senna preparations, in the form of powdered leaf, powdered fruit, or extracts, are typically standardized to a given sennoside content. Non-standardized preparations have unpredictable action and should be avoided. Senna is a stimulant laxative and acts on the wall of the large intestine, increasing peristaltic movement. After oral administration, the sennosides are transformed by intestinal flora into rhein anthrone (Figure 3.33), which appears to be the ultimate purgative principle. The glycoside residues in the active constituents are necessary for water-solubility and subsequent transportation to the site of action. Although purgative action is provided by the aglycones, including anthraquinones, these materials are conjugated and excreted in the urine after oral administration rather than being transported to the colon. Senna is a purgative drug suitable for either habitual constipation, or for occasional use, and is widely prescribed.

Cascara

Cascara is the dried bark of the cascara buckthorn *Rhamnus purshianus* (Rhamnaceae), a small tree native to the forests of the Pacific coast of North America. Most of the drug material is gathered from wild trees in Oregon, Washington, and British Columbia. Trees are felled and the bark is stripped from the trunk and branches, then dried. The fresh bark is unsuitable for drug use, causing griping and nausea, and thus the bark is stored for at least a year before being processed. During this time, enzymic hydrolysis and oxidation modify the anthraquinone-based constituents and thus the cathartic activity. Cascara preparations are mainly formulated from extracts of the bark.

Cascara bark contains about 6–9% of anthracene derivatives, 80–90% of which are anthrone *C*-glycosides. The major constituents are cascarosides A and B (Figure 3.33), which contain both *O*- and *C*-glucoside linkages, and represent a pair of optical isomers differing only in the stereochemistry of the *C*-glucoside bond. These have a substitution pattern analogous to aloe-emodin (Figure 3.30) and oxidative hydrolysis (e.g. aq HNO_3 or H_2O_2/HCl) liberates aloe-emodin. Acid hydrolysis does not cleave the *C*-glucose linkage, and instead generates barbaloin (Figure 3.33), a mixture of two diastereoisomeric forms, which have been named aloin A and aloin B. It is likely that during any chemical manipulation, the two forms may interconvert via the anthranol tautomer (Figure 3.33). Similar components in the bark, though usually present in smaller amounts than cascarosides A and B, are cascarosides C and D (Figure 3.33). These are also a pair of diastereoisomers, and have a substitution pattern analogous to chrysophanol (Figure 3.30). Hydrolysis of the *O*-glucose linkage yields chrysaloin, sometimes referred to as deoxybarbaloin. Barbaloin and chrysaloin are also found in the bark, and are thought to be breakdown products formed by enzymic hydrolysis of the cascarosides. Other compounds identified in the bark include simple anthraquinones and their *O*-glycosides, and some dianthrone derivatives.

The principal purgative activity originates from the cascarosides, the *C*-glycosides barbaloin and chrysaloin being less active when taken orally. As with the sennosides, the actual purgative

(Continued)

agent is produced by the action of intestinal flora, and the cascarosides are transformed into aloe-emodin anthrone (Figure 3.33). Cascara has a similar pharmacological action to senna, i.e. it stimulates peristalsis of the large intestine, and has found major use in the correction of habitual constipation. It has a stronger effect than senna, however, and its routine usage is not now recommended.

Frangula

The bark of the alder buckthorn, *Rhamnus frangula* (Rhamnaceae) is used in a similar way to cascara, and is preferred to cascara in many European countries, though not in the UK. *Rhamnus frangula* is a small tree of European origin, and supplies of the bark come from South-Eastern Europe. The bark is also stored for a year before use. **Frangula** bark contains about 6% anthracene derivatives, mainly anthraquinone *O*-glycosides. These are derivatives of emodin (Figure 3.30) and comprise glucofrangulins A and B, and frangulins A and B (Figure 3.33). Free aglycones emodin, physcion and chrysophanol are also present.

Allied Drugs

Aloes and rhubarb have found considerable use as purgatives in the past, but they both have a rather drastic action and their use for this purpose has largely been abandoned.

Aloes consists of the dried juice from the leaves of various *Aloe* species (Liliaceae/Aloeaceae), including *A. ferox* (Cape aloes), *A. barbadensis* (Curacao aloes), and *A. perryi* (Socotrine aloes). The dark brown–black solid extract is extremely bitter, and contains 10–30% anthracene derivatives, the main component of which is barbaloin (Figure 3.33). Aloinosides A and B (Figure 3.33) are present in some varieties. Large amounts of resinous material form the bulk of the extract. Aloes is still used as a pharmaceutical aid in Compound Benzoin Tincture. The fresh mucilaginous gel obtained from *Aloe* species, particularly *Aloe vera* (= *A. barbadensis*), is held to assist wound healing, and is also widely used in skin cosmetics for its moisturizing and emollient properties. This material, mainly carbohydrate in nature (pectins and glucomannans), does not contain anthraquinone derivatives.

Rhubarb consists of the dried rhizome and root of *Rheum officinale*, *R. palmatum*, and other *Rheum* species (Polygonaceae). This contains 3–7.5% anthracene derivatives, mainly in the form of mono- and di-*O*-glucosides of rhein, physcion, and chrysophanol. Aglycones, especially rhein, are also present, and dianthrone derivatives have also been characterized. A high proportion of tannin-like materials gives rhubarb astringent as well as purgative properties. The common rhubarb cultivated for culinary use is *Rheum rhaponticum*, a species containing similar anthraquinone derivatives to the drug material, but which was not officially acceptable. In common with other *Rheum* species, this plant is considered poisonous due to the high concentration of oxalic acid present in the leaf (though not in the stem, which is edible). Toxic effects result from hypocalcaemia caused by removal of calcium from the bloodstream by formation of the insoluble calcium oxalate.

Dantron (**danthron**; 1,8-dihydroxyanthraquinone) (Figure 3.34) is known as a natural product, but for drug use is produced synthetically. It is prescribed to relieve constipation in geriatric and terminally ill patients. **Dithranol** (1,8-dihydroxyanthrone) is used as topical agent to treat troublesome cases of psoriasis. **Diacetylrhein** is marketed in some countries for the treatment of osteoarthritis.

(Continues)

(*Continued*)

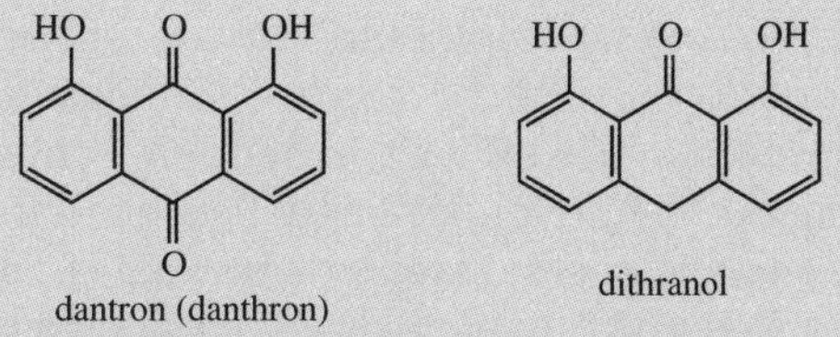

dantron (danthron)

dithranol

Figure 3.34

Hypericum/St John's Wort

The dried flowering tops of St John's Wort (*Hypericum perforatum*; Guttiferae/Hypericaceae) have been used as a herbal remedy for many years, an extract in vegetable oil being used for its antiseptic and wound healing properties. St John's Wort is now a major crop marketed as an antidepressant, that is claimed to be as effective in its action as the widely prescribed antidepressants of the selective serotonin re-uptake inhibitor (SSRI) class such as fluoxetine (Prozac ®), and with fewer side-effects. There is considerable clinical evidence that extracts of St John's Wort are effective in treating mild to moderate depression and improving mood. However, to avoid potentially dangerous side-effects, St John's Wort should not be used at the same time as prescription antidepressants. St John's Wort is a small to medium height herbaceous perennial plant with numerous yellow flowers characteristic of this genus. It is widespread throughout Europe, where it is generally considered a weed, and has also become naturalized in North America. The tops, including flowers at varying stages of development, which contain considerable amounts of the active principles, are harvested and dried in late summer.

The dried herb contains significant amounts of phenolic derivatives, including 4–5% of flavonoids, though the antidepressant activity is considered to derive principally from naphthodianthrone structures such as hypericin (about 0.1%) and pseudohypericin (about 0.2%), and a prenylated phloroglucinol derivative hyperforin (Figure 3.35). The fresh plant also contains significant levels of protohypericin and protopseudohypericin, which are converted into hypericin and pseudophypericin during drying and processing, as a result of irradiation with visible light. Hyperforin is a major lipophilic constituent in the leaves and flowers (2–3%),

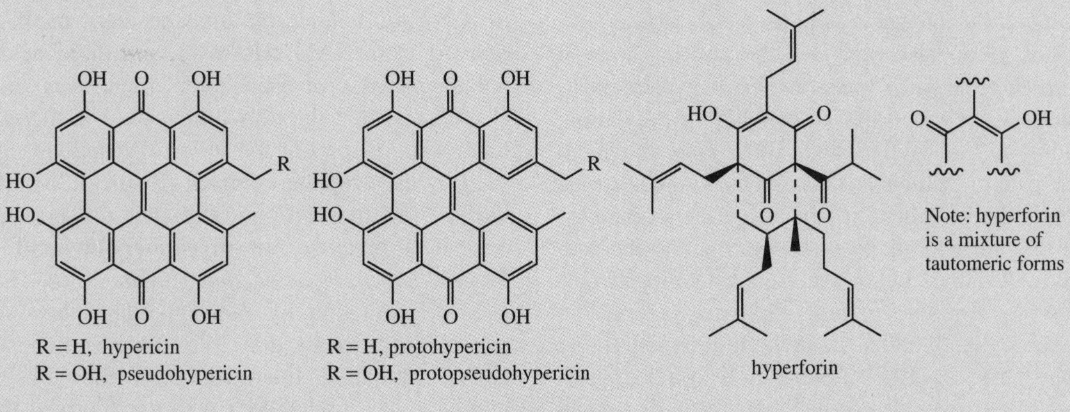

R = H, hypericin
R = OH, pseudohypericin

R = H, protohypericin
R = OH, protopseudohypericin

hyperforin

Note: hyperforin is a mixture of tautomeric forms

Figure 3.35

(*Continues*)

(Continued)

and is now thought to be the major contributor to the antidepressant activity, as well as to the antibacterial properties of the oil extract. Studies show clinical effects of St John's Wort on depression correlate well with hyperforin content. Standardized aqueous ethanolic extracts containing 0.15% hypericin and 5% hyperforin are usually employed. The aqueous solubility of hypericin and pseudohypericin is markedly increased by the presence of flavonoid derivatives in the crude extract, particularly procyanidin B_2, a dimer of epicatechin (see page 151). *Hypericum* extracts have been demonstrated to increase levels of serotonin, noradrenaline, and dopamine, which may be responsible for the antidepressant activity.

Hypericin also possesses extremely high toxicity towards certain viruses, a property that requires light and may arise via photo-excitation of the polycyclic quinone system. It is currently under investigation as an antiviral agent against HIV and hepatitis C. Antiviral activity appears to arise from an inhibition of various protein kinases, including those of the protein kinase C family. Hypericin and pseudohypericin are potent photosensitizers initiating photochemical reactions, and are held responsible for hypericism, a photodermatosis seen in cattle after consuming *Hypericum* plants present in pasture. Patients using St John's Wort as an antidepressant should also be warned to avoid over-exposure to sunlight. There is also considerable evidence that St John's Wort interacts with a number of prescription drugs including the anticoagulant warfarin, the cardiac glycoside digoxin, the bronchodilator theophylline, the HIV protease inhibitor indinavir, the immunosuppressive drug cyclosporin, and oral contraceptives. In some cases, it is known to promote the cytochrome P-450-dependent metabolism of the co-administered drugs.

C-Alkylation Reactions

A common feature of many natural products containing phenolic rings is the introduction of alkyl groups at nucleophilic sites. Obviously, the phenol groups themselves are nucleophilic, and with a suitable alkylating agent, *O*-alkyl derivatives may be formed (see page 12), e.g. the *O*-methylation of emodin to physcion (Figure 3.30). However, a phenol group also activates the ring carbons at the *ortho* and *para* positions, so that these positions similarly become susceptible to alkylation, leading to *C*-alkyl derivatives. The *meta* oxygenation pattern, which is a characteristic feature of acetate-derived phenolics, has the effect of increasing this nucleophilicity considerably, and the process of *C*-alkylation is very much facilitated (see page 12). Suitable natural alkylating agents are *S*-adenosylmethionine (SAM), and dimethylallyl diphosphate (DMAPP). Other polyprenyl diphosphate esters may also be encountered in biological alkylation reactions (e.g. see vitamin K, page 159). A minor inconsistency has been discovered, in

that, while *C*-alkylation with dimethylallyl and higher diphosphates is mediated *after* the initial polyketide cyclization product is liberated from the enzyme, there are several examples where *C*-methylation undoubtedly occurs *before* release of any aromatic compound from the enzyme. **5-methylorsellinic acid** (Figure 3.36) is a simple *C*-methylated analogue of orsellinic acid found in *Aspergillus flaviceps*, and the extra methyl is derived from SAM. However, orsellinic acid is not a precursor of 5-methylorsellinic acid and it is proposed that the poly-β-keto ester is therefore methylated as part of the series of reactions catalysed by the synthase complex (Figure 3.36). Similarly, 5-methylorsellinic acid, but not orsellinic acid is a precursor of **mycophenolic acid*** in *Penicillium brevicompactum* (Figure 3.36). However, *C*-alkylation by farnesyl diphosphate (see page 191) proceeds *after* the aromatization step, and a phthalide intermediate is the substrate involved. The phthalide is a lactone derived from 5-methylorsellinic acid by hydroxylation of its starter methyl group and reaction with the end-

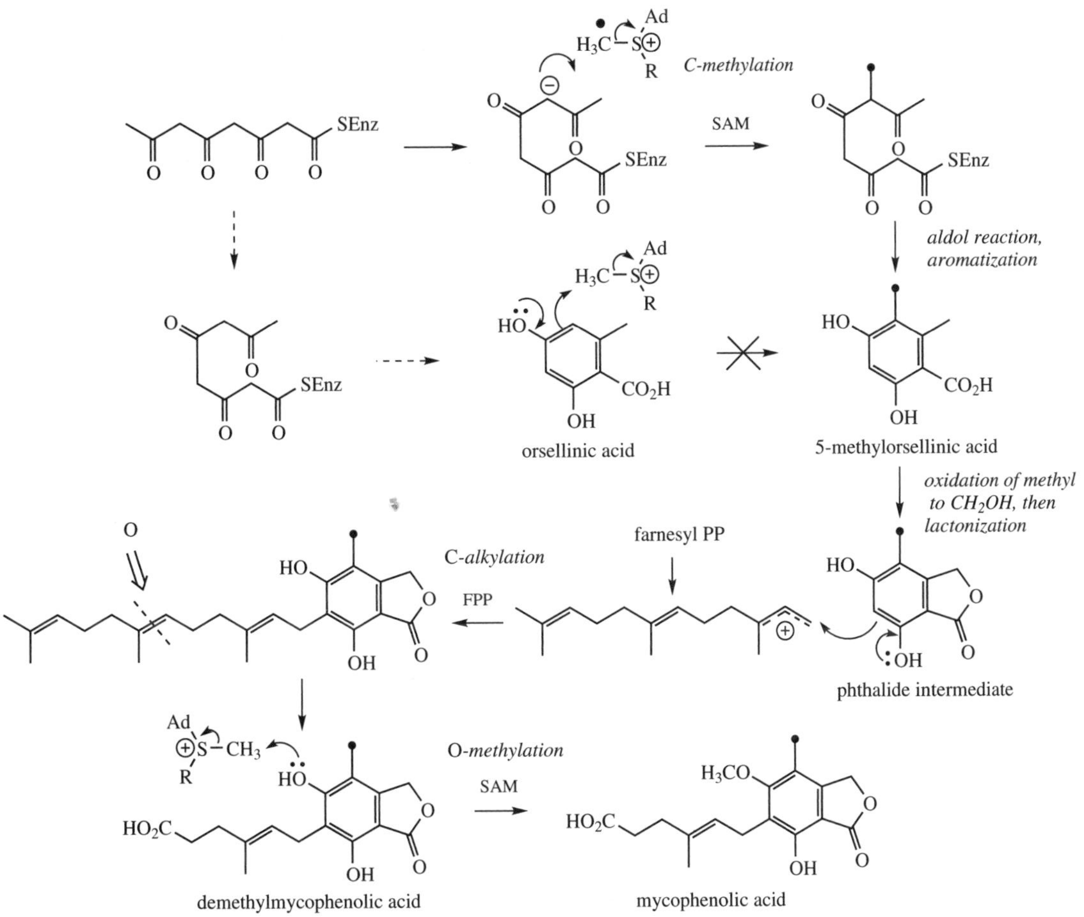

Figure 3.36

Mycophenolic Acid

Mycophenolic acid (Figure 3.36) is produced by fermentation cultures of the fungus *Penicillium brevicompactum*. It has been known for many years to have antibacterial, antifungal, antiviral, and antitumour properties. It has recently been introduced into medicine as an immunosuppressant drug, to reduce the incidence of rejection of transplanted organs, particularly kidney and heart transplants. It is formulated as the *N*-morpholinoethyl ester **mycophenolate mofetil** (Figure 3.37), which is metabolized after ingestion to mycophenolic

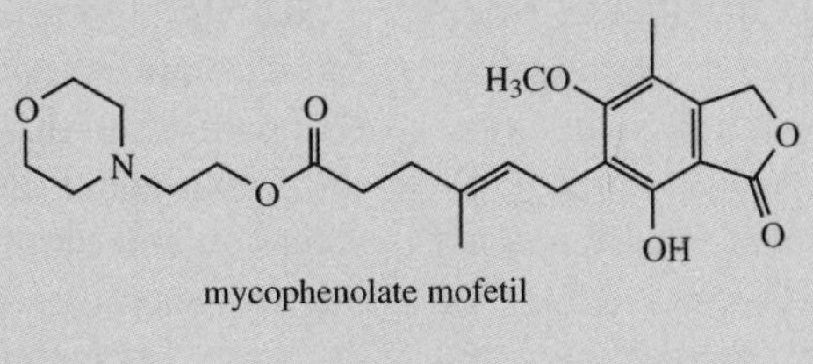

mycophenolate mofetil

Figure 3.37

(Continues)

(Continued)

acid, and is usually administered in combination with cyclosporin (see page 429). The drug is a specific inhibitor of mammalian inosine monophosphate dehydrogenase and has an antiproliferative activity on cells due to inhibition of guanosine nucleotide biosynthesis. This enzyme catalyses the NAD$^+$-dependent oxidation of inosine monophosphate (IMP) to xanthosine monophosphate (XMP), a key transformation in the synthesis of guanosine triphosphate (GTP) (see also caffeine biosynthesis, page 394). Rapidly growing cells have increased levels of the enzyme, so this forms an attractive target for anticancer, antiviral, and immunosuppressive therapy.

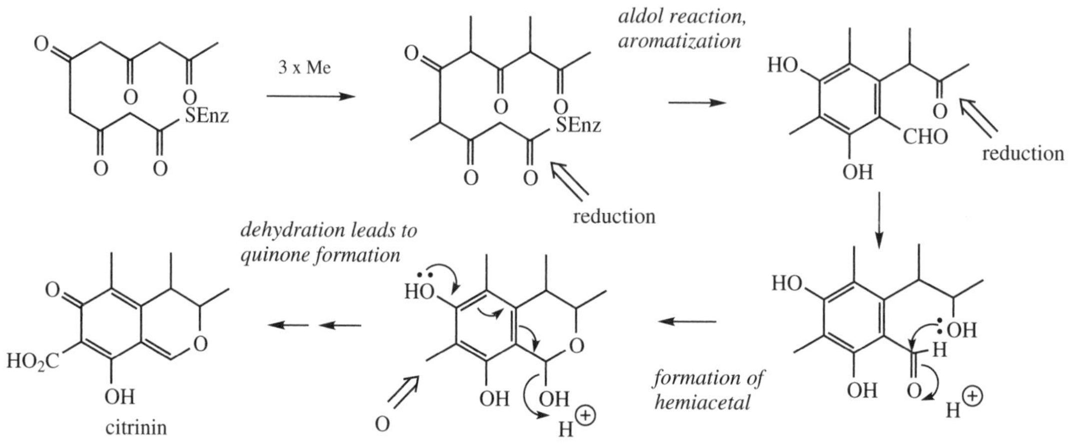

Figure 3.38

of-chain carboxyl. The chain length of the farnesyl alkyl group is then shortened by oxidation of a double bond giving demethylmycophenolic acid, which is then *O*-methylated, again involving SAM, to produce mycophenolic acid. Note that the *O*-methylation step only occurs after the *C*-alkylations, so that the full activating benefit of two *meta* positioned phenols can be utilized for the *C*-alkylation.

Three *C*-methyl substituents are inserted into the acetate-derived skeleton of **citrinin** (Figure 3.38), an antimicrobial metabolite from *Penicillium citrinum* and several *Aspergillus* species, which also displays potentially dangerous carcinogenic and nephrotoxic (kidney-damaging) activity. One of these introduced methyls has undergone oxidation to a carboxyl, adding to the difficulties in immediately recognizing the biosynthetic origins of this compound which contains a quinonemethide system rather than the

simpler aromatic ring. The methyls are probably introduced into the polyketide prior to release of the first aromatic intermediate, which could well be an aldehyde rather than the corresponding acid if a reductase component also forms part of the synthase complex. The hemiacetal can be produced after reduction of the side-chain carbonyl, and then in the later stages, oxidation of one methyl to a carboxyl will follow. The quinonemethide system in citrinin is simply the result of a dehydration reaction on the hemiacetal (Figure 3.38).

Khellin* and **visnagin** (Figure 3.39) are furochromones found in the fruits of *Ammi visnaga* (Umbelliferae/Apiaceae), and the active principles of a crude plant drug which has a long history of use as an antiasthmatic agent. Figure 3.39 presents the sequence of steps utilized in the biosynthesis of these compounds, fully consistent with the biosynthetic rationale developed above. The two carbons C-2′ and C-3′ forming part of the furan

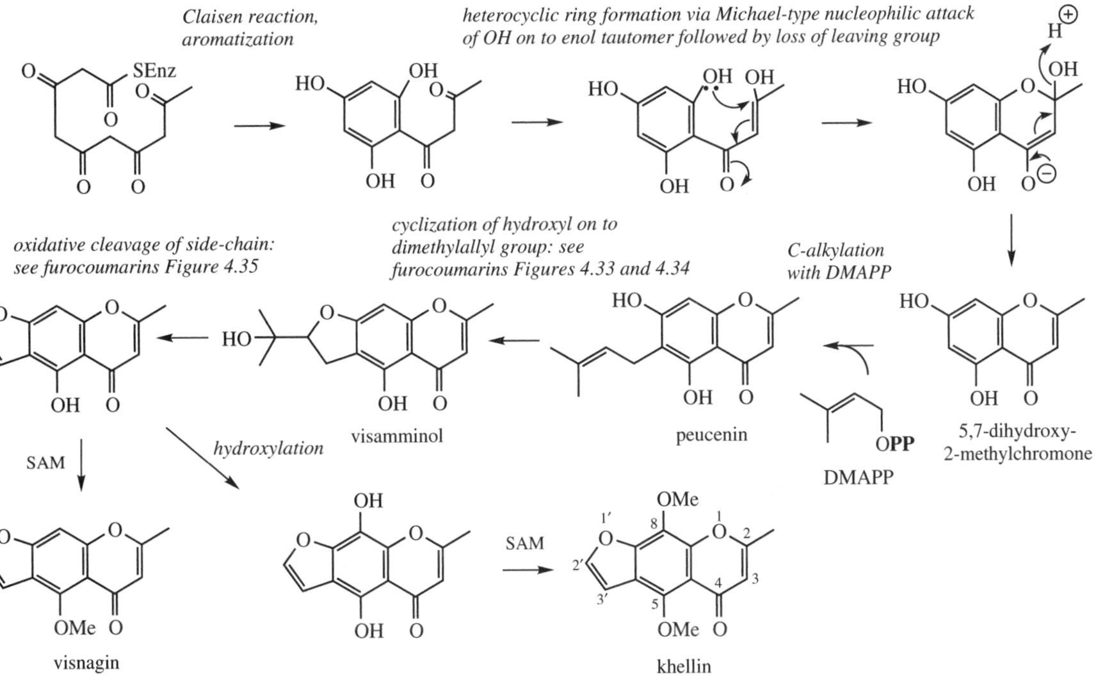

Claisen reaction, aromatization

heterocyclic ring formation via Michael-type nucleophilic attack of OH on to enol tautomer followed by loss of leaving group

oxidative cleavage of side-chain: see furocoumarins Figure 4.35

cyclization of hydroxyl on to dimethylallyl group: see furocoumarins Figures 4.33 and 4.34

C-alkylation with DMAPP

visamminol

peucenin

DMAPP

5,7-dihydroxy-2-methylchromone

SAM

hydroxylation

SAM

visnagin

khellin

Figure 3.39

Khellin and Cromoglicate

The dried ripe fruits of *Ammi visnaga* (Umbelliferae/Apiaceae) have a long history of use in the Middle East as an antispasmodic and for the treatment of angina pectoris. The drug contains small amounts of coumarin derivatives, e.g. visnadin (Figure 3.40) (compare *Ammi majus*, a rich source of furocoumarins, page 146), but the major constituents (2–4%) are

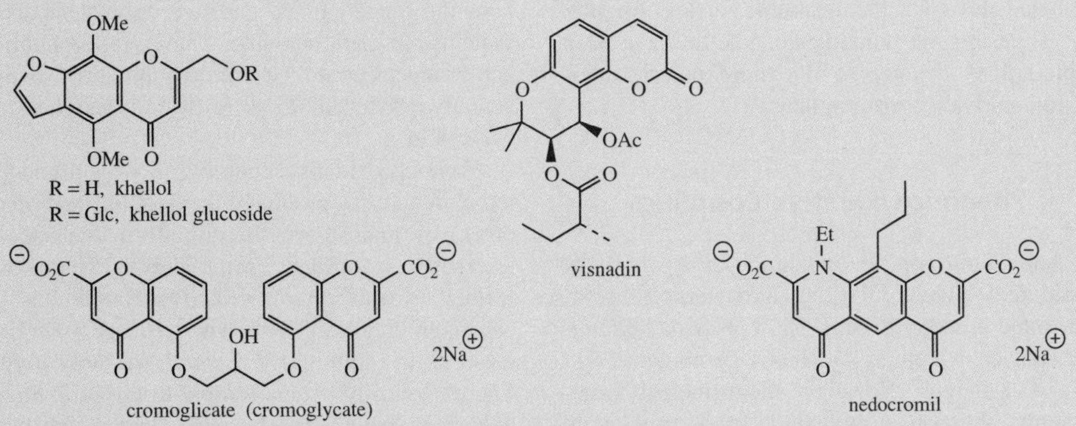

R = H, khellol
R = Glc, khellol glucoside

visnadin

cromoglicate (cromoglycate)

nedocromil

Figure 3.40

(Continues)

(Continued)

furochromones, including khellin and visnagin (Figure 3.34), and khellol and khellol glucoside (Figure 3.40). Both khellin and visnadin are coronary vasodilators and spasmolytic agents, with visnadin actually being the more potent agent. Khellin has been used in the treatment of angina pectoris and bronchial asthma. The synthetic analogue **cromoglicate** (**cromoglycate**) (Figure 3.40) is a most effective and widely used agent for the treatment and prophylaxis of asthma, hay fever, and allergic rhinitis. Cromoglicate contains two chromone systems containing polar carboxylic acid functions, joined by a glycerol linker. The mode of action is not fully established. It was believed to prevent the release of bronchospasm mediators by stabilizing mast cell membranes, but an alternative suggestion is that it may act by inhibiting the effect of sensory nerve activation, thus interfering with bronchoconstriction. It is poorly absorbed orally and is thus administered as inhalation or nasal spray. Eyedrops for relief of allergic conjunctivitis are also available. The more potent **nedocromil** (Figure 3.40) has also been introduced.

ring originate by metabolism of a five-carbon dimethylallyl substituent attached to C-6 (for a full discussion, see furocoumarins, page 145). The 8-methoxy group in khellin is absent from visnagin, so must be introduced late in the sequence. The key intermediate is thus 5,7-dihydroxy-2-methylchromone. On inspection, this has the alternate acetate-derived oxygenation pattern and a methyl chain starter, so is formed from a poly-β-keto chain through Claisen condensation then heterocyclic ring formation by an overall dehydration reaction. After formation of the furan ring via the C-dimethylallyl derivative peucenin and then visamminol, **visnagin** can be obtained by O-methylation. Alternatively, further hydroxylation *para* to the free phenol, followed by two methylations, yields **khellin**. The antiasthmatic properties of khellin have been exploited by developing the more polar, water-soluble derivative **cromoglicate***.

Phenolic Oxidative Coupling

C-Methylation also features in the biosynthesis of **usnic acid** (Figure 3.41), an antibacterial metabolite found in many lichens, e.g. *Usnea* and *Cladonia* species, which are symbiotic combinations of alga and fungus. However, the principal structural modification encountered involves phenolic oxidative coupling (see page 28). Two molecules of **methylphloracetophenone** are incorporated, and these are known to derive from a pre-aromatization methylation reaction and not by

methylation of phloracetophenone (Figure 3.41). The two molecules are joined together by an oxidative coupling mechanism which can be rationalized via the one-electron oxidation of a phenol group in methylphloracetophenone giving free radical A, for which resonance forms B and C can be written. Coupling of B and C occurs. Only the left-hand ring can subsequently be restored to aromaticity by keto–enol tautomerism, this state being denied to the right-hand ring because coupling occurred on to the methyl-containing position *para* to the original phenol. Instead, a heterocyclic ring is formed by attack of the phenol on to the enone system (see khellin, above). The outcome of this reaction is enzyme controlled, since two equivalent phenol groups are present as potential nucleophiles, and two equivalent enone systems are also available. Therefore, four different products could be formed, but only one is actually produced. Loss of water then leads to usnic acid.

Phenolic oxidative coupling is widely encountered in natural product biosynthesis, and many other examples are described in subsequent sections. A further acetate-derived metabolite formed as a result of oxidative coupling is the antifungal agent **griseofulvin*** (Figure 3.42) synthesized by cultures of *Penicillium griseofulvin*. The sequence of events leading to griseofulvin has now been established in detail, and the pathway also includes O-methylation steps and the introduction of a halogen (chlorine) atom at one of the nucleophilic sites, which is represented as involving the electrophile Cl^+ (Figure 3.42).

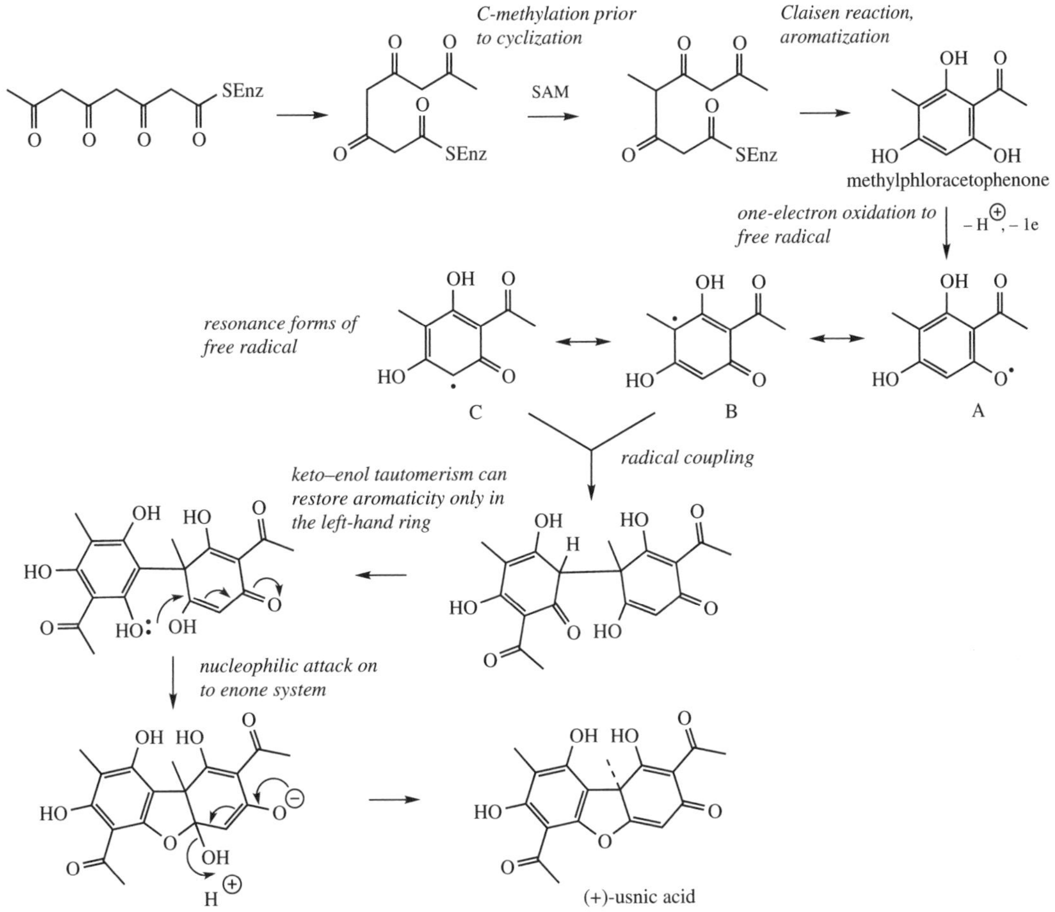

Figure 3.41

Griseofulvin

Griseofulvin is an antifungal agent produced by cultures of *Penicillium griseofulvum* and a number of other *Penicillium* species, including *P. janczewski*, *P. nigrum*, and *P. patulum*. Griseofulvin is the drug of choice for widespread or intractable dermatophyte infections, but is ineffective when applied topically. However, it is well absorbed from the gut and selectively concentrated into keratin, so may be used orally to control dermatophytes such as *Epidermophyton*, *Microsporium*, and *Trichophyton*. Treatment for some conditions, e.g. infections in fingernails, may have to be continued for several months, but the drug is generally free of side-effects. The antifungal action appears to be through disruption of the mitotic spindle, thus inhibiting fungal mitosis.

Initial inspection of the structure of griseofulvin shows the alternate oxygenation pattern, and also a methyl group which identifies the start of the polyketide chain. Cyclization of the C_{14} poly-β-keto chain folded as shown allows both

Claisen (left-hand ring) and aldol (right-hand ring) reactions to occur giving a benzophenone intermediate. Two selective methylations lead to griseophenone C, which is the substrate for chlorination to griseophenone B; both these

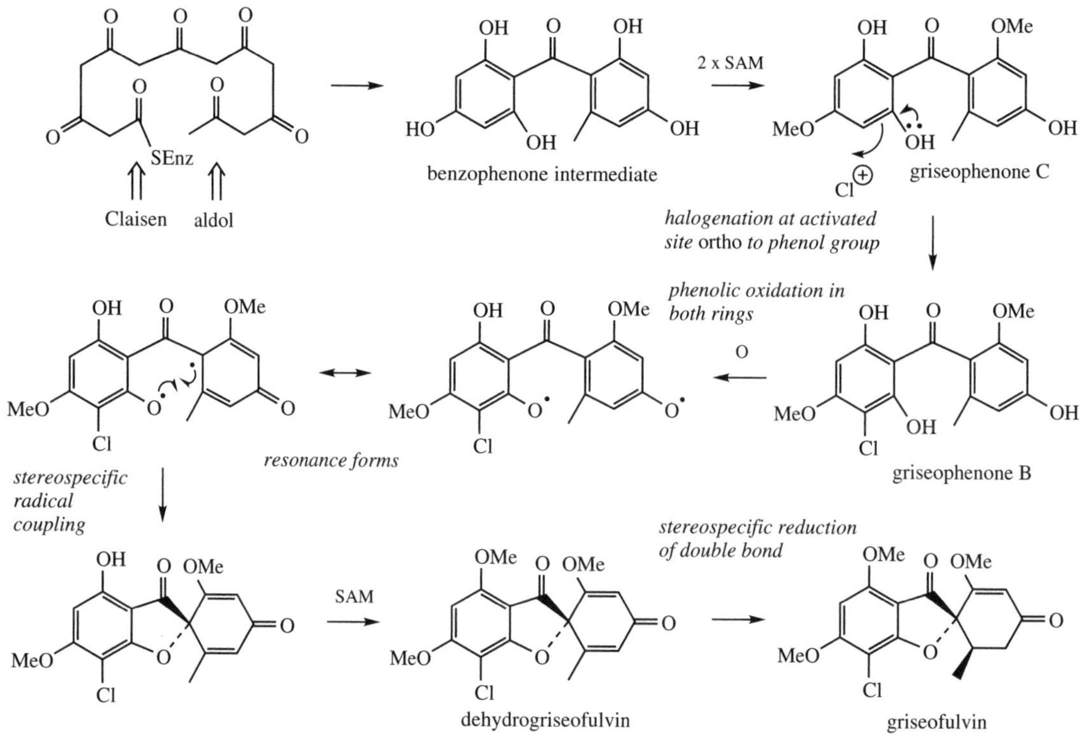

Figure 3.42

compounds appear as minor metabolites in *P. griseofulvin* cultures. One-electron oxidations on a phenolic group in each ring give a diradical and its mesomer, the latter allowing radical coupling to the basic grisan skeleton. **Griseofulvin** is then the result of methylation of the remaining phenol group and stereospecific reduction of the double bond in dehydrogriseofulvin.

Oxidative Cleavage of Aromatic Rings

Perhaps the most drastic modification which can happen to an aromatic ring is ring cleavage brought about by oxidative enzymes called dioxygenases (see page 27). These enzymes typically use catechol (1,2-dihydroxy) or quinol (1,4-dihydroxy) substrates, require molecular oxygen and Fe^{2+} cofactors, and incorporate both the oxygen atoms into the ring-cleaved product. In the case of catechols, cleavage may be between or adjacent to the two hydroxyls, giving products containing aldehyde and/or carboxylic acid functionalities (Figure 3.43). These groups are then able to react with other substituents in the molecule creating

compounds in which the characteristic acetate-derived features are probably no longer apparent. Shikimate-derived aromatic rings can suffer similar oxidative cleavage reactions.

Patulin is an excellent example of an acetate-derived structure synthesized from an aromatic substrate via oxidative cleavage and subsequent modifications (Figure 3.44). Patulin is a potent carcinogen produced by *Penicillium patulum*, a common contaminant on apples. If mould-infected apples find their way into food products, e.g.

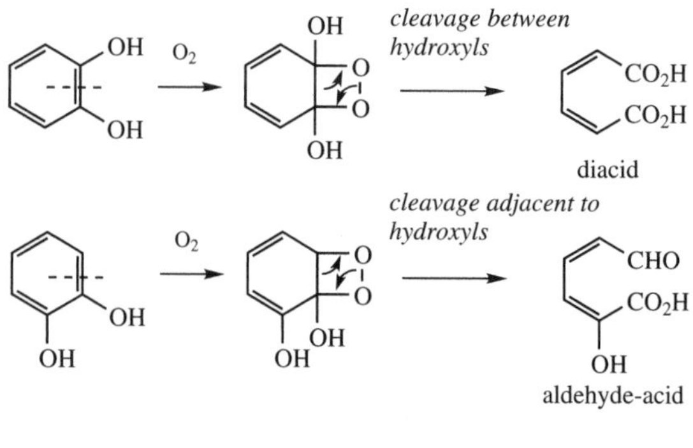

Figure 3.43

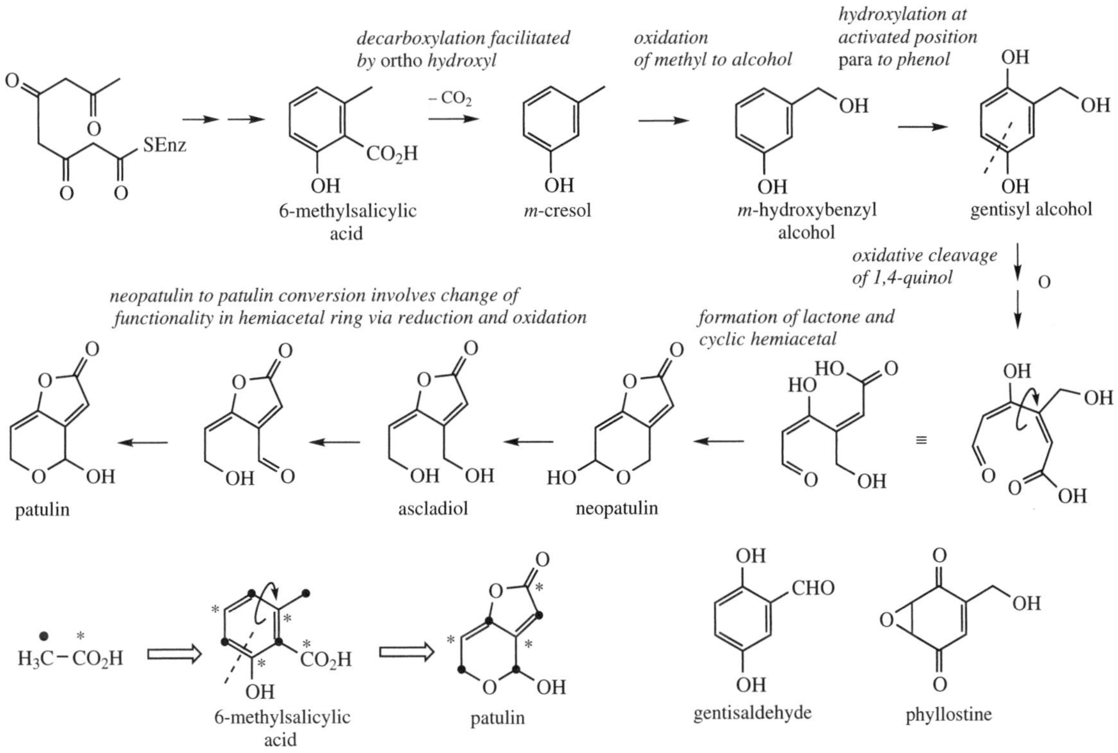

Figure 3.44

apple juice, fruit pies, etc, then these products may contain unacceptable and dangerous levels of patulin. Such food materials are routinely screened for patulin content, with a tolerance level set at 50 μg kg⁻¹. Patulin is derived from acetate via **6-methylsalicylic acid** (Figure 3.26). Decarboxylation and hydroxylation reactions then lead to **gentisyl alcohol** (Figure 3.44), which may suffer oxidative cleavage as shown. Cleavage of the aromatic ring would generate aldehyde and carboxylic acid functions. By rotating the molecule around the carbon–carbon single bond as shown, it is easy to see that **neopatulin** can result by formation of hemiacetal and lactone groups. The reversal of functionality in the hemiacetal ring to produce **patulin** is achieved by reduction and oxidation reactions involving aldehyde and alcohol components of the hemiacetal. The sequence shown in Figure 3.44 has been deliberately simplified to rationalize the oxidative cleavage. The true sequence involves gentisaldehyde and the epoxyquinone phyllostine as intermediates between gentisyl alcohol and neopatulin.

Penicillic acid (Figure 3.45), another microbially produced food contaminant with carcinogenic properties, is synthesized by cultures of *Penicillium cyclopium* and *P. baarnense*, and also features oxidative ring fission of an aromatic compound. This time **orsellinic acid** (Figure 3.25) is a precursor, and ring fission appears to proceed via a quinone, which is the result of decarboxylation, oxidation, and methylation reactions. Figure 3.45 also represents an over-simplistic rationalization of the ring fission process.

Starter Groups Other Than Acetate

In the examples so far discussed, the basic carbon skeleton has been derived from an acetate starter group, with malonate acting as the chain extender. The molecule has then, in some cases, been made more elaborate by the inclusion of other carbon atoms, principally via alkylation reactions. However, the range of natural product structures that are at least partly derived from acetate is increased enormously by altering the nature of the

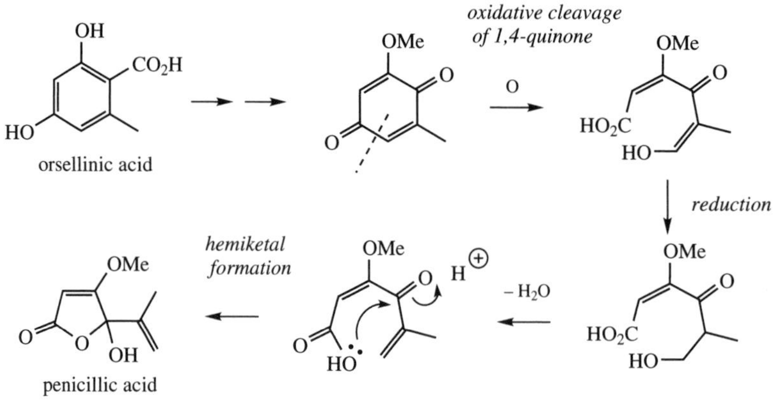

Figure 3.45

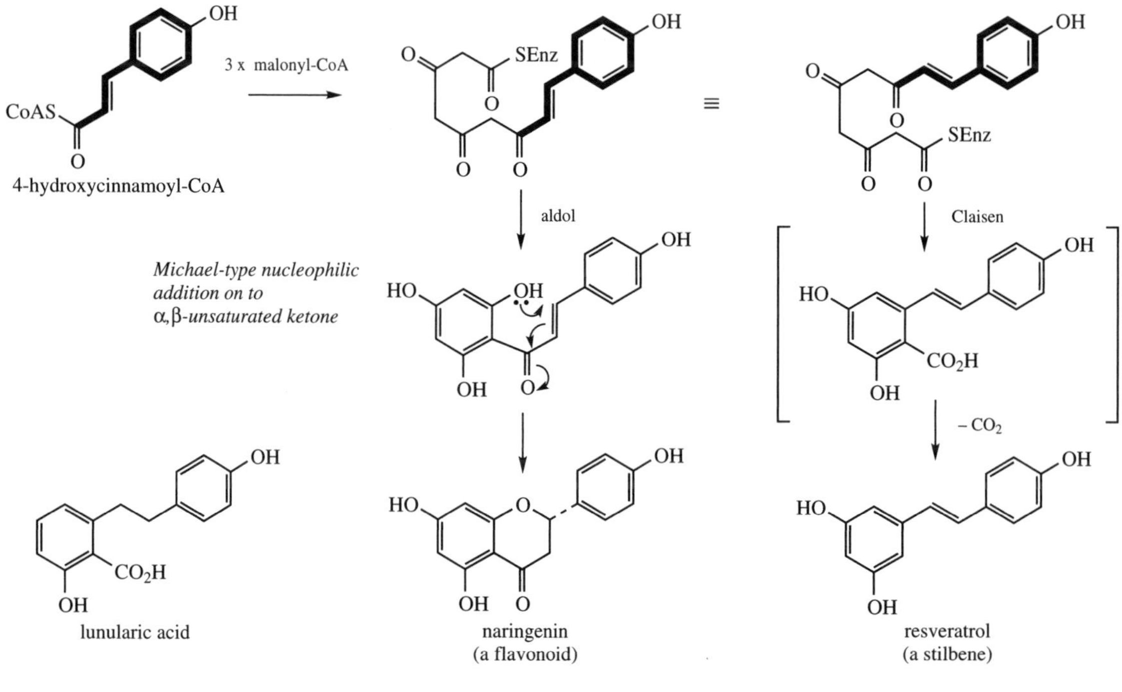

Figure 3.46

starter group from acetate to a different carboxylate system, as its coenzyme A ester, with malonyl-CoA again providing the chain extender. There is less detailed knowledge here about the precise nature of how substrates are bound to the enzyme, and whether coenzyme A esters are initially transformed into thio esters of the ACP type.

Flavonoids and **stilbenes** are simple examples of molecules in which a suitable cinnamoyl-CoA C_6C_3 precursor from the shikimate pathway (see

page 130) has acted as a starter group. Thus, if **4-hydroxycinnamoyl-CoA** (Figure 3.46) is chain extended with three malonyl-CoA units, the poly-β-keto chain can then be folded in two ways, allowing aldol or Claisen-type cyclizations to occur, respectively. The six-membered heterocyclic ring characteristic of most flavonoids, e.g. **naringenin**, is formed by nucleophilic attack of a phenol group from the acetate-derived ring on to the α,β-unsaturated ketone. Stilbenes, such as

resveratrol, incorporate the carbonyl carbon of the cinnamoyl unit into the aromatic ring, and typically lose the end-of-chain carboxyl by a decarboxylation reaction. Although some related structures, e.g. **lunularic acid** from the liverwort *Lunularia cruciata*, still contain this carboxyl, in general it is lost in a pre-cyclization modification, and intermediates of the type shown in brackets are not produced. Flavonoids and stilbenes are discussed in more detail in Chapter 4 (see page 149).

Anthranilic acid (2-aminobenzoic acid) (see page 126) is another shikimate-derived compound which, as its CoA ester anthraniloyl-CoA, can act as a starter unit for malonate chain extension. Aromatization of the acetate-derived portion then leads to **quinoline** or **acridine** alkaloids, according to the number of acetate units incorporated (Figure 3.47). These products are similarly discussed elsewhere, under alkaloids (Chapter 6, page 376).

Fatty acyl-CoA esters are similarly capable of participating as starter groups. Fatty acid biosynthesis and aromatic polyketide biosynthesis are distinguished by the sequential reductions as the chain length increases in the former, and by the stabilization of a reactive poly-β-keto chain in the latter, with little or no reduction involved. It is thus interesting to see natural product structures containing both types of acetate–malonate-derived chains. In plants of the Anacardiaceae, e.g. poison ivy* (*Rhus radicans*) and poison oak* (*Rhus toxicodendron*), contact allergens called **urushiols** are encountered, which derive from just such a pathway. Thus, **palmitoleoyl-CoA** (Δ^9-hexadecenoyl-CoA) can act as starter group for extension by three malonyl-CoA units, with a reduction step during chain extension (Figure 3.48). Aldol cyclization then gives **anacardic acid**, which is likely to be the precursor of **urushiol** by decarboxylation/hydroxylation. It is likely that different fatty acyl-CoAs can participate in this sequence, since urushiols from poison ivy can contain up to three double bonds in the C_{15} side-chain, whilst those from poison oak also have variable unsaturation

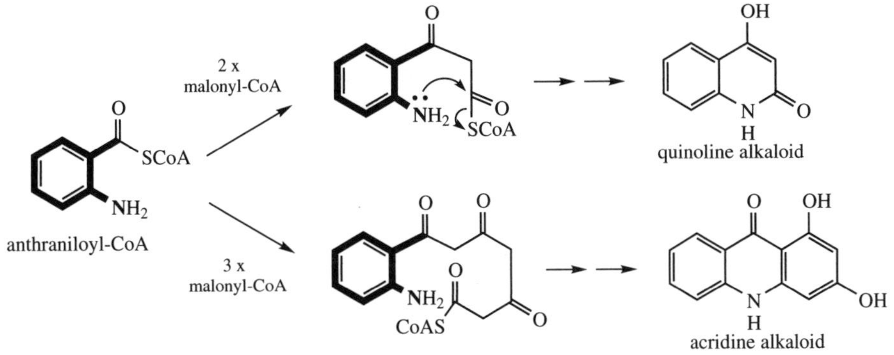

Figure 3.47

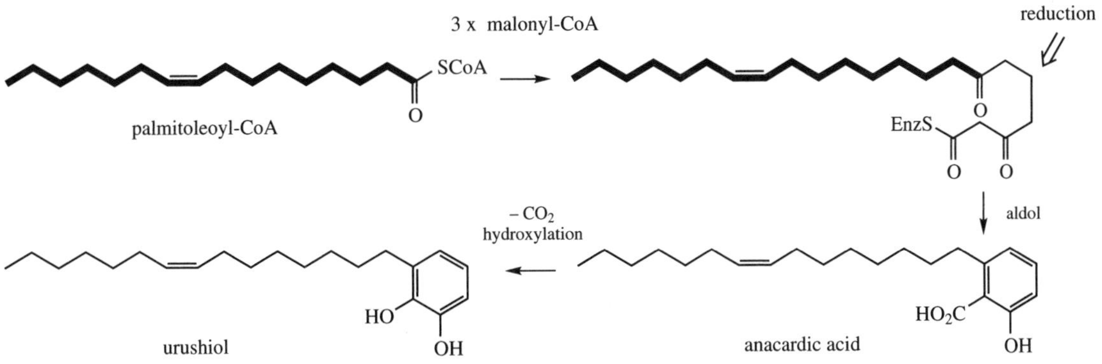

Figure 3.48

Poison Ivy and Poison Oak

Poison ivy (*Rhus radicans* or *Toxicodendron radicans*; Anacardiaceae) is a woody vine with three-lobed leaves that is common in the USA. The plant may be climbing, shrubby, or may trail over the ground. It presents a considerable hazard to humans should the sap, which exudes from damaged leaves or stems, come into contact with the skin. The sap sensitizes most individuals, producing delayed contact dermatitis after a subsequent encounter. This results in watery blisters that break open, the fluid quickly infecting other parts of the skin. The allergens may be transmitted from one person to another on the hands, on clothing, or by animals. The active principles are urushiols, a mixture of alkenyl polyphenols. In poison ivy, these are mainly pentadecylcatechols with varying degrees of unsaturation (Δ^8, $\Delta^{8,11}$, $\Delta^{8,11,14}$) in the side-chain. Small amounts of C_{17} side-chain analogues are present. These catechols become oxidized to an *ortho*-quinone, which is then attacked by nucleophilic groups in proteins to yield an antigenic complex.

Poison oak (*Rhus toxicodendron* or *Toxicodendron toxicaria*: Anacardiaceae) is nearly always found as a low-growing shrub, and has lobed leaflets similar to those of oak. It is also common throughout North America. There appears considerable confusion over nomenclature, and *Rhus radicans* may also be termed poison oak, and *R. toxicodendron* oakleaf poison ivy. Poison oak contains similar urushiol structures in its sap as poison ivy, though heptadecylcatechols (i.e. C_{17} side-chains) predominate over pentadecylcatechols (C_{15} side-chains).

Related species of *Rhus*, e.g. *R. diversiloba* (Pacific poison oak) and *R. vernix* (poison sumach, poison alder, poison dogwood) are also allergenic with similar active constituents. The allergen-containing species of *Rhus* have been reclassified under the genus *Toxicodendron*, though this nomenclature is not commonly employed. Dilute purified extracts containing urushiols may be employed to stimulate antibody production and thus build up immunity to the allergens.

in a C_{17} side-chain. Large quantities of anacardic acids containing C_{15} side-chains with one, two, and three double bonds are also found in the shells of cashew nuts (*Anacardium occidentale*; Anacardiaceae).

A saturated C_6 **hexanoate** starter unit is used in the formation of the **aflatoxins***, a group of highly toxic metabolites produced by *Aspergillus flavus*, and probably responsible for the high incidence of liver cancer in some parts of Africa. These compounds were first detected following the deaths of young turkeys fed on mould-contaminated peanuts (*Arachis hypogaea*; Leguminosae/Fabaceae). Peanuts still remain one of the crops most likely to represent a potential risk to human health because of contamination with fungal toxins. These and other food materials must be routinely screened to ensure levels of aflatoxins do not exceed certain set limits. The aflatoxin structures contain a bisfuran unit fused to an aromatic

ring, e.g. **aflatoxin B_1** and **aflatoxin G_1**, and their remarkably complex biosynthetic origin begins with a poly-β-keto chain derived from a hexanoyl-CoA starter and seven malonyl-CoA extender units (Figure 3.49). This gives an anthraquinone **norsolorinic acid** by now-familiar condensation reactions, but the folding of the chain is rather different from that seen with simpler anthraquinones (see page 64). The six-carbon side-chain of norsolorinic acid is cyclized to give, in several steps, the ketal **averufin**. **Versiconal acetate** is another known intermediate, and its formation involves a Baeyer–Villiger oxidation (see page 28), resulting principally in transfer of a two-carbon fragment (the terminal ethyl of hexanoate) to become an ester function. These two carbons can then be lost in formation of **versicolorin B**, now containing the tetrahydrobisfuran moiety, oxidized in **versicolorin A** to a dihydrobisfuran system. **Sterigmatocystin** is derived from versicolorin A by

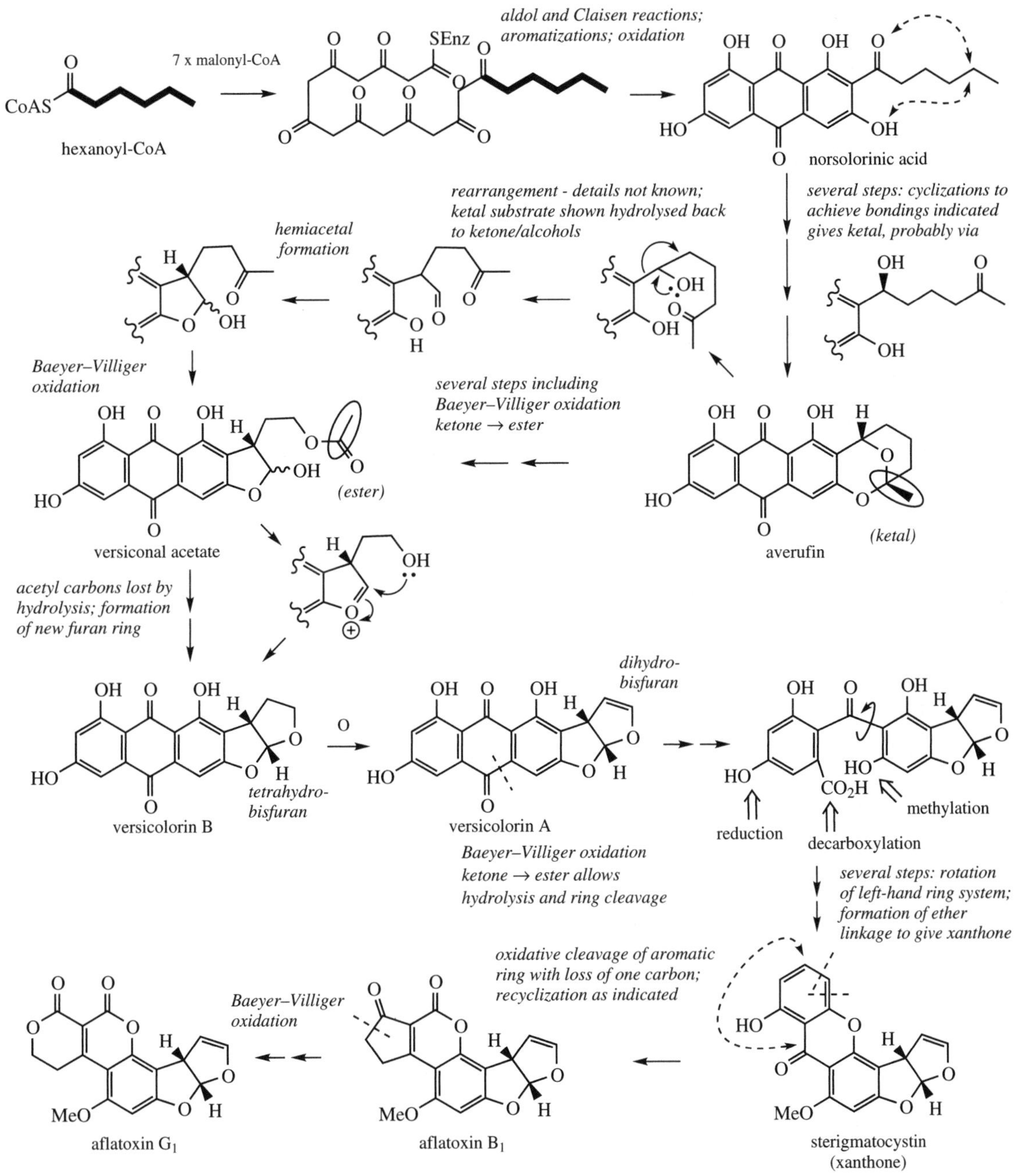

Figure 3.49

oxidative cleavage of the anthraquinone system involving a second Baeyer–Villiger oxidation, and recyclization through phenol groups to give a xanthone skeleton. Rotation of an intermediate leads to the angular product as opposed to a linear product. One phenol group is methylated,

and, quite unusually, another phenol group is lost (contrast loss of oxygen functions via reduction/dehydration prior to cyclization, see page 62). **Aflatoxin B₁** formation requires oxidative cleavage of an aromatic ring in sterigmatocystin, loss of one carbon and recyclization exploiting the

Aflatoxins

Aflatoxins are potent mycotoxins produced by the fungi *Aspergillus flavus* and *A. parasiticus*. Four main naturally occurring aflatoxins, aflatoxins B_1, B_2, G_1, and G_2 (Figure 3.50), are recognized, but these can be metabolized by microorganisms and animals to other aflatoxin structures, which are also toxic. Aflatoxin B_1 is the most commonly encountered member of the group, and is also the most acutely toxic and carcinogenic example. Aflatoxin B_2 is a dihydro derivative of aflatoxin B_1, whilst aflatoxins G_1 and G_2 are an analogous pair with a six-membered lactone rather than a five-membered cyclopentenone ring. These toxins are most commonly associated with peanuts (groundnuts), maize, rice, pistachio nuts, and Brazil nuts, though other crops can be affected, and, although found world-wide, they are partic-ularly prevalent in tropical and subtropical regions. Aflatoxin M_1 (Figure 3.50) is a hydroxy derivative of aflatoxin B_1 and equally toxic. It may occur in cow's milk as a result of mammalian metabolism of aflatoxin B_1 originally contaminating the animal's food. Because these com-pounds fluoresce strongly under UV light, they are relatively easily detected and monitored.

The aflatoxins primarily affect the liver, causing enlargement, fat deposition, and necrosis, at the same time causing cells of the bile duct to proliferate, with death resulting from irreversible loss of liver function. In the case of aflatoxin B_1, this appears to be initiated by cytochrome P-450-dependent metabolism in the body to the epoxide (Figure 3.50). The epoxide intercalates with DNA, and in so doing becomes orientated towards nucleophilic attack from guanine residues. This leads to inhibition of DNA replication and of RNA synthesis, and initiates mutagenic activity. Aflatoxins are also known to cause hepatic carcinomas, this varying with the species of animal. The above normal incidence of liver cancer in parts of Africa and Asia has been suggested to be linked to the increased amounts of aflatoxins found in foodstuffs, and a tolerance level of 30 ppb has been recommended. Acute hepatitis may result from food containing aflatoxin B_1 at levels of the order of 0.1 ppm, and levels of more than 1 ppm are frequently encountered.

The biosynthesis of aflatoxins proceeds through intermediates sterigmatocystin and versicolorin (see Figure 3.49). Toxins related to these structures but differing in aromatic substituents are also produced by various fungi. The sterigmatocystins are synthesized by species of *Aspergillus* and *Bipolaris*, and contain a reduced bifuran fused to a xanthone, whilst the versicolorins from *Aspergillus versicolor* contain the same type of reduced bisfuran system but fused to an anthraquinone. Like the aflatoxins, the sterigmatocystins are acutely toxic and carcinogenic. The versicolorins are less toxic though still carcinogenic.

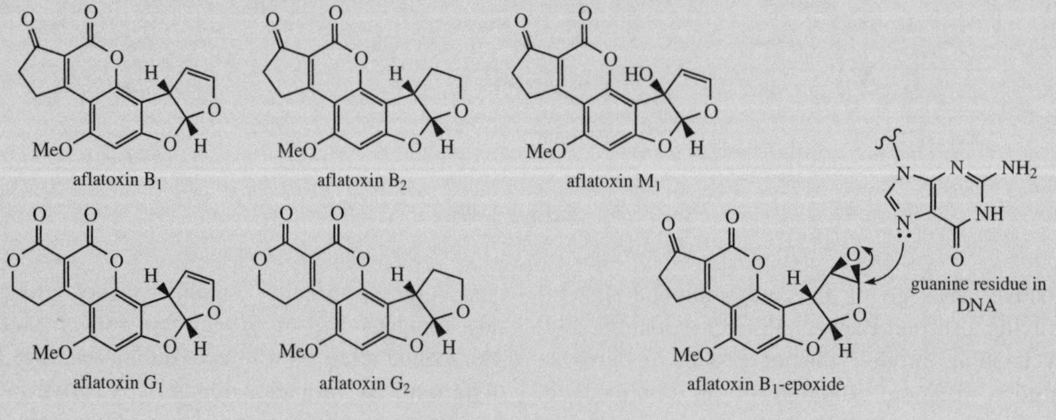

aflatoxin B_1 aflatoxin B_2 aflatoxin M_1

guanine residue in DNA

aflatoxin G_1 aflatoxin G_2 aflatoxin B_1-epoxide

Figure 3.50

carbonyl functionality. **Aflatoxin G$_1$** is derived by further modification of aflatoxin B$_1$, cleaving the cyclopentenone ring and forming a lactone, perhaps via a further Baeyer–Villiger reaction.

Hexanoate is also likely to feature as a starter unit in the formation of the **cannabinoids**, a group of **terpenophenolics** found in Indian hemp (*Cannabis sativa*; Cannabaceae). This plant, and preparations from it, known under a variety of names including hashish, marihuana, pot, bhang, charas, and dagga, have been used for centuries for the pleasurable sensations and mild euphoria experienced after its consumption, usually by smoking.

The principal psychoactive component is **tetrahydrocannabinol** (**THC**) (Figure 3.51), whilst structurally similar compounds such as **cannabinol** (**CBN**) and **cannabidiol** (**CBD**), present in similar or larger amounts, are effectively inactive. In recent years, the beneficial effects of cannabis*, and especially THC, in alleviating nausea and vomiting in cancer patients undergoing chemotherapy, and in the treatment of glaucoma and multiple sclerosis, has led to a study of cannabinoid analogues for potentially useful medicinal activity. All the cannabinoid structures contain a monoterpene C$_{10}$ unit attached to a phenolic ring having a C$_5$ alkyl chain. The aromatic ring/C$_5$ chain

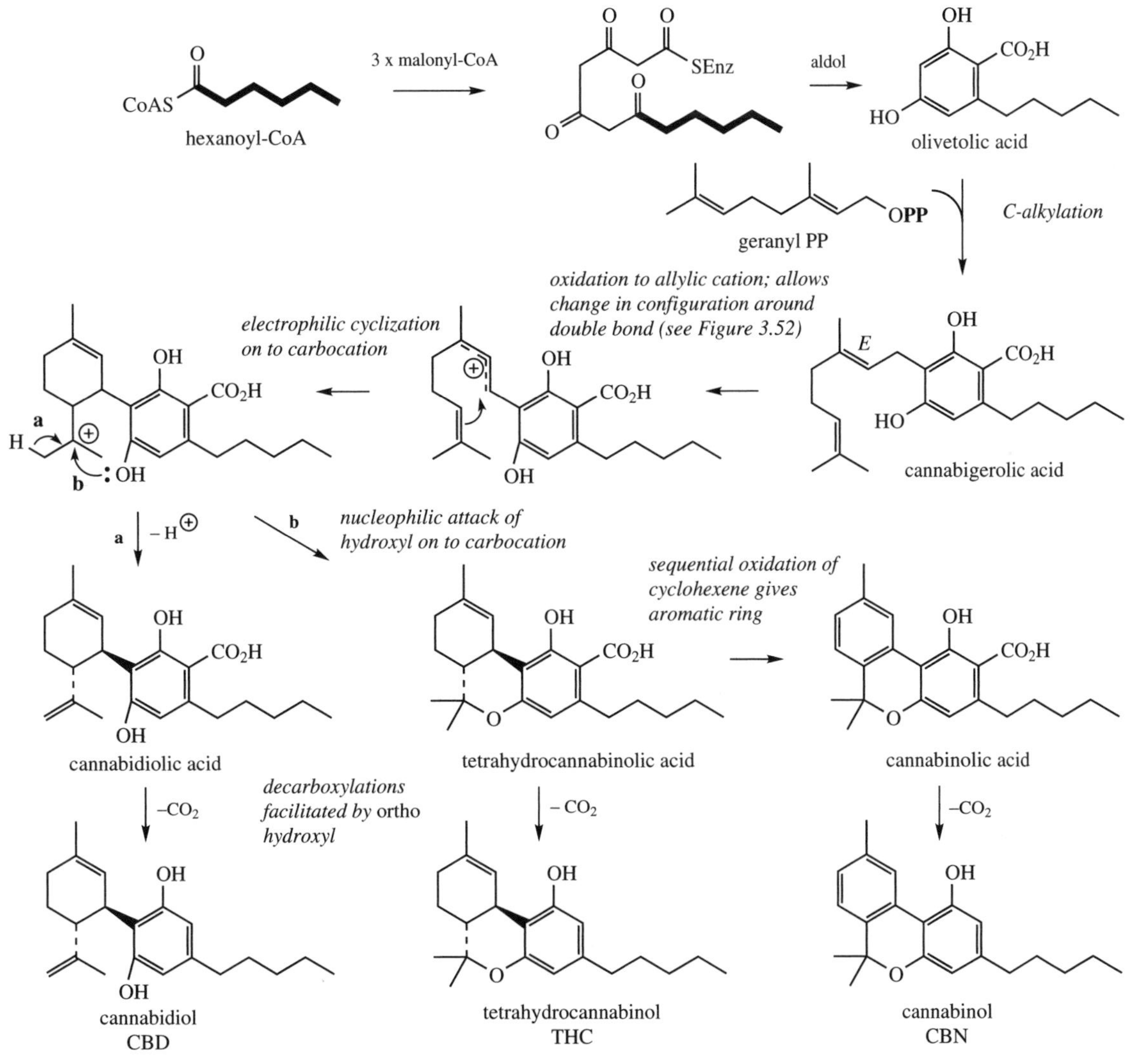

Figure 3.51

is likely to originate from hexanoate and malonate, cyclization to a polyketide giving **olivetolic acid**, from which **cannabigerolic acid** can be obtained by *C*-alkylation with the monoterpene unit geranyl diphosphate (Figure 3.51). Cyclization in the monoterpene unit necessitates a change in configuration of the double bond, and this may be rationalized as involving the allylic cation, which will then also allow electrophilic cyclization to proceed (for further detail see Figure 3.52,

and compare terpenoid cyclization mechanisms, page 173). **Cannabidiolic acid** is the result of proton loss, whilst **tetrahydrocannabinolic acid** is the product from heterocyclic ring formation. **CBD** and **THC** are then the respective decarboxylation products from these two compounds. The aromatic terpenoid derived ring in **cannabinolic acid** and **cannabinol** can arise via a dehydrogenation process (compare thymol, page 186).

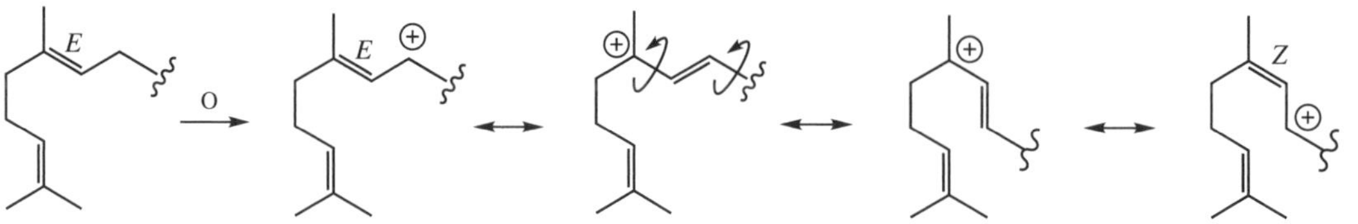

Figure 3.52

Cannabis

Indian hemp, *Cannabis sativa* (Cannabaceae) is an annual herb indigenous to Central and Western Asia, cultivated widely in India and many tropical and temperate regions for its fibre (hemp) and seed (for seed oil). The plant is also grown for its narcotic and mild intoxicant properties, and in most countries of the world its possession and consumption is illegal. Over many years, cannabis plants have been selected for either fibre production or drug use, the former resulting in tall plants with little pharmacological activity, whilst the latter tend to be short, bushy plants. Individual plants are almost always male or female, though the sex is not distinguishable until maturity and flowering. Seeds will produce plants of both sexes in roughly equal proportions. The active principles are secreted as a resin by glandular hairs, which are more numerous in the upper parts of female plants, and resin is produced from the time flowers first appear until the seeds reach maturity. However, all parts of the plant, both male and female, contain cannabinoids. In a typical plant, the concentration of cannabinoids increases in the following order: large leaves, small leaves, flowers, and bracts (which surround the ovaries), with stems containing very little. Material for drug use (ganja) is obtained by collecting the flowering tops (with little leaf) from female plants, though lower quality material (bhang) consisting of leaf from both female and male plants may be employed. By rubbing the flowering tops, the resin secreted by the glandular hairs can be released and subsequently scraped off to provide cannabis resin (charas) as an amorphous brown solid or semi-solid. A potent form of cannabis, called cannabis oil, is produced by alcoholic extraction of cannabis resin. A wide variety of names are used for cannabis products according to their nature and the geographical area. In addition to the Indian words above, the names hashish (Arabia), marihuana (Europe, USA), kief and dagga (Africa) are frequently used. The term 'assassin' is a corruption of 'hashishin', a group of 13th century murderous Persians who were said to have been rewarded for their activities with hashish. The names grass, dope, pot, hash, weed, and wacky backy are more likely to be in current usage.

(Continues)

(*Continued*)

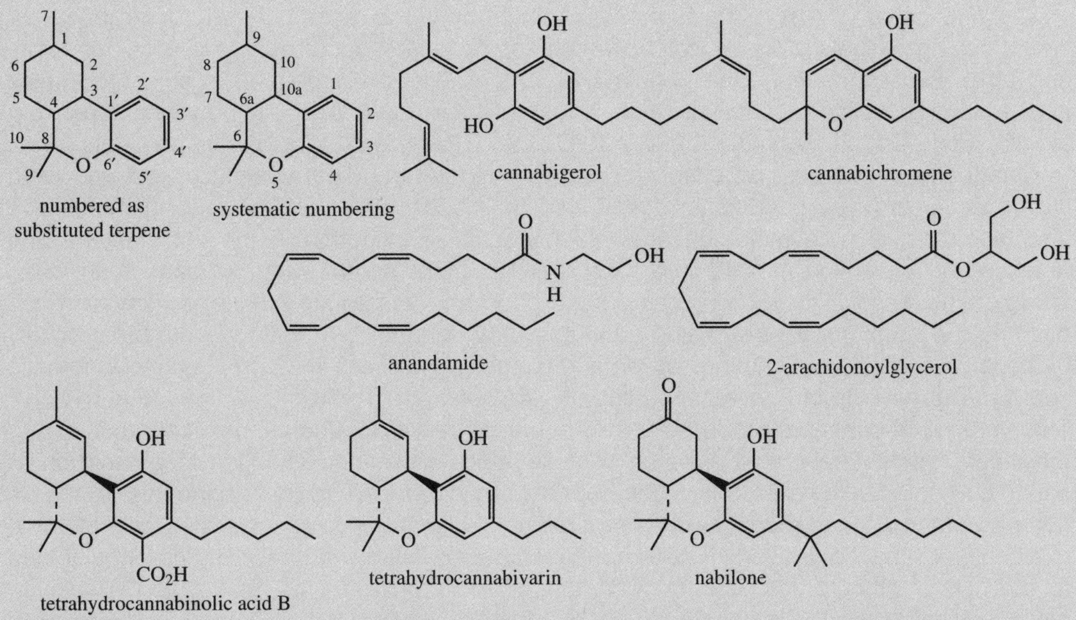

Figure 3.53

The quantity of resin produced by the flowering tops of high quality Indian cannabis is about 15–20%. The amount produced by various plants is dependent on several features, however, and this will markedly alter biological properties. Thus, in general, plants grown in a tropical climate produce more resin than those grown in a temperate climate. The tall fibre-producing plants are typically low resin producers, even in tropical zones. However, the most important factor is the genetic strain of the plant, and the resin produced may contain high levels of psychoactive compounds, or mainly inactive constituents. The quality of any cannabis drug is potentially highly variable.

The major constituents in cannabis are termed cannabinoids, a group of more than 60 structurally related terpenophenolics. The principal psychoactive agent is tetrahydrocannabinol (THC) (Figure 3.51). This is variously referred to as Δ^1-THC or Δ^9-THC according to whether the numbering is based on the terpene portion, or as a systematic dibenzopyran (Figure 3.53). Both systems are currently in use. Also found, often in rather similar amounts, are cannabinol (CBN) and cannabidiol (CBD) (Figure 3.51), which have negligible psychoactive properties. These compounds predominate in the inactive resins. Many other cannabinoid structures have been characterized, including cannabigerol and cannabichromene (Figure 3.53). A range of cannabinoid acids, e.g. cannabidiolic acid, tetrahydrocannabinolic acid, and tetrahydrocannabinolic acid B (Figure 3.53) are also present, as are some analogues of the other compounds mentioned, where a propyl side-chain replaces the pentyl group, e.g. tetrahydrocannabivarin (Figure 3.53). The latter compounds presumably arise from the use of butyrate rather than hexanoate as starter unit in the biosynthetic sequence.

(*Continues*)

(Continued)

The THC content of high quality cannabis might be in the range 0.5–1% for large leaves, 1–3% for small leaves, 3–7% for flowering tops, 5–10% for bracts, 14–25% for resin, and up to 60% in cannabis oil. Higher amounts of THC are produced in selected strains known as skunk cannabis, so named because of their powerful smell; flowering tops from skunk varieties might contain 10–15% THC. The THC content in cannabis products tends to deteriorate on storage, an effect accelerated by heat and light. Cannabis leaf and resin stored under ordinary conditions rapidly lose their activity and can be essentially inactive after about 2 years. A major change which occurs is oxidation in the cyclohexene ring resulting in conversion of THC into CBN. THC is more potent when smoked than when taken orally, its volatility allowing rapid absorption and immediate effects, so smoking has become the normal means of using cannabis. Any cannabinoid acids will almost certainly be decarboxylated upon heating, and thus the smoking process will also effectively increase somewhat the levels of active cannabinoids available, e.g. THC acid → THC (Figure 3.51). The smoking of cannabis produces a mild euphoria similar to alcohol intoxication, inducing relaxation, contentment, and a sense of well-being, with some changes in perception of sound and colour. However, this is accompanied by a reduced ability to concentrate and do complicated tasks, and a loss of short-term memory. Users claim cannabis is much preferable to alcohol or tobacco, insisting it does not cause dependence, withdrawal symptoms, or lead to the use of other drugs, and they campaign vociferously for its legalization. However, psychological dependence does occur, and cannabis can lead to hallucinations, depression, anxiety, and panic, with the additional risk of bronchitis and lung cancer if the product is smoked.

Cannabis has been used medicinally, especially as a mild analgesic and tranquillzer, but more effective and reliable agents replaced it, and even controlled prescribing was discontinued. In recent times, cannabis has been shown to have valuable anti-emetic properties, which help to reduce the side-effects of nausea and vomiting caused by cancer chemotherapeutic agents. This activity stems from THC, and has resulted in some use of **THC (dronabinol)** and the prescribing of cannabis for a small number of patients. A synthetic THC analogue, **nabilone** (Figure 3.53), has been developed as an anti-emetic drug for reducing cytotoxic-induced vomiting. Some of the psychoactive properties of THC, e.g. euphoria, mild hallucinations, and visual disturbances, may be experienced as side-effects of nabilone treatment. Cannabis has also been shown to possess properties which may be of value in other medical conditions. There is now ample evidence that cannabis can give relief to patients suffering from chronic pain, multiple sclerosis, glaucoma, asthma, migraine, epilepsy, and other conditions. Many sufferers who cannot seem to benefit from any of the current range of drugs are obtaining relief from their symptoms by using cannabis, but are breaking the law to obtain this medication. Current thinking is that cannabis offers a number of beneficial pharmacological responses and that there should be legal prescribing of cannabinoids or derivatives. Clinical trials have already confirmed the value of cannabis and/or THC taken orally for the relief of chronic pain and the painful spasms characteristic of multiple sclerosis, and in reducing intraocular pressure in glaucoma sufferers. In general, cannabis is only able to alleviate the symptoms of these diseases, and does not provide a cure. The non-psychoactive CBD has been shown to have anti-inflammatory properties potentially useful in arthritis treatment.

Recently, the ethanolamide of arachidonic acid (anandamide; ananda is the Sanskrit word for bliss) (Figure 3.53) has been isolated from animal brain tissue, and has been shown

(Continued)

to mimic several of the pharmacological properties of THC. This appears to be a natural ligand which interacts with central receptors (CB1) to which cannabinoids also bind. Two other polyunsaturated fatty acid ethanolamides, namely dihomo-γ-linolenoyl- (20:3) and adrenoyl- (22:4) ethanolamides have also been isolated from mammalian brain, and shown to have THC-like properties. Another type of cannabinoid receptor (CB2), expressed mainly in the immune system, has been identified; its natural ligand is 2-arachidonoylglycerol (Figure 3.53). Since this compound also interacts with the anandamide receptor, and levels of 2-arachidonoylglycerol in the brain are some 800 times higher than those of anandamide, it is now thought to be the physiological ligand for both receptors, rather than anandamide. The identification of these endogenous materials may open up other ways of exploiting some of the desirable pharmacological features of cannabis.

Table 3.3 Tetracyclines

	R^1	R^2	R^3	R^4	R^5	
	5	6α	6β	7		
tetracycline	H	Me	OH	H	H	natural
chlortetracycline	H	Me	OH	Cl	H	
oxytetracycline	OH	Me	OH	H	H	
demeclocycline	H	H	OH	H	H	
methacycline	OH	=CH$_2$		H	H	semi-synthetic
doxycycline	OH	Me	H	H	H	
minocycline	H	H	H	NMe$_2$	H	
lymecycline	H	Me	OH	H		

The **tetracyclines*** (Table 3.3) are a group of broad spectrum, orally active antibiotics produced by species of *Streptomyces*, and several natural and semi-synthetic members are used clinically. They contain a linear tetracyclic skeleton of polyketide origin in which the starter group is **malonamyl-CoA** (Figure 3.54), i.e. the coenzyme A ester of malonate semi-amide. Thus, in contrast to most acetate-derived compounds, malonate supplies all carbon atoms of the tetracycline skeleton, the starter group as well as the chain extenders. The main features of the pathway (Figure 3.54) were deduced from extensive studies of mutant strains of *Streptomyces aureofaciens* with genetic blocks causing accumulation of mutant metabolites or production of abnormal tetracyclines. This organism typically produces **chlortetracycline**, whilst the parent compound **tetracycline** (Table 3.3) is in fact an aberrant product synthesized in mutants blocked in the chlorination step. The use of mutants with genetic blocks has also enabled the shikimate pathway (Chapter 4) to be delineated. In that case, since a primary metabolic pathway was affected, mutants tended to accumulate intermediates and could not grow unless later components of the pathway were supplied. With the tetracyclines, a secondary metabolic pathway is involved, and the relatively broad specificity of some of the

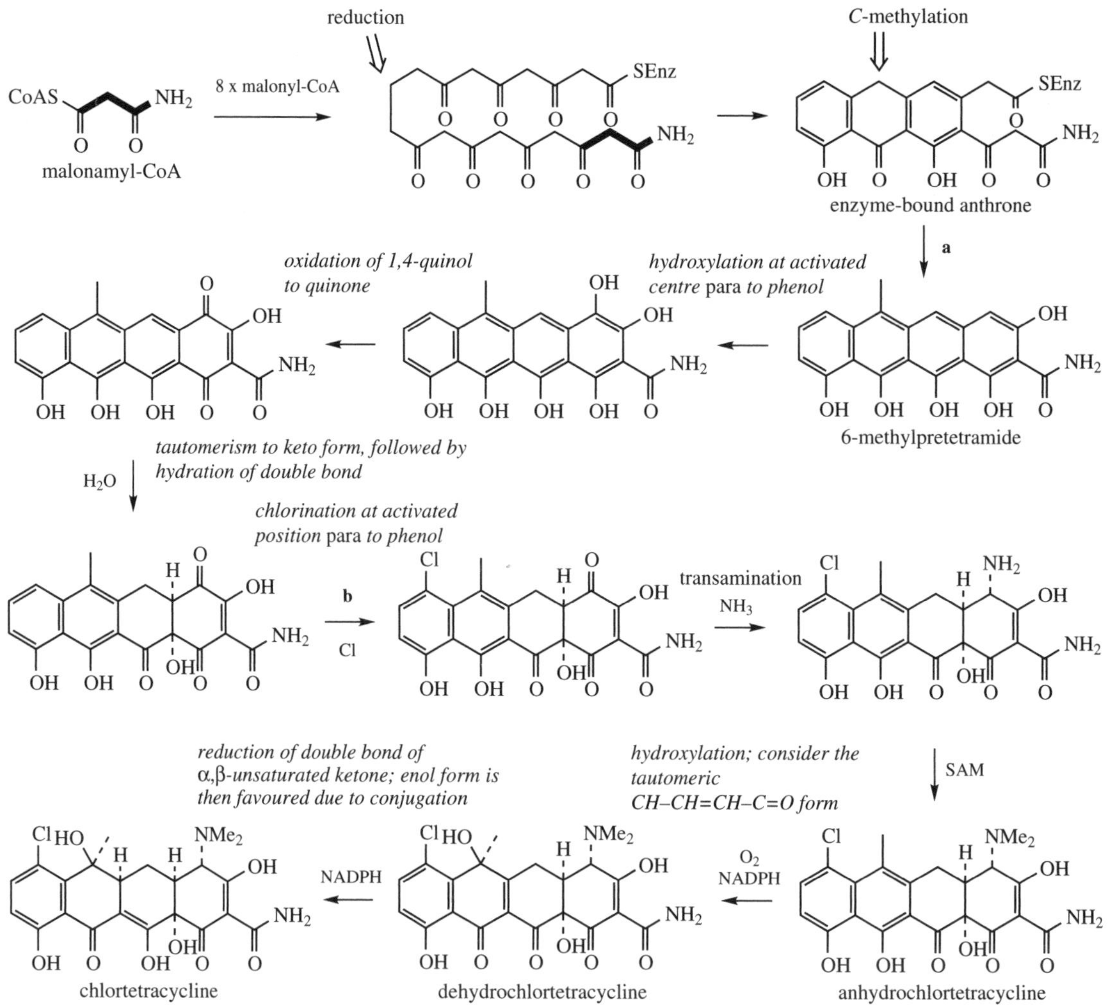

Figure 3.54

Tetracyclines

The tetracyclines (Table 3.3) are a group of broad spectrum, orally active antibiotics produced by cultures of *Streptomyces* species. **Chlortetracycline** isolated from *Streptomyces aureofaciens* was the first of the group to be discovered, closely followed by **oxytetracycline** from cultures of *S. rimosus*. **Tetracycline** was found as a minor antibiotic in *S. aureofaciens*, but may be produced in quantity by utilizing a mutant strain blocked in the chlorination step *b* (Figure 3.54). Similarly, the early C-6 methylation step (included in *a*) can also be blocked, and such mutants accumulate 6-demethyltetracyclines, e.g. **demeclocycline** (**deme**thyl**chloro**tetra**cycline**). These reactions can also be inhibited in the normal strain of *S. aureofaciens* by supplying cultures with either aminopterin (which inhibits C-6 methylation) or mercaptothiazole (which inhibits C-7 chlorination). Oxytetracycline from *S. rimosus* lacks

(*Continues*)

the chlorine substituent, but has an additional 5α-hydroxyl group, probably introduced at a late stage. Only minor alterations can be made to the basic tetracycline structure to modify the antibiotic activity, and these are at positions 5, 6, and 7. Other functionalities in the molecule are all essential to retain activity. Semi-synthetic tetracyclines used clinically include **methacycline**, obtained by a dehydration reaction from oxytetracycline, and **doxycycline**, via reduction of the 6-methylene in methacycline. **Minocycline** contains a 7-dimethylamino group and is produced by a sequence involving aromatic nitration. **Lymecycline** is an example of an antibiotic developed by chemical modification of the primary amide function at C-2.

Having both amino and phenolic functions, tetracyclines are amphoteric compounds, and are more stable in acid than under alkaline conditions. They are thus suitable for oral administration, and are absorbed satisfactorily. However, because of the sequence of phenol and carbonyl substituents in the structures, they act as chelators and complex with metal ions, especially calcium, aluminium, iron, and magnesium. Accordingly, they should not be administered with foods such as milk and dairy products (which have a high calcium content), aluminium- and magnesium-based antacid preparations, iron supplements, etc, otherwise erratic and unsatisfactory absorption will occur. A useful feature of doxycycline and minocycline is that their absorptions are much less affected by metal ions. Chelation of tetracyclines with calcium also precludes their use in children developing their adult teeth, and in pregnant women, since the tetracyclines become deposited in the growing teeth and bone. In children, this would cause unsightly and permanent staining of teeth with the chelated yellow tetracycline.

Although the tetracycline antibiotics have a broad spectrum of activity spanning Gram-negative and Gram-positive bacteria, their value has decreased as bacterial resistance has developed in pathogens such as *Pneumococcus*, *Staphylococcus*, *Streptococcus*, and *E. coli*. These organisms appear to have evolved mechanisms of resistance involving decreased cell permeability; a membrane-embedded transport protein exports the tetracycline out of the cell before it can exert its effect. Nevertheless, tetracyclines are the antibiotics of choice for infections caused by *Chlamydia*, *Mycoplasma*, *Brucella*, and *Rickettsia*, and are valuable in chronic bronchitis due to activity against *Haemophilus influenzae*. They are also used systemically to treat severe cases of acne, helping to reduce the frequency of lesions by their effect on skin flora. There is little significant difference in the antimicrobial properties of the various agents, except for **minocycline**, which has a broader spectrum of activity, and being active against *Neisseria meningitidis* is useful for prophylaxis of meningitis. The individual tetracyclines do have varying bioavailabilities, however, which may influence the choice of agent. **Tetracycline** and **oxytetracycline** are probably the most commonly prescribed agents. Tetracyclines are formulated for oral application or injection, as ear and eye drops, and for topical use on the skin. **Doxycycline** also finds use as a prophylactic against malaria in areas where there is widespread resistance to chloroquine and mefloquine (see page 363).

Their antimicrobial activity arises by inhibition of protein synthesis. This is achieved by interfering with the binding of aminoacyl-tRNA to acceptor sites on the ribosome by disrupting the codon–anticodon interaction (see page 407). Evidence points to a single strong binding site on the smaller 30S subunit of the ribosome. Although tetracyclines can also bind to mammalian ribosomes, there appears to be preferential penetration into bacterial cells, and there are few major side-effects from using these antibiotics.

A series of tetracycline derivatives has recently been isolated from species of *Dactylosporangium*. These compounds, the dactylocyclines (Figure 3.55), are glycosides and have the opposite configuration at C-6 to the natural tetracyclines. Importantly, these compounds are active towards tetracycline-resistant bacteria.

(Continues)

(Continued)

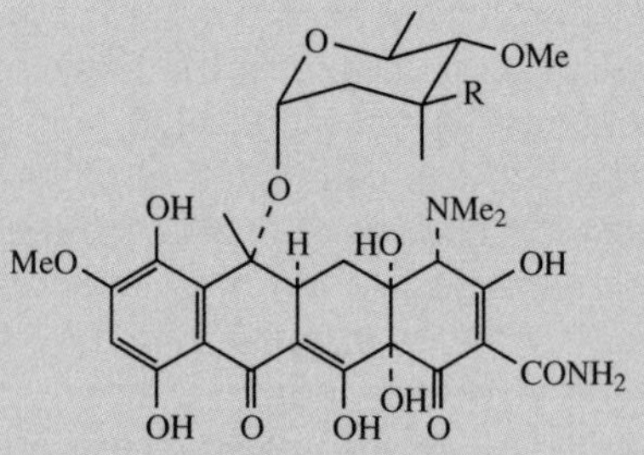

R = NHOH, dactylocycline-A
R = NO$_2$, dactylocycline-B
R = NHOAc, dactylocycline-C
R = OH, dactylocycline-E

Figure 3.55

enzymes concerned allows many of the later steps to proceed even if one step, e.g. the chlorination, is not achievable. This has also proved valuable for production of some of the clinical tetracycline antibiotics.

One of the early intermediates in the pathway to chlortetracycline is 6-methylpretetramide (Figure 3.54). This arises from the poly-β-keto ester via an enzyme-bound anthrone (compare Figure 3.30). Reduction of one carbonyl will occur during chain extension, whilst the methylation must be a later modification. Hydroxylation in ring A followed by oxidation gives a quinone, the substrate for hydration at the A/B ring fusion. The product now features the keto tautomer in ring B, since its aromaticity has been destroyed. Chlorination of ring D at the nucleophilic site *para* to the phenol follows, and an amine group is then introduced stereospecifically into ring A by a transamination reaction. This amino function is then di-*N*-methylated using SAM as the methylating agent yielding **anhydrochlortetracycline**. In the last two steps, C-6 is hydroxylated via an O$_2$-, NADPH-, and flavin-dependent oxygenase giving the enone **dehydrochlortetracycline**, and NADPH reduction of the C-5a/11a double bond generates **chlortetracycline**.

A number of **anthracycline antibiotics***, e.g. **doxorubicin** (Figure 3.56) from *Streptomyces peuceticus* and **daunorubicin** from *S. coeruleorubicus*, have structurally similar tetracyclic skeletons and would appear to be related to the tetracyclines. There are similarities in that the molecules are essentially acetate derived, but for

the anthracyclines the starter group is **propionate** rather than malonamide, and labelling studies have demonstrated a rather different folding of the poly-β-keto chain (Figure 3.56). As a result, the end-of-chain carboxyl is ultimately lost through decarboxylation. This carboxyl is actually retained for a considerable portion of the pathway, and is even protected against decarboxylation by methylation to the ester, until no longer required. Most of the modifications which occur during the biosynthetic pathway are easily predictable. Thus, the anthraquinone portion is likely to be formed first, then the fourth ring can be elaborated by a aldol reaction (Figure 3.56). A feature of note in molecules such as doxorubicin and daunorubicin is the amino sugar L-daunosamine which originates from TDPglucose (thymidine diphosphoglucose; compare UDPglucose, page 29) and is introduced in the latter stages of the sequence. Hydroxylation of daunorubicin to doxorubicin is the very last step. Doxorubicin and daunorubicin are used as antitumour drugs rather than antimicrobial agents. They act primarily at the DNA level and so also have cytotoxic properties. Doxorubicin in particular is a highly successful and widely used antitumour agent, employed in the treatment of leukaemias, lymphomas, and a variety of solid tumours.

MACROLIDES AND POLYETHERS

Extender Groups other than Malonate

The use of propionate as a starter group as in the formation of the anthracyclines is perhaps

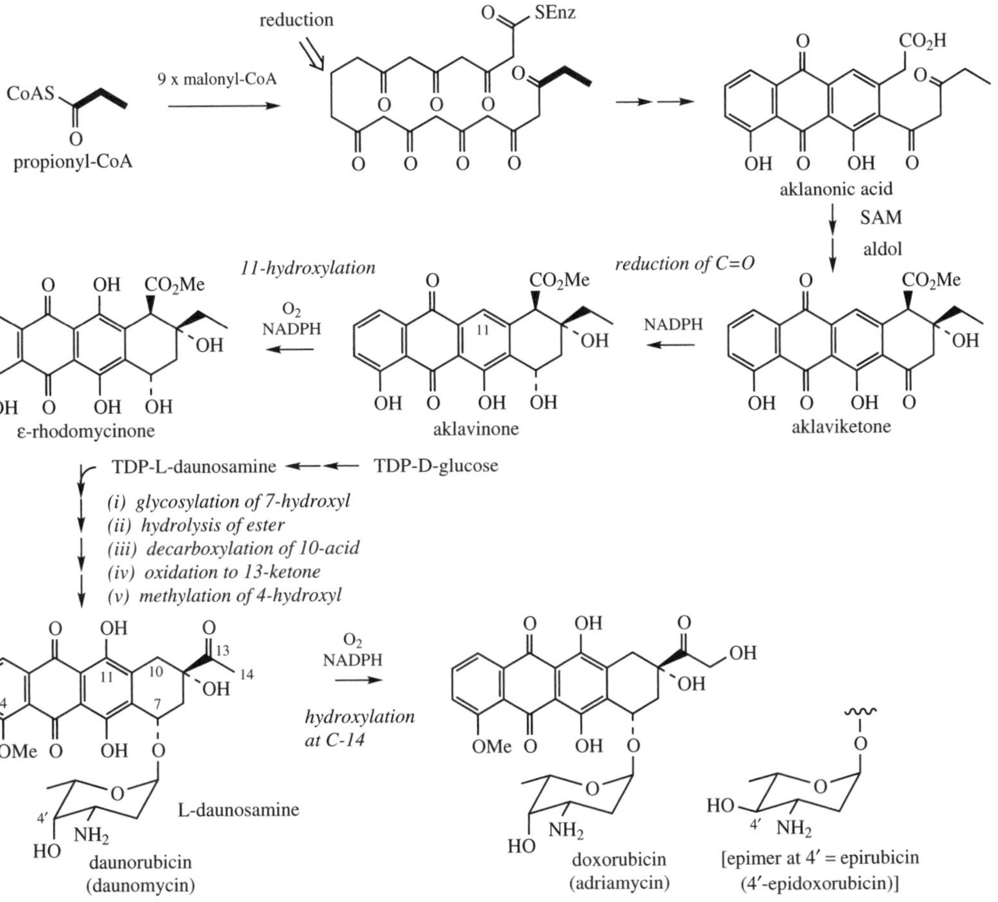

Figure 3.56

Anthracycline Antibiotics

Doxorubicin (adriamycin) (Figure 3.56) is produced by cultures of *Streptomyces peucetius* var *caesius* and is one of the most successful and widely used antitumour drugs. The organism is a variant of *S. peucetius*, a producer of daunorubicin (see below), in which mutagen treatment resulted in expression of a latent hydroxylase enzyme and thus synthesis of doxorubicin by 14-hydroxylation of daunorubicin. Doxorubicin has one of the largest spectra of antitumour activity shown by antitumour drugs and is used to treat acute leukaemias, lymphomas, and a variety of solid tumours. It is administered by intravenous injection and largely excreted in the bile. It inhibits the synthesis of RNA copies of DNA by intercalation of the planar molecule between base pairs on the DNA helix. The sugar unit provides further binding strength and also plays a major role in sequence-recognition for the binding. Doxorubicin also exerts some of its cytotoxic effects by inhibition of the enzyme topoisomerase II, which is responsible for cleaving and resealing of double-stranded DNA during replication (see page 137). Common toxic effects include nausea and vomiting, bone marrow suppression, hair loss, and local tissue necrosis, with cardiotoxicity at higher dosage. **Daunorubicin** (Figure 3.56) is produced by *Streptomyces coeruleorubidus* and *S. peucetius*,

(Continues)

(Continued)

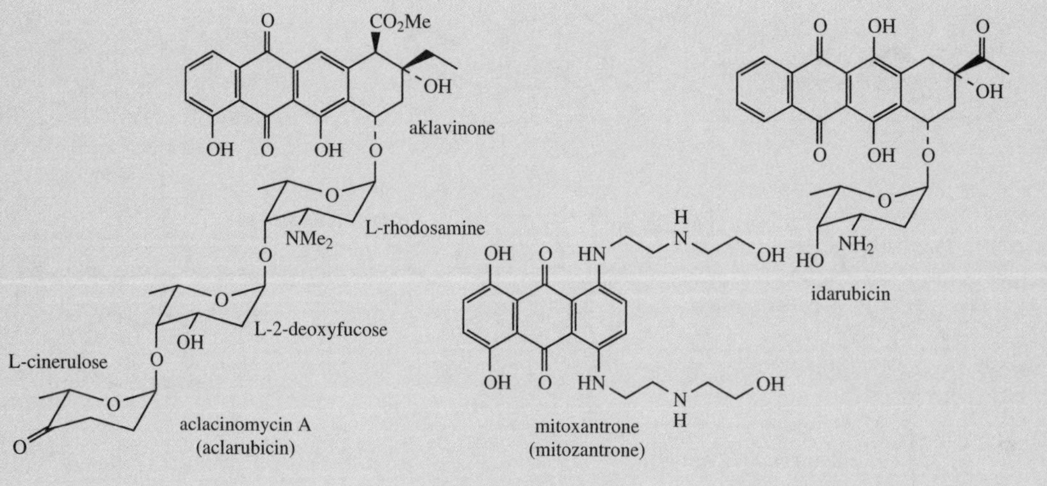

Figure 3.57

and, though similar to doxorubicin in its biological and chemical properties, it is no longer used therapeutically to any extent. It has a much less favourable therapeutic index than doxorubicin, and the markedly different effectiveness as an antitumour drug is not fully understood, though differences in metabolic degradation may be responsible. **Epirubicin** (Figure 3.56), the 4′-epimer of doxorubicin, is particularly effective in the treatment of breast cancer, producing lower side-effects than doxorubicin. The antileukaemics **aclarubicin** from *Streptomyces galilaeus*, a complex glycoside of aklavinone (Figure 3.56), and the semi-synthetic **idarubicin** are shown in Figure 3.57. These compounds are structurally related to doxorubicin but can show increased activity with less cardiotoxicity. The principal disadvantage of all of these agents is their severe cardiotoxicity which arises through inhibition of cardiac Na⁺,K⁺-ATPase.

Mitoxantrone (mitozantrone) (Figure 3.57) is a synthetic analogue of the anthracyclinones in which the non-aromatic ring and the aminosugar have both been replaced with aminoalkyl side-chains. This agent has reduced toxicity compared with doxorubicin, and is effective in the treatment of solid tumours and leukaemias.

less common than incorporating it as a chain extender via methylmalonyl-CoA. We have already encountered this process in the formation of some branched-chain fatty acids with methyl substituents on the basic chain (see page 49). Of course, methyl groups can also be added to a fatty acid chain via SAM (see page 49), and there are also many examples for the methylation of poly-β-keto chains, several of which have already been discussed. Accordingly, methylation using SAM, and incorporation of propionate via methylmalonyl-CoA, provide two different ways of synthesizing a methylated polyketide (Figure 3.58). The former process is the more common in fungi, whilst Actinomycetes (e.g. *Streptomyces*) tend to

employ propionate by the latter route. The incorporation of propionate by methylmalonate extender units can frequently be interrupted and normal malonate extenders are added, thus giving an irregular sequence of methyl side-chains.

The **macrolide antibiotics*** provide us with excellent examples of natural products conforming to the acetate pathway, but composed principally of propionate units, or mixtures of propionate and acetate units. The macrolides are a large family of compounds, many with antibiotic activity, characterized by a macrocyclic lactone ring, typically 12, 14, or 16 membered, reflecting the number of units utilized. **Zearalenone** (Figure 3.59), a toxin produced by the

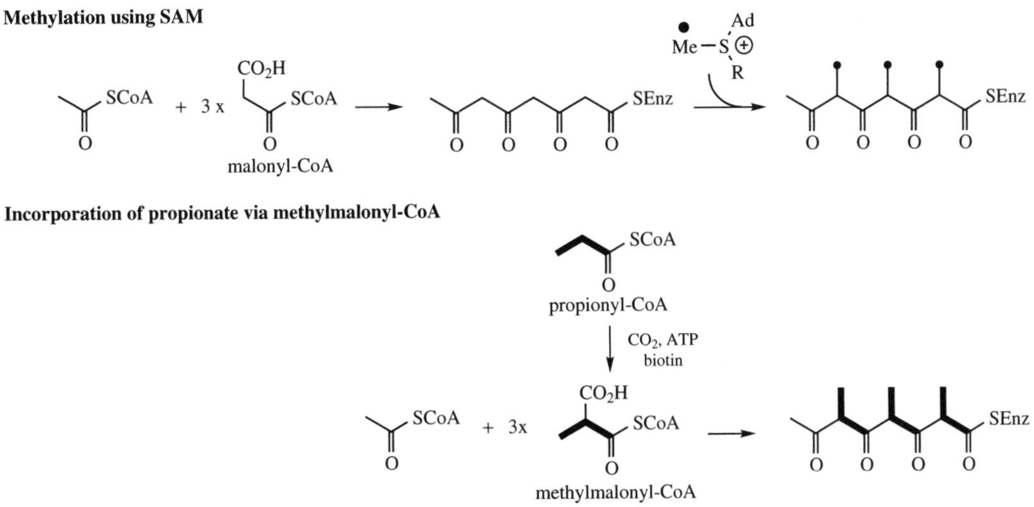

Methylation using SAM

Incorporation of propionate via methylmalonyl-CoA

Figure 3.58

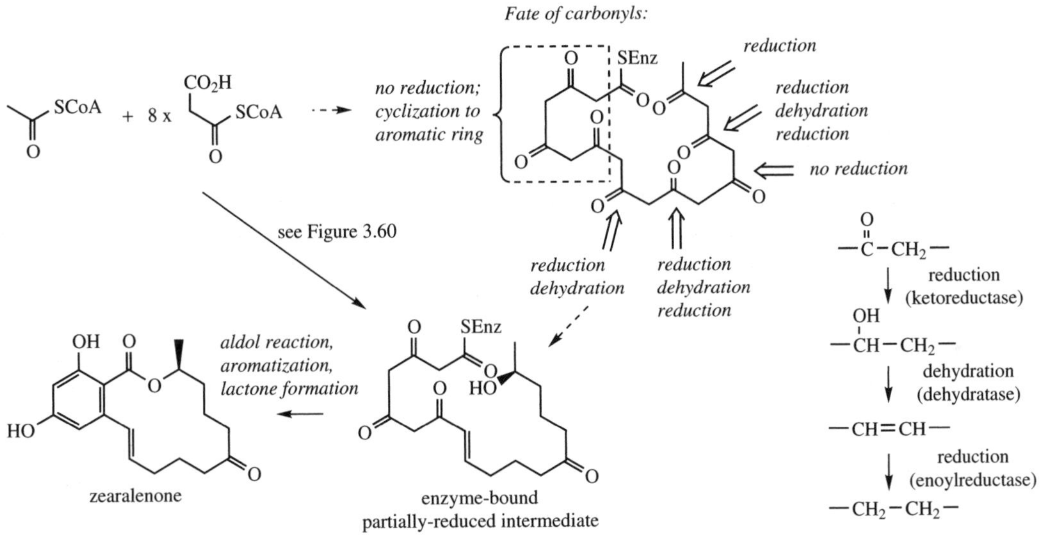

Figure 3.59

fungus *Gibberella zeae* and several *Fusarium* species, has a relatively simple structure which is derived entirely from acetate–malonate units. It could be envisaged as a cyclization product from a poly-β-keto ester, requiring a variety of reduction processes and formation of an aromatic ring by aldol condensation near the carboxyl terminus (Figure 3.59). However, the poly-β-keto ester shown in Figure 3.59 would not be produced, since its reactivity might tend to favour formation of a polycyclic aromatic system (compare anthraquinones, page 63, and

tetracyclines, page 89). Instead, appropriate reductions, dehydrations, etc, involving the β-carbonyl group are achieved during the chain extension process as in the fatty acid pathway (see page 36), and *before* further malonyl-CoA extender units are added (Figure 3.60). In contrast to fatty acid biosynthesis, where there is total reduction of each carbonyl group before further chain extension, macrolide biosynthesis frequently involves partial reduction, with the enzymic machinery being accurately controlled to leave the units at the right oxidation level before further chain extension

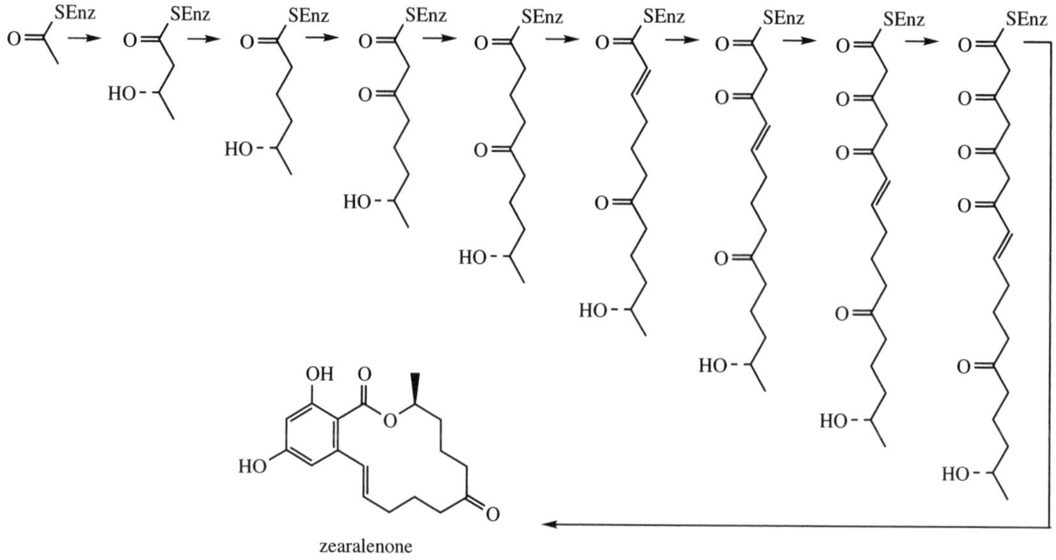

Figure 3.60

occurs. This then provides an enzyme-bound inter-mediate, which leads on to the final product (Figure 3.59). As a result, zearalenone is a remark-able example of an acetate-derived metabolite con-taining all types of oxidation level seen during the fatty acid extension cycle, i.e. carbonyl, secondary alcohol (eventually forming part of the lactone), alkene, and methylene, as well as having a portion which has cyclized to an aromatic ring because no reduction processes occurred in that fragment of the chain. There is now extensive genetic evidence from a variety of **polyketide synthase** systems to show that macrolide assembly is accomplished on a biological production line of multifunctional pro-teins organised as discrete modules, in which the developing polyketide chain attached to an acyl carrier protein is modified according to the appro-priate enzyme activities encoded genetically, and is then passed on to another ACP prior to the next condensation and modification (see page 115 for more details).

Erythromycin A (Figure 3.61) from *Saccha-ropolyspora erythraea* is a valuable antibacte-rial drug and contains a 14-membered macrocy-cle composed entirely of propionate units, both as starter and extension units, the latter via methylmalonyl-CoA. In common with many anti-bacterial macrolides, sugar units, including amino sugars, are attached through glycoside linkages. These unusual 6-deoxy sugars are frequently

restricted to this group of natural products. In erythromycin A, the sugars are L-cladinose and D-desosamine. Chain extension and appropriate reduction processes lead to an enzyme-bound polyketide in which one carbonyl group has suf-fered total reduction, four have been reduced to alcohols, whilst one carbonyl is not reduced, and remains throughout the sequence. These processes ultimately lead to release of the modified polyke-tide as the macrolide ester **deoxyerythronolide**, a demonstrated intermediate in the pathway to erythromycins (Figure 3.61; see also page 115). The stereochemistry in the chain is controlled by the condensation and reduction steps during chain extension, but a reassuring feature is that there appears to be a considerable degree of stereochemical uniformity throughout the known macrolide antibiotics. In the later stages of the biosynthesis of erythromycin, hydroxylations at carbons 6 and 12, and addition of sugar units, are achieved.

A combination of propionate and acetate units is used to produce the 14-membered macrocyclic ring of **oleandomycin** (Figure 3.62) from *Strep-tomyces antibioticus*, but otherwise many of the structural features and the stereochemistry of ole-andomycin resemble those of erythromycin A. One acetate provides the starter unit, whilst seven propi-onates, via methylmalonyl-CoA, supply the exten-sion units (Figure 3.62). One methyl group derived

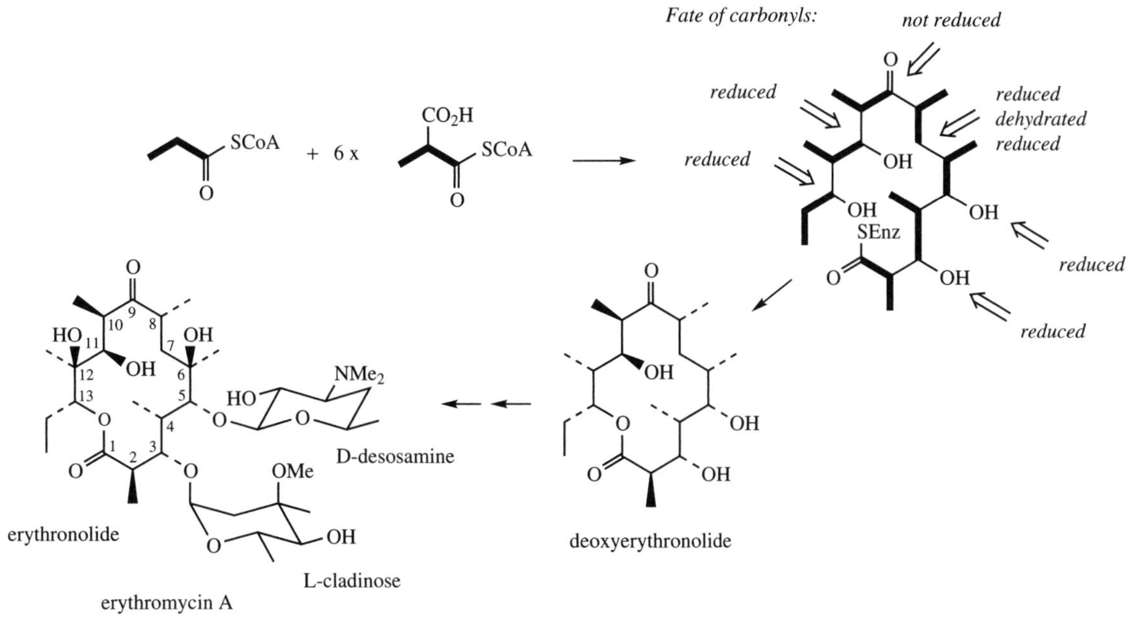

Figure 3.61

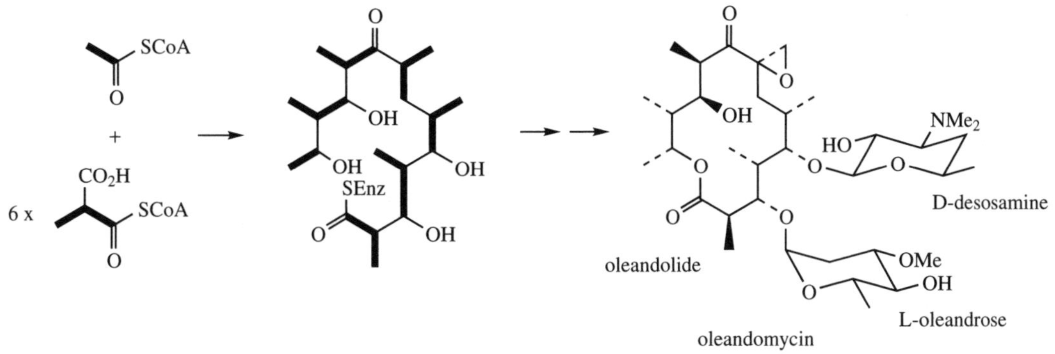

Figure 3.62

from propionate has been modified to give an epoxide function. The sugar units in oleandomycin are L-oleandrose and D-desosamine. **Spiramycin I** (Figure 3.63) from *Streptomyces ambofaciens* has a 16-membered lactone ring, and is built up from a combination of six acetate units (one as starter), one propionate extender, together with a further variant, butyrate as chain extender. Butyrate will be incorporated via ethylmalonyl-CoA and yield an extension unit having an ethyl side-chain. This is outlined in Figure 3.63. In due course, this ethyl group is oxidized generating an aldehyde. Spiramycin I also contains a conjugated diene,

the result of carbonyl reductions being followed by dehydration during chain assembly. **Tylosin** (Figure 3.64) from *Streptomyces fradiae* has many structural resemblances to the spiramycins, but can be analysed as a propionate starter with chain extension from two malonyl-CoA, four methylmalonyl-CoA, and one ethylmalonyl-CoA.

The **avermectins*** (Figure 3.67) have no antibacterial activity, but possess anthelmintic, insecticidal, and acaricidal properties, and these are exploited in human and veterinary medicine. The avermectins are also 16-membered macrolides, but their structures are made up from a much longer

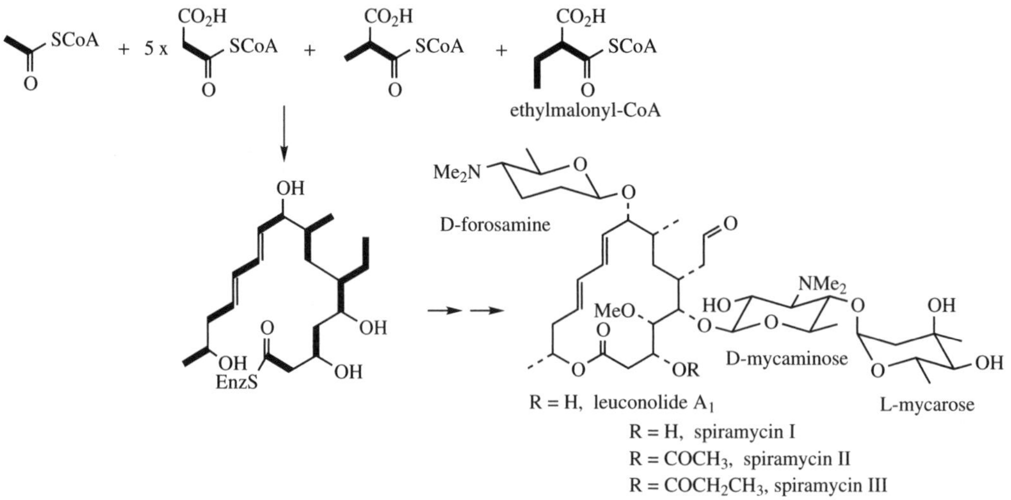

Figure 3.63

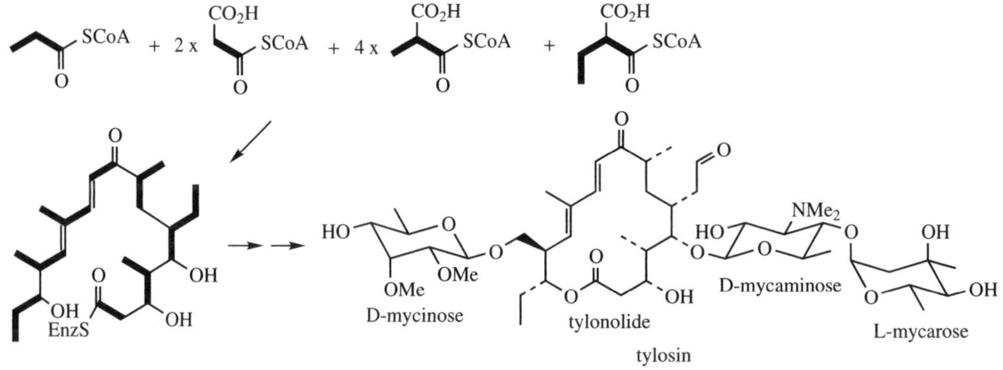

Figure 3.64

Macrolide Antibiotics

The macrolide antibiotics are macrocyclic lactones with a ring size typically 12–16 atoms, and with extensive branching through methyl substituents. Two or more sugar units are attached through glycoside linkages, and these sugars tend to be unusual 6-deoxy structures often restricted to this class of compounds. Examples include L-cladinose, L-mycarose, D-mycinose, and L-oleandrose. At least one sugar is an amino sugar, e.g. D-desosamine, D-forosamine, and D-mycaminose. These antibiotics have a narrow spectrum of antibacterial activity, principally against Gram-positive microorganisms. Their antibacterial spectrum resembles, but is not identical to, that of the penicillins, so they provide a valuable alternative for patients allergic to the penicillins. Erythromycin is the principal macrolide antibacterial currently used in medicine.

The erythromycins (Figure 3.65) are macrolide antibiotics produced by cultures of *Saccharopolyspora erythraea* (formerly *Streptomyces erythreus*). The commercial product

(Continues)

(Continued)

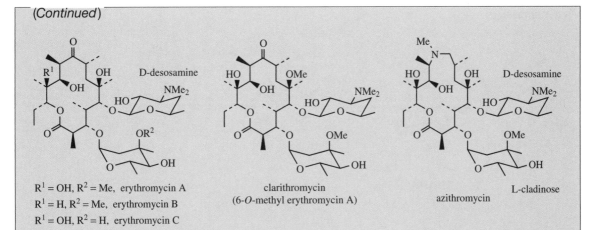

R^1 = OH, R^2 = Me, erythromycin A
R^1 = H, R^2 = Me, erythromycin B
R^1 = OH, R^2 = H, erythromycin C

clarithromycin
(6-O-methyl erythromycin A)

azithromycin

Figure 3.65

erythromycin is a mixture containing principally erythromycin A, plus small amounts of erythromycins B and C (Figure 3.65). Erythromycin activity is predominantly against Gram-positive bacteria, and the antibiotic is prescribed for penicillin-allergic patients. It is also used against penicillin-resistant *Staphylococcus* strains, in the treatment of respiratory tract infections, and systemically for skin conditions such as acne. It is the antibiotic of choice for infections of *Legionella pneumophila*, the cause of legionnaire's disease. Erythromycin exerts its antibacterial action by inhibiting protein biosynthesis in sensitive organisms. It binds reversibly to the larger 50S subunit of bacterial ribosomes and blocks the translocation step in which the growing peptidyl-tRNA moves from the aminoacyl acceptor site to the peptidyl donor site on the ribosome (see page 408). The antibiotic is a relatively safe drug with few serious side-effects. Nausea and vomiting may occur, and if high doses are prescribed, a temporary loss of hearing might be experienced. Hepatotoxicity may also occur at high dosage.

Erythromycin is unstable under acidic conditions, undergoing degradation to inactive compounds by a process initiated by the 6-hydroxyl attacking the 9-carbonyl to form a hemiketal. Dehydration then follows (Figure 3.66). The 14-membered ring in erythromycin A adopts a modified version of the diamond lattice chairlike conformation shown in Figure 3.66. Studies have indicated that carbon 6 is displaced from this conformation to reduce the 1,3-diaxial interactions at C-4 and C-6, and the two relatively large sugar units attached to the hydroxyls at C-3 and C-5 also distort the ring system further. The distortion of the chairlike conformation brings the 6-hydroxyl sufficiently close to react with the 9-carbonyl. A similar reaction may occur between the C-12 hydroxyl and the 9-carbonyl. Thus, to protect oral preparations of erythromycin against gastric acid, they are formulated as enteric-coated tablets, or as insoluble esters (e.g. ethyl succinate esters), which are then hydrolysed in the intestine. Esterification typically involves the hydroxyl of the amino sugar desosamine. To reduce this acid instability, semi-synthetic analogues of erythromycin have also been developed. **Clarithromycin** (Figure 3.65) is a 6-O-methyl derivative of erythromycin A; this modification blocks hemiketal formation as in Figure 3.66. **Azithromycin** (Figure 3.65) is a ring-expanded aza-macrolide in which the carbonyl function has been reduced. In both analogues, the changes enhance activity compared with that of erythromycin.

Bacterial resistance to erythromycin has become significant and has limited its therapeutic use against many strains of *Staphylococcus*. Several mechanisms of resistance have been

(Continues)

(Continued)

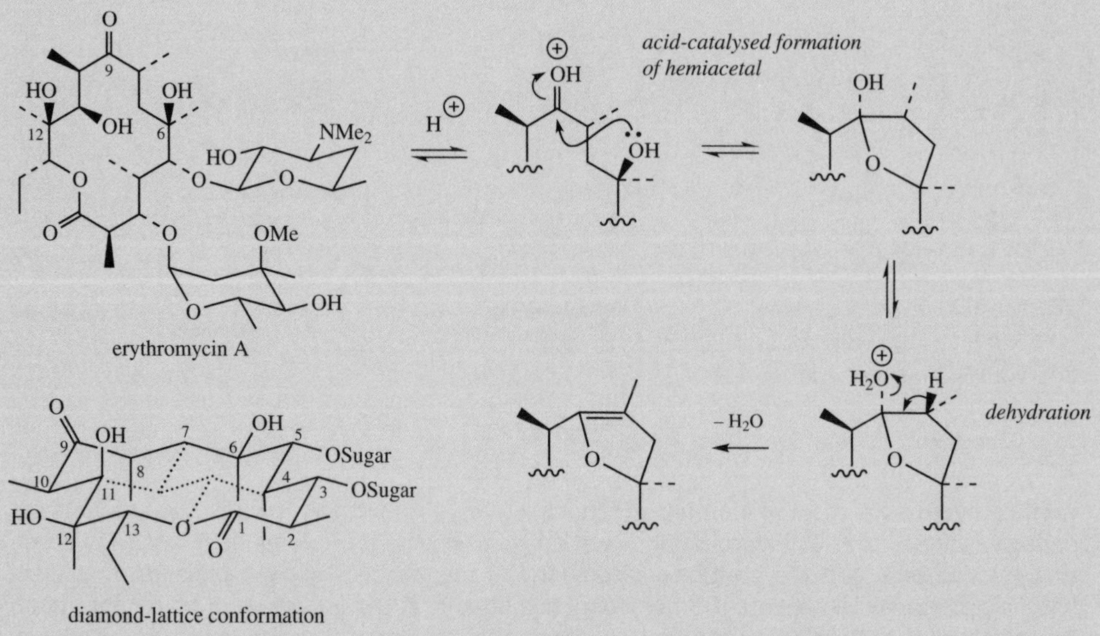

erythromycin A

acid-catalysed formation of hemiacetal

dehydration

diamond-lattice conformation

Figure 3.66

implicated, one of which is a change in permeability of the bacterial cell wall. Differences in permeability also appear to explain the relative insensitivity of Gram-negative bacteria to erythromycin when compared to Gram-positive bacteria. Resistant bacteria may also modify the chemical nature of the binding site on the ribosome, thus preventing antibiotic binding, and some organisms are now known to metabolize the macrolide ring to yield inactive products.

Oleandomycin (Figure 3.62) is produced by fermentation cultures of *Streptomyces antibioticus* and has been used medicinally as its triacetyl ester **troleandomycin** against Gram-positive bacterial infections. The **spiramycins** (Figure 3.63) are macrolides produced by cultures of *Streptomyces ambofaciens*. The commercial antibiotic is a mixture containing principally spiramycin I, together with smaller amounts (10–15% each) of the acetyl ester spiramycin II and the propionyl ester spiramycin III. This antibiotic has recently been introduced into medicine for the treatment of toxoplasmosis, infections caused by the protozoan *Toxoplasma gondii*.

Tylosin (Figure 3.64) is an important veterinary antibiotic. It is produced by *Streptomyces fradiae*, and is used to control chronic respiratory diseases caused by *Mycoplasma galliseptum* in poultry, and to treat Gram-positive infections in pigs.

polyketide chain, which is also used to form oxygen heterocycles fused to the macrolide. **Avermectin B_{1a}** exemplifies a typical structure and the basic carbon skeleton required to produce this can be postulated as in Figure 3.67. The starter unit in this case would be 2-methylbutyryl-CoA, which is derived from the amino acid L-isoleucine (compare necic acids, page 305, and tiglic acid, page 197).

Both malonyl-CoA and methylmalonyl-CoA are then utilized as extender units. The heterocyclic rings are easily accounted for: the spiro system is merely a ketal, though the tetrahydrofuran ring requires further hydroxylations of the basic skeleton for its construction. Avermectins are usually isolated as a mixture in which the main *a* component has a 2-methylpropyl group

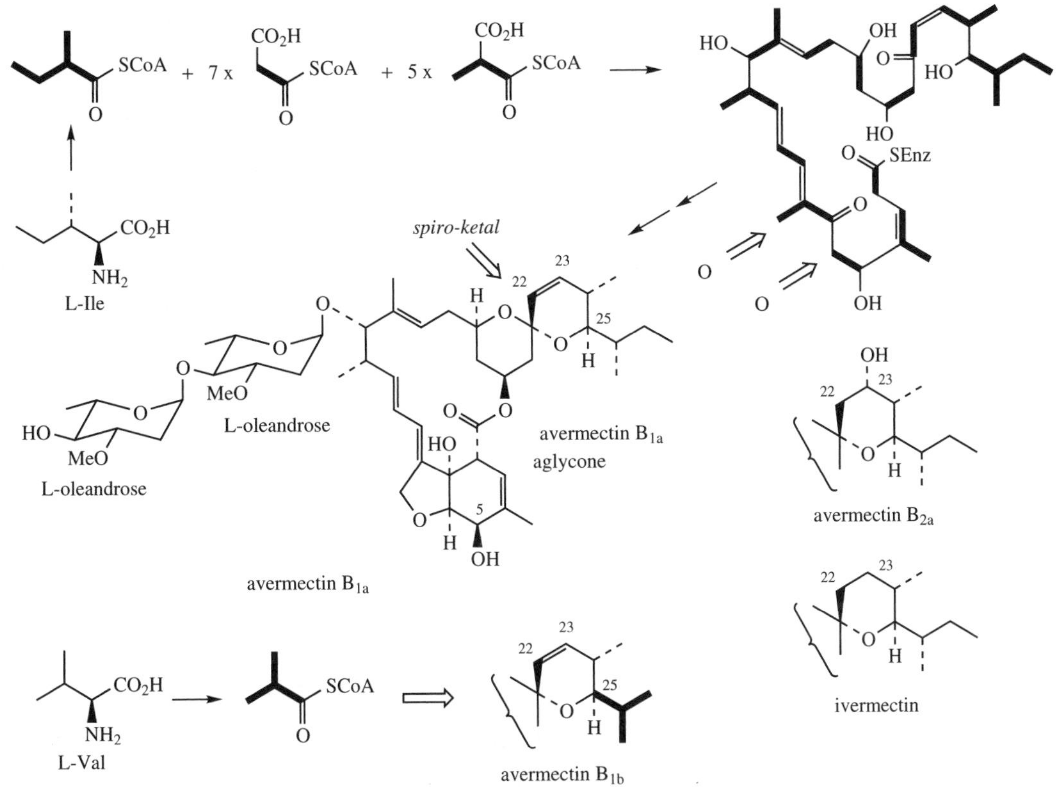

Figure 3.67

Avermectins

The avermectins (Figure 3.67) are a group of macrolides with strong anthelmintic, insecticidal, and acaricidal properties, but with low toxicity to animals and humans. They are produced by cultures of *Streptomyces avermectilis*. Some eight closely related structures have been identified, with avermectins B_{1a} and B_{2a} being the most active antiparasitic compounds. **Abamectin** (a mixture of about 85% avermectin B_{1a} and about 15% avermectin B_{1b}) is used on agricultural crops to control mites and insects. **Ivermectin** (Figure 3.67) is a semi-synthetic 22,23-dihydro derivative of avermectin B_{1a} and was first used in veterinary practice against insects, ticks, mites, and roundworms. Although it is a broad spectrum nematocide against roundworms, it is inactive against tapeworms and flatworms, or against bacteria and fungi. It is an extremely potent agent, and is effective at very low dosages. It has now been introduced for use against filarial and several other worm parasites in humans. Avermectins act by blocking neuromuscular transmission in sensitive organisms by acting on GABA (γ-aminobutyric acid) receptors.

(derived from isoleucine) at C-25, whilst the minor *b* component has an isopropyl group instead, e.g. **avermectin B_{1b}**. In this case, the starter group is 2-methylpropionyl-CoA, derived from the amino acid L-valine. The A-series of avermectins are the 5-methoxy analogues of the B-series.

Even larger macrolides are encountered in the **polyene macrolides***, most of which have antifungal properties, but not antibacterial activity. The macrolide ring size ranges from 26 to 38 atoms, and this also accommodates a conjugated polyene of up to seven *E* double bonds. Relatively few methyl groups are attached to the ring, and thus malonyl-CoA is utilized more frequently than methylmalonyl-CoA as chain extender. Typical examples are **amphotericin B** (Figure 3.68) from *Streptomyces nodosus* and **nystatin A₁** from *Streptomyces noursei*. These have very similar structures and are derived from the same basic precursors (Figure 3.68). The ring size is contracted due to cross-linking by formation of a hemiketal. They have slightly different hydroxylation patterns, part

of which is introduced by hydroxylation, and the two areas of conjugation in nystatin A₁ are extended into a heptaene system in amphotericin B. Both compounds are glycosylated with the amino sugar D-mycosamine, and both are carboxylic acids, a result of oxidation of a propionate-derived methyl group.

An unusual and clinically significant macrolide isolated from *Streptomyces tsukubaensis* is **FK-506 (tacrolimus)*** (Figure 3.69), which contains a

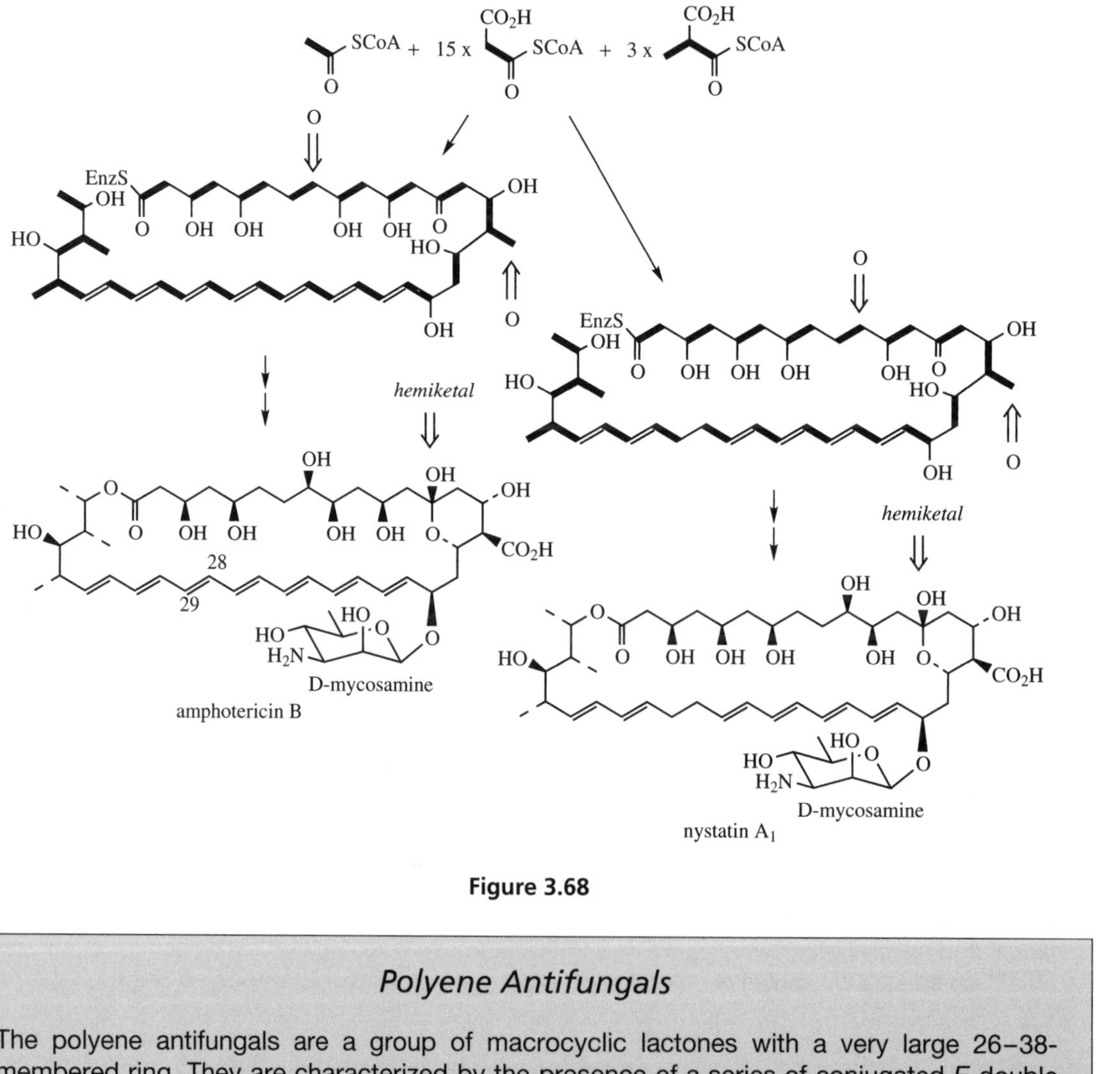

Figure 3.68

Polyene Antifungals

The polyene antifungals are a group of macrocyclic lactones with a very large 26–38-membered ring. They are characterized by the presence of a series of conjugated *E* double bonds and are classified according to the longest conjugated chain present. Medicinally important ones include the heptaene amphotericin B, and the tetraene nystatin. There are relatively few methyl branches in the macrocyclic chain. The polyenes have no antibacterial

(Continues)

(Continued)

activity but are useful antifungal agents. Their activity is a result of binding to sterols in the eukaryotic cell membrane, which action explains the lack of antibacterial activity because bacterial cells do not contain sterol components. Fungal cells are also attacked rather than mammalian cells, since the antibiotics bind much more strongly to ergosterol, the major fungal sterol (see page 253), than to cholesterol, the main animal sterol component (see page 236). This binding modifies the cell wall permeability and leads to formation of transmembrane pores allowing K^+ ions, sugars, and proteins to be lost from the microorganism. Though binding to cholesterol is less than to ergosterol, it is responsible for the observed toxic side-effects of these agents on humans. The polyenes are relatively unstable, undergoing light-catalysed decomposition, and are effectively insoluble in water. This insolubility actually protects the antibiotic from gastric decomposition, allowing oral treatment of infections in the intestinal tract.

Amphotericin is an antifungal polyene produced by cultures of *Streptomyces nodosus* and contains principally the heptaene amphotericin B (Figure 3.68) together with structurally related compounds, e.g. the tetraene amphotericin A (about 10%), which is the 28,29-dihydro analogue of amphotericin B. Amphotericin A is much less active than amphotericin B. Amphotericin is active against most fungi and yeasts, but it is not absorbed from the gut, so oral administration is restricted to the treatment of intestinal candidiasis. It is administered intravenously for treating potentially life-threatening systemic fungal infections. However, it then becomes highly protein bound resulting in poor penetration and slow elimination from the body. After parenteral administration, toxic side-effects, including nephrotoxicity, are relatively common. Close supervision and monitoring of the patient is thus necessary, especially since the treatment may need to be prolonged. A liposome-encapsulated formulation of amphotericin has been shown to be much less toxic and may prove a significant advance. *Candida* infections in the mouth or on the skin may be treated with appropriate formulations.

Nystatin is a mixture of tetraene antifungals produced by cultures of *Streptomyces noursei*. The principal component is nystatin A_1 (Figure 3.68), but the commercial material also contains nystatin A_2 and A_3, which have additional glycoside residues. Nystatin is too toxic for intravenous use, but has value for oral treatment of intestinal candidiasis, as lozenges for oral infections, and as creams for topical control of *Candida* species.

23-membered macrolactone that also incorporates an *N*-heterocyclic ring. This compound is known to be derived from acetate and propionate, the fragments of which can readily be identified in the main chain. The starter unit is cyclohexanecarboxylic acid, a reduction product from shikimate, and the piperidine ring and adjacent carbonyl are incorporated as pipecolic acid (see page 310) via an amide linkage on to the end of the growing chain. An unusual pentanoic acid unit is also incorporated to provide the propenyl side-chain. FK-506 is a particularly effective immunosuppressant, and is proving valuable in organ transplant surgery. Although **rapamycin (sirolimus)*** contains a very large 31-membered macrocycle, several portions of the structure are identical to those of FK-506. Cyclohexanecarboxylic acid and pipecolic acid are again utilized in its formation, whilst the rest of the skeleton is supplied by simple acetate and propionate residues (Figure 3.69).

Attracting considerable interest at the present time are the **epothilones** (Figure 3.70), a group of macrolides produced by cultures of the bacterium *Sorangium cellulosum*. These compounds employ an unusual starter unit containing a thiazole ring, which is almost certainly constructed from the amino acid cysteine and an acetate unit (see also thiazole rings in bleomycin, page 429). The macrolide ring also contains an extra methyl group at C-4, the result of methylation after or during polyketide chain assembly. The other interesting feature is that this bacterium produces epothilone A and epothilone B in the ratio of about 2:1. These compounds differ in the nature

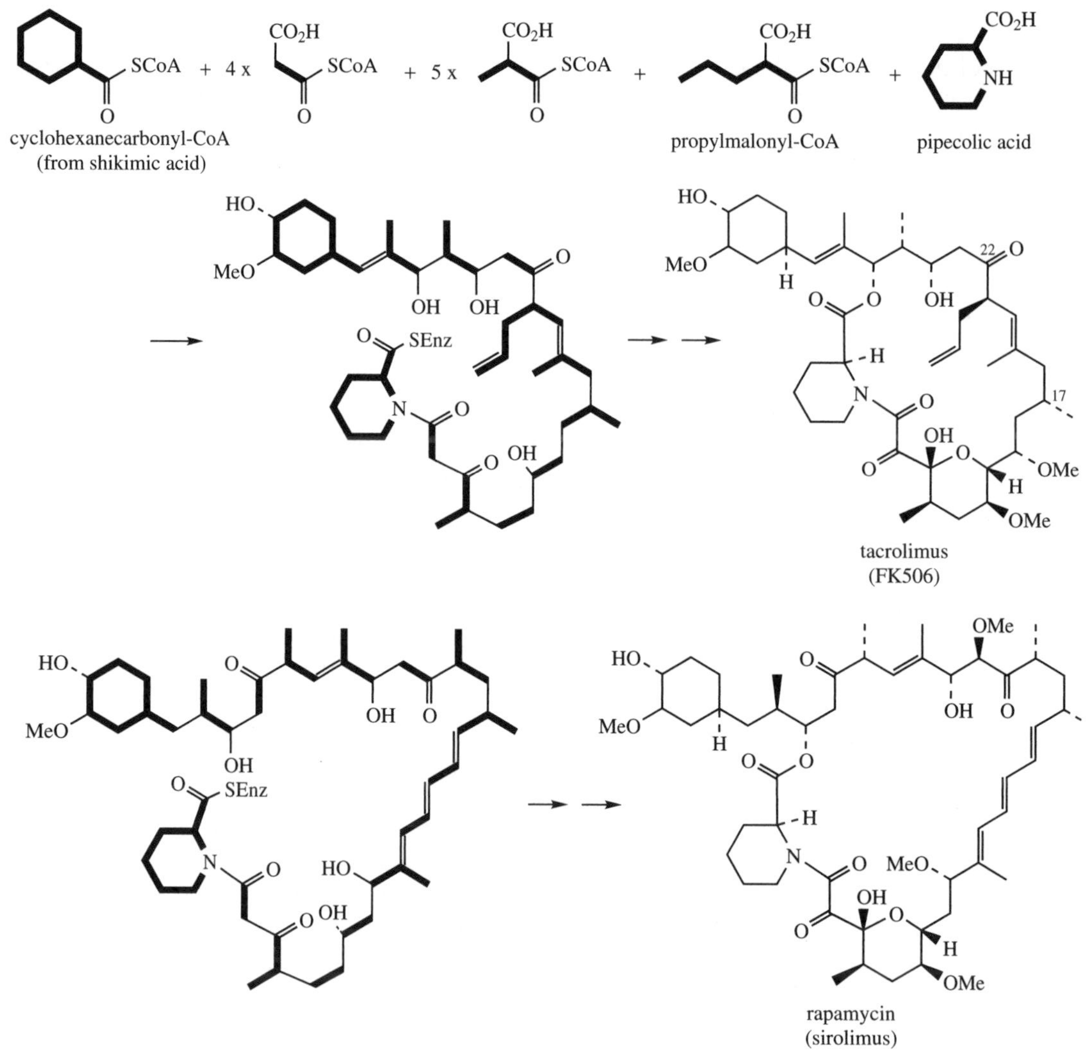

Figure 3.69

Tacrolimus and Sirolimus

Tacrolimus (FK-506) (Figure 3.69) is a macrolide immunosuppressant isolated from cultures of *Streptomyces tsukubaensis*. It is used in liver and kidney transplant surgery. Despite the significant structural differences between tacrolimus and the cyclic peptide cyclosporin A (ciclosporin; see page 429), these two agents have a similar mode of action. They both inhibit T-cell activation in the immunosuppressive mechanism by binding first to a receptor protein giving a complex, which then inhibits a phosphatase enzyme called calcineurin. The resultant aberrant phosphorylation reactions prevent appropriate gene transcription and subsequent T-cell activation. Structural similarities between the region C-17 to C-22 and fragments of the cyclosporin A peptide chain have been postulated to account for this binding. Tacrolimus is

(Continues)

(Continued)

up to 100 times more potent than cyclosporin A, but produces similar side-effects including neurotoxicity and nephrotoxicity.

Rapamycin (sirolimus) (Figure 3.69) is produced by cultures of *Streptomyces hygroscopicus* and is also being investigated as an immunosuppressant drug. Although tacrolimus and rapamycin possess a common structural unit, and both inhibit T-cell activation, they appear to achieve this by somewhat different mechanisms. The first-formed rapamycin–receptor protein binds not to calcineurin, but to a different protein. Rapamycin suppresses lymphocyte production. Rapamycin also possesses pronounced antifungal activity, but is not active against bacteria.

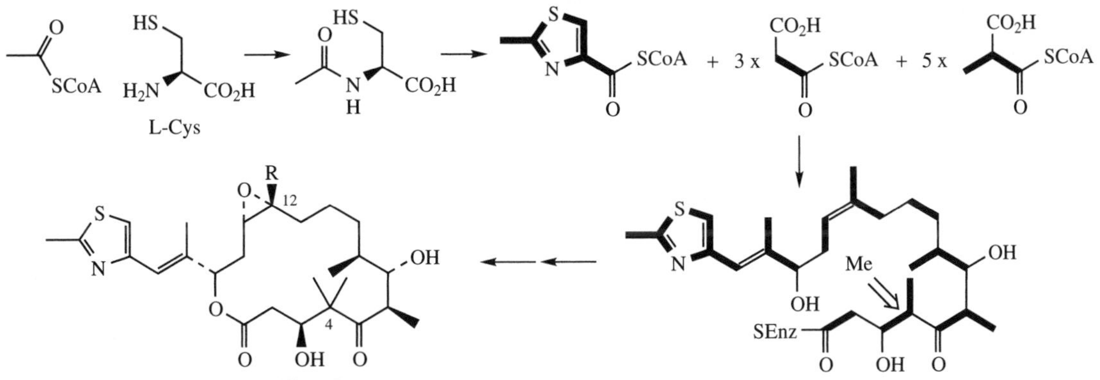

Figure 3.70

of the substituent at C-12, which is hydrogen in epothilone A but a methyl group in epothilone B. Genetic evidence shows that the polyketide synthase enzyme can accept either malonyl-CoA or methylmalonyl-CoA extender units for this position. Thus, epothilone B is constructed from three malonate and five methylmalonate extender units as shown in Figure 3.70, whilst epothilone A requires four units of each type. The epothilones display marked antitumour properties with a mode of action paralleling that of the highly successful anticancer drug taxol (see page 205). However, the epothilones have a much higher potency (2000–5000 times) and are active against cell lines which are resistant to taxol and other drugs. There appears to be considerable potential for developing the epothilones or analogues into valuable anticancer drugs.

A further group of macrolides in which nonadjacent positions on an aromatic ring are bridged by the long aliphatic chain is termed **ansa macrolides***. These are actually lactams rather than lactones, and the nitrogen atom originates from **3-amino-5-hydroxybenzoic acid**, which acts as the starter unit for chain extension with malonyl-CoA or methylmalonyl-CoA. 3-Amino-5-hydroxybenzoic acid (Figure 3.71) is a simple phenolic acid derivative produced by an unusual variant of the shikimate pathway (see Chapter 4), in which aminoDAHP is formed in the initial step, and then the pathway continues with amino analogues. This proceeds through to aminodehydroshikimic acid which yields 3-amino-5-hydroxybenzoic acid on dehydration. In the biosynthesis of **rifamycin B** (Figure 3.71) in *Amycolatopsis mediterranei*, this starter unit, plus two malonyl-CoA and eight methylmalonyl-CoA extenders, are employed to fabricate **proansamycin X** as the first product released from the enzyme. The enzyme-bound intermediate shown in Figure 3.71 is not strictly correct, in that the naphthoquinone ring system is now

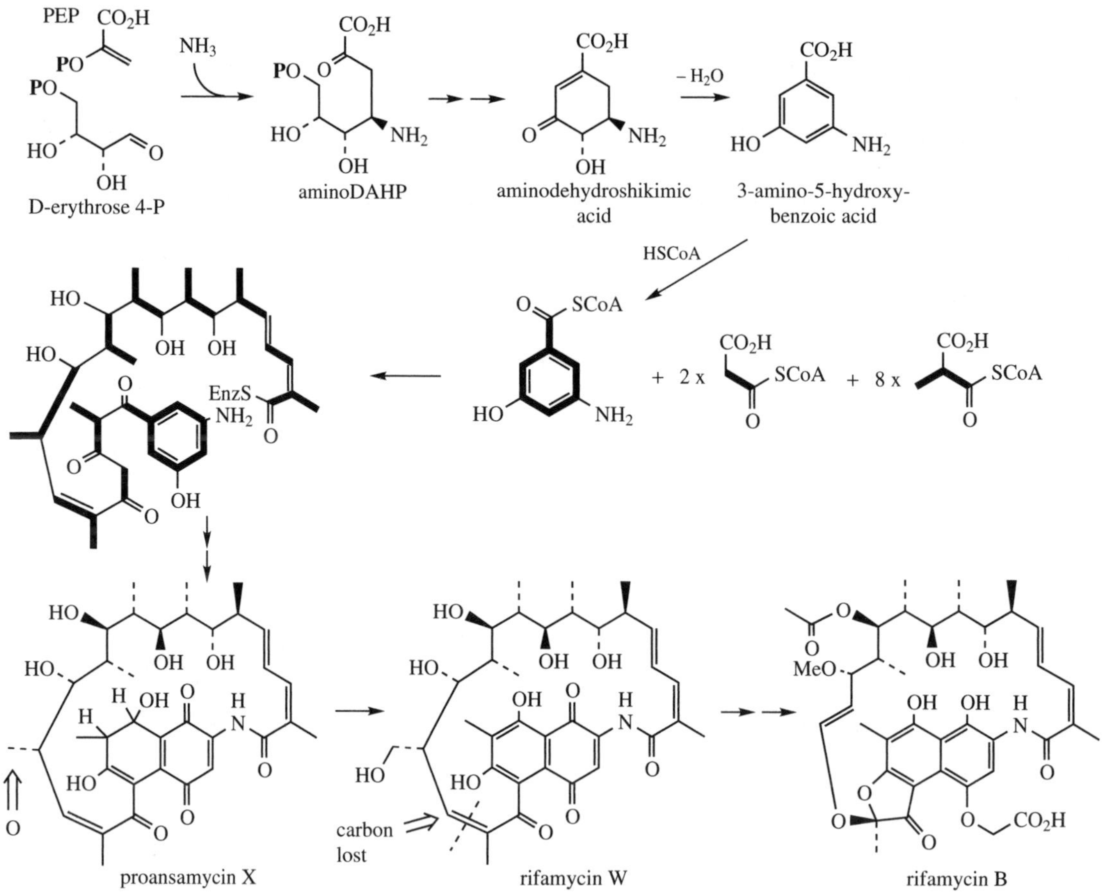

Figure 3.71

known to be constructed during, not after, chain assembly. **Rifamycin W** and then the antibiotic rifamycin B are the result of further modifications including cleavage of the double bond, loss of one carbon, then formation of the ketal. **Maytansine** (Figure 3.72) is a plant-derived ansa macrolide from *Maytenus serrata* (Celastraceae), though other esters of the parent alcohol, maytansinol, are produced by species of the fungus *Nocardia*. Maytansine has been extensively investigated for its potential antitumour activity.

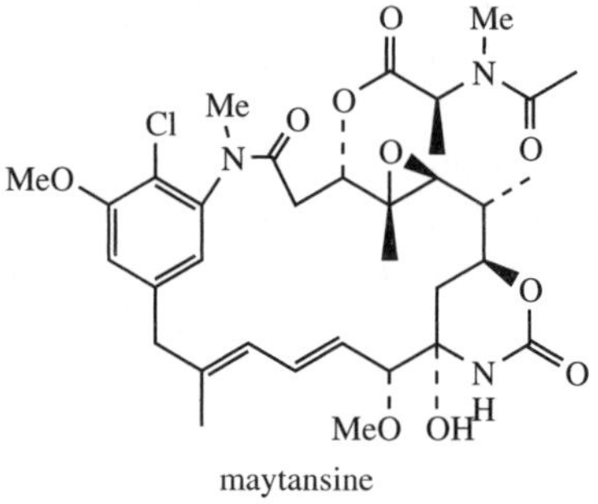

The macrolide systems described above are produced by formation of an intramolecular ester or amide linkage, utilizing appropriate functionalities in the growing polyketide chain. Macrolide formation does not always occur, and similar acetate–propionate precursors might also be expected to yield molecules which are essentially linear in nature. Good examples of such molecules

maytansine

Figure 3.72

are **lasalocid A** (Figure 3.74) from *Streptomyces lasaliensis* and **monensin A** (Figure 3.75) from *Streptomyces cinnamonensis*, representatives of a large group of compounds called **polyether antibiotics**. These, and other examples, are of value in veterinary medicine, being effective in preventing

Ansa Macrolides

Ansamycins are a class of macrocyclic compounds in which non-adjacent positions on an aromatic ring system are spanned by the long aliphatic bridge (Latin: *ansa* = handle). The aromatic portion may be a substituted naphthalene or naphthaquinone, or alternatively a substituted benzene ring. The macrocycle in the ansamycins is closed by an amide rather than an ester linkage, i.e. ansamycins are lactams. The only ansamycins currently used therapeutically are semi-synthetic naphthalene-based macrocycles produced from rifamycin B.

The rifamycins are ansamycin antibiotics produced by cultures of *Amycolatopsis mediterranei* (formerly *Nocardia mediterranei* or *Streptomyces mediterranei*). The crude antibiotic mixture was found to contain five closely related substances rifamycins A–E, but if the organism was cultured in the presence of sodium diethyl barbiturate (barbitone or barbital), the product was almost entirely rifamycin B (Figure 3.71). Rifamycin B has essentially no antibacterial activity, but on standing in aqueous solution in the presence of air, it is readily transformed by oxidation and intramolecular nucleophilic addition into rifamycin O, which

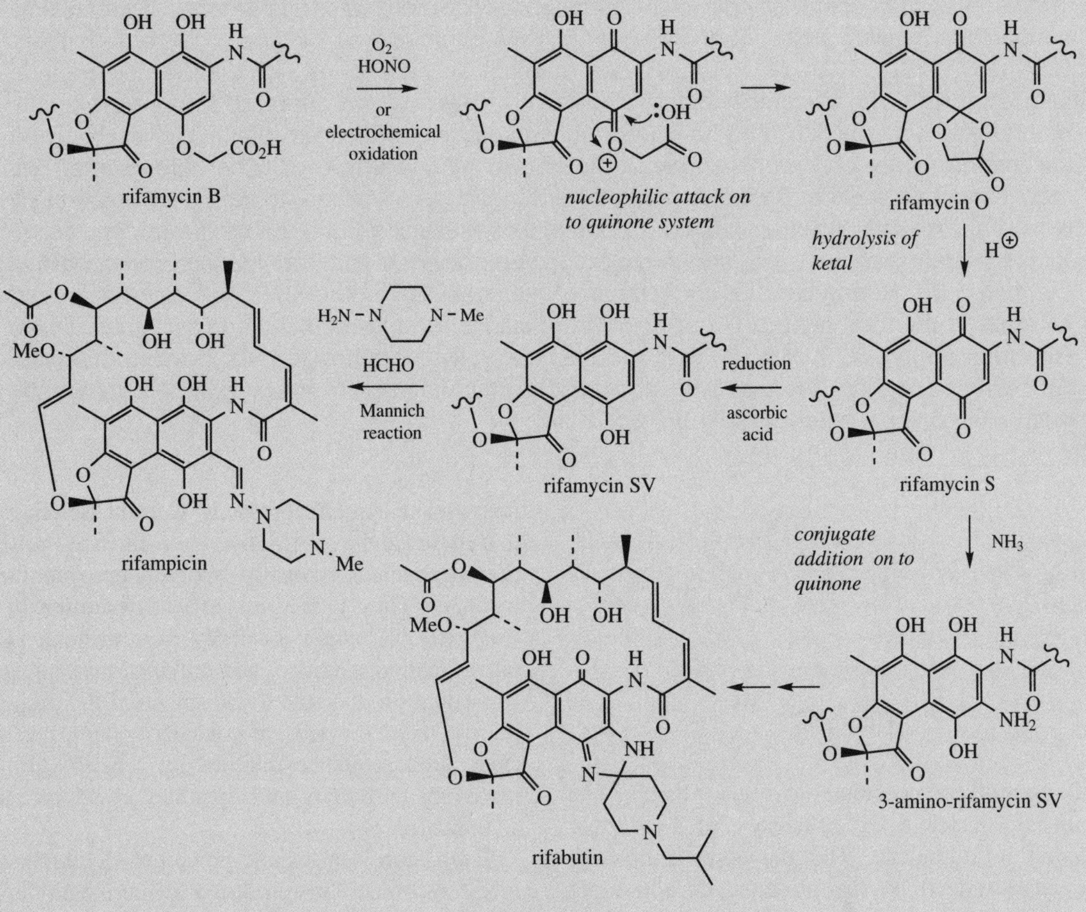

Figure 3.73

(*Continues*)

(*Continued*)

under acidic conditions then hydrolyses and gives rifamycin S, a highly active antibacterial agent (Figure 3.73). Chemical reduction of rifamycin S using ascorbic acid (vitamin C) converts the quinone into a quinol and provides a further antibacterial, rifamycin SV. Rifamycins O, S, and SV can all be obtained by fermentation using appropriate strains of *A. mediterranei*. Rifamycin SV is actually the immediate biosynthetic precursor of rifamycin B under normal conditions, so this conversion can be genetically blocked and lead to accumulation of rifamycin SV. Several other rifamycin analogues have also been characterized. Rifamycin O is usually produced by chemical or electrochemical oxidation of rifamycin B, and converted into rifamycin SV as in Figure 3.73.

The most useful rifamycin employed clinically is **rifampicin** (Figure 3.73), a semi-synthetic derivative produced from rifamycin SV via a Mannich reaction (see page 18) using formaldehyde and *N*-amino-*N*′-methylpiperazine. Rifampicin has a wide antibacterial spectrum, with high activity towards Gram-positive bacteria and a lower activity towards Gram-negative organisms. Its most valuable activity is towards *Mycobacterium tuberculosis* and rifampicin is a key agent in the treatment of tuberculosis, usually in combination with at least one other drug to reduce the chances for development of resistant bacterial strains. It is also useful in control of meningococcal meningitis and leprosy. Rifampicin's antibacterial activity arises from inhibition of RNA synthesis by binding to DNA-dependent RNA polymerase. RNA polymerase from mammalian cells does not contain the peptide sequence to which rifampicin binds, so RNA synthesis is not affected. In contrast to the natural rifamycins which tend to have poor absorption properties, rifampicin is absorbed satisfactorily after oral administration, and is also relatively free of toxic side-effects. The most serious side-effect is disturbance of liver function. A trivial, but to the patient potentially worrying, side-effect is discoloration of body fluids, including urine, saliva, sweat, and tears, to a red–orange colour, a consequence of the naphthalene/naphthoquinone chromophore in the rifamycins. **Rifamycin**, the sodium salt of rifamycin SV (Figure 3.73), has also been used clinically in the treatment of Gram-positive infections, and particularly against tuberculosis. **Rifabutin** (Figure 3.73) is a newly introduced derivative, synthesized via 3-amino-rifamycin SV, which also has good activity against the *Mycobacterium avium* complex frequently encountered in patients with AIDS.

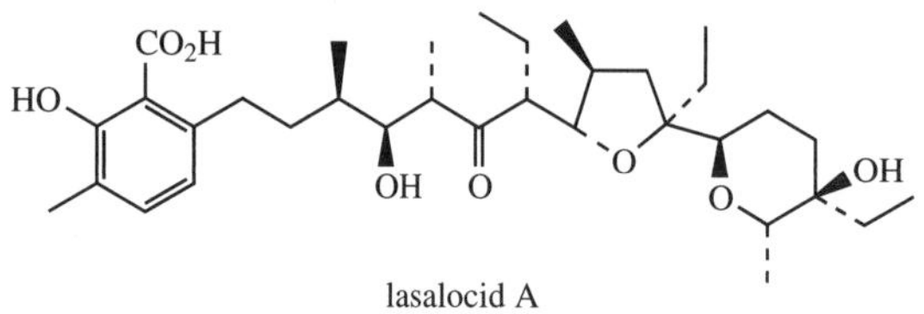

lasalocid A

Figure 3.74

and controlling coccidiae and also having the ability to improve the efficiency of food conversion in ruminants. The polyether antibiotics are characterized by the presence of a number of tetrahydrofuran and/or tetrahydropyran rings along the basic chain. The polyether acts as an ionophore, increasing influx of sodium ions into the parasite, causing a resultant and fatal

increase in osmotic pressure. Current thinking is that these ring systems arise via a cascade cyclization mechanism, probably involving epoxide intermediates. Thus, in the biosynthesis of **monensin A** (Figure 3.75), chain assembly from acetate, malonate, methylmalonate, and ethylmalonate precursors could produce the triene shown. If the triepoxide is then formed, a concerted stereospecific cyclization sequence initiated by a hydroxyl and involving carbonyls and epoxides could proceed as indicated.

Even more remarkable polyether structures are found in some toxins produced by marine dinoflagellates, which are in turn taken up by shellfish and pass on their toxicity to the shellfish. **Okadaic acid** (Figure 3.76) and related polyether structures from *Dinophysis* species are responsible for

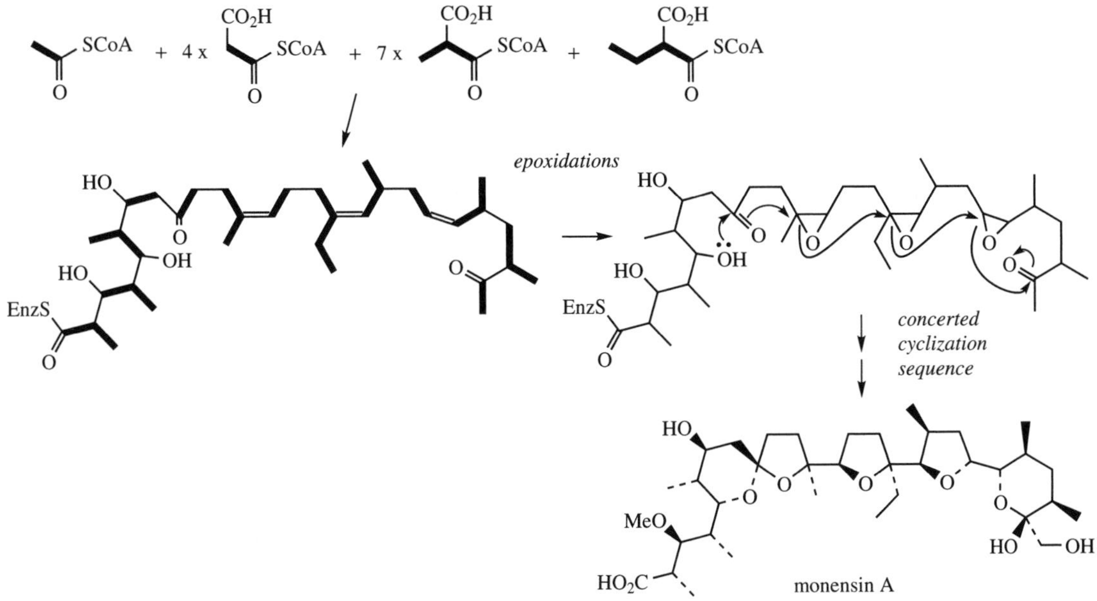

Figure 3.75

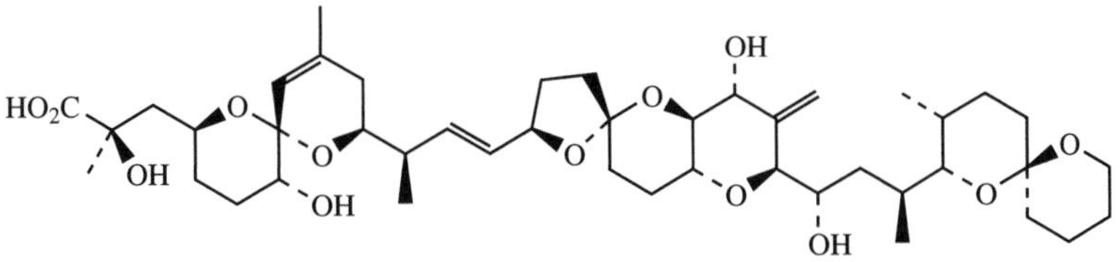

okadaic acid

Figure 3.76

diarrhoeic shellfish poisoning in mussels, causing severe diarrhoea in consumers of contaminated shellfish in many parts of the world. **Brevetoxin A** (Figure 3.77) is an example of the toxins associated with 'red tide' blooms of dinoflagellates, which affect fishing and also tourism especially in Florida and the Gulf of Mexico. The red tide toxins are derived from *Gymnodimium breve* and are the causative agents of neurotoxic shellfish poisoning, leading to neurological disorders as well as gastrointestinal troubles. The toxins are known to bind to sodium channels, keeping them in an open state. Fatalities among marine life, e.g. fish, dolphins, whales, and in humans, are associated with these toxins synthesized by organisms at the base of the marine food chain. These compounds are postulated to be produced from a polyunsaturated fatty acid by epoxidation of the double bonds, and then a concerted sequence of epoxide ring openings leads to the extended polyether structure (Figure 3.77). The carbon skeleton does not conform to a simple polyketide chain, and biosynthetic studies have shown that fragments from the citric acid cycle and a four-carbon starter unit from mevalonate are also involved, and that some of the methyls originate from methionine. **Ciguatoxin** (Figure 3.78) is one of the most complex examples of a polyether structure found in nature. This is found in the moray eel (*Gymnothorax javanicus*) and in a variety of coral reef fish, such as red snapper (*Lutjanus bohar*). Ciguatoxin is remarkably toxic even at microgram levels, causing widespread food poisoning (ciguatera) in tropical and subtropical regions, characterized by vomiting, diarrhoea, and neurological problems. Most sufferers slowly recover, and few cases are fatal, due principally to the very low

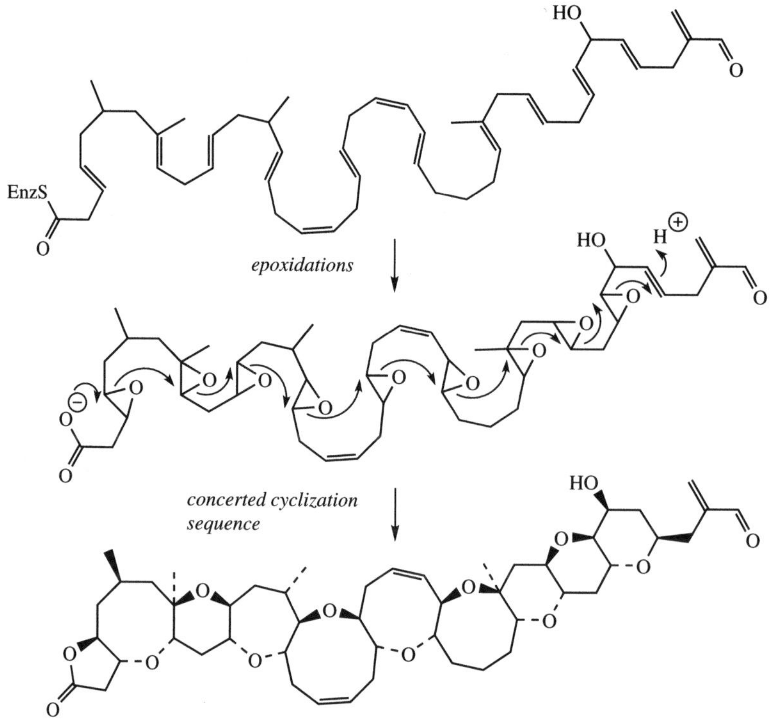

brevetoxin-A

Figure 3.77

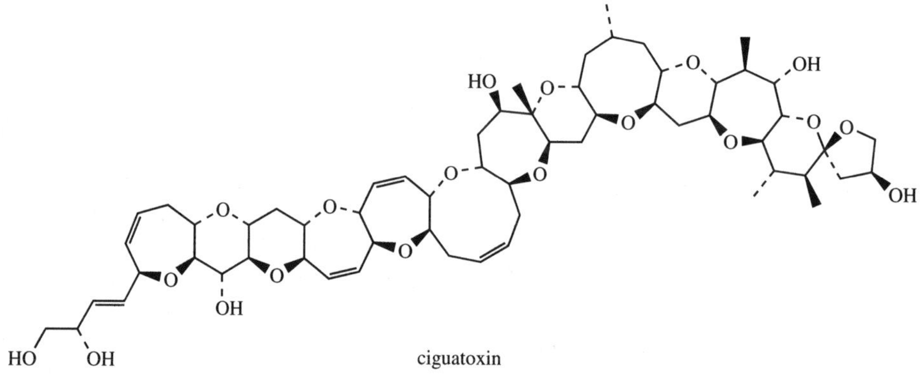

ciguatoxin

Figure 3.78

levels of toxin actually present in the fish. A dinoflagellate *Gambierdiscus toxicus* is ultimately responsible for polyether production, synthesizing a less toxic analogue, which is passed through the food chain and eventually modified into the very toxic ciguatoxin by the fish.

The **zaragozic acids** (**squalestatins**) are not macrolides, but they are primarily acetate derived, and the central ring system is suggested to be

formed by an epoxide-initiated process resembling the polyether derivatives just described. Thus, **zaragozic acid A** (Figure 3.79) is known to be constructed from two acetate-derived chains and a C_4 unit such as the Krebs cycle intermediate oxaloacetate (see Figure 2.1). One chain has a benzoyl-CoA starter (from the shikimate pathway, see page 141), and both contain two methionine-derived side-chain substituents (Figure 3.79). The

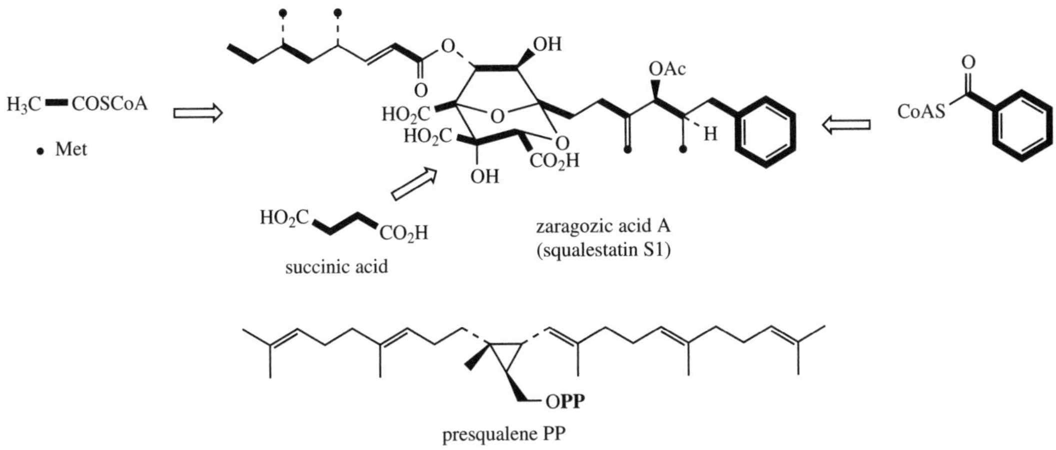

Figure 3.79

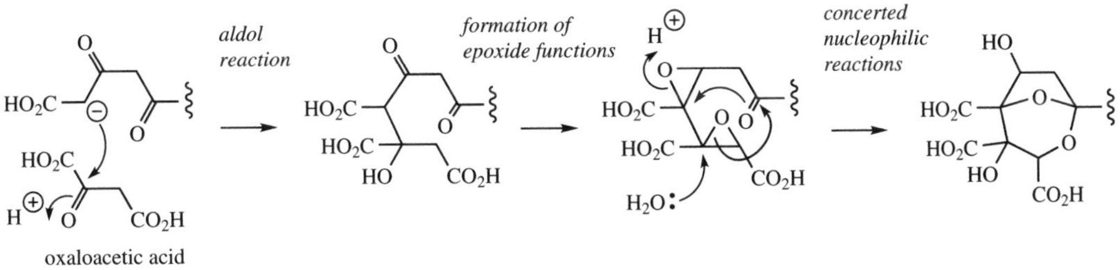

Figure 3.80

heterocyclic ring system can be envisaged as arising via nucleophilic attack on to oxaloacetic acid, formation of a diepoxide, then a concerted sequence of reactions as indicated (Figure 3.80). The zaragozic acids are produced by a number of fungi, including *Sporomiella intermedia* and *Leptodontium elatius*, and are attracting considerable interest since they are capable of reducing blood cholesterol levels in animals by acting as potent inhibitors of the enzyme squalene synthase (see page 212). This is achieved by mimicking the steroid precursor presqualene PP (Figure 3.79) and irreversibly inactivating the enzyme. They thus have considerable medical potential for reducing the incidence of coronary-related deaths (compare the statins, below).

CYCLIZATION THROUGH DIELS–ALDER REACTIONS

A number of cyclic structures, typically containing cyclohexane rings, are known to be formed via the acetate pathway, but experimental evidence supports cyclization processes different from the aldol and Claisen reactions seen in the biosynthesis of aromatic compounds. They can, however, be rationalized in terms of an enzymic Diels–Alder reaction, represented as the electrocyclic sequence shown in Figure 3.81. Thus, **lovastatin** can be formulated as arising from two polyketide chains with *C*-methylation as outlined in Figure 3.82, with relatively few of the oxygen functions being retained in the final product. Accordingly, it is possible that lovastatin is formed by cyclization of the trienoic acid (Figure 3.82), which is likely to arise by a variant of the macrolide

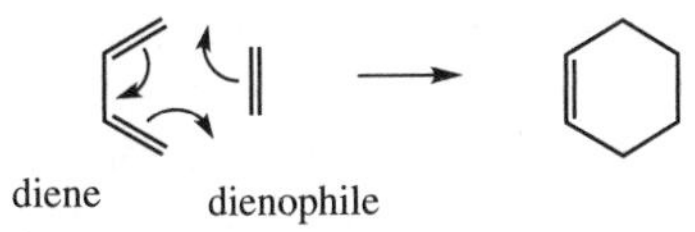

diene dienophile

Diels–Alder reaction

Figure 3.81

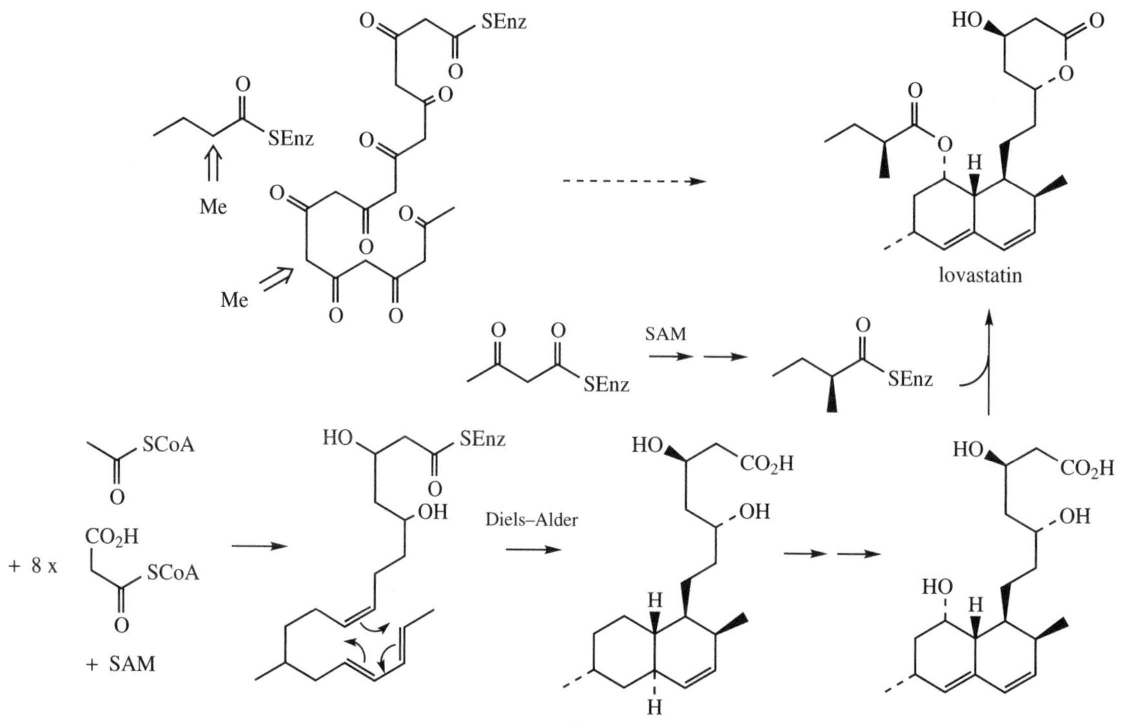

Figure 3.82

Mevastatin and other Statins

Mevastatin (formerly compactin) (Figure 3.83) is produced by cultures of *Penicillium citrinum* and *P. brevicompactum*, and was shown to be a reversible competitive inhibitor of HMG-CoA reductase, dramatically lowering sterol biosynthesis in mammalian cell cultures and animals, and reducing total and low density lipoprotein cholesterol levels (see page 236). Mevastatin in its ring-opened form (Figure 3.84) mimics the half-reduced substrate mevaldate hemithioacetal during the two-stage reduction of HMG-CoA to mevalonate (see page 170), and the affinity of this agent towards HMG-CoA reductase is 10 000-fold more than the normal substrate. High blood cholesterol levels contribute to the incidence of coronary heart disease (see page 236), so mevastatin, or analogues, are of potential value in treating high risk coronary patients, and some agents are already in use. Although lowering of cholesterol levels reduces the risk of heart attacks, there is evidence that the beneficial effects of statins may extend beyond simply cholesterol reduction.

  Lovastatin (formerly called mevinolin or monacolin K) (Figure 3.83) is produced by *Monascus ruber* and *Aspergillus terreus* and is slightly more active than mevastatin, but has been superseded by more active agents. **Simvastatin** is obtained from lovastatin by ester hydrolysis and then re-esterification, and is two to three times as potent as lovastatin. **Pravastatin** is prepared from mevastatin by microbiological hydroxylation using *Streptomyces carbophilus* and is consequently more hydrophilic than the other drugs, with an activity similar to lovastatin. Lovastatin and simvastatin are both lactones, and are inactive until metabolized

(Continues)

(Continued)

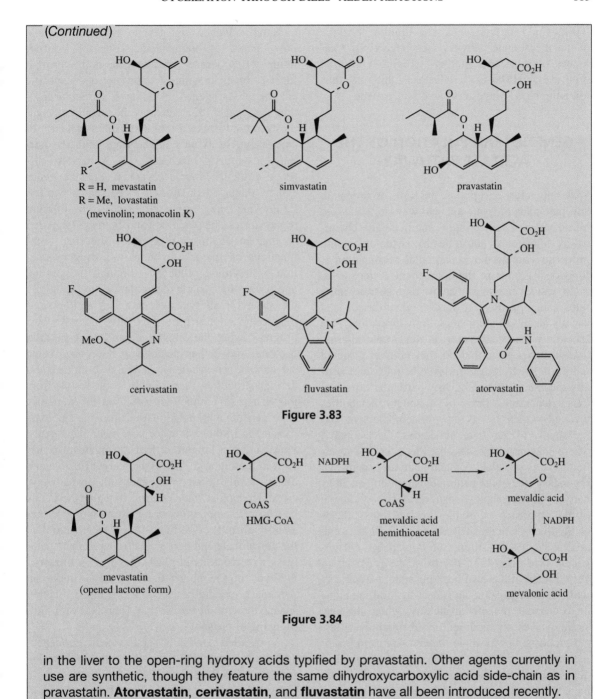

R = H, mevastatin
R = Me, lovastatin
(mevinolin; monacolin K)

simvastatin

pravastatin

cerivastatin

fluvastatin

atorvastatin

Figure 3.83

mevastatin
(opened lactone form)

HMG-CoA

mevaldic acid
hemithioacetal

mevaldic acid

mevalonic acid

Figure 3.84

in the liver to the open-ring hydroxy acids typified by pravastatin. Other agents currently in use are synthetic, though they feature the same dihydroxycarboxylic acid side-chain as in pravastatin. **Atorvastatin, cerivastatin,** and **fluvastatin** have all been introduced recently.

biosynthetic processes, though *C*-methylation must occur during chain assembly whilst activating carbonyl groups are available. The Diels–Alder reaction can then account for formation of the decalin system and further reactions will allow the other functional groups in lovastatin to be produced. The ester side-chain is derived as a separate unit from two acetates with a methyl from methionine, again with *C*-methylation preceding reduction processes. Lovastatin was isolated from cultures of *Aspergillus terreus* and was found to be a potent inhibitor of hydroxymethylglutaryl-CoA

(HMG-CoA) reductase, a rate-limiting enzyme in the mevalonate pathway (see page 169). Analogues of lovastatin (statins*) (Figure 3.83) find drug use as HMG-CoA reductase inhibitors, thus lowering blood cholesterol levels in patients.

GENETIC MANIPULATION OF THE ACETATE PATHWAY

With only a few exceptions, the transformations in any particular biosynthetic pathway are catalysed by enzymes. These proteins facilitate the chemical modification of substrates by virtue of binding properties conferred by a particular combination of functional groups in the constituent amino acids. As a result, enzymes tend to demonstrate quite remarkable specificity towards their substrates, and usually catalyse only a single transformation. This specificity means enzymes do not accept alternative substrates, or, if they do, they convert a limited range of structurally similar substrates and usually much less efficiently. Any particular organism thus synthesizes a range of secondary metabolites dictated largely by its enzyme complement and the supply of substrate molecules. Occasionally, where enzymes do possess broader substrate specificities, it is possible to manipulate an organism's secondary metabolite pattern by supplying an alternative, but acceptable, substrate. A good example of this approach is in the directed biosynthesis of modified penicillins by the use of phenylacetic acid analogues in cultures of *Penicillium chrysogenum* (see page 437), but its scope is generally very limited. It has also been possible, particularly with microorganisms, to select natural mutants, or to generate mutants artificially, where the new strain synthesizes modified or substantially different products. For example, mutant strains of *Streptomyces aureofaciens* synthesize tetracycline or demeclocycline rather than chlortetracycline (see page 90). Such mutants are usually deficient in a single enzyme and are thus unable to carry out a single transformation, but the broader specificity of later enzymes in the sequence means subsequent modifications may still occur. However, as exemplified throughout this book, the vast bulk of modified natural products of medicinal importance are currently obtained by chemical synthesis or semi-synthesis.

Rapid advances in genetic engineering have now opened up tremendous scope for manipulating the processes of biosynthesis by providing an organism with, or depriving it of, specific enzymes. The genes encoding a particular protein (see page 407) can now be identified, synthesized, and inserted into a suitable organism for expression; to avoid complications with the normal biosynthetic machinery, this is usually different from the source organism. Specific genes can be damaged or deleted to prevent a particular enzyme being expressed. Genes from different organisms can be combined and expressed together so that an organism synthesizes abnormal combinations of enzyme activities, allowing production of modified products. Although the general approaches for genetic manipulation are essentially the same for all types of organism and/or natural product, it has proved possible to make best progress using the simpler organisms, especially bacteria, and in particular there have been some substantial achievements in the area of acetate-derived structures. Accordingly, some results from this group of compounds are used to exemplify how genetic manipulation may provide an extra dimension in the search for new medicinal agents. However, it is important that an organism is not viewed merely as a sackful of freely diffusible and always available enzymes; biosynthetic pathways are under sophisticated controls in which there may be restricted availability or localization of enzymes and/or substrates (see the different localizations of the mevalonate and deoxyxylulose phosphate pathways to terpenoids in plants, page 172). Enzymes involved in the biosynthesis of many important secondary metabolites are often grouped together as enzyme complexes, or may form part of a multifunctional protein.

A detailed study of amino acid sequences and mechanistic similarities in various **polyketide synthase** (PKS) enzymes has led to two main types being distinguished. **Type I enzymes** consist of one or more large multifunctional proteins that possess a distinct active site for every enzyme-catalysed step. On the other hand, **Type II enzymes** are multienzyme complexes that carry out a single set of repeating activities. Like fatty acid synthases, PKSs catalyse the condensation of coenzyme A esters of simple carboxylic acids. However, the variability at each step in the biosynthetic pathway gives rise to much more structural diversity than encountered with fatty acids. The usual starter units employed are acetyl-CoA or propionyl-CoA, whilst malonyl-CoA or methylmalonyl-CoA are the main extender units. At each cycle of chain extension, Type I PKSs may retain the β-ketone, or modify it to a hydroxyl, methenyl, or methylene, according to the presence of ketoreductase, dehydratase, or enoylreductase activities (see page 95). The enzyme activities for each extension cycle with its subsequent modification is considered a 'module'. The linear sequence of modules in the enzyme corresponds to the generated sequence of extender units in the polyketide product. The β-ketone groups are predominantly left intact by Type II PKSs, and the highly reactive polyketide backbone undergoes further enzyme-catalysed intramolecular cyclization reactions, which are responsible for generating a range of aromatic structures (see page 61).

6-Deoxyerythronolide B synthase (DEBS) is a modular Type I PKS involved in **erythromycin** biosynthesis (see page 96) and its structure and function are illustrated in Figure 3.85. The enzyme contains three subunits (DEBS-1, 2, and 3), each encoded by a gene (*eryA*-I, II, and III). It has a linear organization of six modules, each of which

contains the activities needed for one cycle of chain extension. A minimal module contains a β-ketoacyl synthase (KS), an acyltransferase (AT), and an acyl carrier protein (ACP), that together would catalyse a two-carbon chain extension. The specificity of the AT for either malonyl-CoA or an alkyl-malonyl-CoA determines which two-carbon chain extender is used. The starter unit used is similarly determined by the specificity of the AT in a loading domain in the first module. After each condensation reaction, the oxidation state of the β-carbon is determined by the presence of a β-ketoacyl reductase (KR), a KR + a dehydratase (DH), or a KR + DH + an enoylreductase (ER) in the appropriate module. The sequence is finally terminated by a thioesterase (TE) activity which releases the polyketide from the enzyme and allows cyclization. Thus in DEBS, module 3 lacks any β-carbon modifying domains, modules 1, 2, 5, and 6 contain KR domains and are responsible for hydroxy substituents, whereas module 4 contains the complete KR, DH, and ER set, and results in complete reduction to a methylene. Overall, the AT specificity and the catalytic domains on each module determine the structure and stereochemistry of each two-carbon extension unit, the order of the modules specifies the sequence of the units,

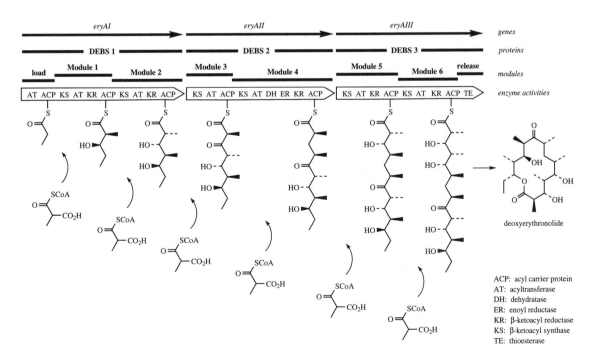

Figure 3.85

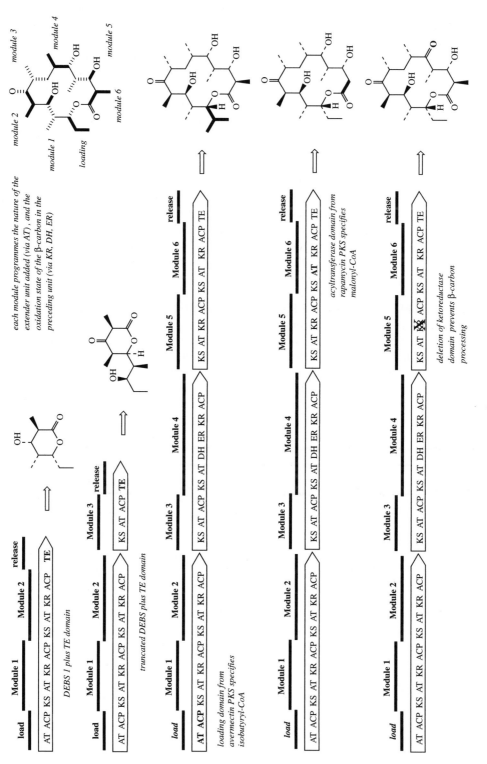

Figure 3.86

and the number of modules determines the size of the polyketide chain. The vast structural diversity of natural polyketides arises from combinatorial possibilities of arranging modules containing the various catalytic domains, the sequence and number of modules, and the post-PKS enzymes which subsequently modify the first-formed product, e.g. 6-deoxyerythronolide B → erythromycin (see page 96). Genetic engineering now offers vast opportunities for rational modification of the resultant polyketide structure.

A few representative examples of successful experiments leading to engineered polyketides are shown in Figure 3.86. Reducing the size of the gene sequence so that it encodes fewer modules results in the formation of smaller polyketides, characterized by the corresponding loss of extender units; in these examples the gene encoding the chain terminating thioesterase also has to be attached to complete the biosynthetic sequence. Replacing the loading domain of DEBS with that from another PKS, e.g. that producing avermectin (see page 97), alters the specificity of the enzyme for the starter unit. The loading module of the avermectin-producing PKS actually has a much broader specificity than that for DEBS; Figure 3.86 shows the utilization of isobutyryl-CoA as features in the natural biosynthesis of avermectin B_{1b}. Other examples include the replacement of an AT domain (in DEBS specifying a methylmalonyl extender) with a malonyl-specific AT domain from the rapamycin-producing PKS (see page 103), and deletion of a KR domain, thus stopping any β-carbon processing for that module with consequent retention of a carbonyl group. Not all experiments in gene modification are successful, and even when they are yields can be disappointingly lower than in the natural system. There is always a fundamental requirement that enzymes catalysing steps after the point of modification need to have sufficiently broad substrate specificities to accept and process the abnormal compounds being synthesized; this becomes more unlikely where two or more genetic changes have been made. Nevertheless, multiple modifications have been successful, and it has also been possible to exploit changes in a combinatorial fashion using different expression vectors for the individual subunits, thus creating a library of polyketides, which may then be screened for potential biological activity.

Non-ribosomal peptide synthases (see page 421) are also modular and lend themselves to similar genetic manipulation as the Type I PKSs. The production of modified aromatic polyketides by genetically engineered Type II PKSs is not quite so 'obvious' as with the modular Type I enzymes, but significant progress has been made in many systems. Each Type II PKS contains a minimal set of three protein subunits, two β-ketoacyl synthase (KS) subunits and an ACP to which the growing chain is attached. Additional subunits, including KRs, cyclases (CYC), and aromatases (ARO), are responsible for modification of the nascent chain to form the final cyclized structure. Novel polyketides have been generated by manipulating Type II PKSs, exchanging KS, CYC, and ARO subunits among different systems. However, because of the highly reactive nature of poly-β-keto chains, the cyclizations that occur with the modified gene product frequently vary from those in the original compound. Compared with Type I PKSs, the formation of new products with predictable molecular structure has proven less controllable.

The polyketide synthases responsible for chain extension of cinnamoyl-CoA starter units leading to flavonoids and stilbenes, and of anthraniloyl-CoA leading to quinoline and acridine alkaloids (see page 377) do not fall into either of the above categories and have now been termed **Type III PKSs**. These enzymes differ from the other examples in that they are homodimeric proteins, they utilize coenzyme A esters rather than acyl carrier proteins, and they employ a single active site to perform a series of decarboxylation, condensation, cyclization, and aromatization reactions.

FURTHER READING

Biosynthesis

Rawlings BJ (1999) Biosynthesis of polyketides (other than actinomycete macrolides). *Nat Prod Rep* **16**, 425–484. Earlier review: (1997) **14**, 523–556.

Fatty Acids and Fats

Gunstone F (1996) *Fatty Acid and Lipid Chemistry*. Blackie, Glasgow.
Harwood JL (1996) Recent advances in the biosynthesis of plant fatty acids. *Biochim Biophys Acta* **1301**, 7–56.

Hasenhuettl GL (1993) Fats and fatty oils. *Kirk–Othmer Encyclopedia of Chemical Technology*, 4th edn, Vol 10. Wiley, New York, 252–267.

Kawaguchi A and Iwamoto-Kihara A (1999) Biosynthesis and degradation of fatty acids. *Comprehensive Natural Products Chemistry*, Vol 1. Elsevier, Amsterdam, pp 23–59.

Lie Ken Jie MSF and Pasha MK (1998) Fatty acids, fatty acid analogues and their derivatives. *Nat Prod Rep* **15**, 607–629. Earlier review: Lie Ken Jie MSF, Pasha MK and Syed-Rahmatullah MSK (1997) **14**, 163–189.

Mason P (2000) Nutrition: fish oils – an update. *Pharm J* **265**, 720–724.

Ohlrogge J and Browse J (1995) Lipid biosynthesis. *Plant Cell* **7**, 957–970.

Rawlings BJ (1998) Biosynthesis of fatty acids and related metabolites. *Nat Prod Rep* **15**, 275–308. Earlier review: (1997), **14**, 335–358.

Shanklin J and Cahoon EB (1998) Desaturation and related modifications of fatty acids. *Annu Rev Plant Physiol Plant Mol Biol* **49**, 611–641.

Wolf WJ (1997) Soybeans and other oilseeds. *Kirk–Othmer Encyclopedia of Chemical Technology*, 4th edn, Vol 22. Wiley, New York, pp 591–619.

Platelet-Activating Factor

Prescott SM, Zimmermann GA and McIntyre TM (1990) Platelet-activating factor. *J Biol Chem* **265**, 17 382–17 384.

Snyder F (1995) Platelet-activating factor: the biosynthetic and catabolic enzymes. *Biochem J* **305**, 689–705.

Echinacea

Houghton P (1994) Herbal products: echinacea. *Pharm J* **253**, 342–343.

Perry NB, van Klink JW, Burgess EJ and Parmenter GA (1997) Alkamide levels in *Echinacea purpurea*: a rapid analytical method revealing differences among roots, rhizomes, stems, leaves and flowers. *Planta Med* **63**, 58–62.

Prostaglandins, Thromboxanes, Leukotrienes

Beuck M (1999) Nonsteroidal antiinflammatory drugs: a new generation of cyclooxygenase inhibitors. *Angew Chem Int Ed* **38**, 631–633.

Clissold D and Thickitt C (1994) Recent eicosanoid chemistry. *Nat Prod Rep* **11**, 621–637.

Collins PW (1996) Prostaglandins. *Kirk–Othmer Encyclopedia of Chemical Technology*, 4th edn, Vol 20. Wiley, New York, 302–351.

Collins PW and Djuric SW (1993) Synthesis of therapeutically useful prostaglandins and prostacyclin analogs. *Chem Rev* **93**, 1533–1564.

Hamanaka N (1999) Eicosanoids in mammals. *Comprehensive Natural Products Chemistry*, Vol 1. Elsevier, Amsterdam, pp 159–206.

Jackson WT and Fleisch JH (1996) Development of novel anti-inflammatory agents: a pharmacologic perspective on leukotrienes and their receptors. *Prog Drug Res* **46**, 115–168.

Lawson JA, Rokach J and FitzGerald GA (1999) Isoprostanes: formation, analysis and use as indices of lipid peroxidation *in vivo*. *J Biol Chem* **274**, 24 441–24 444.

Marnett LJ, Goodwin DC, Rowlinson SW, Kalgutkar AS and Landino LM (1999) Structure, function, and inhibition of prostaglandin endoperoxide synthases. *Comprehensive Natural Products Chemistry*, Vol 5. Elsevier, Amsterdam, pp 225–261.

Marnett LJ and Kalgutker AS (1999) Cyclooxygenase 2 inhibitors: discovery, selectivity and the future. *Trends in Pharmacological Sciences* **20**, 465–469.

Marnett LJ, Rowlinson SW, Goodwin DC, Kalgutkar AS and Lanzo CA (1999) Arachidonic acid oxygenation by COX-1 and COX-2. Mechanisms of catalysis and inhibition. *J Biol Chem* **274**, 22 903–22 906.

Masferrer JL and Needleman P (2000) Anti-inflammatories for cardiovascular disease. *Proc Natl Acad Sci USA* **97**, 12 400–12 401.

Mueller MJ (1998) Radically novel prostaglandins in animals and plants: the isoprostanes. *Chem Biol* **5**, R323–333.

Peters-Golden M and Brock TG (2001) Intracellular compartmentalization of leukotriene synthesis: unexpected nuclear secrets. *FEBS Lett* **487**, 323–326.

Smith WL, Garavito RM and DeWitt DL (1996) Prostaglandin endoperoxide H synthases (cyclooxygenases)-1 and -2. *J Biol Chem* **271**, 33 157–33 160.

Talley JJ (1999) Selective inhibitors of cyclooxygenase-2 (COX-2). *Prog Med Chem* **36**, 201–234.

Ullrich V and Brugger R (1994) Prostacyclin and thromboxane synthase: new aspects of hemethiolate catalysis. *Angew Chem Int Ed Engl* **33**, 1911–1919.

Versteeg HH, van Bergen en Henegouwen PMP, van Deventer SJH and Peppelenbosch MP (1999) Cyclooxygenase-dependent signalling: molecular events and consequences. *FEBS Lett* **445**, 1–5.

Yamamoto S (1999) Biosynthesis and metabolism of eicosanoids. *Comprehensive Natural Products Chemistry*, Vol 1. Elsevier, Amsterdam, pp 255–271.

Anthraquinones

Atherton P (1998) First aid plant. *Chem Brit* **34** (5), 33–36 (aloes).

Butler AR and Moffett J (1995) Pass the rhubarb. *Chem Brit*, 462–465.

Falk H (1999) From the photosensitizer hypericin to the photoreceptor stentorin – the chemistry of phenanthroperylene quinones. *Angew Chem Int Ed* **38**, 3116–3136.

Khellin/Cromoglicate

Bernstein PR (1992) *Antiasthmatic agents. Kirk–Othmer Encyclopedia of Chemical Technology*, 4th edn, Vol 2. Wiley, New York, 830–854.

Griseofulvin

Cauwenbergh G (1992) *Antiparasitic agents (antimycotics). Kirk–Othmer Encyclopedia of Chemical Technology*, 4th edn, Vol 3. Wiley, New York, 473–489.

Aflatoxins

Minto RE and Townsend CA (1997) Enzymology and molecular biology of aflatoxin biosynthesis. *Chem Rev* **97**, 2537–2555.

Townsend CA and Minto RE (1999) Biosynthesis of aflatoxins. *Comprehensive Natural Products Chemistry*, Vol 1. Elsevier, Amsterdam, pp 443–471.

Cannabinoids

Devane WA (1994) New dawn of cannabinoid pharmacology. *Trends in Pharmacological Sciences* **15**, 40–41.

Mechoulam R and Ben-Shabat S (1999) From *gan-zi-gun-nu* to anandamide and 2-arachidonoylglycerol: the ongoing story of cannabis. *Nat Prod Rep* **16**, 131–143.

Metchoulam R, Hanus L and Fride E (1998) Towards cannabinoid drugs – revisited. *Prog Med Chem* **35**, 199–243.

O'Driscoll C (2000) High hopes for cannabis. *Chem Brit* **36** (5), 27.

Piomelli D, Giuffrida A, Calignano A and de Fonseca FR (2000) The endocannabinoid system as a target for therapeutic drugs. *Trends in Pharmacological Sciences* **21**, 218–224.

Seth R and Sinha S (1991) Chemistry and pharmacology of cannabis. *Prog Drug Res* **36**, 71-115.

Stevenson R (1998) Cannabis: proscribed or prescribed? *Chem Brit* **34** (7), 34–36.

Straus SE (2000) Immunoactive cannabinoids: therapeutic prospects for marijuana constituents. *Proc Natl Acad Sci USA* **97**, 9363–9364.

Williamson EM and Evans FJ (2000) Cannabinoids in clinical practice. *Drugs* **60**, 1303–1314.

Tetracyclines

Hlavka JJ, Ellestad GA and Chopra I (1992) Antibiotics (tetracyclines). *Kirk–Othmer Encyclopedia of Chemical Technology*, 4th edn, Vol 3. Wiley, New York, 331–346.

Anthracyclines

Fujii I and Ebizuka Y (1997) Anthracycline biosynthesis in *Streptomyces galilaeus*. *Chem Rev* **97**, 2511–2523.

Hutchinson CR (1997) Biosynthetic studies of daunorubicin and tetracenomycin C. *Chem Rev* **97**, 2525–2535.

Lown JW (1993) Discovery and development of anthracycline antitumour antibiotics. *Chem Soc Rev* **22**, 165–176.

Macrolides

Kirst HA (1992) Antibiotics (macrolides). *Kirk–Othmer Encyclopedia of Chemical Technology*, 4th edn, Vol 3. Wiley, New York, 169–213.

Kirst HA (1993) Semi-synthetic derivatives of erythromycin. *Prog Med Chem* **30**, 57–88.

Nicolau KC, Roschangar F and Vourloumis D (1998) Chemical biology of epothilones. *Angew Chem Int Ed* **37**, 2014–2045.

Pieper R, Kao C, Khosla C, Luo G and Cane DE (1996) Specificity and versatility in erythromycin biosynthesis. *Chem Soc Rev* **25**, 297–302.

Staunton J and Wilkinson B (1997) Biosynthesis of erythromycin and rapamycin. *Chem Rev* **97**, 2611–2629.

Staunton J and Wilkinson B (1999) Biosynthesis of erythromycin and related macrolides. *Comprehensive Natural Products Chemistry*, Vol 1. Elsevier, Amsterdam, pp 495–532.

Polyene Antifungals

Cauwenbergh G (1992) Antiparasitic agents (antimycotics). *Kirk–Othmer Encyclopedia of Chemical*

Technology, 4th edn, Vol 3. Wiley, New York, 473–489.

Hoeprich PD (1995) Antifungal chemotherapy. *Prog Drug Res* **44**, 87–127.

Avermectins

Campbell WC (1991) Ivermectin as an antiparasitic agent for use in humans. *Annu Rev Microbiol* **45**, 445–474.

Davies HG and Green RH (1991) Avermectins and milbemycins. *Chem Soc Rev* **20**, 211–269; 271–339.

Fisher MH and Mrozik H (1992) Antiparasitic agents (avermectins). *Kirk–Othmer Encyclopedia of Chemical Technology*, 4th edn, Vol 3. Wiley, New York, 526–540.

Ikeda H and Omura S (1997) Avermectin biosynthesis. *Chem Rev* **97**, 2591–2609.

Ansa Macrolides

Antosz FJ (1992) Antibiotics (ansamacrolides). *Kirk–Othmer Encyclopedia of Chemical Technology*, 4th edn, Vol 2. Wiley, New York, 926-961.

Tacrolimus

Clardy J (1995) The chemistry of signal transduction. *Proc Natl Acad Sci USA* **92**, 56–61.

Rosen MK and Schreiber SL (1992) Natural products as probes of cellular function: studies of immunophilins. *Angew Chem Int Ed Engl* **31**, 384–400.

Wong S (1995) *Immunotherapeutic agents. Kirk–Othmer Encyclopedia of Chemical Technology*, 4th edn, Vol 14. Wiley, New York, 64–86.

Polyethers

Crandall LW and Hamill RL (1992) Antibiotics (polyethers). *Kirk–Othmer Encyclopedia of Chemical Technology*, 4th edn, Vol 3. Wiley, New York, 306–331.

Dutton CJ, Banks BJ and Cooper CB (1995) Polyether ionophores. *Nat Prod Rep* **12**, 165–181.

Riddell FG (1992) Ionophoric antibiotics. *Chem Brit*, 533–537.

Robinson JA (1991) Chemical and biochemical aspects of polyether-ionophore antibiotic biosynthesis. *Prog Chem Org Nat Prod* **58**, 1–81.

Shimizu Y (1993) Microalgal metabolites. *Chem Rev* **93**, 1685–1698.

Yasumoto T and Murata M (1993) Marine toxins. *Chem Rev* **93**, 1897–1909.

Zaragozic Acids (Squalestatins)

Bergstrom JD, Dufresne C, Bills GF, Nallin-Omstead M and Byrne K (1995) Discovery, biosynthesis and mechanism of action of the zaragozic acids: potent inhibitors of squalene synthase. *Annu Rev Microbiol* **49**, 607–639.

Nadin A and Nicolaou KC (1996) Chemistry and biology of the zaragozic acids (squalestatins). *Angew Chem Int Ed Engl* **35**, 1623–1656.

Watson NS and Procopiou PA (1996) Squalene synthase inhibitors: their potential as hypocholesterolaemic agents. *Prog Med Chem* **33**, 331–378.

Statins

Cervoni P, Crandall DL and Chan PS (1993) Cardiovascular agents – atherosclerosis and antiatherosclerosis agents. Kirk–Othmer Encyclopedia of Chemical Technology, 4th edn, Vol 5. Wiley, New York, 257–261.

Endo A and Hasumi K (1993) HMG-CoA reductase inhibitors. *Nat Prod Rep* **10**, 541–550.

Gopinath L (1996) Cholesterol drug dilemma. *Chem Brit* **32** (11), 38–41.

Laschat S (1996) Pericyclic reactions in biological systems – does nature know about the Diels–Alder reaction? *Angew Chem Int Ed Engl* **35**, 289–291.

Yalpani M (1996) Cholesterol-lowering drugs. *Chem Ind*, 85–89.

Polyketide Synthases/Genetic Manipulation

Cane DE, Walsh CT and Khosla C (1998) Harnessing the biosynthetic code: combinations, permutations, and mutations. *Science* **262**, 63–68.

Hutchinson CR (1999) Microbial polyketide synthases: more and more prolific. *Proc Natl Acad Sci USA* **96**, 3336–3338.

Katz L (1997) Manipulation of modular polyketide synthases. *Chem Rev* **97**, 2557–2575.

Khosla C (1997) Harnessing the potential of modular polyketide synthases. *Chem Rev* **97**, 2577–2590.

Khosla C, Gokhale RS, Jacobsen JR and Cane DE (1999) Tolerance and specificity of polyketide synthases. *Annu Rev Biochem* **68**, 219–253.

Richardson M and Khosla C (1999) Structure, function, and engineering of bacterial polyketide synthases. *Comprehensive Natural Products Chemistry*, Vol 1. Elsevier, Amsterdam, pp 473–494.

4

THE SHIKIMATE PATHWAY: AROMATIC AMINO ACIDS AND PHENYLPROPANOIDS

Shikimic acid and its role in the formation of aromatic amino acids, benzoic acids, and cinnamic acids is described, along with further modifications leading to lignans and lignin, phenylpropenes, and coumarins. Combinations of the shikimate pathway and the acetate pathway are responsible for the biosynthesis of styrylpyrones, flavonoids and stilbenes, flavonolignans, and isoflavonoids. Terpenoid quinones are formed by a combination of the shikimate pathway with the terpenoid pathway. Monograph topics giving more detailed information on medicinal agents include folic acid, chloramphenicol, podophyllum, volatile oils, dicoumarol and warfarin, psoralens, kava, *Silybum marianum*, phyto-oestrogens, derris and lonchocarpus, vitamin E, and vitamin K.

The shikimate pathway provides an alternative route to aromatic compounds, particularly the aromatic amino acids L-**phenylalanine**, L-**tyrosine** and L-**tryptophan**. This pathway is employed by microorganisms and plants, but not by animals, and accordingly the aromatic amino acids feature among those essential amino acids for man which have to be obtained in the diet. A central intermediate in the pathway is **shikimic acid** (Figure 4.1), a compound which had been isolated from plants of *Illicium* species (Japanese 'shikimi') many years before its role in metabolism had been discovered. Most of the intermediates in the pathway were identified by a careful study of a series of *Escherichia coli* mutants prepared by UV irradiation. Their nutritional requirements for growth, and any by-products formed, were then characterized. A mutant strain capable of growth usually differs from its parent in only a single gene, and the usual effect is the impaired synthesis of a single enzyme. Typically, a mutant blocked in the transformation of compound A into compound B will require B for growth whilst accumulating A in its culture medium. In this way, the pathway from phosphoenolpyruvate (from

glycolysis) and D-erythrose 4-phosphate (from the pentose phosphate cycle) to the aromatic amino acids was broadly outlined. Phenylalanine and tyrosine form the basis of C_6C_3 phenylpropane units found in many natural products, e.g. cinnamic acids, coumarins, lignans, and flavonoids, and along with tryptophan are precursors of a wide range of alkaloid structures. In addition, it is found that many simple benzoic acid derivatives, e.g. gallic acid (Figure 4.1) and *p*-aminobenzoic acid (4-aminobenzoic acid) (Figure 4.4) are produced via branchpoints in the shikimate pathway.

AROMATIC AMINO ACIDS AND SIMPLE BENZOIC ACIDS

The shikimate pathway begins with a coupling of phosphoenolpyruvate (PEP) and D-erythrose 4-phosphate to give the seven-carbon 3-deoxy-D-*arabino*-heptulosonic acid 7-phosphate (DAHP) (Figure 4.1). This reaction, shown here as an aldol-type condensation, is known to be mechanistically more complex in the enzyme-catalysed version; several of the other transformations in the pathway have also been found to be surprisingly

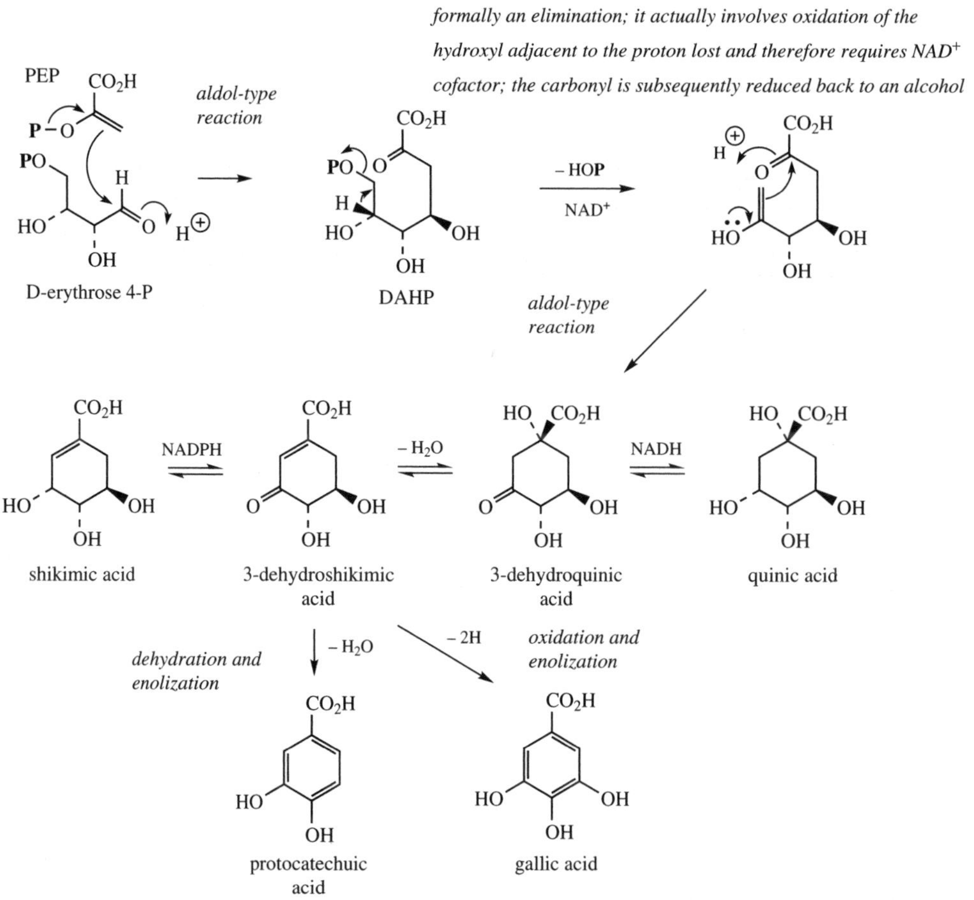

Figure 4.1

complex. Elimination of phosphoric acid from DAHP followed by an intramolecular aldol reaction generates the first carbocyclic intermediate **3-dehydroquinic acid**. However, this also represents an oversimplification. The elimination of phosphoric acid actually follows an NAD^+-dependent oxidation of the central hydroxyl, and this is then re-formed in an NADH-dependent reduction reaction on the intermediate carbonyl compound prior to the aldol reaction occurring. All these changes occur in the presence of a single enzyme. Reduction of 3-dehydroquinic acid leads to **quinic acid**, a fairly common natural product found in the free form, as esters, or in combination with alkaloids such as quinine (see page 362). **Shikimic acid** itself is formed from 3-dehydroquinic acid via **3-dehydroshikimic acid** by dehydration and reduction steps. The simple phenolic acids **protocatechuic acid** (3,4-dihydroxybenzoic acid) and

gallic acid (3,4,5-trihydroxybenzoic acid) can be formed by branchpoint reactions from 3-dehydroshikimic acid, which involve dehydration and enolization, or, in the case of gallic acid, dehydrogenation and enolization. **Gallic acid** features as a component of many tannin materials (**gallotannins**), e.g. pentagalloylglucose (Figure 4.2), found in plants, materials which have been used for thousands of years in the tanning of animal hides to make leather, due to their ability to cross-link protein molecules. Tannins also contribute to the astringency of foods and beverages, especially tea, coffee and wines (see also condensed tannins, page 151).

A very important branchpoint compound in the shikimate pathway is **chorismic acid** (Figure 4.3), which has incorporated a further molecule of PEP as an enol ether side-chain. PEP combines with shikimic acid 3-phosphate produced in a

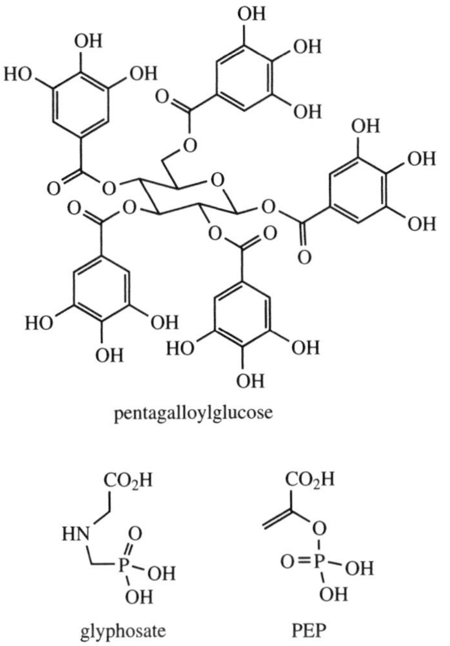

pentagalloylglucose

glyphosate   PEP

Figure 4.2

simple ATP-dependent phosphorylation reaction. This combines with PEP via an addition–elimination reaction giving **3-enolpyruvylshikimic acid 3-phosphate (EPSP)**. This reaction is catalysed by the enzyme EPSP synthase. The synthetic *N*-(phosphonomethyl)glycine derivative **glyphosate** (Figure 4.2) is a powerful inhibitor of this

enzyme, and is believed to bind to the PEP binding site on the enzyme. Glyphosate finds considerable use as a broad spectrum herbicide, a plant's subsequent inability to synthesize aromatic amino acids causing its death. The transformation of EPSP to **chorismic acid** (Figure 4.3) involves a 1,4-elimination of phosphoric acid, though this is probably not a concerted elimination.

4-hydroxybenzoic acid (Figure 4.4) is produced in bacteria from chorismic acid by an elimination reaction, losing the recently introduced enolpyruvic acid side-chain. However, in plants, this phenolic acid is formed by a branch much further on in the pathway via side-chain degradation of cinnamic acids (see page 141). The three phenolic acids so far encountered, 4-hydroxybenzoic, protocatechuic, and gallic acids, demonstrate some of the hydroxylation patterns characteristic of shikimic acid-derived metabolites, i.e. a single hydroxy *para* to the side-chain function, dihydroxy groups arranged *ortho* to each other, typically 3,4- to the side-chain, and trihydroxy groups also *ortho* to each other and 3,4,5- to the side-chain. The single *para*-hydroxylation and the *ortho*-polyhydroxylation patterns contrast with the typical *meta*-hydroxylation patterns characteristic of phenols derived via the acetate pathway (see page 62), and in most cases allow the biosynthetic origin (acetate or shikimate) of an aromatic ring to be deduced by inspection.

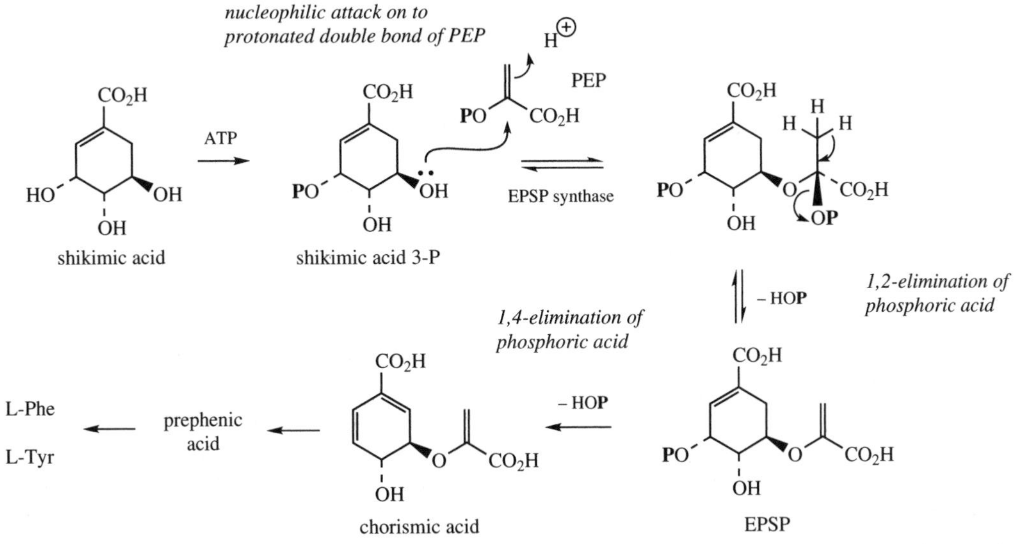

Figure 4.3

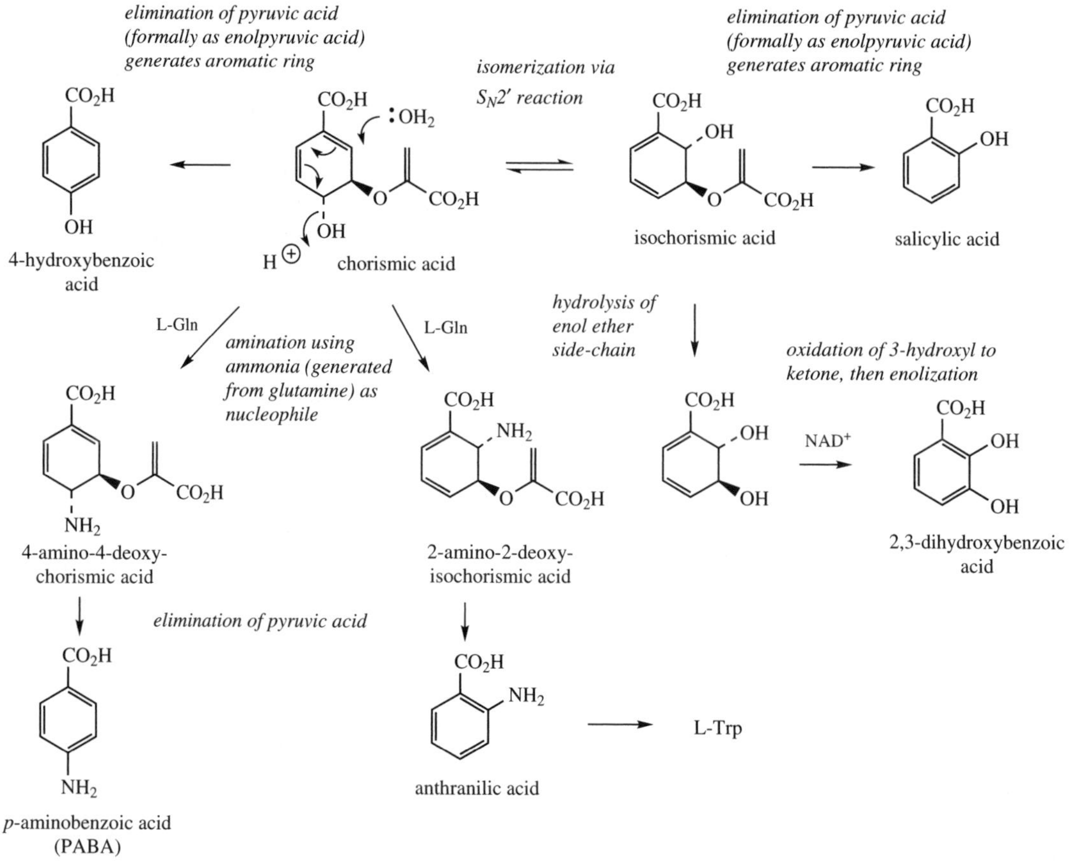

Figure 4.4

2,3-dihydroxybenzoic acid, and **salicylic acid** (2-hydroxybenzoic acid) (in microorganisms, but not in plants, see page 141), are derived from chorismic acid via its isomer **isochorismic acid** (Figure 4.4). The isomerization involves an S_N2'-type of reaction, an incoming water nucleophile attacking the diene system and displacing the hydroxyl. Salicyclic acid arises by an elimination reaction analogous to that producing 4-hydroxybenzoic acid from chorismic acid. In the formation of 2,3-dihydroxybenzoic acid, the side-chain of isochorismic acid is first lost by hydrolysis, then dehydrogenation of the 3-hydroxy to a 3-keto allows enolization and formation of the aromatic ring. 2,3-Dihydroxybenzoic acid is a component of the powerful iron chelator (siderophore) **enterobactin** (Figure 4.5) found in *Escherichia coli* and many other Gram-negative bacteria. Such compounds play an important role in bacterial growth by making

available sufficient concentrations of essential iron. Enterobactin comprises three molecules of 2,3-dihydroxybenzoic acid and three of the amino acid L-serine, in cyclic triester form.

Simple amino analogues of the phenolic acids are produced from chorismic acid by related transformations in which ammonia, generated from glutamine, acts as a nucleophile (Figure 4.4). Chorismic acid can be aminated at C-4 to give 4-amino-4-deoxychorismic acid and then *p*-aminobenzoic (4-aminobenzoic) acid, or at C-2 to give the isochorismic acid analogue which will yield 2-aminobenzoic (anthranilic) acid. Amination at C-4 has been found to occur with retention of configuration, so perhaps a double inversion mechanism is involved. ***p*-Aminobenzoic acid (PABA)** forms part of the structure of **folic acid** (vitamin B$_9$)* (Figure 4.6). The folic acid structure is built up (Figure 4.6) from a dihydropterin diphosphate which reacts with *p*-aminobenzoic

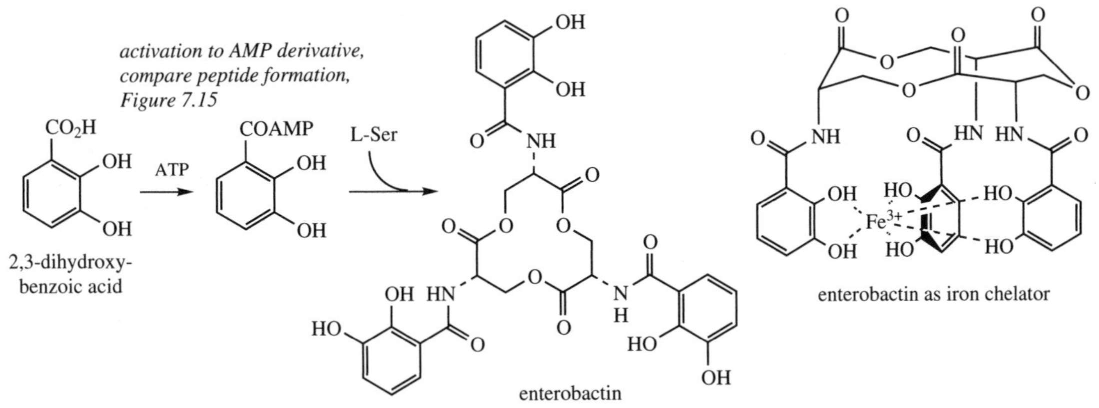

Figure 4.5

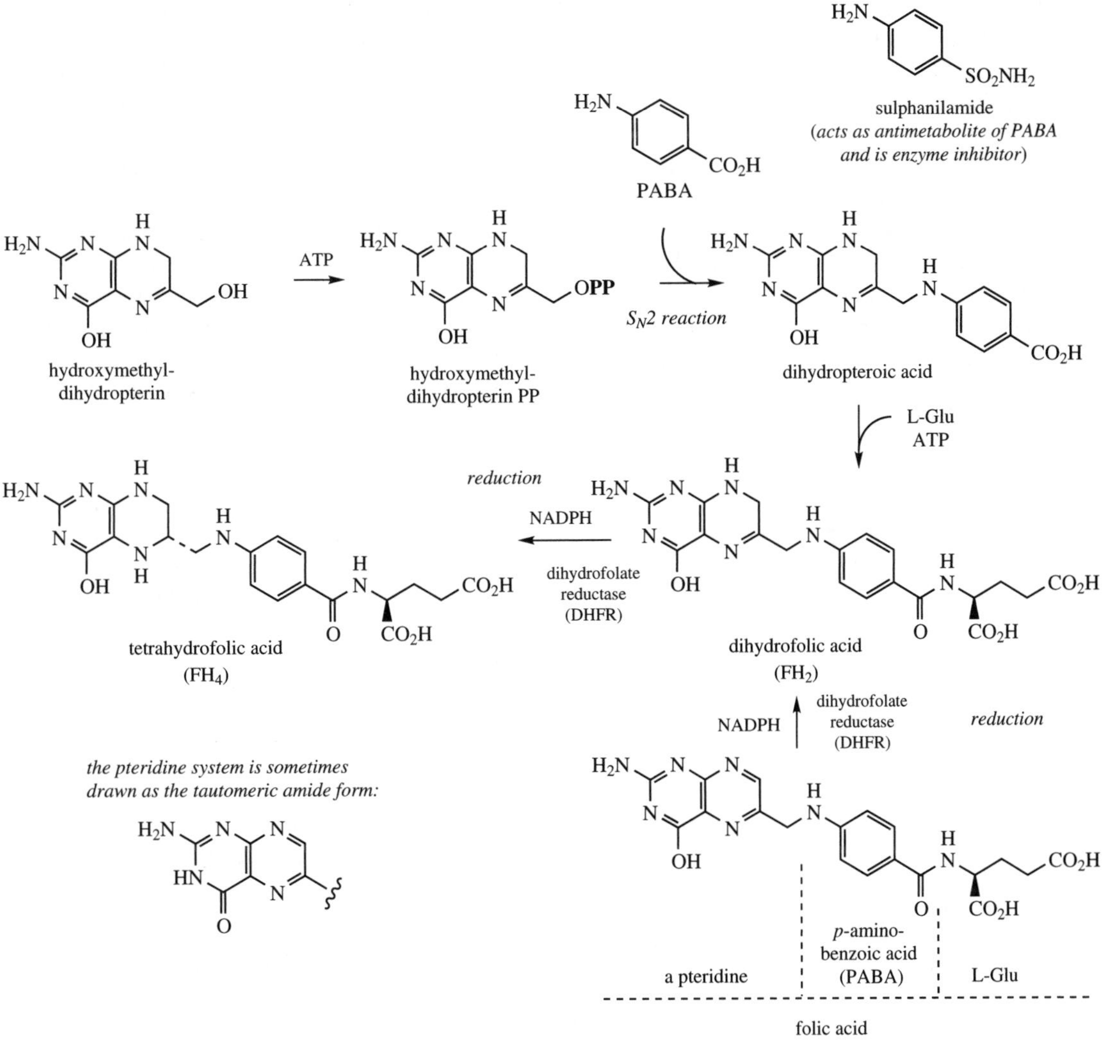

Figure 4.6

acid to give dihydropteroic acid, an enzymic step for which the sulphonamide antibiotics are inhibitors. **Dihydrofolic acid** is produced from the dihydropteroic acid by incorporating glutamic acid, and reduction yields **tetrahydrofolic acid**. This reduction step is also necessary for the continual regeneration of tetrahydrofolic acid, and forms an important site of action for some antibacterial, anti-malarial, and anticancer drugs.

Anthranilic acid (Figure 4.4) is an intermediate in the biosynthetic pathway to the indole-containing aromatic amino acid L-**tryptophan** (Figure 4.10).

Folic Acid (Vitamin B₉)

Folic acid (vitamin B₉) (Figure 4.6) is a conjugate of a pteridine unit, p-aminobenzoic acid, and glutamic acid. It is found in yeast, liver, and green vegetables, though cooking may destroy up to 90% of the vitamin. Deficiency gives rise to anaemia, and supplementation is often necessary during pregnancy. Otherwise, deficiency is not normally encountered unless there is malabsorption, or chronic disease. Folic acid used for supplementation is usually synthetic, and it becomes sequentially reduced in the body by the enzyme dihydrofolate reductase to give dihydrofolic acid and then tetrahydrofolic acid (Figure 4.6). Tetrahydrofolic acid then functions as a carrier of one-carbon groups, which may be in the form of methyl, methylene, methenyl, or formyl groups, by the reactions outlined in Figure 4.7. These groups are involved in amino acid and nucleotide metabolism. Thus a methyl group is transferred in the regeneration of methionine from homocysteine, purine biosynthesis involves methenyl and formyl transfer, and pyrimidine biosynthesis utilizes methylene transfer. Tetrahydrofolate derivatives also serve as acceptors of one-carbon units in degradative pathways.

Mammals must obtain their tetrahydrofolate requirements from their diet, but microorganisms are able to synthesize this material. This offers scope for selective action and led to the use of sulphanilamide and other antibacterial sulpha drugs, compounds which competitively inhibit dihydropteroate synthase, the biosynthetic enzyme incorporating p-aminobenzoic acid into the structure. These sulpha drugs thus act as antimetabolites of p-aminobenzoate. Specific dihydrofolate reductase inhibitors have also become especially useful as antibacterials,

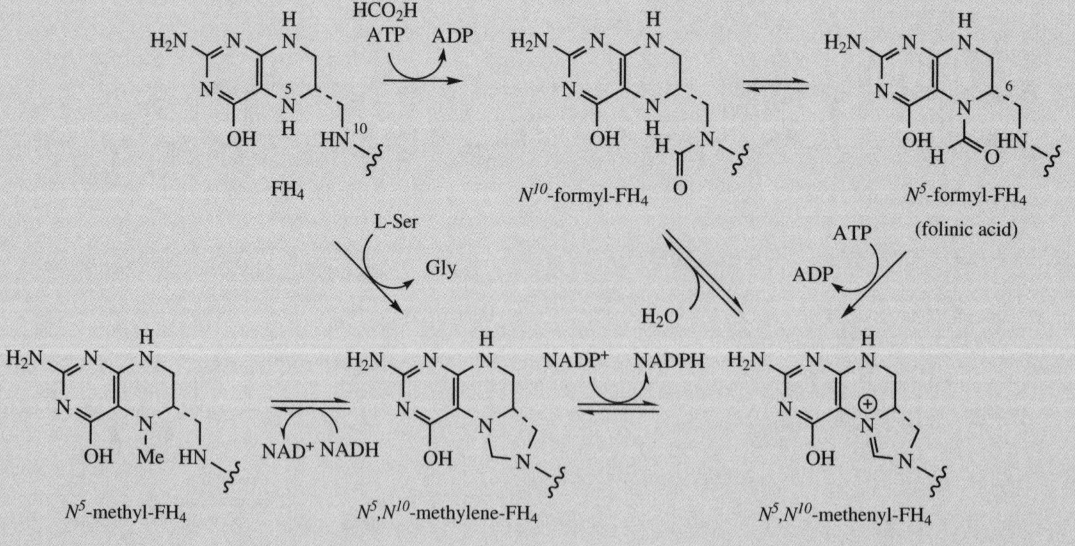

Figure 4.7

(Continues)

(Continued)

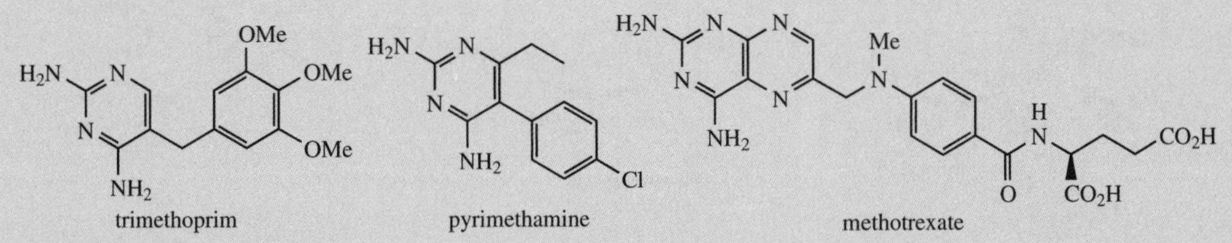

Figure 4.8

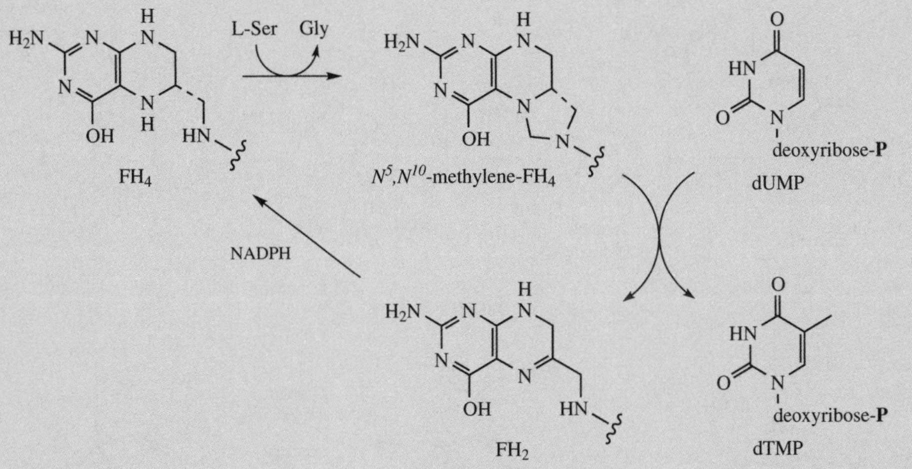

Figure 4.9

e.g. **trimethoprim** (Figure 4.8), and antimalarial drugs, e.g. **pyrimethamine**, relying on the differences in susceptibility between the enzymes in humans and in the infective organism. Anticancer agents based on folic acid, e.g. **methotrexate** (Figure 4.8), primarily block pyrimidine biosynthesis, but are less selective than the antimicrobial agents, and rely on a stronger binding to the enzyme than the natural substrate has. Regeneration of tetrahydrofolate from dihydrofolate is vital for DNA synthesis in rapidly proliferating cells. The methylation of deoxyuridylate (dUMP) to deoxythymidylate (dTMP) requires N^5,N^{10}-methylenetetrahydrofolate as the methyl donor, which is thereby transformed into dihydrofolate (Figure 4.9). N^5-Formyl-tetrahydrofolic acid (**folinic acid, leucovorin**) (Figure 4.7) is used to counteract the folate-antagonist action of anticancer agents like methotrexate. The natural 6S isomer is termed **levofolinic acid** (**levoleucovorin**); **folinic acid** in drug use is usually a mixture of the 6R and 6S isomers.

In a sequence of complex reactions, which will not be considered in detail, the indole ring system is formed by incorporating two carbons from phosphoribosyl diphosphate, with loss of the original anthranilate carboxyl. The remaining ribosyl carbons are then removed by a reverse aldol reaction, to be replaced on a bound form of indole by those from L-serine, which then becomes the side-chain of L-tryptophan. Although a precursor of L-tryptophan, anthranilic acid may also be produced by metabolism of tryptophan. Both compounds feature as building blocks for a variety of alkaloid structures (see Chapter 6).

Returning to the main course of the shikimate pathway, a singular rearrangement process occurs transforming **chorismic acid** into **prephenic acid**

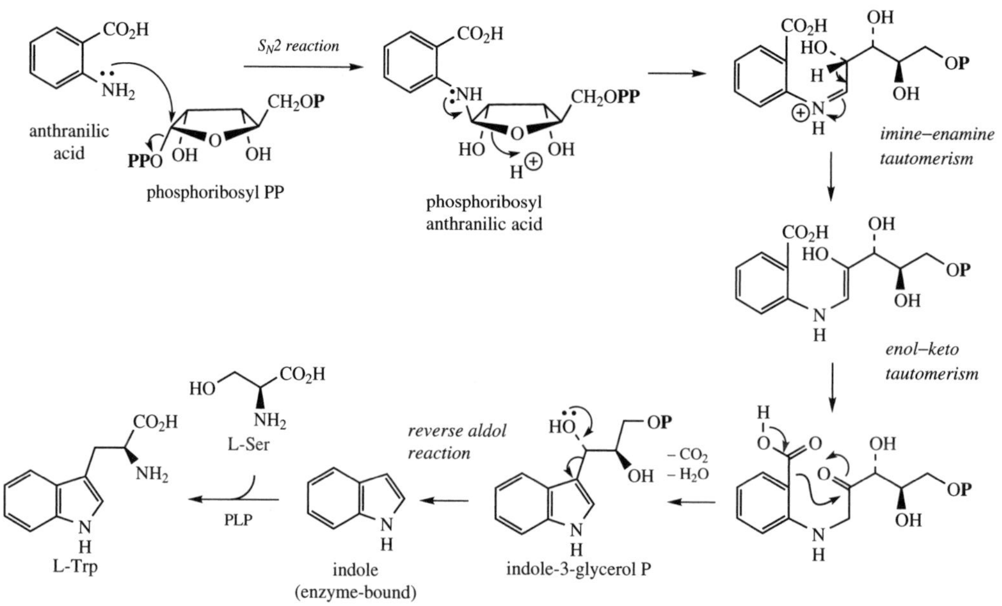

Figure 4.10

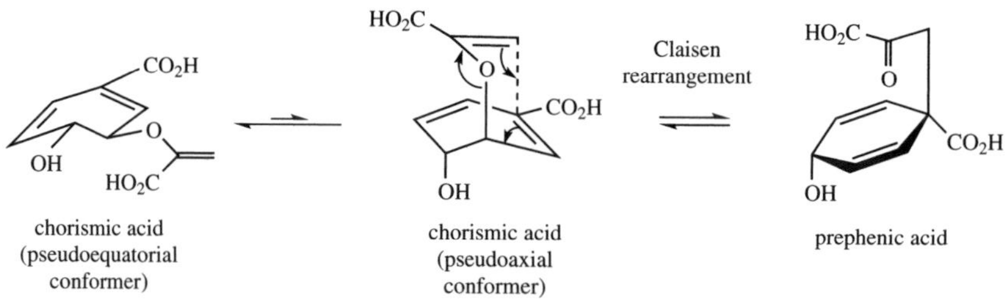

Figure 4.11

(Figure 4.11). This reaction, a Claisen rearrangement, transfers the PEP-derived side-chain so that it becomes directly bonded to the carbocycle, and so builds up the basic carbon skeleton of phenylalanine and tyrosine. The reaction is catalysed in nature by the enzyme chorismate mutase, and, although it can also occur thermally, the rate increases some 10^6-fold in the presence of the enzyme. The enzyme achieves this by binding the pseudoaxial conformer of chorismic acid, allowing a transition state with chairlike geometry to develop.

Pathways to the aromatic amino acids L-**phenylalanine** and L-**tyrosine** via prephenic acid may vary according to the organism, and often more than one route may operate in a particular species according to the enzyme activities that are available (Figure 4.12). In essence, only three reactions are involved, decarboxylative aromatization,

transamination, and in the case of tyrosine biosynthesis an oxidation, but the order in which these reactions occur differentiates the routes. Decarboxylative aromatization of prephenic acid yields **phenylpyruvic acid**, and PLP-dependent transamination leads to L-phenylalanine. In the presence of an NAD$^+$-dependent dehydrogenase enzyme, decarboxylative aromatization occurs with retention of the hydroxyl function, though as yet there is no evidence that any intermediate carbonyl analogue of prephenic acid is involved. Transamination of the resultant **4-hydroxyphenylpyruvic acid** subsequently gives L-tyrosine. L-**Arogenic acid** is the result of transamination of prephenic acid occurring prior to the decarboxylative aromatization, and can be transformed into both L-phenylalanine and L-tyrosine depending on the absence or presence of a suitable enzymic dehydrogenase activity. In some organisms, broad activity enzymes are known to be capable of accepting both prephenic acid and arogenic acid as substrates. In microorganisms and plants, L-phenylalanine and L-tyrosine tend to be synthesized separately as in Figure 4.12, but in animals, which lack the shikimate pathway, direct hydroxylation of L-phenylalanine to L-tyrosine, and of L-tyrosine to L-**DOPA** (dihydroxyphenylalanine), may be achieved (Figure 4.13). These reactions are catalysed by tetrahydropterin-dependent hydroxylase enzymes, the hydroxyl oxygen being derived from molecular oxygen. L-DOPA is a precursor of the **catecholamines**, e.g. the neurotransmitter noradrenaline and the hormone adrenaline (see page 316). Tyrosine and DOPA are also converted

by oxidation reactions into a heterogeneous polymer **melanin**, the main pigment in mammalian skin, hair, and eyes. In this material, the indole system is not formed from tryptophan, but arises from DOPA by cyclization of DOPAquinone, the nitrogen of the side-chain then attacking the *ortho*-quinone (Figure 4.13).

Some organisms are capable of synthesizing an unusual variant of L-phenylalanine, the aminated derivative L-*p*-aminophenylalanine (L-PAPA) (Figure 4.14). This is known to occur by a series of reactions paralleling those in Figure 4.12, but utilizing the PABA precursor 4-amino-4-deoxychorismic acid (Figure 4.4) instead of chorismic acid. Thus, amino derivatives of prephenic acid and pyruvic acid are elaborated. One important metabolite known to be formed from L-PAPA is the antibiotic

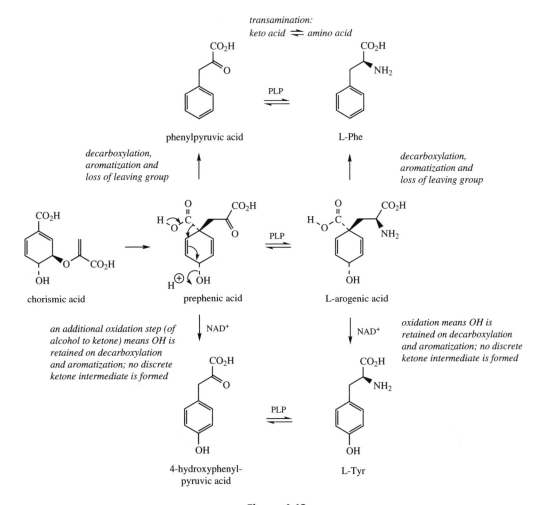

Figure 4.12

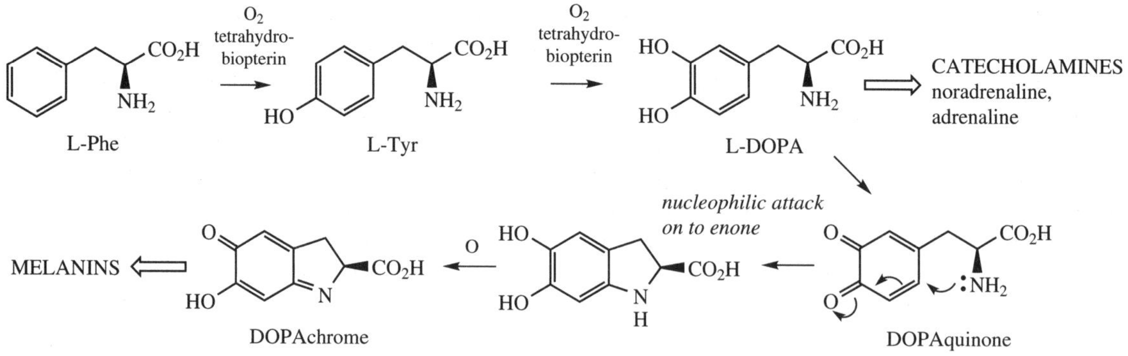

Figure 4.13

chloramphenicol*, produced by cultures of *Streptomyces venezuelae*. The late stages of the pathway (Figure 4.14) have been formulated to involve hydroxylation and *N*-acylation in the side-chain, the latter reaction probably requiring a coenzyme A ester of dichloroacetic acid. Following reduction of the carboxyl group, the final reaction is oxidation of the 4-amino group to a nitro, a fairly rare substituent in natural product structures.

CINNAMIC ACIDS

L-Phenylalanine and L-tyrosine, as C_6C_3 building blocks, are precursors for a wide range of natural products. In plants, a frequent first step is the elimination of ammonia from the side-chain to generate the appropriate *trans* (*E*) cinnamic acid. In the case of phenylalanine, this would give **cinnamic acid**, whilst tyrosine could yield **4-coumaric acid** (*p*-coumaric acid) (Figure 4.15). All plants appear to have the ability to deaminate phenylalanine via the enzyme phenylalanine ammonia lyase (PAL), but the corresponding transformation of tyrosine is more restricted, being mainly limited to members of the grass family (the Graminae/Poaceae). Whether a separate enzyme tyrosine ammonia lyase (TAL) exists, or whether grasses merely have a broad specificity PAL also

Chloramphenicol

Chloramphenicol (chloromycetin) (Figure 4.14) was initially isolated from cultures of *Streptomyces venezuelae*, but is now obtained for drug use by chemical synthesis. It was one of the first broad spectrum antibiotics to be developed, and exerts its antibacterial action by inhibiting protein biosynthesis. It binds reversibly to the 50S subunit of the bacterial ribosome, and in so doing disrupts peptidyl transferase, the enzyme that catalyses peptide bond formation (see page 408). This reversible binding means that bacterial cells not destroyed may resume protein biosynthesis when no longer exposed to the antibiotic. Some microorganisms have developed resistance to chloramphenicol by an inactivation process involving enzymic acetylation of the primary alcohol group in the antibiotic. The acetate binds only very weakly to the ribosomes, so has little antibiotic activity. The value of chloramphenicol as an antibacterial agent has been severely limited by some serious side-effects. It can cause blood disorders including irreversible aplastic anaemia in certain individuals, and these can lead to leukaemia and perhaps prove fatal. Nevertheless, it is still the drug of choice for some life-threatening infections such as typhoid fever and bacterial meningitis. The blood constitution must be monitored regularly during treatment to detect any abnormalities or adverse changes. The drug is orally active, but may also be injected. Eye-drops are useful for the treatment of bacterial conjunctivitis.

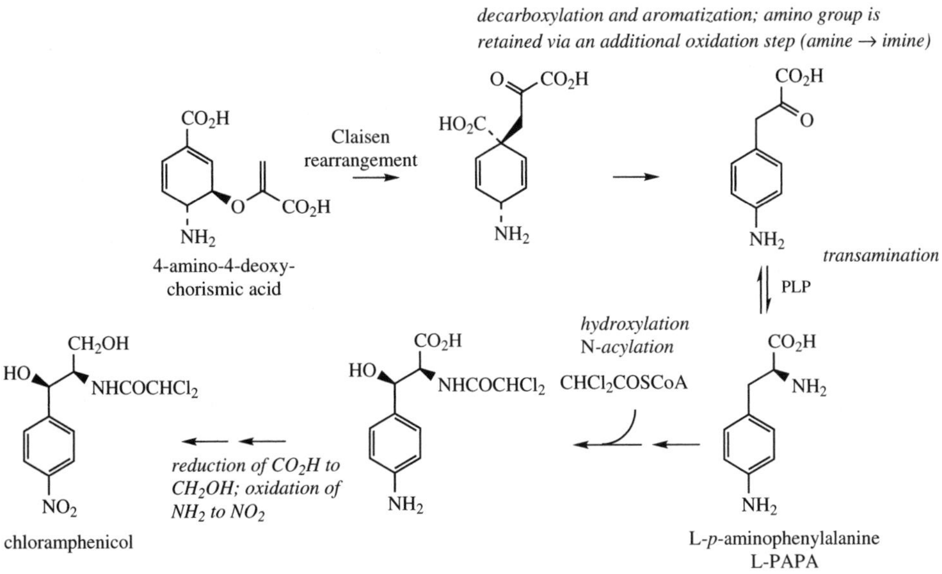

Figure 4.14

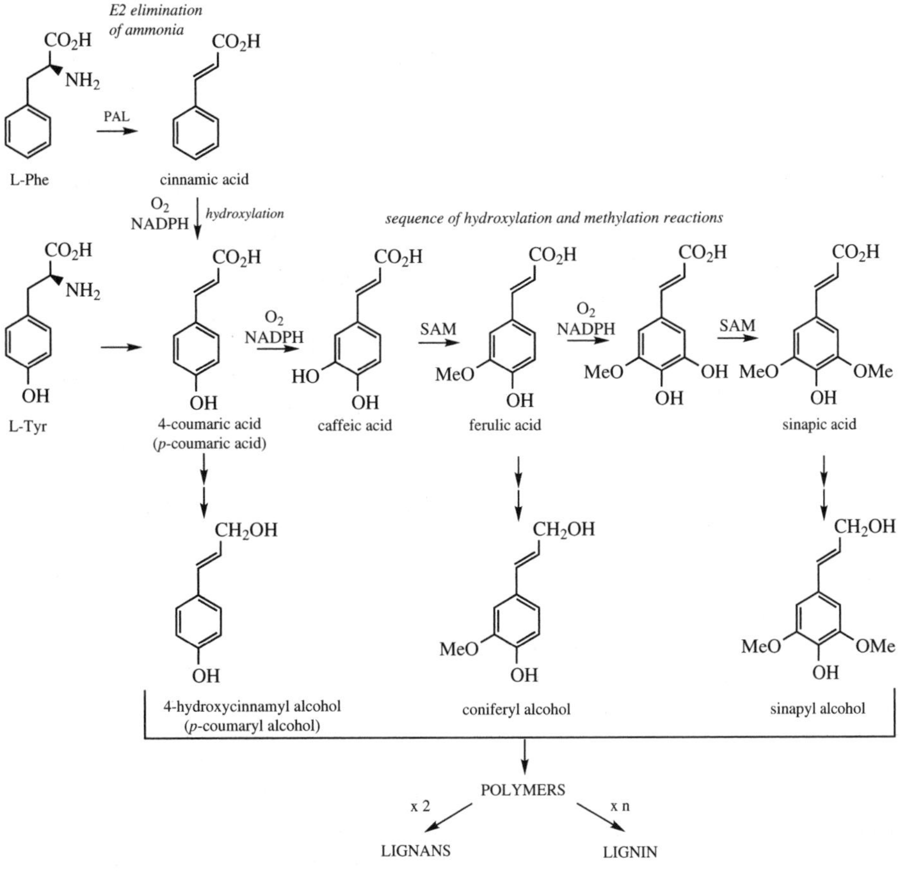

Figure 4.15

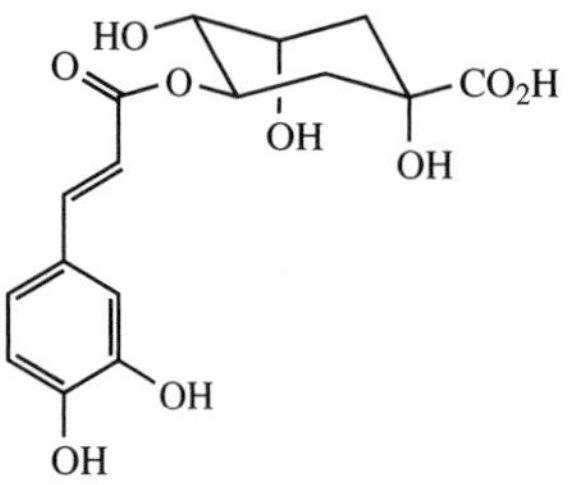

chlorogenic acid
(5-*O*-caffeoylquinic acid)

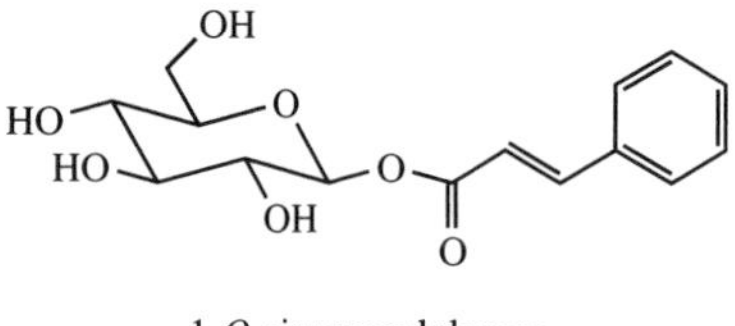

1-*O*-cinnamoylglucose

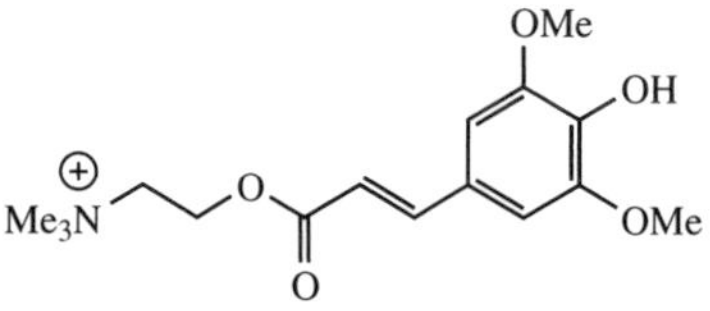

sinapine

Figure 4.16

capable of deaminating tyrosine, is still debated. Those species that do not transform tyrosine synthesize 4-coumaric acid by direct hydroxylation of cinnamic acid, in a cytochrome P-450-dependent reaction, and tyrosine is often channelled instead into other secondary metabolites, e.g. alkaloids. Other cinnamic acids are obtained by further hydroxylation and methylation reactions, sequentially building up substitution patterns typical of shikimate pathway metabolites, i.e. an *ortho* oxygenation pattern (see page 123). Some of the more common natural cinnamic acids are 4-coumaric, **caffeic, ferulic**, and **sinapic acids** (Figure 4.15). These can be found in plants in free form and in a range of esterified forms, e.g. with quinic acid as in **chlorogenic acid** (5-*O*-caffeoylquinic acid) (see coffee, page 395), with glucose as in **1-*O*-cinnamoylglucose**, and with choline as in **sinapine** (Figure 4.16).

LIGNANS AND LIGNIN

The cinnamic acids also feature in the pathways to other metabolites based on C_6C_3 building blocks.

Pre-eminent amongst these, certainly as far as nature is concerned, is the plant polymer **lignin**, a strengthening material for the plant cell wall which acts as a matrix for cellulose microfibrils (see page 473). Lignin represents a vast reservoir of aromatic materials, mainly untapped because of the difficulties associated with release of these metabolites. The action of wood-rotting fungi offers the most effective way of making these useful products more accessible. Lignin is formed by phenolic oxidative coupling of hydroxycinnamyl alcohol monomers, brought about by peroxidase enzymes (see page 28). The most important of these monomers are **4-hydroxycinnamyl alcohol** (*p*-coumaryl alcohol), **coniferyl alcohol**, and **sinapyl alcohol** (Figure 4.15), though the monomers used vary according to the plant type. Gymnosperms polymerize mainly coniferyl alcohol, dicotyledonous plants coniferyl alcohol and sinapyl alcohol, whilst monocotyledons use all three alcohols. The alcohols are derived by reduction of cinnamic acids via coenzyme A esters and aldehydes (Figure 4.17), though the substitution patterns are not necessarily elaborated completely at the cinnamic acid stage, and coenzyme A esters and aldehydes may also be substrates for aromatic hydroxylation and methylation. Formation of the coenzyme A ester facilitates the first reduction step by introducing a better leaving group (CoAS$^-$) for the NADPH-dependent reaction. The second reduction step, aldehyde to alcohol, utilizes a further molecule of NADPH and is reversible. The peroxidase enzyme then achieves one-electron oxidation of the phenol group. One-electron oxidation of a simple phenol allows delocalization of the unpaired electron, giving resonance forms in which the free electron resides at positions *ortho* and *para* to the oxygen function (see page 29). With cinnamic acid derivatives, conjugation allows the unpaired electron to be delocalized also into the side-chain (Figure 4.18). Radical pairing of resonance structures can then provide a range of dimeric systems containing reactive quinonemethides, which are susceptible to nucleophilic attack from hydroxyl groups in the same system, or by external water molecules. Thus, **coniferyl alcohol** monomers can couple, generating linkages as exemplified by **guaiacylglycerol β-coniferyl ether** (β-arylether linkage), **dehydrodiconiferyl alcohol** (phenylcoumaran linkage), and

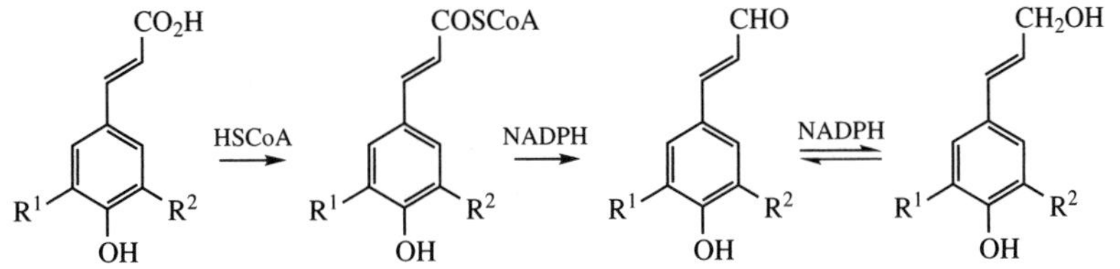

Figure 4.17

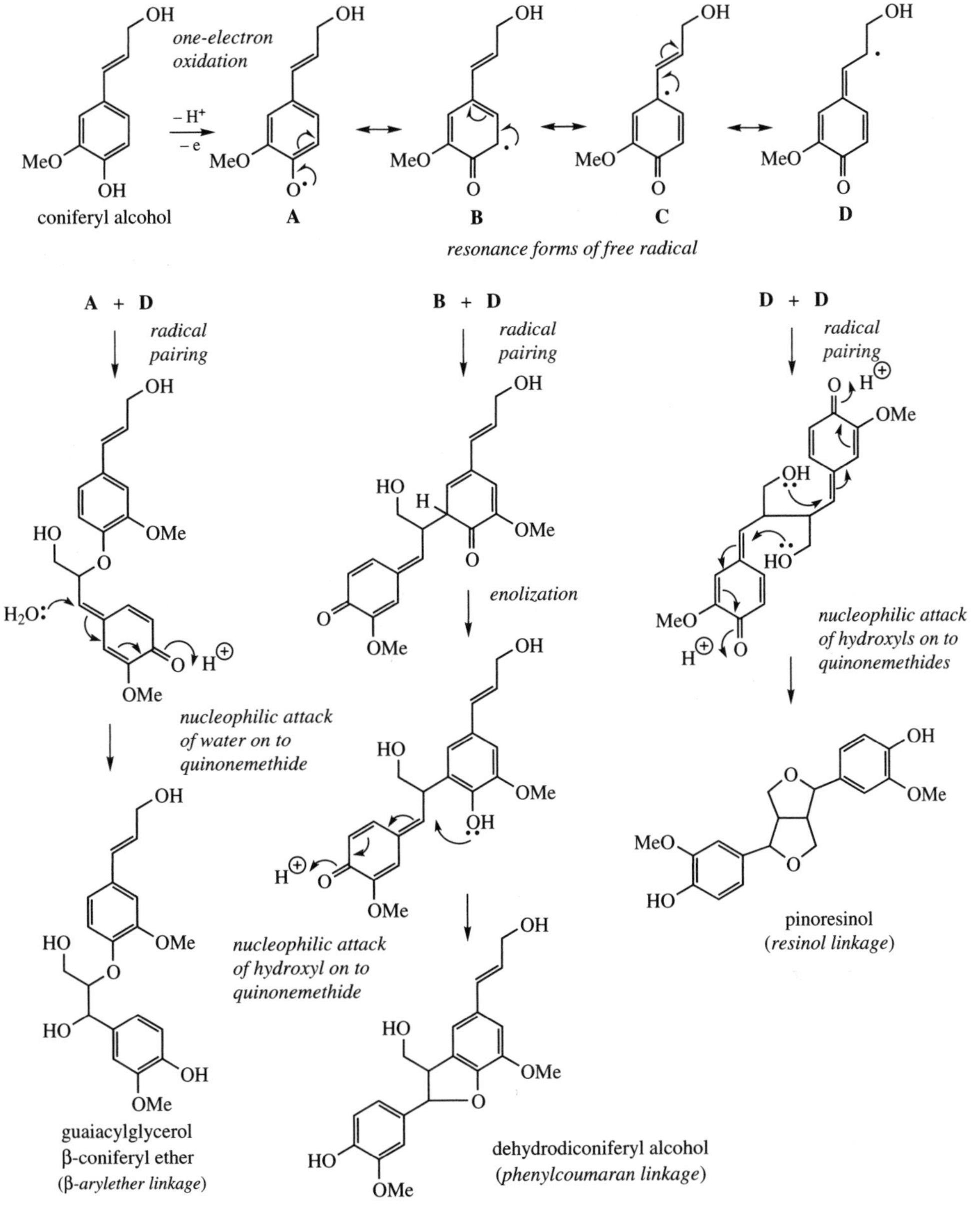

Figure 4.18

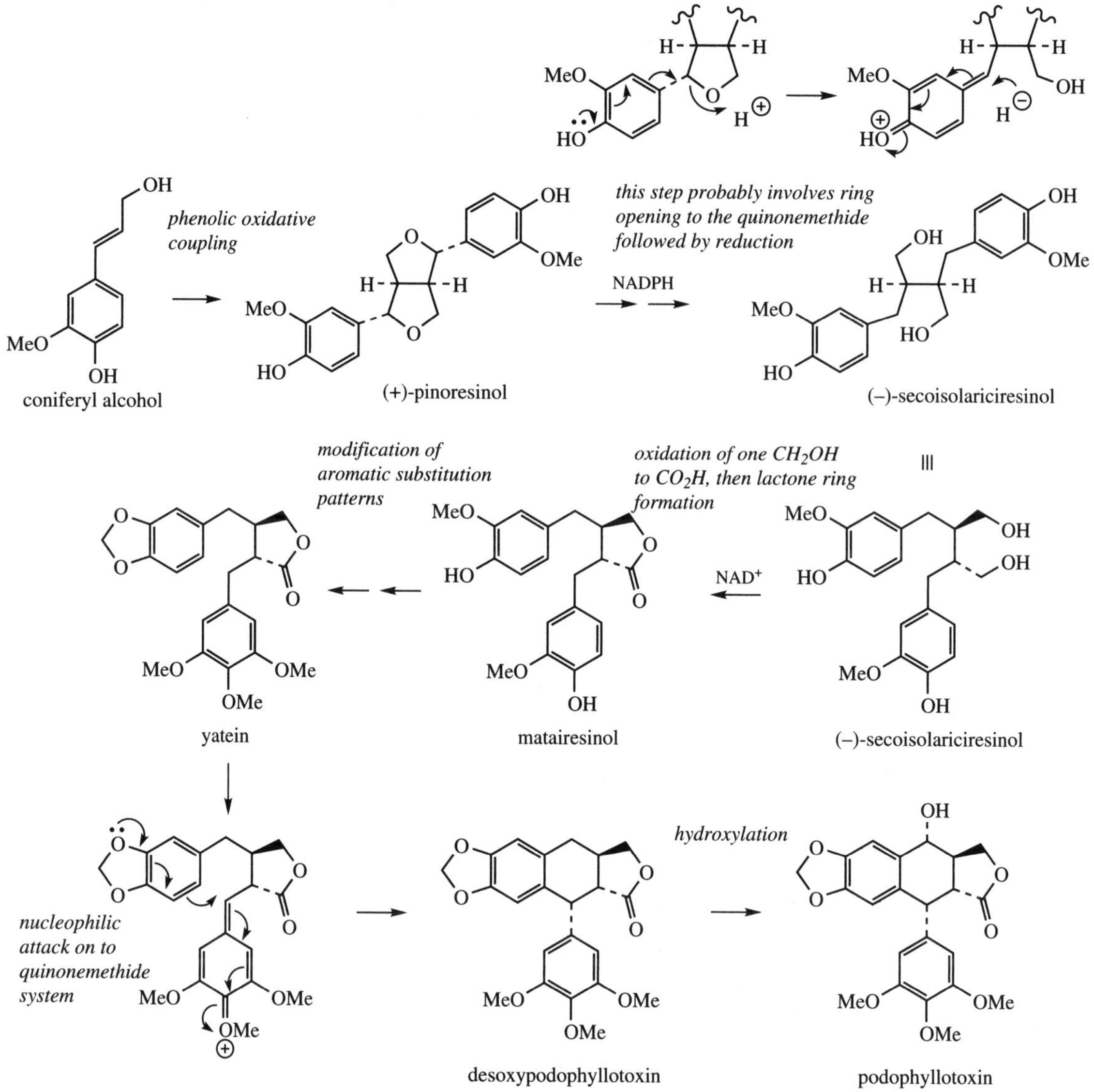

Figure 4.19

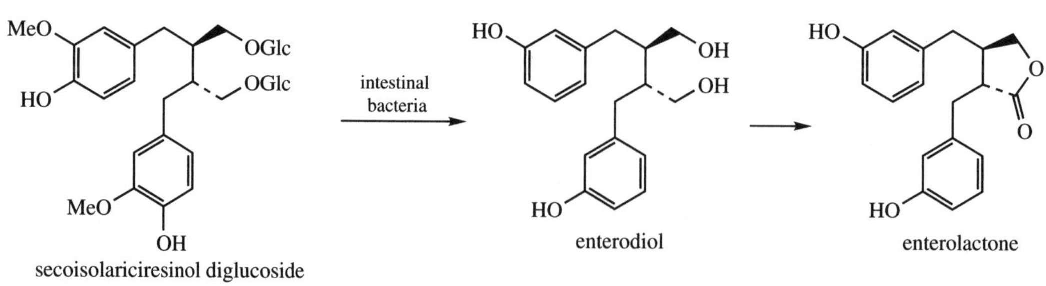

Figure 4.20

pinoresinol (resinol linkage). These dimers can react further by similar mechanisms to produce a lignin polymer containing a heterogeneous series of inter-molecular bondings as seen in the various dimers. In contrast to most other natural polymeric materials, lignin appears to be devoid of ordered repeating units, though some 50–70% of the linkages are of the β-arylether type. The dimeric materials are also found in nature and are called **lignans**. Some authorities like to restrict the term lignan specifically to molecules in which the two phenylpropane units are coupled at the central carbon of the side-chain, e.g. pinoresinol, whilst compounds containing other types of coupling, e.g. as in guaiacylglycerol β-coniferyl ether and dehydrodiconiferyl alcohol, are then referred to as **neolignans**. Lignan/neolignan formation and lignin biosynthesis are catalysed by different enzymes, and a consequence is that natural lignans/neolignans are normally enantiomerically pure because they arise from stereochemically controlled coupling. The control mechanisms for lignin biosynthesis are less well defined, but the enzymes appear to generate products lacking optical activity.

Further cyclization and other modifications can create a wide range of lignans of very different structural types. One of the most important of the natural lignans having useful biological activity is the aryltetralin lactone **podophyllotoxin** (Figure 4.19), which is derived from coniferyl alcohol via the dibenzylbutyrolactones **matairesinol** and **yatein**, cyclization probably occurring as shown in Figure 4.19. Matairesinol

is known to arise by reductive opening of the furan rings of **pinoresinol**, followed by oxidation of a primary alcohol to the acid and then lactonization. The substitution pattern in the two aromatic rings is built up further during the pathway, i.e. matairesinol → yatein, and does not arise by initial coupling of two different cinnamyl alcohol residues. The methylenedioxy ring system, as found in many shikimate-derived natural products, is formed by an oxidative reaction on an *ortho*-hydroxymethoxy pattern (see page 27). Podophyllotoxin and related lignans are found in the roots of *Podophyllum** species (Berberidaceae), and have clinically useful cytotoxic and anticancer activity. The lignans **enterolactone** and **enterodiol** (Figure 4.20) were discovered in human urine, but were subsequently shown to be derived from dietary plant lignans, especially secoisolariciresinol diglucoside, by the action of intestinal microflora. Enterolactone and enterodiol have oestrogenic activity and have been implicated as contributing to lower levels of breast cancer amongst vegetarians (see phyto-oestrogens, page 156).

PHENYLPROPENES

The reductive sequence from an appropriate cinnamic acid to the corresponding cinnamyl alcohol is not restricted to lignin and lignan biosynthesis, and is utilized for the production of various phenylpropene derivatives. Thus **cinnamaldehyde** (Figure 4.23) is the principal component in the

---

Podophyllum

Podophyllum consists of the dried rhizome and roots of *Podophyllum hexandrum* (*P. emodi*) or *P. peltatum* (Berberidaceae). *Podophyllum hexandrum* is found in India, China, and the Himalayas and yields Indian podophyllum, whilst *P. peltatum* (May apple or American mandrake) comes from North America and is the source of American podophyllum. Plants are collected from the wild. Both plants are large-leafed perennial herbs with edible fruits, though other parts of the plant are toxic. The roots contain cytotoxic lignans and their glucosides, *P. hexandrum* containing about 5%, and *P. peltatum* about 1%. A concentrated form of the active principles is obtained by pouring an ethanolic extract of the root into water, and drying the precipitated podophyllum resin or 'podophyllin'. Indian podophyllum yields about 6–12% of resin containing 50–60% lignans, and American podophyllum 2–8% of resin containing 14–18% lignans.

(Continues)

(Continued)

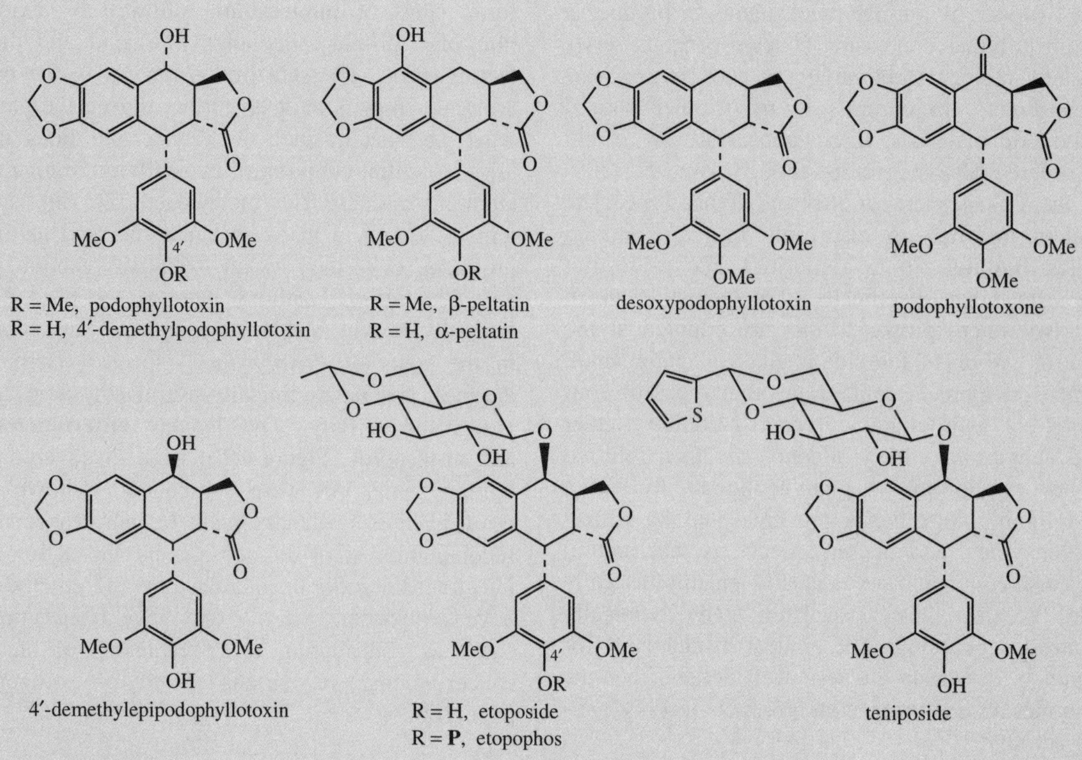

R = Me, podophyllotoxin
R = H, 4'-demethylpodophyllotoxin

R = Me, β-peltatin
R = H, α-peltatin

desoxypodophyllotoxin

podophyllotoxone

4'-demethylepipodophyllotoxin

R = H, etoposide
R = P, etopophos

teniposide

Figure 4.21

The lignan constituents of the two roots are the same, but the proportions are markedly different. The Indian root contains chiefly podophyllotoxin (Figure 4.21) (about 4%) and 4'-demethylpodophyllotoxin (about 0.45%). The main components in the American root are podophyllotoxin (about 0.25%), β-peltatin (about 0.33%) and α-peltatin (about 0.25%). Desoxypodophyllotoxin and podophyllotoxone are also present in both plants, as are the glucosides of podophyllotoxin, 4'-demethylpodophyllotoxin, and the peltatins, though preparation of the resin results in considerable losses of the water-soluble glucosides.

Podophyllum resin has long been used as a purgative, but the discovery of the cytotoxic properties of podophyllotoxin and related compounds has now made podophyllum a commercially important drug plant. Preparations of **podophyllum resin** (the Indian resin is preferred) are effective treatments for warts, and pure **podophyllotoxin** is available as a paint for venereal warts, a condition which can be sexually transmitted. The antimitotic effect of podophyllotoxin and the other lignans is by binding to the protein tubulin in the mitotic spindle, preventing polymerization and assembly into microtubules (compare vincristine, page 356, and colchicine, page 343). During mitosis, the chromosomes separate with the assistance of these microtubules, and after cell division the microtubules are transformed back to tubulin. Podophyllotoxin and other *Podophyllum* lignans were found to be unsuitable for clinical use as anticancer agents due to toxic side-effects, but the semi-synthetic derivatives etoposide and teniposide (Figure 4.21), which are manufactured from natural podophyllotoxin, have proved excellent antitumour agents. They were developed as modified forms (acetals) of the natural

(Continues)

(Continued)

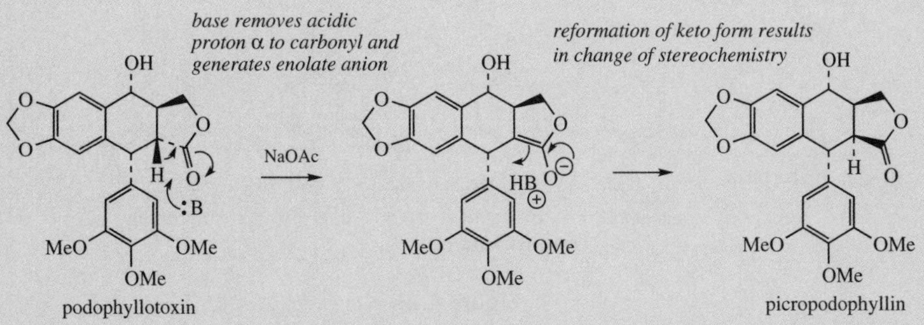

base removes acidic proton α to carbonyl and generates enolate anion

reformation of keto form results in change of stereochemistry

NaOAc

podophyllotoxin

picropodophyllin

Figure 4.22

4′-demethylpodophyllotoxin glucoside. Attempted synthesis of the glucoside inverted the stereochemistry at the sugar–aglycone linkage, and these agents are thus derivatives of 4′-demethylepipodophyllotoxin (Figure 4.21). **Etoposide** is a very effective anticancer agent, and is used in the treatment of small cell lung cancer, testicular cancer and lymphomas, usually in combination therapies with other anticancer drugs. It may be given orally or intravenously. The water-soluble pro-drug **etopophos** (etoposide 4′-phosphate) is also available. **Teniposide** has similar anticancer properties, and, though not as widely used as etoposide, has value in paediatric neuroblastoma.

Remarkably, the 4′-demethylepipodophyllotoxin series of lignans do not act via a tubulin-binding mechanism as does podophyllotoxin. Instead, these drugs inhibit the enzyme topoisomerase II, thus preventing DNA synthesis and replication. Topoisomerases are responsible for cleavage and resealing of the DNA strands during the replication process, and are classified as type I or II according to their ability to cleave one or both strands. Camptothecin (see page 365) is an inhibitor of topoisomerase I. Etoposide is believed to inhibit strand-rejoining ability by stabilizing the topoisomerase II–DNA complex in a cleavage state, leading to double-strand breaks and cell death. Development of other topoisomerase inhibitors based on podophyllotoxin-related lignans is an active area of research. Biological activity in this series of compounds is very dependent on the presence of the *trans*-fused five-membered lactone ring, this type of fusion producing a highly-strained system. Ring strain is markedly reduced in the corresponding *cis*-fused system, and the natural compounds are easily and rapidly converted into these *cis*-fused lactones by treatment with very mild bases, via enol tautomers or enolate anions (Figure 4.22). Picropodophyllin is almost devoid of cytotoxic properties.

Podophyllotoxin is also found in significant amounts in the roots of other *Podophyllum* species, and in closely related genera such as *Diphylleia* (Berberidaceae).

oil from the bark of cinnamon (*Cinnamomum zeylanicum*; Lauraceae), widely used as a spice and flavouring. Fresh bark is known to contain high levels of **cinnamyl acetate**, and cinnamaldehyde is released from this by fermentation processes which are part of commercial preparation of the bark, presumably by enzymic hydrolysis and participation of the reversible aldehyde–alcohol oxidoreductase. Cinnamon leaf, on the other hand,

contains large amounts of **eugenol** (Figure 4.23) and much smaller amounts of cinnamaldehyde. Eugenol is also the principal constituent in oil from cloves (*Syzygium aromaticum*; Myrtaceae), used for many years as a dental anaesthetic, as well as for flavouring. The side-chain of eugenol is derived from that of the cinnamyl alcohols by reduction, but differs in the location of the double bond. This change is accounted for by resonance

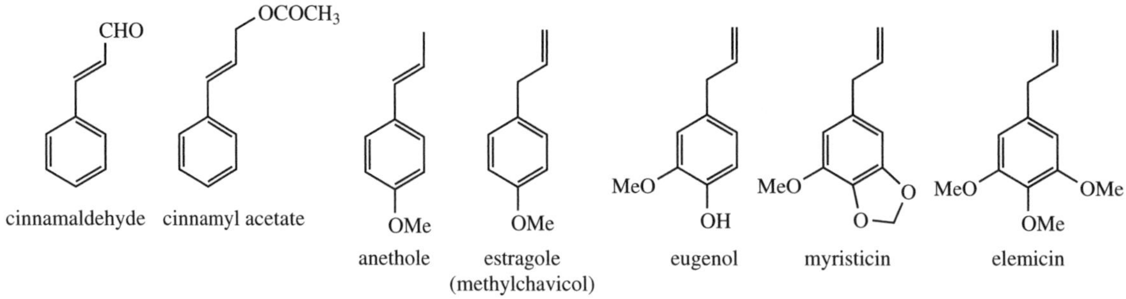

Figure 4.23

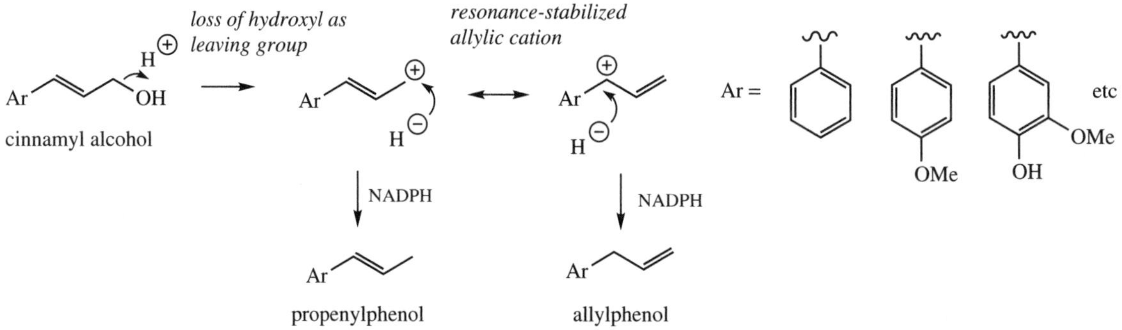

Figure 4.24

forms of the allylic cation (Figure 4.24), and addition of hydride (from NADPH) can generate either allylphenols, e.g. eugenol, or propenylphenols, e.g. **anethole** (Figure 4.23). Loss of hydroxyl from a cinnamyl alcohol may be facilitated by protonation, or perhaps even phosphorylation, though there is no evidence for the latter. **Myristicin** (Figure 4.23) from nutmeg (*Myristica fragrans*; Myristicaceae) is a further example of an allylphenol found in flavouring materials. Myristicin also has a history of being employed as a mild hallucinogen via ingestion of ground nutmeg. Myristicin is probably metabolized in the body via an amination reaction to give an amfetamine-like derivative (see page 385). **Anethole** is the main component in oils from aniseed (*Pimpinella anisum*; Umbelliferae/Apiaceae), star anise (*Illicium verum*; Illiciaceae), and fennel (*Foeniculum vulgare*; Umbelliferae/Apiaceae). The propenyl components of flavouring materials such as cinnamon, star anise, nutmeg, and sassafras (*Sassafras albidum*; Lauraceae) have reduced their commercial use somewhat since these constituents

have been shown to be weak carcinogens in laboratory tests on animals. In the case of **safrole** (Figure 4.25), the main component of sassafras oil, this has been shown to arise from hydroxylation in the side-chain followed by sulphation, giving an agent which binds to cellular macromolecules. Further data on volatile oils containing aromatic constituents isolated from these and other plant materials are given in Table 4.1. Volatile oils in which the main components are terpenoid in nature are listed in Table 5.1, page 177.

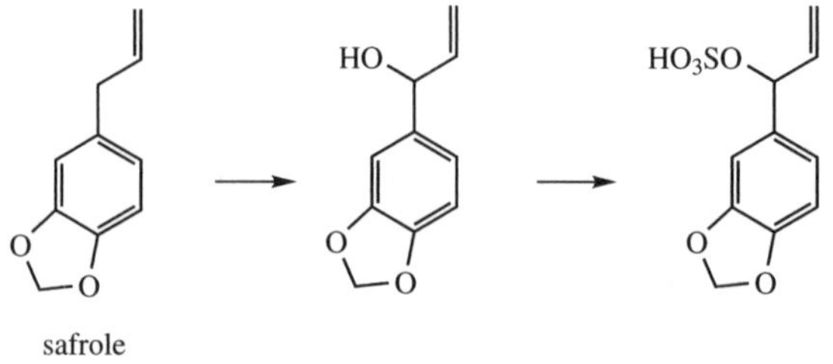

Figure 4.25

Table 4.1 Volatile oils containing principally aromatic compounds

Volatile or essential oils are usually obtained from the appropriate plant material by steam distillation, though if certain components are unstable at these temperatures, other less harsh techniques such as expression or solvent extraction may be employed. These oils, which typically contain a complex mixture of low boiling components, are widely used in flavouring, perfumery, and aromatherapy. Only a small number of oils have useful therapeutic properties, e.g. clove and dill, though a wide range of oils is now exploited for aromatherapy. Most of those employed in medicines are simply added for flavouring purposes. Some of the materials are commercially important as sources of chemicals used industrially, e.g. turpentine.

For convenience, the major oils listed are divided into two groups. Those which contain principally chemicals which are aromatic in nature and which are derived by the shikimate pathway are given in Table 4.1 below. Those oils which are composed predominantly of terpenoid compounds are listed in Table 5.1 on page 177, since they are derived via the deoxyxylulose phosphate pathway. It must be appreciated that many oils may contain aromatic and terpenoid components, but usually one group predominates. The oil yields, and the exact composition of any sample of oil will be variable, depending on the particular plant material used in its preparation. The quality of an oil and its commercial value is dependent on the proportion of the various components.

Oil	Plant source	Plant part used	Oil content (%)	Major constituents with typical (%) composition	Uses, notes
Aniseed (Anise)	*Pimpinella anisum* (Umbelliferae/ Apiaceae)	ripe fruit	2–3	anethole (80–90) estragole (1–6)	flavour, carminative, aromatherapy
Star anise	*Illicium verum* (Illiciaceae)	ripe fruit	5–8	anethole (80–90) estragole (1–6)	flavour, carminative fruits contain substantial amounts of shikimic and quinic acids
Cassia	*Cinnamomum cassia* (Lauraceae)	dried bark, or leaves and twigs	1–2	cinnamaldehyde (70–90) 2-methoxycinnamal- dehyde (12)	flavour, carminative known as cinnamon oil in USA
Cinnamon bark	*Cinnamomum zeylanicum* (Lauraceae)	dried bark	1–2	cinnamaldehyde (70–80) eugenol (1–13) cinnamyl acetate (3–4)	flavour, carminative, aromatherapy
Cinnamon leaf	*Cinnamomum zeylanicum* (Lauraceae)	leaf	0.5–0.7	eugenol (70–95)	flavour

(Continued overleaf)

Table 4.1 (*Continued*)

Oil	Plant source	Plant part used	Oil content (%)	Major constituents with typical (%) composition	Uses, notes
Clove	*Syzygium aromaticum* (*Eugenia caryophyllus*) (Myrtaceae)	dried flower buds	15–20	eugenol (75–90) eugenyl acetate (10–15) β-caryophyllene (3)	flavour, aromatherapy, antiseptic
Fennel	*Foeniculum vulgare* (Umbelliferae/ Apiaceae)	ripe fruit	2–5	anethole (50–70) fenchone (10–20) estragole (3–20)	flavour, carminative, aromatherapy
Nutmeg	*Myristica fragrans* (Myristicaceae)	seed	5–16	sabinene (17–28) α-pinene (14–22) β-pinene (9–15) terpinen-4-ol (6–9) myristicin (4–8) elemicin (2)	flavour, carminative, aromatherapy although the main constituents are terpenoids, most of the flavour comes from the minor aromatic constituents, myristicin, elemicin, etc myristicin is hallucinogenic (see page 385)
Wintergreen	*Gaultheria procumbens* (Ericacae) or *Betula lenta* (Betulaceae)	leaves bark	0.7–1.5 0.2–0.6	methyl salicylate (98%)	flavour, antiseptic, antirheumatic prior to distillation, plant material is macerated with water to allow enzymic hydrolysis of glycosides methyl salicylate is now produced synthetically

BENZOIC ACIDS FROM C$_6$C$_3$ COMPOUNDS

Some of the simple hydroxybenzoic acids (C$_6$C$_1$ compounds) such as 4-hydroxybenzoic acid and gallic acid can be formed directly from intermediates early in the shikimate pathway, e.g. 3-dehydroshikimic acid or chorismic acid (see page 121), but alternative routes exist in which cinnamic acid derivatives (C$_6$C$_3$ compounds) are cleaved at the double bond and lose two carbons from the side-chain. Thus, 4-coumaric acid may act as a precursor of **4-hydroxybenzoic acid**, and ferulic acid may give **vanillic acid** (4-hydroxy-3-methoxybenzoic acid) (Figure 4.26). A sequence analogous to that involved in the β-oxidation of fatty acids (see page 18) is possible, so that the double bond in the coenzyme A ester would be hydrated, the hydroxyl group oxidized to a ketone, and the β-ketoester would then lose acetyl-CoA by a reverse Claisen reaction, giving the coenzyme A ester of 4-hydroxybenzoic acid. Whilst this sequence has been generally accepted, newer evidence supports another side-chain cleavage mechanism, which is different from the fatty acid β-oxidation pathway (Figure 4.26).

Coenzyme A esters are not involved, and though a similar hydration of the double bond occurs, chain shortening features a reverse aldol reaction, generating the appropriate aromatic aldehyde. The corresponding acid is then formed via an NAD$^+$-dependent oxidation step. Thus, aromatic aldehydes such as **vanillin**, the main flavour compound in vanilla (pods of the orchid *Vanilla planiflora*; Orchidaceae) would be formed from the correspondingly substituted cinnamic acid without proceeding through intermediate benzoic acids or esters. Whilst the substitution pattern in these C$_6$C$_1$ derivatives is generally built up at the C$_6$C$_3$ cinnamic acid stage, prior to chain shortening, there exists the possibility of further hydroxylations and/or methylations occurring at the C$_6$C$_1$ level, and this is known in certain examples. **Salicylic acid** (Figure 4.27) is synthesized in microorganisms directly from isochorismic acid (see page 124), but can arise in plants by two other mechanisms. It can be produced by hydroxylation of benzoic acid, or by side-chain cleavage of 2-coumaric acid, which itself is formed by an *ortho*-hydroxylation of cinnamic acid. **Methyl salicylate** is the principal component of oil of wintergreen from *Gaultheria procumbens* (Ericaceae), used for many years for pain relief. It is derived by

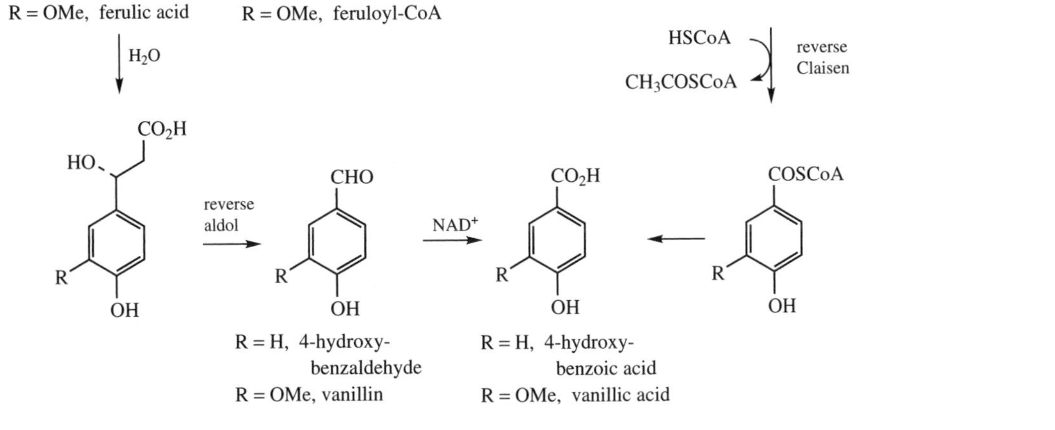

β-*oxidation pathway, as in fatty acid metabolism (Figure 2.11)*

R = H, 4-coumaric acid
R = OMe, ferulic acid

R = H, 4-coumaroyl-CoA
R = OMe, feruloyl-CoA

R = H, 4-hydroxy-benzaldehyde
R = OMe, vanillin

R = H, 4-hydroxy-benzoic acid
R = OMe, vanillic acid

Figure 4.26

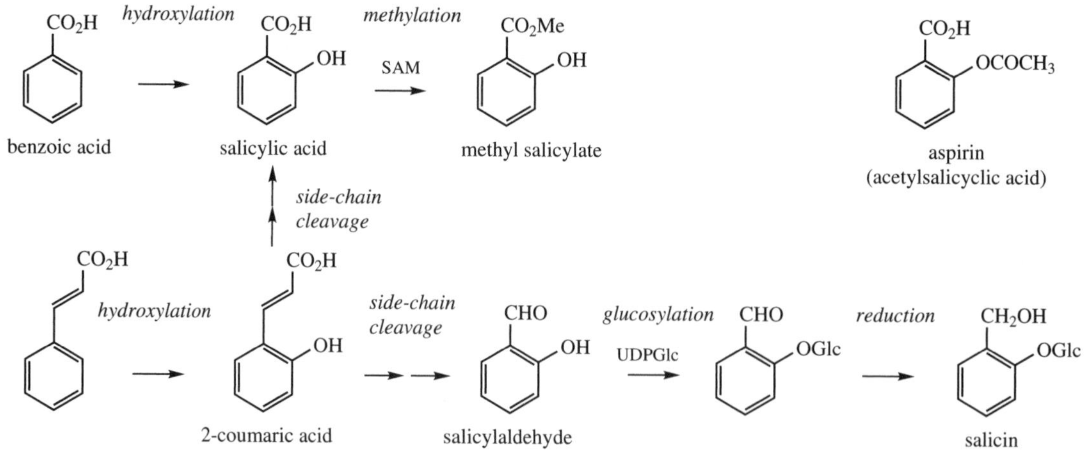

Figure 4.27

SAM-dependent methylation of salicylic acid. The salicyl alcohol derivative **salicin**, found in many species of willow (*Salix* species; Salicaceae), is not derived from salicylic acid, but probably via glucosylation of salicylaldehyde and then reduction of the carbonyl (Figure 4.27). Salicin is responsible for the analgesic and antipyretic effects of willow barks, widely used for centuries, and the template for synthesis of acetylsalicyclic acid (**aspirin**) (Figure 4.27) as a more effective analogue.

COUMARINS

The hydroxylation of cinnamic acids *ortho* to the side-chain as seen in the biosynthesis of salicylic acid is a crucial step in the formation of a group of cinnamic acid lactone derivatives, the **coumarins**. Whilst the direct hydroxylation of the aromatic ring of the cinnamic acids is common, hydroxylation generally involves initially the 4-position *para* to the side-chain, and subsequent hydroxylations then proceed *ortho* to this substituent (see page 132). In contrast, for the coumarins, hydroxylation of cinnamic acid or 4-coumaric acid can occur *ortho* to the side-chain (Figure 4.28). In the latter case, the 2,4-dihydroxycinnamic acid produced confusingly seems to possess the *meta* hydroxylation pattern characteristic of phenols derived via the acetate pathway. Recognition of the C_6C_3 skeleton should help to avoid this confusion. The two 2-hydroxycinnamic acids then suffer a change in configuration in the side-chain, from

the *trans* (*E*) to the less stable *cis* (*Z*) form. Whilst *trans–cis* isomerization would be unfavourable in the case of a single isolated double bond, in the cinnamic acids the fully conjugated system allows this process to occur quite readily, and UV irradiation, e.g. daylight, is sufficient to produce equilibrium mixtures which can be separated (Figure 4.29). The absorption of energy promotes an electron from the π-orbital to a higher energy state, the π^*-orbital, thus temporarily destroying the double bond character and allowing rotation. Loss of the absorbed energy then results in re-formation of the double bond, but in the *cis*-configuration. In conjugated systems, the π–π^* energy difference is considerably less than with a non-conjugated double bond. Chemical lactonization can occur on treatment with acid. Both the *trans–cis* isomerization and the lactonization are enzyme-mediated in nature, and light is not necessary for coumarin biosynthesis. Thus, cinnamic acid and 4-coumaric acid give rise to the coumarins **coumarin** and **umbelliferone** (Figure 4.28). Other coumarins with additional oxygen substituents on the aromatic ring, e.g. **aesculetin** and **scopoletin**, appear to be derived by modification of umbelliferone, rather than by a general cinnamic acid to coumarin pathway. This indicates that the hydroxylation *meta* to the existing hydroxyl, discussed above, is a rather uncommon occurrence.

Coumarins are widely distributed in plants, and are commonly found in families such as the Umbelliferae/Apiaceae and Rutaceae, both in the free form and as glycosides. **Coumarin** itself is

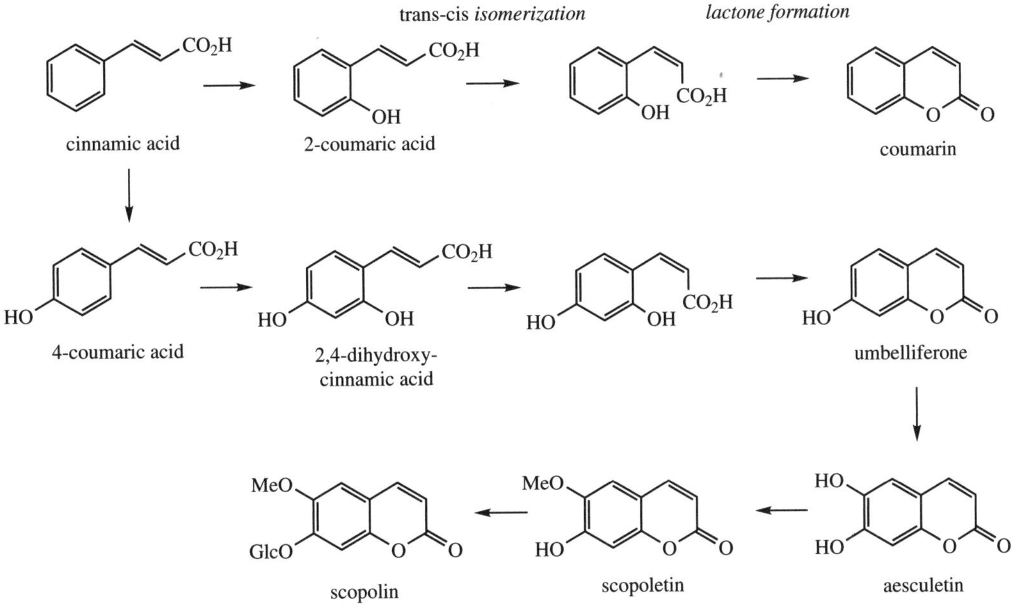

Figure 4.28

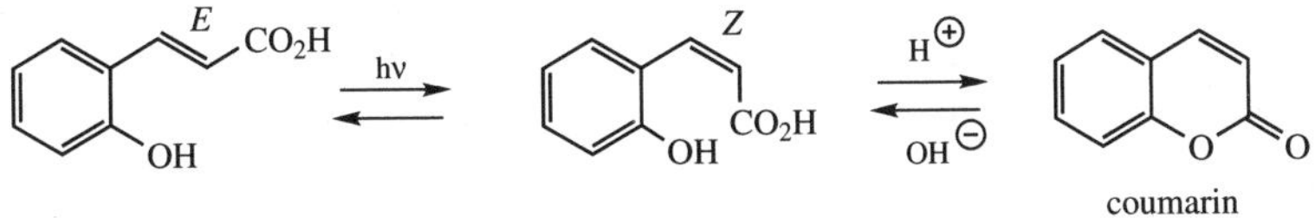

Figure 4.29

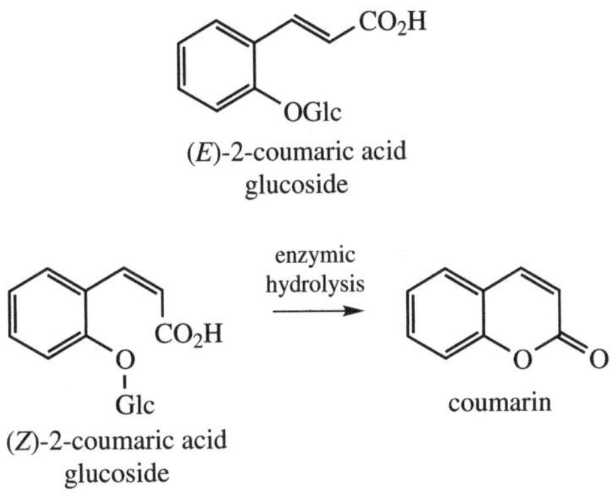

Figure 4.30

found in sweet clover (*Melilotus* species; Leguminosae/Fabaceae) and contributes to the smell of new-mown hay, though there is evidence that the plants actually contain the glucosides of (*E*)- and (*Z*)-2-coumaric acid (Figure 4.30), and coumarin is only liberated as a result of enzymic

hydrolysis and lactonization through damage to the plant tissues during harvesting and processing (Figure 4.30). If sweet clover is allowed to ferment, 4-hydroxycoumarin is produced by the action of microorganisms on 2-coumaric acid (Figure 4.31) and this can react with formaldehyde, which is usually present due to microbial degradative reactions, combining to give **dicoumarol**. Dicoumarol* is a compound with pronounced blood anticoagulant properties, which can cause the deaths of livestock by internal bleeding, and is the forerunner of the warfarin* group of medicinal anticoagulants.

Many other natural coumarins have a more complex carbon framework and incorporate extra carbons derived from an isoprene unit (Figure 4.33). The aromatic ring in umbelliferone is activated at positions *ortho* to the hydroxyl group and can thus be alkylated by a suitable alkylating agent, in this case dimethylallyl diphosphate. The newly introduced dimethylallyl

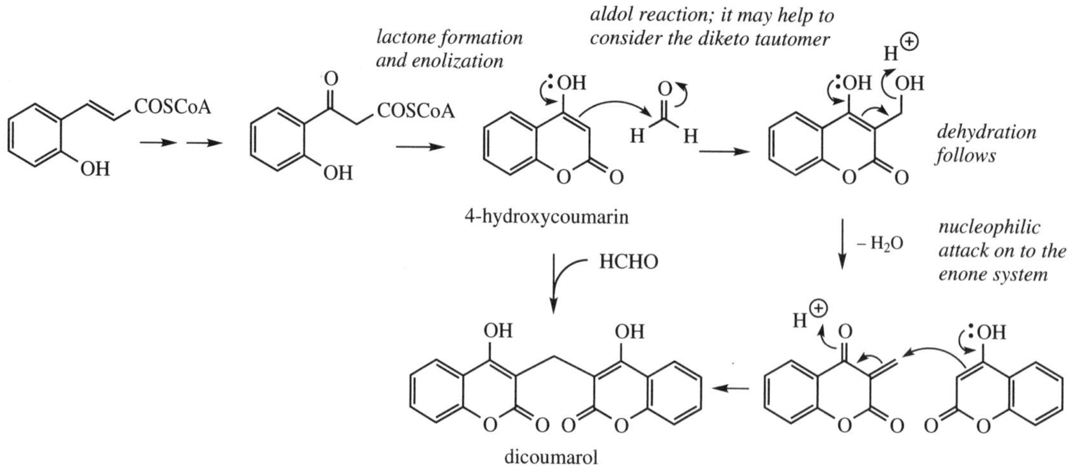

Figure 4.31

Dicoumarol and Warfarin

The cause of fatal haemorrhages in animals fed spoiled sweet clover (*Melilotus officinalis*; Leguminosae/Fabaceae) was traced to dicoumarol (bishydroxycoumarin) (Figure 4.31). This agent interferes with the effects of vitamin K in blood coagulation (see page 163), the blood loses its ability to clot, and thus minor injuries can lead to severe internal bleeding. Synthetic dicoumarol has been used as an oral blood anticoagulant in the treatment of thrombosis, where the risk of blood clots becomes life threatening. It has been superseded by salts of **warfarin** and **acenocoumarol** (**nicoumalone**) (Figure 4.32), which are synthetic developments from the natural product. An overdose of warfarin may be countered by injection of vitamin K_1.

Warfarin was initially developed as a rodenticide, and has been widely employed for many years as the first choice agent, particularly for destruction of rats. After

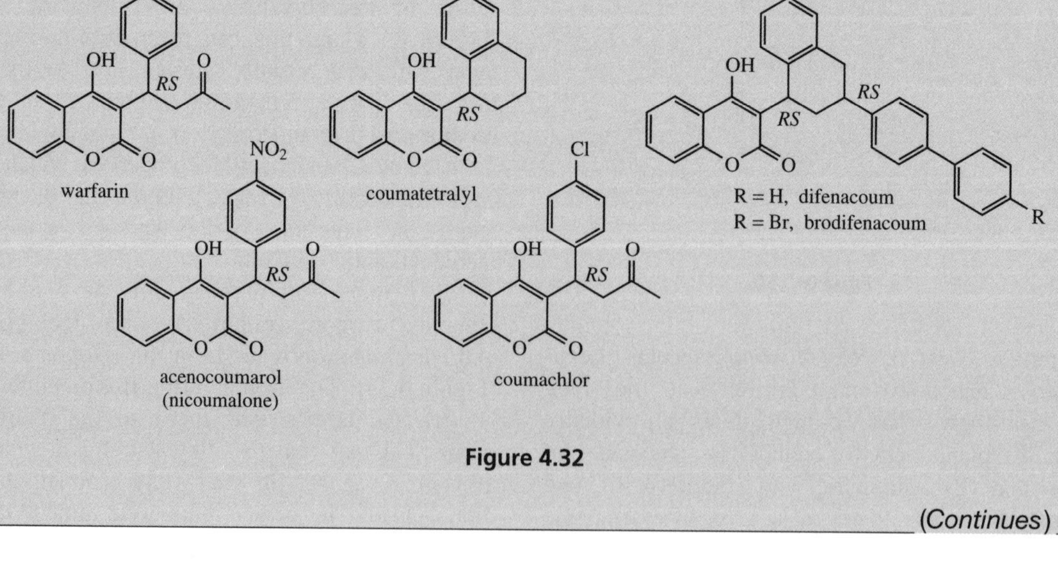

Figure 4.32

(Continues)

(*Continued*)

consumption of warfarin-treated bait, rats die from internal haemorrhage. Other coumarin derivatives employed as rodenticides include **coumachlor** and **coumatetralyl** (Figure 4.32). In an increasing number of cases, rodents are becoming resistant to warfarin, an ability which has been traced to elevated production of vitamin K by their intestinal microflora. Modified structures **defenacoum** and **brodifenacoum** have been found to be more potent than warfarin, and are also effective against rodents that have become resistant to warfarin.

group in **demethylsuberosin** is then able to cyclize with the phenol group giving **marmesin**. This transformation is catalysed by a cytochrome P-450-dependent mono-oxygenase, and requires cofactors NADPH and molecular oxygen. For many years, the cyclization had been postulated to involve an intermediate epoxide, so that nucleophilic attack of the phenol on to the epoxide group might lead to formation of either five-membered furan or six-membered pyran heterocycles as commonly encountered in natural products (Figure 4.34). Although the reactions of Figure 4.34 offer a convenient rationalization for cyclization, epoxide intermediates have not been demonstrated in any of the enzymic systems so far investigated, and therefore some direct oxidative cyclization mechanism must operate. A second cytochrome P-450-dependent mono-oxygenase enzyme then cleaves off the hydroxyisopropyl fragment (as acetone) from **marmesin** giving the furocoumarin **psoralen** (Figure 4.35). This does not involve any hydroxylated intermediate, and cleavage is believed to be initiated by a radical abstraction process. Psoralen can act as a precursor for the further substituted furocoumarins **bergapten, xanthotoxin**, and **isopimpinellin** (Figure 4.33), such modifications occurring late in the biosynthetic sequence rather than at the cinnamic acid stage. Psoralen, bergapten, etc are termed 'linear' furocoumarins. 'Angular' furocoumarins, e.g. **angelicin** (Figure 4.33), can arise by a similar sequence of reactions, but these involve dimethylallylation at the alternative position *ortho* to the phenol. An isoprene-derived furan ring system has already been noted in the formation of khellin (see page 74), though the

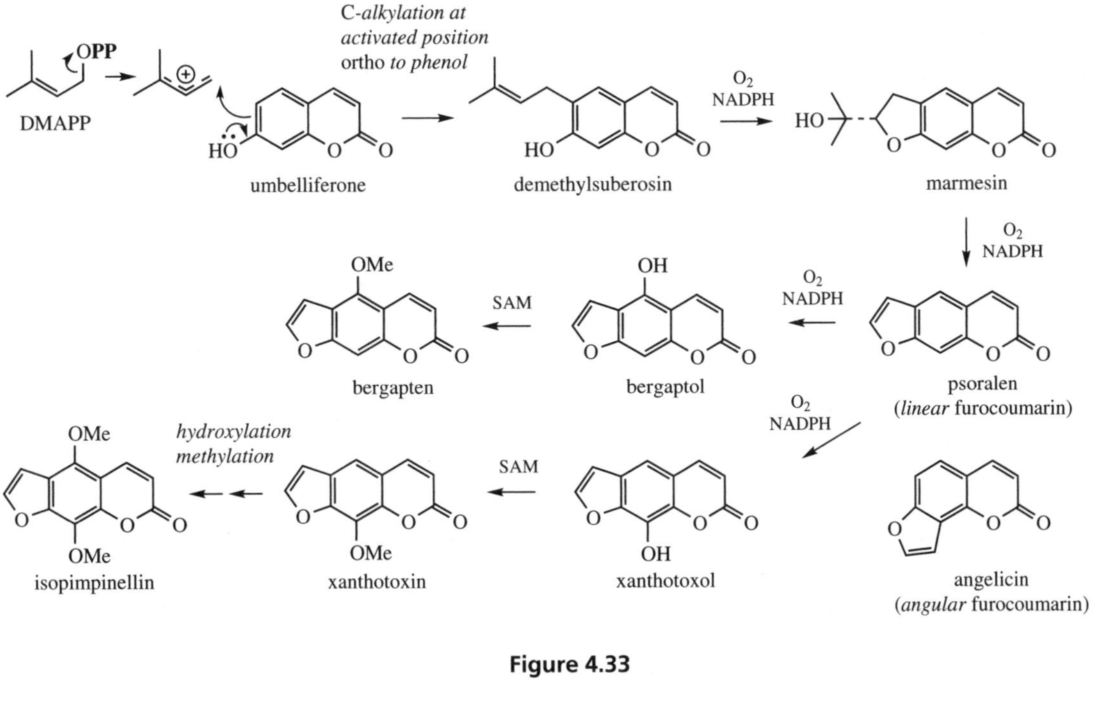

Figure 4.33

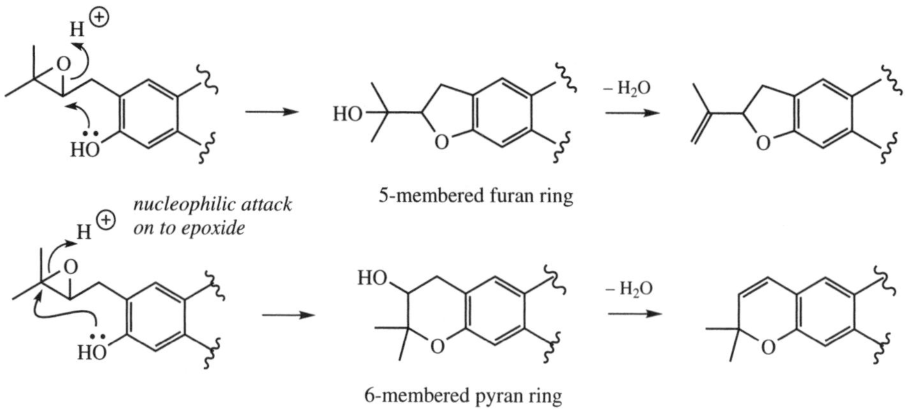

*nucleophilic attack
on to epoxide*

5-membered furan ring

6-membered pyran ring

Figure 4.34

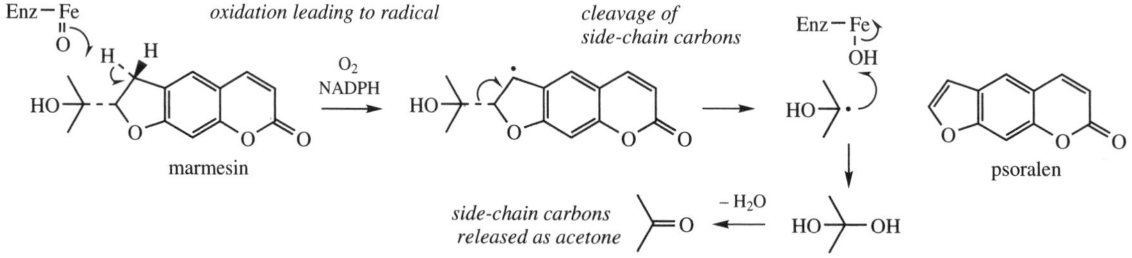

oxidation leading to radical   *cleavage of
side-chain carbons*

marmesin   psoralen

*side-chain carbons
released as acetone*

Figure 4.35

aromatic ring to which it was fused was in that case a product of the acetate pathway. Linear furocoumarins (**psoralens**)* can be troublesome to humans since they can cause photosensitization towards UV light, resulting in sunburn or serious blistering. Used medicinally, this effect may be valuable in promoting skin pigmentation and treating psoriasis.

Psoralens

Psoralens are linear furocoumarins which are widely distributed in plants, but are particularly abundant in the Umbelliferae/Apiaceae and Rutaceae. The most common examples are psoralen, bergapten, xanthotoxin, and isopimpinellin (Figure 4.33). Plants containing psoralens have been used internally and externally to promote skin pigmentation and sun-tanning. Bergamot oil obtained from the peel of *Citrus aurantium* ssp. *bergamia* (Rutaceae) (see page 179) can contain up to 5% bergapten, and is frequently used in external suntan preparations. The psoralen, because of its extended chromophore, absorbs in the near UV and allows this radiation to stimulate formation of melanin pigments (see page 129).

Methoxsalen (xanthotoxin; 8-methoxypsoralen) (Figure 4.36), a constituent of the fruits of *Ammi majus* (Umbelliferae/Apiaceae), is used medically to facilitate skin repigmentation where severe blemishes exist (vitiligo). An oral dose of methoxsalen is followed by long wave UV irradiation, though such treatments must be very carefully regulated to

(Continues)

(Continued)

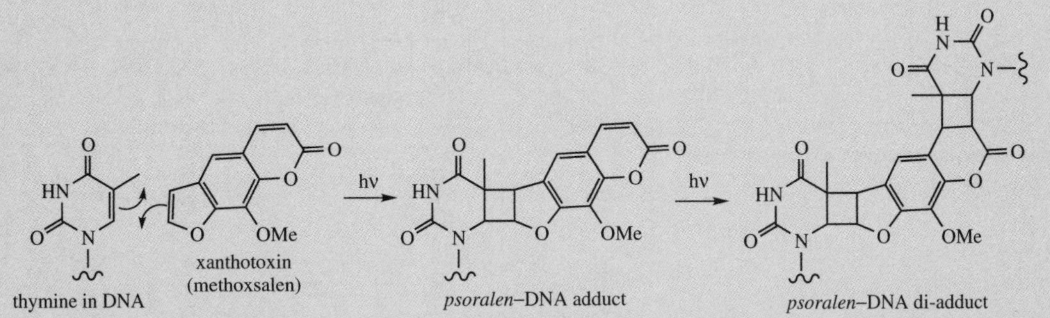

Figure 4.36

minimize the risk of burning, cataract formation, and the possibility of causing skin cancer. The treatment is often referred to as PUVA (psoralen + UV-A). PUVA is also of value in the treatment of psoriasis, a widespread condition characterized by proliferation of skin cells. Similarly, methoxsalen is taken orally, prior to UV treatment. Reaction with psoralens inhibits DNA replication and reduces the rate of cell division. Because of their planar nature, psoralens intercalate into DNA, and this enables a UV-initiated cycloaddition reaction between pyrimidine bases (primarily thymine) in DNA and the furan ring of psoralens (Figure 4.36). In some cases, di-adducts can form involving further cycloaddition via the pyrone ring, thus cross-linking the nucleic acid.

A troublesome extension of these effects can arise from the handling of plants that contain significant levels of furocoumarins. Celery (*Apium graveolens*; Umbelliferae/Apiaceae) is normally free of such compounds, but fungal infection with the natural parasite *Sclerotinia sclerotiorum* induces the synthesis of furocoumarins (xanthotoxin and others) as a response to the infections. Some field workers handling these infected plants have become very sensitive to UV light and suffer from a form of sunburn termed photophytodermatitis. Infected parsley (*Petroselinum crispum*) can give similar effects. Handling of rue (*Ruta graveolens*; Rutaceae) or giant hogweed (*Heracleum mantegazzianum*; Umbelliferae/Apiaceae), which naturally contain significant amounts of psoralen, bergapten, and xanthotoxin, can cause similar unpleasant reactions, or more commonly rapid blistering by direct contact with the sap. The giant hogweed can be particularly dangerous. Individuals vary in their sensitivity towards furocoumarins; some are unaffected whilst others tend to become sensitized by an initial exposure and then develop the allergic response on subsequent exposures.

STYRYLPYRONES

Cinnamic acids, as their coenzyme A esters, may also function as starter units for chain extension with malonyl-CoA units, thus combining elements of the shikimate and acetate pathways (see page 80). Most commonly, three C_2 units are added via malonate giving rise to flavonoids and stilbenes, as described in the next section (page 149). However, there are several examples of products formed from a cinnamoyl-CoA starter plus one or two C_2 units from malonyl-CoA. The short poly-β-keto chain frequently cyclizes to form a lactone derivative (compare triacetic acid lactone, page 62). Thus, Figure 4.37 shows the proposed derivation of **yangonin** via cyclization of the di-enol tautomer of the polyketide formed from 4-hydroxycinnamoyl-CoA and two malonyl-CoA extender units. Two methylation reactions complete the sequence. Yangonin and a series of related structures form the active principles of kava root (*Piper methysticum*; Piperaceae), a herbal remedy popular for its anxiolytic activity.

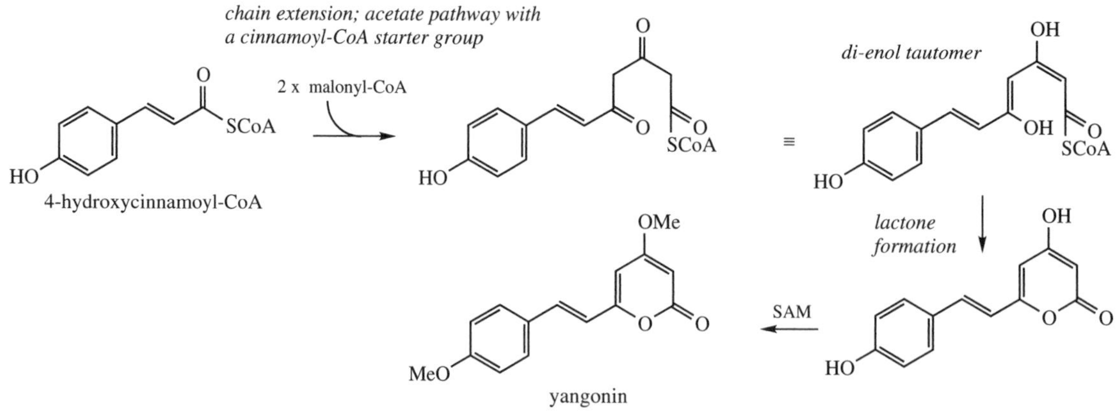

Figure 4.37

Kava

Aqueous extracts from the root and rhizome of *Piper methysticum* (Piperaceae) have long been consumed as an intoxicating beverage by the peoples of Pacific islands comprising Polynesia, Melanesia and Micronesia, and the name kava or kava-kava referred to this drink. In herbal medicine, the dried root and rhizome is now described as kava, and it is used for the treatment of anxiety, nervous tension, agitation and insomnia. The pharmacological activity is associated with a group of styrylpyrone derivatives termed kavapyrones or kavalactones, good quality roots containing 5–8% kavapyrones. At least 18 kavapyrones have been characterized, the six major ones being the enolides kawain, methysticin, and their dihydro derivatives reduced in the cinnamoyl side-chain, and the dienolides yangonin and demethoxyyangonin (Figure 4.38). Compared with the dienolides, the enolides have a reduced pyrone ring and a chiral centre. Clinical trials have indicated kava extracts to be effective as an anxiolytic, the kavapyrones also displaying anticonvulsive, analgesic, and central muscle relaxing action. Several of these compounds have been shown to have an effect on neurotransmitter systems including those involving glutamate, GABA, dopamine, and serotonin.

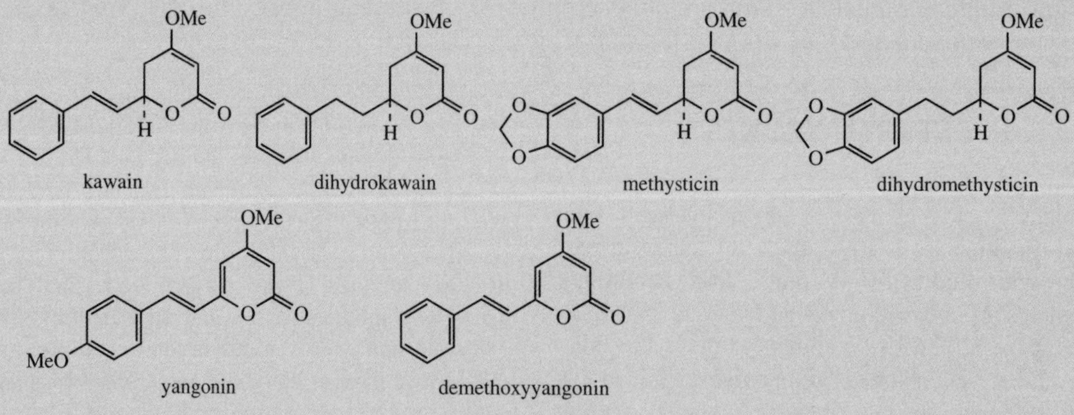

Figure 4.38

FLAVONOIDS AND STILBENES

Flavonoids and stilbenes are products from a cinnamoyl-CoA starter unit, with chain extension using three molecules of malonyl-CoA. This initially gives a polyketide (Figure 4.39), which, according to the nature of the enzyme responsible, can be folded in two different ways. These allow aldol or Claisen-like reactions to occur, generating aromatic rings as already seen in Chapter 3 (see page 80). Enzymes stilbene synthase and chalcone synthase couple a cinnamoyl-CoA unit with three malonyl-CoA units giving stilbenes, e.g. **resveratrol** or chalcones, e.g. **naringenin-chalcone** respectively.

Both structures nicely illustrate the different characteristic oxygenation patterns in aromatic rings derived from the acetate or shikimate pathways. With the **stilbenes**, it is noted that the terminal ester function is no longer present, and therefore hydrolysis and decarboxylation have also taken place during this transformation. No intermediates, e.g. carboxylated stilbenes, have been detected, and the transformation from cinnamoyl-CoA/malonyl-CoA to stilbene is catalysed by the single enzyme. **Resveratrol** has assumed greater relevance in recent years as a constituent of grapes and wine, as well as other food products, with antioxidant, anti-inflammatory, anti-platelet, and cancer preventative properties. Coupled with

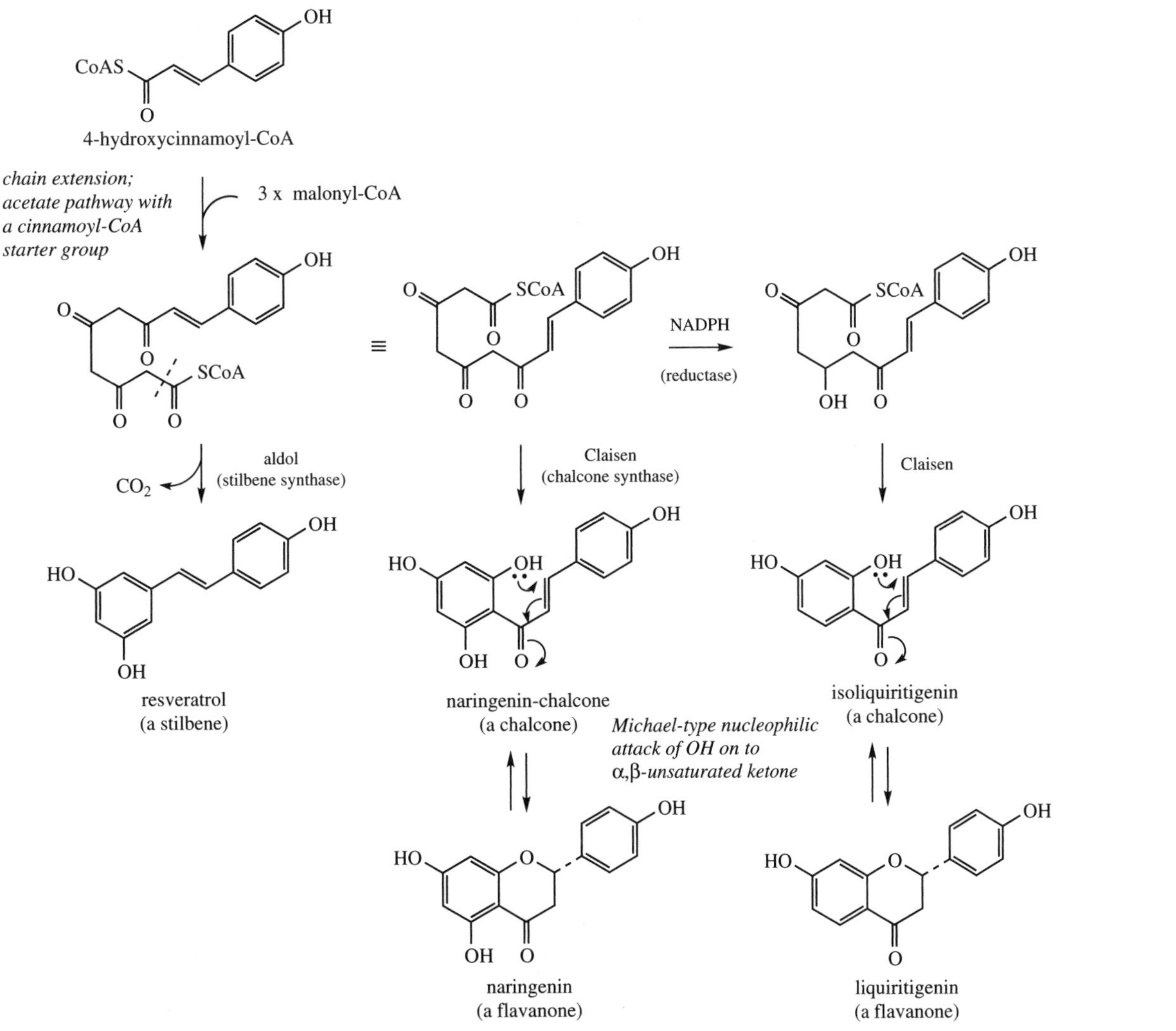

Figure 4.39

the cardiovascular benefits of moderate amounts of alcohol, and the beneficial antioxidant effects of flavonoids (see page 151), red wine has now emerged as an unlikely but most acceptable medicinal agent.

Chalcones act as precursors for a vast range of **flavonoid** derivatives found throughout the plant kingdom. Most contain a six-membered heterocyclic ring, formed by Michael-type nucleophilic attack of a phenol group on to the unsaturated ketone giving a **flavanone**, e.g. **naringenin** (Figure 4.39). This isomerization can occur chemically, acid conditions favouring the flavanone and basic conditions the chalcone, but in nature the reaction is enzyme catalysed and stereospecific, resulting in formation of a single flavanone enantiomer. Many flavonoid structures, e.g. **liquiritigenin**, have lost one of the hydroxyl groups, so that the acetate-derived aromatic ring

has a resorcinol oxygenation pattern rather than the phloroglucinol system. This modification has been tracked down to the action of a reductase enzyme concomitant with the chalcone synthase, and thus **isoliquiritigenin** is produced rather than naringenin-chalcone. Flavanones can then give rise to many variants on this basic skeleton, e.g. **flavones, flavonols, anthocyanidins**, and **catechins** (Figure 4.40). Modifications to the hydroxylation patterns in the two aromatic rings may occur, generally at the flavanone or dihydroflavonol stage, and methylation, glycosylation, and dimethylallylation are also possible, increasing the range of compounds enormously. A high proportion of flavonoids occur naturally as water-soluble glycosides. Considerable quantities of flavonoids are consumed daily in our vegetable diet, so adverse biological effects on man are not particularly intense. Indeed, there is growing belief that some

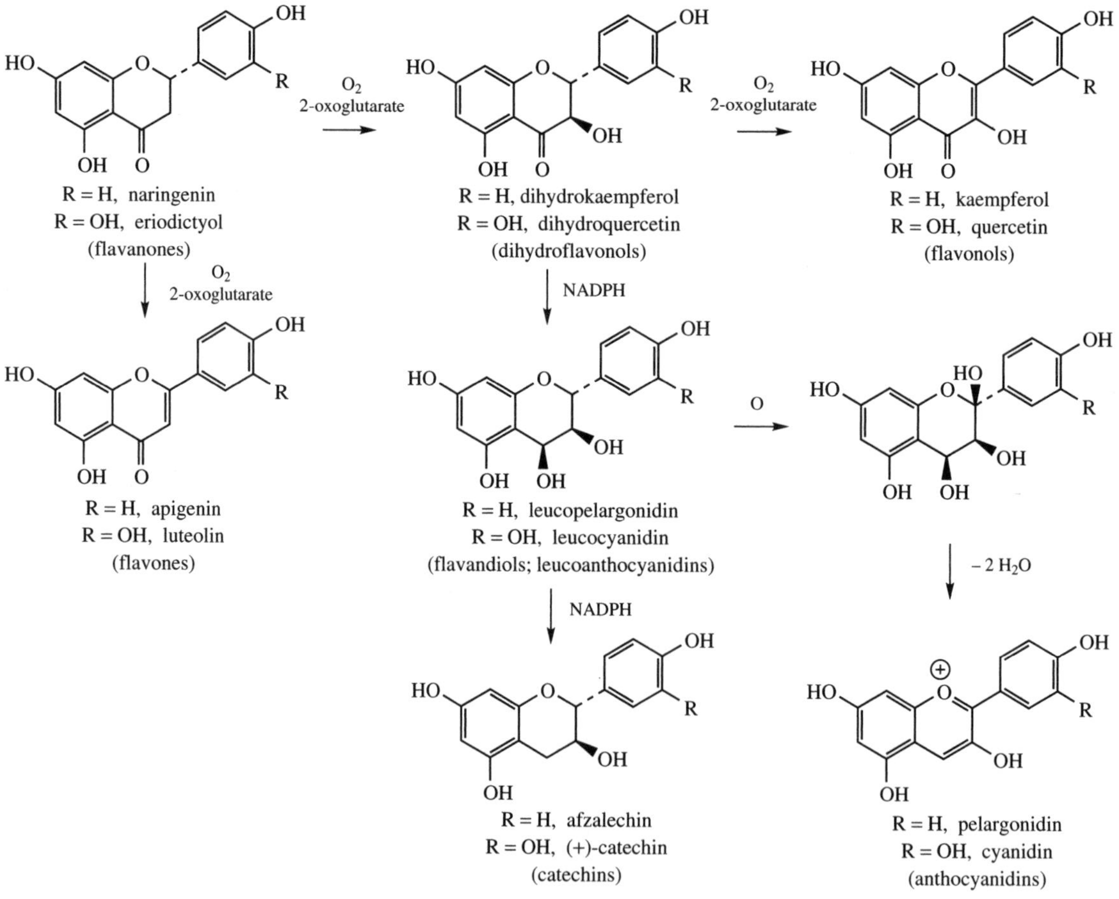

Figure 4.40

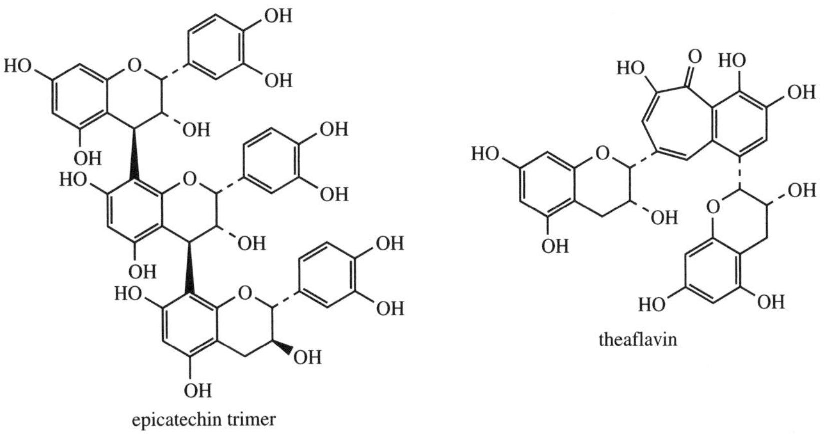

epicatechin trimer

theaflavin

Figure 4.41

flavonoids are particularly beneficial, acting as antioxidants and giving protection against cardiovascular disease, certain forms of cancer, and, it is claimed, age-related degeneration of cell components. Their polyphenolic nature enables them to scavenge injurious free radicals such as superoxide and hydroxyl radicals. **Quercetin** in particular is almost always present in substantial amounts in plant tissues, and is a powerful antioxidant, chelating metals, scavenging free radicals, and preventing oxidation of low density lipoprotein. Flavonoids in red wine (**quercetin, kaempferol**, and anthocyanidins) and in tea (**catechins** and catechin gallate esters) are also demonstrated to be effective antioxidants. Flavonoids contribute to plant colours, yellows from chalcones and flavonols, and reds, blues, and violets from anthocyanidins. Even the colourless materials, e.g. flavones, absorb strongly in the UV and are detectable by insects, probably aiding flower pollination. Catechins form small polymers (oligomers), the **condensed tannins**, e.g. the epicatechin trimer (Figure 4.41) which contribute astringency to our foods and drinks, as do the simpler gallotannins (see page 122), and are commercially important for tanning leather. **Theaflavins**, antioxidants found in fermented tea (see page 395), are dimeric catechin structures in which oxidative processes have led to formation of a seven-membered tropolone ring.

The flavonol glycoside **rutin** (Figure 4.42) from buckwheat (*Fagopyrum esculentum*; Polygonaceae) and rue (*Ruta graveolens*; Rutaceae), and the flavanone glycoside **hesperidin** from *Citrus*

peels have been included in dietary supplements as vitamin P, and claimed to be of benefit in treating conditions characterized by capillary bleeding, but their therapeutic efficacy is far from conclusive. **Neohesperidin** (Figure 4.42) from bitter orange (*Citrus aurantium*; Rutaceae) and **naringin** from grapefruit peel (*Citrus paradisi*) are intensely bitter flavanone glycosides. It has been found that conversion of these compounds into **dihydrochalcones** by hydrogenation in alkaline solution (Figure 4.43) produces a remarkable change to their taste, and the products are now intensely sweet, being some 300–1000 times as sweet as sucrose. These and other dihydrochalcones have been investigated as non-sugar sweetening agents.

FLAVONOLIGNANS

An interesting combination of flavonoid and lignan structures is found in a group of compounds called **flavonolignans**. They arise by oxidative coupling processes between a flavonoid and a phenylpropanoid, usually coniferyl alcohol. Thus, the dihydroflavonol **taxifolin** through one-electron oxidation may provide a free radical, which may combine with the free radical generated from **coniferyl alcohol** (Figure 4.44). This would lead to an adduct, which could cyclize by attack of the phenol nucleophile on to the quinone methide system provided by coniferyl alcohol. The product would be **silybin**, found in *Silybum marianum* (Compositae/Asteraceae) as a mixture of two *trans* diastereoisomers, reflecting a lack of stereospecificity for the original radical coupling. In addition,

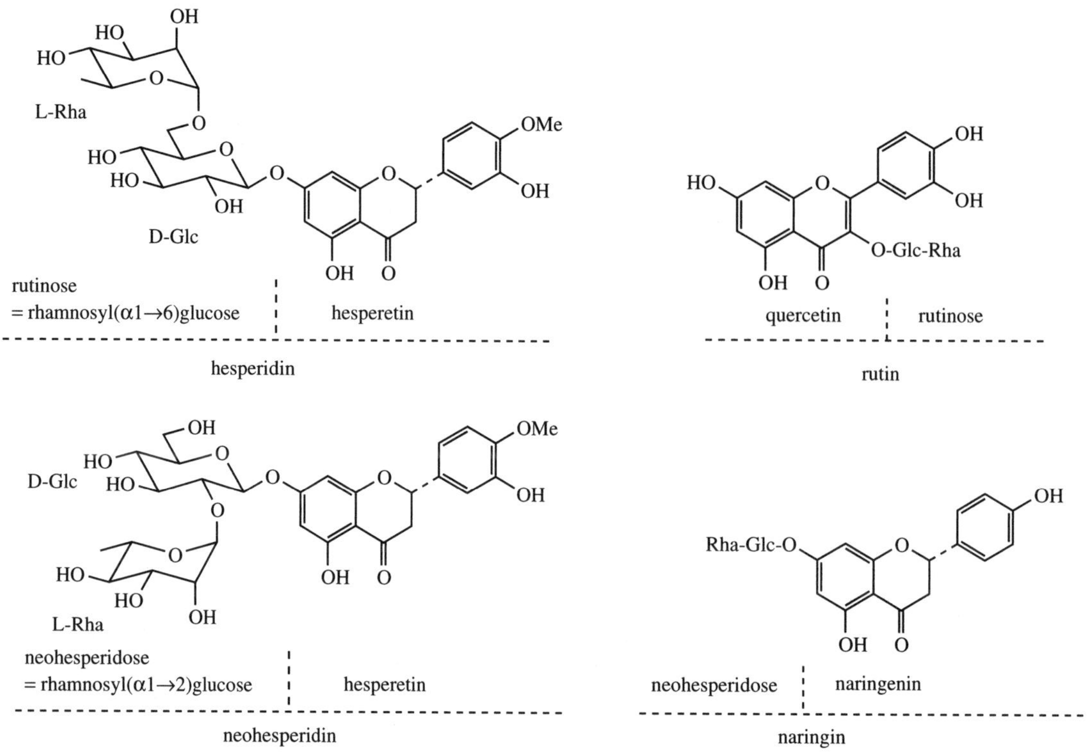

Figure 4.42

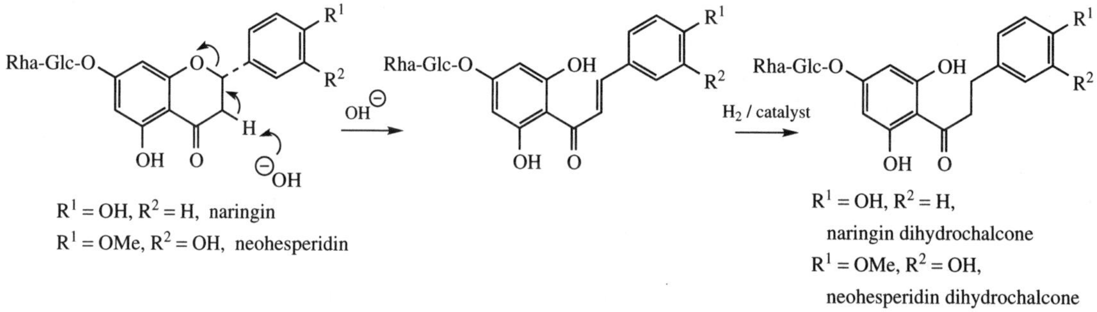

Figure 4.43

the regioisomer **isosilybin** (Figure 4.45), again a mixture of *trans* diastereoisomers, is also found in *Silybum*. **Silychristin** (Figure 4.45) demonstrates a further structural variant which can be seen to originate from a mesomer of the taxifolin-derived free radical, in which the unpaired electron is localized on the carbon *ortho* to the original 4-hydroxyl function. The more complex structure in **silydianin** is accounted for by the mechanism shown in Figure 4.46, in which the initial coupling product cyclizes further by intramolecular attack

of an enolate nucleophile on to the quinonemethide. Hemiketal formation finishes the process. The flavonolignans from *Silybum** (milk thistle) have valuable antihepatotoxic properties, and can provide protection against liver-damaging agents. Coumarinolignans, which are products arising by a similar oxidative coupling mechanism which combines a coumarin with a cinnamyl alcohol, may be found in other plants. The benzodioxane ring as seen in silybin and isosilybin is a characteristic feature of many such compounds.

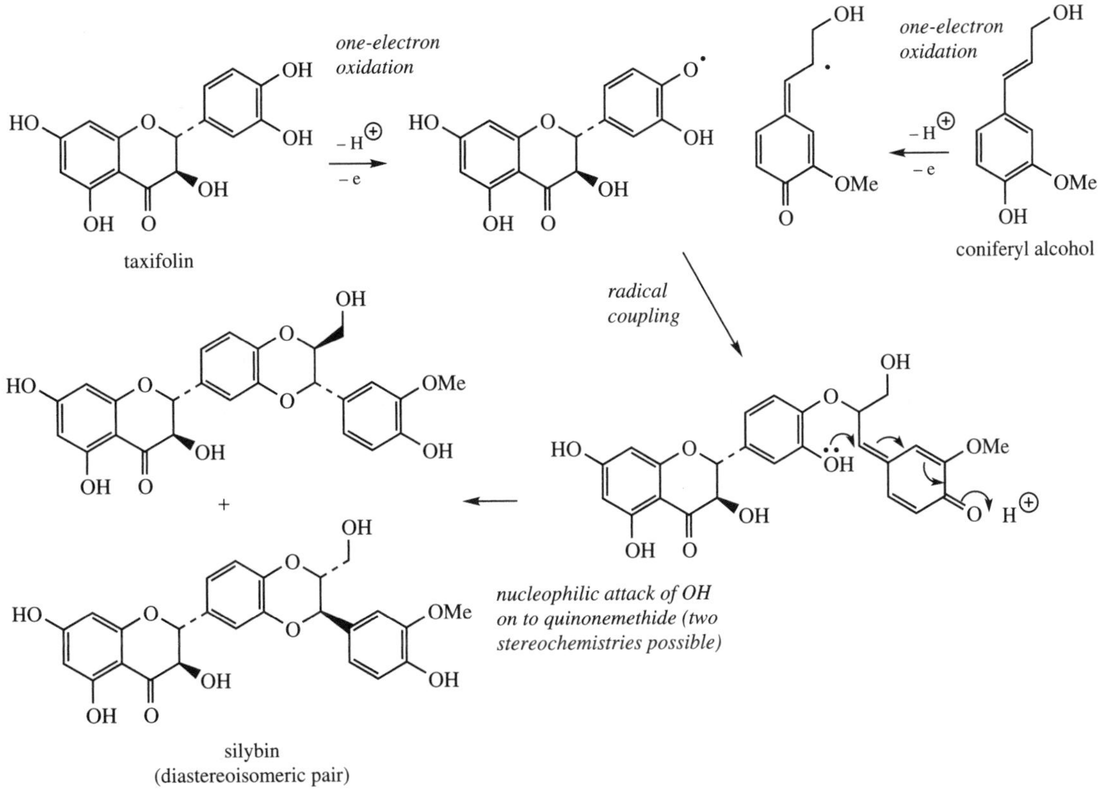

one-electron oxidation

$-H^{\oplus}$
$-e$

taxifolin

one-electron oxidation

$-H^{\oplus}$
$-e$

coniferyl alcohol

radical coupling

nucleophilic attack of OH on to quinonemethide (two stereochemistries possible)

silybin
(diastereoisomeric pair)

Figure 4.44

Silybum marianum

Silybum marianum (Compositae/Asteraceae) is a biennial thistle-like plant (milk thistle) common in the Mediterranean area of Europe. The seeds yield 1.5–3% of flavonolignans collectively termed silymarin. This mixture contains mainly silybin (Figure 4.44), together with silychristin (Figure 4.45), silydianin (Figure 4.46), and small amounts of isosilybin (Figure 4.45). Both silybin and isosilybin are equimolar mixtures of two *trans* diastereoisomers. *Silybum marianum* is widely used in traditional European medicine, the fruits being used to treat a variety of hepatic and other disorders. Silymarin has been shown to protect animal livers against the damaging effects of carbon tetrachloride, thioacetamide, drugs such as paracetamol, and the toxins α-amanitin and phalloin found in the death cap fungus (*Amanita phalloides*) (see page 433). Silymarin may be used in many cases of liver disease and injury, though it still remains peripheral to mainstream medicine. It can offer particular benefit in the treatment of poisoning by the death cap fungus. These agents appear to have two main modes of action. They act on the cellular membrane of hepatocytes inhibiting absorption of toxins, and secondly, because of their phenolic nature, they can act as antioxidants and scavengers for free radicals which cause liver damage originating from liver detoxification of foreign chemicals. Derivatives of silybin with improved water-solubility and/or bioavailability have been developed, e.g. the bis-hemisuccinate and a phosphatidylcholine complex.

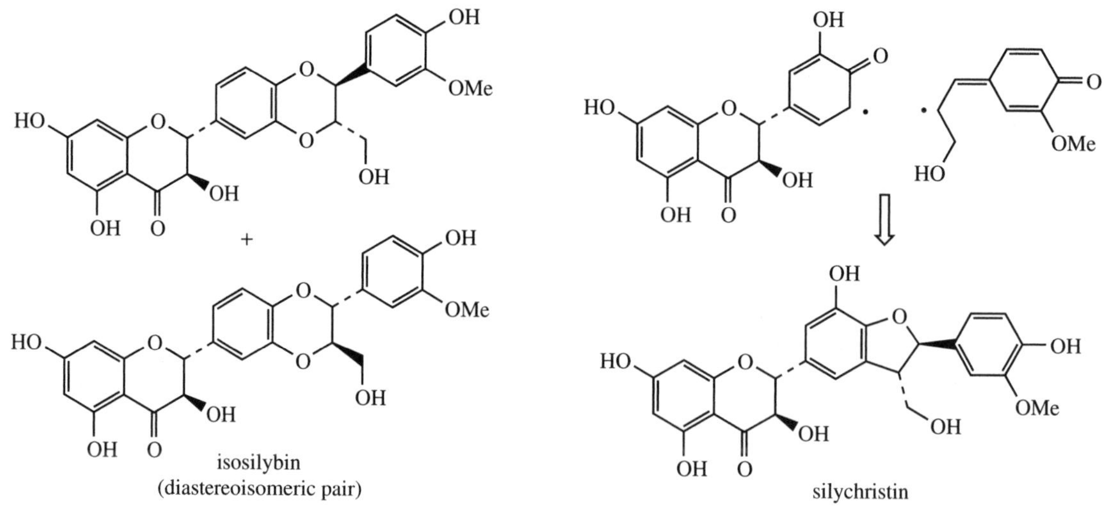
isosilybin
(diastereoisomeric pair)
silychristin

Figure 4.45

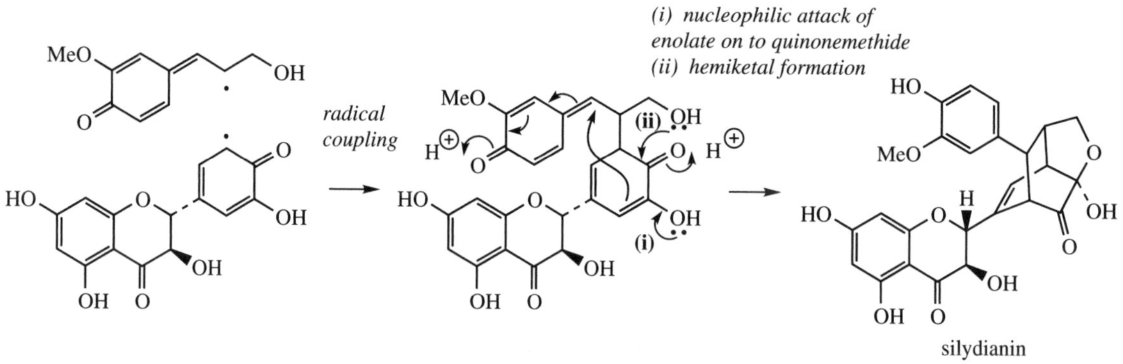
radical
coupling
(i) nucleophilic attack of
enolate on to quinonemethide
(ii) hemiketal formation
silydianin

Figure 4.46

ISOFLAVONOIDS

The **isoflavonoids** form a quite distinct subclass of flavonoid compound, being structural variants in which the shikimate-derived aromatic ring has migrated to the adjacent carbon of the heterocycle. This rearrangement process is brought about by a cytochrome P-450-dependent enzyme requiring NADPH and O_2 cofactors, which transforms the flavanones **liquiritigenin** or **naringenin** into the isoflavones **daidzein** or **genistein** respectively via intermediate hydroxyisoflavanones (Figure 4.47). A radical mechanism has been proposed. This rearrangement is quite rare in nature, and isoflavonoids are almost entirely restricted to the plant family the Leguminosae/Fabaceae. Nevertheless, many hundreds of different isoflavonoids have been identified, and structural complexity is brought about by hydroxylation and alkylation reactions, varying the oxidation level of the heterocyclic ring, or forming additional heterocyclic rings. Some of the many variants are shown in Figure 4.48. **Pterocarpans**, e.g. **medicarpin** from lucerne (*Medicago sativa*), and **pisatin** from pea (*Pisum sativum*), have antifungal activity and form part of these plants' natural defence mechanism against fungal attack. Simple **isoflavones** such as **daidzein** and **coumestans** such as **coumestrol** from lucerne and clovers (*Trifolium* species), have sufficient oestrogenic activity to seriously affect the reproduction of grazing animals, and are termed **phyto-oestrogens***. These planar molecules undoubtedly mimic the shape and polarity of the steroid hormone estradiol

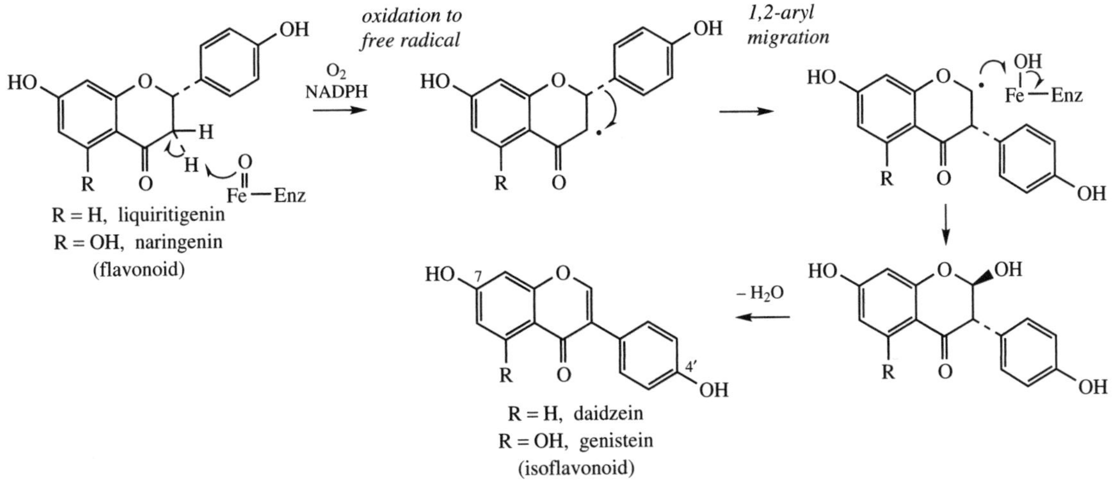

Figure 4.47

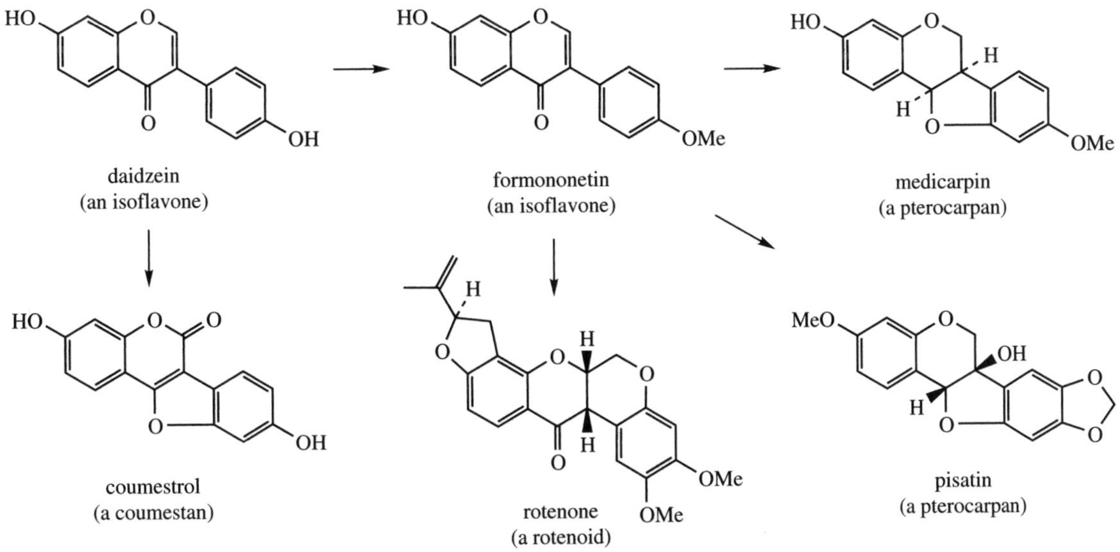

Figure 4.48

(see page 276). The consumption of legume fodder crops by animals must therefore be restricted, or low isoflavonoid producing strains have to be selected. Isoflavonoids in the human diet, e.g. from soya (*Glycine max*) products, are believed to give some protection against oestrogen-dependent cancers such as breast cancer, by restricting the availability of the natural hormone. In addition, they can feature as dietary oestrogen supplements in the reduction of menopausal symptoms, in a similar way to hormone replacement therapy

(see page 279). The **rotenoids** take their name from the first known example **rotenone**, and are formed by ring cyclization of a methoxyisoflavone (Figure 4.49). Rotenone itself contains a C_5 isoprene unit (as do virtually all the natural rotenoids) introduced via dimethylallylation of **demethylmunduserone**. The isopropenylfurano system of rotenone, and the dimethylpyrano of **deguelin**, are formed via rotenonic acid (Figure 4.49) without any detectable epoxide or hydroxy intermediates (compare furocoumarins, page 145). Rotenone

oxidation of OMe group (hydroxylation and loss of hydroxide, compare Figure 2.21)

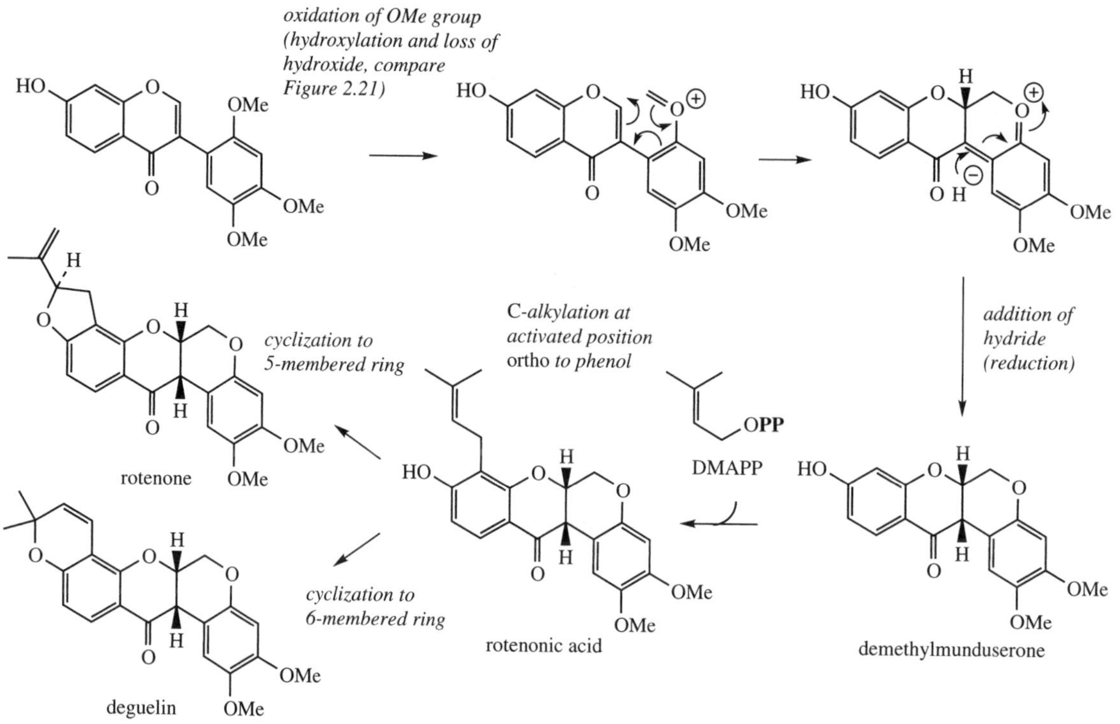

addition of hydride (reduction)

C-alkylation at activated position ortho to phenol

cyclization to 5-membered ring

cyclization to 6-membered ring

rotenone

deguelin

rotenonic acid

demethylmunduserone

Figure 4.49

and other rotenoids are powerful insecticidal and piscicidal (fish poison) agents, interfering with oxidative phosphorylation. They are relatively harmless to mammals unless they enter the blood stream, being metabolized rapidly upon ingestion. Rotenone thus provides an excellent biodegradable insecticide, and is used as such either in pure or powdered plant form. Roots of *Derris elliptica** or *Lonchocarpus** species are rich sources of rotenone.

Phyto-oestrogens

Phyto-oestrogen (phytoestrogen) is a term applied to non-steroidal plant materials displaying oestrogenic properties. Pre-eminent amongst these are isoflavonoids. These planar molecules mimic the shape and polarity of the steroid hormone estradiol (see page 279), and are able to bind to an oestrogen receptor, though their activity is less than that of estradiol. In some tissues, they stimulate an oestrogenic response, whilst in others they can antagonize the effect of oestrogens. Such materials taken as part of the diet therefore influence overall oestrogenic activity in the body by adding their effects to normal levels of steroidal oestrogens (see page 282). Foods rich in isoflavonoids are valuable in countering some of the side-effects of the menopause in women, such as hot flushes, tiredness, and mood swings. In addition, there is mounting evidence that phyto-oestrogens also provide a range of other beneficial effects, helping to prevent heart attacks and other cardiovascular diseases, protecting against osteoporosis, lessening the risk of breast and uterine cancer, and in addition displaying significant antioxidant activity which may reduce the risk of Alzheimer's disease. Whilst some of these benefits may be obtained by the use of steroidal

(Continues)

(Continued)

oestrogens, particularly via hormone replacement therapy (HRT; see page 279), phyto-oestrogens offer a dietary alternative.

The main food source of isoflavonoids is the soya bean (*Glycine max*; Leguminosae/Fabaceae) (see also page 256), which contains significant levels of the isoflavones daidzein, and genistein (Figure 4.47), in free form and as their 7-*O*-glucosides. Total isoflavone levels fall in the range 0.1–0.4%, according to variety. Soya products such as soya milk, soya flour, tofu, and soya-based textured vegetable protein may all be used in the diet for their isoflavonoid content. Breads in which wheat flour is replaced by soya flour are also popular. Extracts from red clover (*Trifolium pratense*; Leguminosae/Fabaceae) are also used as a dietary supplement. Red clover isoflavones are predominantly formononetin (Figure 4.48) and daidzein, together with their 7-*O*-glucosides.

The lignans enterodiol and enterolactone (see page 135) are also regarded as phyto-oestrogens. These compounds are produced by the action of intestinal microflora on lignans such as secoisolariciresinol or matairesinol ingested in the diet. A particularly important precursor is secoisolariciresinol diglucoside from flaxseed (*Linum usitatissimum*; Linaceae), and flaxseed may be incorporated into foodstuffs along with soya products. Enterolactone and enterodiol were first detected in human urine, and their origins were traced back to dietary fibre-rich foods. Levels in the urine were much higher in vegetarians, and have been related to a lower incidence of breast cancer in vegetarians.

Derris and Lonchocarpus

Species of *Derris* (e.g. *D. elliptica*, *D. malaccensis*) and *Lonchocarpus* (e.g. *L. utilis, L. urucu*) (Leguminosae/Fabaceae) have provided useful insecticides for many years. Roots of these plants have been employed as a dusting powder, or extracts have been formulated for sprays. *Derris* plants are small shrubs cultivated in Malaysia and Indonesia, whilst *Lonchocarpus* includes shrubs and trees, with commercial material coming from Peru and Brazil. The insecticidal principles are usually supplied as a black, resinous extract. Both *Derris* and *Lonchocarpus* roots contain 3–10% of rotenone (Figure 4.49) and smaller amounts of other rotenoids, e.g. deguelin (Figure 4.49). The resin may contain rotenone (about 45%) and deguelin (about 20%).

Rotenone and other rotenoids interfere with oxidative phosphorylation, blocking transfer of electrons to ubiquinone (see page 159) by complexing with NADH:ubiquinone oxidoreductase of the respiratory electron transport chain. However, they are relatively innocuous to mammals unless they enter the blood stream, being metabolized rapidly upon ingestion. Insects and also fish seem to lack this rapid detoxification. The fish poison effect has been exploited for centuries in a number of tropical countries, allowing lazy fishing by the scattering of powdered plant material on the water. The dead fish were collected, and when subsequently eaten produced no ill effects on the consumers. More recently, rotenoids have been used in fish management programmes to eradicate undesirable fish species prior to restocking with other species. As insecticides, the rotenoids still find modest use, and are valuable for their selectivity and rapid biodegradability. However, they are perhaps inactivated too rapidly in the presence of light and air to compete effectively with other insecticides such as the modern pyrethrin derivatives (see page 188).

TERPENOID QUINONES

Quinones are potentially derivable by oxidation of suitable phenolic compounds, catechols (1,2-dihydroxybenzenes) giving rise to *ortho*-quinones and quinols (1,4-dihydroxybenzenes) yielding *para*-quinones (see page 25). Accordingly, quinones can be formed from phenolic systems generated by either the acetate or shikimate pathways, provided a catechol or quinol

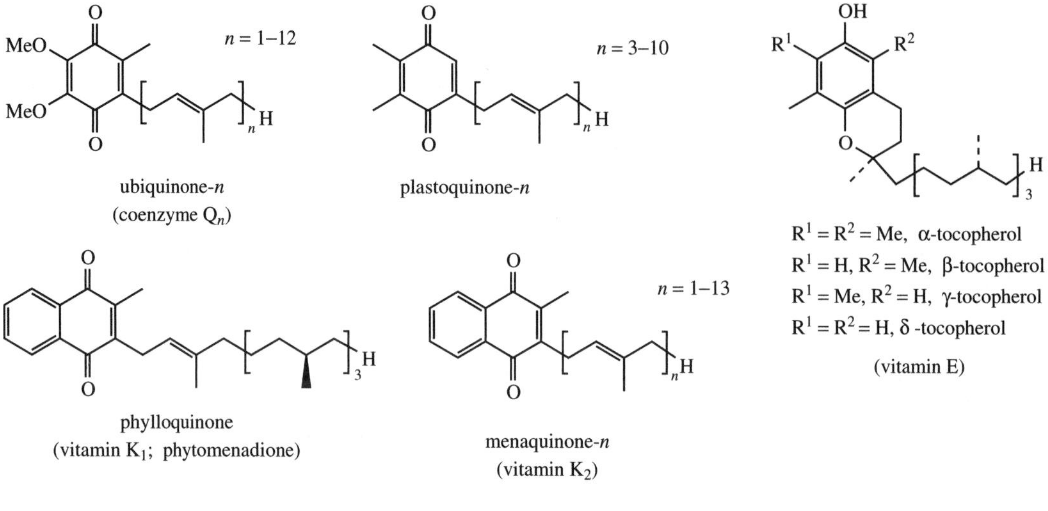

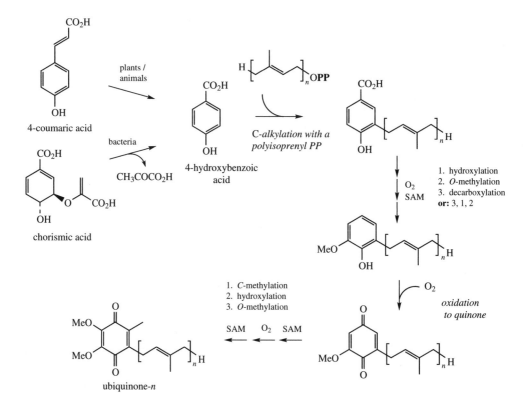

Figure 4.50

Figure 4.51

system has been elaborated, and many examples are found in nature. A range of quinone derivatives and related structures containing a terpenoid fragment as well as a shikimate-derived portion are also widely distributed. Many of these have important biochemical functions in electron transport systems for respiration or photosynthesis, and some examples are shown in Figure 4.50.

Ubiquinones (coenzyme Q) (Figure 4.50) are found in almost all organisms and function as electron carriers for the electron transport chain in mitochondria. The length of the terpenoid chain is variable ($n = 1-12$), and dependent on species, but most organisms synthesize a range of compounds, of which those where $n = 7-10$ usually predominate. The human redox carrier is coenzyme Q_{10}. They are derived from **4-hydroxybenzoic acid** (Figure 4.51), though the origin of this compound varies according to organism (see pages 123, 141). Thus, bacteria are known to transform chorismic acid by enzymic elimination of pyruvic acid, whereas plants and animals utilize a route from phenylalanine or tyrosine via 4-hydroxycinnamic acid (Figure 4.51). 4-Hydroxybenzoic acid is the substrate for *C*-alkylation *ortho* to the phenol group with a polyisoprenyl diphosphate of appropriate chain length (see page 231). The product then undergoes further elaboration, the exact sequence of modifications, i.e. hydroxylation, *O*-methylation, and decarboxylation, varying in eukaryotes and prokaryotes. Quinone formation follows in an O_2-dependent combined hydroxylation–oxidation process, and ubiquinone production then involves further hydroxylation, and *O*- and *C*-methylation reactions.

Plastoquinones (Figure 4.50) bear considerable structural similarity to ubiquinones, but are not derived from 4-hydroxybenzoic acid. Instead, they are produced from **homogentisic acid**, a phenylacetic acid derivative formed from **4-hydroxyphenylpyruvic acid** by a complex reaction involving decarboxylation, O_2-dependent hydroxylation, and subsequent migration of the $-CH_2CO_2H$ side-chain to the adjacent position on the aromatic ring (Figure 4.52). *C*-Alkylation of homogentisic acid *ortho* to a phenol group follows, and involves a polyisoprenyl diphosphate with $n = 3-10$, but most commonly with $n = 9$, i.e. **solanesyl diphosphate**. However, during the alkylation reaction, the $-CH_2CO_2H$ side-chain of homogentisic acid

suffers decarboxylation, and the product is thus an alkyl methyl *p*-quinol derivative. Further aromatic methylation (via *S*-adenosylmethionine) and oxidation of the *p*-quinol to a quinone follow to yield the plastoquinone. Thus, only one of the two methyl groups on the quinone ring of the plastoquinone is derived from SAM. Plastoquinones are involved in the photosynthetic electron transport chain in plants.

Tocopherols are also frequently found in the chloroplasts and constitute members of the vitamin E* group. Their biosynthesis shares many of the features of plastoquinone biosynthesis, with an additional cyclization reaction involving the *p*-quinol and the terpenoid side-chain to give a chroman ring (Figure 4.52). Thus, the tocopherols, e.g. **α-tocopherol** and **γ-tocopherol**, are not in fact quinones, but are indeed structurally related to plastoquinones. The isoprenoid side-chain added, from **phytyl diphosphate**, contains only four isoprene units, and three of the expected double bonds have suffered reduction. Again, decarboxylation of homogentisic acid cooccurs with the alkylation reaction. *C*-Methylation steps using SAM, and the cyclization of the *p*-quinol to γ-tocopherol, have been established as in Figure 4.52. Note once again that one of the nuclear methyls is homogentisate-derived, whilst the others are supplied by SAM.

The **phylloquinones** (vitamin K_1) and **menaquinones** (vitamin K_2) are shikimate-derived naphthoquinone derivatives found in plants and algae (vitamin K_1*) or bacteria and fungi (vitamin K_2). The most common phylloquinone structure (Figure 4.50) has a diterpenoid side-chain, whereas the range of menaquinone structures tends to be rather wider with 1–13 isoprene units. These quinones are derived from chorismic acid via its isomer **isochorismic acid** (Figure 4.55). Additional carbons for the naphthoquinone skeleton are provided by 2-oxoglutaric acid, which is incorporated by a mechanism involving the coenzyme thiamine diphosphate (TPP). 2-Oxoglutaric acid is decarboxylated in the presence of TPP to give the TPP anion of succinic semialdehyde, which attacks isochorismic acid in a Michael-type reaction. Loss of the thiamine cofactor, elimination of pyruvic acid, and then dehydration yield the intermediate *o*-succinylbenzoic acid (OSB). This is activated by formation of a coenzyme A ester, and a Dieckmann-like condensation allows ring formation. The dihydroxynaphthoic acid is the

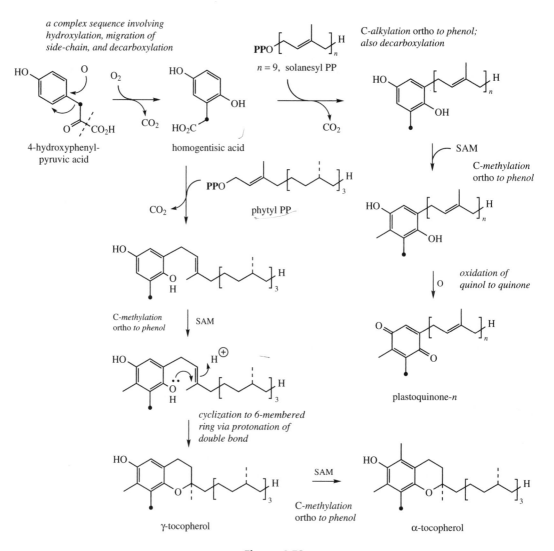

Figure 4.52

Vitamin E

Vitamin E refers to a group of fat-soluble vitamins, the tocopherols, e.g. α-, β-, γ-, and δ-tocopherols (Figure 4.53), which are widely distributed in plants, with high levels in cereal seeds such as wheat, barley, and rye. Wheat germ oil is a particularly good source. The proportions of the individual tocopherols vary widely in different seed oils, e.g. principally β- in wheat oil, γ- in corn oil, α- in safflower oil, and γ- and δ- in soybean oil. Vitamin E deficiency is virtually unknown, with most of the dietary intake coming from food oils and margarine, though much can be lost during processing and cooking. Rats deprived of the vitamin display reproductive abnormalities. α-Tocopherol has the highest activity (100%), with the relative activities of β-, γ-, and δ-tocopherols being 50%, 10%, and 3% respectively. **α-Tocopheryl acetate** is the main commercial form used for food supplementation and

(Continues)

(Continued)

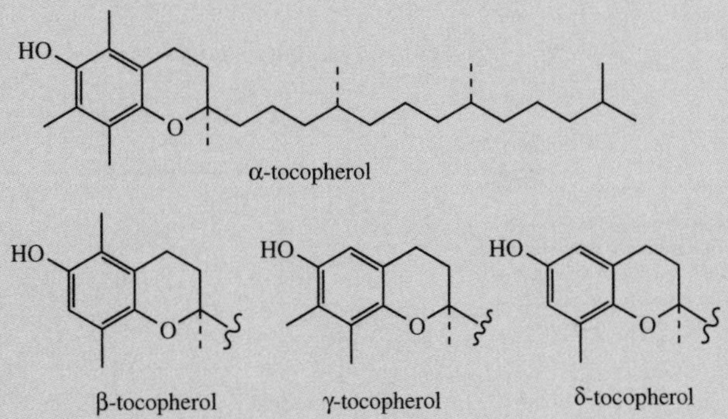

α-tocopherol

β-tocopherol γ-tocopherol δ-tocopherol

Figure 4.53

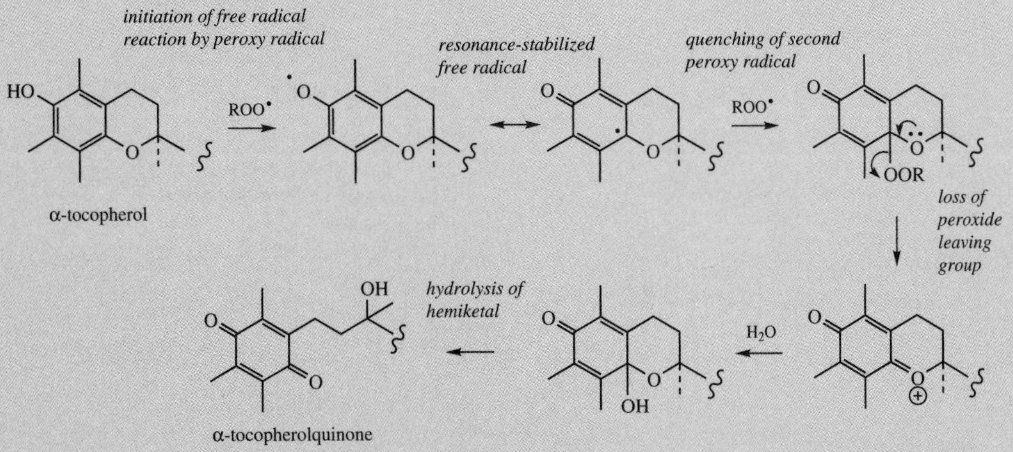

initiation of free radical reaction by peroxy radical

resonance-stabilized free radical

quenching of second peroxy radical

α-tocopherol

loss of peroxide leaving group

hydrolysis of hemiketal

α-tocopherolquinone

Figure 4.54

for medicinal purposes. The vitamin is known to provide valuable antioxidant properties, probably preventing the destruction by free radical reactions of vitamin A and unsaturated fatty acids in biological membranes. It is used commercially to retard rancidity in fatty materials in food manufacturing, and there are also claims that it can reduce the effects of ageing and help to prevent heart disease. Its antioxidant effect is likely to arise by reacting with peroxyl radicals, generating by one-electron phenolic oxidation a resonance-stabilized free radical that does not propagate the free radical reaction, but instead mops up further peroxyl radicals (Figure 4.54). In due course, the tocopheryl peroxide is hydrolysed to the tocopherolquinone.

more favoured aromatic tautomer from the hydrolysis of the coenzyme A ester. This compound is now the substrate for alkylation and methylation as seen with ubiquinones and plastoquinones. However, the terpenoid fragment is found to replace the carboxyl group, and the decarboxylated analogue is not involved. The transformation of **1,4-dihydroxynaphthoic acid** to the isoprenylated naphthoquinone appears to be catalysed by a single enzyme, and can be rationalized by the mechanism in Figure 4.56. This involves alkylation (shown in Figure 4.56 using the diketo tautomer), decarboxylation of the resultant β-keto acid, and finally an oxidation to the *p*-quinone.

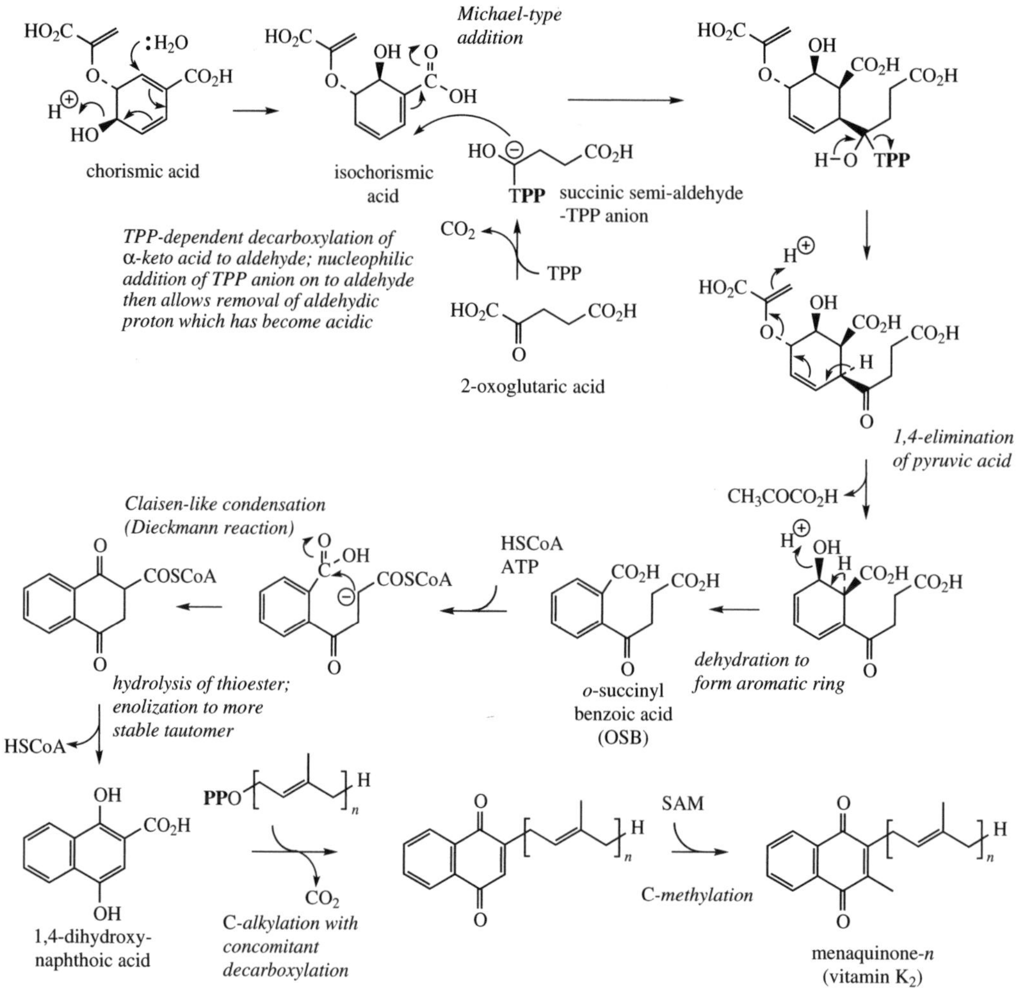

Figure 4.55

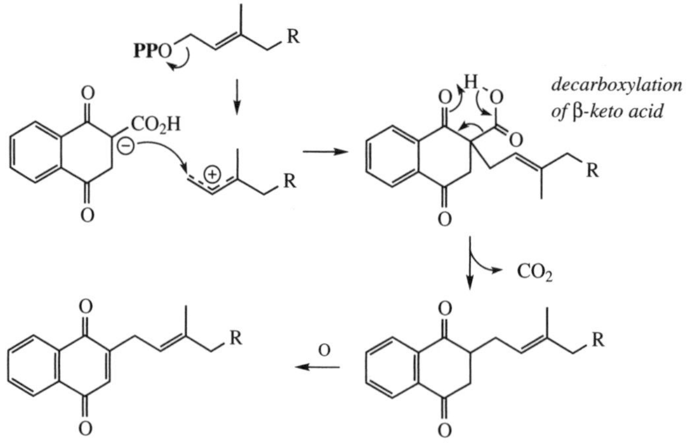

Figure 4.56

Vitamin K

Vitamin K comprises a number of fat-soluble naphthoquinone derivatives, with vitamin K_1 (phylloquinone) (Figure 4.50) being of plant origin whilst the vitamins K_2 (menaquinones) are produced by microorganisms. Dietary vitamin K_1 is obtained from almost any green vegetable, whilst a significant amount of vitamin K_2 is produced by the intestinal microflora. As a result, vitamin K deficiency is rare. Deficiencies are usually the result of malabsorption of the vitamin, which is lipid soluble. Vitamin K_1 (**phytomenadione**) or the water-soluble **menadiol phosphate** (Figure 4.57) may be employed as supplements. Menadiol is oxidized in the body to the quinone, which is then alkylated, e.g. with geranylgeranyl diphosphate, to yield a metabolically active product.

Vitamin K is involved in normal blood clotting processes, and a deficiency would lead to haemorrhage. Blood clotting requires the carboxylation of glutamate residues in the protein prothrombin, generating bidentate ligands that allow the protein to bind to other factors. This carboxylation requires carbon dioxide, molecular oxygen, and the reduced quinol form of vitamin K (Figure 4.57). During the carboxylation, the reduced vitamin K suffers epoxidation, and vitamin K is subsequently regenerated by reduction. Anticoagulants such as dicoumarol and warfarin (see page 144) inhibit this last reduction step. However, the polysaccharide anticoagulant heparin (see page 477) does not interfere with vitamin K metabolism, but acts by complexing with blood clotting enzymes.

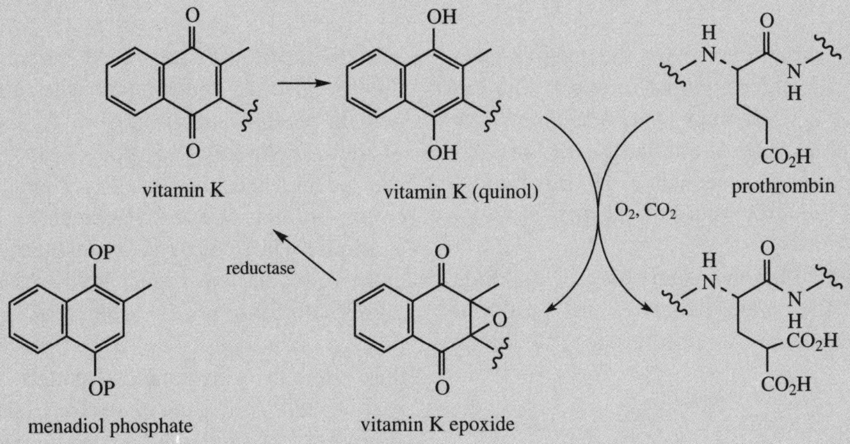

Figure 4.57

OSB, and 1,4-dihydroxynaphthoic acid, or its diketo tautomer, have been implicated in the biosynthesis of a wide range of plant naphthoquinones and anthraquinones. There are parallels with the later stages of the menaquinone sequence shown in Figure 4.55, or differences according to the plant species concerned. Some of these pathways are illustrated in Figure 4.58. Replacement of the carboxyl function by an isoprenyl substituent is found to proceed via a disubstituted intermediate in *Catalpa* (Bignoniaceae) and

Streptocarpus (Gesneriaceae), e.g. **catalponone** (compare Figure 4.56), and this can be transformed to **deoxylapachol** and then **menaquinone-1** (Figure 4.58). **Lawsone** is formed by an oxidative sequence in which hydroxyl replaces the carboxyl. A further interesting elaboration is the synthesis of an anthraquinone skeleton by effectively cyclizing a dimethylallyl substituent on to the naphthaquinone system. Rather little is known about how this process is achieved but many examples are known from the results of labelling studies.

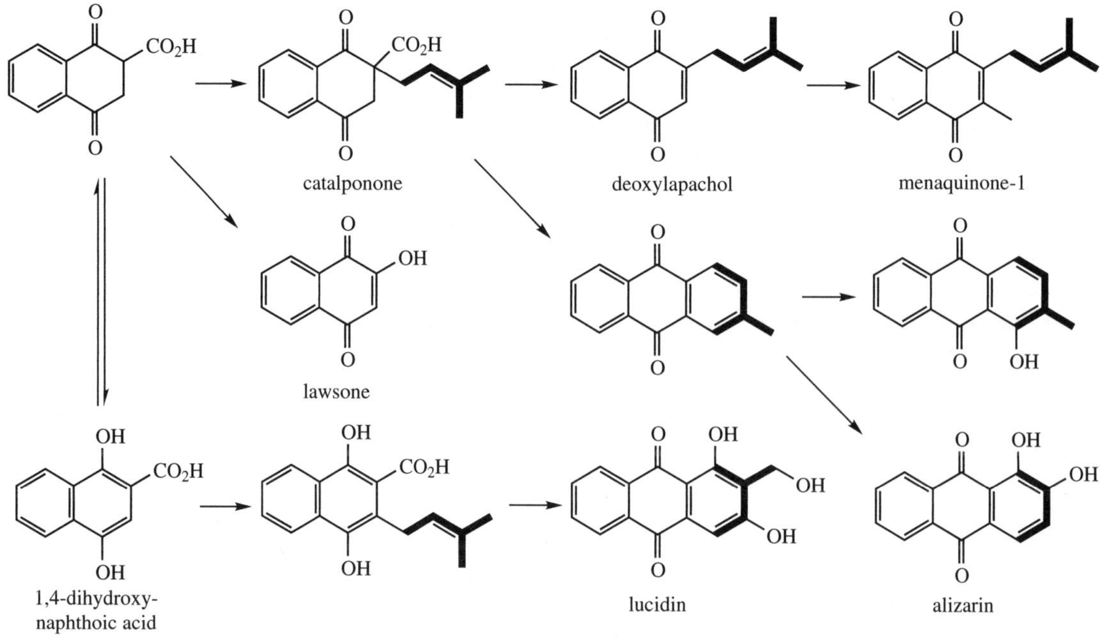

Figure 4.58

Some of these structures retain the methyl from the isoprenyl substituent, whilst in others this has been removed, e.g. **alizarin** from madder (*Rubia tinctorum*; Rubiaceae), presumably via an oxidation–decarboxylation sequence. Hydroxylation, particularly in the terpenoid-derived ring, is also a frequent feature.

Some other quinone derivatives, although formed from the same pathway, are produced by dimethylallylation of 1,4-dihydroxynaphthoic acid at the non-carboxylated carbon. Obviously, this is also a nucleophilic site and alkylation here is mechanistically sound. Again, cyclization of the dimethylallyl to produce an anthraquinone can occur, and the potently mutagenic **lucidin** from *Galium* species (Rubiaceae) is a typical example. The hydroxylation patterns seen in the anthraquinones in Figure 4.58 should be compared with those noted earlier in acetate/malonate-derived structures (see page 63). Remnants of the alternate oxygenation pattern are usually very evident in acetate-derived anthraquinones (Figure 4.59), whereas such a pattern cannot easily be incorporated into typical shikimate/2-oxoglutarate/isoprenoid structures. Oxygen substituents are not usually present in positions fitting the polyketide hypothesis.

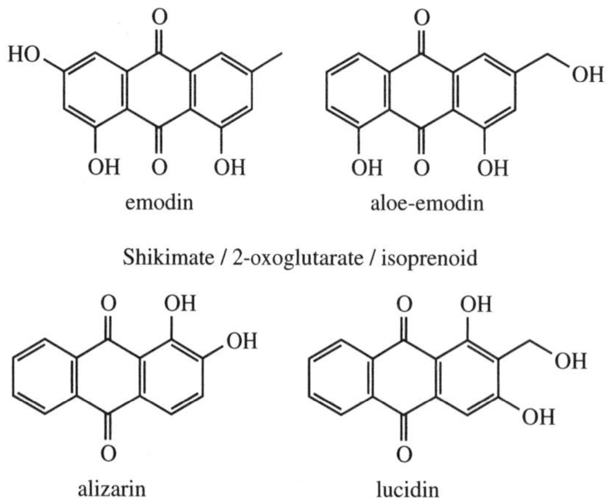

Figure 4.59

FURTHER READING

Shikimate Pathway

Abell C (1999) Enzymology and molecular biology of the shikimate pathway. *Comprehensive Natural Products Chemistry*, Vol 1. Elsevier, Amsterdam, pp 573–607.

Floss HG (1997) Natural products derived from unusual variants of the shikimate pathway. *Nat Prod Rep* **14**, 433–452.

Haslam E (1996) Aspects of the enzymology of the shikimate pathway. *Prog Chem Org Nat Prod* **69**, 157–240.

Herrmann KM and Weaver LM (1999) The shikimate pathway. *Annu Rev Plant Physiol Plant Mol Biol* **50**, 473–503.

Knaggs AR (2000) The biosynthesis of shikimate metabolites. *Nat Prod Rep* **17**, 269–292. Earlier reviews: 1999, **16**, 525–560; Dewick PM (1998) **15**, 17–58.

Folic acid

Cossins EA and Chen L (1997) Folates and one-carbon metabolism in plants. *Phytochemistry* **45**, 437–452.

Maden BEH (2000) Tetrahydrofolate and tetrahydromethanopterin compared: Functionally distinct carriers in C_1 metabolism. *Biochem J* **350**, 609–629 (see errata **352**, 935–936).

Mason P (1999) Nutrition: folic acid – new roles for a well known vitamin. *Pharm J* **263**, 673–677.

Pufulete M (1999) Eat your greens. *Chem Brit* **35** (6), 26–28.

Rawalpally TR (1998) Vitamins (folic acid). *Kirk–Othmer Encyclopedia of Chemical Technology*, 4th edn, Vol 25. Wiley, New York, pp 64–82.

Young DW (1994) Studies on thymidylate synthase and dihydrofolate reductase – two enzymes involved in the synthesis of thymidine. *Chem Soc Rev* **23**, 119–128.

Chloramphenicol

Nagabhushan T, Miller GH and Varma KJ (1992) Antibiotics (chloramphenicol and analogues). *Kirk–Othmer Encyclopedia of Chemical Technology*, 4th edn, Vol 2. Wiley, New York, pp 961–978.

Melanins

Prota G (1995) The chemistry of melanins and melanogenesis. *Prog Chem Org Nat Prod* **64**, 93–148.

Volatile Oils

Mookherjee BD and Wilson RA (1996) Oils, essential. *Kirk–Othmer Encyclopedia of Chemical Technology*, 4th edn, Vol 17. Wiley, New York, pp 603–674.

Lignin

Lewis NG and Yamamoto E (1990) Lignin: occurrence, biogenesis and biodegradation. *Annu Rev Plant Physiol Plant Mol Biol* **41**, 455–496.

Lignans

Canel C, Moraes RM, Dayan FE and Ferreira D (2000) Molecules of interest: podophyllotoxin. *Phytochemistry* **54**, 115–120.

Davin LB and Lewis NG (2000) Dirigent proteins and dirigent sites explain the mystery of specificity of radical precursor coupling in lignan and lignin biosynthesis. *Plant Physiol* **123**, 453–461.

Lewis NG and Davin LB (1999) Lignans: biosynthesis and function. *Comprehensive Natural Products Chemistry*, Vol 1. Elsevier, Amsterdam, pp 639–712.

Stähelin HF and von Wartburg A (1991) The chemical and biological route from podophyllotoxin glucoside to etoposide. *Cancer Res* **51**, 5–15.

Ward RS (1999) Lignans, neolignans and related compounds. *Nat Prod Rep* **16**, 75–96. Earlier reviews: 1997, **14**, 43–74; 1995, **12**, 183–205.

Coumarins

Bell WR (1992) Blood, coagulants and anticoagulants. *Kirk–Othmer Encyclopedia of Chemical Technology*, 4th edn, Vol 4. Wiley, New York, pp 333–360.

Estévez-Braun A and González AG (1997) Coumarins. *Nat Prod Rep* **14**, 465–475. Earlier review: Murray RDH (1995) **12**, 477–505.

Matern U, Lüer P and Kreusch D (1999) Biosynthesis of coumarins. *Comprehensive Natural Products Chemistry*, Vol 1. Elsevier, Amsterdam, pp 623–637.

Murray RDH (1997) Naturally occurring plant coumarins. *Prog Chem Org Nat Prod* **72**, 1–119.

Styrylpyrones

Häberlein H, Boonen G and Beck M-A (1997) *Piper methysticum*: enantiomeric separation of kavapyrones by high performance liquid chromatography. *Planta Med* **63**, 63–65.

Flavonoids

Das A, Wang JH and Lien EJ (1994) Carcinogenicity, mutagenicity and cancer preventing activities of flavonoids: a structure–system–activity relationship (SSAR) analysis. *Prog Drug Res* **42**, 133–166.

Forkmann G and Heller W (1999) Biosynthesis of flavonoids. *Comprehensive Natural Products Chemistry*, Vol 1. Elsevier, Amsterdam, pp 713–748.

Gordon MH (1996) Dietary antioxidants in disease prevention. *Nat Prod Rep* **13**, 265–273.

Harborne JB and Williams CA (1998) Anthocyanins and other flavonoids. *Nat Prod Rep* **15**, 631–652. Earlier review: 1995, **12**, 639–657.

Harborne JB and Williams CA (2000) Advances in flavonoid research since 1992. *Phytochemistry* **55**, 481–504.

Waterhouse AL (1995) Wine and heart disease. *Chem Ind* 338–341.

Isoflavonoids

Crombie L and Whiting DA (1998) Biosynthesis in the rotenoid group of natural products: applications of isotope methodology. *Phytochemistry*, **49**, 1479–1507.

Davis SR, Dalais FS, Simpson ER and Murkies AL (1999) Phytoestrogens in health and disease. *Rec Prog Hormone Res* **54**, 185–210.

Dixon RA (1999) Isoflavonoids: biochemistry, molecular biology, and biological function. *Comprehensive Natural Products Chemistry*, Vol 1. Elsevier, Amsterdam, pp 773–823.

Donnelly DMX and Boland GM (1998) Isoflavonoids and related compounds. *Nat Prod Rep* **15**, 241–260. Earlier review: 1995, **12**, 321–338.

Mason P (2001) Nutrition: isoflavones. *Pharm J* **266**, 16–19.

Metcalf RL (1995) Insect control technology. *Kirk–Othmer Encyclopedia of Chemical Technology*, 4th edn, Vol 14. Wiley, New York, pp 524–602.

Tannins

Ferreira D, Brandt EV, Coetzee J and Malan E (1999) Condensed tannins. *Prog Chem Org Nat Prod* **77**, 21–67.

Ferreira D and Li X-C (2000) Oligomeric proanthocyanidins: naturally occurring *O*-heterocycles. *Nat Prod Rep* **17**, 193–212. Earlier review: Ferreira D and Bekker R (1996) **13**, 411–433.

Ferreira D, Nel RJJ and Bekker R (1999) Condensed tannins. *Comprehensive Natural Products Chemistry*, Vol 3. Elsevier, Amsterdam, pp 747–797.

Gross GG (1999) Biosynthesis of hydrolyzable tannins. *Comprehensive Natural Products Chemistry*, Vol 3. Elsevier, Amsterdam, pp 799–826.

Haslam E (1996) Natural polyphenols (vegetable tannins) as drugs: possible modes of action. *J Nat Prod* **59**, 205–215.

Haslam E and Cai Y (1994) Plant polyphenols (vegetable tannins): gallic acid metabolism. *Nat Prod Rep* **11**, 41–66.

Okuda T, Yoshida T and Hatano T (1995) Hydrolyzable tannins and related polyphenols. *Prog Chem Org Nat Prod* **66**, 1–117.

Vitamin E

Casani R (1998) Vitamins (vitamin E). *Kirk–Othmer Encyclopedia of Chemical Technology*, 4th edn, Vol 25. Wiley, New York, pp 256–268.

Gordon MH (1996) Dietary antioxidants in disease prevention. *Nat Prod Rep* **13**, 265–273.

Scott G (1995) Antioxidants – the modern elixir? *Chem Brit* 879–882.

Vitamin K

Dowd P, Hershline R, Ham SW and Naganathan S (1994) Mechanism of action of vitamin K. *Nat Prod Rep* **11**, 251–264.

Van Arnum SD (1998) Vitamins (vitamin K). *Kirk–Othmer Encyclopedia of Chemical Technology*, 4th edn, Vol 25. Wiley, New York, pp 269–283.

5

THE MEVALONATE AND DEOXYXYLULOSE PHOSPHATE PATHWAYS: TERPENOIDS AND STEROIDS

The two pathways leading to terpenoids are described: the mevalonate pathway and the recently discovered mevalonate-independent pathway via deoxyxylulose phosphate. Terpenoids may be classified according to the number of isoprenoid units incorporated, and hemiterpenes, monoterpenes and the variants irregular monoterpenes and iridoids, sesquiterpenes, diterpenes, sesterterpenes, triterpenes, tetraterpenes, and higher terpenoids are described in turn, representing groups with increasing numbers of isoprene units. Structures are rationalized through extensive use of carbocation mechanisms and subsequent Wagner–Meerwein rearrangements. Steroids as examples of modified triterpenoids are discussed in detail, including stereochemistry and molecular shape. There follows specific consideration of cholesterol, steroidal saponins, cardioactive glycosides, phytosterols, vitamin D, bile acids, corticosteroids and their semi-synthesis, progestogens, oestrogens, and androgens. Monograph topics giving more detailed information on medicinal agents include volatile oils, pyrethrins, valerian, feverfew, chamomile and matricaria, *Artemisia annua* and artemisinin, gossypol, trichothecenes, *Taxus brevifolia* and taxol, *Ginkgo biloba*, forskolin, liquorice, quillaia, ginseng, vitamin A, cholesterol, dioscorea, fenugreek, sisal, sarsaparilla, yucca, *Digitalis purpurea, Digitalis lanata*, strophanthus, convallaria, squill, soya bean sterols, fusidic acid, vitamin D, bile acids, corticosteroid drugs, progestogen drugs, oestrogen drugs, aromatase inhibitors, oestrogen receptor antagonists, and androgen drugs.

The terpenoids form a large and structurally diverse family of natural products derived from C_5 **isoprene units** (Figure 5.1) joined in a head-to-tail fashion. Typical structures contain carbon skeletons represented by $(C_5)_n$, and are classified as **hemiterpenes** (C_5), **monoterpenes** (C_{10}), **sesquiterpenes** (C_{15}), **diterpenes** (C_{20}), **sesterterpenes** (C_{25}), **triterpenes** (C_{30}) and **tetraterpenes** (C_{40}) (Figure 5.2). Higher polymers are encountered in materials such as rubber. Isoprene itself (Figure 5.1) had been characterized as a decomposition product from various natural cyclic hydrocarbons, and was suggested as the fundamental building block for these compounds, also referred to as 'isoprenoids'. Isoprene is produced naturally but is not involved in the formation of these compounds, and the biochemically active isoprene units were identified as the diphosphate (pyrophosphate) esters **dimethylallyl diphosphate (DMAPP)** and **isopentenyl diphosphate (IPP)** (Figure 5.2). Relatively few of the natural terpenoids conform exactly to the simple concept of a linear head-to-tail combination of isoprene units as seen with **geraniol** (C_{10}), **farnesol** (C_{15}), and **geranylgeraniol** (C_{20}) (Figure 5.3). **Squalene** (C_{30}) and **phytoene** (C_{40}), although formed entirely of isoprene units, display a tail-to-tail linkage at the centre of the molecules. Most terpenoids are modified further by cyclization reactions, but the head-to-tail arrangement of the units can usually still be recognized, e.g. **menthol, bisabolene**, and **taxadiene**. The linear arrangement of

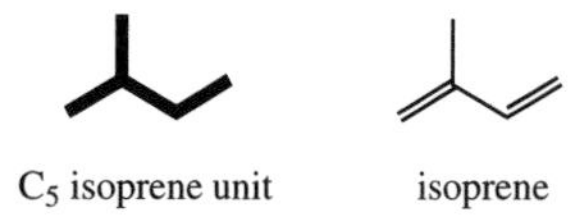

C$_5$ isoprene unit   isoprene

Figure 5.1

isoprene units can be more difficult to appreciate in many other structures when rearrangement reactions have taken place, e.g. steroids, where in addition, several carbons have been lost. Nevertheless, such compounds are formed via regular terpenoid precursors.

Many other natural products contain terpenoid elements in their molecules, in combination with carbon skeletons derived from other sources, such as the acetate and shikimate pathways. Many alkaloids, phenolics, and vitamins discussed in other chapters are examples of this. A particularly common terpenoid fragment in such cases is a

single C$_5$ unit, usually a dimethylallyl substituent, and molecules containing these isolated isoprene units are sometimes referred to as '**meroterpenoids**'. Some examples include furocoumarins (see page 145), rotenoids (see page 155), and ergot alkaloids (see page 368). One should also note that the term '**prenyl**' is in general use to indicate the dimethylallyl substituent. Even macromolecules like proteins can be modified by attaching terpenoid chains. Cysteine residues are alkylated with farnesyl or geranylgeranyl groups, thereby increasing the lipophilicity of the protein and its ability to associate with membranes.

The biochemical isoprene units may be derived by two pathways, by way of intermediates **mevalonic acid** (**MVA**) (Figure 5.4) or 1-deoxy-D-xylulose 5-phosphate (**deoxyxylulose phosphate; DXP**) (Figure 5.6). Mevalonic acid, itself a product of acetate metabolism, had been established as a precursor of the animal sterol cholesterol, and

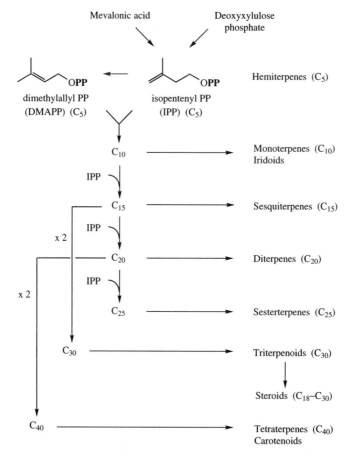

Figure 5.2

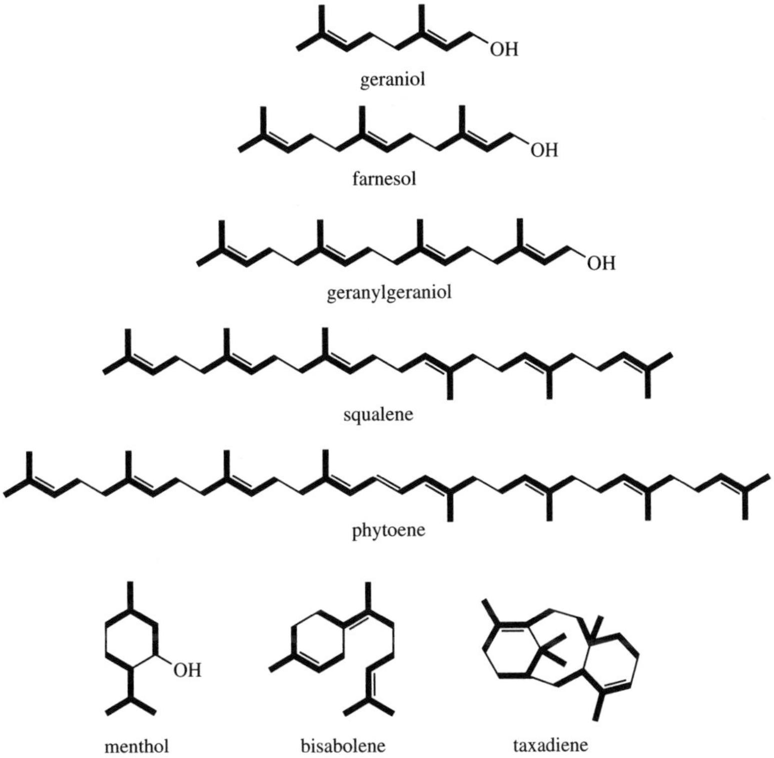

geraniol

farnesol

geranylgeraniol

squalene

phytoene

menthol   bisabolene   taxadiene

Figure 5.3

the steps leading to and from mevalonic acid were gradually detailed in a series of painstakingly executed experiments. For many years, the early parts of the mevalonate pathway were believed to be common to the whole range of natural terpenoid derivatives, but it has since been discovered that an alternative pathway to IPP and DMAPP exists, via deoxyxylulose phosphate, and that this pathway is probably more widely utilized in nature than is the mevalonate pathway. This pathway is also referred to as the **mevalonate-independent pathway** or the **methylerythritol phosphate pathway**.

Three molecules of acetyl-coenzyme A are used to form **mevalonic acid**. Two molecules combine initially in a Claisen condensation to give acetoacetyl-CoA, and a third is incorporated via a stereospecific aldol addition giving the branched-chain ester **β-hydroxy-β-methylglutaryl-CoA** (**HMG-CoA**) (Figure 5.4). This third acetyl-CoA molecule appears to be bound to the enzyme via a thiol group, and this linkage is subsequently hydrolysed to form the free acid group of HMG-CoA. In the acetate pathway, an acetoacetic acid thioester (bound to the acyl carrier protein) would have been formed using the more nucleophilic thioester of malonic acid. The mevalonate pathway does not use malonyl derivatives and it thus diverges from the acetate pathway at the very first step. In the second step, it should be noted that, on purely chemical grounds, acetoacetyl-CoA is the more acidic substrate, and might be expected to act as the nucleophile rather than the third acetyl-CoA molecule. The enzyme thus achieves what is a less favourable reaction. The conversion of HMG-CoA into (3R)-MVA involves a two-step reduction of the thioester group to a primary alcohol, and provides an essentially irreversible and rate-limiting transformation. Drug-mediated inhibition of this enzyme (**HMG-CoA reductase**) can be used to regulate the biosynthesis of mevalonate and ultimately of the steroid cholesterol (see statins, page 112).

The six-carbon compound MVA is transformed into the five-carbon phosphorylated isoprene units in a series of reactions, beginning with

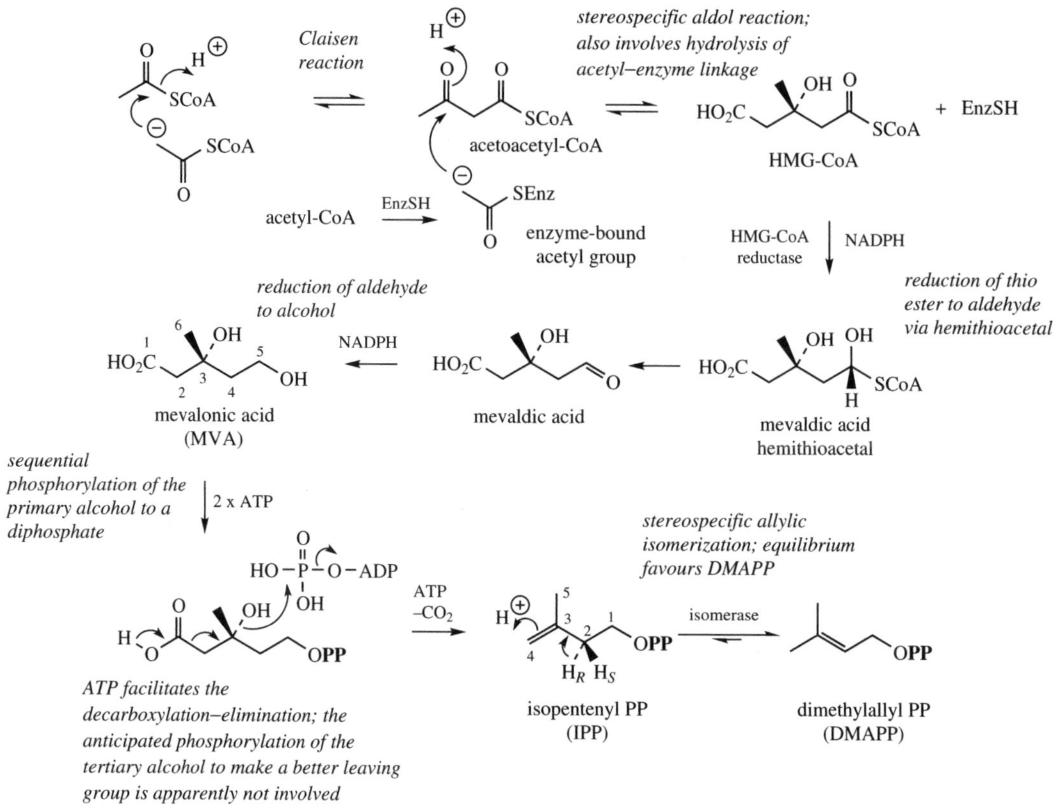

Figure 5.4

phosphorylation of the primary alcohol group. Two different ATP-dependent enzymes are involved, resulting in mevalonic acid diphosphate, and decarboxylation/dehydration then follow to give **IPP**. Whilst a third molecule of ATP is required for this last transformation, there is no evidence for phosphorylation of the tertiary hydroxyl, though this would convert the hydroxyl into a better leaving group. Perhaps ATP assists the loss of the hydroxyl as shown in Figure 5.4. IPP is isomerized to the other isoprene unit, **DMAPP**, by an isomerase enzyme which stereospecifically removes the *pro-R* proton (H_R) from C-2, and incorporates a proton from water on to C-4. Whilst the isomerization is reversible, the equilibrium lies heavily on the side of DMAPP. This conversion generates a reactive electrophile and therefore a good alkylating agent. DMAPP possesses a good leaving group, the diphosphate, and can yield via an S_N1 process an allylic carbocation which is stabilized by charge delocalization (Figure 5.5). In contrast, IPP with its terminal double bond

is more likely to act as a nucleophile, especially towards the electrophilic DMAPP. These differing reactivities are the basis of terpenoid biosynthesis, and carbocations feature strongly in mechanistic rationalizations of the pathways.

1-Deoxy-D-xylulose 5-phosphate is formed from the glycolytic pathway intermediates pyruvic acid and glyceraldehyde 3-phosphate with the loss of the pyruvate carboxyl (Figure 5.6). Thiamine diphosphate-mediated decarboxylation of pyruvate

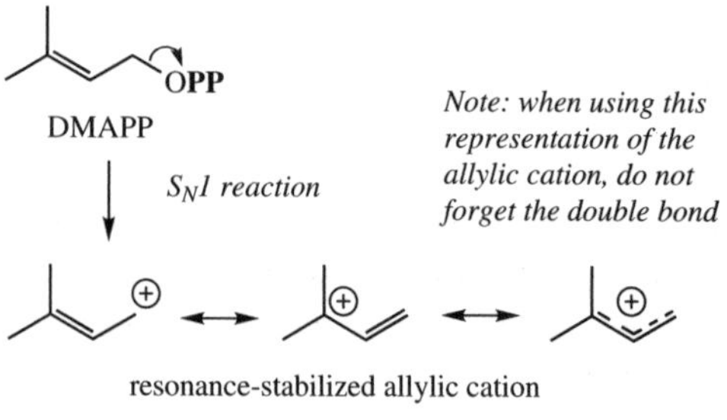

Note: when using this representation of the allylic cation, do not forget the double bond

resonance-stabilized allylic cation

Figure 5.5

(compare page 21) produces an acetaldehyde equivalent bound in the form of an enamine, which reacts as a nucleophile in an addition reaction with the glyceraldehyde 3-phosphate. Subsequent release from the TPP carrier generates **deoxyxylulose phosphate**, which is transformed into **2-C-methyl-D-erythritol 4-phosphate** by a rearrangement reaction, conveniently rationalized as a pinacol-like rearrangement (Figure 5.6), coupled

with a reduction. The expected aldehyde product from the rearrangement step is not detectable, and the single enzyme catalyses the rearrangement and reduction reactions without release of any intermediate. Analogous rearrangements are seen in the biosynthesis of the amino acids valine, leucine, and isoleucine. The methylerythritol phosphate contains the branched-chain system equivalent to the isoprene unit, but the complete sequence of steps

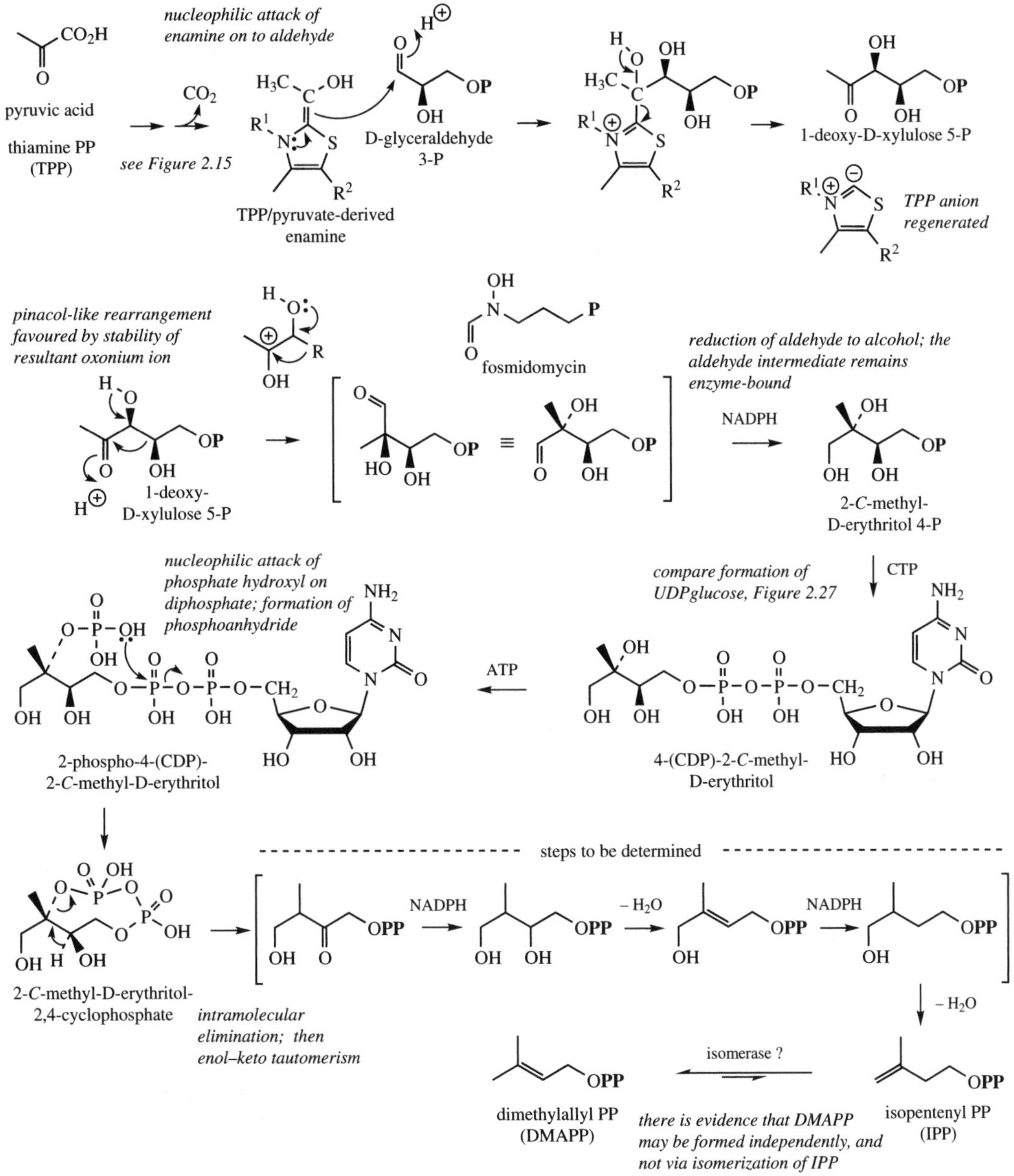

Figure 5.6

leading to the intermediate **isopentenyl phosphate** has yet to be elucidated. Reaction of methylerythritol phosphate with cytidine triphosphate (CTP) produces a cytidine diphospho derivative (compare uridine diphosphoglucose in glycosylation, page 29), which is then phosphorylated via ATP. The resultant 2-phosphate is converted into a cyclic phosphoanhydride with loss of cytidine phosphate. This cyclophosphate, by steps not yet known (a possible sequence is proposed in Figure 5.6), leads to **IPP**, and links the deoxyxylulose pathway with the mevalonate pathway. **DMAPP** may then be derived by isomerism of IPP, or may be produced independently; this also remains to be clarified. Deoxyxylulose phosphate also plays an important role as a precursor of thiamine (vitamin B_1, page 30) and pyridoxol phosphate (vitamin B_6, page 33).

Whether the mevalonate pathway or the deoxyxylulose phosphate pathway supplies isoprene units for the biosynthesis of a particular terpenoid has to established experimentally. Animals appear to lack the deoxyxylulose phosphate pathway, so utilize the mevalonate pathway exclusively. Many other organisms, including plants, are equipped to employ both pathways, often concurrently. In plants, the two pathways appear to be compartmentalized, so that the mevalonate pathway enzymes are localized in the cytosol, whereas the deoxyxylulose phosphate pathway enzymes are found in chloroplasts. Accordingly, triterpenoids and steroids (cytosolic products) are formed by the mevalonate pathway, whilst most other terpenoids are formed in the chloroplasts and are deoxyxylulose phosphate derived. Of course there are exceptions. There are also examples where the two pathways can supply different portions of a molecule, or where there is exchange of late-stage common intermediates between the two pathways resulting in a contribution of isoprene units from each pathway. In the following part of this chapter, these complications will not be considered further, and in most cases there is no need to consider the precise source of the isoprene units. The only area of special pharmacological interest where the early pathway is of particular concern is steroid biosynthesis, which appears to be from mevalonate in the vast majority of organisms. Thus, inhibitors of the mevalonate pathway enzyme HMG-CoA reductase will reduce steroid production, but will

not affect the formation of terpenoids derived via deoxyxylulose phosphate. Equally, it is possible to inhibit terpenoid production without affecting steroid formation by the use of deoxyxyxlulose phosphate pathway inhibitors, such as the antibiotic **fosmidomycin** from *Streptomyces lavendulae*. This acts as an analogue of the rearrangement intermediate (Figure 5.6). Regulation of cholesterol production in humans is an important health concern (see page 236).

HEMITERPENES (C$_5$)

IPP and DMAPP are reactive hemiterpene intermediates in the pathways leading to more complex terpenoid structures. They are also used as alkylating agents in the formation of meroterpenoids as indicated above, but examples of these structures are discussed under the section appropriate to the major substructure, e.g. alkaloids, shikimate, acetate. Relatively few true hemiterpenes are produced in nature, with **isoprene**, a volatile compound which is released by many species of plants, especially trees, being the notable example. Isoprene is formed by loss of a proton from the allylic cation (Figure 5.7).

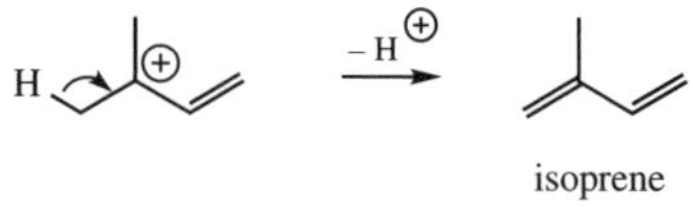

isoprene

Figure 5.7

MONOTERPENES (C$_{10}$)

Combination of DMAPP and IPP via the enzyme prenyl transferase yields **geranyl diphosphate** (**GPP**) (Figure 5.8). This is believed to involve ionization of DMAPP to the allylic cation, addition to the double bond of IPP, followed by loss of a proton. Stereochemically, the proton lost (H_R) is analogous to that lost on the isomerization of IPP to DMAPP. This produces a monoterpene diphosphate, geranyl PP, in which the new double bond is *trans* (*E*). **Linalyl PP** and **neryl PP** are isomers of geranyl PP, and are likely to be formed from geranyl PP by ionization to the allylic cation, which can thus allow a change in attachment of the diphosphate group (to the tertiary carbon in linalyl

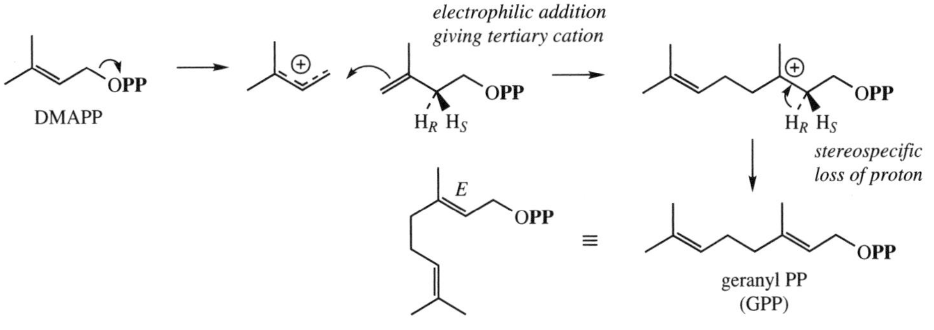

Figure 5.8

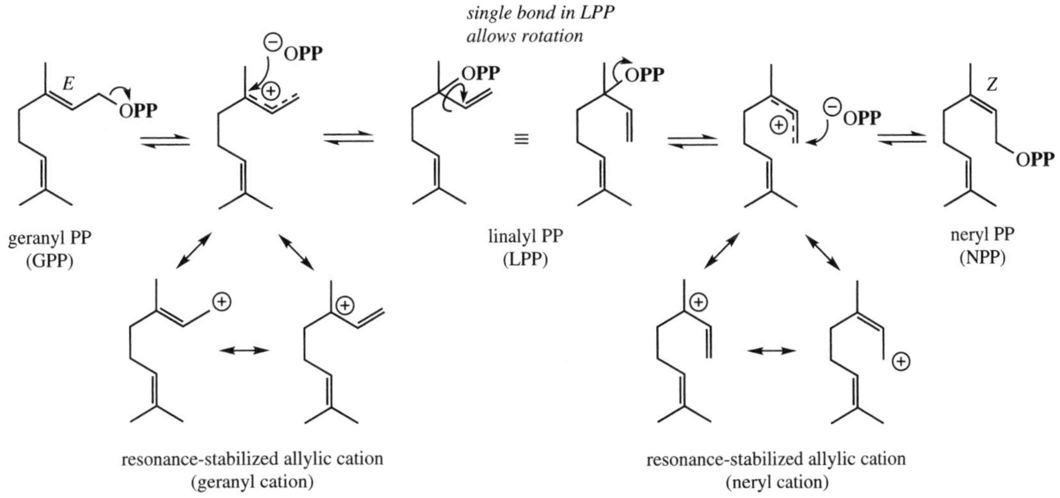

Figure 5.9

PP) or a change in stereochemistry at the double bond (to *Z* in neryl PP) (Figure 5.9). These three compounds, by relatively modest changes, can give rise to a range of linear monoterpenes found as components of volatile oils used in flavouring and perfumery (Figure 5.10). The resulting compounds may be hydrocarbons, alcohols, aldehydes, or perhaps esters, especially acetates.

The range of monoterpenes encountered is extended considerably by cyclization reactions, and monocyclic or bicyclic systems can be created. Some of the more important examples of these ring systems are shown in Figure 5.11. Such cyclizations would not be expected to occur with the precursor geranyl diphosphate, the *E* stereochemistry of the double bond being unfavourable for ring formation (Figure 5.9). Neryl PP or linalyl PP, however, do have favourable stereochemistry, and either or both

of these would seem more immediate precursors of the monocyclic menthane system, formation of which could be represented as shown in Figure 5.12, generating a carbocation (termed menthyl or α-terpinyl) having the menthane skeleton. It has been found that monoterpene cyclase enzymes are able to accept all three diphosphates, with linalyl PP being the best substrate, and it appears they have the ability to isomerize the substrates initially as well as to cyclize them. It is convenient therefore to consider the species involved in the cyclization as the delocalized allylic cation tightly bound to the diphosphate anion, and bond formation follows due to the proximity of the π-electrons of the double bond (Figure 5.12).

In Chapter 2, the possible fates of carbocations were discussed. These include quenching with nucleophiles (especially water), loss of a proton,

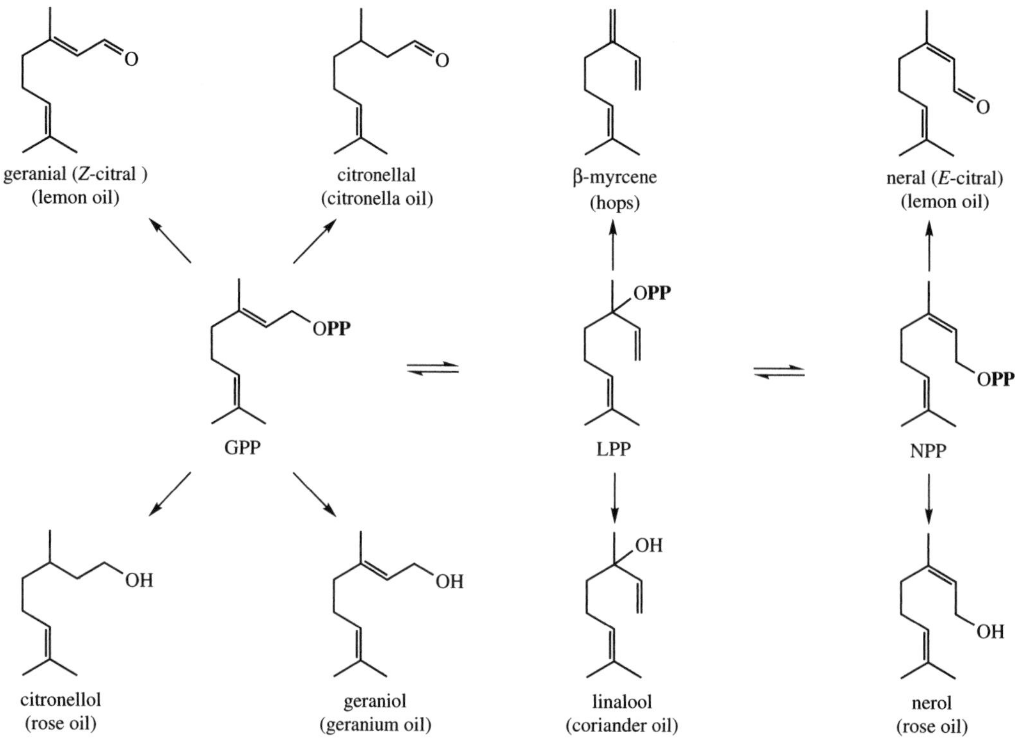

Figure 5.10

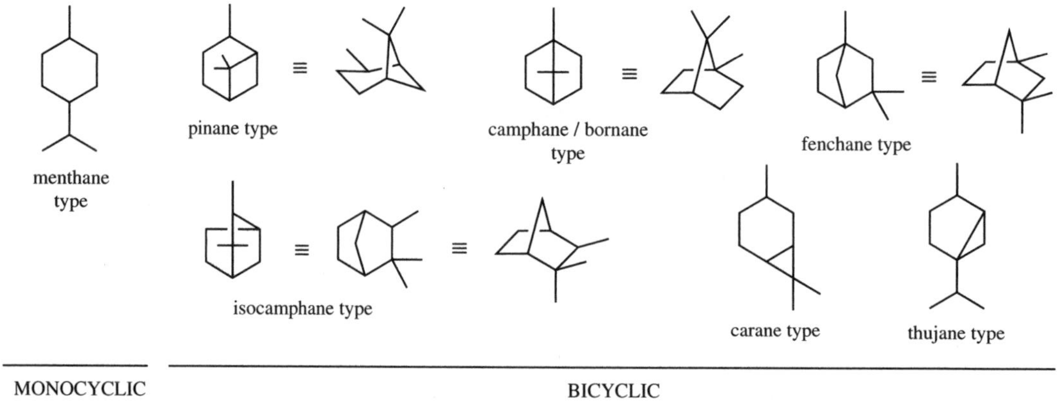

MONOCYCLIC   BICYCLIC

Figure 5.11

cyclization, and the possibility that Wagner–Meerwein rearrangements might occur (see page 15). All feature strongly in terpenoid biosynthesis. The newly generated menthyl cation could be quenched by attack of water, in which case the alcohol **α-terpineol** would be formed, or it could lose a proton to give **limonene** (Figure 5.13). Alternatively, folding the cationic side-chain towards the double bond (via the surface characteristics of the enzyme) would allow a repeat of the cyclization mechanism, and produce bicyclic bornyl and pinyl cations, according to which end of the double bond was involved in forming the new bonds (Figure 5.14). **Borneol** would result from quenching of the bornyl cation with water, and then oxidation of the secondary

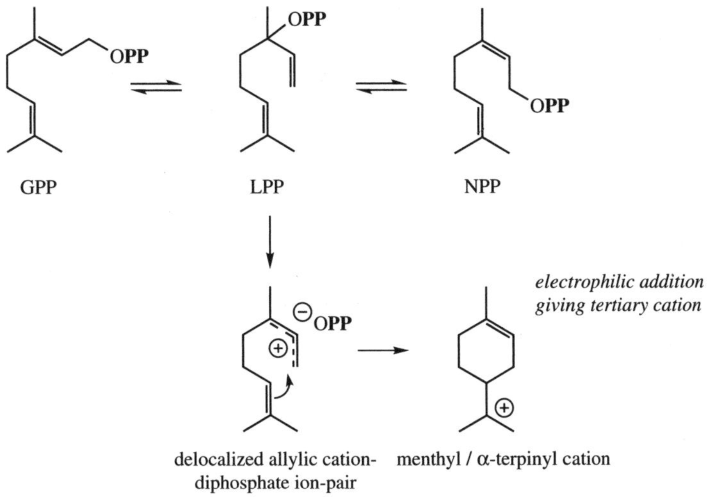

Figure 5.12

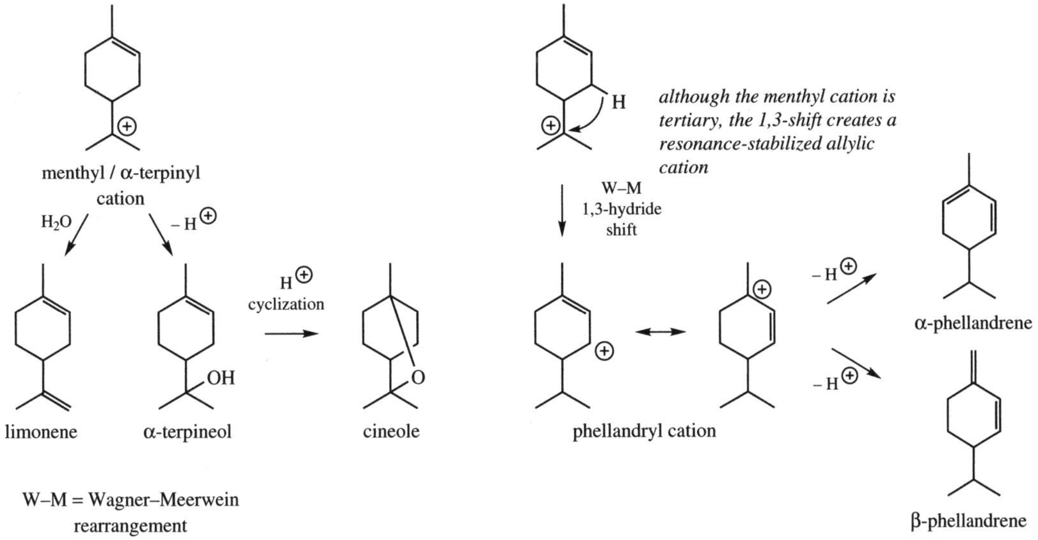

Figure 5.13

alcohol could generate the ketone **camphor**. As an alternative to discharging the positive charge by adding a nucleophile, loss of a proton would generate an alkene. Thus **α-pinene** and **β-pinene** arise by loss of different protons from the pinyl cation, producing the double bonds as cyclic or exocyclic respectively. A less common termination step involving loss of a proton is the formation of a cyclopropane ring as exemplified by **3-carene** and generation of the carane skeleton.

The chemistry of terpenoid formation is essentially based on the reactivity of carbocations,

even though, in nature, these cations may not exist as such discrete species, but rather as tightly bound ion pairs with a counter-anion, e.g. diphosphate. The analogy with carbocation chemistry is justified, however, since a high proportion of natural terpenoids have skeletons which have suffered rearrangement processes. Rearrangements of the Wagner–Meerwein type (see page 15), in which carbons or hydride migrate to achieve enhanced stability for the cation via tertiary against secondary character, or by reduction of ring strain, give a mechanistic rationalization for the biosynthetic pathway. The menthyl cation, although it is a tertiary, may be converted by a 1,3-hydride shift into a favourable resonance-stabilized allylic cation (Figure 5.13). This allows the formation of α- and β-phellandrenes by loss of a proton from the phellandryl carbocation. The bicyclic pinyl cation, with a strained four-membered ring, rearranges to the less strained five-membered fenchyl cation (Figure 5.14), a change which presumably more than makes up for the unfavourable tertiary to secondary carbocation transformation. This produces the fenchane skeleton, exemplified by **fenchol** and **fenchone**. The isocamphyl tertiary carbocation is formed from the bornyl secondary carbocation by a Wagner–Meerwein rearrangement, and so leads to **camphene**. A hydride shift converting the menthyl cation into the terpinen-4-yl cation

only changes one tertiary carbocation system for another, but allows formation of **α-terpinene**, **γ-terpinene**, and the α-terpineol isomer, **terpinen-4-ol**. A further cyclization reaction on the terpinen-4-yl cation generates the thujane skeleton, e.g. **sabinene** and **thujone**. Terpinen-4-ol is the primary antibacterial component of tea tree oil from *Melaleuca alternifolia* (Myrtaceae); thujone has achieved notoriety as the neurotoxic agent in wormwood oil from *Artemisia absinthium* (Compositae/Asteraceae) used in preparation of the drink absinthe, now banned in most countries.

So far, little attention has been given to the stereochemical features of the resultant monoterpene. Individual enzyme systems present in a particular organism will, of course, control the folding of the substrate molecule and thus define the stereochemistry of the final product. Most

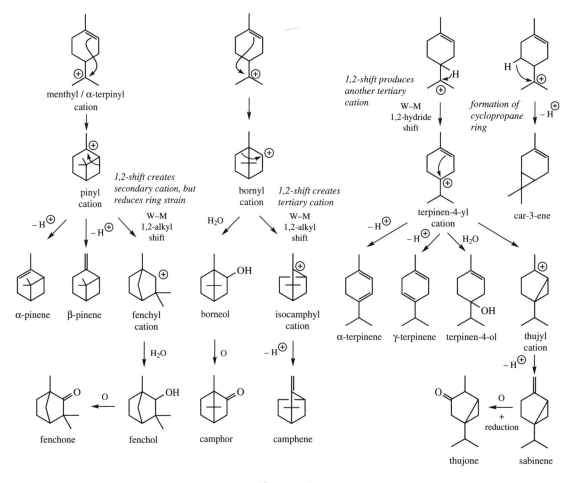

Figure 5.14

monoterpenes are optically active, and there are many examples known where enantiomeric forms of the same compound can be isolated from different sources, e.g. (+)-**camphor** in sage (*Salvia officinalis*; Labiatae/Lamiaceae) and (−)-camphor in tansy (*Tanacetum vulgare*; Compositae/Asteraceae), or (+)-**carvone** in caraway (*Carum carvi*; Umbelliferae/Apiaceae) and (−)-carvone in spearmint (*Mentha spicata*; Labiatae/Lamiaceae). There are also examples of compounds found in both enantiomeric forms in the same organism, examples being (+)- and (−)-**limonene** in peppermint (*Mentha* x *piperita*; Labiatae/Lamiaceae) and (+)- and (−)-**α-pinene** in pine (*Pinus* species; Pinaceae). The individual enantiomers can produce different biological responses, especially towards olfactory receptors in the nose. Thus the characteristic caraway odour is

due to (+)-carvone whereas (−)-carvone smells of spearmint. (+)-Limonene smells of oranges whilst (−)-limonene resembles the smell of lemons. The origins of the different enantiomeric forms of limonene and α-pinene are illustrated in Figure 5.15. This shows the precursor geranyl PP being folded in two mirror image conformations, leading to formation of the separate enantiomers of linalyl PP. Analogous carbocation reactions will then explain production of the optically active monoterpenes. Where a single plant produces both enantiomers, it appears to contain two separate enzyme systems each capable of elaborating a single enantiomer. Furthermore, a single enzyme typically accepts geranyl PP as substrate, catalyses the isomerization to linalyl PP, and converts this into a final product without the release of free intermediates. Sometimes, multiple products in varying

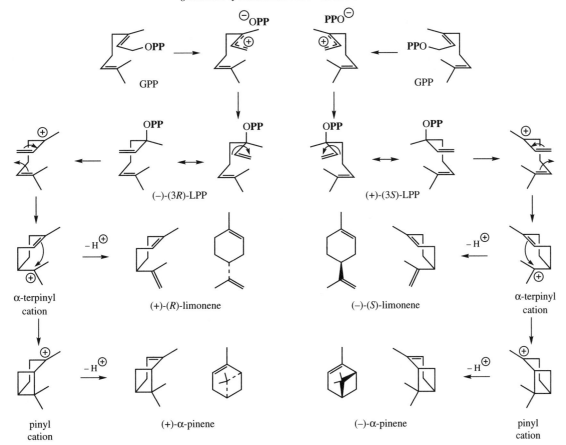

GPP can be folded in two different ways, thus allowing generation of enantiomeric LPP molecules

Figure 5.15

amounts, e.g. limonene, myrcene, α-pinene, and β-pinene, are synthesized by a single enzyme, reflecting the common carbocation chemistry involved in these biosyntheses, and suggesting the enzyme is predominantly providing a suitable environment for the folding and cyclization of the substrate. Subsequent reactions such as oxidation of an alcohol to a ketone, e.g. borneol to **camphor** (Figure 5.14), or heterocyclic ring formation in the conversion of α-terpineol into **cineole** (Figure 5.13), require additional enzyme systems.

In other systems, a particular structure may be found as a mixture of diastereoisomers. Peppermint (*Mentha* x *piperita*; Labiatae/Lamiaceae) typically produces (−)-**menthol**, with smaller amounts of the stereoisomers (+)-**neomenthol**, (+)-**isomenthol**, and (+)-**neoisomenthol**, covering four of the possible eight stereoisomers (Figure 5.16). Oils from various *Mentha* species also contain significant amounts of ketones, e.g. (−)-**menthone**, (+)-**isomenthone**, (−)-**piperitone**, or (+)-**pulegone**. The metabolic relationship of

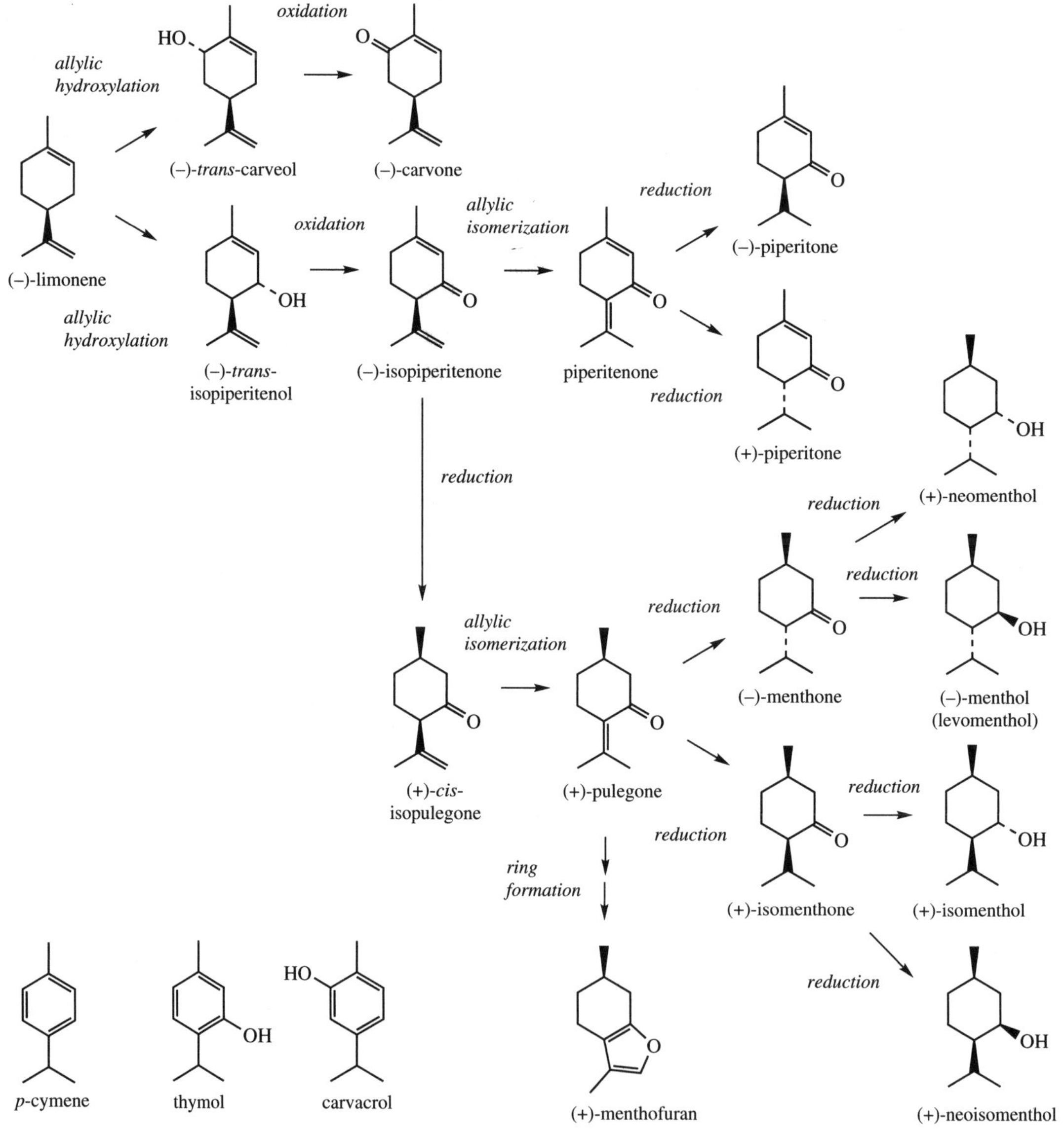

Figure 5.16

Table 5.1 Volatile oils containing principally terpenoid compounds

Major volatile oils have been divided into two groups. Those oils containing principally chemicals which are terpenoid in nature and which are derived by the deoxyxylulose phosphate pathway are given in Table 5.1 below. Oils which are composed predominantly of aromatic compounds which are derived via the shikimate pathway are listed in Table 4.1 on page 139. The introductory remarks to Table 4.1 are also applicable to Table 5.1.

Oils	Plant source	Plant part used	Oil content (%)	Major constituents with typical (%) composition	Uses, notes
Bergamot	*Citrus aurantium* ssp. *bergamia* (Rutaceae)	fresh fruit peel (expression)	0.5	limonene (42) linalyl acetate (27) γ-terpinene (8) linalool (7)	flavouring, aromatherapy, perfumery also contains the furocoumarin bergapten (up to 5%) and may cause severe photosensitization (see page 146)
Camphor oil	*Cinnamomum camphora* (Lauraceae)	wood	1–3	camphor (27–45) cineole (4–21) safrole (1–18)	soaps
Caraway	*Carum carvi* (Umbelliferae/Apiaceae)	ripe fruit	3–7	(+)-carvone (50–70) limonene (47)	flavour, carminative, aromatherapy
Cardamom	*Elettaria cardamomum* (Zingiberaceae)	ripe fruit	3–7	α-terpinyl acetate (25–35) cineole (25–45) linalool (5)	flavour, carminative, ingredient of curries, pickles

(Continued overleaf)

Table 5.1 (*Continued*)

Oils	Plant source	Plant part used	Oil content (%)	Major constituents with typical (%) composition	Uses, notes
Chamomile (Roman chamomile)	*Chamaemelum nobile* (*Anthemis nobilis*) (Compositae/Asteraceae)	dried flowers	0.4–1.5	aliphatic esters of angelic, tiglic, isovaleric, and isobutyric acids (75–85) small amounts of monoterpenes	flavouring, aromatherapy blue colour of oil is due to chamazulene (see page 196)
Citronella	*Cymbopogon winterianus C. nardus* (Graminae/Poaceae)	fresh leaves	0.5–1.2	(+)-citronellal (25–55) geraniol (+)-citronellol (10–15) geranyl acetate (8) (20–40)	perfumery, aromatherapy, insect repellent
Coriander	*Coriandrum sativum* (Umbelliferae/Apiaceae)	ripe fruit	0.3–1.8	(+)-linalool (60–75) γ-terpinene (5) α-pinene (5) camphor (5)	flavour, carminative
Dill	*Anethum graveolens* (Umbelliferae/Apiaceae)	ripe fruit	3–4	(+)-carvone (40–65)	flavour, carminative
Eucalyptus	*Eucalyptus globulus E. smithii E. polybractea* (Myrtaceae)	fresh leaves	1–3	cineole (= eucalyptol) (70–85) α-pinene (14)	flavour, antiseptic, aromatherapy
Eucalyptus (lemon-scented)	*Eucalyptus citriodora* (Myrtaceae)	fresh leaves	0.8	citronellal (65–85)	perfumery

Ginger	*Zingiber officinale* (Zingiberaceae)	dried rhizome	1.5–3	zingiberene (34) β-sesquiphellandrene (12) β-phellandrene (8) β-bisabolene (6)	flavouring the main pungent principles in ginger (gingerols) are not volatile
Juniper	*Juniperus communis* (Cupressaceae)	dried ripe berries	0.5–2	α-pinene (45–80) myrcene (10–25) limonene (1–10) sabinene (0–15)	flavouring, antiseptic, diuretic, aromatherapy juniper berries provide the flavouring for gin
Lavender	*Lavandula angustifolia* L. officinalis (Labiatae/Lamiaceae)	fresh flowering tops	0.3–1	linalyl acetate (25–45) linalool (25–38)	perfumery, aromatherapy inhalation produces mild sedation and facilitates sleep
Lemon	*Citrus limon* (Rutaceae)	dried peel from fruit (expression)	0.1–3	(+)-limonene (60–80) β-pinene (8–12) γ-terpinene (8–10) citral (= geranial + neral) (2–3)	flavouring, perfumery, aromatherapy terpeneless lemon oil is obtained by removing much of the terpenes under reduced pressure; this oil is more stable and contains 40–50% citral

(Continued overleaf)

Table 5.1 (*Continued*)

Oils	Plant source	Plant part used	Oil content (%)	Major constituents with typical (%) composition	Uses, notes
Lemon-grass	*Cymbopogon citratus* (Graminae/Poaceae)	fresh leaves	0.1–0.3	citral (= geranial + neral) (50–85)	perfumery, aromatherapy
Matricaria (German chamomile)	*Matricaria chamomilla* (*Chamomilla recutica*) (Compositae/Asteraceae)	dried flowers	0.3–1.5	(−)-α-bisabolol (10–25%) bisabolol oxides A and B (10–25%) chamazulene (1–15%)	flavouring dark blue colour of oil is due to chamazulene
Orange (bitter)	*Citrus aurantium* ssp. *amara* (Rutaceae)	dried peel from fruit (expression)	0.5–2.5	(+)-limonene (92–94) myrcene (2)	flavouring, aromatherapy the main flavour and odour comes from the minor oxygenated components terpeneless orange oil is obtained by removing much of the terpenes under reduced pressure; this oil contains about 20% aldehydes, mainly decanal
Orange (sweet)	*Citrus sinensis* (Rutaceae)	dried peel from fruit (expression)	0.3	(+)-limonene (90–95) myrcene (2)	flavouring, aromatherapy

Name	Botanical source	Plant part	Yield (%)	Main constituents	Uses
Orange flower (Neroli)	*Citrus aurantium* ssp. *amara* (Rutaceae)	fresh flowers	0.1	linalool (36) β-pinene (16) limonene (12) linalyl acetate (6)	flavour, perfumery, aromatherapy the main flavour and odour comes from the minor oxygenated components terpeneless orange oil is obtained by removing much of the terpenes under reduced pressure; this oil contains about 20% aldehydes, mainly octanal and decanal.
Peppermint	*Mentha* x *piperita* (Labiatae/Lamiaceae)	fresh leaf	1–3	menthol (30–50) menthone (15–32) menthyl acetate (2–10), menthofuran (1–9)	flavouring, carminative, aromatherapy
Pine	*Pinus palustris* or other *Pinus* species (Pinaceae)	needles, twigs		α-terpineol (65)	antiseptic, disinfectant, aromatherapy
Pumilio pine	*Pinus mugo* ssp. *pumilio* (Pinaceae)	needles	0.3–0.4	α- and β-phellandrene (60) α- and β-pinene (10–20) bornyl acetate (3–10)	inhalant the minor components bornyl acetate and borneol are mainly responsible for the aroma

(*Continued overleaf*)

Table 5.1 (*Continued*)

Oils	Plant source	Plant part used	Oil content (%)	Major constituents with typical (%) composition	Uses, notes
Rose (attar of rose, otto of rose)	*Rosa damascena, gallica, R. alba,* and *R. centifolia* (Rosaceae)	fresh flowers	0.02–0.03	citronellol (36) geraniol (17) 2-phenylethanol (3) C_{14}–C_{23} straight chain hydrocarbons (25)	perfumery, aromatherapy
Rosemary	*Rosmarinus officinalis* (Labiatae/Lamiaceae)	fresh flowering tops	1–2	cineole (15–45) α-pinene (10–25) camphor (10–25) β-pinene (8)	perfumery, aromatherapy
Sage	*Salvia officinalis* (Labiatae/Lamiaceae)	fresh flowering tops	0.7–2.5	thujone (40–60) camphor (5–22) cineole (5–14) β-caryophyllene (10) limonene (6)	aromatherapy, food flavouring
Sandalwood	*Santalum album* (Santalaceae)	heartwood	4.5–6.3	sesquiterpenes: α-santalol (50) β-santalol (21)	perfumery, aromatherapy

Spearmint	*Mentha spicata* (Labiatae/Lamiaceae)	fresh leaf	1–2	(−)-carvone (50–70) (−)-limonene (2–25)	flavouring, carminative, aromatherapy
Tea tree	*Melaleuca alternifolia* (Myrtaceae)	fresh leaf	1.8	terpinen-4-ol (30–45) γ-terpinene (10–28) α-terpinene (5–13) *p*-cymene (0.5–12) cineole (0.5–10) α-terpineol (1.5–8)	antiseptic, aromatherapy an effective broad spectrum antiseptic widely used in creams, cosmetics, toiletries
Thyme	*Thymus vulgaris* (Labiatae/Lamiaceae)	fresh flowering tops	0.5–2.5	thymol (40) *p*-cymene (30) linalool (7) carvacrol (1)	antiseptic, aromatherapy, food flavouring
Turpentine oil	*Pinus palustris* and other *Pinus* species (Pinaceae)	distillation of the resin (turpentine) secreted from bark		(+)- and (−)-α-pinene (35:65) (60–70) β-pinene (20–25)	counter-irritant, important source of industrial chemicals residue from distillation is colophony (rosin), composed chiefly of diterpene acids (abietic acids, see page 209)

these various compounds has been established as in Figure 5.16, which illustrates how the stereochemistry at each centre can be established by stereospecific reduction processes on double bonds or carbonyl groups. The pathway also exemplifies that oxygen functions can be introduced into the molecule at positions activated by adjacent double bonds (allylic oxidation), as well as being introduced by quenching of carbocations with water. Thus limonene is a precursor of **carvone** (the main constituent of spearmint oil from *Mentha spicata*) as well as menthone and piperitone, initial hydroxylation occurring at an alternative allylic site on the ring. **Menthofuran** exemplifies a further oxidative modification generating a heterocyclic ring. Both pulegone and menthofuran are considered hepatotoxic. Pulegone is a major constituent of oil of pennyroyal from *Mentha pulegium*, which has a folklore history as an abortifacient. Pulegone is metabolized in humans first to menthofuran, and then to electrophilic metabolites that form adducts with cellular proteins (compare pyrrolizidine alkaloids, page 305).

p-**Cymene**, and the phenol derivatives **thymol** and **carvacrol** (Figure 5.16) found in thyme (*Thymus vulgaris*; Labiatae/Lamiaceae), are representatives of a small group of aromatic compounds that are produced in nature from isoprene units, rather than by the much more common routes to aromatics involving acetate or shikimate (see also cannabinol, page 85, and gossypol, page 200). These compounds all possess the carbon skeleton typical of monocyclic monoterpenes, and their structural relationship to limonene and the more common oxygenated monoterpenes such as menthone or carvone suggests pathways in which additional dehydrogenation reactions are involved.

Data on volatile oils containing terpenoid constituents isolated from these and other plant materials are given in Table 5.1. Volatile oils in which the main components are aromatic and derived from the shikimate pathway are listed in Table 4.1, page 139.

IRREGULAR MONOTERPENES

A number of natural monoterpene structures contain carbon skeletons which, although obviously derived from isoprene C_5 units, do not seem to

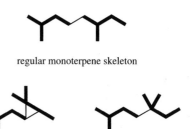

regular monoterpene skeleton

irregular monoterpene skeletons

Figure 5.17

fit the regular head-to-tail coupling mechanism, e.g. those in Figure 5.17. These structures are termed irregular monoterpenes and seem to be limited almost exclusively to members of the plant family the Compositae/Asteraceae. Allowing for possible rearrangements, the two isoprene units appear to have coupled in another manner and this is borne out by information available on their biosynthesis, though this is far from fully understood. Thus, although DMAPP and IPP are utilized in their biosynthesis, geranyl PP and neryl PP do not appear to be involved. Pre-eminent amongst these structures are **chrysanthemic acid** and **pyrethric acid** (Figure 5.18), found in ester form as the **pyrethrins*** (pyrethrins, cinerins, and jasmolins, Figure 5.18), which are valuable insecticidal components in pyrethrum flowers, the flower heads of *Chrysanthemum cinerariaefolium* (Compositae/Asteraceae). These cyclopropane structures are readily recognizable as derived from two isoprene units, and a mechanism for the derivation of chrysanthemic acid is given in Figure 5.19 (compare this mechanism with that involved in the formation of presqualene PP during steroid biosynthesis, page 214). This invokes two DMAPP units joining by a modification of the standard mechanism, with termination achieved by cyclopropane ring formation. Little is known about the origins of **pyrethrolone**, **cinerolone**, and **jasmolone** (Figure 5.18), the alcohol portions of the pyrethrins, though it is possible that these are cyclized and modified fatty acid derivatives, the cyclization resembling the biosynthetic pathway to prostaglandins (see page 53). Thus, α-linolenic acid via 12-oxophytodienoic acid could be the precursor of jasmolone, with β-oxidation and then decarboxylation accounting for the chain shortening (Figure 5.20). Certainly, this type of

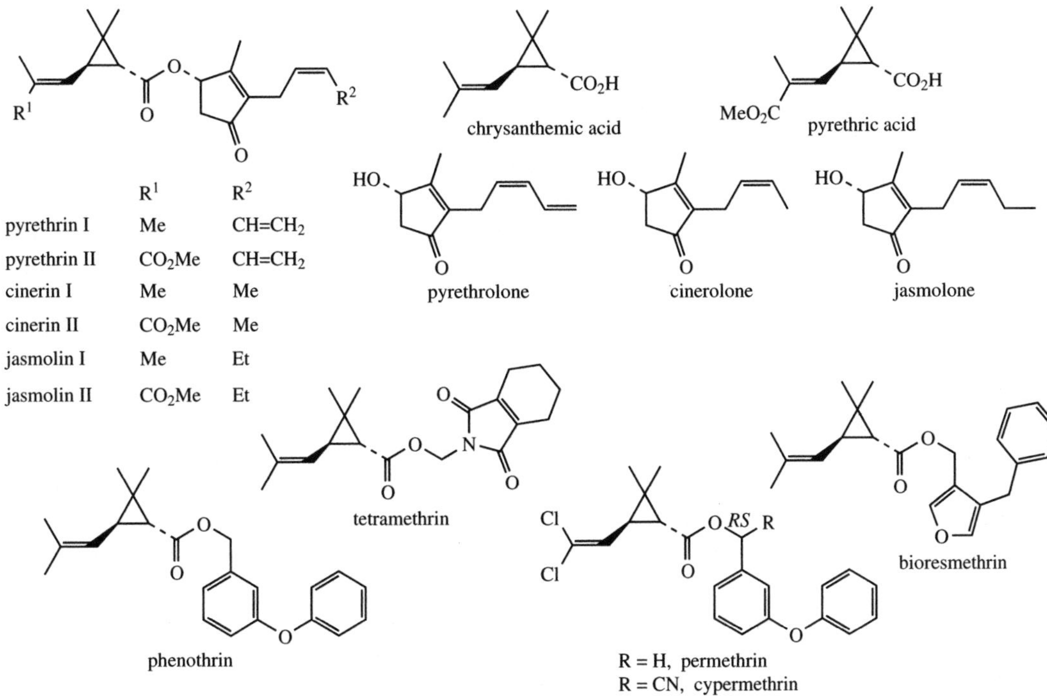

Figure 5.18

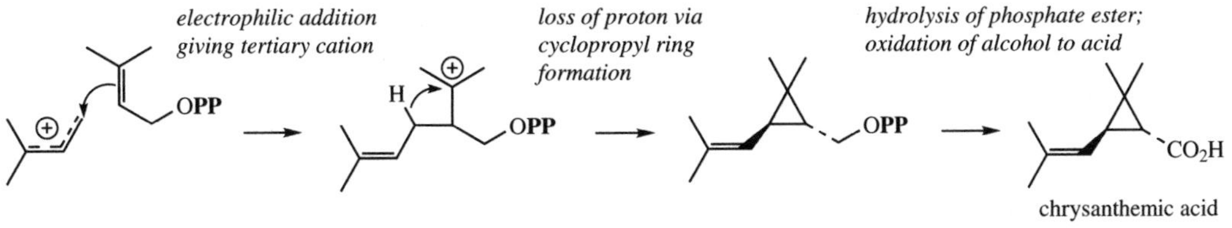

electrophilic addition giving tertiary cation

loss of proton via cyclopropyl ring formation

hydrolysis of phosphate ester; oxidation of alcohol to acid

chrysanthemic acid

Figure 5.19

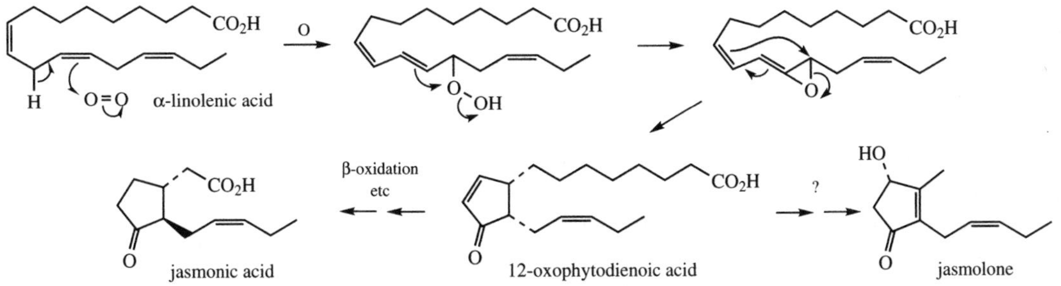

α-linolenic acid

β-oxidation etc

jasmonic acid

12-oxophytodienoic acid

jasmolone

Figure 5.20

pathway operates in the formation of **jasmonic acid** (Figure 5.20), which forms part of a general signalling system in plants, particularly the synthesis of secondary metabolites in response to wounding or microbial infection.

IRIDOIDS (C$_{10}$)

The **iridane** skeleton (Figure 5.21) found in **iridoids** is monoterpenoid in origin and contains a cyclopentane ring which is usually fused

Pyrethrins

The **pyrethrins** are valuable insecticidal components of pyrethrum flowers, *Chrysanthemum cinerariaefolium* (= *Tanacetum cinerariifolium*) (Compositae/Asteraceae). The flowers are harvested just before they are fully expanded, and usually processed to an extract. Pyrethrum cultivation is conducted in East Africa, especially Kenya, and more recently in Ecuador and Australia. The natural pyrethrins are used as a constituent of insect sprays for household use and as post-harvest insecticides, having a rapid action on the nervous system of insects, whilst being biodegradable and non-toxic to mammals, though they are toxic to fish and amphibians. This biodegradation, initiated by air and light, means few insects develop resistance to the pyrethrins, but it does limit the lifetime of the insecticide under normal conditions to just a few hours.

The flowers may contain 0.7–2% of pyrethrins, representing about 25–50% of the extract. A typical pyrethrin extract contains pyrethrin I (35%), pyrethrin II (32%), cinerin I (10%), cinerin II (14%), jasmolin I (5%), and jasmolin II (4%), which structures represent esters of chrysanthemic acid or pyrethric acid with the alcohols pyrethrolone, cinerolone, and jasmolone (Figure 5.18). Pyrethrin I is the most insecticidal component, with pyrethrin II providing much of the rapid knock-down (paralysing) effect. A wide range of synthetic pyrethroid analogues, e.g. **bioresmethrin**, **tetramethrin**, **phenothrin**, **permethrin**, and **cypermethrin** (Figure 5.18), have been developed, which have increased lifetimes up to several days and greater toxicity towards insects. These materials have become widely used household and agricultural insecticides. Tetramethrin, bioresmethrin, and phenothrin are all esters of chrysanthemic acid but with a modified alcohol portion, providing improvements in knock-down effect and in insecticidal activity. Replacement of the terminal methyls of chrysanthemic acid with chlorine atoms, e.g. permethrin, conferred greater stability towards air and light, and opened up the use of pyrethroids in agriculture. Inclusion of a cyano group in the alcohol portion as in cypermethrin improved insecticidal activity several-fold. Modern pyrethroids now have insecticidal activities over a thousand times that of pyrethrin I, whilst maintaining extremely low mammalian toxicity. Permethrin and phenothrin are employed against skin parasites such as head lice.

to a six-membered oxygen heterocycle, e.g. **nepetalactone** from catmint *Nepeta cataria* (Labiatae/Lamiaceae), a powerful attractant and stimulant for cats. The iridoid system arises from geraniol by a type of folding (Figure 5.22) which is different from that already encountered with monoterpenoids, and also different is the lack of phosphorylated intermediates and subsequent carbocation mechanism in its formation. The fundamental cyclization to **iridodial** is formulated as attack of hydride on the dialdehyde, produced by a series of hydroxylation and oxidation reactions on geraniol. Further oxidation gives **iridotrial**, in which hemiacetal formation then leads to production of the heterocyclic ring. In iridotrial, there is an equal chance that the original methyls from the head of geraniol end up as

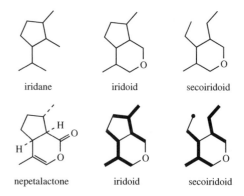

iridane   iridoid   secoiridoid

nepetalactone   iridoid   secoiridoid

Figure 5.21

the aldehyde or in the heterocyclic ring. A large number of iridoids are found as glycosides, e.g. **loganin**, glycosylation effectively transforming the hemiacetal linkage into an acetal. The pathway to loganin involves, in addition, a sequence of reactions in which the remaining aldehyde group is oxidized to the acid and methylated, giving **deoxyloganin**, and the final step is a hydroxylation reaction. Loganin is a key intermediate in the biosynthesis of many other iridoid structures, and also features in the pathway to a range of complex terpenoid indole alkaloids (see page 350) and tetrahydroisoquinoline alkaloids (see page 343). Fundamental in this further metabolism is cleavage of the simple monoterpene

skeleton still recognizable in loganin to give **secologanin**, representative of the **secoiridoids** (Figure 5.21). This is catalysed by a cytochrome P-450-dependent mono-oxygenase, and a free radical mechanism is proposed in Figure 5.22. Secologanin now contains a free aldehyde group, together with further aldehyde and enol groups, these latter two fixed as an acetal by the presence of the glucose. As we shall see with some of the complex alkaloids, these functionalities can be released again by hydrolysing off the glucose and reopening the hemiacetal linkage. **Gentiopicroside**

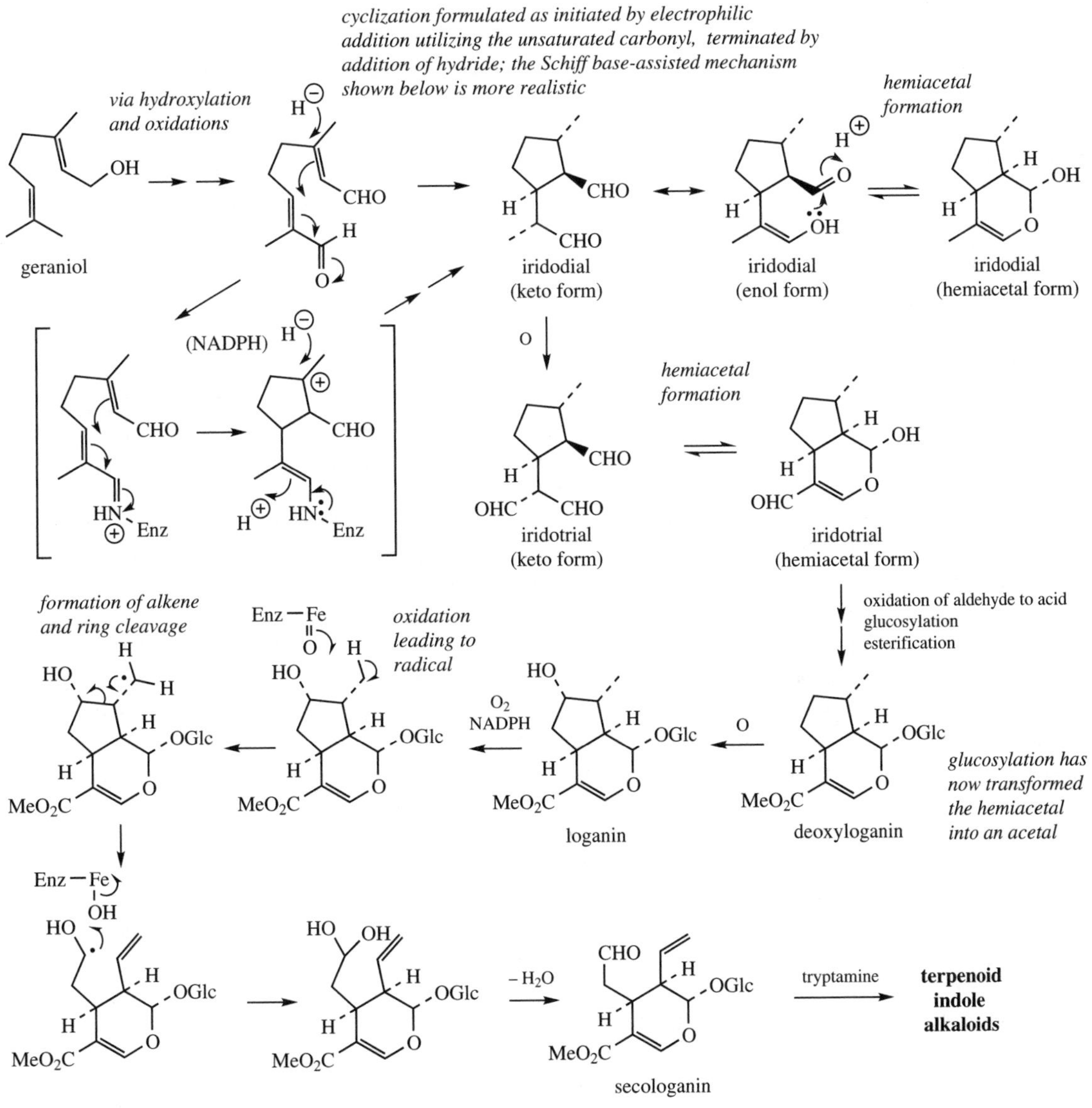

Figure 5.22

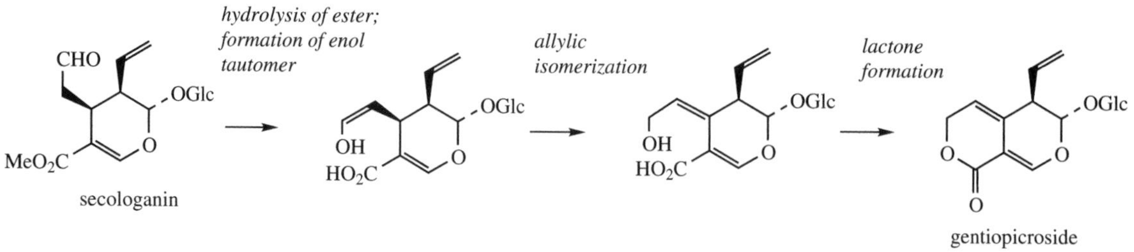

Figure 5.23

(Figure 5.23) is another example of a secoiridoid, and is found in Gentian root (*Gentiana lutea*; Gentianaceae), contributing to the bitter taste of this herbal drug. Its relationship to secologanin is suggested in Figure 5.23. The alkaloid gentianine (see page 386) is also found in Gentian root, and represents a nitrogen analogue of the secoiridoids, in which the pyran oxygen has been replaced by nitrogen.

A range of epoxyiridoid esters has been identified in the drug valerian* (*Valeriana officinalis*; Valerianaceae). These materials, responsible for the sedative activity of the crude drug, are termed **valepotriates**. **Valtrate** (Figure 5.24) is a typical example, and illustrates the structural relationship to loganin, though these compounds contain additional ester functions, frequently isovaleryl. The hemiacetal is now fixed as an ester, rather than as a glycoside.

Valerian

Valerian root consists of the dried underground parts of *Valeriana officinalis* (Valerianaceae), a perennial herb found throughout Europe. Drug material comes from wild and cultivated plants, and is carefully dried at low temperature (less than 40 °C) to minimize decomposition of constituents. Valerian preparations are widely used as herbal tranquillizers to relieve nervous tension, anxiety, and insomnia. Valerian was especially popular during the First World War, when it was used to treat shell-shock. The drug does possess mild sedative and tranquillizing properties, but for maximum activity the roots need to be freshly harvested and carefully dried. The major active principles are generally held to be a number of epoxyiridoid esters called valepotriates (0.5–1.6%), the principal component of which is valtrate (about 80%) (Figure 5.24). Minor valepotriates have the same parent iridoid alcohol as valtrate, but differ with respect to esterifying acids, e.g. isovaltrate (Figure 5.24), or are based on the reduced iridoid seen in didrovaltrate, again with various ester functionalities. Acid entities characterized in this group of compounds are mainly isovaleric (3-methylbutyric) and acetic (as in valtrate/isovaltrate/didrovaltrate), though more complex diester groups involving 3-acetoxyisovaleric and isovaleroxyisovaleric acids are encountered. During drying and storage, some of the valepotriate content may decompose by hydrolysis to liberate quantities of isovaleric acid, giving a characteristic odour, and structures such as baldrinal (Figure 5.24) (from valtrate) and homobaldrinal (from isovaltrate). Samples of old or poorly prepared valerian may contain negligible amounts of valepotriates. Standardized mixtures of valepotriates, containing didrovaltrate (80%), valtrate (15%), and acevaltrate (Figure 5.24) (5%), are available in some countries. These materials are usually extracted from the roots of other species of *Valeriana*, which produce higher amounts of valepotriates than *V. officinalis*, e.g. *V. mexicana* contains up to about 8%. Some other species of *Valeriana* that contain similar valepotriate constituents are used medicinally, including *V. wallichi* (Indian valerian) and *V. edulis* (Mexican valerian).

(Continues)

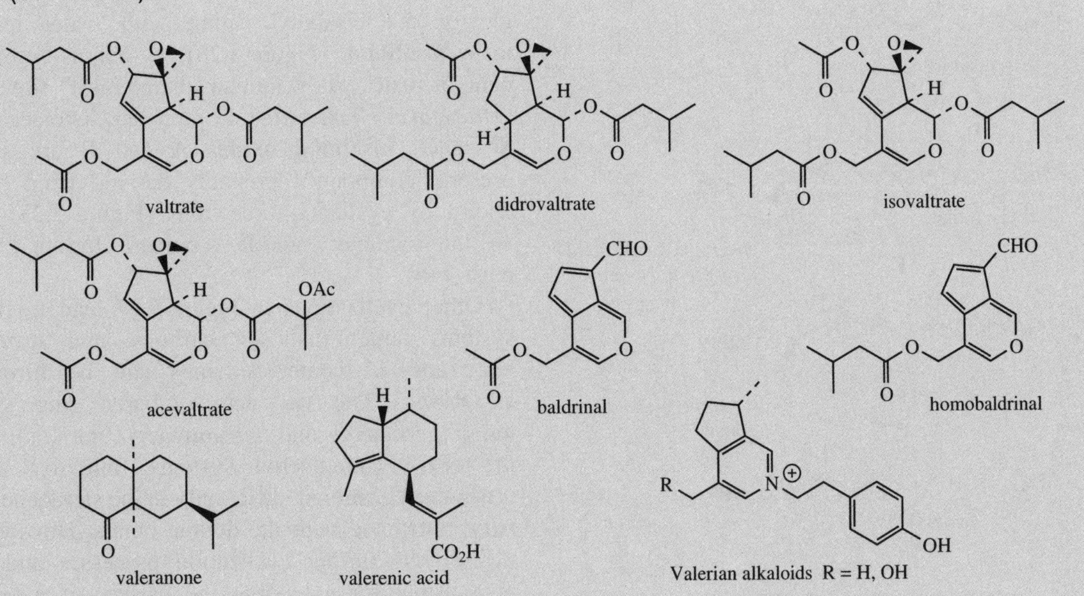

Figure 5.24

Despite the information given above, many workers believe the sedative activity of valerian cannot be due to the valepotriates, which are very unstable and not water soluble. Some of the sedative activity is said to arise from sesquiterpene derivatives such as valerenic acid (about 0.3%) and those constituting the volatile oil content (0.5–1.3%), e.g. valeranone (Figure 5.24), which have been shown to be physiologically active. GABA (γ-aminobutyric acid) and glutamine have also been identified in aqueous extracts of valerian, and these have been suggested to contribute to the sedative properties. Small amounts of some iridoid-related alkaloids (Figure 5.24) have been isolated from valerian root (see page 386). The valepotriates valtrate and didrovaltrate are reported to be cytotoxic *in vitro*, and this may restrict future use of valerian. The reactive epoxide group is likely to be responsible for these cytotoxic properties.

SESQUITERPENES (C$_{15}$)

Addition of a further C$_5$ IPP unit to geranyl diphosphate in an extension of the prenyl transferase reaction leads to the fundamental sesquiterpene precursor, **farnesyl diphosphate (FPP)** (Figure 5.25). Again, an initial ionization of GPP seems likely, and the proton lost from C-2 of IPP is stereochemically analogous to that lost in the previous isoprenylation step. FPP can then give rise to linear and cyclic sesquiterpenes. Because of the increased chain length and additional double bond, the number of possible cyclization modes is also increased, and a huge range of mono-, bi-, and tri-cyclic structures can result. The

stereochemistry of the double bond nearest the diphosphate can adopt an *E* configuration (as in FPP), or a *Z* configuration via ionization, as found with geranyl/neryl PP (Figure 5.26). In some systems, the tertiary diphosphate **nerolidyl PP** (compare linalyl PP, page 172) has been implicated as a more immediate precursor than farnesyl PP (Figure 5.26). This allows different possibilities for folding the carbon chain, dictated of course by the enzyme involved, and cyclization by electrophilic attack on to an appropriate double bond. As with the monoterpenes, standard reactions of carbocations rationally explain most of the common structural skeletons encountered, and a representative selection of these is given in

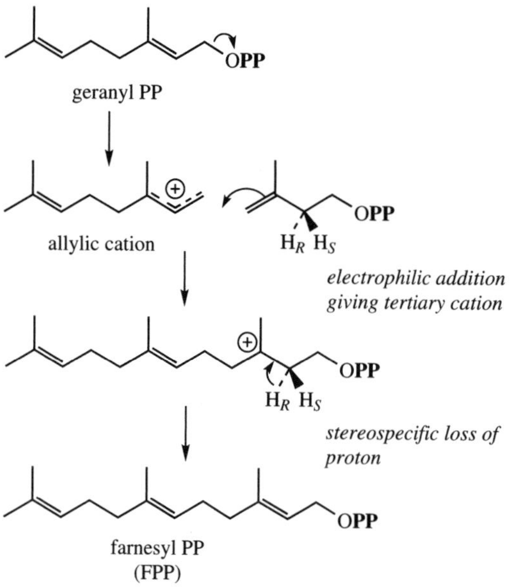

geranyl PP

allylic cation

electrophilic addition giving tertiary cation

stereospecific loss of proton

farnesyl PP
(FPP)

Figure 5.25

Figure 5.27. One of these cyclized systems, the bisabolyl cation, is analogous to the monoterpene menthane system, and further modifications in the six-membered ring can take place to give essentially monoterpene variants with an extended hydrocarbon substituent, e.g. **γ-bisabolene** (Figure 5.28), which contributes to the aroma of ginger (*Zingiber officinale*; Zingiberaceae) along with the related structures such as **zingiberene** and **β-sesquiphellandrene** (Figure 5.29). Sesquiterpenes will in general be

less volatile than monoterpenes. Simple quenching of the bisabolyl cation with water leads to α-**bisabolol** (Figure 5.28), a major component of matricaria (German chamomile)* flowers (*Matricaria chamomilla*; Compositae/Asteraceae). So-called **bisabolol oxides** A and B are also present, compounds probably derived from bisabolol by cyclization reactions (Figure 5.28) on an intermediate epoxide (compare Figure 4.34, page 146).

Other cyclizations in Figure 5.27 lead to ring systems larger than six carbons, and seven-, ten-, and 11-membered rings can be formed as shown. The two ten-membered ring systems (germacryl and *cis*-germacryl cations), or the two 11-membered systems (humulyl and *cis*-humulyl cations), differ only in the stereochemistry associated with the double bonds. However, this affects further cyclization processes and is responsible for extending the variety of natural sesquiterpene derivatives. The germacryl cation, without further cyclization, is a precursor of the germacrane class of sesquiterpenes, as exemplified by **parthenolide** (Figure 5.30), the antimigraine agent in feverfew* (*Tanacetum parthenium*; Compositae/Asteraceae). Parthenolide is actually classified as a germacranolide, the suffix 'olide' referring to the lactone group. Whilst the details of the pathway are not known, a series of simple oxidative transformations (Figure 5.30) can produce the α,β-unsaturated lactone and epoxide groupings.

The α,β-unsaturated lactone functionality is a common feature of many of the biologically

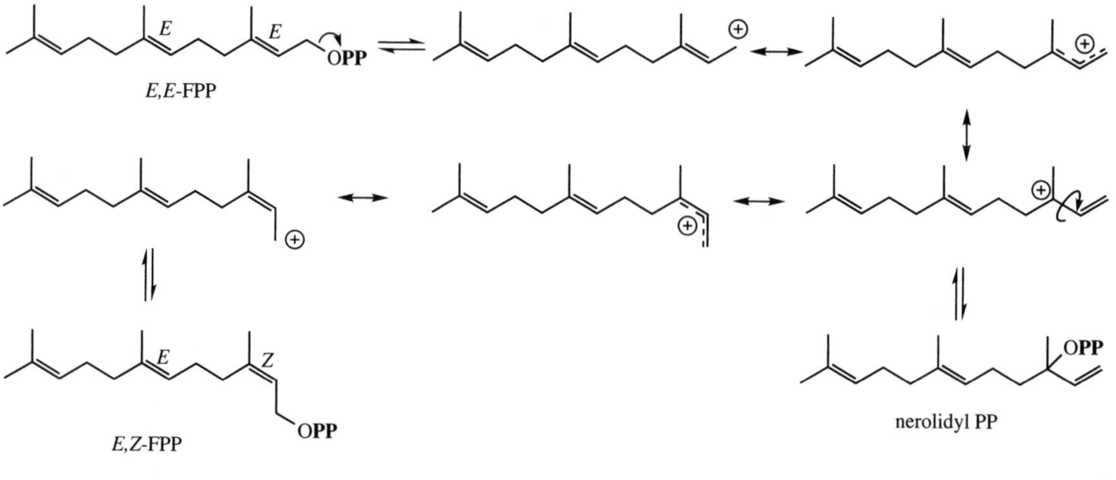

E,E-FPP

E,Z-FPP

nerolidyl PP

Figure 5.26

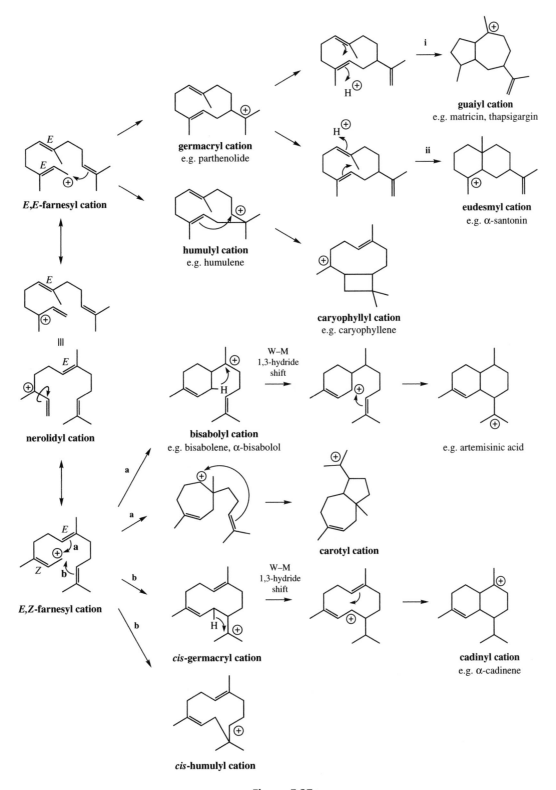

Figure 5.27

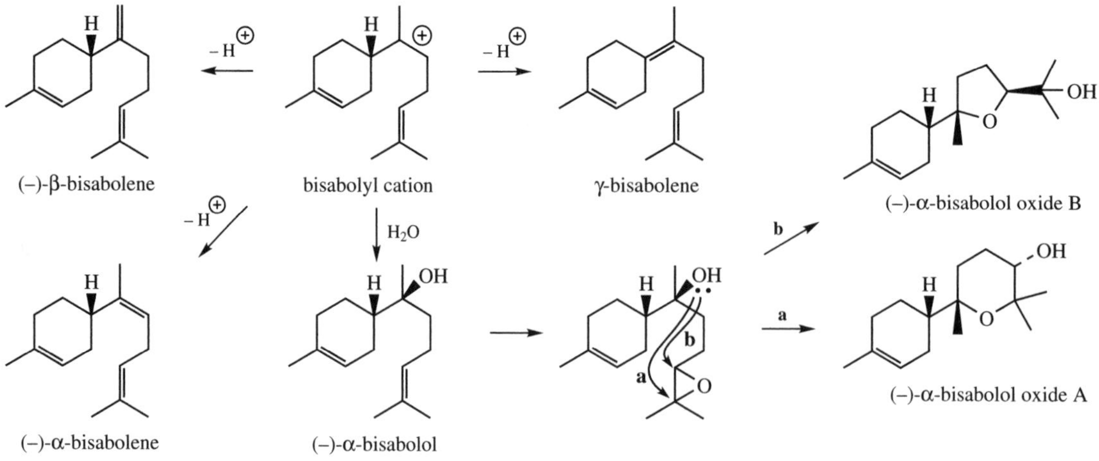

Figure 5.28

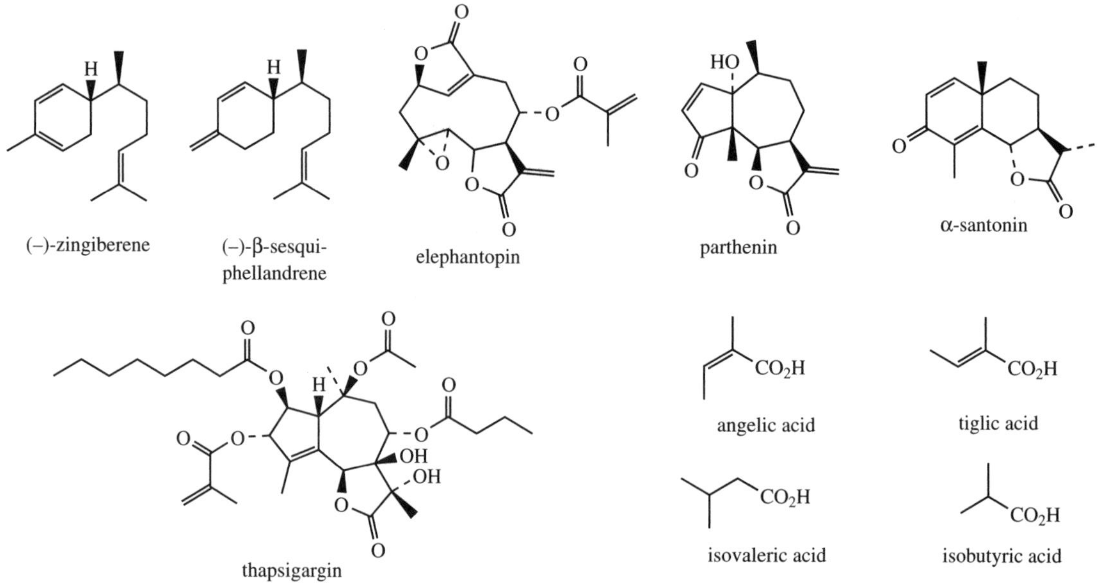

Figure 5.29

active terpenoids. The activity frequently manifests itself as a toxicity, especially cytotoxicity as seen with the germacranolide **elephantopin** (Figure 5.29) from *Elephantopus elatus* (Compositae/Asteraceae), or skin allergies, as caused by the pseudoguaianolide (a rearranged guaianolide) **parthenin** (Figure 5.29) from *Parthenium hysterophorus* (Compositae/Asteraceae), a highly troublesome weed in India. These compounds can be considered as powerful alkylating agents by a

Michael-type addition of a suitable nucleophile, e.g. thiols, on to the α,β-unsaturated lactone. Such alkylation reactions are believed to explain biological activity, and, indeed, activity is typically lost if either the double bond or the carbonyl group is chemically reduced. In some structures, additional electrophilic centres offer further scope for alkylation reactions. In **parthenolide** (Figure 5.31), an electrophilic epoxide group is also present, allowing transannular cyclization and generation of a

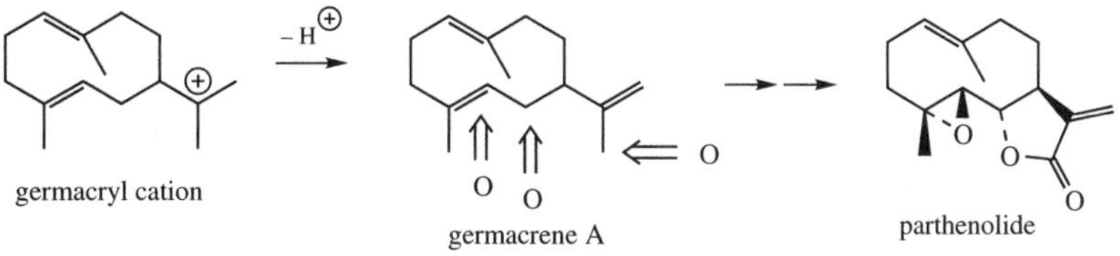

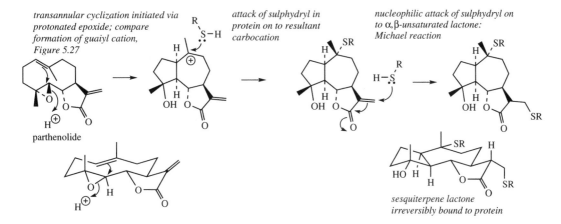

Figure 5.30

Figure 5.31 content:

transannular cyclization initiated via protonated epoxide; compare formation of guaiyl cation, Figure 5.27

parthenolide

attack of sulphydryl in protein on to resultant carbocation

nucleophilic attack of sulphydryl on to α,β-unsaturated lactone: Michael reaction

sesquiterpene lactone irreversibly bound to protein

Figure 5.31

second alkylation site. Cytotoxic agents may irreversibly alkylate critical enzymes that control cell division, whilst allergenic compounds may conjugate with proteins to form antigens which trigger the allergic response. The beneficial effects of parthenolide and structurally related compounds in feverfew have been demonstrated to relate to alkylation of thiol groups.

α-**Santonin** (Figure 5.29) has been identified as the principal anthelmintic component of

Feverfew

Feverfew is a traditional herbal remedy for the relief of arthritis, migraine, toothache, and menstrual difficulties. The plant is a perennial, strongly aromatic herb of the Compositae/Asteraceae family, and has been classified variously as *Tanacetum parthenium*, *Chrysanthemum parthenium*, *Leucanthemum parthenium*, or *Pyrethrum parthenium*, the former name being currently favoured. Studies have confirmed that feverfew is an effective prophylactic treatment in about 70% of migraine sufferers. It reduces the frequency of attacks, the vomiting associated with attacks, and the severity of attacks. The herb has been shown to inhibit blood platelet aggregation, the release of 5-hydroxytryptamine (serotonin) from platelets, the release of histamine from mast cells, and the production of prostaglandins, thromboxanes and leukotrienes. Of a range of sesquiterpene lactones of the germacrane and guianane groups characterized in the leaf material, the principal constituent and major active component is parthenolide (Figure 5.30) (up to about 1% in dried leaves). The powerful pungent odour of the plant arises from the volatile oil constituents, of which the monoterpene camphor (Figure 5.14) is a major constituent. Feverfew may be taken as the fresh leaf, often eaten with bread in the form of a sandwich to minimize the bitter taste, or it can be obtained

(Continues)

(Continued)

in dosage forms as tablets or capsules of the dried powdered leaf. The parthenolide content of dried leaf deteriorates on storage, and many commercial preparations of feverfew have been shown to contain little parthenolide, or to be well below the stated content. This may be a consequence of complexation with plant thiols via Michael addition. Consumers of fresh leaf can be troubled by sore mouth or mouth ulcers, caused by the sesquiterpenes. Parthenolide is also known to be capable of causing some allergic effects, e.g. contact dermatitis. The proposed mechanism of action of parthenolide via alkylation of thiol groups in proteins is shown in Figure 5.31.

---

various *Artemisia* species, e.g. wormseed (*A. cinia*; Compositae/Asteraceae), and has found considerable use for removal of roundworms, although potential toxicity limits its application. Structurally, α-santonin bears much similarity to parthenolide, and the most marked difference lies in the presence of the bicyclic decalin ring system. This basic skeleton, the eudesmane system, is formed from the germacryl cation by protonation and cyclization via the eudesmyl cation (Figure 5.27, route ii), whereas protonation at the more substituted end of a double bond (anti-Markovnikov addition, route i), could generate the guaiyl cation and guaiane skeleton. This latter skeleton is found in **matricin** (Figure 5.32),

again from matricaria* flowers. This compound degrades on heating, presumably by elimination of acetic acid and water, and then decarboxylation to the azulene derivative **chamazulene**, responsible for the blue coloration of oil distilled from the flowers. **Thapsigargin** (Figure 5.29) from *Thapsia garganica* (Umbelliferae/Apiaceae) provides a further example of a guaianolide, esterified with a variety of acid groups. This compound is of considerable pharmacological interest as a tumour promoter, and as a potent activator of cells involved in the inflammatory response.

Another type of decalin-containing sesquiterpene is seen in the structures of α-cadinene and amorpha-4,11-diene. **α-Cadinene** (Figure 5.33) is

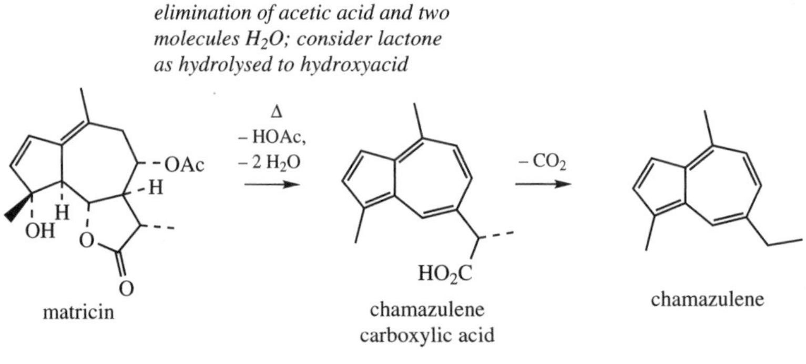

elimination of acetic acid and two molecules H$_2$O; consider lactone as hydrolysed to hydroxyacid

matricin

chamazulene carboxylic acid

chamazulene

Figure 5.32

Chamomile and Matricaria

Two types of **chamomile (camomile)** are commonly employed in herbal medicine, Roman chamomile *Chamaemelum nobile* (formerly *Anthemis nobilis*) (Compositae/Asteraceae), and German chamomile *Matricaria chamomilla* (*Chamomilla recutica*) (Compositae/Asteraceae). German chamomile, an annual plant, is the more important commercially, and is often called **matricaria** to distinguish it from the perennial Roman chamomile. Both plants are cultivated

(Continues)

in various European countries to produce the flowerheads, which are then dried for drug use. Volatile oils obtained by steam distillation or solvent extraction are also available.

Roman chamomile is usually taken as an aqueous infusion (chamomile tea) to aid digestion, curb flatulence, etc, but extracts also feature in mouthwashes, shampoos, and many pharmaceutical preparations. It has mild antiseptic and anti-inflammatory properties. The flowerheads yield 0.4–1.5% of volatile oil, which contains over 75% of aliphatic esters of angelic, tiglic, isovaleric, and isobutyric acids (Figure 5.29), products of isoleucine, leucine, and valine metabolism (see pages 100, 295, 306), with small amounts of monoterpenes and sesquiterpenes. Matricaria is also used as a digestive aid, but is mainly employed for its anti-inflammatory and spasmolytic properties. Extracts or the volatile oil find use in creams and ointments to treat inflammatory skin conditions, and as an antibacterial and antifungal agent. Taken internally, matricaria may help in the control of gastric ulcers. The flowers yield 0.5–1.5% volatile oil containing the sesquiterpenes α-bisabolol (10–25%), bisabolol oxides A and B (10–25%) (Figure 5.28), and chamazulene (0–15%) (Figure 5.32). Chamazulene is a thermal decomposition product from matricin, and is responsible for the dark blue coloration of the oil (Roman chamomile oil contains only trace amounts of chamazulene). α-Bisabolol has some anti-inflammatory, antibacterial, and ulcer-protective properties, but chamazulene is probably a major contributor to the anti-inflammatory activity of matricaria preparations. It has been found to block the cyclooxygenase enzyme in prostaglandin biosynthesis (see page 55) and the anti-inflammatory activity may result from the subsequent inhibition of leukotriene formation.

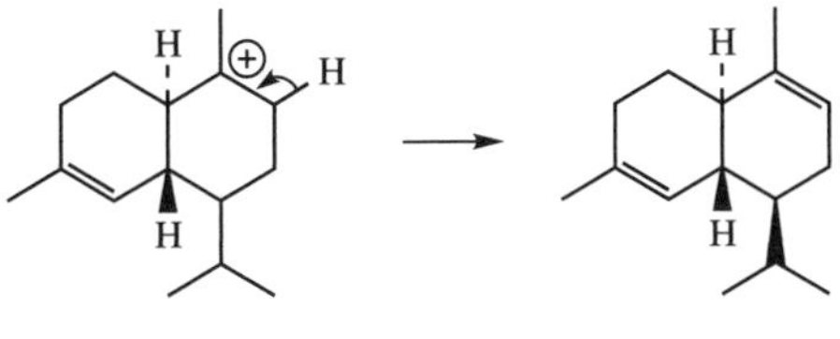

cadinyl cation   α-cadinene

Figure 5.33

one of the many terpenoids found in juniper berries (*Juniperus communis*; Cupressaceae) used in making gin, and this compound is derived from the ten-carbon ring-containing *cis*-germacryl cation. The double bonds in the *cis*-germacryl cation are unfavourably placed for a cyclization reaction as observed with the germacryl cation, and available evidence points to an initial 1,3-shift of hydride to the isopropyl side-chain generating a new cation, and thus allowing cyclization (Figure 5.27). **Amorpha-4,11-diene** (Figure 5.34) is structurally related to α-cadinene, but the different stereochemistry of ring fusion and site of the second double bond is a consequence of a different cyclization mechanism operating

to produce the decalin ring system. In this case, a six-membered ring is most likely formed first giving the bisabolyl cation, and, again, a 1,3-hydride shift is implicated prior to forming the decalin system (Figure 5.27). Amorpha-4,11-diene is an intermediate in the pathway leading to **artemisinin** in *Artemisia annua* (Compositae/Asteraceae) (Figure 5.34). This proceeds through **artemisinic acid** and **dihydroartemisinic acid** via modest oxidation and reduction processes. Dihydroartemisinic acid may be converted chemically into artemisinin by an oxygen-mediated photochemical oxidation under conditions that might normally be present in the plant, suggesting that all further transformations may in fact be non-enzymic. An intermediate in this process also found naturally in *A. annua* is the hydroperoxide of dihydroartemisinic acid. The further modifications postulated in Figure 5.34 include ring expansion by cleavage of this hydroperoxide and a second oxygen-mediated hydroperoxidation. The 1,2,4-trioxane system in artemisinin can be viewed more simply as a combination of hemiketal, hemiacetal, and lactone functions, and the later stages of the pathway merely reflect their construction. Artemisinin* is an important

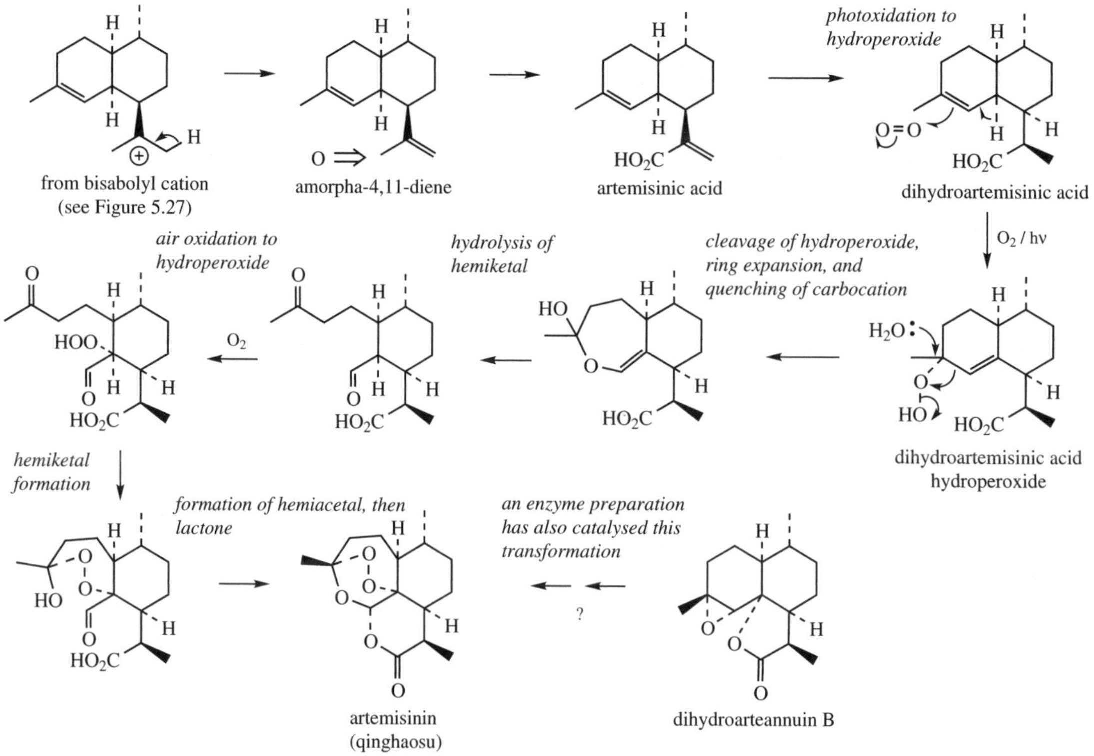

Figure 5.34

antimalarial component in *Artemisia annua**, a Chinese herbal drug. There is currently strong research effort to produce artemisinin or analogues as new antimalarial drugs, since many of the current drugs have become less satisfactory due to resistance (see quinine, page 362).

The 11-carbon ring of the humulyl carbocation may be retained, as in the formation of **humulene** (Figure 5.36), or modified to give the caryophyllyl cation containing a nine-membered ring fused to a four-membered ring, as in **β-caryophyllene** (Figure 5.36). Humulene is found

Artemisia annua and Artemisinin

Artemisia annua (Compositae/Asteraceae) is known as qinghao in Chinese traditional medicine, where it has been used for centuries in the treatment of fevers and malaria. The plant is sometimes called annual or sweet wormwood, and is quite widespread, being found in Europe, North and South America, as well as China. Artemisinin (qinghaosu) (Figure 5.34) was subsequently extracted and shown to be responsible for the antimalarial properties, being an effective blood schizontocide in humans infected with malaria, and showing virtually no toxicity. Malaria is caused by protozoa of the genus *Plasmodium*, especially *P. falciparum*, entering the blood system from the salivary glands of mosquitoes, and world-wide is responsible for 2–3 million deaths each year. Established antimalarial drugs such as chloroquine (see page 363) are proving less effective in the treatment of malaria due to the appearance of drug-resistant strains of *P. falciparum*. Artemisinin is currently effective against these drug-resistant strains.

(Continues)

(*Continued*)

Artemisinin is a sesquiterpene lactone containing a rare peroxide linkage which appears essential for activity. Some plants of *Artemisia annua* have been found to produce as much as 1% artemisinin, but the yield is normally very much less, typically 0.05–0.2%. Apart from one or two low-yielding species, the compound has not been found in any other species of the genus *Artemisia* (about 400 species). Small amounts (about 0.01%) of the related peroxide structure artemisitene (Figure 5.35) are also present in *A. annua*, though this has a lower antimalarial activity. The most abundant sesquiterpenes in the plant are artemisinic acid (arteannuic acid, qinghao acid) (typically 0.2–0.8%) (Figure 5.34), and lesser amounts (0.1%) of arteannuin B (qinghaosu-II) (Figure 5.35). Fortunately, arteannuic acid may be converted chemically into artemisinin by a relatively simple and efficient process. Artemisinin may be reduced to the lactol (hemiacetal) dihydroartemisinin (Figure 5.35), and this has been used for the semi-synthesis of a range of analogues, of which the acetals artemether and arteether (Figure 5.35), and the water-soluble sodium salts of artelinic acid and artesunic acid (Figure 5.35), appear very promising antimalarial agents. These materials have increased activity compared with artemisinin and the chances of infection recurring are also reduced. **Artemether** has rapid action against chloroquinine-resistant *P. falciparum* malaria, and is currently being used as injection formulations. Arteether has similar activity. Being acetals, artemether and arteether are both extensively decomposed in acidic conditions, but are stable in alkali. The ester **artesunic acid** is also used in injection form, but is rather unstable in alkaline solution, hydrolysing to dihydroartemisinin. The ether artelinic acid is considerably more stable. These two compounds have a rapid action and particular application in the treatment of potentially fatal cerebral malaria. Dihydroartemisinin is a more active antimalarial than artemisinin and appears to be the main metabolite of these drugs in the body. They rapidly clear the blood of parasites, but do not have a prophylactic effect. Chemically, these agents are quite unlike any other class of current antimalarial agent, and when thoroughly evaluated, they may well become an important group of drugs in the fight against this life-threatening disease.

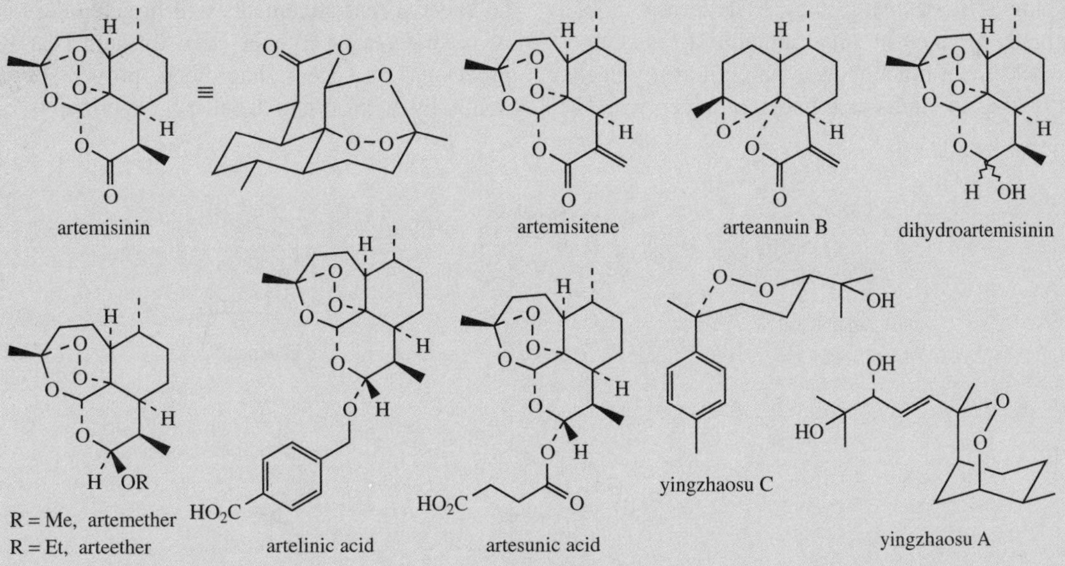

artemisinin   artemisitene   arteannuin B   dihydroartemisinin

R = Me, artemether   artelinic acid   artesunic acid   yingzhaosu C
R = Et, arteether

yingzhaosu A

Figure 5.35

(*Continues*)

(Continued)

The relationship between a peroxide linkage and antimalarial activity is strengthened by the isolation of other sesquiterpene peroxides which have similar levels of activity as artemisinin. Thus, roots of the vine yingzhao (*Artabotrys uncinatus*; Annonaceae), which is also used as a traditional remedy for malaria, contain the bisabolyl derivatives yingzhaosu A and yingzhaosu C (Figure 5.35), the latter containing an aromatic ring of isoprenoid origin (compare the monoterpenes thymol and carvacrol, page 186). Artemisinin, and other peroxide-containing antimalarial agents, appear to complex with haemin, which is a soluble iron–porphyrin material released from haemoglobin as a result of proteolytic digestion by the malarial parasite. This material is toxic to *Plasmodium*, so is normally converted into an insoluble non-toxic form haemozoin (malarial pigment) by enzymic polymerization. Agents like chloroquine (see page 363) interfere with the polymerization process. Complexation of haemin with artemisinin by coordination of the peroxide bridge with the iron atom interrupts the detoxification process and leads to the generation of free radical species through homolytic cleavage of the peroxide. The resulting radicals ultimately damage proteins in *Plasmodium*.

---

in hops (*Humulus lupulus*; Cannabaceae), and β-caryophyllene is found in a number of plants, e.g. in the oils from cloves (*Syzygium aromaticum*; Myrtaceae) and cinnamon (*Cinnamomum zeylanicum*; Lauraceae).

Gossypol* (Figure 5.37) is an interesting and unusual example of a dimeric sesquiterpene in which loss of hydrogen has led to an aromatic system (compare the phenolic monoterpenes thymol and carvacrol, page 186). This material is found in immature flower buds and seeds of the cotton plant (*Gossypium* species; Malvaceae), though originally isolated in small amounts from cottonseed oil. It can function as a male infertility agent, and is used in China as a male contraceptive. The

cadinyl carbocation via **δ-cadinene** is involved in generating the basic aromatic sesquiterpene unit hemigossypol, and then dimerization is simply an example of phenolic oxidative coupling *ortho* to the phenol groups (Figure 5.37).

The formation of sesquiterpenes by a carbocation mechanism means that there is considerable scope for rearrangements of the Wagner–Meerwein type. So far, only occasional hydride migrations have been invoked in rationalizing the examples considered. Obviously, fundamental skeletal rearrangements will broaden the range of natural sesquiterpenes even further. That such processes do occur has been proven beyond doubt by appropriate labelling experiments, and

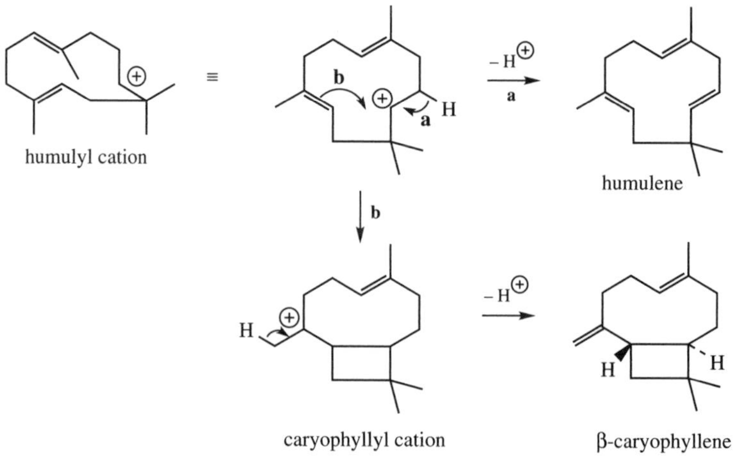

Figure 5.36

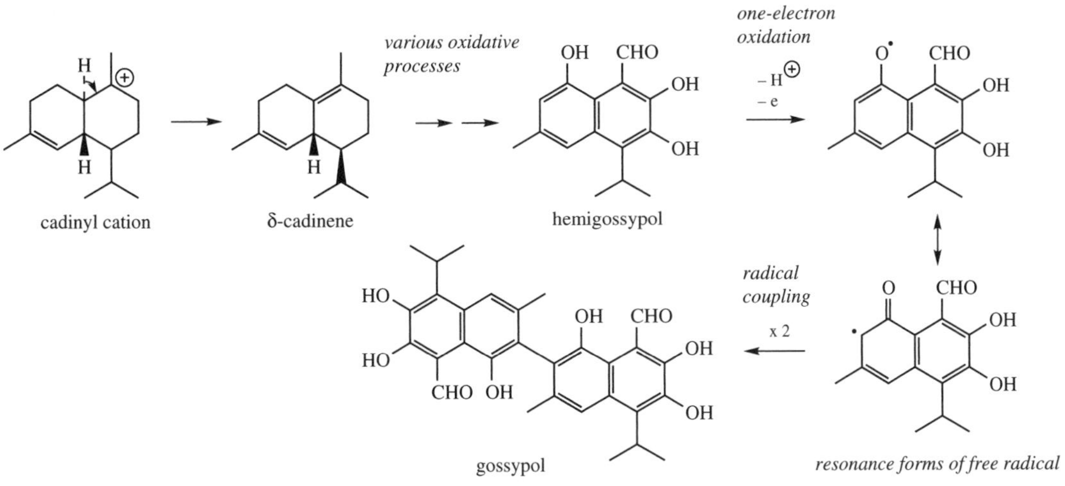

Figure 5.37

Gossypol

Gossypol occurs in the seeds of cotton (*Gossypium* species, e.g. *G. hirsutum*, *G. herbaceum*, *G. arboreum*, *G. barbadense*; Malvaceae) in amounts of 0.1–0.6%. Its contraceptive effects were discovered when subnormal fertility in some Chinese rural communities was traced back to the presence of gossypol in dietary cottonseed oil. Gossypol acts as a male contraceptive, altering sperm maturation, spermatozoid motility, and inactivation of sperm enzymes necessary for fertilization. Extensive clinical trials in China have shown the antifertility effect is reversible after stopping the treatment provided consumption has not been too prolonged. Cases of irreversible infertility have resulted from longer periods of drug use. The molecule is chiral due to restricted rotation, and can thus exist as two atropisomers which do not easily racemize (Figure 5.38). Only the (−)-isomer is pharmacologically active as a contraceptive, whereas most of the toxic symptoms appear to be associated with the (+)-isomer, which also displays antitumour and antiviral activities. Most species of *Gossypium* (except *G. barbadense*) produce gossypol where the (+)-isomer predominates over the (−)-isomer, with amounts varying according to species and cultivar. Racemic (±)-gossypol (but neither of the enantiomers) complexes with acetic acid, so that suitable treatment of cotton seed extracts actually separates the racemate from the excess of (+)-isomer. The racemic form can then be resolved. Other plants in the Gossypieae tribe of the Malvaceae also produce gossypol, with the barks of *Thespia populnea* (3.3%) and *Montezuma speciosissima* (6.1%) being particularly rich sources. Unfortunately, gossypol from these is almost entirely the inactive (+)-form.

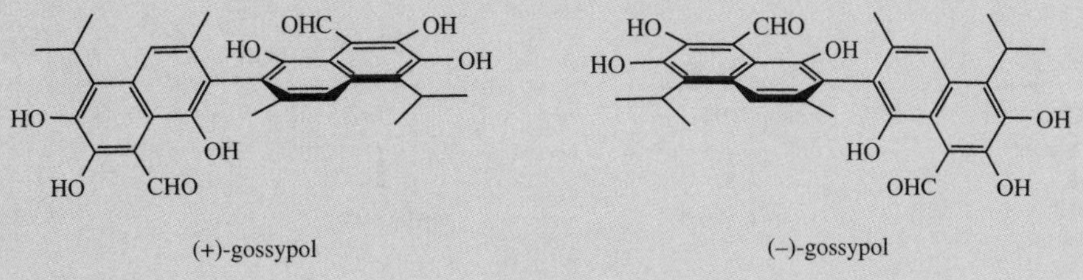

Figure 5.38

a single example will be used as illustration. The **trichothecenes*** are a group of fungal toxins found typically in infected grain foodstuffs. Their name comes from the fungal genus *Trichothecium*, but most of the known structures are derived from cultures of *Fusarium* species. A particularly prominent trichothecene contaminant is **deoxynivalenol** (vomitoxin), which is produced from the less substituted trichothecene **isotrichodermol** by a sequence of oxygenation reactions (Figure 5.39). The trichothecenes have their origins in **nerolidyl diphosphate**, and ring closure of the bisabolyl cation derived from it generates a new carbocation with a five-membered ring (Figure 5.39). At this stage, a series of one hydride and two methyl migrations occur to give a cation, which loses a proton to produce the specific trichothecene precursor **trichodiene**. These migrations are fully backed up by experimental data, and although not immediately predictable, can be rationalized satisfactorily by consideration of the cation suitably bound to the enzyme surface as shown in Figure 5.39. The sequence is initiated by a 1,4-hydride shift which is spatially allowed by the relative proximity of the centres. Two

1,2-methyl shifts then follow, and it is important to note that each migrating group attacks the opposite side of the centre from which the previous group is departing, i.e. inverting the configuration at these centres. Accordingly, a concerted sequence of migrations is feasible, such a process being seen more vividly in the formation of triterpenoids and steroids (see page 216). Loss of a proton and generation of a double bond terminates the process giving trichodiene. Oxygenation of trichodiene gives, in several steps, **isotrichotriol**. Two of the hydroxylations are at activated allylic positions; hydroxylation on the five-membered ring will therefore occur before the epoxidation. Ether formation, involving perhaps protonation, loss of water and generation of an allylic cation, completes the pathway to the basic trichothecene structure as in **isotrichodermol**.

Finally, it is worth noting how many of the sesquiterpene derivatives described above are found in plants belonging to the daisy family, the Compositae/Asteraceae. Whilst sesquiterpenes are by no means restricted to this family, the Compositae/Asteraceae undoubtedly provides a very rich source.

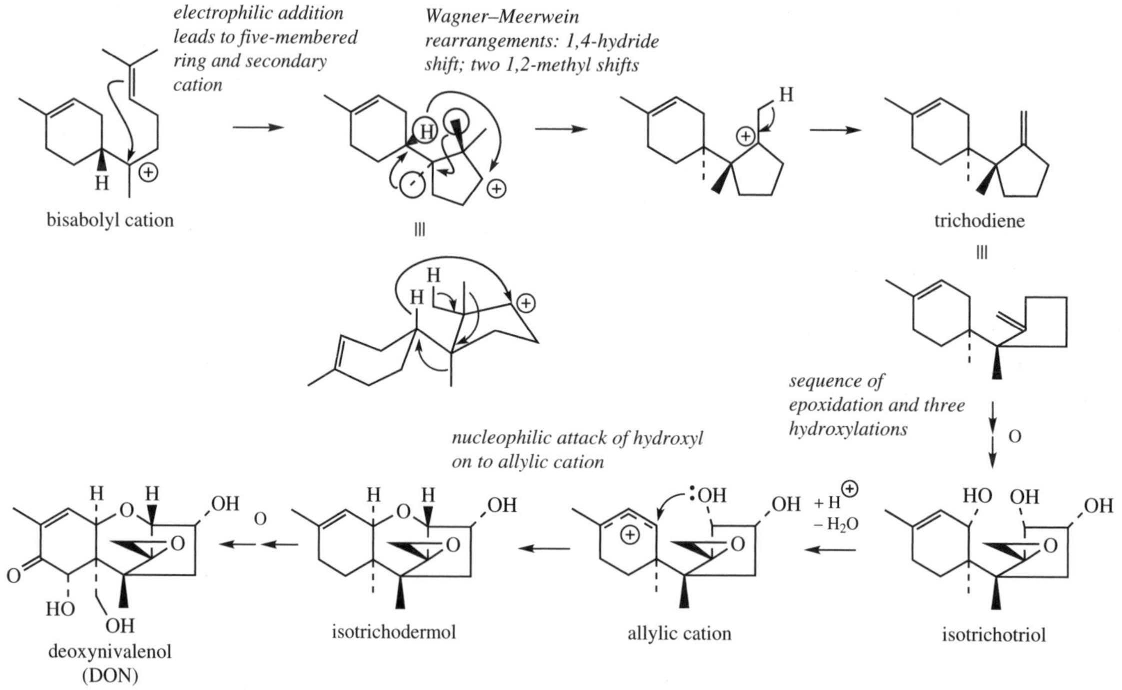

Figure 5.39

Trichothecenes

The trichothecenes are a group of sesquiterpene toxins produced by several fungi of the genera *Fusarium*, *Myrothecium*, *Trichothecium*, and *Trichoderma*, which are parasitic on cereals such as maize, wheat, rye, barley, and rice. About 150 different structures have been identified, with some of these being isolated from plants of the genus *Baccharis* (Compositae/Asteraceae), where a symbiotic plant–fungus relationship may account for their production. Examples of trichothecene structures commonly encountered as food contaminants include deoxynivalenol (DON) (Figure 5.39), and diacetoxyscirpenol (DAS), T-2 toxin, and verrucarin A (Figure 5.40). The double bond and the epoxide group in the basic trichothecene skeleton are essential for toxicity, and the number of oxygen substituents and ester functions also contribute. Macrocyclic ester functions as seen in verrucarin A tend to produce the most toxic examples. Although these compounds are more toxic when injected, oral toxicity is relatively high, and lethal amounts can easily be consumed because of the nature of the host plants. They are sufficiently toxic to warrant routine analysis of foodstuffs such as wheat and flour, and also flour-derived products, e.g. bread, since they survive processing and the high temperatures used in baking. DON levels above 1 ppm are considered hazardous for human consumption. It is relevant to note that when mammals ingest these compounds, a degree of de-epoxidation can occur, ascribed to gut microflora, thus providing some detoxification by removing a structural feature necessary for toxicity.

As their main mechanism of action, these compounds inhibit protein biosynthesis by binding to the ribosome and inhibiting peptidyl transferase activity (see page 407). They also inhibit DNA biosynthesis. A major human condition known to be caused by trichothecenes is alimentary toxic aleukia (ATA), characterized by destruction of the skin, haemorrhaging, inflammation, sepsis, a decrease in red and white blood corpuscles, bone marrow atrophy, and a high mortality rate. A severe outbreak of ATA was recorded in the former Soviet Union shortly after the Second World War when food shortages necessitated the consumption of grain that had overwintered in the field. This had become badly contaminated with *Fusarium sporotrichioides* and hence T-2 toxin. It is estimated that tens of thousands died as a result.

Many trichothecene derivatives have been tested as potential anticancer agents but have proved too toxic for clinical use.

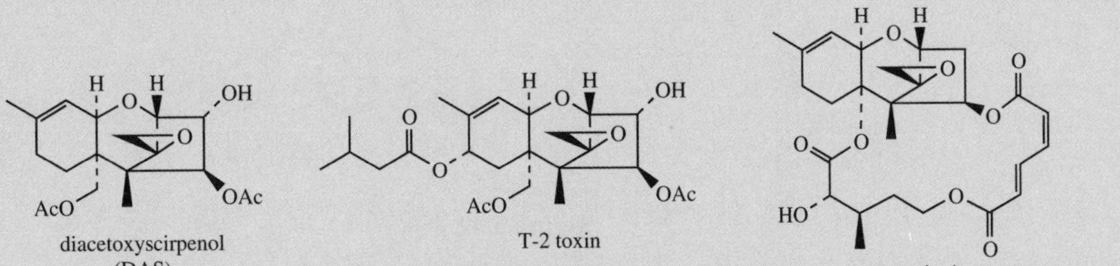

diacetoxyscirpenol
(DAS)

T-2 toxin

verrucarin A

Figure 5.40

DITERPENES (C$_{20}$)

The diterpenes arise from **geranylgeranyl diphosphate** (**GGPP**), which is formed by addition of a further IPP molecule to farnesyl diphosphate in the same manner as described for the lower terpenoids (Figure 5.41). One of the simplest and most important of the diterpenes is **phytol** (Figure 5.42), a reduced form of geranylgeraniol, which forms the lipophilic side-chain of the chlorophylls, e.g. **chlorophyll *a*** (Figure 5.42). Related haem molecules, porphyrin components

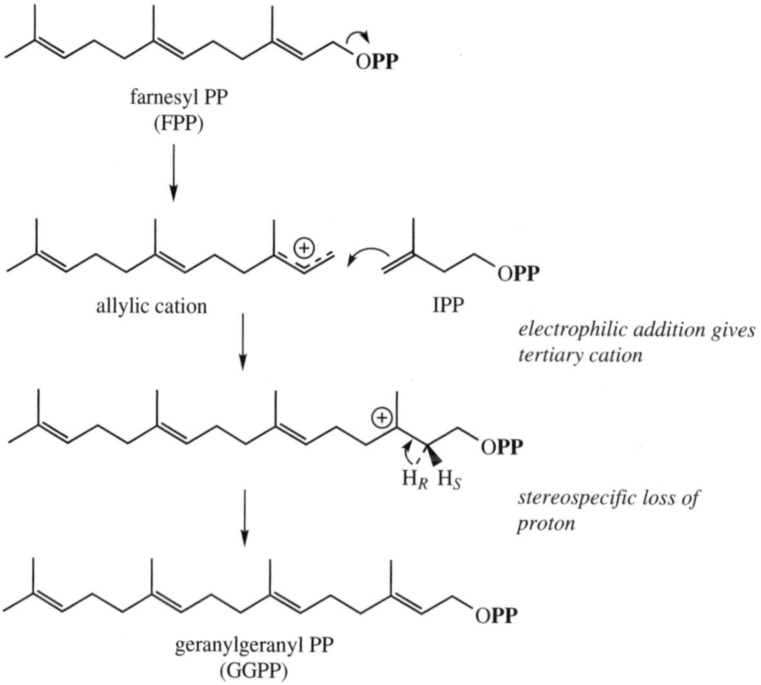
OPP
farnesyl PP
(FPP)
allylic cation
IPP
OPP
electrophilic addition gives tertiary cation
OPP
H_R H_S
stereospecific loss of proton
OPP
geranylgeranyl PP
(GGPP)

Figure 5.41

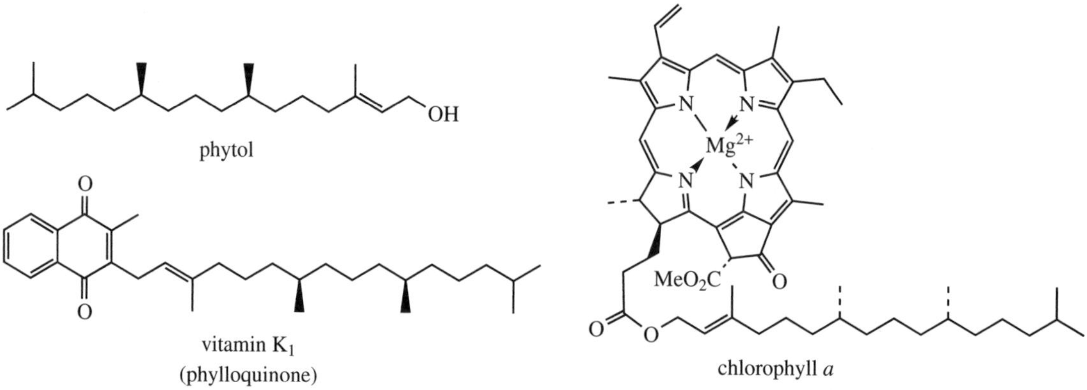
OH
phytol
O
O
vitamin K_1
(phylloquinone)
N
N
Mg^{2+}
N
N
MeO_2C
O
O
O
chlorophyll *a*

Figure 5.42

of haemoglobin, lack such lipophilic side-chains. Available evidence suggests that geranylgeranyl diphosphate is involved in forming the ester linkage, and the three reduction steps necessary to form the phytol ester occur after attachment to the chlorophyll molecule. A phytyl substituent is also found in **vitamin K_1 (phylloquinone)** (Figure 5.42), a naphthoquinone derivative found in plants, though other members of the vitamin K group (**menaquinones**) from bacteria have unsaturated terpenoid side-chains of variable length. The phytyl group of phylloquinone is introduced by alkylation of dihydroxynaphthoic acid with phytyl diphosphate and a similar phytylation of homogentisic acid features in the formation of the E group vitamins (tocopherols). These compounds are discussed further under shikimate derivatives (see page 158).

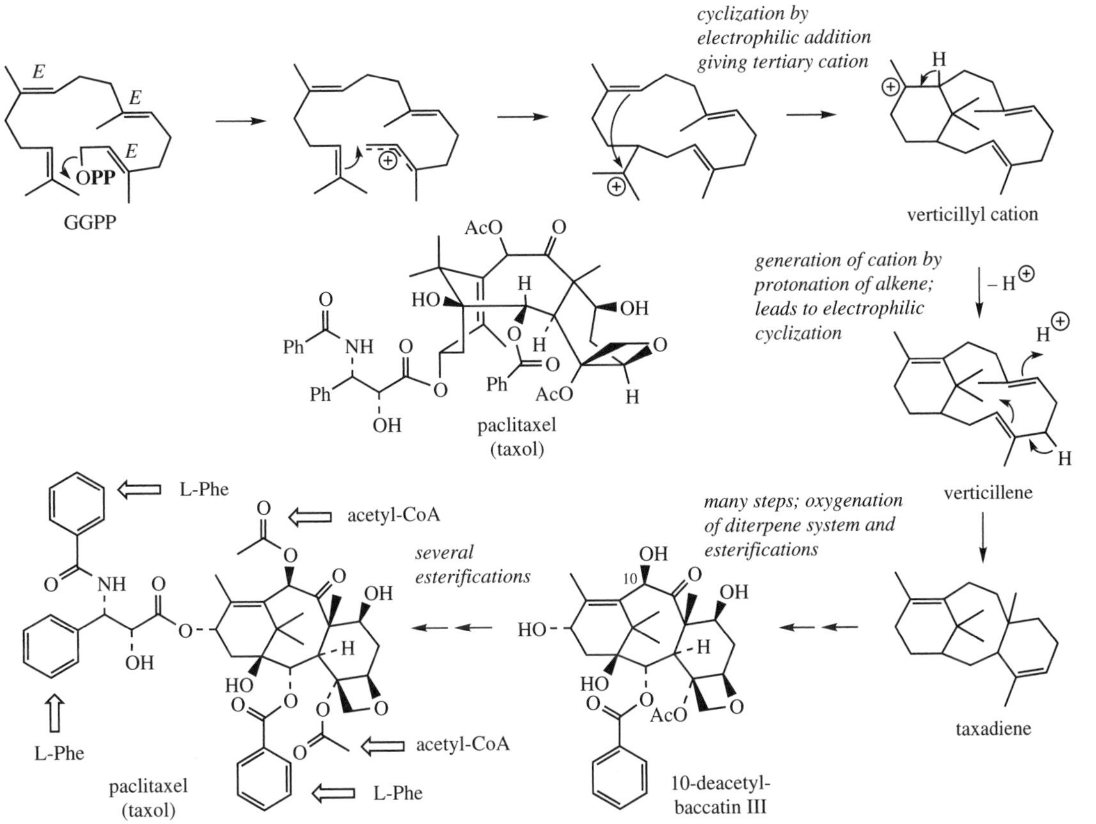

Figure 5.43

Cyclization reactions of GGPP mediated by carbocation formation, plus the potential for Wagner–Meerwein rearrangements, will allow many structural variants of diterpenoids to be produced. The toxic principle 'taxine' from common yew (*Taxus baccata*; Taxaceae) has been shown to be a mixture of at least eleven compounds based on the **taxadiene** skeleton which can be readily rationalized as in Figure 5.43, employing the same mechanistic principles as seen with mono- and sesqui-terpenes.

Although these compounds are sometimes classified as diterpenoid alkaloids, the nitrogen atom is not incorporated into the diterpene skeleton, as exemplified by **taxol (paclitaxel)*** (Figure 5.43) from Pacific yew (*Taxus brevifolia*)*. The side-chains in taxol containing aromatic rings are derived from shikimate via phenylalanine. Taxol is an important new anticancer agent, with a broad spectrum of activity against some cancers which do not respond to other agents.

Taxus brevifolia and Taxol (Paclitaxel)

A note on nomenclature: the name taxol was given to a diterpene ester with anticancer properties when it was first isolated in 1971 from *Taxus brevifolia*. When the compound was subsequently exploited commercially as a drug, Taxol was registered as a trademark. Accordingly, the generic name paclitaxel has been assigned to the compound. The literature now contains an unhappy mixture of the two names, though the original name taxol is most often employed.

(Continues)

(Continued)

The anticancer drug taxol (Figure 5.43) is extracted from the bark of the Pacific yew, *Taxus brevifolia* (Taxaceae), a slow growing shrub/tree found in the forests of North-West Canada (British Columbia) and the USA (Washington, Oregon, Montana, Idaho, and North California). Although the plant is not rare, it does not form thick populations, and needs to be mature (about 100 years old) to be large enough for exploitation of its bark. The wood of *T. brevifolia* is not suitable for timber, and in some areas, plants have been systematically destroyed to allow cultivation of faster-growing commercially exploitable conifers. Harvesting is now strictly regulated, but it is realized that this will not provide a satisfactory long term supply of the drug. The bark from about three mature 100-year-old trees is required to provide one gram of taxol, and a course of treatment may need 2 grams of the drug. Current demand for taxol is in the region of 100–200 kg per annum.

All parts of *Taxus brevifolia* contain a wide range of diterpenoid derivatives termed taxanes, which are structurally related to the toxic constituents found in other *Taxus* species, e.g. the common yew, *Taxus baccata*. Over a hundred taxanes have been characterized from various *Taxus* species, and taxol is a member of a small group of compounds possessing a four-membered oxetane ring and a complex ester side-chain in their structures, both of which are essential for antitumour activity. Taxol is found predominantly in the bark of *T. brevifolia*, but in relatively low amounts (about 0.01–0.02%). Up to 0.033% of taxol has been recorded in some samples of leaves and twigs, but generally the taxol content is much lower than in the bark. The content of some other taxane derivatives in the bark is considerably higher, e.g. up to 0.2% baccatin III (Figure 5.44). Other taxane derivatives characterized include 10-deacetyltaxol, 10-deacetylbaccatin III, cephalomannine and 10-deacetylcephalomannine. A more satisfactory solution currently exploited for the supply of taxol and derivatives for drug use is to produce these compounds by semi-synthesis from more accessible structurally related materials. Both baccatin III and 10-deacetylbaccatin III (Figure 5.44) have been efficiently transformed into taxol. 10-Deacetylbaccatin III is readily extracted from the leaves and twigs of *Taxus baccata*, and, although the content is variable, it is generally present at much higher levels (up to 0.2%) than taxol can be found in *T. brevifolia*. *Taxus baccata,* the common yew, is widely planted as an ornamental tree in Europe and the USA and is much faster growing than the Pacific yew. Cell cultures of *T. baccata* also offer excellent potential for production of taxol or 10-deacetylbaccatin III but are not yet economic; taxol yields of up to 0.2% dry weight

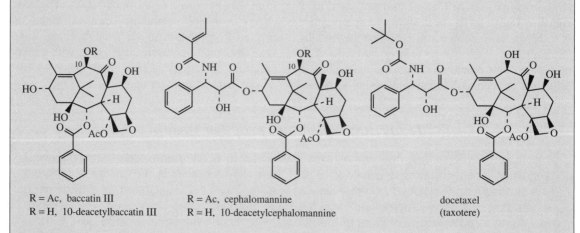

R = Ac, baccatin III
R = H, 10-deacetylbaccatin III

R = Ac, cephalomannine
R = H, 10-deacetylcephalomannine

docetaxel
(taxotere)

Figure 5.44

(Continues)

(Continued)

cultured cells have been obtained. The use of microorganisms and enzymes to specifically hydrolyse ester groups from the mixture of structurally related taxanes in crude extracts and thus improve the yields of 10-deacetylbaccatin III has also been reported.

There is further optimism for new methods of obtaining taxol by microbial culture. Thus, a fungus, *Taxomyces adreanae*, isolated from the inner bark of *Taxus brevifolia* appears to have inherited the necessary genes from the tree (or vice versa) and is able to synthesize taxol and other taxanes in culture, though at only very low levels (20–50 ng l^{-1}). A fungus, *Pestalotiopsis microspora*, recently isolated from the inner bark of the Himalayan yew (*Taxus wallachiana*) produces higher levels (60–70 μg l^{-1}), and if this could be optimized further it might form the basis for commercial production.

Paclitaxel (Taxol®) is being used clinically in the treatment of ovarian and breast cancers, non-small-cell lung cancer, small-cell lung cancer, and cancers of the head and neck. **Docetaxel (Taxotere®)** (Figure 5.44) is a side-chain analogue of taxol, which has also been produced by semi-synthesis from 10-deacetylbaccatin III. It has improved water-solubility compared with taxol, and is being used clinically against ovarian and breast cancers. Taxol acts as an antimitotic by binding to microtubules, promoting their assembly from tubulin, and stabilizing them against depolymerization during cell division. The resultant abnormal tubulin–microtubule equilibrium disrupts the normal mitotic spindle apparatus and blocks cell proliferation. Taxol thus has a different mechanism of action to other antimitotics such as vincristine (see page 356) or podophyllotoxin (see page 136), which bind to the protein tubulin in the mitotic spindle, preventing polymerization and assembly into microtubules. Taxol has also been shown to bind to a second target, a protein which normally blocks the process of apoptosis (cell death). Inhibition of this protein allows apoptosis to proceed.

The latex of some plants in the genus *Euphorbia* (Euphorbiaceae) is toxic, and can cause poisoning in humans and animals, skin dermatitis, cell proliferation, and tumour promotion (co-carcinogen activity). Many species of *Euphorbia* are regarded as potentially toxic, and the latex can produce severe irritant effects, especially on mucous membranes and the eye. Most of the biological effects are due to diterpene esters, e.g. esters of **phorbol** (Figure 5.45), which activate protein kinase C, an important and widely distributed enzyme responsible for phosphorylating many biochemical entities. The permanent activation of protein kinase C is thought to lead to the uncontrolled cancerous growth. The most commonly encountered ester of phorbol is 12-*O*-myristoylphorbol 13-acetate (Figure 5.45). The origins of phorbol are not fully delineated, but may be rationalized as in Figure 5.45. Cyclization of GGPP generates a cation containing a 14-membered ring system. Loss of a proton via cyclopropane ring formation leads to **casbene**, an antifungal metabolite produced by the castor oil plant, *Ricinus communis* (Euphorbiaceae). Casbene, via the ring closures

shown in Figure 5.45, is then likely to be the precursor of the phorbol ring system.

In contrast to the cyclization mechanisms shown in Figures 5.43 and 5.45, where loss of diphosphate generates the initial carbocation, many of the natural diterpenes have arisen by a different mechanism. Carbocation formation is initiated by protonation of the double bond at the head of the chain leading to a first cyclization sequence. Loss of the diphosphate later on also produces a carbocation and facilitates further cyclization. The early part of the sequence resembles that involved in hopanoid biosynthesis (see page 218), and to some extent triterpenoid and steroid biosynthesis (see page 214), though in the latter cases opening of the epoxide ring of the precursor squalene oxide is responsible for generation of the cationic intermediates. Protonation of GGPP can initiate a concerted cyclization sequence, terminated by loss of a proton from a methyl, yielding **copalyl PP** (Figure 5.46, a). The stereochemistry in this product is controlled by the folding of the substrate on the enzyme surface, though an alternative folding can lead to **labdadienyl PP**,

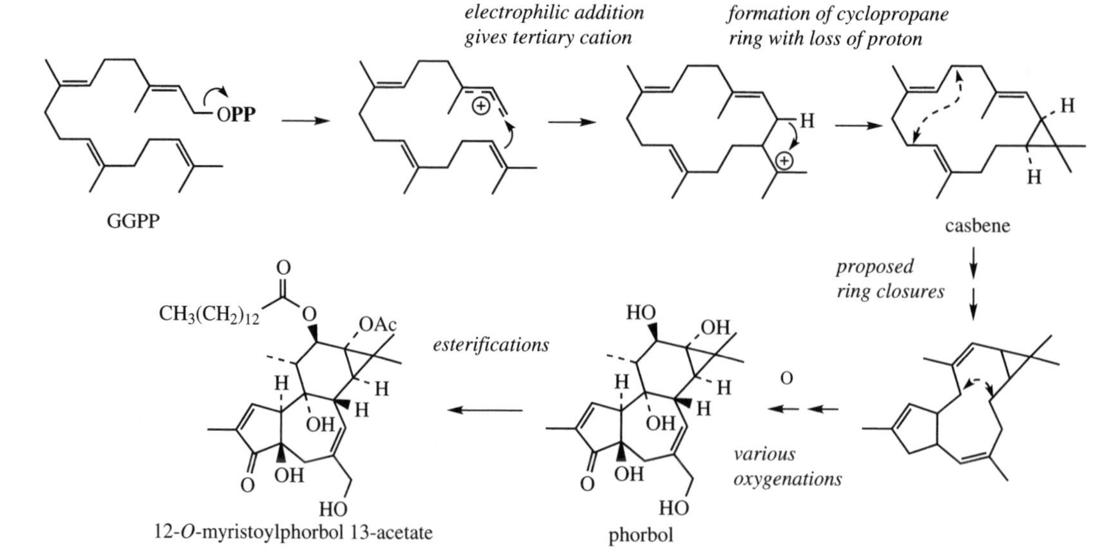

Figure 5.45

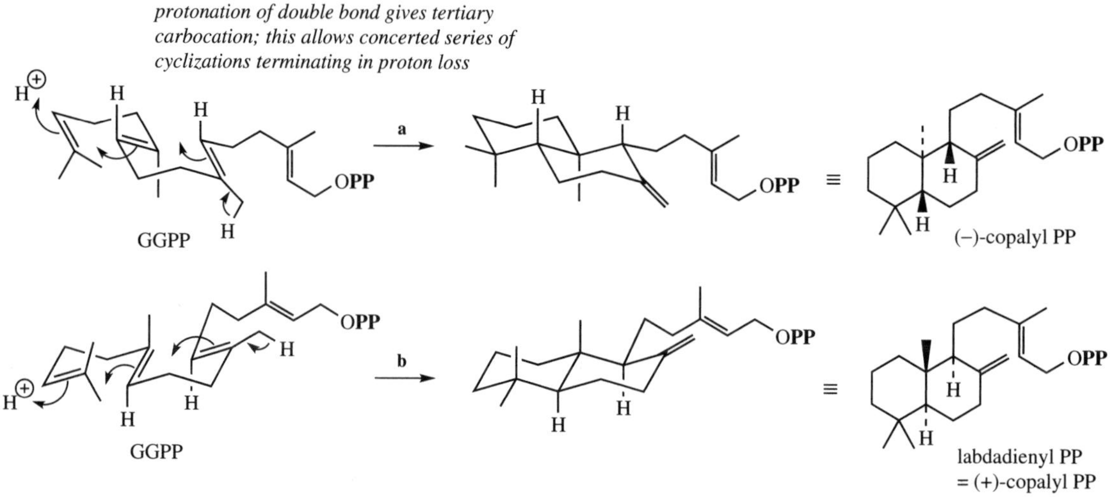

Figure 5.46

the enantiomeric product having opposite configurations at the newly generated chiral centres (Figure 5.46, b). From copalyl PP, a sequence of cyclizations and a rearrangement, all catalysed by a single enzyme, leads to ***ent*-kaurene** (Figure 5.47). As shown, this involves loss of the diphosphate leaving group enabling carbocation-mediated formation of the third ring system, and subsequent production of the fourth ring. Then follows a Wagner–Meerwein migration, effectively contracting the original six-membered ring to a five-membered

one, whilst expanding the five-membered ring to give a six-membered ring. The driving force is transformation of a secondary carbocation to give a tertiary one, but this also results in the methyl group no longer being at a bridgehead, and what appears at first glance to be merely a confusing change in stereochemistry. Loss of a proton from this methyl generates the exocyclic double bond of ***ent*-kaurene** and provides an exit from the carbocationic system. The prefix *ent* is used to indicate enantiomeric; the most common stereochemistry

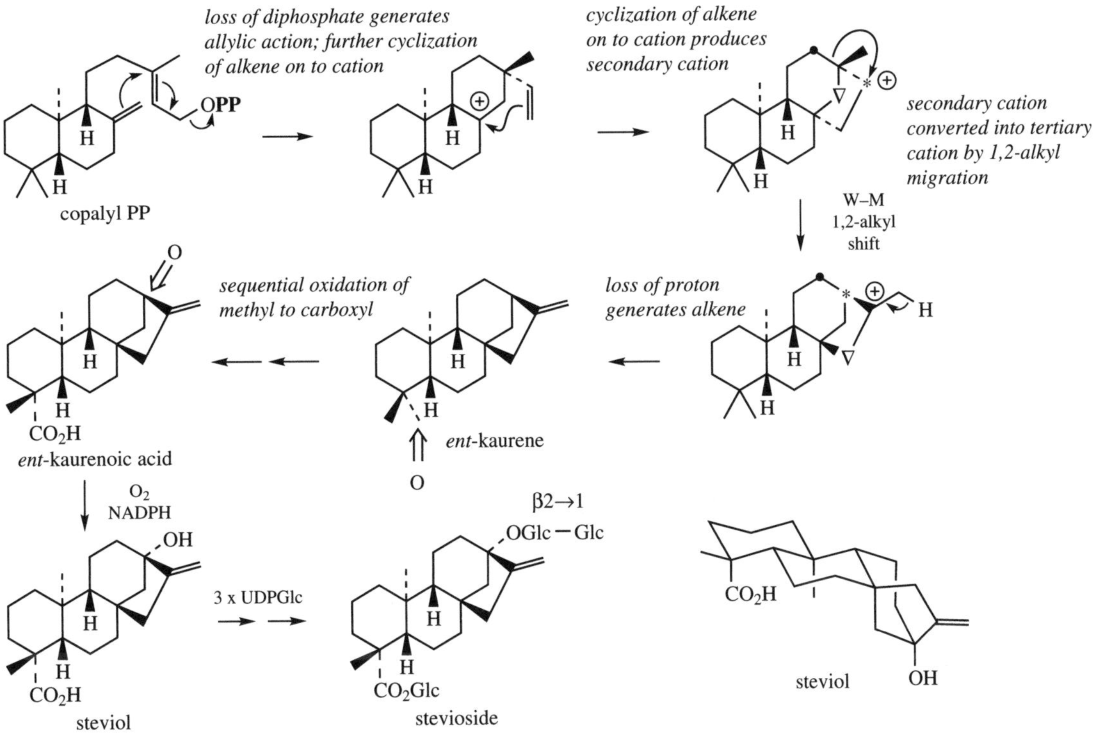

Figure 5.47

is that found in labdadienyl PP (Figure 5.46) and derivatives, so the kaurene series is termed enantiomeric.

ent-Kaurene is the precursor of **stevioside** (Figure 5.47) in the plant *Stevia rebaudiana* (Compositae/Asteraceae) by relatively simple hydroxylation, oxidation, and glucosylation reactions. Both glucosyl ester and glucoside linkages are present in stevioside, which help to confer an intensely sweet taste to this and related compounds. Stevioside is present in the plant leaf in quite large amounts (3–10%), is some 100–300 times as sweet as sucrose, and is being used commercially as a sweetening agent.

The alternative stereochemistry typified by labdadienyl PP can be seen in the structure of **abietic acid** (Figure 5.48), the major component of the rosin fraction of turpentine from pines and other conifers (Table 5.1). Initially, the tricyclic system is built up as in the pathway to *ent*-kaurene (Figure 5.47), via the same mechanism, but generating the enantiomeric series of compounds. The cation loses a proton to give **sandaracopimaradiene** (Figure 5.48), which undergoes a methyl

migration to modify the side-chain, and further proton loss to form the diene **abietadiene**. **Abietic acid** results from sequential oxidation of the 4α-methyl. Wounding of pine trees leads to an accumulation at the wound site of both monoterpenes and diterpenes, which may be fractionated by distillation to give turpentine oil and rosin (Table 5.1). The volatile monoterpenes seem to act as a solvent to allow deposition of the rosin layer to seal the wound. The diterpenes in rosin have both antifungal and insecticidal properties.

Extensive modification of the labdadienyl diterpene skeleton is responsible for generation of the **ginkgolides**, highly oxidized diterpene trilactones which are the active principles of *Ginkgo biloba** (Ginkgoaceae). Several rearrangements, ring cleavage, and formation of lactone rings can broadly explain its origin (Figure 5.49), though this scheme is highly speculative and likely to be incorrect. Although detailed evidence is lacking, it is known that labdadienyl PP is a precursor, and most probably dehydroabietane also. The unusual *tert*-butyl substituent arises as a consequence of the A ring cleavage. **Bilobalide**

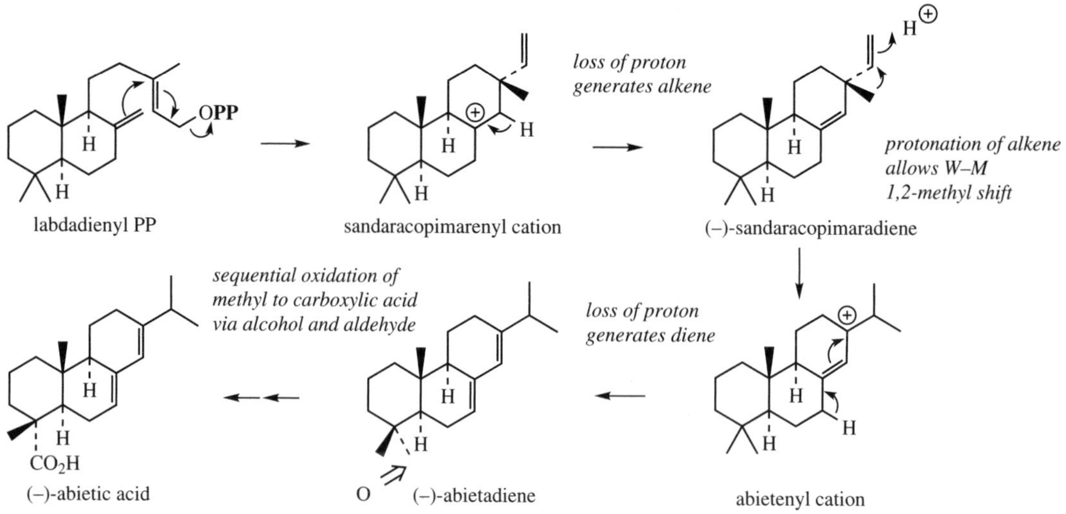

Figure 5.48

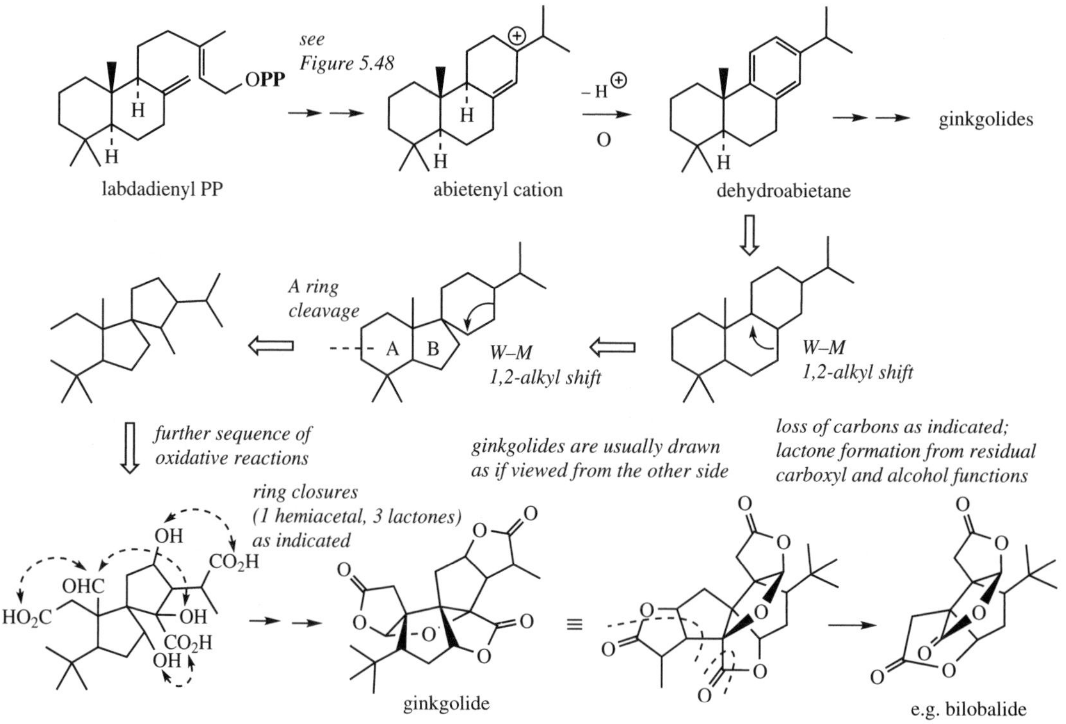

Figure 5.49

(Figure 5.49) contains a related C_{15}-skeleton, and is most likely a partially degraded ginkgolide. *Ginkgo* is the world's oldest tree species, and its leaves are now a currently fashionable health supplement, taken in the hope that it can delay

some of the degeneration of the faculties normally experienced in old age.

In **forskolin** (Figure 5.51), the third ring is heterocyclic rather than carbocyclic. The basic skeleton of forskolin can be viewed as the result

Ginkgo biloba

Ginkgo biloba is a primitive member of the gymnosperms and the only survivor of the Ginkgoaceae, all other species being found only as fossils. It is a small tree native to China, but widely planted as an ornamental, and cultivated for drug use in Korea, France, and the United States. Standardized extracts of the leaves are marketed against cerebral vascular disease and senile dementia. Extracts have been shown to improve peripheral and cerebrovascular circulation. The decline in cognitive function and memory processes in old age can be due to disturbances in brain blood circulation, and thus **ginkgo** may exert beneficial effects by improving this circulation, and assist with other symptoms such as vertigo, tinnitus, and hearing loss.

The active constituents have been characterized as mixtures of terpenoids and flavonoids. The dried leaves contain 0.1–0.25% terpene lactones, comprising five ginkgolides (A, B, C, J, and M) and bilobalide (Figure 5.50). Bilobalide comprises about 30–40% of the mixture, whilst ginkgolide A is the predominant ginkgolide (about 30%). The ginkgolides are diterpenoid in nature, whilst bilobalide is described as sesquiterpenoid. However, bilobalide bears such a structural similarity to the ginkgolides, it is most probably a degraded ginkgolide. The ginkgolides have been shown to have potent and selective antagonistic activity towards platelet-activating factor (PAF, see page 39), which is implicated in many physiological processes. The flavonoid content of the dried leaves is 0.5–1.0%, and consists of a mixture of mono-, di-, and tri-glycosides of the flavonols kaempferol and quercetin (Figure 5.50; see also page 151) and some biflavonoids. These probably also contribute to the activity of ginkgo, and may act as radical scavengers.

Extracts of ginkgo for drug use are usually standardized to contain flavonoid glycosides and terpene lactones in a ratio of 24% to 6%, or 27% to 7%. Ginkgo may be combined with ginseng (see page 222) in the treatment of geriatric disorders. Ginkgo and the ginkgolides are undergoing extensive investigation in conditions where there are high PAF levels, e.g. shock, burns, ulceration, and inflammatory skin disease.

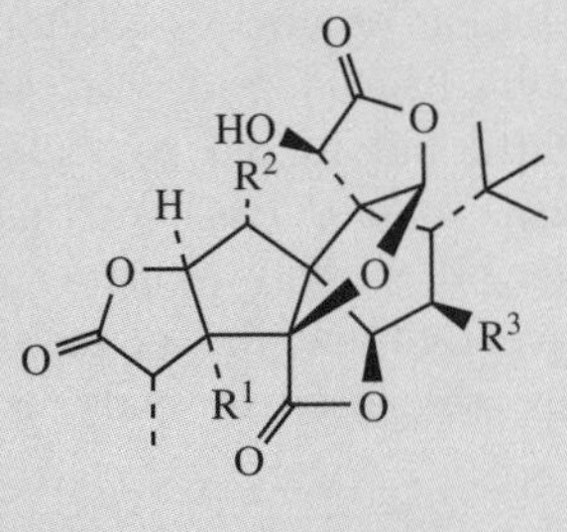

	R^1	R^2	R^3
ginkgolide A	OH	H	H
ginkgolide B	OH	OH	H
ginkgolide C	OH	OH	OH
ginkgolide J	OH	H	OH
ginkgolide M	H	OH	OH

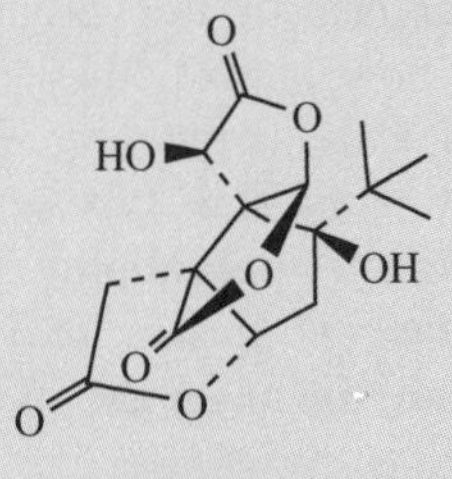

bilobalide

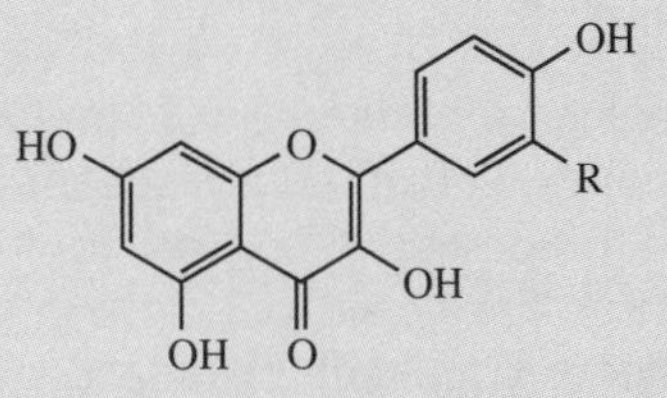

R = H, kaempferol
R = OH, quercetin

Figure 5.50

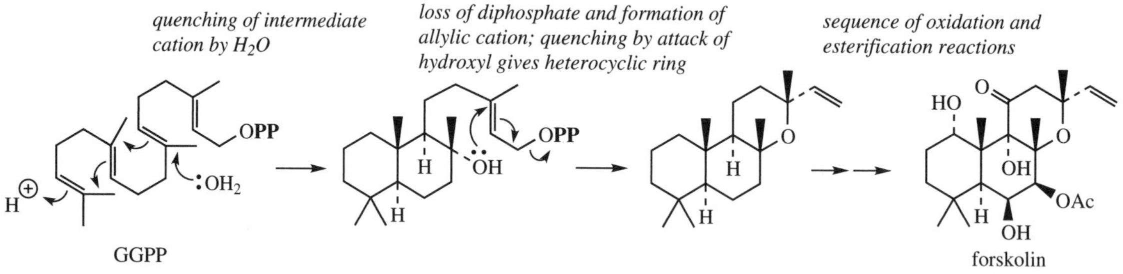

quenching of intermediate cation by H₂O

loss of diphosphate and formation of allylic cation; quenching by attack of hydroxyl gives heterocyclic ring

sequence of oxidation and esterification reactions

GGPP

forskolin

Figure 5.51

of quenching of the cation by water as opposed to proton loss, followed by S_N2' nucleophilic substitution on to the allylic diphosphate (or nucleophilic substitution on to the allylic cation generated by loss of diphosphate) (Figure 5.51). A series of oxidative modifications will then lead to forskolin. This compound has been isolated from *Coleus forskohlii* (Labiatae/Lamiaceae), a plant used in Indian traditional medicine and shown to have quite pronounced hypotensive and antispasmodic activities. Forskolin* is a valuable pharmacological tool as a potent stimulator of adenylate cyclase activity, and it is being investigated for its cardiovascular and bronchospasmolytic effects.

shown in Figure 5.52, and provides no novel features except for an experimentally demonstrated 1,5-hydride shift. GFPP arises by a continuation of the chain extension process, adding a further IPP unit to GGPP. Ophiobolin A shows a broad spectrum of biological activity against bacteria, fungi, and nematodes. The most common type of marine sesterterpenoid is exemplified by **sclarin** and this structure can be envisaged as the result of a concerted cyclization sequence (Figure 5.53) analogous to that seen with GGPP in the diterpenoids, and with squalene oxide in the triterpenoids (see below).

SESTERTERPENES (C₂₅)

Although many examples of this group of natural terpenoids are now known, they are found principally in fungi and marine organisms, and span relatively few structural types. The origin of **ophiobolene** and **ophiobolin A** in the plant pathogen *Helminthosporium maydis* from cyclization of **geranylfarnesyl PP** (**GFPP**) is

TRITERPENES (C₃₀)

Triterpenes are not formed by an extension of the now familiar process of adding IPP to the growing chain. Instead, two molecules of farnesyl PP are joined tail to tail to yield the hydrocarbon **squalene** (Figure 5.54), originally isolated from the liver oil of shark (*Squalus* sp.). Squalene was subsequently found in rat liver and yeast, and these systems were used to study its biosynthetic role

Forskolin

In a screening programme of Indian medicinal plants, extracts from the roots of *Coleus forskohlii* (Labiatae/Lamiaceae) were discovered to lower blood pressure and have cardioactive properties. This led to the isolation of the diterpene forskolin (= coleonol) (Figure 5.51) as the active principle in yields of about 0.1%. Forskolin has been shown to exert its effects by direct stimulation of adenylate cyclase, and has become a valuable pharmacological tool in the study of this enzyme and its functions. It has shown promising potential for the treatment of glaucoma, congestive heart failure, hypertension, and bronchial asthma, though drug use is limited by poor water-solubility, and derivatives or analogues will need to be developed.

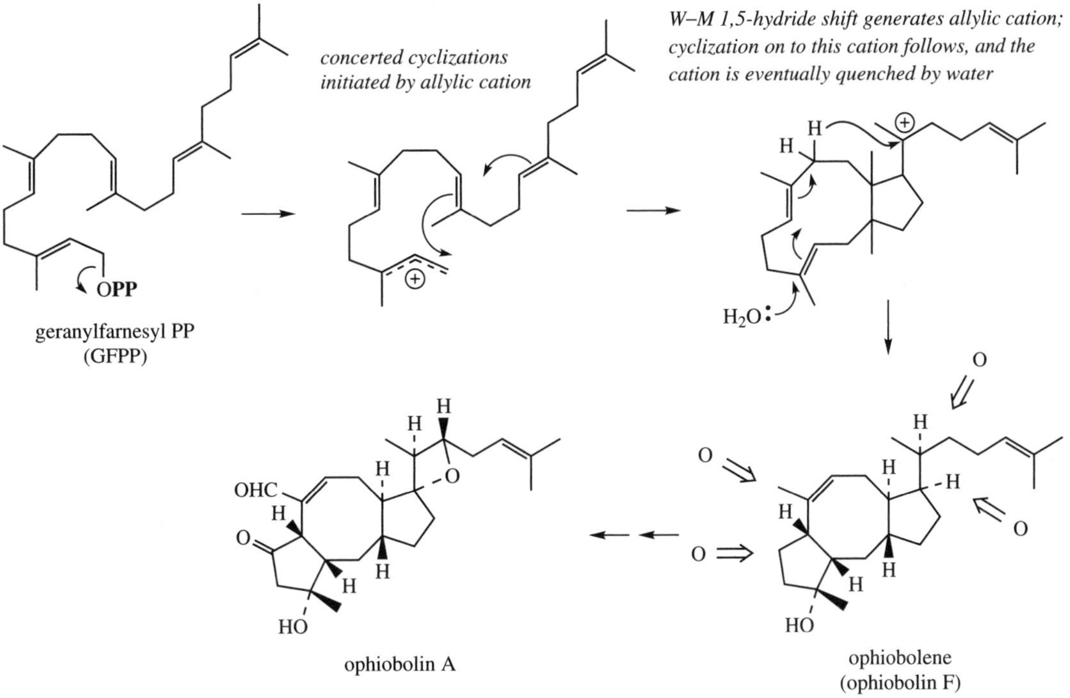

concerted cyclizations
initiated by allylic cation

W–M 1,5-hydride shift generates allylic cation;
cyclization on to this cation follows, and the
cation is eventually quenched by water

geranylfarnesyl PP
(GFPP)

H$_2$O:

ophiobolin A

ophiobolene
(ophiobolin F)

Figure 5.52

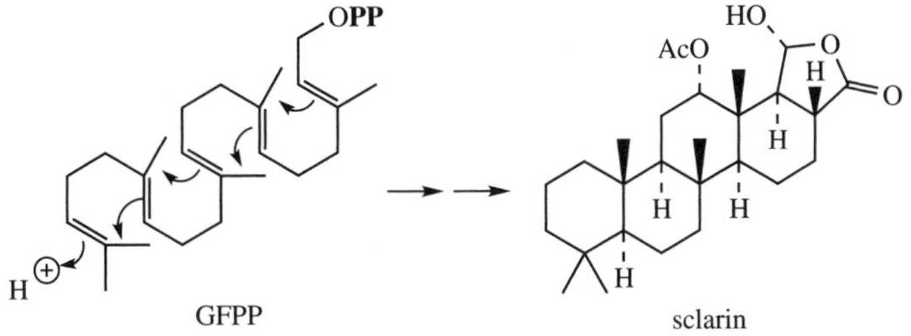

GFPP

sclarin

Figure 5.53

as a precursor of triterpenes and steroids; several seed oils are now recognized as quite rich sources of squalene, e.g. *Amaranthus cruentus* (Amaranthaceae). During the coupling process, which on paper merely requires removal of the two diphosphate groups, a proton from a C-1 position of one molecule of FPP is lost, and a proton from NADPH is inserted. Difficulties with formulating a plausible mechanism for this unlikely reaction were resolved when an intermediate in the process, **presqualene diphosphate**, was isolated from rat liver. Its characterization as a cyclopropane derivative immediately ruled out all the hypotheses current at the time.

The formation of presqualene PP is represented in Figure 5.54 as attack of the 2,3-double bond of FPP on to the farnesyl cation, analogous to the chain extension using IPP (see also the proposal for the origins of irregular monoterpenes, page 186). The resultant tertiary cation is discharged by loss of a proton and formation of the cyclopropane ring, giving presqualene PP. Obviously, to form squalene, carbons-1 of the two FPP units must eventually be coupled, whilst presqualene PP formation has actually joined C-1 of one molecule to C-2 of the other. To account for the subsequent change in bonding of the two FPP units, a further cyclopropane cationic intermediate is proposed. Loss of

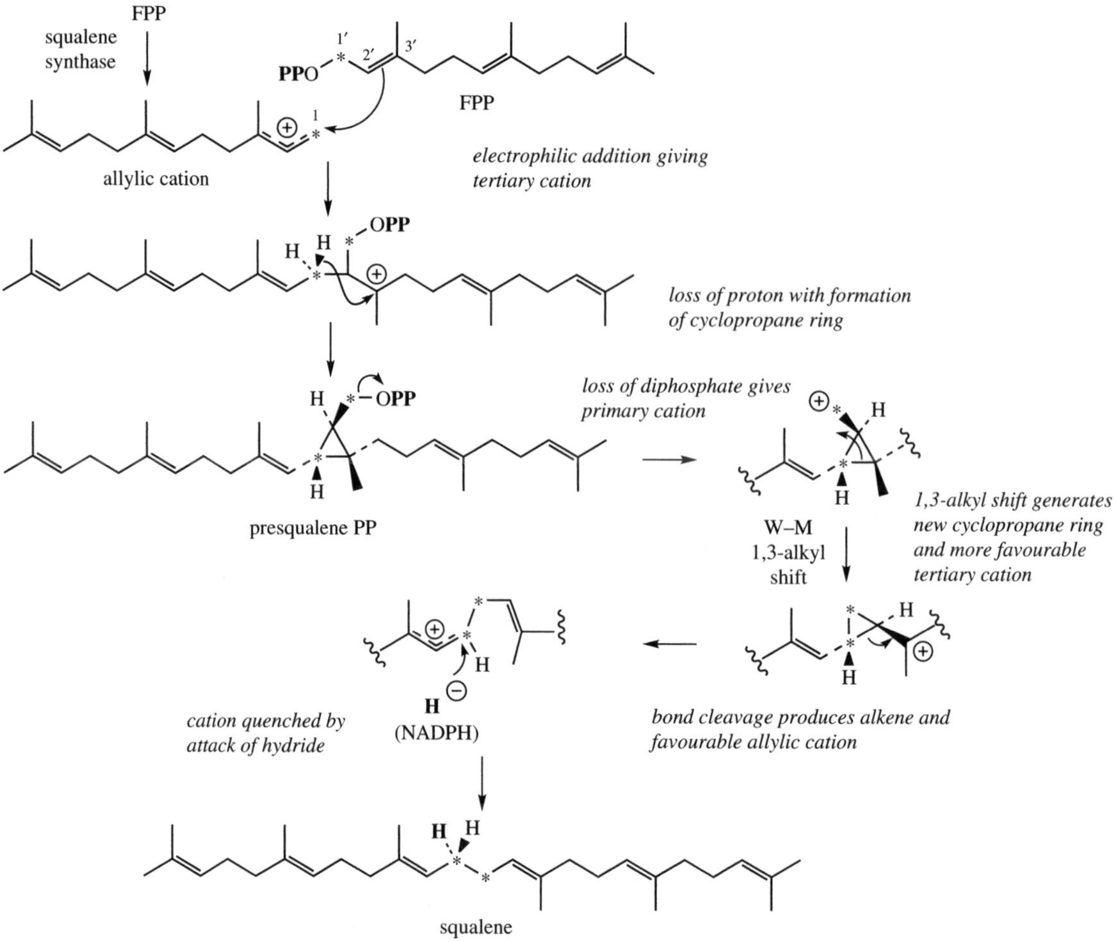

Figure 5.54

diphosphate from presqualene PP would give the unfavourable primary cation, which via Wagner–Meerwein rearrangement can generate a tertiary carbocation and achieve the required C-1–C-1′ bond. Breaking the original but now redundant C-1–C-2′ bond can give an allylic cation, and the generation of **squalene** is completed by supply of hydride from NADPH.

Cyclization of squalene is via the intermediate **squalene-2,3-oxide** (Figure 5.55), produced in a reaction catalysed by a flavoprotein requiring O_2 and NADPH cofactors. If squalene oxide is suitably positioned and folded on the enzyme surface, the polycyclic triterpene structures formed can be rationalized in terms of a series of cyclizations, followed by a sequence of concerted Wagner–Meerwein migrations of methyls and hydrides

(Figure 5.55). The cyclizations are carbocation mediated and proceed in a step-wise sequence (Figure 5.56). Thus, protonation of the epoxide group will allow opening of this ring and generation of the preferred tertiary carbocation, suitably placed to allow electrophilic addition to a double bond, formation of a six-membered ring and production of a new tertiary carbocation. This process continues twice more, generating the preferred tertiary carbocation (Markovnikov addition) after each ring formation, though the third ring formed is consequently a five-membered one. This is expanded to a six-membered ring via a Wagner–Meerwein 1,2-alkyl shift, resulting in some relief of ring strain, though sacrificing a tertiary carbocation for a secondary one. A further electrophilic addition generates the tertiary protosteryl

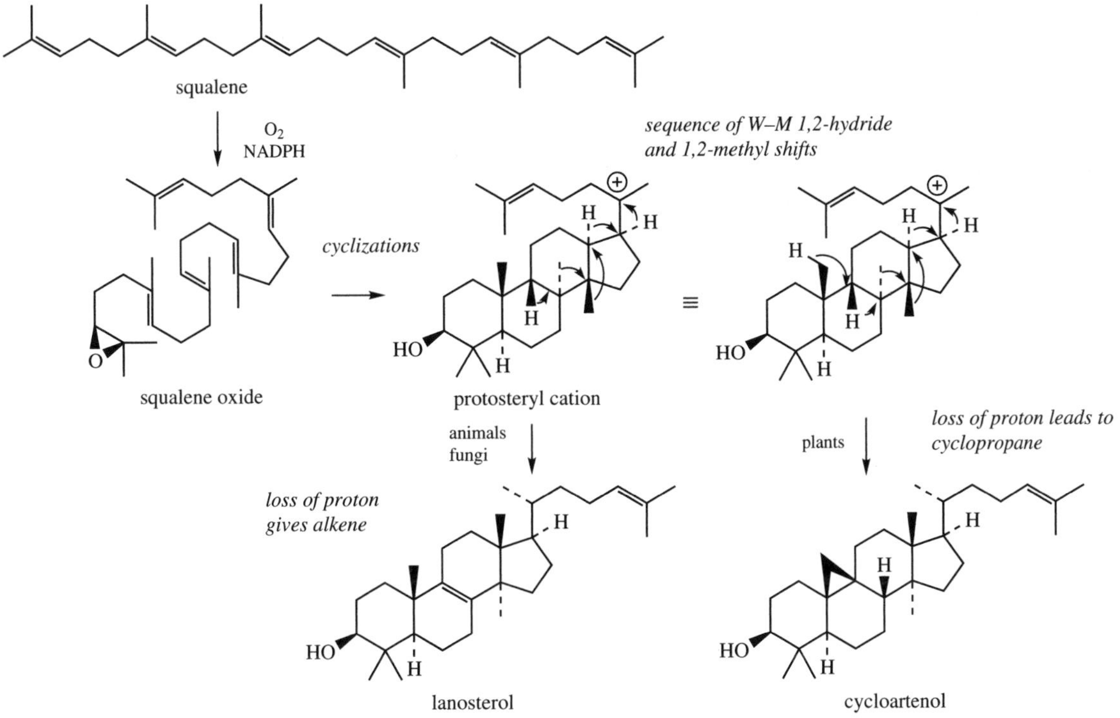

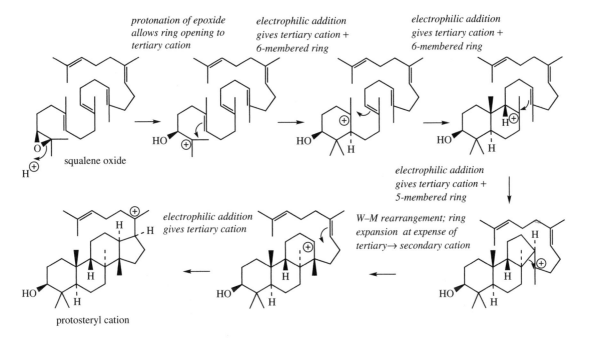

squalene

O₂
NADPH

cyclizations

*sequence of W–M 1,2-hydride
and 1,2-methyl shifts*

squalene oxide

protosteryl cation

≡

HO

HO

*animals
fungi*

plants

*loss of proton leads to
cyclopropane*

*loss of proton
gives alkene*

HO   HO

lanosterol   cycloartenol

Figure 5.55

*protonation of epoxide
allows ring opening to
tertiary cation*

*electrophilic addition
gives tertiary cation +
6-membered ring*

*electrophilic addition
gives tertiary cation +
6-membered ring*

squalene oxide

HO   HO   HO

H⁺

*electrophilic addition
gives tertiary cation +
5-membered ring*

*electrophilic addition
gives tertiary cation*

*W–M rearrangement; ring
expansion at expense of
tertiary→ secondary cation*

HO   HO   HO

protosteryl cation

Figure 5.56

cation (Figure 5.56). The stereochemistries in this cation are controlled by the type of folding achieved on the enzyme surface, and this probably also limits the extent of the cyclization process. Thus, if the folded squalene oxide approximates to a chair–boat–chair–boat conformation (Figure 5.57), the transient **protosteryl cation** will

be produced with these conformational characteristics. This cation then undergoes a series of Wagner–Meerwein 1,2-shifts, firstly migrating a hydride and generating a new cation, migrating the next hydride, then a methyl and so on until a proton is lost forming a double bond and thus creating **lanosterol** (Figure 5.57). The stereochemistry

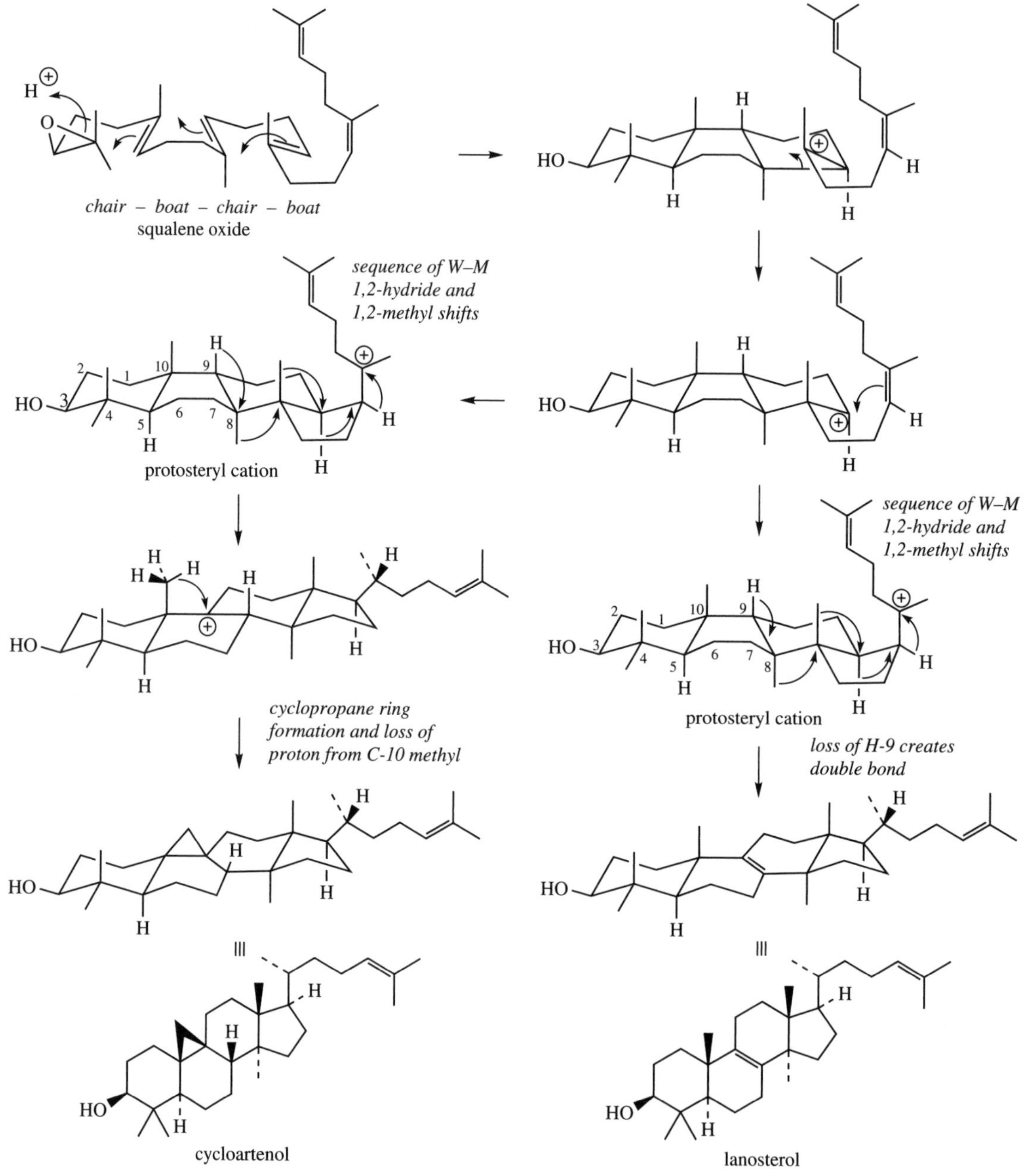

Figure 5.57

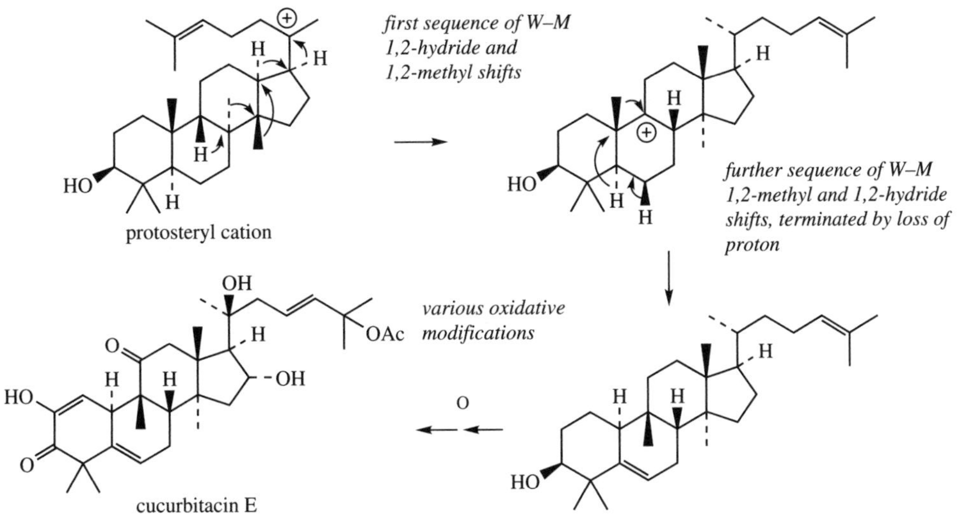

Figure 5.58

of the protosteryl cation in Figure 5.57 shows how favourable this sequence will be, and emphasizes that in the ring system, the migrating groups are positioned *anti* to each other, one group entering whilst the other leaves from the opposite side of the stereocentre. This, of course, inverts configurations at each appropriate centre. No *anti* group is available to migrate to C-9 (steroid numbering), and the reaction terminates by loss of proton H-9. Lanosterol is a typical animal triterpenoid, and the precursor for cholesterol and other sterols in animals (see page 233) and fungi (see page 254). In plants, its intermediate role is taken by **cycloartenol** (Figure 5.57), which contains a cyclopropane ring, generated by inclusion of carbon from the methyl at C-10. For cycloartenol, H-9 is not lost, but migrates to C-8, and the carbocation so formed is quenched by cyclopropane formation and loss of one of the methyl protons. For many plant steroids, this cyclopropane ring has then to be reopened (see page 235). Most natural triterpenoids and steroids contain a 3-hydroxyl group, the original epoxide oxygen from squalene oxide.

An additional feature of the protosteryl cation is that the C-10 methyl and H-5 also share an *anti*-axial relationship, and are also susceptible to Wagner–Meerwein rearrangements, so that the C-9 cation formed in the cycloartenol sequence may then initiate further migrations. This can be terminated by formation of a 5,6-double

bond (Figure 5.58), as in the pathway to the **cucurbitacins**, a group of highly oxygenated triterpenes encountered in the cucumber/melon/marrow family, the Cucurbitaceae. These compounds are characteristically bitter tasting, purgative, and extremely cytotoxic.

Should **squalene oxide** be folded on to another type of cyclase enzyme, this time in a roughly chair–chair–chair–boat conformation (Figure 5.59), then an identical carbocation mechanism ensues, and the transient **dammarenyl cation** formed now has different stereochemical features to the protosteryl cation. Whilst a series of Wagner–Meerwein migrations can occur, there is relatively little to be gained on purely chemical grounds, since these would invert stereochemistry and destroy the already favourable conformation. Instead, the dammarenyl cation typically undergoes further carbocation promoted cyclizations, without any major changes to the ring system already formed. There are occasions in which the migrations do occur, however, and **euphol** from *Euphorbia* species (Euphorbiaceae) is a stereoisomer of lanosterol (Figure 5.55).

Should the Wagner–Meerwein rearrangements not occur, the dammarenyl cation could be quenched with water, giving the epimeric **dammarenediols**, as found in Dammar resin from *Balanocarpus heimii* (Dipterocarpaceae) and ginseng* (*Panax ginseng*; Araliaceae) (Figure 5.60). Alternatively, the migration shown to give the

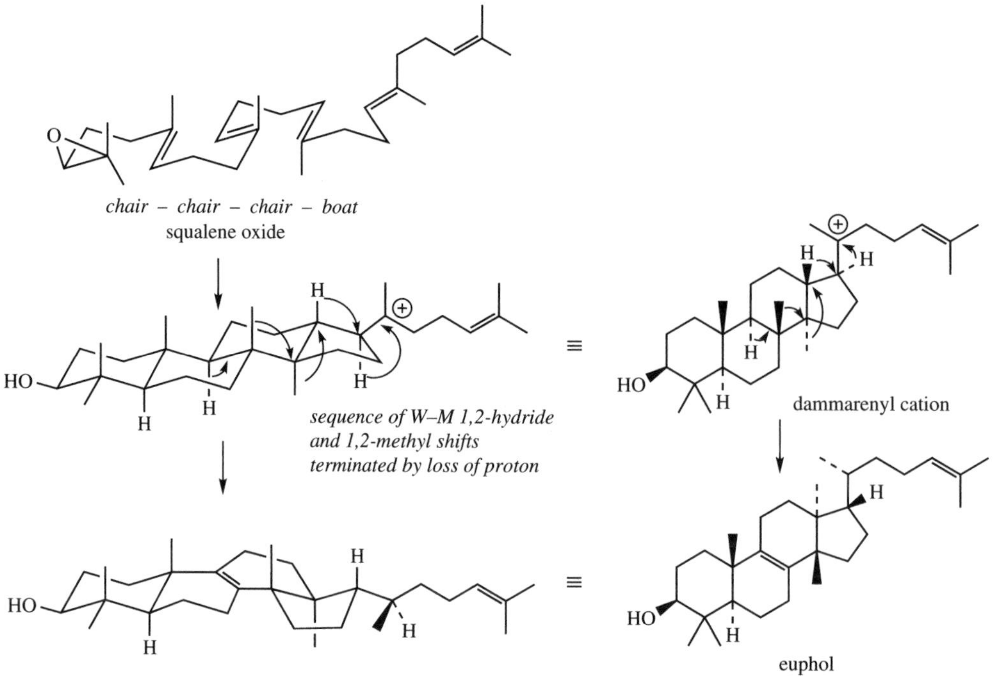

Figure 5.59

baccharenyl cation relieves some ring strain by creating a six-membered ring, despite sacrificing a tertiary carbocation for a secondary one. A pentacyclic ring system can now be formed by cyclization on to the double bond, giving a new five-membered ring and the tertiary lupenyl cation. Although this appears to contradict the reasoning used above for the dammarenyl → baccharenyl transformation, the contribution of the enzyme involved must also be considered in each case. A five-membered ring is not highly strained as evidenced by all the natural examples encountered. Loss of a proton from the lupenyl cation gives **lupeol**, found in lupin (*Lupinus luteus*; Leguminosae/Fabaceae). Ring expansion in the lupenyl cation by bond migration gives the oleanyl system, and labelling studies have demonstrated this ion is discharged by hydride migrations and loss of a proton, giving the widely distributed **β-amyrin**. Formation of the isomeric **α-amyrin** involves first the migration of a methyl in the oleanyl cation, then discharge of the new taraxasteryl cation by three hydride migrations and loss of a proton. Loss of a proton from the non-migrated methyl in the taraxasteryl cation is an alternative way of achieving a neutral molecule, and yields

taraxasterol found in dandelion (*Taraxacum officinale*; Compositae/Asteraceae). Comparison with α-amyrin shows the subtly different stereochemistry present because the inversions of configuration caused by hydride migrations have not occurred. Where evidence is available, these extensive series of cyclizations and Wagner–Meerwein rearrangements appear to be catalysed by a single enzyme, which converts squalene oxide into the final product, e.g. lanosterol, cycloartenol, α-amyrin, or β-amyrin.

Bacterial membranes frequently contain **hopanoids** (Figure 5.61), triterpenoid compounds that appear to take the place of the sterols that are typically found in the membranes of higher organisms, helping to maintain the structural integrity and to control permeability. Hopanoids arise from squalene by a similar carbocation cyclization mechanism, but do not involve the initial epoxidation to squalene oxide. Instead, the carbocation is produced by protonation (compare the cyclization of GGPP to labdadienyl PP, page 207), and the resultant compounds tend to lack the characteristic 3-hydroxyl group, e.g. **hopan-22-ol** from *Alicyclobacillus acidocaldarius* (Figure 5.61). On the other hand, **tetrahymanol** from the protozoan

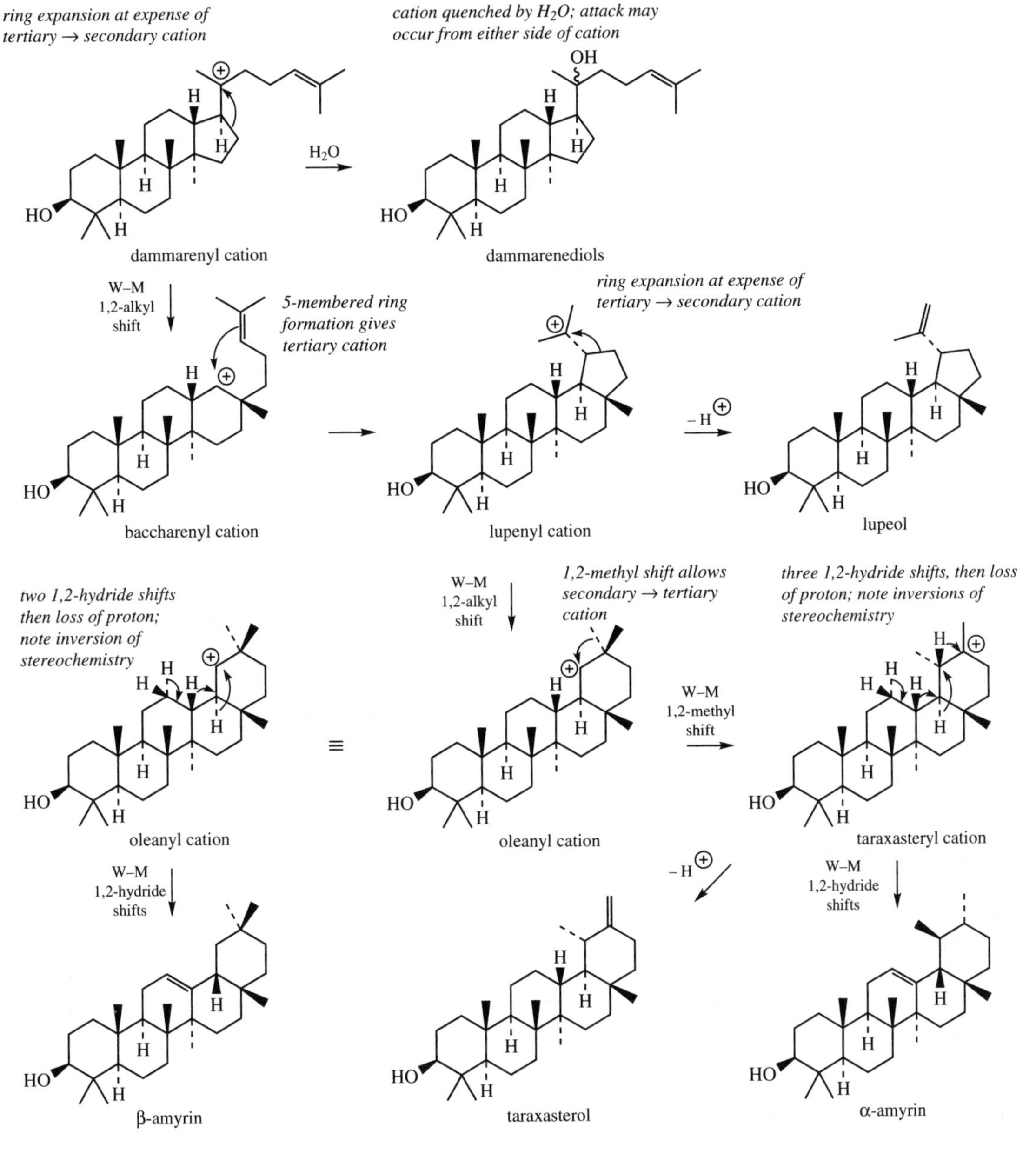

ring expansion at expense of tertiary → secondary cation
cation quenched by H_2O; attack may occur from either side of cation
H_2O
dammarenyl cation
dammarenediols
W–M 1,2-alkyl shift
5-membered ring formation gives tertiary cation
ring expansion at expense of tertiary → secondary cation
– H⊕
baccharenyl cation
lupenyl cation
lupeol
two 1,2-hydride shifts then loss of proton; note inversion of stereochemistry
W–M 1,2-alkyl shift
1,2-methyl shift allows secondary → tertiary cation
three 1,2-hydride shifts, then loss of proton; note inversions of stereochemistry
W–M 1,2-methyl shift
oleanyl cation
oleanyl cation
taraxasteryl cation
W–M 1,2-hydride shifts
– H⊕
W–M 1,2-hydride shifts
β-amyrin
taraxasterol
α-amyrin

Figure 5.60

Tetrahymena pyriformis, because of its symmetry, might appear to have a 3-hydroxyl group, but this is derived from water, and not molecular oxygen as would be the case if squalene oxide were involved. As in formation of the protosteryl cation (page 214), Wagner–Meerwein ring expansions occur during the cyclization mechanisms shown in Figure 5.61 so that the first-formed tertiary carbocation/five-membered ring (Markovnikov addition) becomes a secondary carbocation/six-membered ring.

Triterpenoid Saponins

The pentacyclic triterpenoid skeletons exemplified by lupeol, α-amyrin, and β-amyrin (Figure 5.60) are frequently encountered in the form of triterpenoid saponin structures. Saponins are

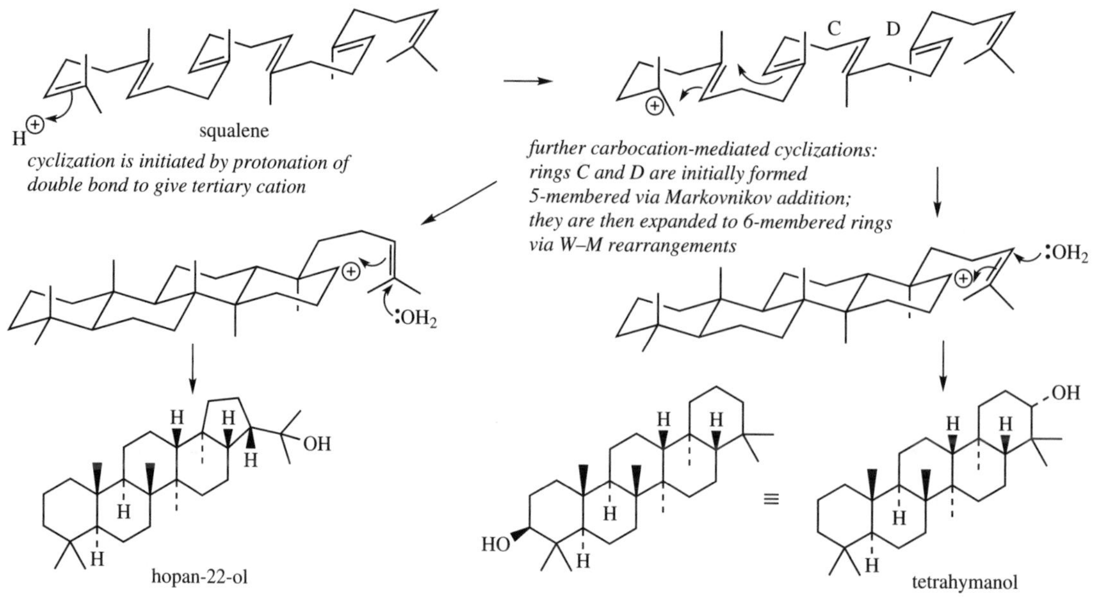

squalene

cyclization is initiated by protonation of double bond to give tertiary cation

further carbocation-mediated cyclizations: rings C and D are initially formed 5-membered via Markovnikov addition; they are then expanded to 6-membered rings via W–M rearrangements

hopan-22-ol

tetrahymanol

Figure 5.61

glycosides which, even at low concentrations, produce a frothing in aqueous solution, because they have surfactant and soaplike properties. The name comes from the Latin *sapo*, soap, and plant materials containing saponins were originally used for cleansing clothes, e.g. soapwort (*Saponaria officinalis*; Caryophyllaceae) and quillaia or soapbark (*Quillaja saponaria*; Rosaceae). These materials also cause haemolysis, lysing red blood cells by increasing the permeability of the plasma membrane, and thus they are highly toxic when injected into the blood stream. Some saponin-containing plant extracts have been used as arrow poisons. However, saponins are relatively harmless when taken orally, and some of our valuable food materials, e.g. beans, lentils, soybeans, spinach, and oats, contain significant amounts. Sarsaparilla (see page 242) is rich in steroidal saponins but is widely used in the manufacture of non-alcoholic drinks. Toxicity is minimized during ingestion by low absorption, and by hydrolysis. Acid-catalysed hydrolysis of saponins liberates sugar(s) and an aglycone (sapogenin), which can be either triterpenoid or steroidal (see page 237) in nature. Some plants may contain exceptionally high amounts of saponins, e.g. about 10% in quillaia bark.

Triterpenoid saponins are rare in monocotyledons, but abundant in many dicotyledonous families. Medicinally useful examples are mainly based on the β-amyrin subgroup (Figure 5.62), and many of these possess carboxylic acid groups derived by oxidation of methyl groups, those at positions 4 (C-23), 17 (C-28), and 20 (C-30) on the aglycone ring system being subject to such oxidation. In some structures, less oxidized formyl (−CHO) or hydroxymethyl (−CH$_2$OH) groups may also be encountered. Positions 11 and 16 may also be oxygenated. Sugar residues are usually attached to the 3-hydroxyl, with one to six monosaccharide units, the most common being glucose, galactose, rhamnose, and arabinose, with uronic acid units (glucuronic acid and galacturonic acid) also featuring (see page 467). Thus, quillaia* bark contains a saponin mixture with **quillaic acid** (Figure 5.62) as the principal aglycone, and the medicinally valuable root of liquorice* (*Glycyrrhiza glabra*; Leguminosae/Fabaceae) contains **glycyrrhizin**, a mixture of potassium and calcium salts of **glycyrrhizic acid** (Figure 5.63), which is composed of the aglycone **glycyrrhetic acid** and two glucuronic acid units. Seeds of horsechestnut (*Aesculus hippocastrum*; Hippocastanaceae), sometimes used in herbal preparations as an anti-inflammatory and anti-bruising remedy, contain a complex mixture of saponins termed **aescin**, based on the polyhydroxylated aglycones **protoaescigenin** and **barringtogenol**

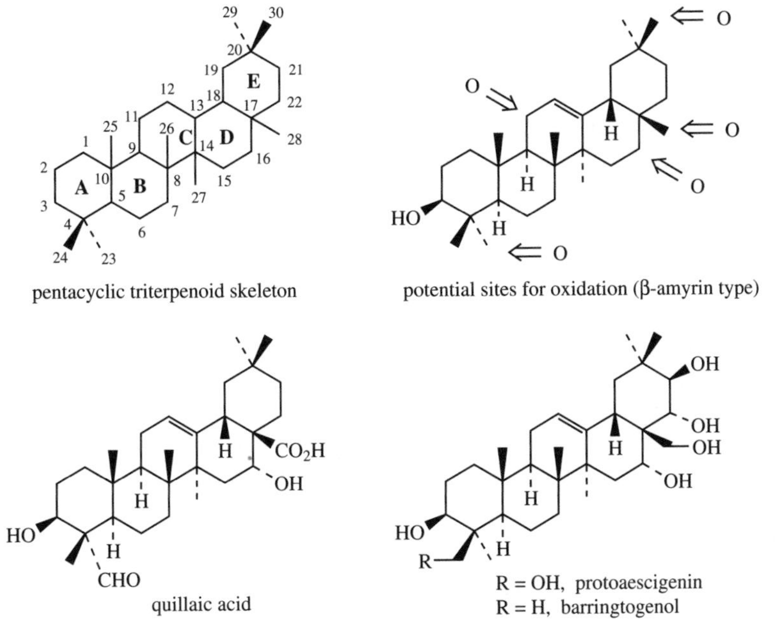

pentacyclic triterpenoid skeleton

potential sites for oxidation (β-amyrin type)

quillaic acid

R = OH, protoaescigenin
R = H, barringtogenol

Figure 5.62

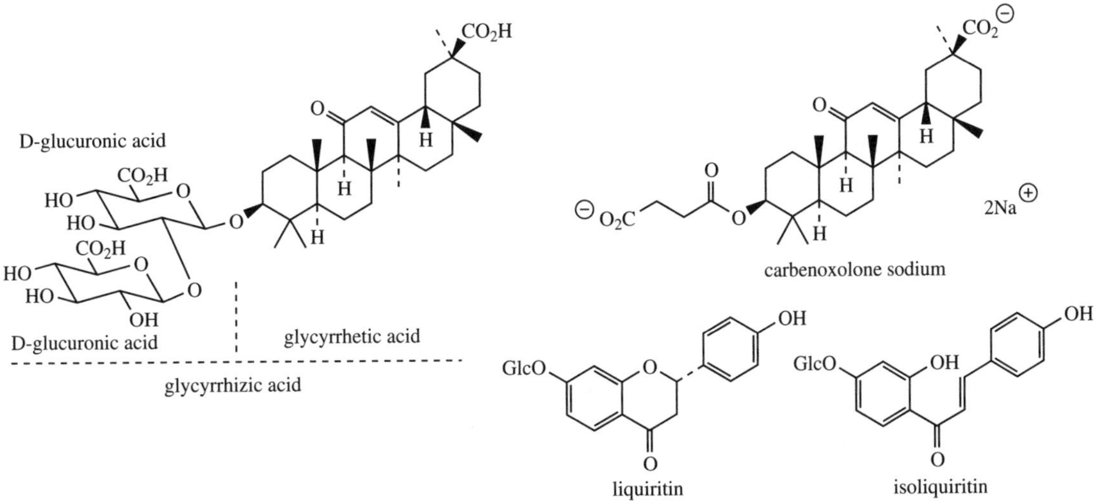

D-glucuronic acid

D-glucuronic acid

glycyrrhetic acid

glycyrrhizic acid

carbenoxolone sodium

liquiritin

isoliquiritin

Figure 5.63

Liquorice

Liquorice (licorice; glycyrrhiza) is the dried unpeeled rhizome and root of the perennial herb *Glycyrrhiza glabra* (Leguminosae/Fabaceae). A number of different varieties are cultivated commercially, including *G. glabra* var. *typica* (Spanish liquorice) in Spain, Italy, and France, and *G. glabra* var. *glandulifera* (Russian liquorice) in Russia. Russian liquorice is usually peeled

(Continues)

(Continued)

before drying. *Glycyrrhiza uralensis* (Manchurian liquorice) from China is also commercially important. Much of the liquorice is imported in the form of an extract, prepared by extraction with water, then evaporation to give a dark black solid. Most of the liquorice produced is used in confectionery and for flavouring, including tobacco, beers, and stouts. Its pleasant sweet taste and foaming properties are due to saponins. Liquorice root contains about 20% of water soluble extractives, and much of this (typically 3–5% of the root, but up to 12% in some varieties) is comprised of glycyrrhizin, a mixture of the potassium and calcium salts of glycyrrhizic (glycyrrhizinic) acid (Figure 5.63). Glycyrrhizic acid is a diglucuronide of the aglycone glycyrrhetic (glycyrrhetinic) acid. The bright yellow colour of liquorice root is provided by flavonoids (1–1.5%) including liquiritin and isoliquiritin (Figure 5.63), and their corresponding aglycones (see page 150). Considerable amounts (5–15%) of sugars (glucose and sucrose) are also present.

Glycyrrhizin is reported to be 50–150 times as sweet as sucrose, and liquorice has thus long been used in pharmacy to mask the taste of bitter drugs. Its surfactant properties have also been exploited in various formulations, as have its demulcent and mild expectorant properties. More recently, some corticosteroid-like activity has been recognized, liquorice extracts displaying mild anti-inflammatory and mineralocorticoid activities. These have been exploited in the treatment of rheumatoid arthritis, Addison's disease (chronic adrenocortical insufficiency), and various inflammatory conditions. Glycyrrhetic acid has been implicated in these activities, and has been found to inhibit enzymes that catalyse the conversion of prostaglandins and glucocorticoids into inactive metabolites. This results in increased levels of prostaglandins, e.g. PGE_2 and $PGF_{2\alpha}$ (see page 54), and of hydrocortisone (see page 268). Perhaps the most important current application is to give systematic relief from peptic ulcers by promoting healing through increased prostaglandin activity. A semi-synthetic derivative of glycyrrhetic acid, the hemisuccinate **carbenoxolone sodium** (Figure 5.63), is widely prescribed for the treatment of gastric ulcers, and also duodenal ulcers. The mineralocorticoid effects (sodium and water retention) may exacerbate hypertension and cardiac problems. Surprisingly, a **deglycyrrhizinized liquorice** preparation is also marketed for treatment of peptic ulcers, but its efficiency has been questioned.

Quillaia

Quillaia bark or soapbark is derived from the tree *Quillaja saponaria* (Rosaceae) and other *Quillaja* species found in Chile, Peru, and Bolivia. The bark contains up to 10% saponins, a mixture known as 'commercial saponin', which is used for its detergent properties. Quillaia's surfactant properties are occasionally exploited in pharmaceutical preparations where it is used in the form of quillaia tincture as an emulsifying agent, particularly for fats, tars, and volatile oils. The bark contains a mixture of saponins which on hydrolysis liberates quillaic acid (Figure 5.62) as the aglycone, together with sugars, uronic acids, and acids from ester functions.

Ginseng

The roots of the herbaceous plants *Panax ginseng* (Araliaceae) from China, Korea, and Russia, and related *Panax* species, e.g. *P. quinquefolium* (American ginseng) from the USA

(Continues)

(Continued)

and Canada, and *P. notoginseng* (Sanchi-ginseng) from China, have been widely used in China and Russia for the treatment of a number of diseases including anaemia, diabetes, gastritis, insomnia, sexual impotence, and as a general restorative. Interest in the drug has increased considerably in recent years and ginseng is widely available as a health food in the form of powders, extracts, and teas. The dried and usually peeled root provides white ginseng, whereas red ginseng is obtained by steaming the root, this process generating a reddish-brown caramel-like colour, and reputedly enhancing biological activity. **Ginseng** is classified as an 'adaptogen', helping the body to adapt to stress, improving stamina and concentration, and providing a normalizing and restorative effect. It is also widely promoted as an aphrodisiac. The Korean root is highly prized and the most expensive. Long term use of ginseng can lead to symptoms similar to those of corticosteroid poisoning, including hypertension, nervousness, and sleeplessness in some subjects, yet hypotension and tranquillizing effects in others.

The benefits of ginseng treatment are by no means confirmed at the pharmacological level, though CNS-stimulating, CNS-sedative, tranquillizing, antifatigue, hypotensive, and hypertensive activities have all been demonstrated. Many of the secondary metabolites present in the root have now been identified. It contains a large number of triterpenoid saponins based on the dammarane subgroup, saponins that have been termed ginsenosides by Japanese investigators, or panaxosides by Russian researchers. These are derivatives of two main aglycones, protopanaxadiol and protopanaxatriol (Figure 5.64), though the aglycones liberated on acid hydrolysis are panaxadiol and panaxatriol respectively. Acid-catalysed cyclization in the side-chain produces an ether ring (Figure 5.64). Sugars are present in the saponins on the 3- and 20-hydroxyls in the diol series, and the 6- and 20-hydroxyls in the triol series. About 30 ginsenosides have been characterized from the different varieties of ginseng, with ginsenoside R_{b-1} (Figure 5.64) of the diol series typically being the most abundant constituent. Ginsenoside R_{g-1} (Figure 5.64) is usually the major component representative of the triol series. Other variants are shown in Figure 5.64. Particularly in white ginseng, many of the ginsenosides are also present as esters with malonic acid. Steaming to prepare red ginseng causes partial hydrolysis of esters and glycosides. Ginsenosides R_{b-1} and R_{g-1} appear to be the main representatives in *Panax ginseng*, ginsenosides R_{b-1}, R_{g-1}, and R_d in *P. notoginseng*, and ginsenosides R_{b-1}, R_e, and malonylated R_{b-1} in *P. quinquefolium*. The pentacyclic triterpenoid sapogenin oleanolic acid (Figure 5.65) is also produced by hydrolysis of the total saponins of *P. ginseng*, and is present in some saponin structures (chikusetsusaponins). The saponin contents of *Panax notoginseng* (about 12%) and *P. quinquefolium* (about 6%) are generally higher than that of *P. ginseng* (1.5–2%).

The root of *Eleutherococcus senticosus* (*Acanthopanax senticosus*) (Araliaceae) is used as an inexpensive substitute for ginseng, and is known as **Russian** or **Siberian ginseng**. This material is held to have similar adaptogenic properties as *Panax ginseng* and a number of eleutherosides have been isolated. However, the term eleutheroside has been applied to compounds of different chemical classes, and the main active anti-stress constituents appear to be lignan glycosides, e.g. eleutheroside E (≡ syringaresinol diglucoside) (Figure 5.65) (see page 132) and phenylpropane glycosides, e.g. eleutheroside B (≡ syringin). The leaves of Russian ginseng contain a number of saponins based on oleanolic acid, but these are quite different to the ginsenosides/panaxosides found in *Panax*. Whilst there is sufficient evidence to support the beneficial adaptogen properties for *Eleutherococcus senticosus*, detailed pharmacological confirmation is not available.

(Continues)

(Continued)

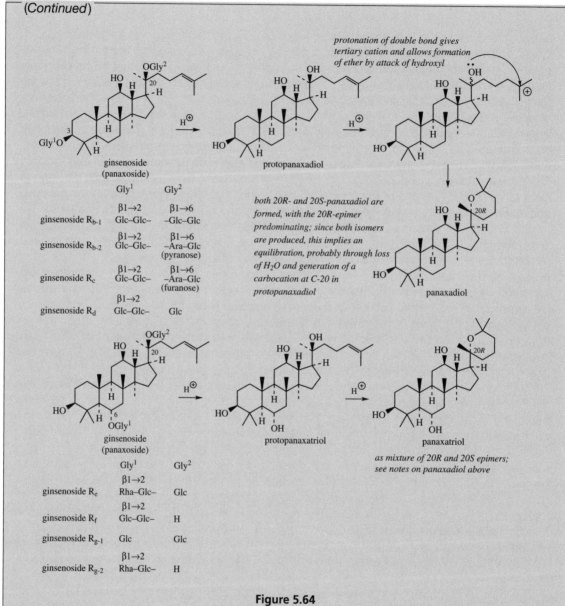

Figure 5.64

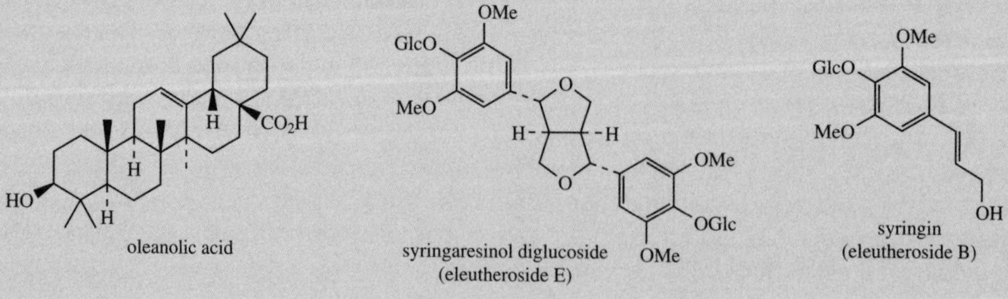

Figure 5.65

(Figure 5.62). Several of these hydroxyls are esterified with aliphatic acids, e.g. acetic, tiglic, and angelic acids.

Modified Triterpenoids

The triterpenoid skeletons may be subjected to a variety of structural modifications as already illustrated. However, the particular modifications considered in this section are those that lead to loss of several skeletal carbon atoms. Pre-eminent amongst such degraded triterpenoids are

the steroids, and these are so important that they are considered separately. Other degraded triterpenoids include the **limonoids** (tetranortriterpenoids), in which four terminal carbons from the side-chain are removed, and the **quassinoids**, which have lost ten carbons, including one of the C-4 methyls. The quassinoids thus have a C_{20} skeleton which could be misinterpreted as a diterpene structure. Biosynthetic information is relatively sparse, but the relationship to precursors of the euphol type is outlined in Figure 5.66. Limonoids are found mainly in plants of the

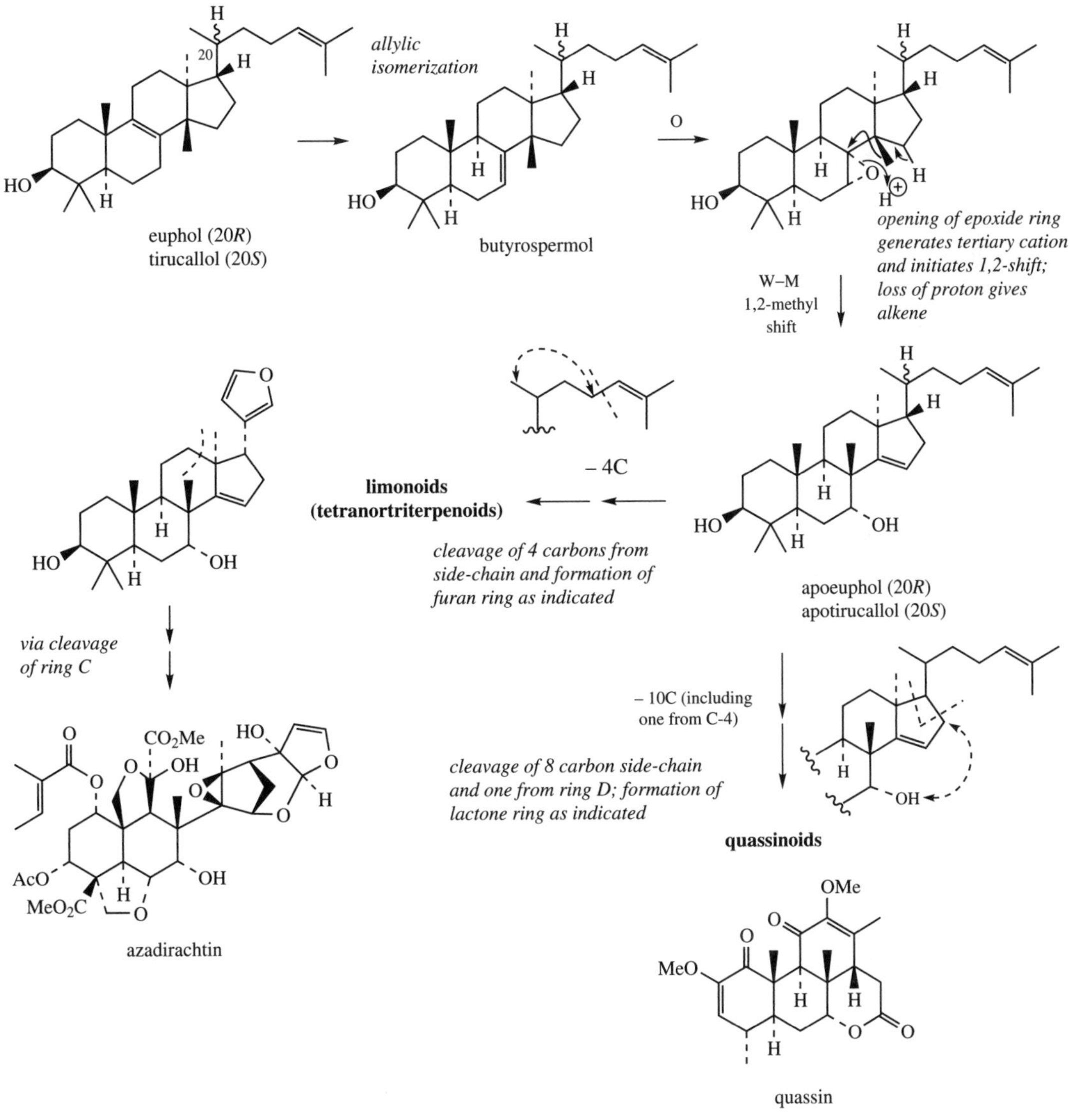

Figure 5.66

families Rutaceae, Meliaceae, and Simaroubaceae. **Azadirachtin** (Figure 5.66) is probably one of the most complex limonoid structures to be encountered, but is currently of considerable interest. This material has potent insect antifeedant properties and is extracted commercially from seeds of the Neem tree (*Azadirachta indica*; Meliaceae) for use as an agricultural pesticide to prevent insect damage to crops. It is a relatively inexpensive and ecologically sound pesticide. Quassinoids are produced by many plants in the Simaroubaceae family, in particular *Quassia*. **Quassin** (Figure 5.66) from *Q. amara* (quassia wood) is a typical example. They have attracted considerable study because of their cytotoxic, antimalarial, and amoebicidal properties.

TETRATERPENES (C₄₀)

The tetraterpenes are represented by only one group of compounds, the **carotenoids**, though several hundred natural structural variants are known. These compounds play a role in photosynthesis, but they are also found in non-photosynthetic plant tissues, in fungi and bacteria. Formation of the tetraterpene skeleton, e.g. **phytoene**, involves tail-to-tail coupling of two molecules of **geranylgeranyl diphosphate** (GGPP) in a sequence essentially analogous to that seen for squalene and triterpenes (Figure 5.67). A cyclopropyl compound, **prephytoene diphosphate** (compare presqualene diphosphate, page 213) is an intermediate in the sequence, and the main difference between the tetraterpene and triterpene pathways is how the resultant allylic cation is discharged. For squalene formation, the allylic cation accepts a hydride ion from NADPH, but for phytoene biosynthesis, a proton is lost, generating a double bond in the centre of the molecule, and thus a short conjugated chain is developed. In plants and fungi, this new double bond has the *Z* (*cis*) configuration, whilst in bacteria, it is *E* (*trans*). This triene system prevents the type of cyclization seen with squalene. Conjugation is extended then by a sequence of desaturation reactions, removing pairs of hydrogens alternately from each side of the triene system, giving eventually **lycopene** (Figure 5.67), which, in common with the majority of carotenoids, has the all-*trans* configuration. This means that in

plants and fungi, an additional isomerization step is involved to change the configuration of the central double bond.

The extended π-electron system confers colour to the carotenoids, and accordingly they contribute yellow, orange, and red pigmentations to plant tissues. Lycopene is the characteristic carotenoid pigment in ripe tomato fruit (*Lycopersicon esculente*; Solanaceae). The orange colour of carrots (*Daucus carota*; Umbelliferae/Apiaceae) is caused by **β-carotene** (Figure 5.68), though this compound is widespread in higher plants. β-Carotene and other natural carotenoids (Figure 5.68) are widely employed as colouring agents for foods, drinks, confectionery, and drugs. β-Carotene displays additional cyclization of the chain ends, which can be rationalized by the carbocation mechanism shown in Figure 5.69. Depending on which proton is lost from the cyclized cation, three different cyclic alkene systems can arise at the end of the chain, described as β-, γ-, or ε-ring systems. **α-Carotene** (Figure 5.68) has a β-ring at one end of the chain, and an ε-type at the other, and is representative of carotenoids lacking symmetry. **γ-Carotene** (a precursor of β-carotene) and **δ-carotene** (a precursor of α-carotene) illustrate carotenoids where only one end of the chain has become cyclized. Oxygenated carotenoids (termed xanthophylls) are also widely distributed, and the biosynthetic origins of the oxygenated rings found in some of these, such as **zeathanthin**, **lutein**, and **violaxanthin** (Figure 5.68), all common green leaf carotenoids, are shown in Figure 5.69. The epoxide grouping in violaxanthin allows further chemical modifications, such as ring contraction to a cyclopentane, exemplified by **capsanthin** (Figure 5.68), the brilliant red pigment of peppers (*Capsicum annuum*; Solanaceae), or formation of an allene as in **fucoxanthin**, an abundant carotenoid in brown algae (*Fucus* species; Fucaceae). **Astaxanthin** (Figure 5.68) is commonly found in marine animals and is responsible for the pink/red coloration of crustaceans, shellfish, and fish such as salmon. These animals are unable to synthesize carotenoids and astaxanthin is produced by modification of plant carotenoids, e.g. β-carotene, obtained in the diet.

Carotenoids function along with chlorophylls in photosynthesis as accessory light-harvesting

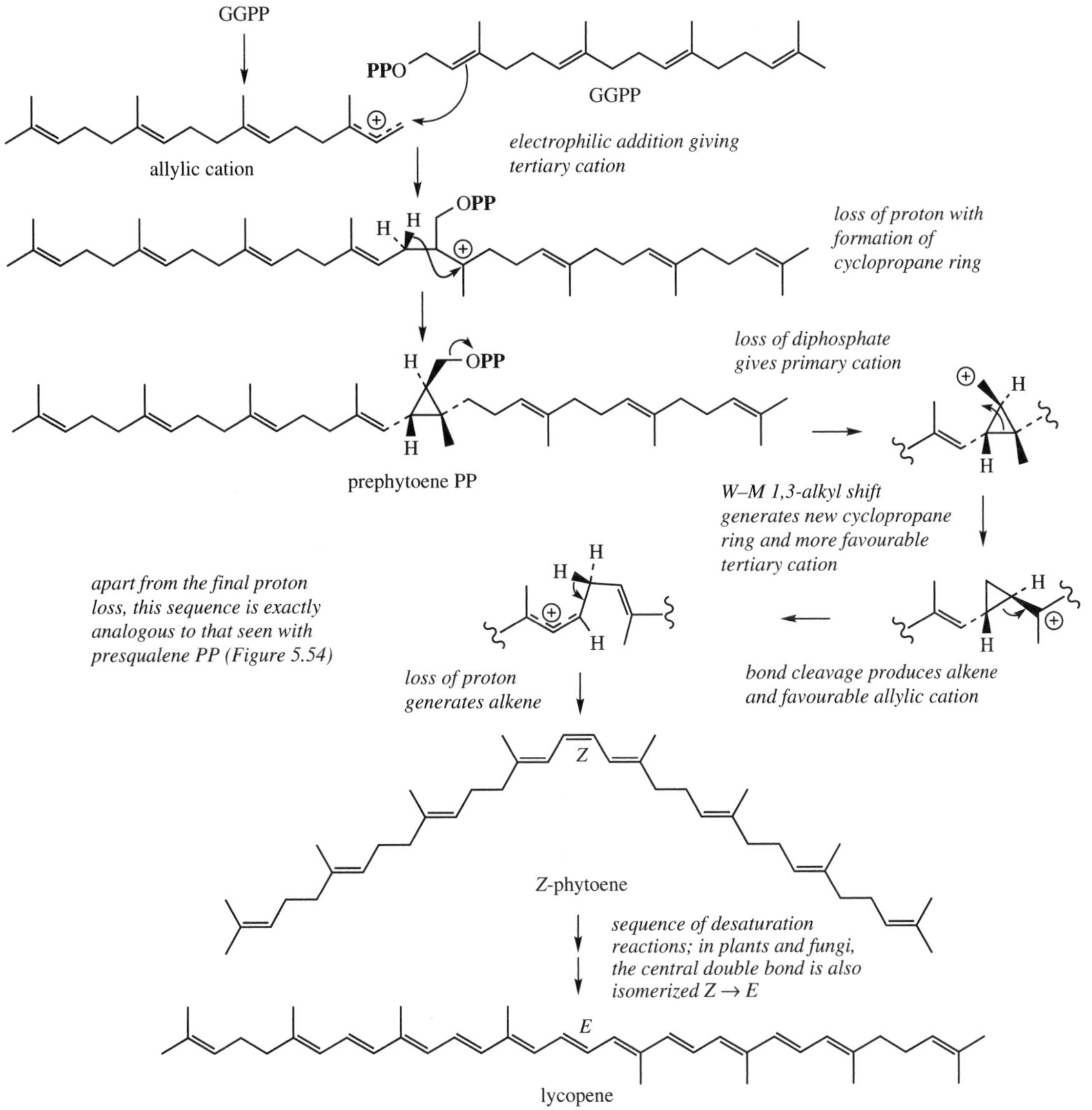

Figure 5.67

pigments, effectively extending the range of light absorbed by the photosynthetic apparatus. They also serve as important protectants for plants and algae against photo-oxidative damage, quenching toxic oxygen species. Some herbicides (bleaching herbicides) act by inhibiting carotenoid biosynthesis, and the unprotected plant is subsequently killed by photo-oxidation. Recent research also suggests carotenoids are important antioxidant molecules in humans, quenching singlet oxygen and scavenging peroxyl radicals, thus minimizing cell damage

and affording protection against some forms of cancer. The most significant dietary carotenoid in this respect is **lycopene**, with tomatoes and processed tomato products featuring as the predominant source. The extended conjugated system allows free radical addition reactions and hydrogen abstraction from positions allylic to this chain.

The A group of vitamins* are important metabolites of carotenoids. **Vitamin A$_1$** (**retinol**) (Figure 5.70) effectively has a diterpene structure, but it is derived in mammals by oxidative

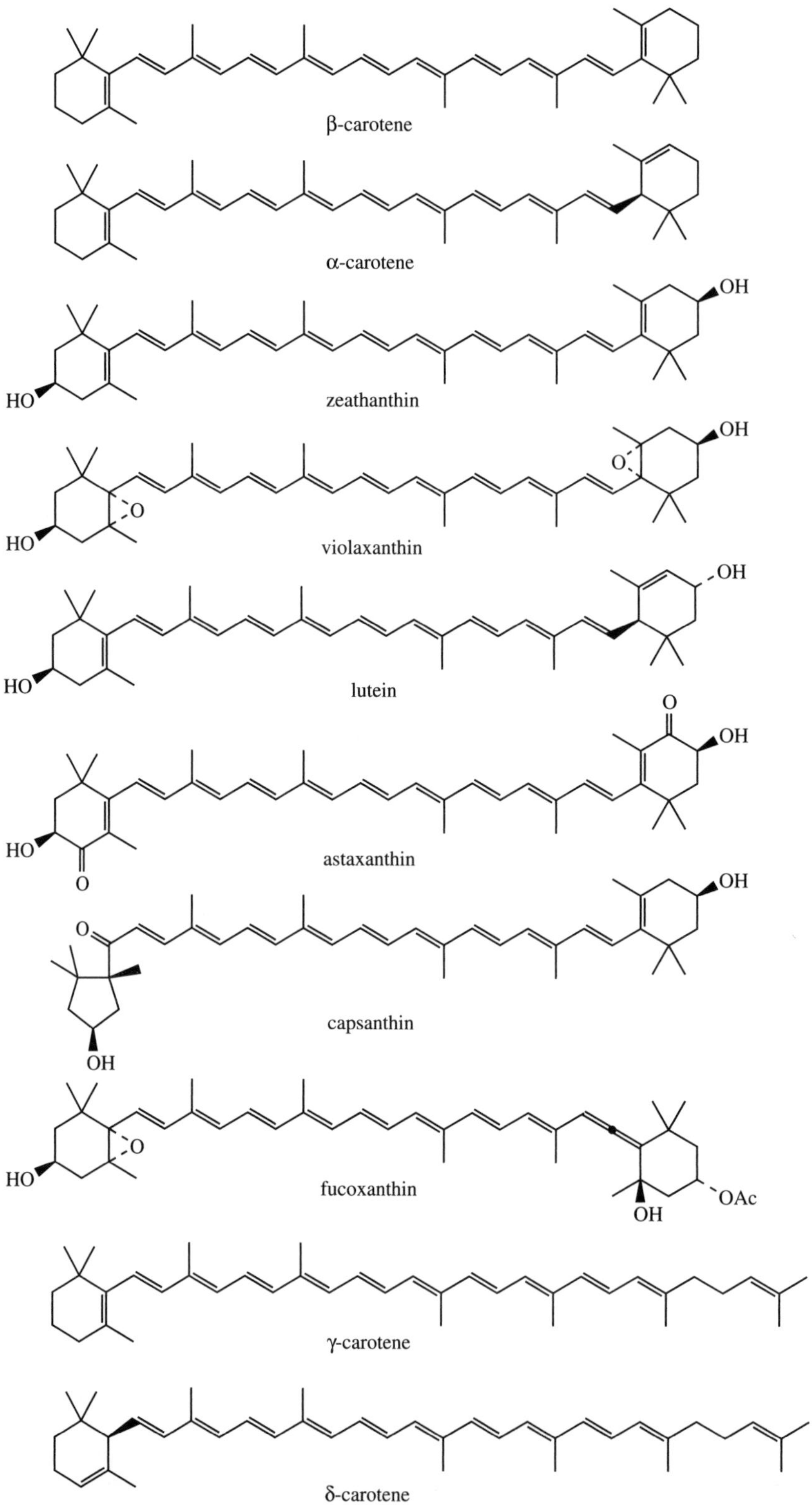

β-carotene

α-carotene

zeathanthin

violaxanthin

lutein

astaxanthin

capsanthin

fucoxanthin

γ-carotene

δ-carotene

Figure 5.68

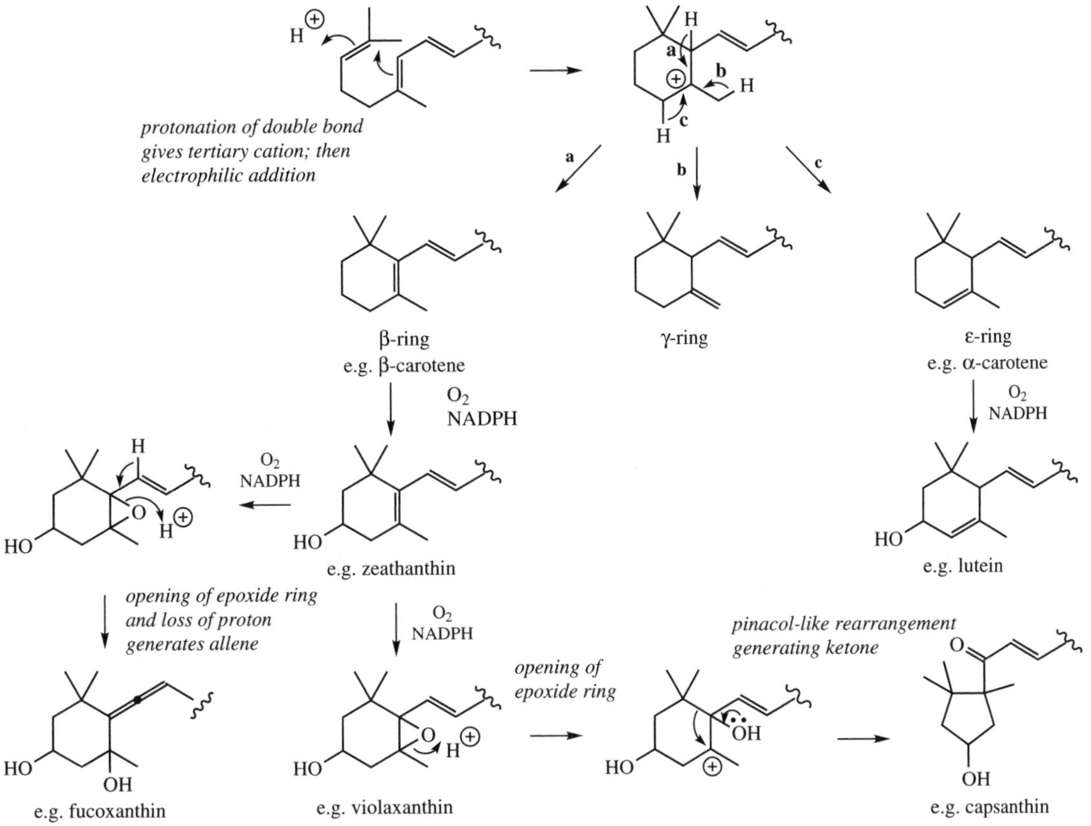

*protonation of double bond
gives tertiary cation; then
electrophilic addition*

β-ring
e.g. β-carotene

O_2
NADPH

γ-ring

ε-ring
e.g. α-carotene

O_2
NADPH

O_2
NADPH

e.g. zeathanthin

e.g. lutein

*opening of epoxide ring
and loss of proton
generates allene*

O_2
NADPH

*opening of
epoxide ring*

*pinacol-like rearrangement
generating ketone*

e.g. fucoxanthin

e.g. violaxanthin

e.g. capsanthin

Figure 5.69

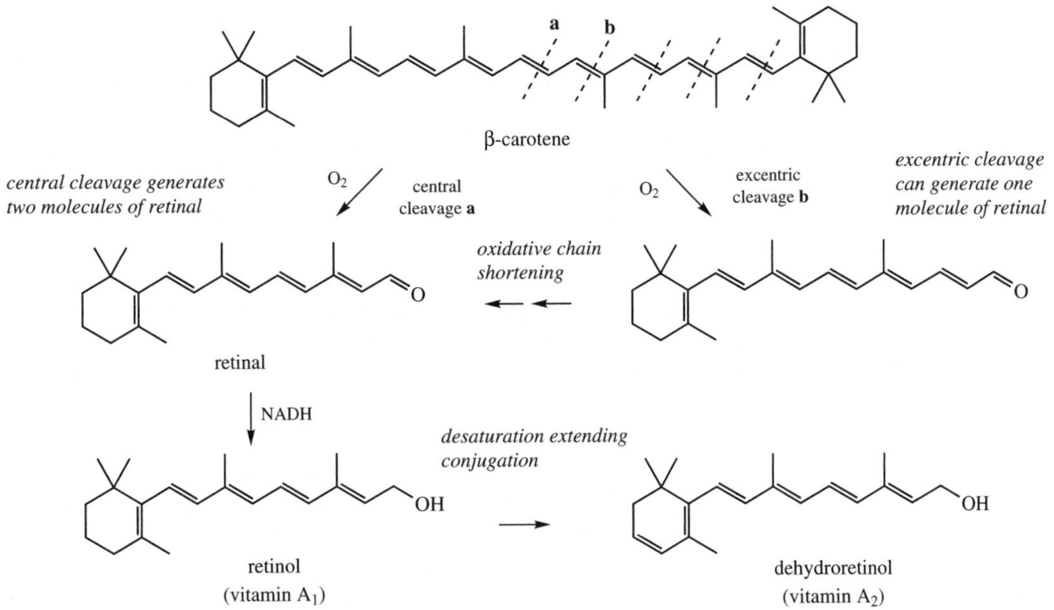

β-carotene

*central cleavage generates
two molecules of retinal*

O_2

central
cleavage **a**

O_2

excentric
cleavage **b**

*excentric cleavage
can generate one
molecule of retinal*

*oxidative chain
shortening*

retinal

NADH

*desaturation extending
conjugation*

retinol
(vitamin A_1)

dehydroretinol
(vitamin A_2)

Figure 5.70

metabolism of a tetraterpenoid, mainly β-carotene, taken in the diet. Cleavage occurs in the mucosal cells of the intestine, and is catalysed by an O_2-dependent dioxygenase, probably via an intermediate peroxide. This can theoretically yield two molecules of the intermediate aldehyde, **retinal**, which is subsequently reduced to the alcohol, retinol (Figure 5.70). Although β-carotene cleaved at the central double bond is capable of giving rise to two molecules of retinol, there is evidence that cleavage can also occur at other double bonds, so-called excentric cleavage (Figure 5.70). Further chain shortening then produces retinal, but only one molecule is produced per molecule of β-carotene. **Vitamin A₂** (**dehydroretinol**) (Figure 5.70) is an analogue of retinol containing a cyclohexadiene ring system; the corresponding aldehyde, and retinal, are also included in the A group of vitamins. Retinol and its derivatives are found only in animal products, and these provide some of our dietary needs. Cod-liver

bixin

crocetin

Figure 5.71

oil and halibut-liver oil are rich sources used as dietary supplements. However, carotenoid sources are equally important. These need to have at least one non-hydroxylated ring system of the β-type, e.g. β-carotene, α-carotene, and γ-carotene.

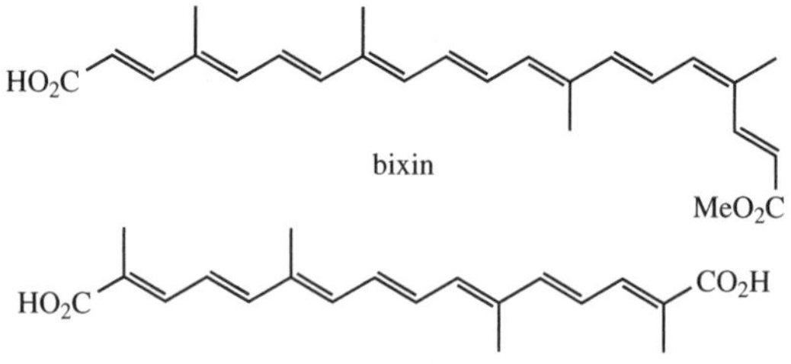

Cleavage of carotenoid precursors is likely to explain the formation of **bixin** and **crocetin** (Figure 5.71) and, indeed, these are classified as apocarotenoids. Large amounts (up to 10%) of the red pigment bixin are found in the seed

---

Vitamin A

Vitamin A₁ (**retinol**) and **vitamin A₂** (dehydroretinol) (Figure 5.72) are fat-soluble vitamins found only in animal products, especially eggs, dairy products, and animal livers and kidneys. Fish liver oils, e.g. cod liver oil and halibut liver oil (see Table 3.2) are particularly rich sources. They exist as the free alcohols, or as esters with acetic and palmitic acid. Vitamin A₂ has about 40% of the activity of vitamin A₁. Carotenoid precursors (provitamins) are widely distributed in plants, and after ingestion, these are subsequently transformed into vitamin A in the liver. Green vegetables and rich plant sources such as carrots help to provide adequate levels. A deficiency of vitamin A leads to vision defects, including impairment at low light levels (night blindness) and a drying and degenerative disease of the cornea. It also is necessary for normal growth of young animals. Retinoids (vitamin A and analogues) are now known to act as signalling molecules which regulate diverse aspects of cell differentiation, embryonic development, growth, and vision. For the processes of vision, retinol needs to be converted first by oxidation into the aldehyde all-*trans*-retinal, and then by enzymic isomerization to *cis*-retinal (Figure 5.73). *cis*-Retinal is then bound to the protein opsin in the retina via a Schiff base linkage to give the red visual pigment rhodopsin, and its sensitivity to light involves isomerization of the *cis*-retinal portion back to the all-*trans* form, thus translating the light energy into molecular change, which triggers a nerve impulse to the brain. The absorption of light energy promotes an electron from a π- to a π*-orbital, thus temporarily destroying the double bond character and allowing rotation. A similar *cis*–*trans* isomerization affecting cinnamic acids was discussed under coumarins (see page 142). All-*trans*-retinal is then subsequently released for the process to continue. Vitamin A is relatively unstable, and sensitive to oxidation and light. Antioxidant stabilizers such as vitamin E and vitamin C are sometimes added. It is more stable in oils such as the fish liver oils, which are thus good vehicles for administering the vitamin. Synthetic material is also used. Excessive intake of

(Continues)

(Continued)

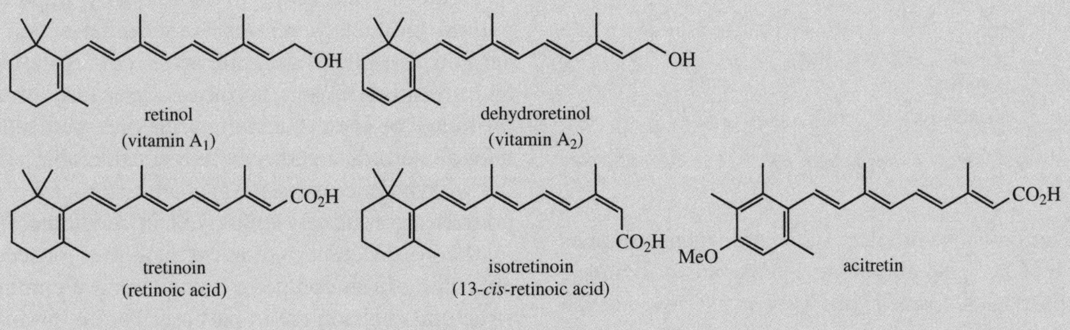

Figure 5.72

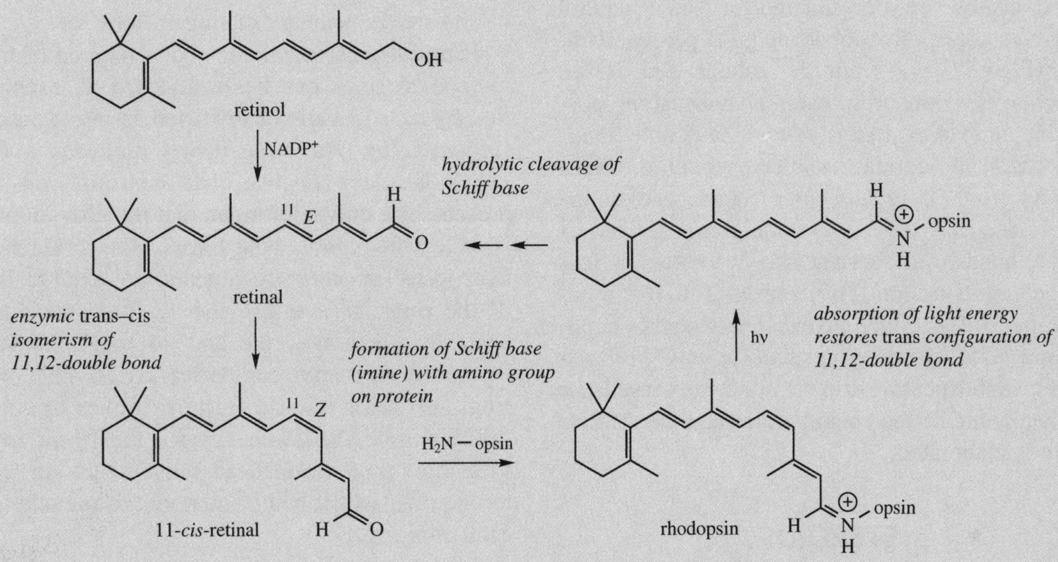

Figure 5.73

vitamin A can lead to toxic effects, including pathological changes in the skin, hair loss, blurred vision, and headaches.

The synthetic retinoic acids **tretinoin** (retinoic acid) and **isotretinoin** (13-*cis*-retinoic acid) (Figure 5.72) are retinoids that are used as topical or oral treatments for acne vulgaris, reducing levels of dehydroretinol and modifying skin keratinization. Dehydroretinol levels in the skin become markedly elevated in conditions such as eczema and psoriasis. **Acitretin** (Figure 5.72) is an aromatic analogue which can give relief in severe cases of psoriasis. All these materials can produce toxic side-effects.

coats of annatto (*Bixa orellana*; Bixaceae), and bixin is widely used as a natural food colorant. Crocetin, in the form of esters with gentiobiose [D-Glc(β1 → 6)D-Glc], is the major pigment in stigmas of *Crocus sativus* (Iridaceae), which comprise the spice saffron.

HIGHER TERPENOIDS

Terpenoid fragments containing several isoprene units are found as alkyl substituents in shikimate-derived quinones (see page 158). Thus ubiquinones typically have $C_{40}-C_{50}$ side-chains,

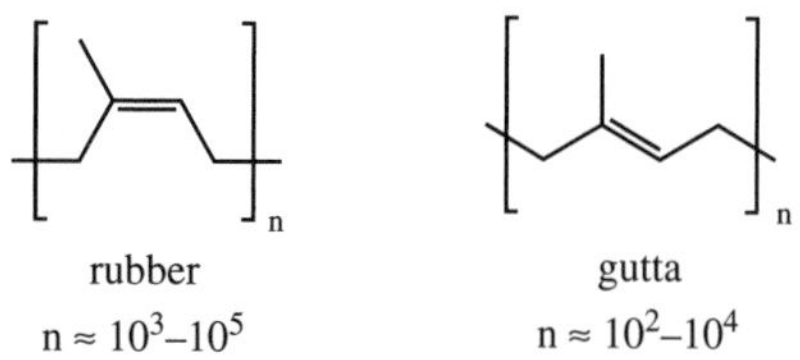

rubber
$n \approx 10^3-10^5$

gutta
$n \approx 10^2-10^4$

Figure 5.74

plastoquinones usually C_{45}, and menaquinones up to C_{65}. The alkylating agents are polyprenyl diphosphates, formed simply by an extension of the prenyltransferase reaction (see page 172), repeatedly adding IPP residues. Even longer polyisoprene chains are encountered in some natural polymers, especially rubber and gutta percha. **Rubber** (Figure 5.74), from the rubber tree *Hevea brasiliensis* (Euphorbiaceae), is unusual in possessing an extended array of *cis* (*Z*) double bonds rather than the normal *trans* configuration. **Gutta percha**, from *Palaquium gutta* (Sapotaceae) on the other hand, has *trans* (*E*) double bonds. The *cis* double bonds in rubber are known to arise by loss of the *pro-S* proton (H_S) from C-2 of IPP (contrast loss of H_R, which gives a *trans* double bond) (Figure 5.75). However, a small (up to C_{20}) *trans*-allylic diphosphate initiator is actually used for the beginning of the chain before the extended *cis* chain is elaborated.

STEROIDS

Stereochemistry

The steroids are modified triterpenoids containing the tetracyclic ring system of lanosterol (Figure 5.55), but lacking the three methyl groups at C-4 and C-14. **Cholesterol** (Figure 5.76) typifies the fundamental structure, but further modifications, especially to the side-chain, help to create a wide range of biologically important natural products, e.g. sterols, steroidal saponins, cardioactive glycosides, bile acids, corticosteroids, and mammalian sex hormones. Because of the profound biological activities encountered, many natural steroids together with a considerable number of synthetic and semi-synthetic steroidal compounds are routinely employed in medicine. The markedly different biological activities observed emanating from compounds containing a common structural skeleton are in part ascribed to the functional groups attached to the steroid nucleus, and in part to the overall shape conferred on this nucleus by the stereochemistry of ring fusions.

Ring systems containing six-membered or five-membered rings can be *trans*-fused as exemplified by *trans*-decalin or *cis*-fused as in *cis*-decalin (Figure 5.76). The *trans*-fusion produces a flattish molecule when two chair conformations are present. The only conformational mobility allowed is to less favourable boat forms. Both bridgehead hydrogens (or other substituents) are axial to both of the rings. In contrast, the *cis*-fused decalin is basically a bent molecule, and is found to be flexible in that alternative conformers are possible, both rings still being in chair form. However, this flexibility will be lost if either ring is then *trans*-fused to a third ring. Bridgehead substituents are axial to one ring, whilst being equatorial to the other, in each conformer.

In natural steroids, there are examples of the A/B ring fusion being *trans* or *cis*, or having unsaturation, either Δ^4 or Δ^5. In some compounds, notably the oestrogens, ring A can even be aromatic; clearly there can then be no bridgehead substituent at C-10 and the normal C-10 methyl (C-19) must therefore be lost. All natural steroids have a *trans* B/C fusion, though *cis* forms can be

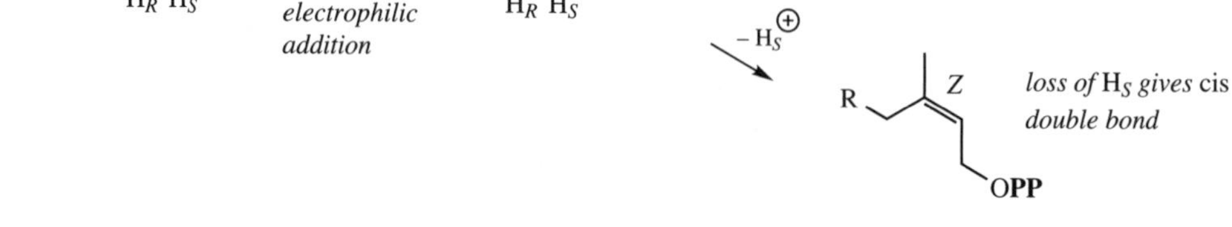

Figure 5.75

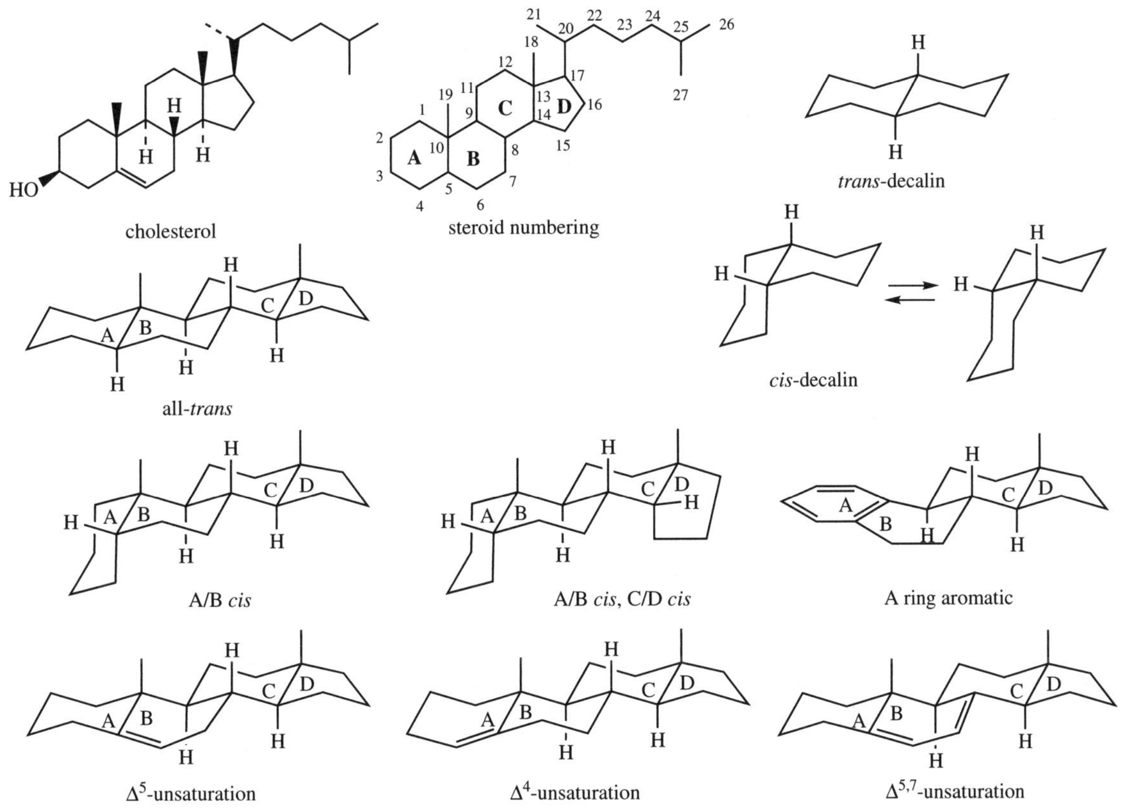

Figure 5.76

made synthetically. The C/D fusion is also usually *trans*, though there are notable exceptions such as the cardioactive glycosides. Of course, such comments apply equally to some of the triterpenoid structures already considered. However, it is in the steroid field that the relationship between stereochemistry and biological activity is most marked. The overall shapes of some typical steroid skeletons are shown in Figure 5.76.

Systematic **steroid nomenclature** is based on a series of parent hydrocarbons, including **estrane**, **androstane**, **pregnane**, **cholane**, **cholestane**, **ergostane**, **campestane**, **stigmastane** and **poriferastane** (Figure 5.77). The triterpenoid hydrocarbons **lanostane** and **cycloartane** are similarly used in systematic nomenclature and are also included in Figure 5.77. It is usual to add only unsaturation (ene/yne) and the highest priority functional group as suffixes to the root name; other groups are added as prefixes. Stereochemistry of substituents is represented by α (on the lower face of the molecule when it is drawn according

to customary conventions as in Figure 5.77), or β (on the upper face). Ring fusions may be designated by using α or β for the appropriate bridgehead hydrogen, particularly those at positions 5 and 14, which will define the A/B and C/D fusions respectively, e.g. 5β-cholestane has the A/B rings *cis*-fused. Since the parent hydrocarbon assumes that ring fusions are *trans*, the stereochemistry for ring fusions is usually only specified where it is *cis*. Cholesterol is thus cholest-5-en-3β-ol or 5-cholesten-3β-ol. The term *nor* is affixed to indicate loss of a carbon atom, e.g. 19-norsteroids (see page 274) lack C-19 (the methyl at C-10).

Cholesterol

In animals, the triterpenoid alcohol **lanosterol** (Figure 5.55) is converted into **cholesterol*** (Figure 5.76), a process requiring, as well as the loss of three methyl groups, reduction of the side-chain double bond, and generation of a $\Delta^{5,6}$

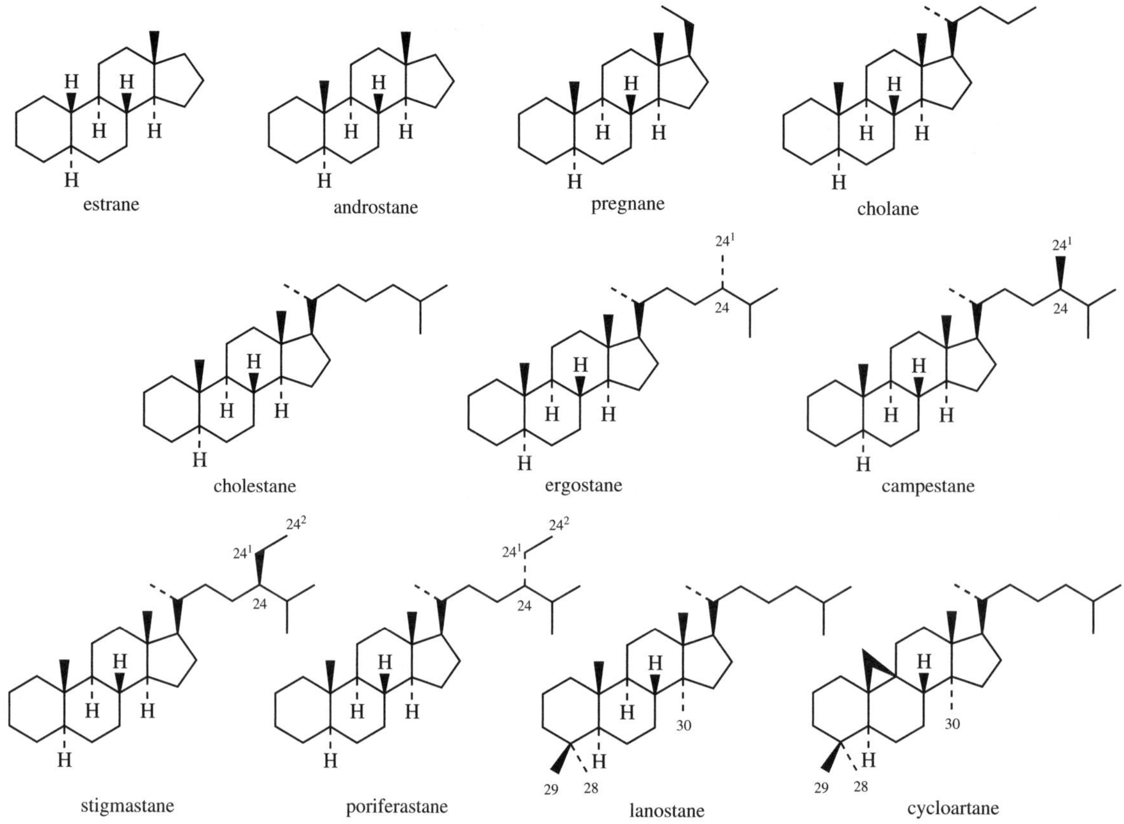

Figure 5.77

double bond in place of the $\Delta^{8,9}$ double bond. The sequence of these steps is to some extent variable, and dependent on the organism involved. Accordingly, these individual transformations are considered rather than the overall pathway.

The methyl at C-14 is usually the one lost first, and this is removed as formic acid. The reaction is catalysed by a cytochrome P-450 monooxygenase, which achieves two oxidation reactions to give the 14α-formyl derivative (Figure 5.78), and loss of this formyl group giving the $\Delta^{8,14}$ diene, most probably via homolytic cleavage of the peroxy adduct as indicated (compare similar peroxy adducts involved in side-chain cleavage from ring D, page 278, and in A ring aromatization, page 278). The 14-demethyl sterol is then obtained by an NADPH-dependent reduction step, the 15-proton being derived from water.

Loss of the C-4 methyls occurs sequentially, usually after removal of the 14α-methyl, with both carbons being cleaved off via a decarboxylation

mechanism (Figure 5.79). This is facilitated by oxidizing the 3-hydroxyl to a ketone, thus generating intermediate β-keto acids. In this sequence, the enolate is restored to a ketone, in which the remaining C-4 methyl takes up the more favourable equatorial (4α) orientation.

The side-chain Δ^{24} double bond is reduced by an NADPH-dependent reductase, hydride from the coenzyme being added at C-25, with H-24 being derived from water (Figure 5.80). The Δ^8 double bond is effectively migrated to Δ^5 via Δ^7 and the $\Delta^{5,7}$ diene (Figure 5.81). This sequence involves an allylic isomerization, a dehydrogenation, and a reduction. Newly introduced protons at C-9 and C-8 originate from water, and that at C-7 from NADPH.

The role of lanosterol in non-photosynthetic organisms (animals, fungi) is taken in photosynthetic organisms (plants, algae) by the cyclopropane triterpenoid **cycloartenol** (Figure 5.55). This cyclopropane feature is found in a number of

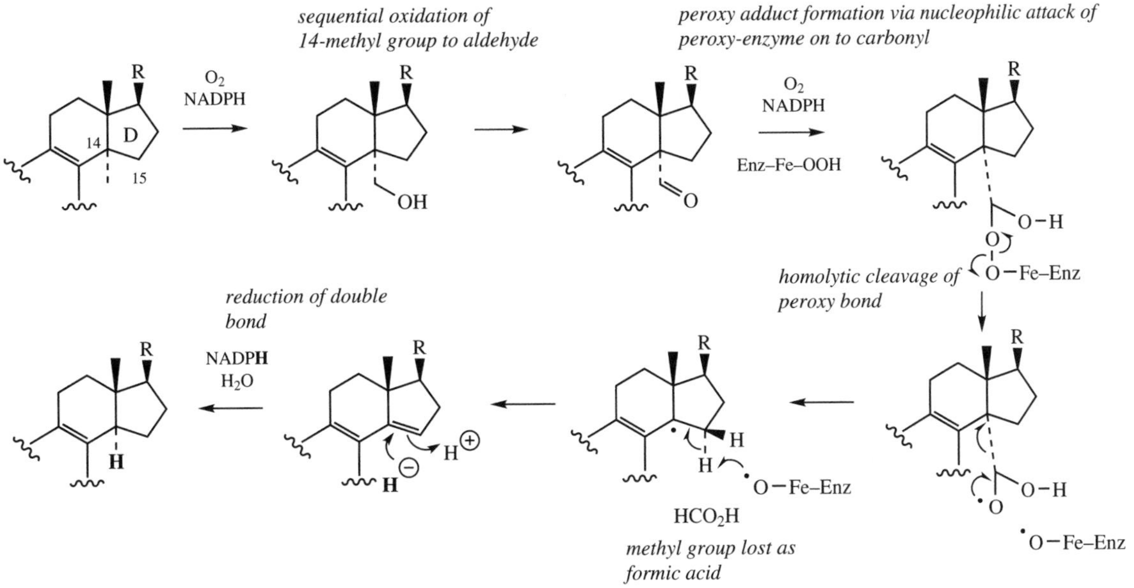
sequential oxidation of
14-methyl group to aldehyde
peroxy adduct formation via nucleophilic attack of
peroxy-enzyme on to carbonyl
O2
NADPH
Enz–Fe–OOH
homolytic cleavage of
peroxy bond
reduction of double
bond
NADPH
H2O
HCO2H
methyl group lost as
formic acid
O–Fe–Enz

Figure 5.78

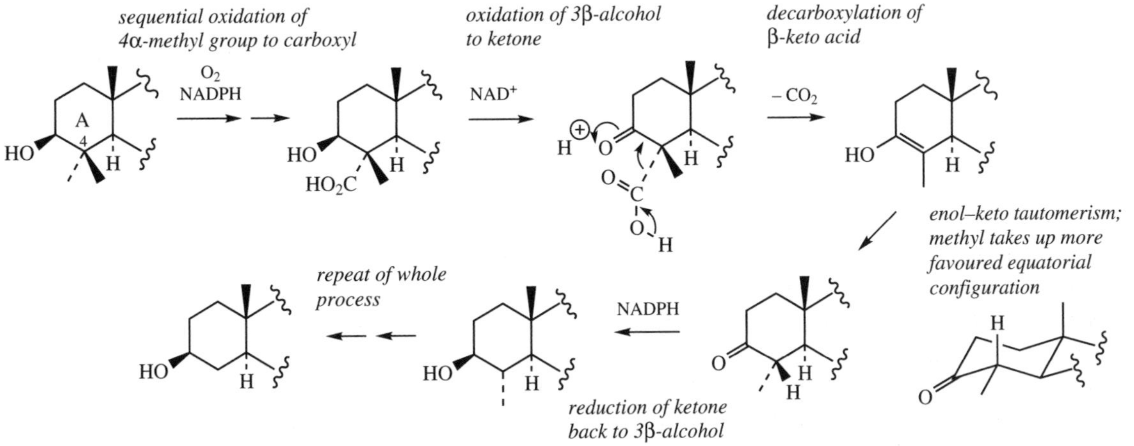
sequential oxidation of
4α-methyl group to carboxyl
oxidation of 3β-alcohol
to ketone
decarboxylation of
β-keto acid
O2
NADPH
NAD+
– CO2
enol–keto tautomerism;
methyl takes up more
favoured equatorial
configuration
repeat of whole
process
NADPH
reduction of ketone
back to 3β-alcohol

Figure 5.79

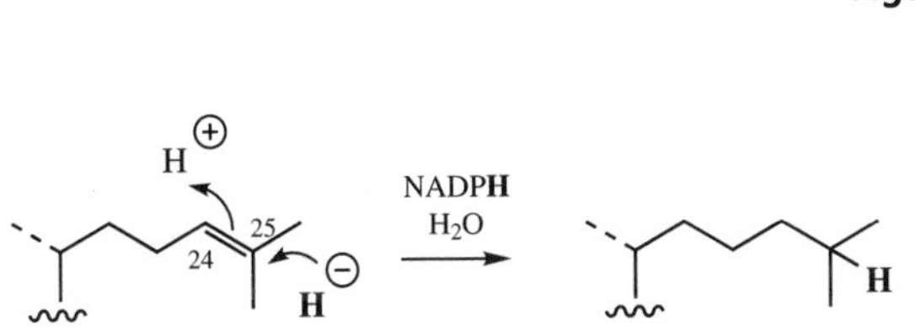
NADPH
H2O

Figure 5.80

plant sterols, but the majority of plant steroids contain the normal methyl at C-10. This means that, in addition to the lanosterol → cholesterol modifications outlined above, a further mechanism to reopen the cyclopropane ring is necessary. This is shown in Figure 5.82. The stereochemistry at C-8 (Hβ) is unfavourable for a concerted mechanism involving loss of H-8 with cyclopropane ring opening. It is suggested therefore that a nucleophilic group from the enzyme attacks C-9, opening the cyclopropane ring and incorporating a proton from water. A *trans* elimination then generates the Δ^8 double bond. The cyclopropane ring-opening process seems specific to 4α-monomethyl sterols. In plants, removal of the first 4-methyl group (4α;

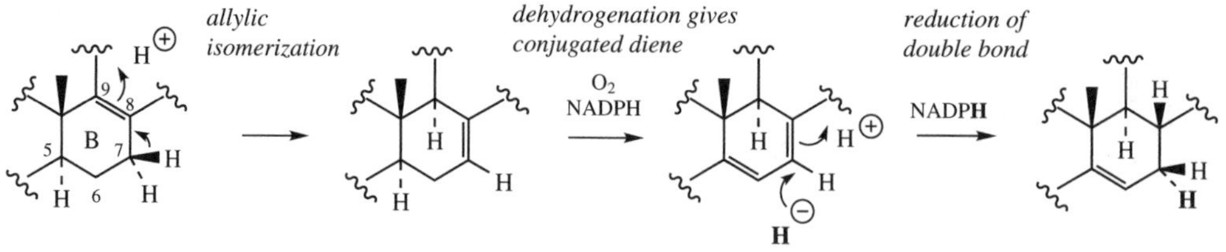

Figure 5.81

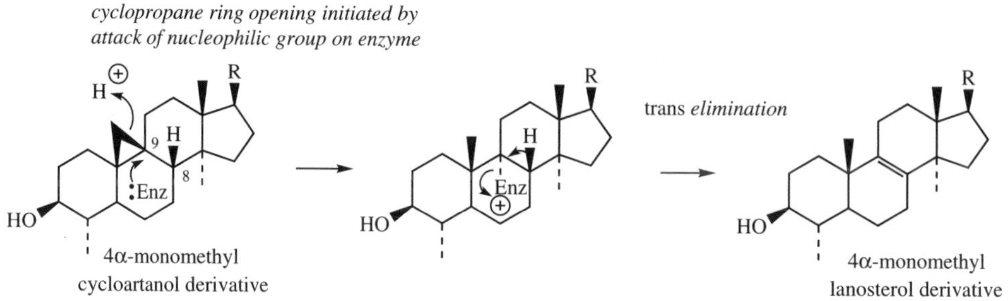

Figure 5.82

note the remaining 4β-methyl group then takes up the α-orientation) is also known to precede loss of the 14α-methyl. Accordingly, the substrate shown in Figure 5.82 has both 4α- and 14α-methyl groups. The specificity of the cyclopropane ring-opening enzyme means cycloartenol is not converted into lanosterol, and lanosterol is thus absent from virtually all plant tissues. Cholesterol is almost always present in plants, though often in only trace amounts, and is formed via cycloartenol.

Cholesterol

Cholesterol (Figure 5.76) is the principal animal sterol and since it is a constituent of cell membranes has been found in all animal tissues. Human gallstones are almost entirely composed of cholesterol precipitated from the bile. Cholesterol is currently available in quantity via the brains and spinal cords of cattle as a by-product of meat production, and these form one source for medicinal steroid semi-synthesis. Large quantities are also extractable from lanolin, the fatty material coating sheep's wool. This is a complex mixture of esters of long chain fatty acids (including straight-chain, branched-chain, and hydroxy acids) with long chain aliphatic alcohols and sterols. Cholesterol is a major sterol component. Saponification of crude lanolin gives an alcohol fraction (lanolin alcohols or wool alcohols) containing about 34% cholesterol and 38% lanosterol/dihydrolanosterol. Wool alcohols are also used as an ointment base.

Although the processes involved are quite complex, there appears to be a clear correlation between human blood cholesterol levels and heart disease. Atherosclerosis is a hardening of the arteries caused by deposition of cholesterol, cholesterol esters, and other lipids in the artery wall, causing a narrowing of the artery and thus an increased risk of forming blood clots (thrombosis). Normally, most of the cholesterol serves a structural element in cell walls, whilst the remainder is transported via the blood and is used for synthesis of steroid hormones, vitamin D (page 259), or bile acids (page 261). Transport of cholesterol is facilitated by formation of lipoprotein carriers, comprising protein and phospholipid shells

(Continues)

surrounding a core of cholesterol, in both free and esterified forms. Risk of atherosclerosis increases with increasing levels of low density lipoprotein (LDL) cholesterol, and is reduced with increasing levels of high density lipoprotein (HDL) cholesterol. Blood LDL cholesterol levels are thus a good statistical indicator of the potential risk of a heart attack. The risks can be lessened by avoiding foods rich in cholesterol, e.g. eggs, reducing the intake of foods containing high amounts of saturated fatty acids such as animal fats, and replacing these with vegetable oils and fish that are rich in polyunsaturated fatty acids (see page 40). Blood LDL cholesterol levels may also be reduced by incorporating into the diet plant sterol esters or plant stanol esters, which reduce the absorption of cholesterol (see page 256). In humans, dietary cholesterol is actually a smaller contributor to LDL cholesterol levels than is dietary saturated fat. Cholesterol biosynthesis may also be inhibited by drug therapy using specific inhibitors of the mevalonate pathway, e.g. lovastatin and related compounds (see page 112).

Steroidal Saponins

Steroidal saponins have similar biological properties to the triterpenoid saponins (see page 219), but are less widely distributed in nature. They are found in many monocotyledon families, especially the Dioscoreaceae (e.g. *Dioscorea*), the Agavaceae (e.g. *Agave, Yucca*) and the Liliaceae (e.g. *Smilax, Trillium*). Their sapogenins are C_{27} sterols in which the side-chain of cholesterol has undergone modification to produce a spiroketal, e.g. **dioscin** (Figure 5.83) from *Dioscorea*. Acid hydrolysis of dioscin liberates the aglycone **diosgenin**. All the

steroidal saponins have the same configuration at the *spiro* centre C-22, but stereoisomers at C-25 exist, e.g. **yamogenin** (Figure 5.84), and often mixtures of the C-25 stereoisomers cooccur in the plant. The sugar moiety is usually at position 3, and typically contains fewer monosaccharide units than found with triterpenoid saponins. One to three monosaccharide units are most common. The three-dimensional shape of diosgenin is indicated in Figure 5.83.

The spiroketal function is derived from the cholesterol side-chain by a series of oxygenation reactions, hydroxylating C-16 and one of the

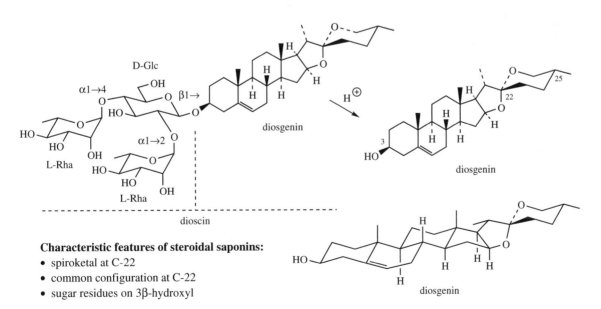

Characteristic features of steroidal saponins:
- spiroketal at C-22
- common configuration at C-22
- sugar residues on 3β-hydroxyl

diosgenin

diosgenin

diosgenin

dioscin

Figure 5.83

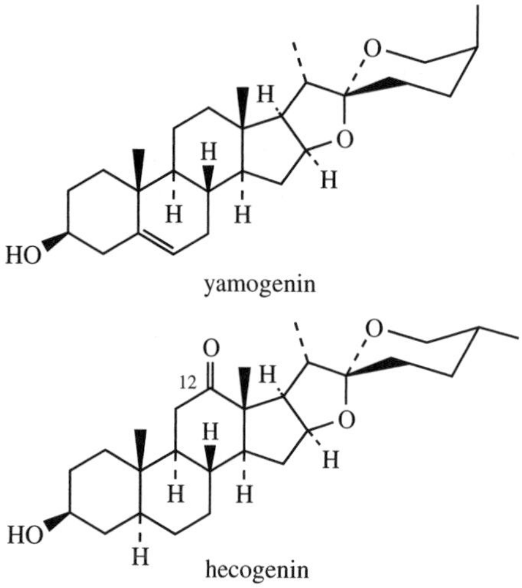

yamogenin

hecogenin

Figure 5.84

terminal methyls, and then producing a ketone function at C-22 (Figure 5.85). This proposed intermediate is transformed into the hemiketal and then the spiroketal. The chirality at C-22 is fixed by the stereospecificity in the formation of the

ketal whilst the different possible stereochemistries at C-25 are dictated by whether C-26 or C-27 is hydroxylated in the earlier step. Glycoside derivatives, e.g. **protodioscin** (Figure 5.86) have been isolated from plants. These are readily hydrolysed, and then spontaneously cyclize to the spiroketal (Figure 5.86). Allowing homogenized fresh plant tissues to stand and autolyse through the action of endogenous glycosidase enzymes not only achieves cyclization of such open-chain saponins, but can hydrolyse off the sugar units at C-3, thus yielding the aglycone or sapogenin. This is a standard approach employed in commercial production of steroidal sapogenins, important starting materials for the semi-synthesis of steroidal drugs. **Diosgenin** is the principal example and is obtained from Mexican yams (*Dioscorea* spp.; Dioscoreaceae)*. Fenugreek* (*Trigonella foenum-graecum*; Leguminosae/Fabaceae) is another potentially useful commercial source. Sisal* (*Agave sisalana*; Agavaceae) is also used commercially, yielding **hecogenin** (Figure 5.84), a 12-keto derivative with *trans*-fused A/B rings, the result of reduction of the Δ^5 double bond.

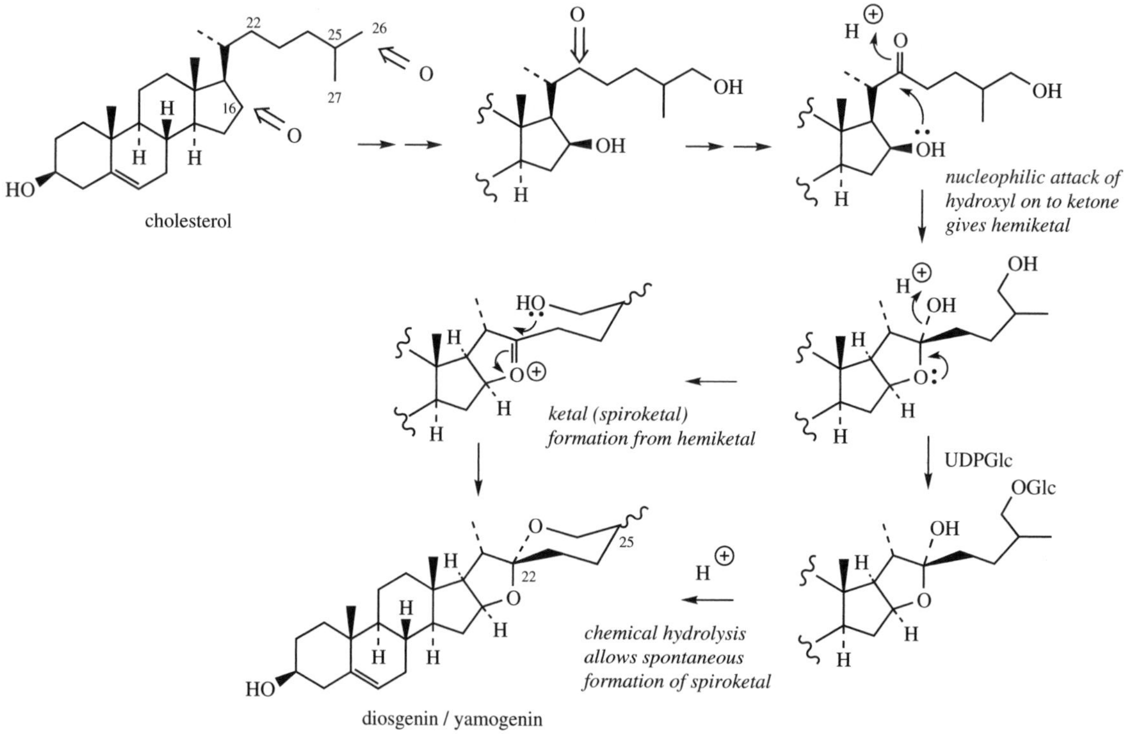

cholesterol

nucleophilic attack of hydroxyl on to ketone gives hemiketal

ketal (spiroketal) formation from hemiketal

UDPGlc

chemical hydrolysis allows spontaneous formation of spiroketal

diosgenin / yamogenin

Figure 5.85

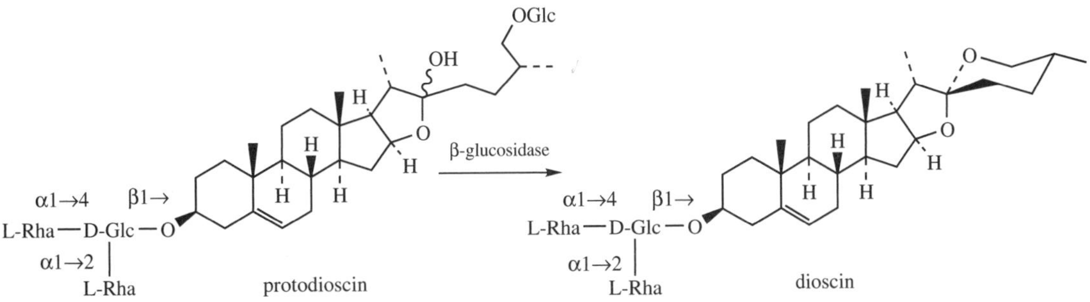

Figure 5.86

Dioscorea

About 600 species of *Dioscorea* (Dioscoreaceae) are known, and a number of these are cultivated for their large starchy tubers, commonly called yams, which are an important food crop in many parts of the world. Important edible species are *Dioscorea alata* and *D. esculenta* (S E Asia), *D. rotundata* and *D. cayenensis* (W Africa) and *D. trifida* (America). A number of species accumulate quite high levels of saponins in their tubers, which make them bitter and inedible, but these provide suitable sources of steroidal material for drug manufacture.

Dioscoreas are herbaceous, climbing, vinelike plants, the tuber being totally buried, or sometimes protruding from the ground. Tubers weigh anything up to 5 kg, but in some species, tubers have been recorded to reach weights as high as 40–50 kg. Drug material is obtained from both wild and cultivated plants, with plants collected from the wild having been exploited considerably more than cultivated ones. Commercial cultivation is less economic, requiring a 4–5 year growing period, and some form of support for the climbing stems. Much of the world's production has come from Mexico, where tubers from *D. composita* (barbasco), *D. mexicana*, and *D. floribunda*, mainly harvested from wild plants, are utilized. The saponin content of the tubers varies, usually increasing as tubers become older. Typically, tubers of *D. composita* may contain 4–6% total saponins, and *D. floribunda* 6–8%. Other important sources of *Dioscorea* used commercially now include India (*D. deltoidea*), South Africa (*D. sylvatica*) and China (*D. collettii*, *D. pathaica*, and *D. nipponica*).

Sapogenins are isolated by chopping the tubers, allowing them to ferment for several days, then completing the hydrolysis of saponins by heating with aqueous acid. The sapogenins can then be solvent extracted. The principal sapogenin in the species given above is diosgenin (Figure 5.83), with small quantities of the 25β-epimer yamogenin (Figure 5.84). Demand for diosgenin for pharmaceuticals is huge, equivalent to 10 000 tonnes of *Dioscorea* tuber per annum, and it is estimated that about 60% of all steroidal drugs are derived from diosgenin.

Powdered Dioscorea (wild yam) root or extract is also marketed to treat the symptoms of menopause as an alternative to hormone replacement therapy (see page 279). Although there is a belief that this increases levels of progesterone, which is then used as a biosynthetic precursor of other hormones, there is no evidence that diosgenin is metabolized in the human body to progesterone, and any beneficial effects may arise from diosgenin itself.

(Continues)

(Continued)

Fenugreek

The seeds of fenugreek (*Trigonella foenum-graecum*; Leguminosae/Fabaceae) are an important spice material, and are ingredients in curries and other dishes. The plant is an annual, and is grown widely, especially in India, both as a spice and as a forage crop. Seeds can yield, after hydrolysis, 1–2% of sapogenins, principally diosgenin (Figure 5.83) and yamogenin (Figure 5.84). Although yields are considerably lower than from *Dioscorea*, the ease of cultivation of fenugreek and its rapid growth make the plant a potentially viable crop for steroid production in temperate countries. Field trials of selected high-yielding strains have been conducted.

Sisal

Sisal (*Agave sisalana*; Agavaceae) has long been cultivated for fibre production, being the source of sisal hemp, used for making ropes, sacking and matting. The plant is a large, rosette-forming succulent with long, tough, spine-tipped leaves containing the very strong fibres. The main area of sisal cultivation is East Africa (Tanzania, Kenya), with smaller plantations in other parts of the world. The sapogenin hecogenin (Figure 5.84) was initially produced from the leaf waste (0.1–0.2% hecogenin) after the fibres had been stripped out. The leaf waste was concentrated, allowed to ferment for several days, then treated with steam under pressure to complete hydrolysis of the saponins. Filtration then produced a material containing about 12% hecogenin, plus other sapogenins. This was refined further in the pharmaceutical industry. Other sapogenins present include tigogenin and neotigogenin (Figure 5.87).

As the demand for natural fibres declined due to the availability of synthetics, so did the supply of sisal waste and thus hecogenin. In due course, hecogenin became a more valuable commodity than sisal, and efforts were directed specifically towards hecogenin production. This has resulted in the cultivation of *Agave* hybrids with much improved hecogenin content.

The fermented sap of several species of Mexican *Agave* provides the alcoholic beverage pulque. Distillation of the fermented sap produces tequila.

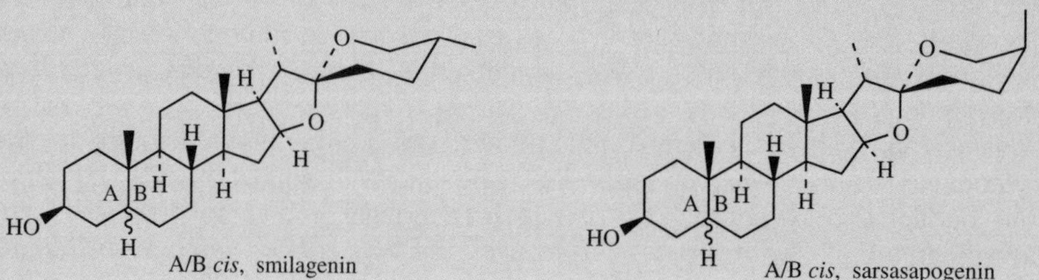

A/B *cis*, smilagenin
A/B *trans*, tigogenin

A/B *cis*, sarsasapogenin
A/B *trans*, neotigogenin

Figure 5.87

Some steroidal alkaloids are nitrogen analogues of steroidal saponins, and display similar properties such as surface activity and haemolytic activity, but these compounds *are* toxic when ingested. These types of compound, e.g. **solasonine** (Figure 5.88) (aglycone **solasodine**), are found in many plants of the genus *Solanum* (Solanaceae), and such plants must thus be regarded as potentially toxic. In contrast to the oxygen analogues, all compounds have the same stereochemistry at C-25 (methyl always equatorial), whilst isomers at C-22 do exist, e.g. **tomatine**

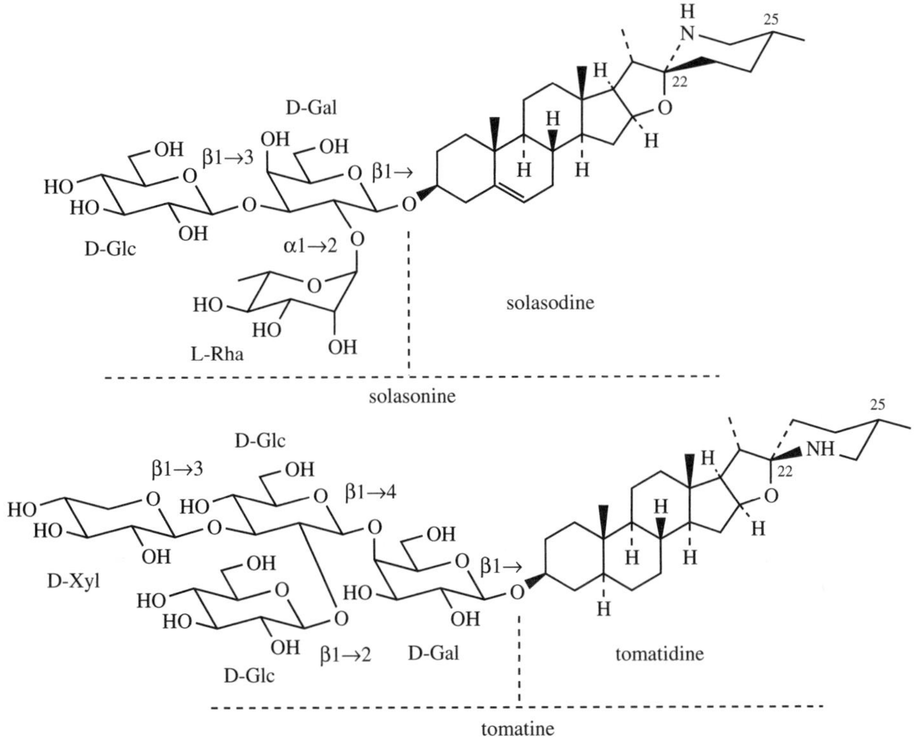

Figure 5.88

(Figure 5.88) (aglycone **tomatidine**) from tomato (*Lycopersicon esculente*; Solanaceae). The nitrogen atom is introduced by a transamination reaction, typically employing an amino acid as donor (see page 20). Since the production of medicinal steroids from steroidal saponins requires preliminary degradation to remove the ring systems containing the original cholesterol side-chain, it is immaterial whether these rings contain oxygen or nitrogen. Thus, plants rich in solasodine or tomatidine could also be employed for commercial steroid production (see page 391).

Smilagenin and **sarsasapogenin** (Figure 5.87) found in sarsaparilla* (*Smilax* spp.; Liliaceae/ Smilacaceae) are reduced forms of diosgenin and yamogenin respectively. These contain *cis*-fused A/B rings, whilst the corresponding *trans*-fused systems are present in **tigogenin** and **neotigogenin** (Figure 5.87) found in *Digitalis purpurea* along with cardioactive glycosides (see page 246). All four stereoisomers are derived from cholesterol, and the stereochemistry of the A/B ring fusion appears to be controlled by the nature of the substrate being reduced. Direct enzymic reduction of

the Δ^5 double bond yields the *trans*-fused system, whereas reduction of a Δ^4 double bond gives the alternative *cis*-fused system (Figure 5.89). Accordingly, to obtain the A/B *cis* fusion, the Δ^5 unsaturation of cholesterol is changed to Δ^4 by oxidation of the 3-hydroxyl and allylic isomerization to the conjugated 4-ene-3-one system, and this is followed by reduction of both functional groups (Figure 5.89) (compare biosynthesis of progesterone, page 243). The sarsaparilla saponins are not present in sufficient quantities to be commercially important for steroid production, but quite large amounts of sarsasapogenin can be extracted from the seeds of *Yucca brevifolia** (Agavaceae).

Cardioactive Glycosides

Many of the plants known to contain cardiac or cardiotonic glycosides have long been used as arrow poisons (e.g. *Strophanthus*) or as heart drugs (e.g. *Digitalis*). They are used to strengthen a weakened heart and allow it to function more efficiently, though the dosage must be controlled very

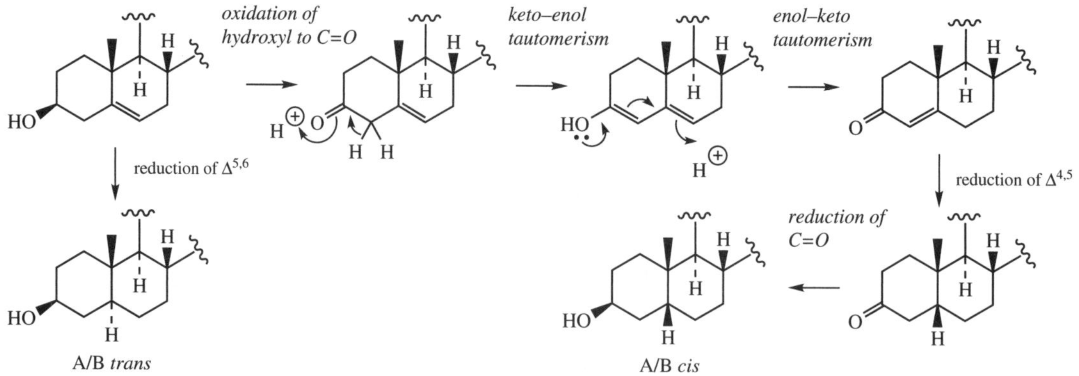

Figure 5.89

Sarsaparilla

Sarsaparilla consists of the dried roots of various *Smilax* species (Liliaceae/Smilacaceae), including *S. aristolochiaefolia*, *S. regelii*, and *S. febrifuga*, known respectively as Mexican, Honduran, and Ecuadorian sarsaparilla. The plants are woody climbers indigenous to Central America. Sarsaparilla has a history of use in the treatment of syphilis, rheumatism, and skin diseases, but is now mainly employed as a flavouring in the manufacture of non-alcoholic drinks. It has some potential as a raw material for the semi-synthesis of medicinal steroids, being a source of sarsasapogenin and smilagenin (Figure 5.87). The roots contain 1.8–2.4% steroidal saponins, including parillin (Figure 5.90).

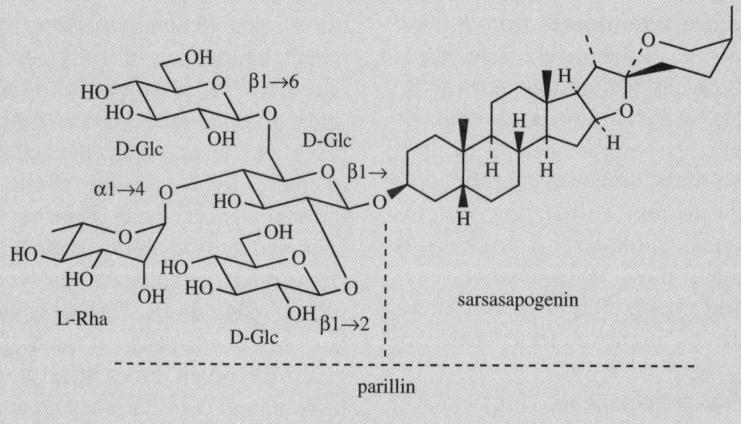

Figure 5.90

Yucca

Yucca brevifolia (Agavaceae) has been explored as a potential source of sarsasapogenin for steroid production, especially at times when market prices of diosgenin from *Dioscorea* became prohibitively expensive. The plant grows extensively in the Mojave desert in California, and high levels of sarsasapogenin (8–13%) are present in the seeds. This means the plants can be harvested regularly without damage. The subsequent stabilization of *Dioscorea* prices in the 1970s stopped further commercial utilization.

carefully since the therapeutic dose is so close to the toxic dose. The cardioactive effects of *Digitalis* were discovered as a result of its application in the treatment of dropsy, an accumulation of water in the body tissues. *Digitalis* alleviated dropsy indirectly by its effect on the heart, improving the blood supply to the kidneys and so removing excess fluid.

The therapeutic action of cardioactive glycosides depends on the structure of the aglycone, and on the type and number of sugar units attached. Two types of aglycone are recognized, **cardenolides**, e.g. **digitoxigenin** from *Digitalis purpurea*, which are C_{23} compounds, and **bufadienolides**, e.g. **hellebrigenin** from *Helleborus niger*, which are C_{24} structures (Figure 5.91). Stereochemistry is very important for activity, and these compounds have *cis* fusions for both the A/B and C/D rings, 3β- and 14β-hydroxyl groups with the glycoside function at C-3, and an α,β-unsaturated lactone grouping at C-17β. This lactone ring is five membered in the cardenolides, and six membered in the bufadienolides. The hellebrigenin structure shows two other modifications not found in the basic steroid skeleton, namely a hydroxyl at the bridgehead carbon C-5, and a formyl group at C-10, being an oxidized form of the normal methyl. The three-dimensional shape of digitoxigenin is shown in Figure 5.91. These basic structures arise biosynthetically by metabolism of cholesterol, in which the side-chain is cleaved to a two-carbon acetyl

group, followed by incorporation of either two or three carbons for cardenolides or bufadienolides respectively (Figure 5.92).

Shortening of the cholesterol side-chain is accomplished by stepwise hydroxylation at C-22 and then C-20, then cleavage of the C-20/22 bond giving **pregnenolone**, which is then oxidized in ring A giving **progesterone** (Figure 5.92). This can be reduced to give the *cis*-fused A/B system as in 3β-hydroxy-5β-pregnan-20-one (compare Figure 5.89) which is the substrate for 14β-hydroxylation, i.e. inverting the stereochemistry at this centre. Inversion is atypical for hydroxylation by mono-oxygenases, which are found to hydroxylate with retention of configuration. Whatever the mechanism of this hydroxylation, no Δ^8 or Δ^{15} double bond intermediates are involved. Hydroxylation in the side-chain at C-21 follows. The lactone ring is created at this stage. An intermediate malonate ester is involved, and ring formation probably occurs via the aldol addition process shown in Figure 5.93 giving the cardenolide **digitoxigenin**, the carboxyl carbon of the malonate ester being lost by decarboxylation during the process (compare malonate in the acetate pathway). Alternatively, three carbons from oxaloacetate can be incorporated by a similar esterification/aldol reaction sequence. This would produce **bufalin** (Figure 5.92), a bufadienolide structure found in the skin of toads (*Bufo* spp.), from which this class of compound was originally isolated and

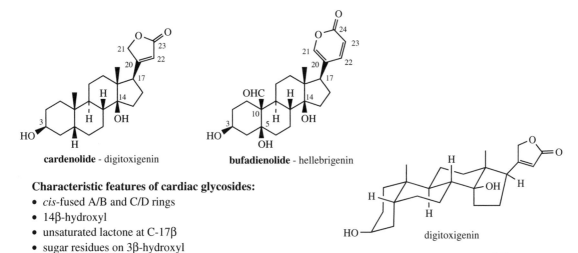

cardenolide - digitoxigenin   **bufadienolide - hellebrigenin**   digitoxigenin

Characteristic features of cardiac glycosides:

- *cis*-fused A/B and C/D rings
- 14β-hydroxyl
- unsaturated lactone at C-17β
- sugar residues on 3β-hydroxyl

Figure 5.91

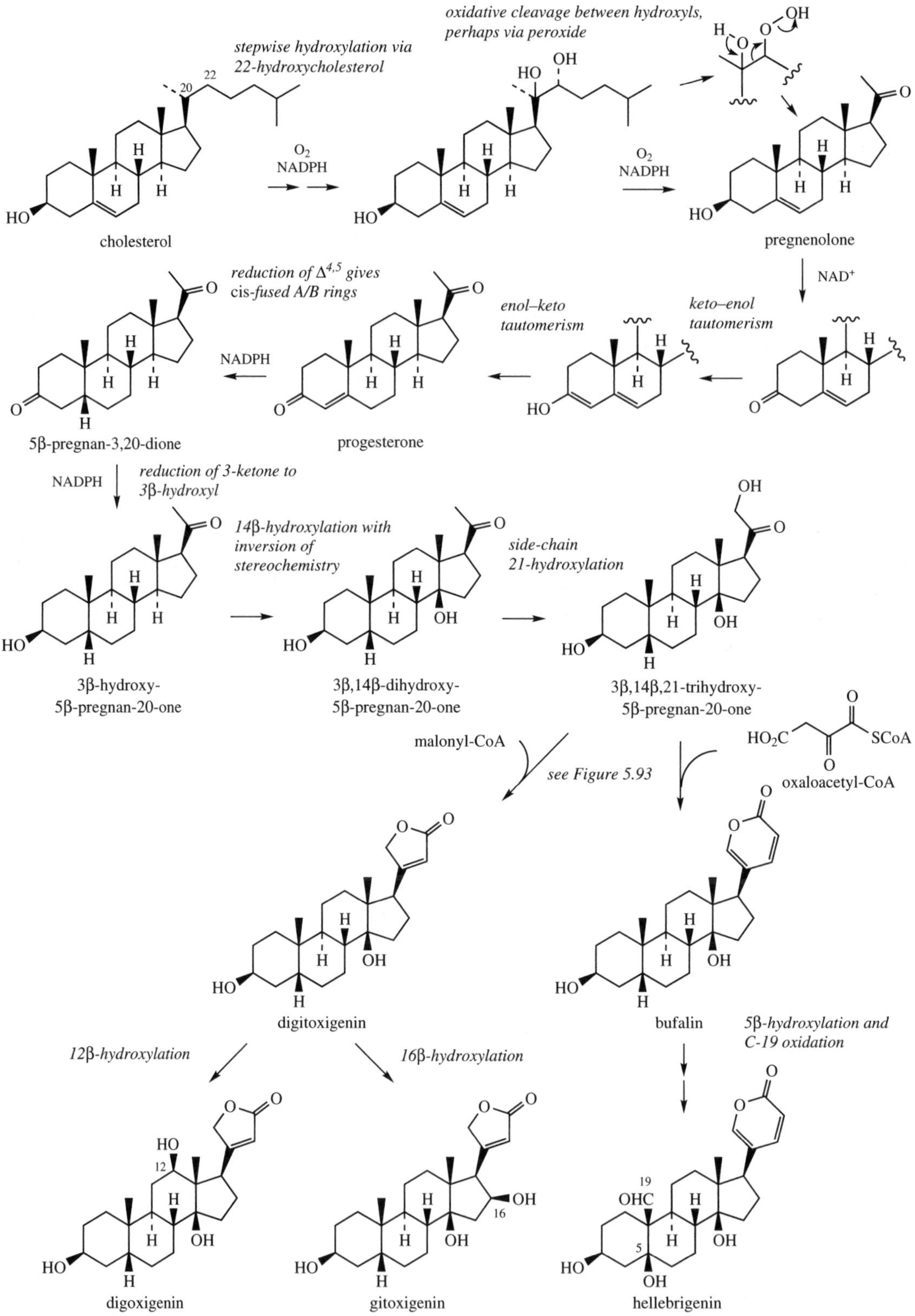
stepwise hydroxylation via 22-hydroxycholesterol
O2 NADPH
oxidative cleavage between hydroxyls, perhaps via peroxide
O2 NADPH
cholesterol
pregnenolone
NAD+
keto–enol tautomerism
enol–keto tautomerism
reduction of Δ4,5 gives cis-fused A/B rings
NADPH
5β-pregnan-3,20-dione
progesterone
NADPH
reduction of 3-ketone to 3β-hydroxyl
14β-hydroxylation with inversion of stereochemistry
side-chain 21-hydroxylation
3β-hydroxy-5β-pregnan-20-one
3β,14β-dihydroxy-5β-pregnan-20-one
3β,14β,21-trihydroxy-5β-pregnan-20-one
malonyl-CoA
see Figure 5.93
oxaloacetyl-CoA
digitoxigenin
bufalin
12β-hydroxylation
16β-hydroxylation
5β-hydroxylation and C-19 oxidation
digoxigenin
gitoxigenin
hellebrigenin

Figure 5.92

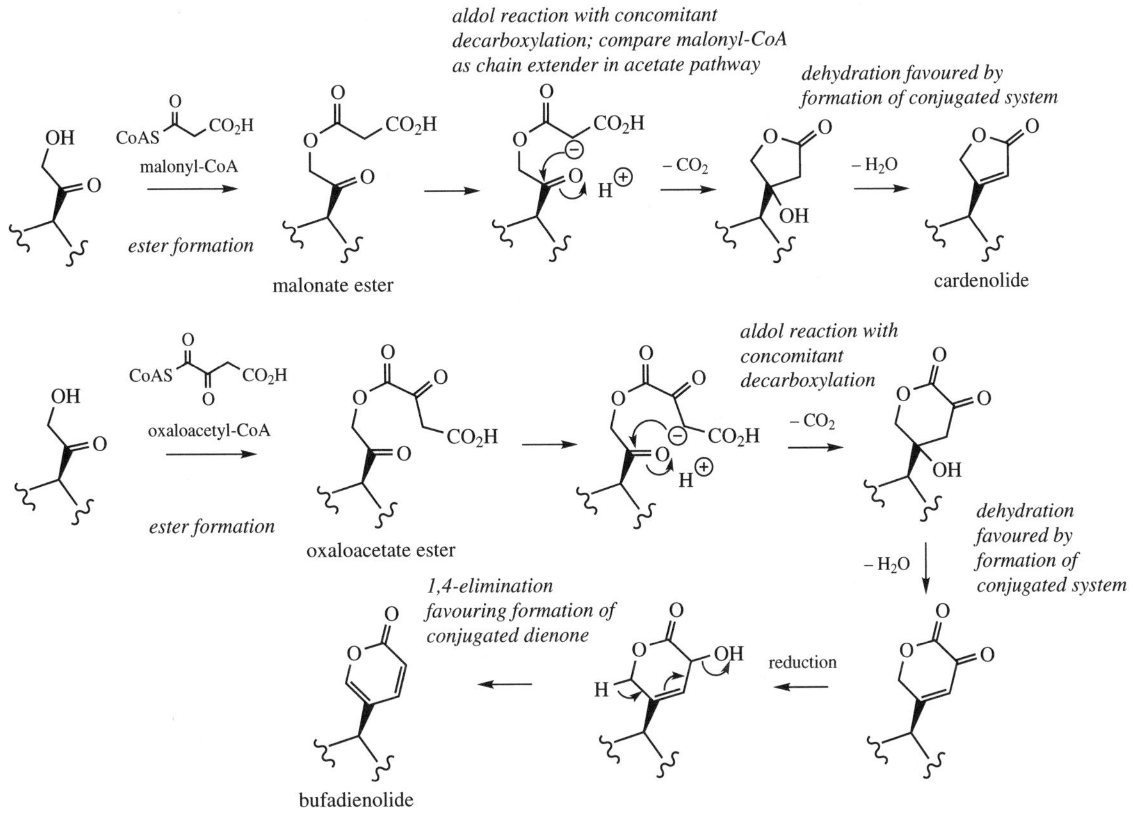

Figure 5.93

has subsequently taken the general name. Note that in the subsequent formation of **hellebrigenin** (Figure 5.92), hydroxylation at C-5 occurs with the expected retention of stereochemistry, and not with inversion as seen at C-14.

The fundamental pharmacological activity resides in the aglycone portion, but is considerably modified by the nature of the sugar at C-3. This increases water solubility and binding to heart muscle. The sugar unit may have one to four monosaccharides; many, e.g. D-**digitoxose** and D-**digitalose** (Figure 5.94), are unique to this group of compounds. About 20 different sugars have been characterized, and with the exception of D-glucose, they are 6-deoxy- (e.g. L-rhamnose) or 2,6-dideoxy- (e.g. D-digitoxose) hexoses, some of which are 3-methyl ethers (e.g. D-digitalose and D-**cymarose** (Figure 5.94)). In plants, cardiac glycosides are confined to the Angiosperms, but are found in both monocotyledons and dicotyledons. The cardenolides are more common, and the plant

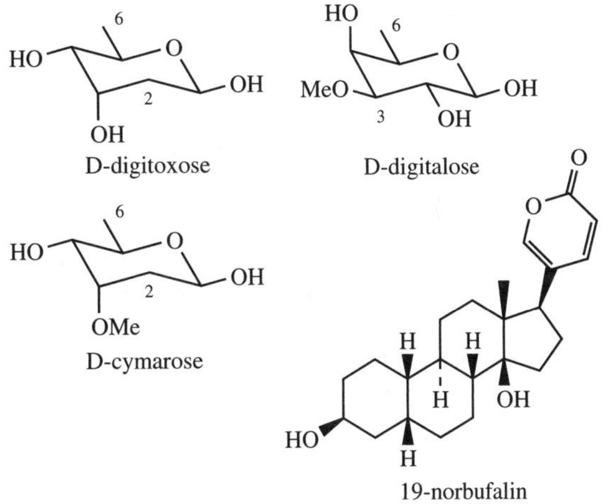

Figure 5.94

families the Apocynaceae (e.g. *Strophanthus*)*, Liliaceae (e.g. *Convallaria*)*, and Scrophulariaceae (e.g. *Digitalis*)* yield medicinal agents. The rarer bufadienolides are found in some members of the Liliaceae (e.g. *Urginea*)* and Ranunculaceae (e.g. *Helleborus*), as well as toads. Monarch butterflies and their larvae are known to accumulate in their bodies a range of cardenolides, which they ingest from their food plant, the common milkweed (*Asclepias syriaca*; Asclepiadaceae). This makes them unpalatable to predators such as birds.

Endogenous *Digitalis*-like compounds have also been detected, albeit in very small quantities, in mammalian tissues. **19-Norbufalin** (Figure 5.94) is found in human eye lenses, at higher levels if these are cataract afflicted, and it is believed to regulate ATPase activity under some physiological and pathological conditions.

Digitalis purpurea

Digitalis leaf consists of the dried leaf of the red foxglove *Digitalis purpurea* (Scrophulariaceae). The plant is a biennial herb, common in Europe and North America, which forms a low rosette of leaves in the first year, and its characteristic spike of purple (occasionally white) bell-shaped flowers in the second year. It is potentially very toxic, but is unlikely to be ingested by humans. *Digitalis purpurea* is cultivated for drug production, principally in Europe, the first year leaves being harvested then rapidly dried at 60°C as soon as possible after collection. This procedure is necessary to inactivate hydrolytic enzymes which would hydrolyse glycoside linkages in the cardioactive glycosides giving rise to less active derivatives. Even so, some partial hydrolysis does occur. Excess heat may also cause dehydration in the aglycone to Δ^{14}-anhydro compounds, which are inactive.

Because of the pronounced cardiac effects of digitalis, the variability in the cardiac glycoside content, and also differences in the range of structures present due to the effects of enzymic hydrolysis, the crude leaf drug is usually assayed biologically rather than chemically. Prepared digitalis is a biologically standardized preparation of powdered leaf, its activity being assessed on cardiac muscle of guinea pig or pigeon and compared against a standard preparation. It may be diluted to the required activity by mixing in powdered digitalis of lower potency, or inactive materials such as lucerne (*Medicago sativa*) or grass. The crude drug is hardly ever used now, having been replaced by the pure isolated glycosides.

The cardioactive glycoside content of *Digitalis purpurea* leaf is 0.15–0.4%, consisting of about 30 different structures. The major components are based on the aglycones digitoxigenin, gitoxigenin, and gitaloxigenin (Figure 5.95), the latter being a formate ester. The glycosides comprise two series of compounds, those with a tetrasaccharide *glucose–(digitoxose)*$_s$–unit and those with a trisaccharide *(digitoxose)*$_3$–unit. The latter group (the secondary glycosides) are produced by partial hydrolysis from the former group (the primary glycosides) during drying by the enzymic action of a β-glucosidase, which removes the terminal glucose. Thus the principal glycosides in the fresh leaves, namely purpureaglycoside A and purpureaglycoside B (Figure 5.95), are partially converted into digitoxin and gitoxin respectively (Figure 5.95), which normally predominate in the dried leaf. These transformations are indicated schematically in Figure 5.96. In the fresh leaf, purpureaglycoside A can constitute about 50% of the glycoside mixture, whilst in the dried leaf, the amounts could be negligible if the plant material is old or poorly stored. The gitaloxigenin-based glycosides are relatively unstable, and the formyl group on the aglycone is readily lost by hydrolysis. Other minor glycosides are present, but neither the fresh nor dried leaf contain any significant quantities of the free aglycones.

Glycosides of the gitoxigenin series are less active than the corresponding members of the digitoxigenin-derived series. **Digitoxin** is the only compound routinely used as a drug, and it is employed in congestive heart failure and treatment of cardiac arrhythmias, particularly atrial fibrillation.

(Continues)

(Continued)

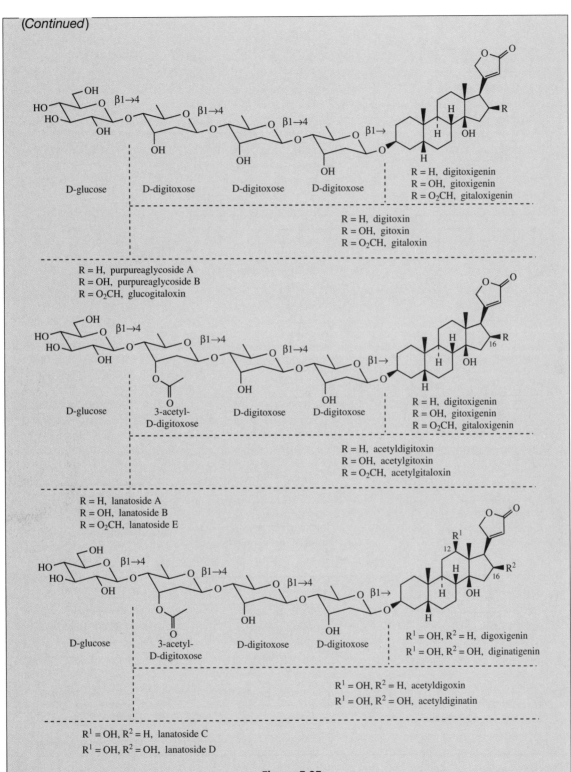

R = H, digitoxigenin
R = OH, gitoxigenin
R = O₂CH, gitaloxigenin

R = H, digitoxin
R = OH, gitoxin
R = O₂CH, gitaloxin

R = H, purpureaglycoside A
R = OH, purpureaglycoside B
R = O₂CH, glucogitaloxin

D-glucose   D-digitoxose   D-digitoxose   D-digitoxose

R = H, digitoxigenin
R = OH, gitoxigenin
R = O₂CH, gitaloxigenin

R = H, acetyldigitoxin
R = OH, acetylgitoxin
R = O₂CH, acetylgitaloxin

R = H, lanatoside A
R = OH, lanatoside B
R = O₂CH, lanatoside E

D-glucose   3-acetyl-D-digitoxose   D-digitoxose   D-digitoxose

R¹ = OH, R² = H, digoxigenin
R¹ = OH, R² = OH, diginatigenin

R¹ = OH, R² = H, acetyldigoxin
R¹ = OH, R² = OH, acetyldiginatin

R¹ = OH, R² = H, lanatoside C
R¹ = OH, R² = OH, lanatoside D

D-glucose   3-acetyl-D-digitoxose   D-digitoxose   D-digitoxose

Figure 5.95

(Continues)

(*Continued*)

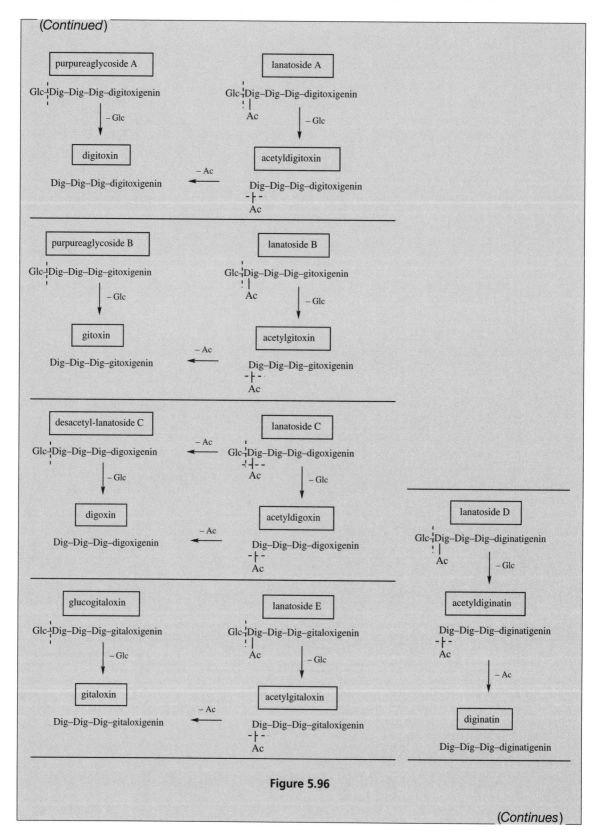

Figure 5.96

(*Continues*)

(Continued)

Digitalis lanata

Digitalis lanata (Scrophulariaceae), the Grecian foxglove, is a perennial or biennial herb from Southern and Central Europe, and differs in appearance from the red foxglove by its long narrow smoother leaves, and its smaller flowers of a yellow-brown colour. It is cultivated in Europe, the United States and South America, and is harvested and dried in a similar manner to *D. purpurea*. It has not featured as a crude drug, but is used exclusively for the isolation of individual cardiac glycosides, principally digoxin and lanatoside C (Figure 5.97).

The total cardenolide content of up to 1% is two to three times that found in *D. purpurea*. The main constituents resemble those of *D. purpurea*, but contain an acetyl ester function on the third digitoxose, that furthest from the aglycone. This acetyl group makes the compounds easier to isolate from the plant material and makes crystallization easier. Drying of the leaf is similarly accompanied by some partial hydrolysis of the original fresh leaf constituents through enzymic action, and both the terminal glucose and the acetyl group may be hydrolysed off, extending the range of compounds isolated. The *D. lanata* cardiac glycosides are based on five aglycones, digitoxigenin, gitoxigenin, and gitaloxigenin, as found in *D. purpurea*, plus digoxigenin and diginatigenin (Figure 5.95), which do not occur in *D. purpurea*. The primary glycosides containing the acetylated tetrasaccharide unit *glucose–acetyldigitoxose–(digitoxose)₂* – are called lanatosides. Lanatosides A and C (Figure 5.95) constitute the major components in the fresh leaf (about 50–70%) and are based on the aglycones digitoxigenin and digoxigenin respectively. Lanatosides B, D, and E (Figure 5.95) are minor components derived from gitoxigenin, diginatigenin, and gitaloxigenin respectively. Enzymic hydrolysis of the lanatosides generally involves loss of the terminal glucose prior to removal of the acetyl function, so that compounds like acetyldigitoxin and acetyldigoxin as well as digitoxin and digoxin are present in the dried leaf as decomposition products from lanatosides A and C respectively. These transformations are also indicated in simplified form in Figure 5.96.

Digoxin (Figure 5.97) has a rapid action and is more quickly eliminated from the body than digitoxin, and is therefore the most widely used of the cardioactive glycosides. Digoxin is more hydrophilic than digitoxin, binds less strongly to plasma proteins and is mainly eliminated by the kidneys, whereas digitoxin is metabolized more slowly by the liver. It is used in congestive

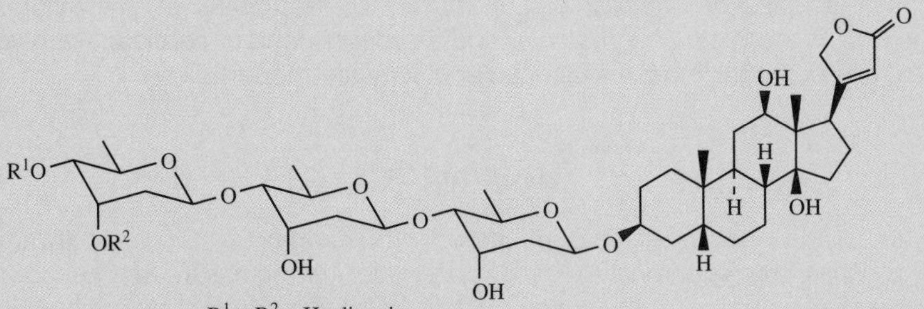

R¹ = R² = H, digoxin
R¹ = Me, R² = H, medigoxin (metildigoxin)
R¹ = Glc, R² = Ac, lanatoside C
R¹ = Glc, R² = H, deslanoside (desacetyl-lanatoside C)

Figure 5.97

(Continues)

(Continued)

heart failure and atrial fibrillation. **Lanatoside C** and **deslanoside (desacetyl-lanatoside C)** (Figure 5.97) have also been employed, though not to the same extent. They have very rapid action and are suited for treatment of cardiac emergencies by injection. A semi-synthetic derivative **medigoxin** or **metildigoxin** (methyl replacing the glucose in lanatoside C) (Figure 5.97) has also been available, being more active via better bioavailability.

The cardioactive glycosides increase the force of contractions in the heart, thus increasing cardiac output and allowing more rest between contractions. The primary effect on the heart appears to be inhibition of Na^+/K^+-ATPase in the cell membranes of heart muscle, specifically inhibiting the Na^+ pump, thereby raising the intracellular Na^+ concentration. The resultant decrease in the Na^+ gradient across the cell membrane reduces the energy available for transport of Ca^{2+} out of the cell, leads to an increase in intracellular Ca^{2+} concentration, and provides the positive ionotropic effect and increased force of contractions. The improved blood circulation also tends to improve kidney function leading to diuresis and loss of oedema fluid often associated with heart disease. However, the diuretic effect, historically important in the treatment of dropsy, is more safely controlled by other diuretic drugs.

To treat congestive heart failure, an initial loading dose of the cardioactive glycoside is followed by regular maintenance doses, the amounts administered depending on drug bioavailability and subsequent metabolism or excretion. Because of the extreme toxicity associated with these compounds (the therapeutic level is 50–60% of the toxic dose; a typical daily dose is only about 1 mg) dosage must be controlled very carefully. Bioavailability has sometimes proved erratic and can vary between different manufacturers' formulations, so patients should not be provided with different preparations during their treatment. Individual patients also excrete the glycosides or metabolize them by hydrolysis to the aglycone at different rates, and ideally these processes should be monitored. Levels of the drug in blood plasma can be measured quite rapidly by radioimmunoassay using a specific antibody. A **digoxin-specific antibody** is available both for assay and also as a means of reversing life-threatening digoxin overdose. It has also successfully reversed digitoxin overdose, thus demonstrating a somewhat broader specificity. The value of digoxin treatment for heart failure where the heartbeat remains regular has recently been called into question. It still remains a recognized treatment for atrial fibrillation.

Many other species of *Digitalis*, e.g. *D. dubia, D. ferruginea, D. grandiflora, D. lutea, D. mertonensis, D. nervosa, D. subalpina*, and *D. thaspi* contain cardioactive glycosides in their leaves, and some have been evaluated and cultivated for drug use.

Strophanthus

Strophanthus is the dried ripe seeds of *Strophanthus kombé* or *S. gratus* (Apocynaceae), which are tall vines from equatorial Africa. *Strophanthus kombé* has a history of use by African tribes as an arrow poison, and the seeds contain 5–10% cardenolides, a mixture known as K-strophanthin. This has little drug use today, though it was formerly used medicinally as a cardiac stimulant. The main glycoside (about 80%) is K-strophanthoside (Figure 5.98) with smaller amounts of K-strophanthin-β and cymarin, related to K-strophanthoside as shown. These are derivatives of the aglycone strophanthidin. *Strophanthus gratus* contains 4–8% of **ouabain** (G-strophanthin) (Figure 5.98), the rhamnoside of ouabigenin. Ouabigenin is rather

(Continues)

(Continued)

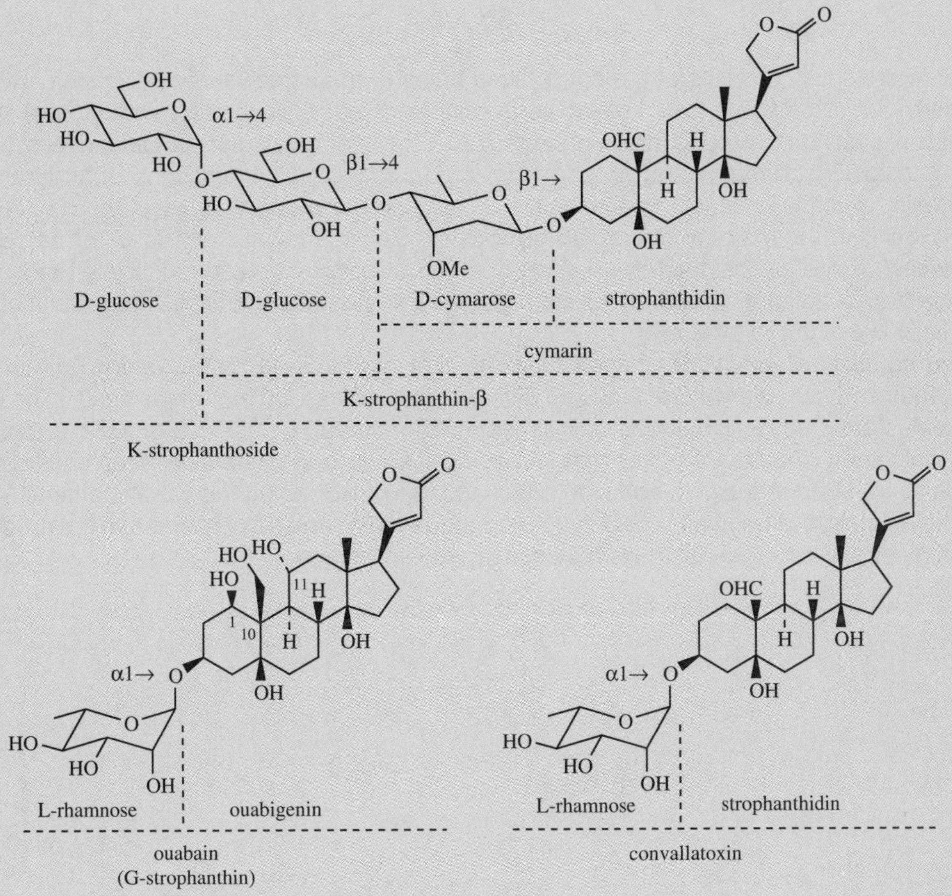

Figure 5.98

unusual in having additional hydroxylation at 1β and 11α, as well as a hydroxymethyl at C-10. Ouabain is a stable, crystalline material, which is often employed as the biological standard in assays for cardiac activity. It is a potent cardiac glycoside and acts quickly, but wears off rapidly. It is very polar with rapid renal elimination and must be injected because it is so poorly absorbed orally. It has been used for emergency treatment in cases of acute heart failure. It is still official in many pharmacopoeias.

Convallaria

The dried roots and tops of lily of the valley, *Convallaria majalis* (Liliaceae/Convallariaceae), contain cardioactive glycosides (0.2–0.3%) and are used in some European countries rather than digitalis. The effects are similar, but the drug is less cumulative. This plant is widely cultivated as an ornamental, particularly for its intensely perfumed small white flowers, and must be considered potentially toxic. The major glycoside (40–50%) is convallatoxin (Figure 5.98), the rhamnoside of strophanthidin.

(Continues)

(Continued)

Squill

Squill (white squill) consists of the dried sliced bulbs of the white variety of *Urginea maritima* (formerly *Scilla maritima*; also known as *Drimia maritima*) (Liliaceae/Hyacinthaceae) which grows on seashores around the Mediterranean. The plant contains bufadienolides (up to 4%), principally scillaren A and proscillaridin A (Figure 5.99). The aglycone of scillaren A is scillarenin, which is unusual in containing a Δ^4 double bond and thus lacks the *cis* A/B ring fusion found in the majority of cardiac glycosides. Squill is not usually used for its cardiac properties, as the glycosides have a short duration of action. Instead, squill is employed for its expectorant action in preparations such as Gee's linctus. Large doses cause vomiting and a digitalis-like action on the heart.

  Red squill is a variety of *Urginea maritima* that contains an anthocyanin pigment (see page 150) and bufadienolides that are different from those of the white squill. The main glycosides are glucoscilliroside and scilliroside (Figure 5.99), glucosides of scillirosidin. This chemical variety should not be present in medicinal squill, and has mainly been employed as a rodenticide. Rodents lack a vomiting reflex and are poisoned by the cardiac effects, whilst in other animals and humans vomiting will occur due to the emetic properties of the drug. The use of red squill as a rodenticide is now considered inhumane.

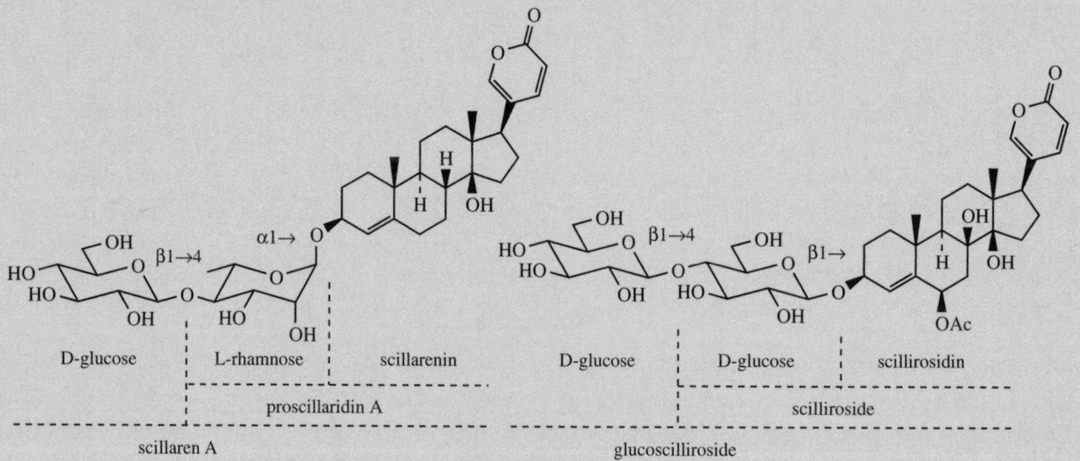

Figure 5.99

Toxic Plants

Many plants containing cardioactive glycosides are widely grown as ornamentals and must be considered toxic and treated with due care and respect. These include *Digitalis* species, *Convallaria majalis*, *Helleborus* species, and oleander (*Nerium oleander*; Apocynaceae).

Phytosterols

The major sterol found in mammals is the C_{27} compound cholesterol, which acts as a precursor for other steroid structures such as sex hormones and corticosteroids. The main sterols in plants, fungi, and algae are characterized by extra one-carbon or two-carbon substituents on the side-chain, attached at C-24. These substituent carbons are numbered 24^1 and 24^2 (Figure 5.77); some older publications

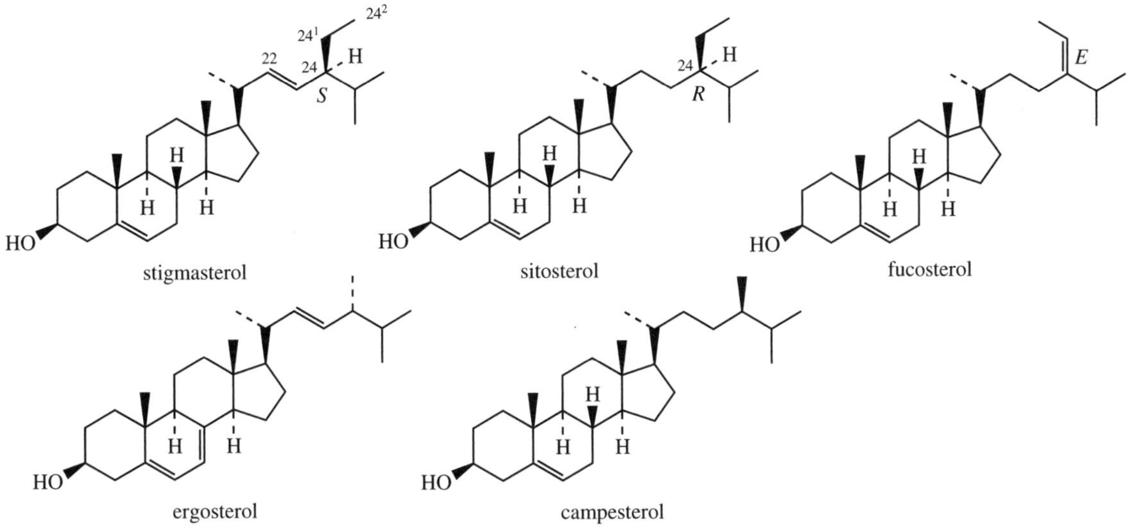

Figure 5.100

may use 28 and 29. The widespread plant sterols **campesterol** and **sitosterol** (Figure 5.100) are respectively 24-methyl and 24-ethyl analogues of cholesterol. **Stigmasterol** contains additional unsaturation in the side-chain, a *trans*-Δ^{22} double bond, a feature seen in many plant sterols, but never in mammalian ones. The introduction of methyl and ethyl groups at C-24 generates a new chiral centre, and the 24-alkyl groups in campesterol, sitosterol, and stigmasterol are designated α. The predominant sterol found in fungi is **ergosterol** (Figure 5.100), which has a β-oriented 24-methyl, as well as a *trans*-Δ^{22} double bond and additional Δ^7 unsaturation. The descriptors α and β unfortunately do not relate to similar terms for the steroid ring system, but are derived from consideration of Fischer projections for the side-chain, substituents to the left being designated α and those to the right as β. Systematic *RS* nomenclature is preferred, but note that this defines sitosterol as 24*R* whilst stigmasterol, because of its extra double bond, is 24*S*. The majority of plant sterols have a 24α-methyl or 24α-ethyl substituent, whilst algal sterols tend to have 24β-ethyls, and fungi 24β-methyls. The most abundant sterol in brown algae (*Fucus* spp.; Fucaceae) is **fucosterol** (Figure 5.100), which demonstrates a further variant, a 24-ethylidene substituent. Such groups can have *E*-configurations as in fucosterol, or the alternative *Z*-configuration. Sterols

are found predominantly in free alcohol form, but also as esters with long chain fatty acids (e.g. palmitic, oleic, linoleic, and α-linolenic acids), as glycosides, and as fatty acylated glycosides. These sterols, termed phytosterols, are structural components of membranes in plants, algae, and fungi, and affect the permeability of these membranes. They also appear to play a role in cell proliferation.

The source of the extra methyl or ethyl side-chain carbons in both cases is *S*-adenosylmethionine (SAM), and to achieve alkylation the side-chain must have a Δ^{24} double bond, i.e. the side-chains seen in lanosterol and cycloartenol. The precise mechanisms involved have been found to vary according to organism, but some of the demonstrated sequences are given in Figure 5.101. Methylation of the Δ^{24} double bond at C-24 via SAM yields a carbocation which undergoes a hydride shift and loss of a proton from C-24^1 to generate the 24-methylene side-chain. This can be reduced to a 24-methyl either directly, or after allylic isomerization. Alternatively, the 24-methylene derivative acts as substrate for a second methylation step with SAM, producing a carbocation. Discharge of this cation by proton loss produces a 24-ethylidene side-chain, and reduction or isomerization/reduction gives a 24-ethyl group. The *trans*-Δ^{22} double bond is introduced only after alkylation at C-24 is completed. No stereochemistry is intended in

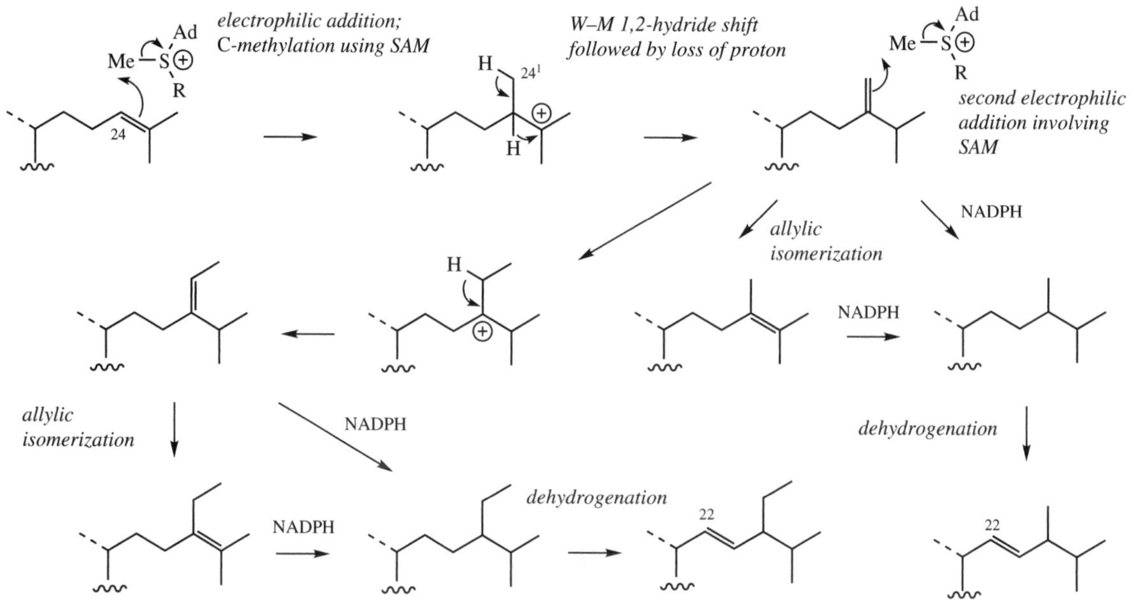

Figure 5.101

Figure 5.101. It is apparent that stereochemistries in the 24-methyl, 24-ethyl, and 24-ethylidene derivatives could be controlled by the reduction processes or by proton loss as appropriate. It is more plausible for different stereochemistries in the 24-methyl and 24-ethyl side-chains to arise from reduction of different double bonds, rather than reduction of the same double bond in two different ways. In practice, other mechanisms involving a 25(26)-double bond are also found to operate.

The substrates for alkylation are found to be cycloartenol in plants and algae, and lanosterol in fungi. The second methylation step in plants and algae usually involves **gramisterol** (24-methylenelophenol) (Figure 5.102). This indicates that the processes of side-chain alkylation and the steroid skeleton modifications, i.e. loss of methyls, opening of the cyclopropane ring, and migration of the double bond, tend to run concurrently rather than sequentially. Accordingly, the range of plant and algal sterol derivatives includes products containing side-chain alkylation, retention of one or more skeletal methyls, and possession of a cyclopropane ring, as well as those more abundant examples such as sitosterol and stigmasterol based on a cholesterol-type skeleton. Most fungal sterols originate from lanosterol, so

less variety is encountered. The most common pathway from lanosterol to **ergosterol** in fungi involves initial side-chain alkylation to **eburicol** (24-methylenedihydrolanosterol), which is the substrate for 14-demethylation (Figure 5.103). Loss of the 4-methyls then gives **fecosterol**, from which ergosterol arises by further side-chain and ring B modifications. Although the transformations are similar to those occurring in the mammalian pathway for lanosterol → cholesterol, the initial side-chain alkylation means the intermediates formed are different. Some useful anti-fungal agents, e.g. ketoconazole and miconazole, are specific inhibitors of the 14-demethylation reaction in fungi, but do not affect cholesterol biosynthesis in humans. Inability to synthesize the essential sterol components of their membranes proves fatal for the fungi. Similarly, 14-demethylation in plants proceeds via **obtusifoliol** (Figure 5.102) and plants are unaffected by azole derivatives developed as agricultural fungicides. The antifungal effect of polyene antibiotics such as amphotericin and nystatin depends on their ability to bind strongly to ergosterol in fungal membranes and not to cholesterol in mammalian cells (see page 102).

Sitosterol and **stigmasterol** (Figure 5.100) are produced commercially from soya beans* (*Glycine max*; Leguminosae/Fabaceae) as raw materials

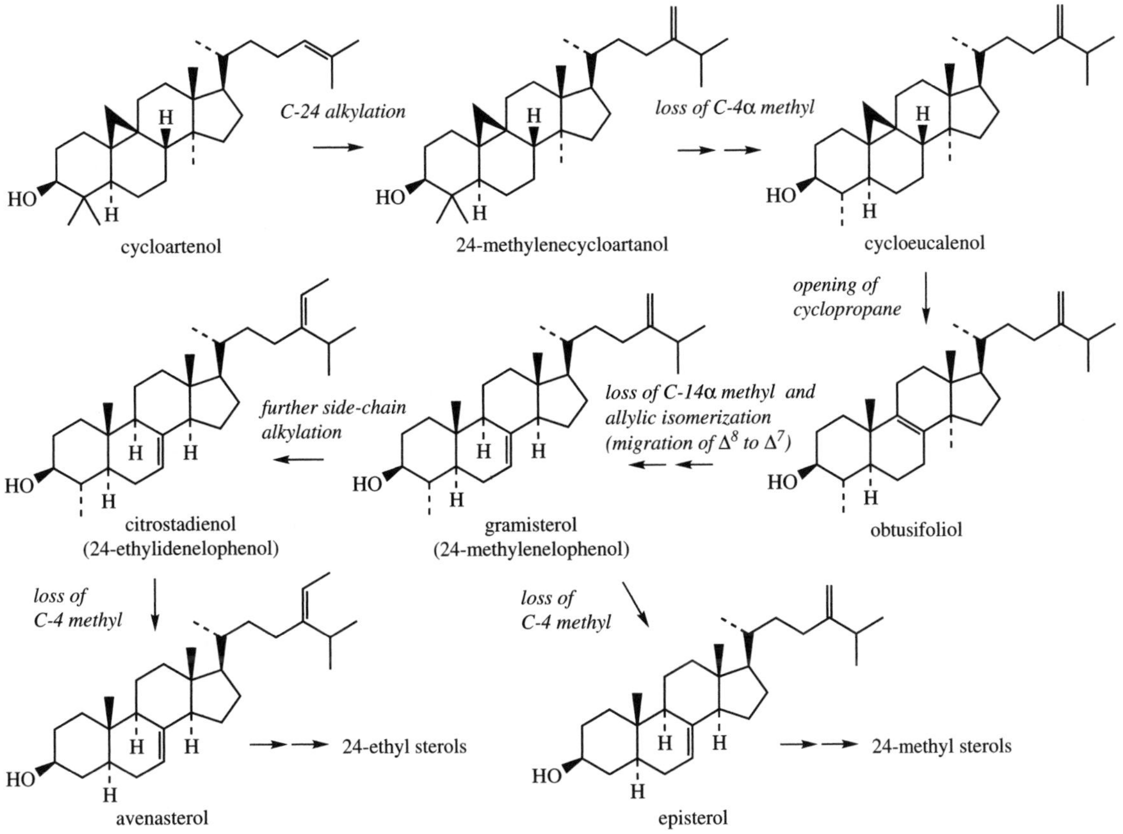

Figure 5.102

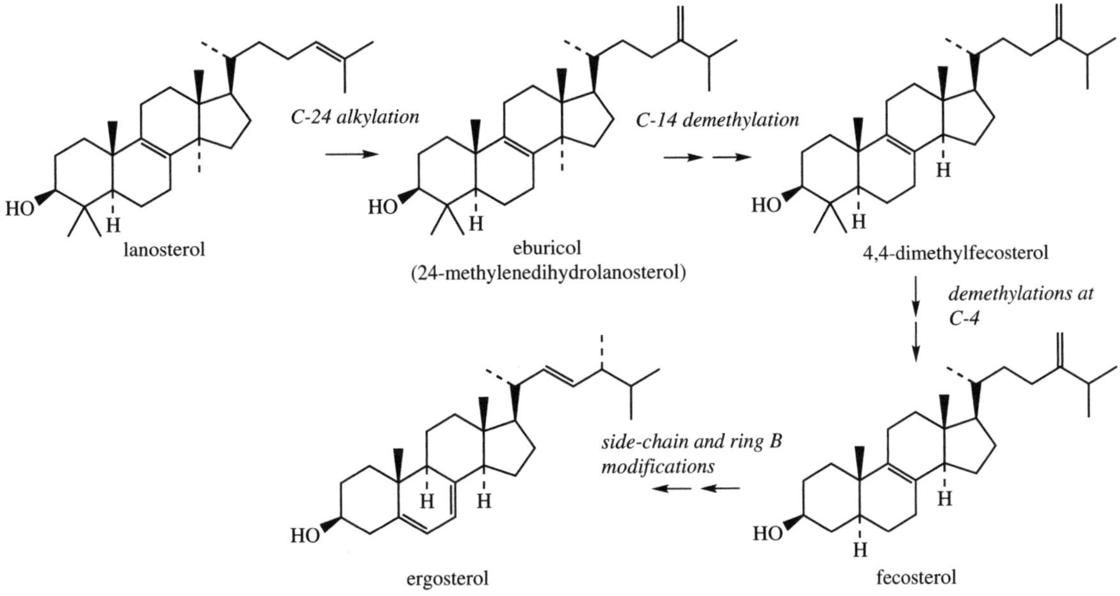

Figure 5.103

Soya Bean Sterols

Soya beans or soybeans (*Glycine max*; Leguminosae/Fabaceae) are grown extensively in the United States, China, Japan, and Malaysia as a food plant. They are used as a vegetable, and provide a high protein flour, an important edible oil (Table 3.2), and an acceptable non-dairy soybean milk. The flour is increasingly used as a meat substitute. Soy sauce is obtained from fermented soybeans and is an indispensable ingredient in Chinese cookery. The seeds also contain substantial amounts (about 0.2%) of sterols. These include stigmasterol (about 20%), sitosterol (about 50%) and campesterol (about 20%) (Figure 5.100), the first two of which are used for the semi-synthesis of medicinal steroids. In the seed, about 40% of the sterol content is in the free form, the remainder being combined in the form of glycosides, or as esters with fatty acids. The oil is usually solvent extracted from the dried flaked seed using hexane. The sterols can be isolated from the oil after basic hydrolysis as a by-product of soap manufacture, and form part of the unsaponifiable matter.

The efficacy of dietary plant sterols in reducing cholesterol levels in laboratory animals has been known for many years. This has more recently led to the introduction of **plant sterol esters** as food additives, particularly in margarines, as an aid to reducing blood levels of low density lipoprotein (LDL) cholesterol, known to be a contributory factor in atherosclerosis and the incidence of heart attacks (see page 236). Plant sterol esters are usually obtained by esterifying sitosterol from soya beans with fatty acids to produce a fat-soluble product. Regular consumption of this material (recommended 1.3 g per day) is shown to reduce blood LDL cholesterol levels by 10–15%. The plant sterols are more hydrophobic than cholesterol and have a higher affinity for micelles involved in fat digestion, effectively decreasing intestinal cholesterol absorption. The plant sterols themselves are not absorbed from the GI tract. Of course, the average diet will normally include small amounts of plant sterol esters. Related materials used in a similar way are **plant stanol esters**. Stanols are obtained by hydrogenation of plant sterols, and will consist mainly of sitostanol (from sitosterol and stigmasterol) and campestanol (from campesterol) (Figure 5.104); these are then esterified with fatty acids. Regular consumption of plant stanol esters (recommended 3.4 g per day) is shown to reduce blood LDL cholesterol levels by an average of 14%. Much of the material used in preparation of plant stanol esters originates from tall oil, a by-product of the wood pulping industry. This contains campesterol, sitosterol, and also sitostanol. The stanols are usually transesterified with rapeseed oil, which is rich in unsaturated fatty acids (see page 43).

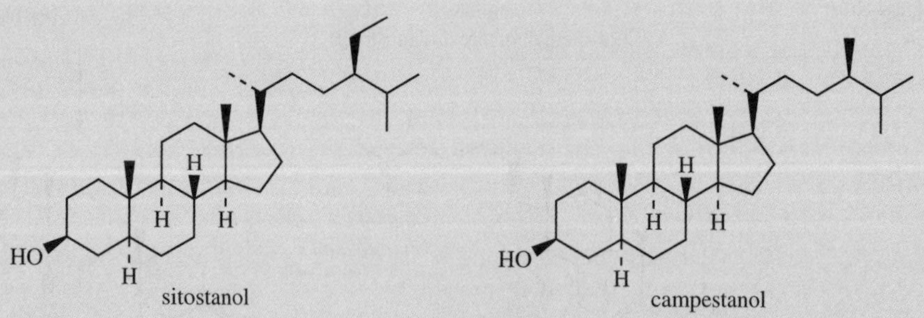

sitostanol   campestanol

Figure 5.104

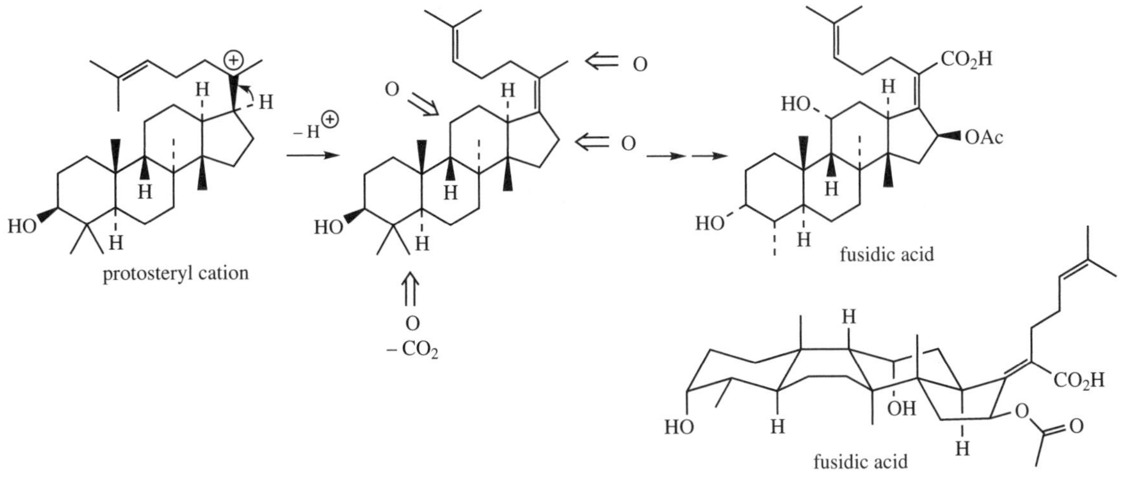

Figure 5.105

for the semi-synthesis of medicinal steroids (see pages 266, 279). For many years, only stigmasterol was utilized, since the Δ^{22} double bond allowed chemical degradation of the side-chain to be effected with ease. The utilization of sitosterol was not realistic until microbiological processes for removal of the saturated side-chain became available.

Fusidic acid[*] (Figure 5.105), an antibacterial agent from *Acremonium fusidioides*, has no additional side-chain alkylation, but has lost one C-4 methyl and undergone hydroxylation and oxidation of a side-chain methyl. Its relationship to the protosteryl cation is shown in Figure 5.105. The stereochemistry in fusidic acid is not typical of most steroids, and ring B adopts a boat conformation; the molecular shape is comparable to the protosteryl cation (Figure 5.57, page 216).

Vitamin D

Vitamin D$_3$ (colecalciferol, cholecalciferol)[*] is a sterol metabolite formed photochemically in animals from **7-dehydrocholesterol** by the sun's irradiation of the skin (Figure 5.106). 7-Dehydrocholesterol is the immediate $\Delta^{5,7}$ diene precursor of cholesterol (see page 234), and a photochemical reaction allows ring opening to precholecalciferol. A thermal 1,7-hydrogen shift follows to give colecalciferol (vitamin D$_3$). Vitamin D$_3$ is also manufactured photosynthetically by the same route. **Vitamin D$_2$ (ergocalciferol)**[*] may be obtained from ergosterol in exactly the same way, and, although found in plants and yeasts, large amounts are obtained semi-synthetically by the sequence shown in Figure 5.106, using ergosterol from yeast

---

Fusidic Acid

Fusidic acid (Figure 5.105) is a steroidal antibiotic produced by cultures of the fungus *Acremonium fusidioides* (formerly *Fusidium coccineum*). It has also been isolated from several *Cephalosporium* species. Fusidic acid and its salts are narrow-spectrum antibiotics active against Gram-positive bacteria. It is primarily used, as its sodium salt, in infections caused by penicillin-resistant *Staphylococcus* species, especially osteomyelitis since fusidic acid concentrates in bone. It is usually administered in combination with another antibiotic to minimize development of resistance. Fusidic acid reversibly inhibits protein biosynthesis at the translocation step by binding to the larger subunit of the ribosome (see page 407).

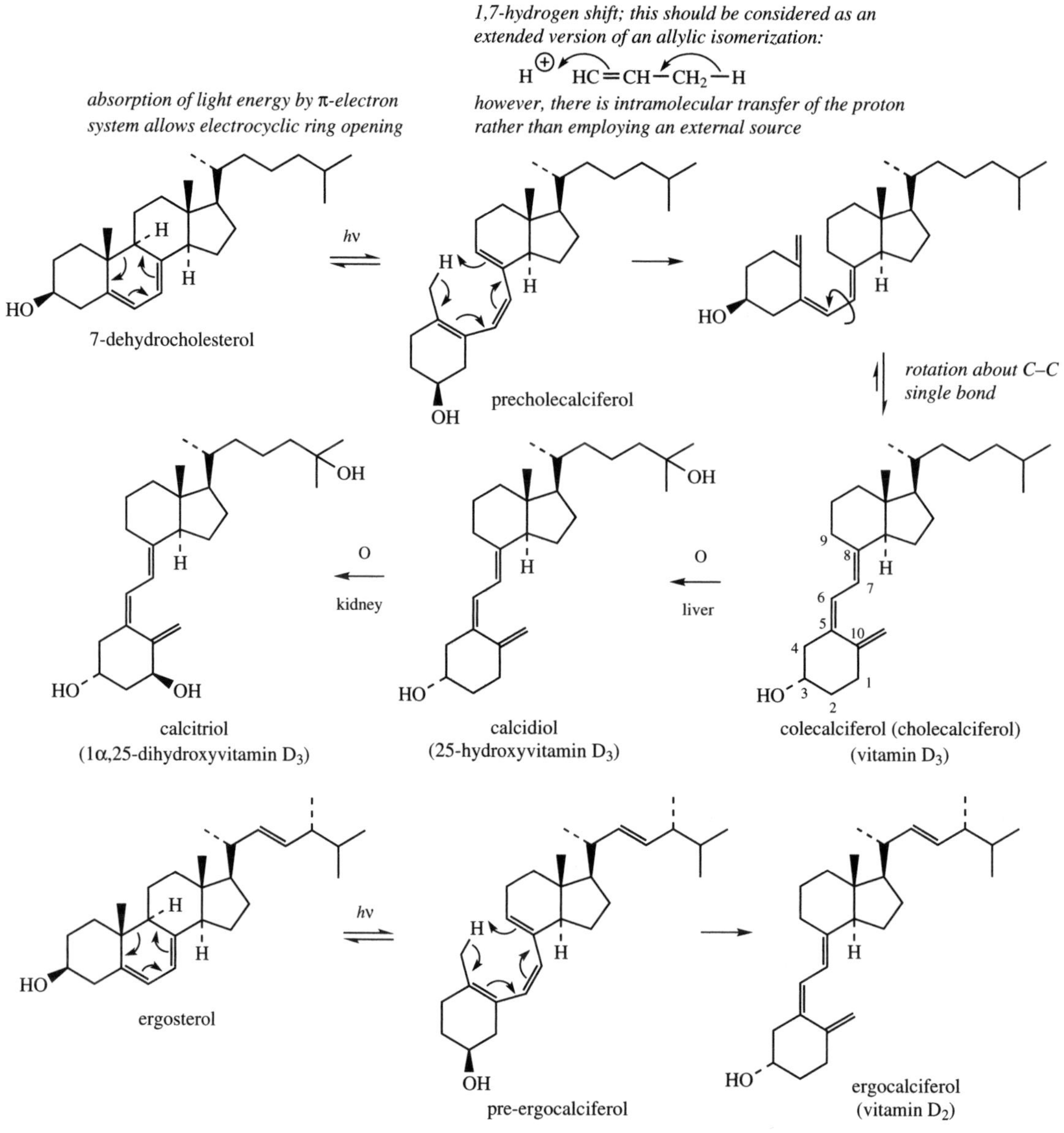

Figure 5.106

(*Saccharomyces cerevisiae*). Vitamin D$_3$ is not itself the active form of the vitamin, and in the body it is hydroxylated first to **calcidiol** and then to **calcitriol** (Figure 5.106). Colecalciferol and calcitriol have also been found in several plant species.

Systematic nomenclature of vitamin D derivatives utilizes the obvious relationship to steroids, and the term *seco* (ring opened) is incorporated into the root name (compare secologanin as a

ring-opened analogue of loganin, page 189). The numbering system for steroids is also retained, and vitamin D$_3$ becomes a derivative of 9,10-seco-cholestane, namely (5Z,7E)-9,10-secocholesta-5, 7,10(19)-trien-3β-ol, '9,10' indicating the site of ring cleavage. Note that it is necessary to indicate the configuration of two of the double bonds, and the somewhat confusing β-configuration for the 3-hydroxy shows it is actually the same as in cholesterol.

Vitamin D

Vitamin D$_3$ (colecalciferol, cholecalciferol) (Figure 5.106) is the main form of the fat-soluble vitamin D found in animals, though **vitamin D$_2$ (ergocalciferol)** (Figure 5.106) is a constituent of plants and yeasts. Vitamin D$_3$ is obtained in the diet from liver and dairy products such as butter, cream, and milk, whilst large amounts can be found in fish liver oils, e.g. cod liver oil and halibut liver oil (Table 3.2). Further requirements are produced naturally when the sterol 7-dehydrocholesterol is converted into colecalciferol by the effects of UV light on the skin. With a proper diet, and sufficient exposure to sunshine, vitamin D deficiency should not occur. Vitamin D deficiency leads to rickets, an inability to calcify the collagen matrix of growing bone, and is characterized by a lack of rigidity in the bones, particularly in children. In adults, osteoporosis may occur. In most countries, foods such as milk and cereals are usually fortified with vitamin D$_3$, obtained commercially by UV irradiation of 7-dehydrocholesterol which is produced in quantity by semi-synthesis from cholesterol. Vitamin D$_2$ has a similar activity in humans and is manufactured by UV irradiation of yeast, thereby transforming the ergosterol content. Other compounds with vitamin D activity have also been produced: vitamin D$_4$ from 22,23-dihydroergosterol, vitamin D$_5$ from 7-dehydrositosterol, vitamin D$_6$ from 7-dehydrostigmasterol, and vitamin D$_7$ from 7-dehydrocampesterol. Vitamin D$_1$ was an early preparation, later shown to be a mixture of vitamin D$_2$ and a photochemical by-product lumisterol (9β,10α-ergosterol). Vitamin D is unstable to heat, light, and air.

Vitamin D$_3$ is not itself the active form of the vitamin, and in the body it is hydroxylated firstly to 25-hydroxyvitamin D$_3$ (calcidiol) (Figure 5.106) by an enzyme in the liver, and then to 1α,25-dihydroxyvitamin D$_3$ (calcitriol) by a kidney enzyme. Calcitriol is then transported to the bones, intestine, and other organs. It stimulates the absorption of calcium and phosphate in the intestine and the mobilization of calcium from bone. **Calcitriol** and other analogues, e.g. **alfacalcidol** and **dihydrotachysterol** (Figure 5.107) are available for use where chronic vitamin D deficiency is due to liver or kidney malfunction. The long term use of calcitriol and alfacalcidol (1α-hydroxyvitamin D$_3$) in the treatment of osteoporosis may lead to toxic effects arising from elevated serum calcium levels. Toxicity is much reduced in the related 1α-hydroxyvitamin D$_2$, and this agent is being investigated further.

Vitamin D is also known to have other physiological functions, including a role in immune suppression, hormone secretion, and the differentiation of both normal and malignant cells.

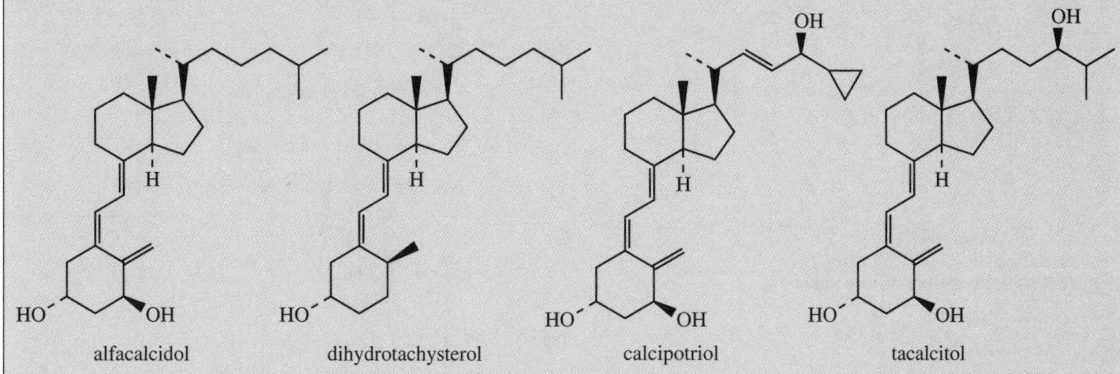

alfacalcidol   dihydrotachysterol   calcipotriol   tacalcitol

Figure 5.107

(Continues)

(Continued)

Two vitamin D derivatives, **calcipotriol** and **tacalcitol** (Figure 5.107) are widely used in the topical treatment of psoriasis, to inhibit the cell proliferation characteristic of this condition.

Vitamin D_2 is also employed as a rodenticide. High doses are toxic to rats and mice, the vitamin causing fatal hypercalcaemia.

Bile Acids

The **bile acids*** are C_{24} steroidal acids that occur in salt form in bile, secreted into the gut to emulsify fats and encourage digestion. They act as detergents by virtue of their relatively non-polar steroid nucleus and the polar side-chain, which contains a carboxylic acid group, that is typically bound via an amide linkage to glycine or taurine. Thus, for example, **cholic acid** (Figure 5.108) is found as **sodium glycocholate** and **sodium taurocholate**. Metabolism to bile acids is also the principal way in which mammals degrade cholesterol absorbed from the diet. These structures are formed in the liver from cholesterol by a sequence which oxidizes off three carbons from the side-chain (Figure 5.109). This is achieved by initial oxidation of one of the side-chain methyls to an acid, followed by a β-oxidation sequence as seen with fatty acids (see Figure 2.11), removing the three-carbon

unit as propionyl-CoA. Other essential features of the molecule are introduced earlier. The A/B ring system is *cis*-fused, and this is achieved by reduction of a Δ^4 rather than a Δ^5 double bond (see page 241). Migration of the double bond is accomplished via the 3-ketone, and when this is reduced back to a hydroxyl the configuration at C-3 is changed to 3α. Both **cholic acid** and **chenodeoxycholic acid** (Figure 5.110) are formed in the liver, though the 7α-hydroxyl functions of these compounds can be removed by intestinal microflora, so that mammalian bile also contains **deoxycholic acid** and **lithocholic acid** (Figure 5.110). The bile salts are then usually reabsorbed and stored in the gall bladder, although they are also excreted as the body's main means of eliminating excess cholesterol. Inability to remove cholesterol by bile acid synthesis and excretion may contribute to atherosclerosis and gallstone disease; gallstones often contain more than 70% of cholesterol.

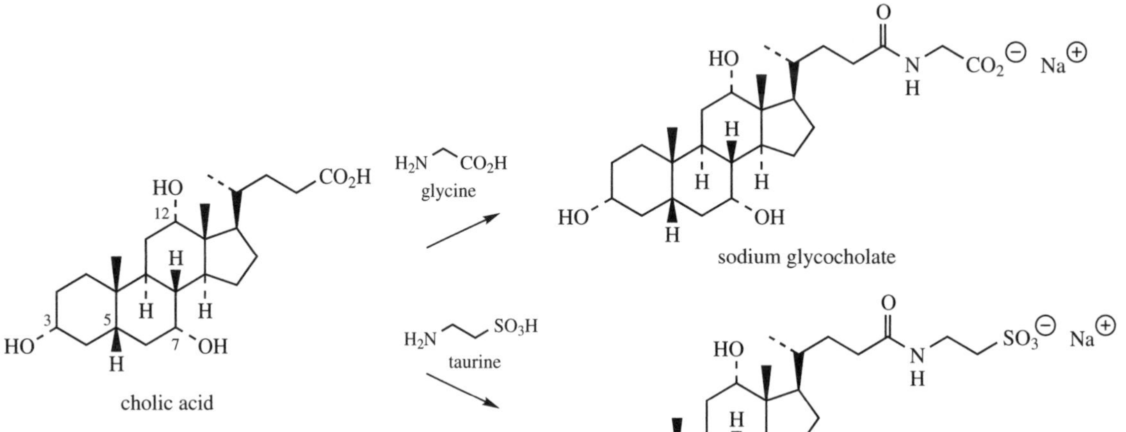

cholic acid

glycine

taurine

sodium glycocholate

sodium taurocholate

Characteristic features of bile acids:

- C_{24} cholane skeleton
- *cis*-fusion of A/B rings
- C_5-carboxylic acid side-chain
- 3α- and 7α-hydroxyls

Figure 5.108

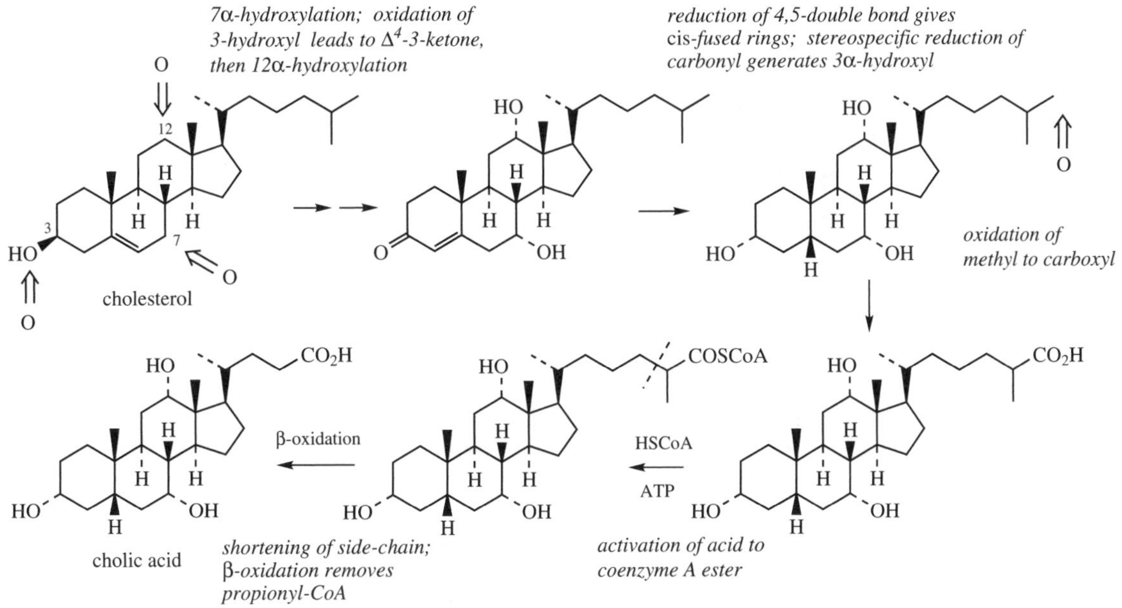

Figure 5.109

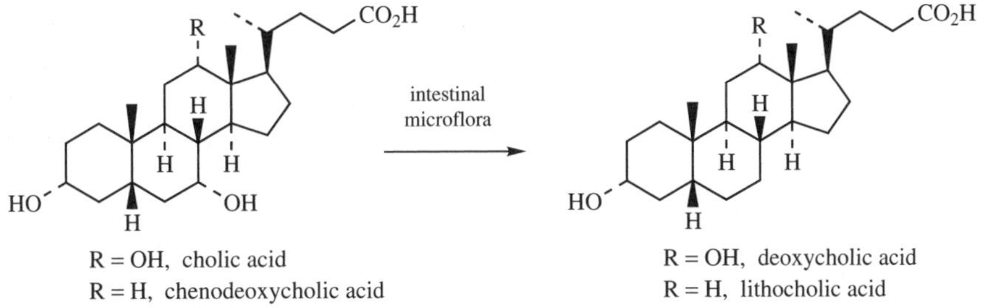

R = OH, cholic acid
R = H, chenodeoxycholic acid

R = OH, deoxycholic acid
R = H, lithocholic acid

Figure 5.110

Bile Acids

Bile acids are obtained by purification from fresh ox bile taken from carcasses as a by-product of the meat trade. **Chenodeoxycholic acid** (Figure 5.110) and **ursodeoxycholic acid** (Figure 5.111) are used to dissolve cholesterol gallstones as an alternative to surgery. By suppressing synthesis of both cholesterol and cholic acid, they contribute to removal of biliary cholesterol and consequently a gradual dissolution of gallstones which may have formed due to supersaturation. Partial or complete dissolution requires treatment over a period of many months, and is not effective for radio-opaque gallstones, which contain appreciable levels of calcium salts. **Dehydrocholic acid** (Figure 5.111) may be used, after surgery, to improve biliary drainage. Anion-exchange resins such as **colestyramine (cholestyramine)** and **colestipol** are used as cholesterol-lowering drugs to bind bile acids and prevent their reabsorption. This promotes hepatic conversion of cholesterol into bile acids, thus increasing breakdown of low density lipoprotein cholesterol, and is of value in treating high risk coronary patients.

(Continues)

(Continued)

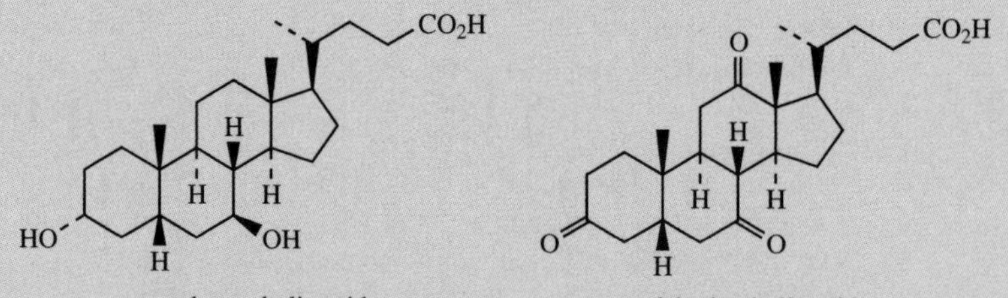

ursodeoxycholic acid   dehydrocholic acid

Figure 5.111

Bile acids are still important as starting materials for the semi-synthesis of other medicinal steroids, being a cheap and readily accessible raw material.

Adrenocortical Hormones/Corticosteroids

A large number of steroid hormones have been isolated and characterized from the adrenal glands. Since they are produced by the adrenal cortex, the outer part of the adrenal glands near the kidneys, they are termed **adrenocortical hormones** or **corticosteroids***. They contain a pregnane C_{21} skeleton and fall into two main activity groups, the **glucocorticoids** and the **mineralocorticoids**, although it is difficult to separate entirely the two types of activity in one molecule. Glucocorticoids are concerned with the synthesis of carbohydrate from protein, and deposition of glycogen in the liver. They also play an important role in inflammatory processes. Mineralocorticoids are concerned with the control of electrolyte balance, active compounds promoting the retention of Na^+ and Cl^-, and the excretion of K^+.

Examples of natural glucocorticoids include **hydrocortisone (cortisol)** and **corticosterone** (Figure 5.112), whilst **aldosterone** and **desoxycorticosterone (cortexone)** (Figure 5.113) typify mineralocorticoids. Desoxycorticosterone has also been found in plants. Some common features of these molecules are the β-$CO.CH_2OH$ side-chain at C-17, and frequently an α-hydroxy also at this position. Ring A usually contains a Δ^4-3-keto functionality. The 11β-hydroxy is essential for glucocorticoid activity. In aldosterone, the principal mineralocorticoid hormone, the methyl

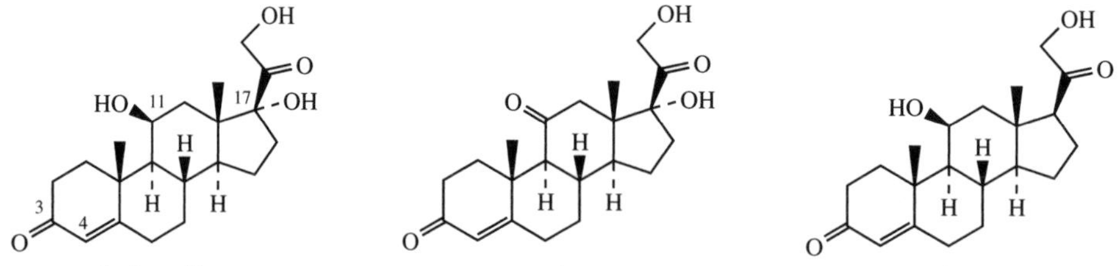

hydrocortisone (cortisol)   cortisone   corticosterone

Characteristic features of glucocorticoids:
- C_{21} pregnane skeleton
- 17β-$CO.CH_2OH$ side-chain
- 11β-hydroxyl
- Δ^4-3-keto (usually)
- 17α-hydroxyl (usually)

Figure 5.112

group (C-18) has been oxidized to an aldehyde, and this is able to react with the 11β-hydroxyl, so that aldosterone exists predominantly in the hemiacetal form (Figure 5.113). This essentially eliminates the glucocorticoid activity.

The corticosteroids are produced from cholesterol via **pregnenolone** and **progesterone**. This involves side-chain cleavage as seen in the biosynthesis of cardioactive glycosides (see page 243) and the same sequence of reactions is operative. From progesterone, the formation of **desoxycorticosterone**, **corticosterone**, and **hydrocortisone** (**cortisol**) (Figure 5.114) requires only a series of hydroxylation steps, catalysed by cytochrome

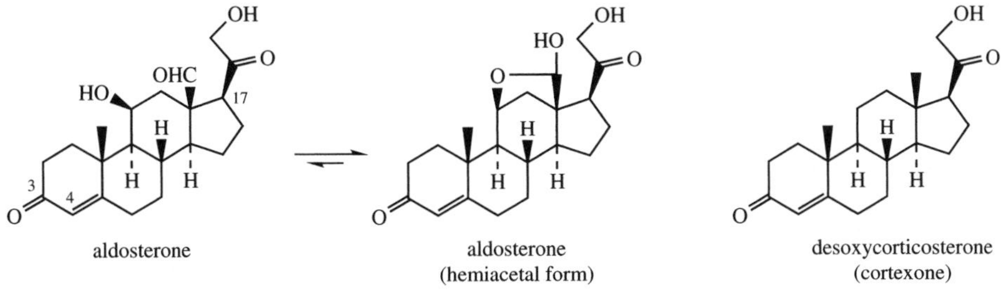

aldosterone　　　　　　　aldosterone　　　　　　　desoxycorticosterone
　　　　　　　　　　　　　(hemiacetal form)　　　　　(cortexone)

Characteristic features of mineralocorticoids:
- C_{21} pregnane skeleton
- 17β-CO.CH$_2$OH side-chain
- Δ^4-3-keto (usually)

Figure 5.113

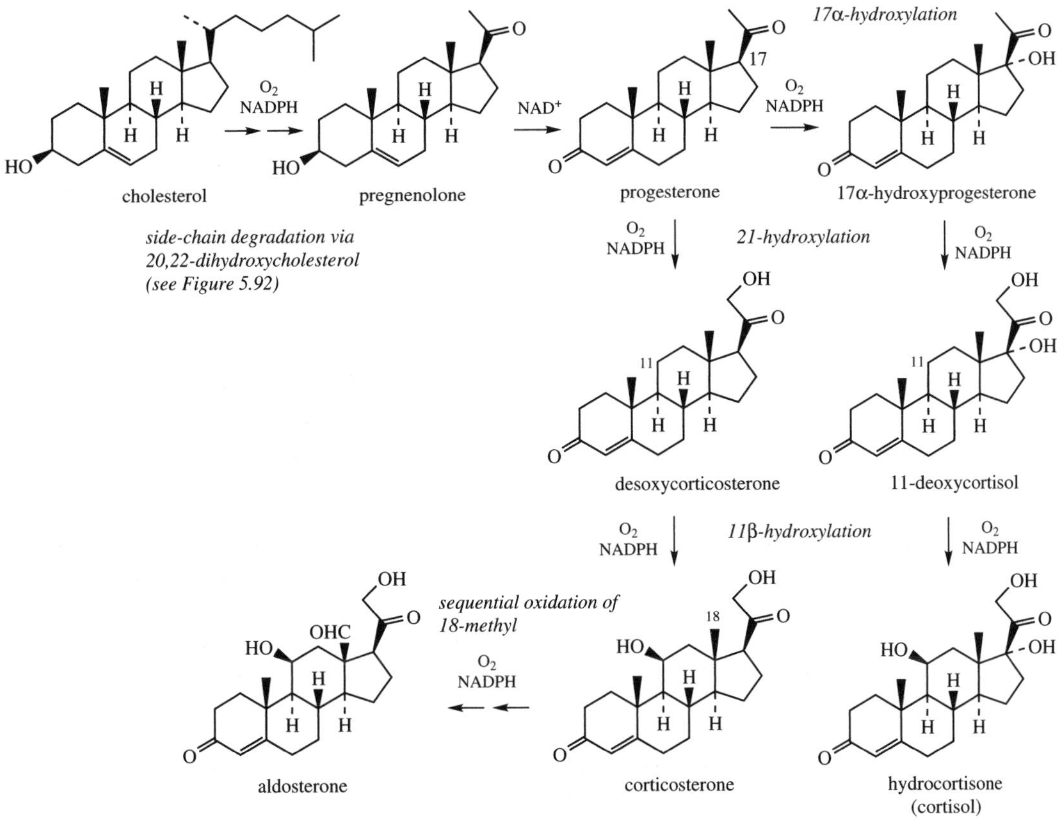

cholesterol　　　pregnenolone　　　progesterone　　　17α-hydroxyprogesterone

side-chain degradation via 20,22-dihydroxycholesterol (see Figure 5.92)

desoxycorticosterone　　　　11-deoxycortisol

aldosterone　　　　corticosterone　　　hydrocortisone (cortisol)

Figure 5.114

P-450-dependent hydroxylases with NADPH and O_2 cofactors. Thus, positions 17, 21 and 11 may be hydroxylated, and the exact order can in fact vary from that shown in Figure 5.114, according to species. It can be seen that production of hydrocortisone from cholesterol actually utilizes cytochrome P-450-dependent enzymes in four of the five steps. The further oxidation of C-18 to an aldehyde via the alcohol allows formation of **aldosterone** from corticosterone, again involving a P-450 system.

Semi-Synthesis of Corticosteroids

The medicinal use of corticosteroids was stimulated by reports of the dramatic effects of **cortisone** on patients suffering from rheumatoid arthritis in the late 1940s and early 1950s. The cortisone employed was isolated from the adrenal glands of cattle, and later was produced semi-synthetically by a laborious process from **deoxycholic acid** (see page 260) isolated from ox bile and necessitating over 30 chemical steps. Increased demand for cortisone and **hydrocortisone** (**cortisol**) (it had been shown that cortisone was reduced in the liver to hydrocortisone as the active agent) led to exploitation of alternative raw materials, particularly plant sterols and saponins. A major difficulty in any semi-synthetic conversion was the need to provide the 11β-hydroxyl group which was essential for glucocorticoid activity.

Sarmentogenin (Figure 5.115) had been identified as a natural 11-hydroxy cardenolide in *Strophanthus sarmentosus* but it was soon appreciated that the amounts present in the seeds, and the limited quantity of plant material available, would not allow commercial exploitation of this compound. As an alternative to using a natural 11-oxygenated substrate, compounds containing a 12-oxygen substituent might be used instead, in that this group activates position 11 and allows chemical modification at the adjacent site. Indeed, this was a feature of the semi-synthesis of cortisone from deoxycholic acid, which contains a 12α-hydroxyl. **Hecogenin** (Figure 5.115) from sisal (*Agave sisalana*; Agavaceae) (see page 240), a steroidal sapogenin with a 12-keto function, made possible the economic production of cortisone on a commercial scale. This material is still used in the semi-synthesis of steroidal drugs, and the critical

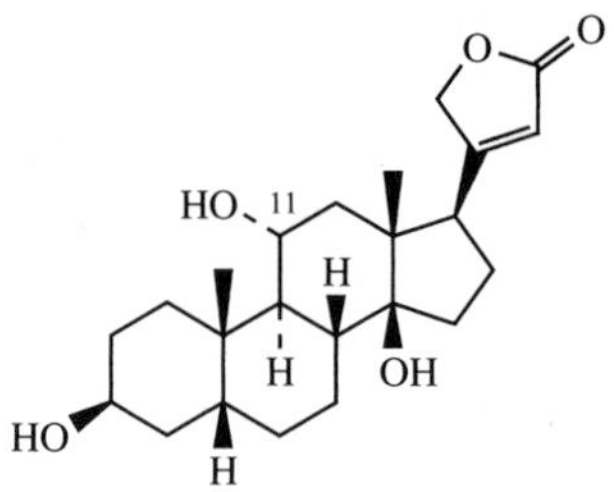

sarmentogenin

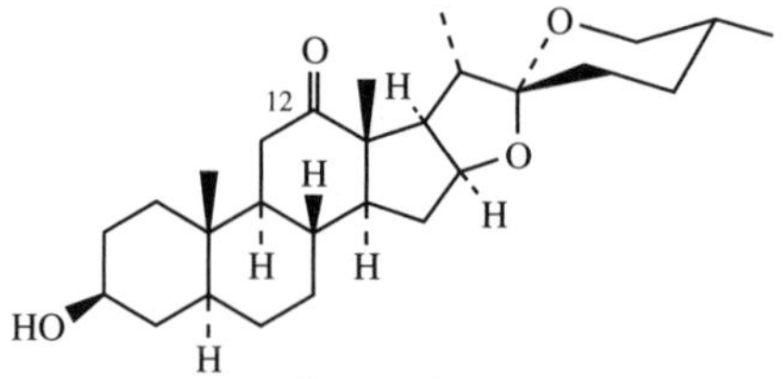

hecogenin

Figure 5.115

modifications in ring C are shown in Figure 5.116. Bromination α to the 12-keto function generates the 11α-bromo derivative, which on treatment with base gives the 12-hydroxy-11-ketone by a base-catalysed keto–enol tautomerism mechanism. The 12-hydroxyl is then removed by hydride displacement of the acetate using calcium in liquid ammonia. The new 11-keto sapogenin is subjected to the side-chain degradation used with other sapogenins, e.g. diosgenin (see Figure 5.119), to the 11-ketopregnane (Figure 5.117) which can then be used for conversion into cortisone, hydrocortisone, and other steroid drugs.

Of much greater importance was the discovery in the mid-1950s that hydroxylation at C-11 could be achieved via a microbial fermentation. **Progesterone** was transformed by *Rhizopus arrhizus* into **11α-hydroxyprogesterone** (Figure 5.118) in yields of up to 85%. More recently, *Rhizopus nigricans* has been employed, giving even higher yields. 11α-Hydroxyprogesterone is then converted into hydrocortisone by chemical means, the 11β configuration being introduced via oxidation to the 11-keto and then a stereospecific reduction step. Progesterone could be obtained in good yields (about 50%) from **diosgenin** extracted from Mexican yams (*Dioscorea* species; Dioscoreaceae) (see page 239) or **stigmasterol** from soya beans (*Glycine max*; Leguminosae/Fabaceae) (see page 256). Steroidal sapogenins such as diosgenin may be degraded by the **Marker degradation**

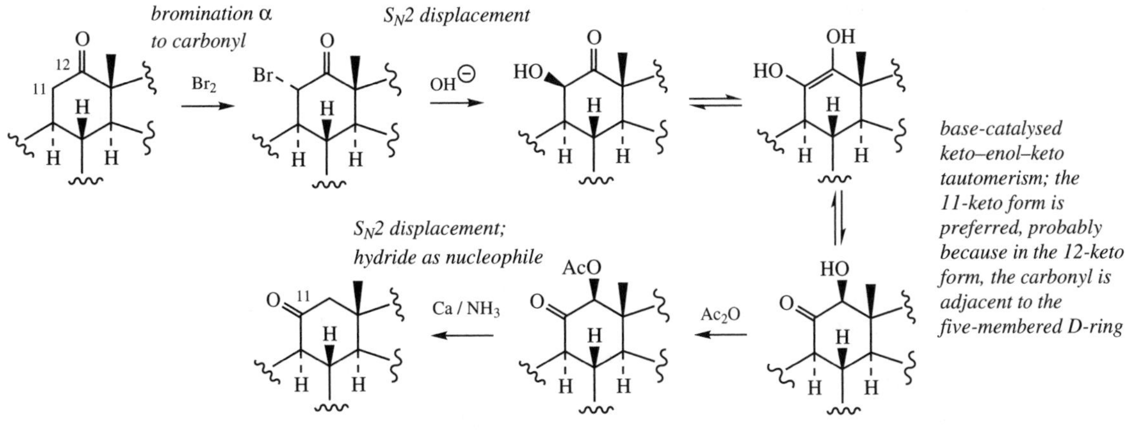

Figure 5.116

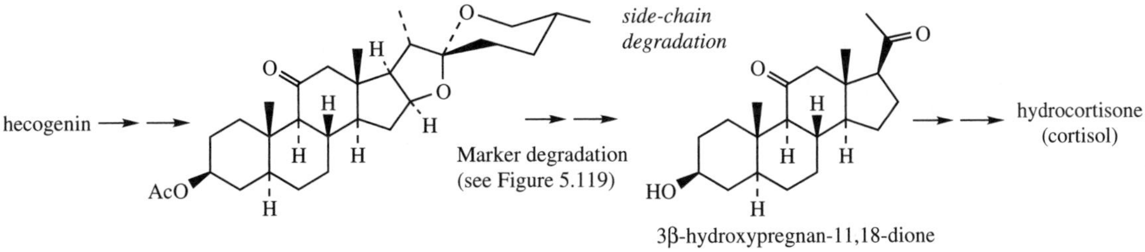

Figure 5.117

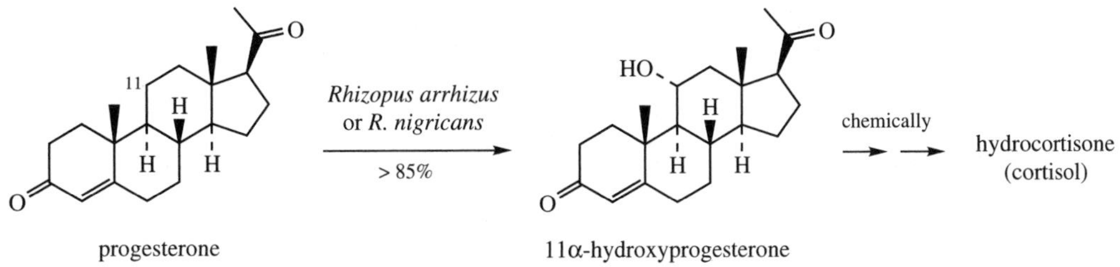

Figure 5.118

(Figure 5.119), which removes the spiroketal portion, leaving carbons C-20 and C-21 still attached to contribute to the pregnane system. Initial treatment with acetic anhydride produces the diacetate, by opening the ketal, dehydrating in ring E, and acetylating the remaining hydroxyls. The double bond in ring E is then selectively oxidized to give a product, which now contains the unwanted side-chain carbons as an ester function, easily removed by hydrolysis. Under the conditions used, the product is the α,β-unsaturated ketone. Hydrogenation of the double bond is achieved in a regioselective

and stereoselective manner, addition of hydrogen being from the less-hindered α-face to give pregnenolone acetate. **Progesterone** is obtained by hydrolysis of the ester function and Oppenauer oxidation to give the preferred α,β-unsaturated ketone (see page 241). It is immediately obvious from Figure 5.119 that, since the objective is to remove the unwanted ring F part of the sapogenin, features like the stereochemistry at C-25 are irrelevant, and the same general degradation procedure can be used for other sapogenins. It is equally applicable to the nitrogen-containing analogues of

Marker Degradation of Diosgenin

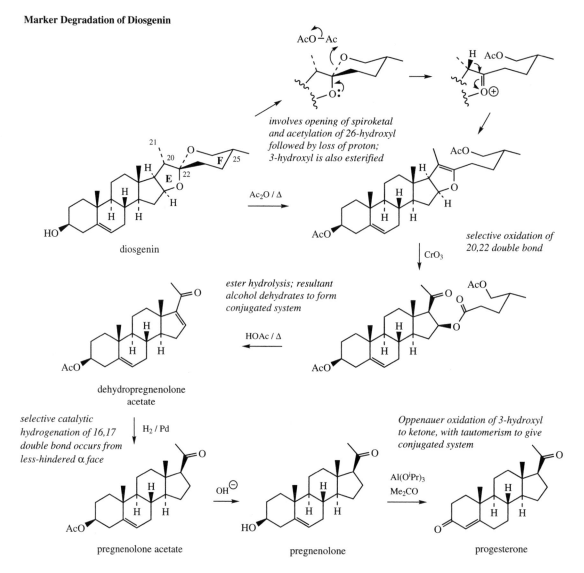

Figure 5.119

sapogenins, e.g. **solasodine** (Figure 5.88). In such compounds, the stereochemistry at C-22 is also quite immaterial.

Degradation of the sterol **stigmasterol** to progesterone is achieved by the sequence shown in Figure 5.120. The double bond in the side-chain allows cleavage by ozonolysis, and the resultant aldehyde is chain shortened via formation of an enamine with piperidine. This can be selectively oxidized to progesterone. In this sequence, the ring A transformations are carried out as the

first reaction. A similar route can be used for the fungal sterol **ergosterol**, though an additional step is required for reduction of the Δ^7 double bond.

An alternative sequence from diosgenin to hydrocortisone has been devised, making use of another microbiological hydroxylation, this time a direct 11β-hydroxylation of the steroid ring system (Figure 5.121). The fungus *Curvularia lunata* is able to 11β-hydroxylate **cortexolone** to **hydrocortisone** in yields of about 60%.

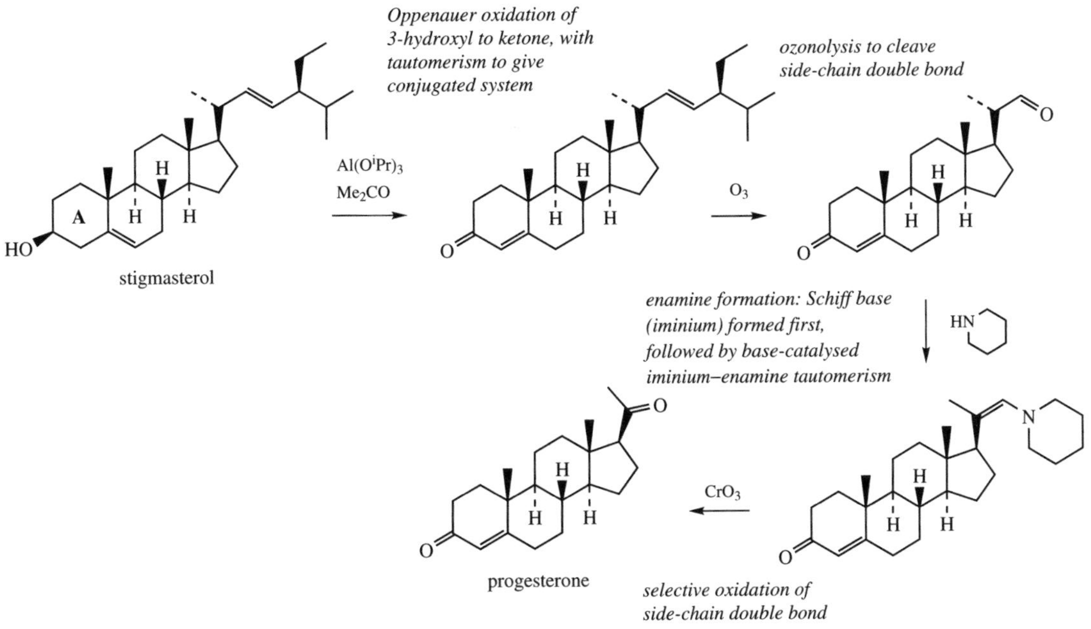

Figure 5.120

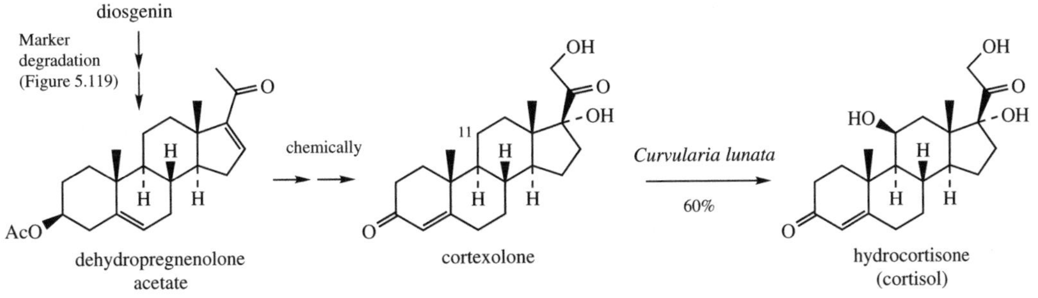

Figure 5.121

Although a natural corticosteroid, cortexolone may be obtained in large amounts by chemical transformation from 16-dehydropregnenolone acetate, an intermediate in the Marker degradation of diosgenin (Figure 5.119).

Some steroid drugs are produced by total synthesis, but, in general, the ability of microorganisms to biotransform steroid substrates has proved invaluable in exploiting inexpensive natural steroids as sources of drug materials. It is now possible via microbial fermentation to hydroxylate the steroid nucleus at virtually any position and with defined stereochemistry. These

processes are in general more expensive than chemical transformations, and are only used commercially when some significant advantage is achieved, e.g. replacement of several chemical steps. The therapeutic properties of cortisone and hydrocortisone can be further improved by the microbial introduction of a 1,2-double bond, giving **prednisone** and **prednisolone** respectively (Figure 5.122). These agents surpass the parent hormones in antirheumatic and antiallergic activity with fewer side effects. As with cortisone, prednisone is converted in the body into the active agent, in this case prednisolone.

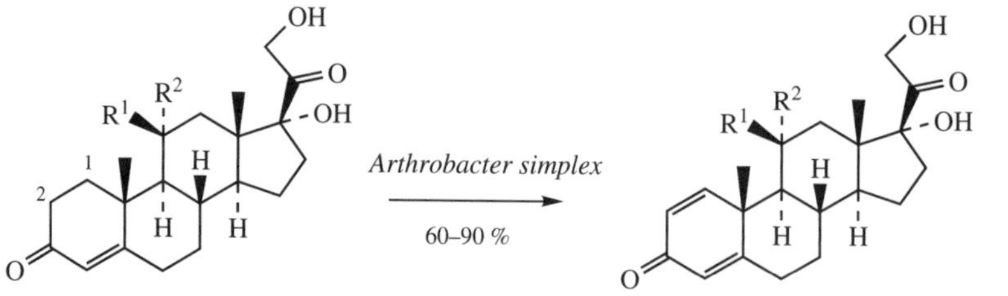

R¹R² = O, cortisone
R¹ = OH, R² = H, hydrocortisone (cortisol)

R¹R² = O, prednisone
R¹ = OH, R² = H, prednisolone

Figure 5.122

Corticosteroid Drugs

Glucocorticoids are primarily used for their antirheumatic and anti-inflammatory activities. They give valuable relief to sufferers of rheumatoid arthritis and osteoarthritis, and find considerable use for the treatment of inflammatory conditions by suppressing the characteristic development of swelling, redness, heat, and tenderness. They exert their action by interfering with prostaglandin biosynthesis, via production of a peptide that inhibits the phospholipase enzyme responsible for release of arachidonic acid from phospholipids (see page 55). However, these agents merely suppress symptoms and they do not provide a cure for the disease. Long term usage may result in serious side-effects, including adrenal suppression, osteoporosis, ulcers, fluid retention, and increased susceptibility to infections. Because of these problems, steroid drugs are rarely the first choice for inflammatory treatment, and other therapies are usually tried first. Nevertheless corticosteroids are widely used for inflammatory conditions affecting the ears, eyes, and skin, and in the treatment of burns. Some have valuable antiallergic properties helping in reducing the effects of hay fever and asthma. In some disease states, e.g. Addison's disease, the adrenal cortex is no longer able to produce these hormones, and replacement therapy becomes necessary. The most common genetic deficiency is lack of the 21-hydroxylase enzyme in the biosynthetic pathway, necessary for both hydrocortisone and aldosterone biosynthesis (Figure 5.114). This can then lead to increased synthesis of androgens (see Figure 5.133).

Mineralocorticoids are primarily of value in maintaining electrolyte balance where there is adrenal insufficiency.

Natural corticosteroid drugs **cortisone** (as **cortisone acetate**) and **hydrocortisone (cortisol)** (Figure 5.112) are valuable in replacement therapies, and hydrocortisone is one of the most widely used agents for topical application in the treatment of inflammatory skin conditions. The early use of natural corticosteroids for anti-inflammatory activity tended to show up some serious side-effects on water, mineral, carbohydrate, protein, and fat metabolism. In particular, the mineralocorticoid activity is usually considered an undesirable effect. In an effort to optimize anti-inflammatory activity, many thousands of chemical modifications to the basic structure were tried. Introduction of a Δ^1 double bond modifies the shape of ring A and was found to increase glucocorticoid over mineralocorticoid activity, e.g. **prednisone** and **prednisolone** (Figure 5.122). A 9α-fluoro substituent increased all activities, whereas 16α- or 16β-methyl groups reduced the mineralocorticoid activity without affecting

(Continues)

(*Continued*)

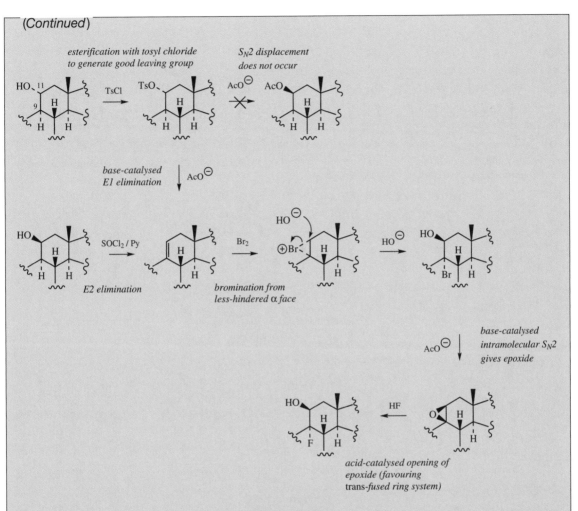

Figure 5.123

the glucocorticoid activity. The discovery that 9α-fluoro analogues had increased activity arose indirectly from attempts to epimerize 11α-hydroxy compounds into the active 11β-hydroxy derivatives (Figure 5.123). Thus, when an 11α-tosylate ester was treated with acetate, a base-catalysed elimination was observed rather than the hoped-for substitution, which is hindered by the methyl groups (Figure 5.123). This *syn* elimination suggests an E1 mechanism is involved. The same $\Delta^{9(11)}$-ene can also be obtained by dehydration of the 11β-alcohol by using thionyl chloride. Addition of HOBr to the 9(11)-double bond proceeds via electrophilic attack from the less-hindered α-face, giving the cyclic bromonium ion, and then ring opening by β-attack of hydroxide at C-11. Attack at C-9 is sterically hindered by the methyl at C-10. 9α-Bromocortisol 21-acetate produced in this way was less active as an anti-inflammatory than cortisol 21-acetate by a factor of three, and 9α-iodocortisol acetate was also less active by a factor of ten. Fluorine must be introduced indirectly by the β-epoxide formed by base treatment of the 9α-bromo-10β-hydroxy analogue (Figure 5.123). The resultant 9α-fluorocortisol 21-acetate (**fluorohydrocortisone** acetate; **fludrocortisone acetate**) (Figure 5.124) was found to be about 11 times more active than cortisol acetate. However, its mineralocorticoid activity was

(*Continues*)

(Continued)

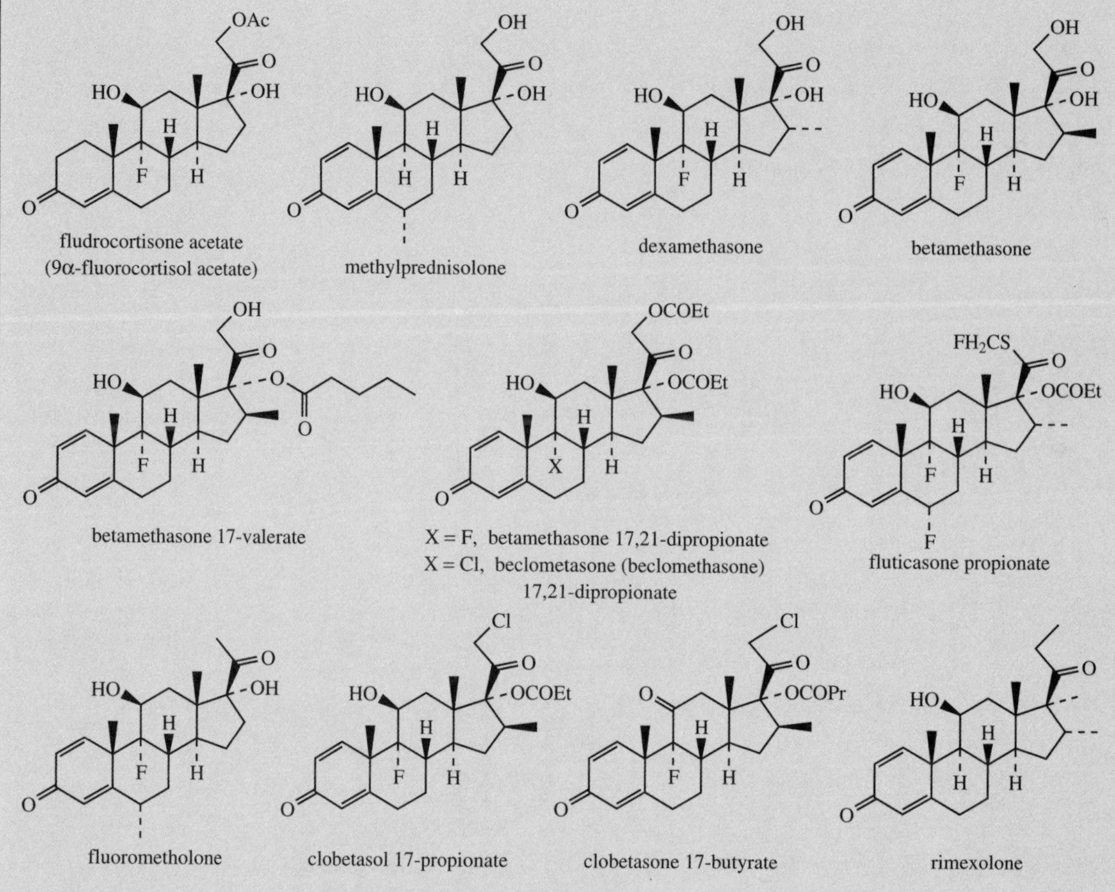

Figure 5.124

also increased some 300-fold, so its anti-inflammatory activity has no clinical relevance, and it is only employed for its mineralocorticoid activity. The introduction of a 9α-fluoro substituent into prednisolone causes powerful Na$^+$ retention. These effects can be reduced (though usually not eliminated entirely) by introducing a substituent at C-16, either a 16α-hydroxy or a 16α/16β-methyl. The 16α-hydroxyl can be introduced microbiologically, e.g. as in the conversion of 9α-fluoroprednisolone into **triamcinolone** (Figure 5.125). The ketal formed from triamcinolone and acetone, **triamcinolone acetonide** (Figure 5.125) provides a satisfactory means of administering this anti-inflammatory by topical application in the treatment of skin disorders such as psoriasis. **Methylprednisolone** (Figure 5.124) is a 6α-methyl derivative of prednisolone showing a modest increase in activity over the parent compound. A 6-methyl group can be supplied by reaction of the Grignard reagent MeMgBr with a suitable 5,6-epoxide derivative. **Dexamethasone** and **betamethasone** (Figure 5.124) exemplify respectively 16α- and 16β-methyl derivatives in drugs with little, if any, mineralocorticoid activity. The 16-methyl group is easily introduced by a similar Grignard reaction with an appropriate α,β-unsaturated Δ^{16}-20-ketone. Betamethasone, for topical application, is typically formulated as a C-17 ester with valeric acid (**betamethasone 17-valerate**), or as

(Continues)

(Continued)

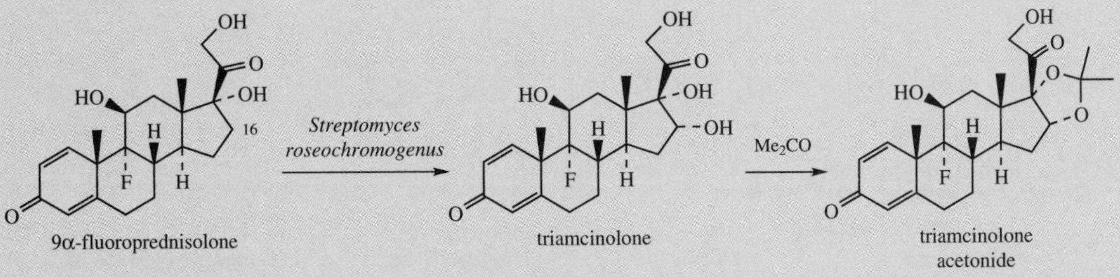

9α-fluoroprednisolone $\xrightarrow{\textit{Streptomyces roseochromogenus}}$ triamcinolone $\xrightarrow{Me_2CO}$ triamcinolone acetonide

Figure 5.125

the 17,21-diester with propionic acid (**betamethasone 17,21-dipropionate**) (Figure 5.124). The 9α-chloro compound **beclometasone 17,21-dipropionate (beclomethasone 17,21-dipropionate**) is also an important topical agent for eczema and psoriasis, and as an inhalant for the control of asthma. **Fluticasone propionate** (Figure 5.124) is also used in asthma treatment, and is representative of compounds where the 17-side-chain has been modified to a carbothiate (sulphur ester).

Although the anti-inflammatory activity of hydrocortisone is lost if the 21-hydroxyl group is not present, considerable activity is restored when a 9α-fluoro substituent is introduced. **Fluorometholone** (Figure 5.124) is a corticosteroid that exploits this relationship and is of value in eye conditions. Other agents are derived by replacing the 21-hydroxyl with a halogen, e.g. **clobetasol 17-propionate** and **clobetasone 17-butyrate** (Figure 5.124), which are effective topical drugs for severe skin disorders. In **rixemolone**, a recently introduced anti-inflammatory for ophthalmic use, neither a 21-hydroxy nor a 9α-fluoro substituent is present, but instead there are 17α- and 16α-methyl substituents. Rimexolone has significant advantages in eye conditions over drugs such as dexamethasone, in that it does not significantly raise intraocular pressure.

Many other corticosteroids are currently available for drug use. Structures of some of these are given in Figure 5.126, grouped according to the most characteristic structural features, namely 16-methyl, 16-hydroxy, and 21-chloro derivatives. The recently introduced **deflazacort** (Figure 5.127) is a drug with high glucocorticoid activity, but does not conveniently fit into any of these groups in that it contains an oxazole ring spanning C-16 and C-17.

Trilostane (Figure 5.127) is an adrenocortical suppressant, which inhibits synthesis of glucocorticoids and mineralocorticoids and has value in treating Cushing's syndrome, characterized by a moon-shaped face and caused by excessive glucocorticoids. This drug is an inhibitor of the dehydrogenase–isomerase that transforms pregnenolone into progesterone (Figures 5.92 and 5.114).

Spironolactone (Figure 5.127) is an antagonist of the endogenous mineralocorticoid aldosterone and inhibits the sodium-retaining action of aldosterone whilst also decreasing the potassium-secreting effect. Classified as a potassium-sparing diuretic, it is employed in combination with other diuretic drugs to prevent excessive potassium loss. Progesterone (page 273) is also an aldosterone antagonist; the spironolactone structure differs from progesterone in its 7α-thioester substituent, and replacement of the 17β side-chain with a 17α-spirolactone.

(Continues)

(Continued)

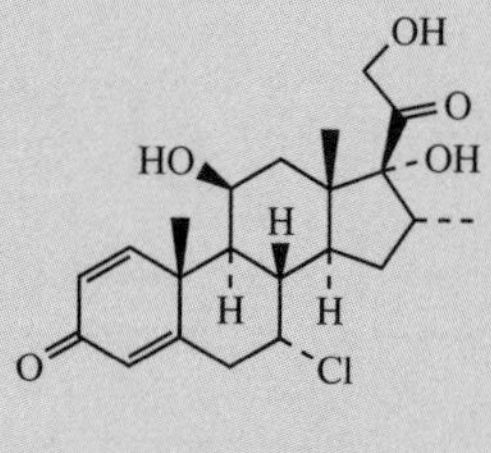

alclometasone
(used as 17,21-dipropionate)

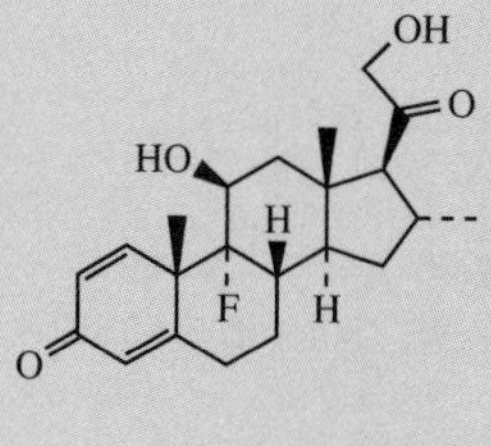

desoximetasone
(desoxymethasone)

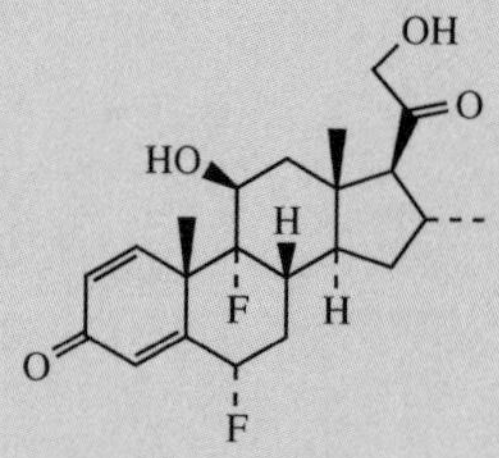

diflucortolone
(used as 21-valerate)

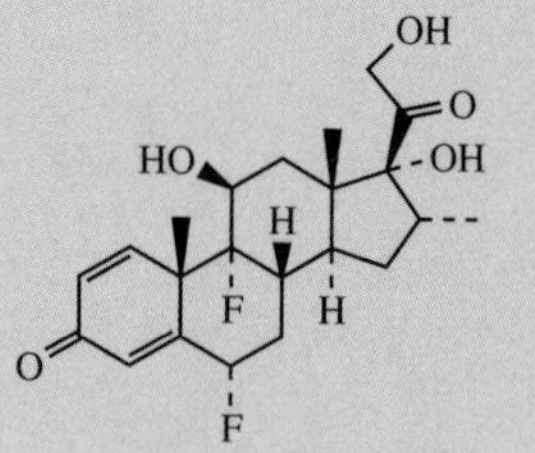

flumetasone (flumethasone)
[used as pivalate (trimethylacetate)]

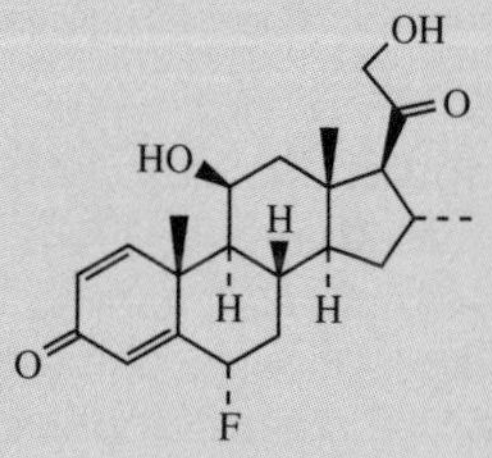

fluocortolone
(used as 21-hexanoate)

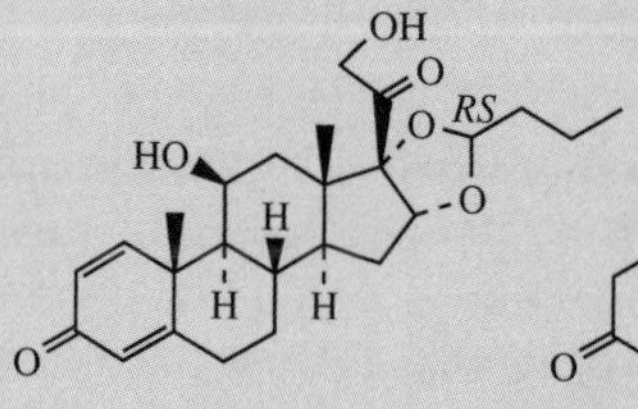

budesonide

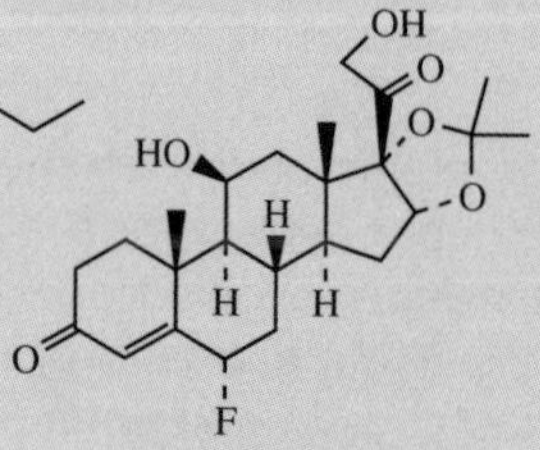

fludroxycortide
(flurandrenolone)

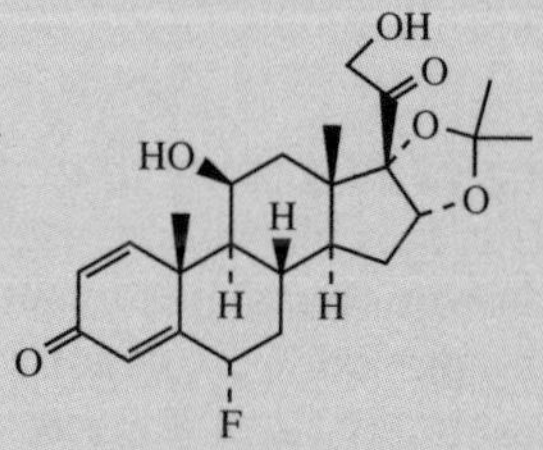

flunisolide

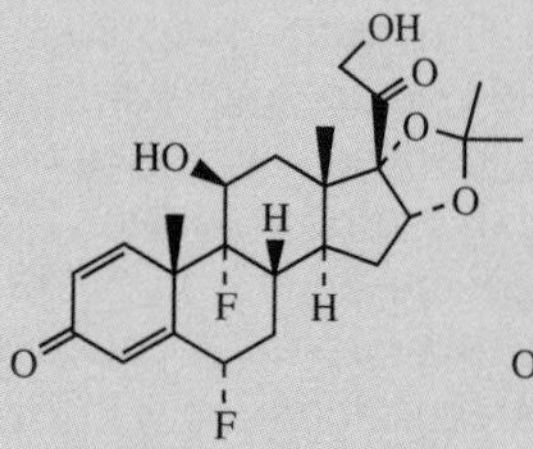

fluocinolone acetonide

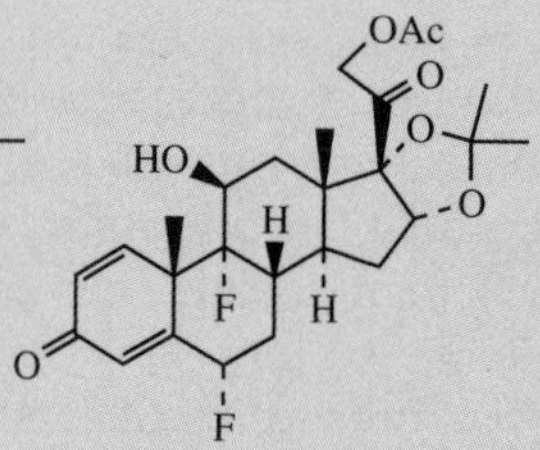

fluocinonide

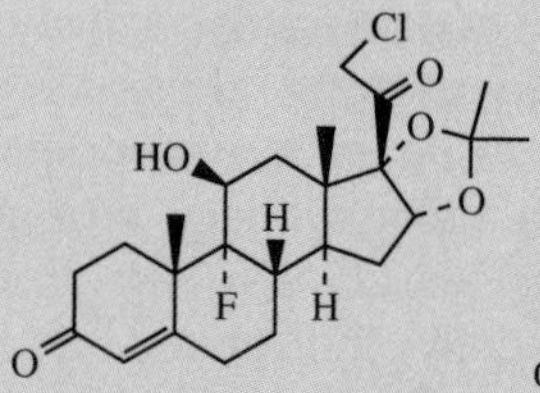

halcinonide

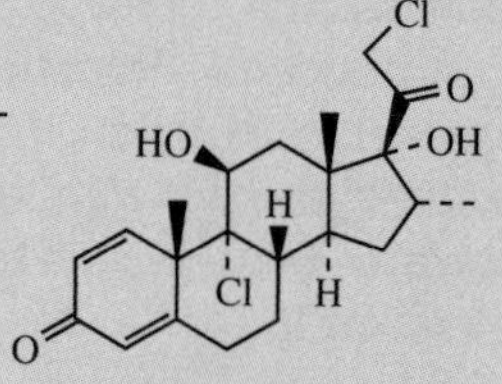

mometasone
(used as 17-furoate)

Figure 5.126

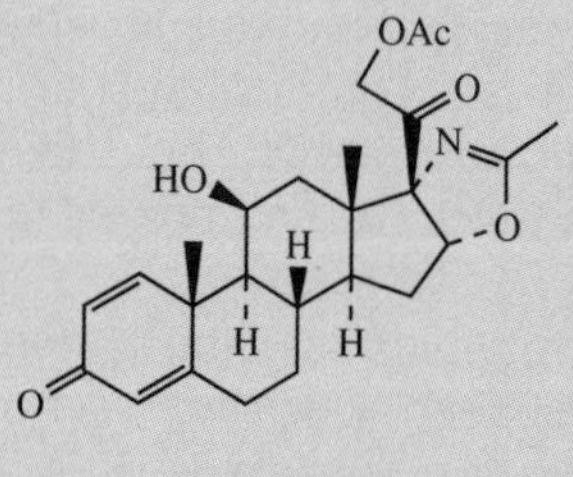

deflazacort

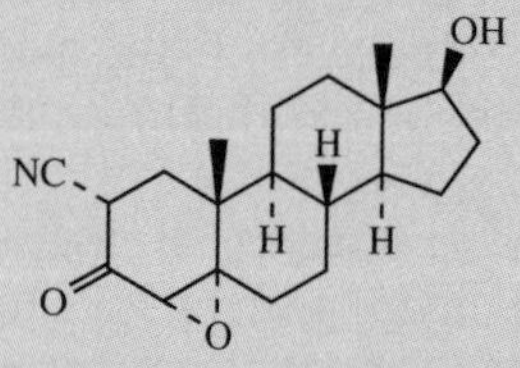

trilostane

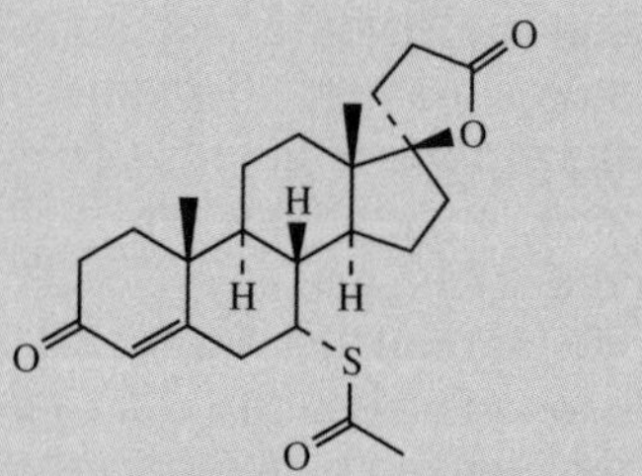

spironolactone

Figure 5.127

Progestogens

Progestogens * (progestins; gestogens) are female sex hormones, concerned with preparing the uterus for pregnancy, and then maintaining the necessary conditions. There is only one naturally occurring progestational steroid and that is **progesterone** (Figure 5.128), which is secreted by the corpus luteum following release of an ovum. Progesterone is also an intermediate in the biosynthesis of the corticosteroids, e.g. hydrocortisone and aldosterone (see page 263), and its derivation from cholesterol via pregnenolone has already been seen in the formation of cardioactive glycosides (see page 243).

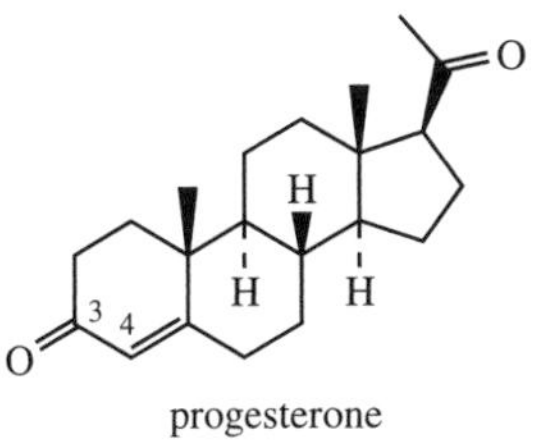

progesterone

Characteristic features of progestogens:

- C_{21} pregnane skeleton
- Δ^4-3-keto

Figure 5.128

Oestrogens

The **oestrogens** (US spelling: **estrogens**) are female sex hormones produced in the ovaries, and also in the placenta during pregnancy. They are responsible for the female sex characteristics, and together with progesterone control the menstrual cycle. Oestrogens were first isolated from the urine of pregnant women, in which levels increase some 50-fold during the pregnancy. In horses, levels rise by as much as 500 times during pregancy. Oestrogens occur both in free form, and as glucuronides

---

Progestogen Drugs

Quantities of **progesterone** (Figure 5.128) for drug use are readily available by semi-synthesis using the Marker degradation (see page 264). However, progesterone is poorly absorbed, and it is not suitable for oral use, being rapidly metabolized in the liver. Many semi-synthetic analogues have been produced, and it was thus appreciated that the α,β-unsaturated ketone system in ring A was essential for activity. The side-chain function at C-17 could be modified, and ethisterone (17α-ethynyltestosterone) (Figure 5.129), developed as a potential androgen, was found to be active orally as a progestational agent. This incorporates an ethynyl side-chain at C-17, a feature of several semi-synthetic steroidal hormones used as drugs. This group, referred to as 'ethinyl' in drug molecules, is introduced by nucleophilic

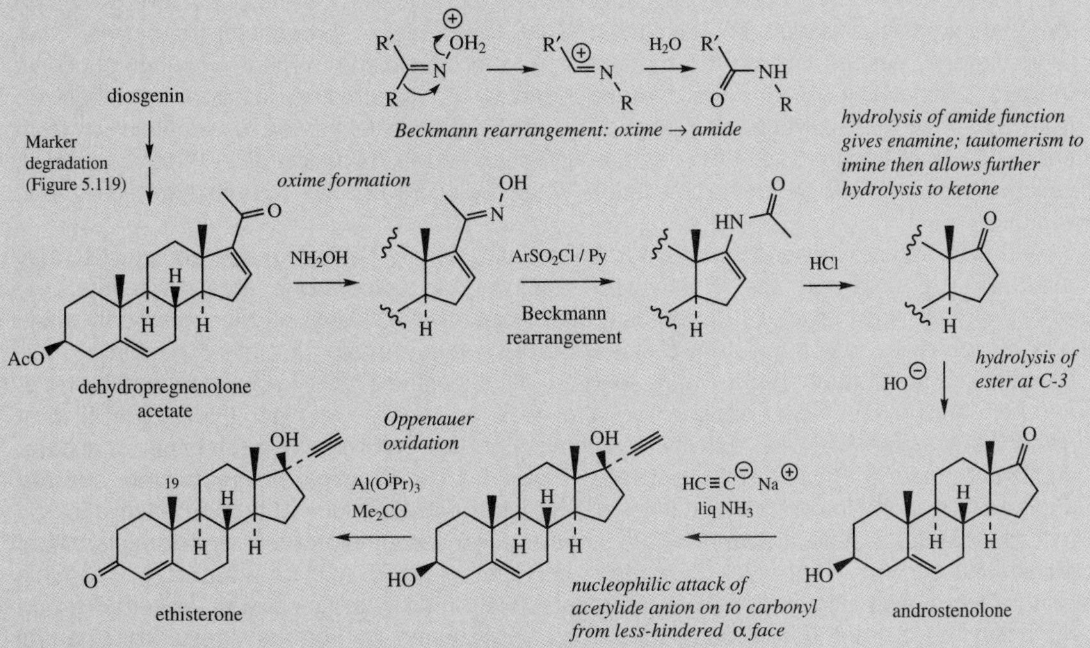

Figure 5.129

(Continues)

(Continued)

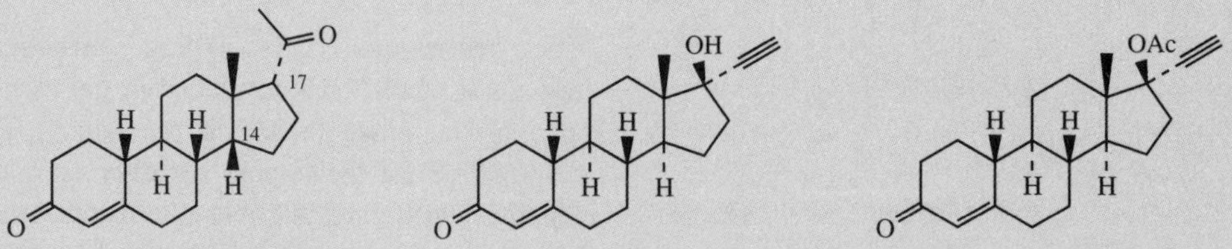

Figure 5.130

attack of acetylide anion on to a C-17 carbonyl (Figure 5.129), attack coming from the α-face, the methyl C-18 hindering approach from the β-face. The substrate androstenolone is readily obtained from the Marker degradation intermediate dehydropregnenolone acetate (Figure 5.119). The oxime (Figure 5.129) is treated with a sulphonyl chloride in pyridine and undergoes a Beckmann rearrangement in which C-17 migrates to the nitrogen giving the amide. This amide is also an enamine and can be hydrolysed to the 17-ketone. Acetylation or other esterification of the 17-hydroxyl in progestogens increases lipid solubility and extends the duration of action by inhibiting metabolic degradation. Examples include norethisterone acetate, medroxyprogesterone acetate, and hydroxyprogesterone caproate, discussed below.

Though considerably better than progesterone, the oral activity of ethisterone is still relatively low, and better agents were required. An important modification from ethisterone was the 19-*nor* analogue, **norethisterone** and its ester **norethisterone acetate** (Figure 5.130). Attention was directed to the 19-norsteroids by the observation that 19-nor-14β,17α-progesterone (Figure 5.130), obtained by degradation of the cardioactive glycoside strophanthidin (see page 250), displayed eight times higher progestational activity than progesterone, despite lacking the methyl C-19, and having the unnatural configurations at the two centres C-14 (C/D rings *cis*-fused) and C-17. Norethisterone can be synthesized from the oestrogen estrone (see page 279) which already lacks the C-9 methyl, or from androstenolone (Figure 5.129) by a sequence which allows oxidation of C-19 to a carboxyl, which is readily lost by decarboxylation when adjacent to the α,β-unsaturated ketone system.

Although ethisterone and norethisterone are structurally C_{21} pregnane derivatives, they may also be regarded as 17-ethynyl derivatives of testosterone (see page 282), the male sex hormone, and 19-nortestosterone respectively. Many of the commonly used progestogens fall into these two classes. Semi-synthetic analogues of progesterone, still containing the 17-acetyl side-chain, tend to be derivatives of 17α-hydroxyprogesterone, another biosynthetic intermediate on the way to hydrocortisone (Figure 5.114) that also has progesterone-like activity. Examples include **hydroxyprogesterone caproate**, and **gestonorone (gestronol) caproate** (Figure 5.131). **Medroxyprogesterone acetate** (Figure 5.131) contains an additional 6α-methyl, introduced to block potential deactivation by metabolic hydroxylation, and is 100–300 times as potent as ethisterone on oral administration. **Megestrol acetate** (Figure 5.131) contains a 6-methyl group and an additional Δ^6 double bond. **Norgestrel** (Figure 5.131) is representative of progestogens with an ethyl group replacing the 13-methyl. Although these can be obtained by semi-synthesis from natural 13-methyl compounds, norgestrel is produced by total synthesis as the racemic compound.

(Continues)

(Continued)

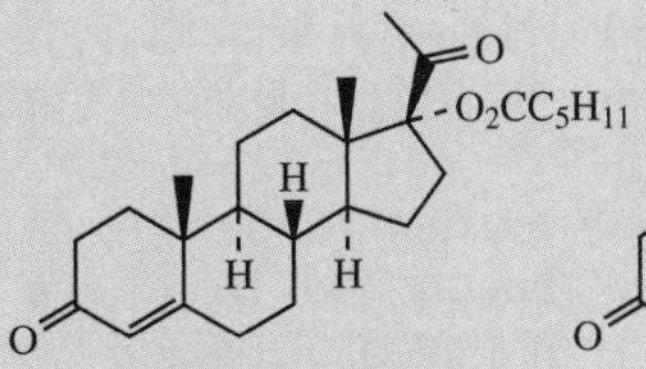

hydroxyprogesterone caproate (hexanoate)

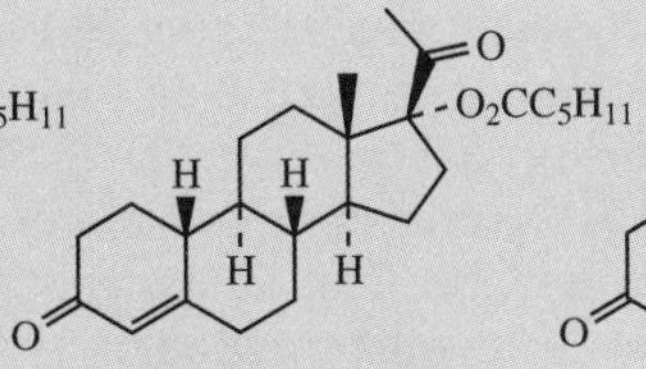

gestonorone (gestronol) caproate (hexanoate)

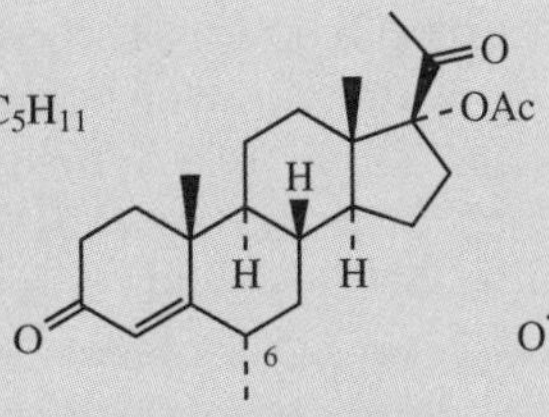

medroxyprogesterone acetate

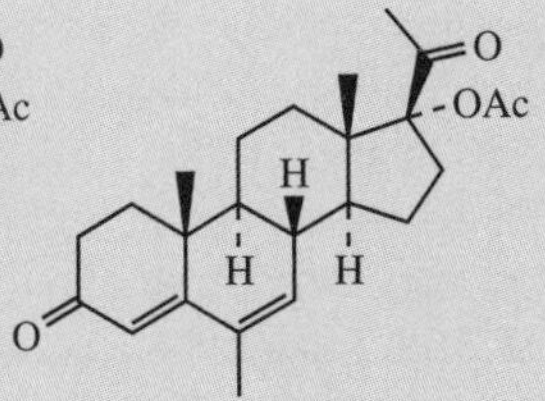

megestrol acetate

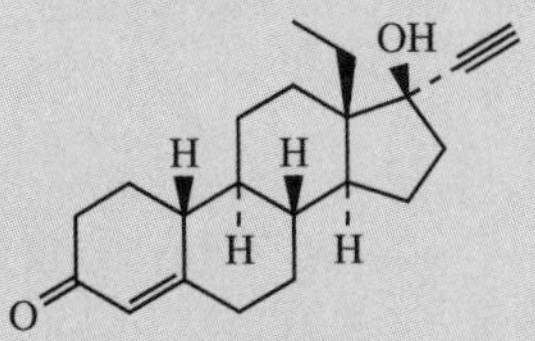

(–): levonorgestrel
(±): norgestrel

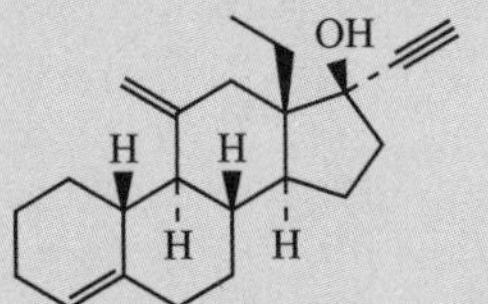

desogestrel

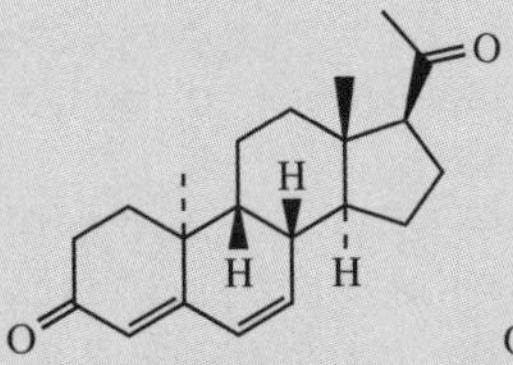

dydrogesterone

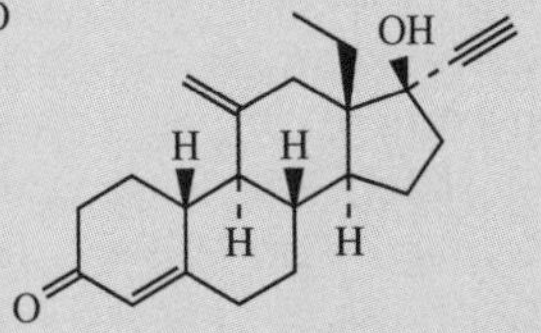

etonogestrel

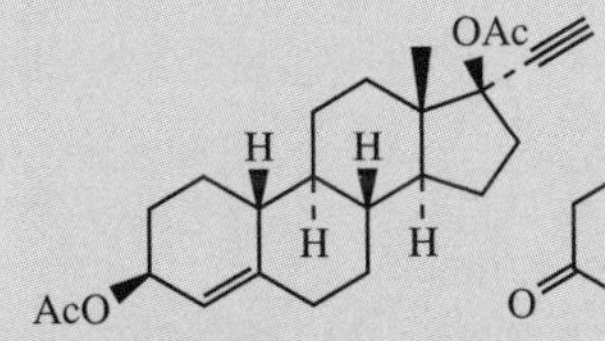

etynodiol (ethynodiol) diacetate

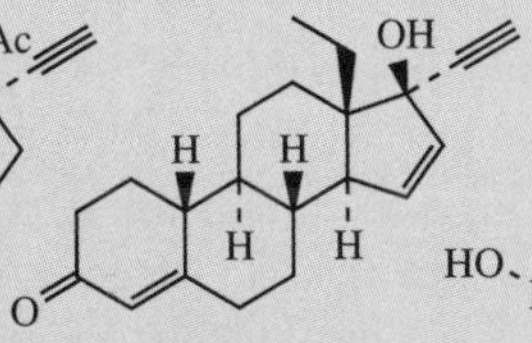

gestodene

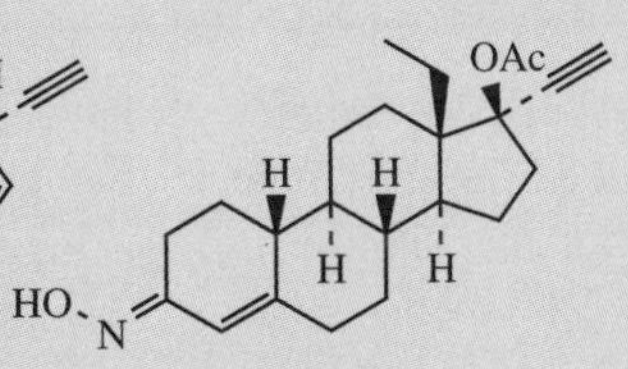

norgestimate

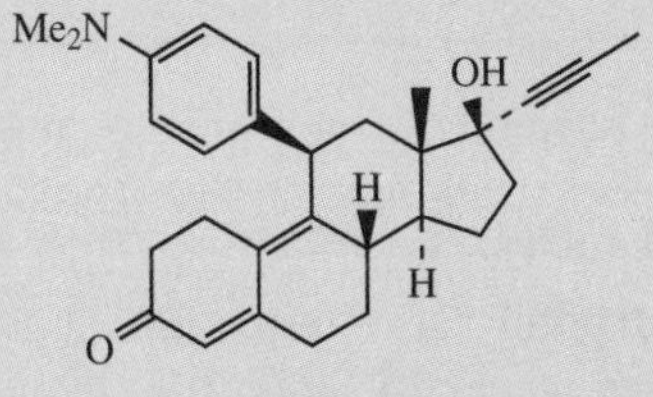

mifepristone

Figure 5.131

Since only the laevorotatory enantiomer which has the natural configuration is biologically active, this enantiomer, **levonorgestrel**, is now replacing the racemic form for drug use. In **desogestrel** (Figure 5.131), further features are the modification of an 11-oxo function to an 11-methylene, and removal of the 3-ketone. Structures of some other currently available progestogen drugs are shown in Figure 5.131.

During pregnancy, the corpus luteum continues to secrete progesterone for the first three months, after which the placenta becomes the supplier of both progesterone and oestrogen. Progesterone prevents further ovulation and relaxes the uterus to prevent the fertilized egg being dislodged. In the absence of pregnancy, a decline in progesterone levels results in shedding of the uterine endometrium and menstruation. Progestogens are useful in many menstrual disorders, and as **oral contraceptives** either alone at low dosage (progestogen-only contraceptives, e.g. norethisterone, levonorgestrel) or in combination with oestrogens (combined oral contraceptives, e.g. ethinylestradiol + norethisterone, ethinylestradiol + levonorgestrel). The combined oestrogen–progestogen preparation inhibits ovulation, but normal menstruation occurs when the drug is withdrawn for several days each month. The low dosage progestogen-only pill appears to interfere with the endometrial lining to inhibit fertilized egg implantation, and thickens cervical mucus making a barrier to sperm movement. The progestogen-only formulation is less likely to cause thrombosis, a serious side-effect sometimes experienced from the use of oral contraceptives. There appears to be a slightly higher risk of thrombosis in patients using the so-called 'third generation' oral

(Continues)

(Continued)

contraceptive pills containing the newer progestogens **desogestrel** and **gestodene**. Current oral contraceptives have a much lower hormone content than the early formulations of the 1960s and 1970s, typically about 10% of the progestogen and 50% of the oestrogen content. Deep muscular injections of medroxyprogesterone or norethisterone esters, and implants of levonorgestrel can be administered to provide long-acting contraception. A high dose of levonorgestrel, alone or in combination with ethinylestradiol, is the drug of choice for emergency contraception after unprotected intercourse, i.e. the 'morning-after' pill. Hormone replacement therapy (HRT) in non-hysterectomized women also uses progestogen–oestrogen combinations (see page 279), whilst progestogens such as norethisterone, megestrol acetate, medroxyprogesterone acetate, and gestonorone caproate also find application in the treatment of breast cancers.

Mifepristone (Figure 5.131) is a progestogen antagonist used orally as an abortifacient to terminate pregnancy. This drug has a higher affinity for the progesterone receptor than does the natural hormone, and prevents normal responses. This leads to loss of integrity of the uterine endometrial lining, and detachment of the implanted fertilized egg.

at position 3; they are not restricted to females since small amounts are produced in the male testis. The principal and most potent example is **estradiol** (also **oestradiol**, but US spelling has been generally adopted), though only low levels are found in urine, and larger amounts of the less active metabolites **estrone (oestrone)** and the 16α-hydroxylated derivative **estriol (oestriol)** are present (Figure 5.132). Estrone has also been found in significant quantities in some plant seeds, e.g. pomegranate and date palm. These compounds have an aromatic A ring, a consequence of which is that C-19, the methyl on C-10, is absent. There is now no carbon side-chain at C-17, and the basic C_{18} skeleton is termed estrane.

The biosynthetic pathway to estradiol and estrone (Figure 5.133) proceeds from cholesterol via pregnenolone and bears a resemblance to the hydrocortisone pathway (Figure 5.114) in the early 17-hydroxylation step. Indeed, the same cytochrome P-450-dependent enzyme catalyses 17-hydroxylation of both pregnenolone and

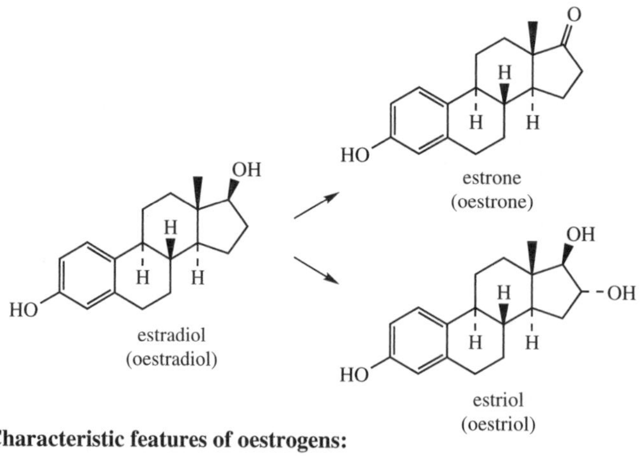

estradiol
(oestradiol)

estrone
(oestrone)

estriol
(oestriol)

Characteristic features of oestrogens:

- C_{18} estrane skeleton
- aromatic A ring (consequent loss of C-10 methyl)
- no side-chain

Figure 5.132

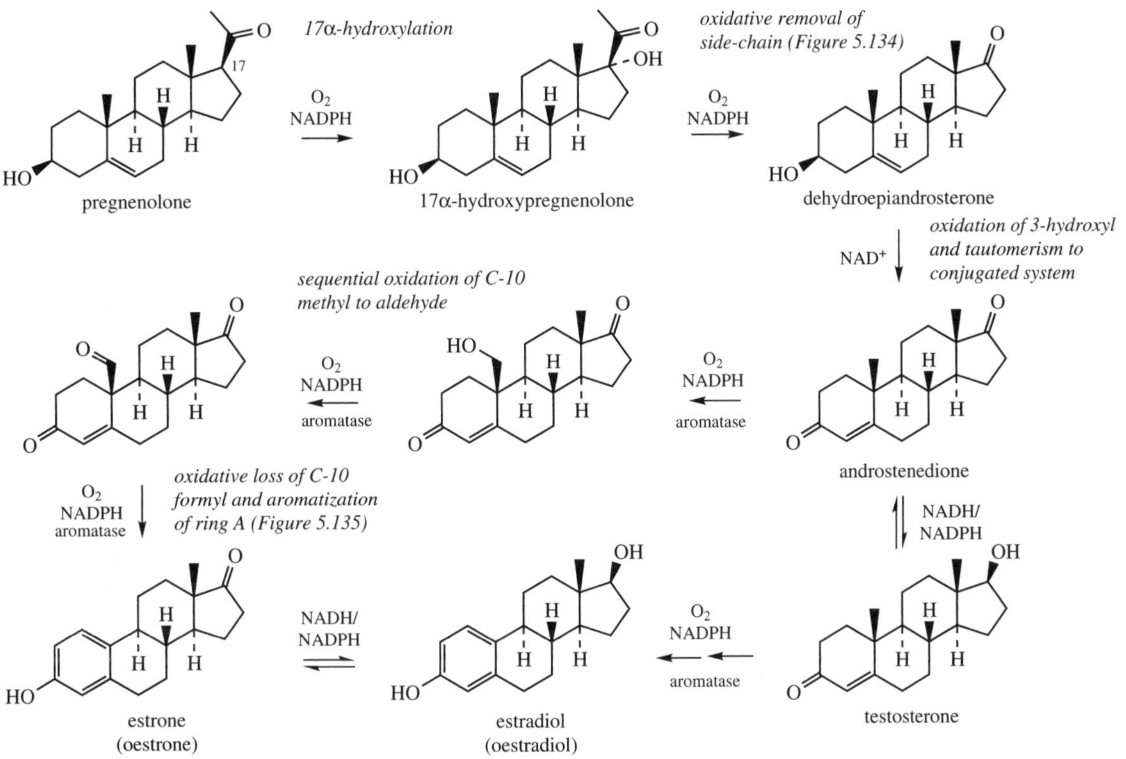

Figure 5.133

progesterone, as well as the next step in oestrogen biosynthesis, and it plays a significant role in controlling the direction of steroid synthesis. Whilst 17α-hydroxyprogesterone is transformed by 21-hydroxylation for hydrocortisone biosynthesis, in oestrogen biosynthesis the α-hydroxyketone function is oxidized from 17α-hydroxypregnenolone, cleaving off the two-carbon side-chain as acetic acid. The product is the 17-ketone **dehydroepiandrosterone**, which is the most abundant steroid in the blood of young adult humans, with levels peaking at about 20 years of age, then declining as the person ages. Apart from its role as a precursor of hormones, it presumably has other physiological functions, though these still remain to be clarified. A mechanism for the side-chain cleavage reaction, initiated by attack of an enzyme-linked peroxide, is shown in Figure 5.134, and is analogous to that proposed for loss of the 14-methyl group during cholesterol biosynthesis (see page 235). Oxidation and tautomerism in rings A/B then give **androstenedione** (Figure 5.133). Either androstenedione, or its reduction product

testosterone, is a substrate for aromatization in ring A, with loss of C-19. This sequence is also catalysed by a single cytochrome P-450-dependent enzyme, called **aromatase**, and the reaction proceeds via sequential oxidation of the methyl, with its final elimination as formic acid (Figure 5.135). The mechanism suggested is analogous to that of the side-chain cleavage reaction. Formation of the aromatic ring is then a result of enolization. As with other steroid hormones, the exact order of some of the steps, including formation of the Δ^4-3-keto function, 17-hydroxylation, reduction of the 17-keto, and aromatization in ring A, can vary according to organism, or site of synthesis in the body. Since breast tumours require oestrogens for growth, the design of **aromatase inhibitors*** has become an important target for anticancer drug research.

The aromatic ring makes the oestrogen molecule almost planar (see page 233) and is essential for activity. Changes which remove the aromaticity, e.g. partial reduction, or alter stereochemistry, give analogues with reduced or no activity.

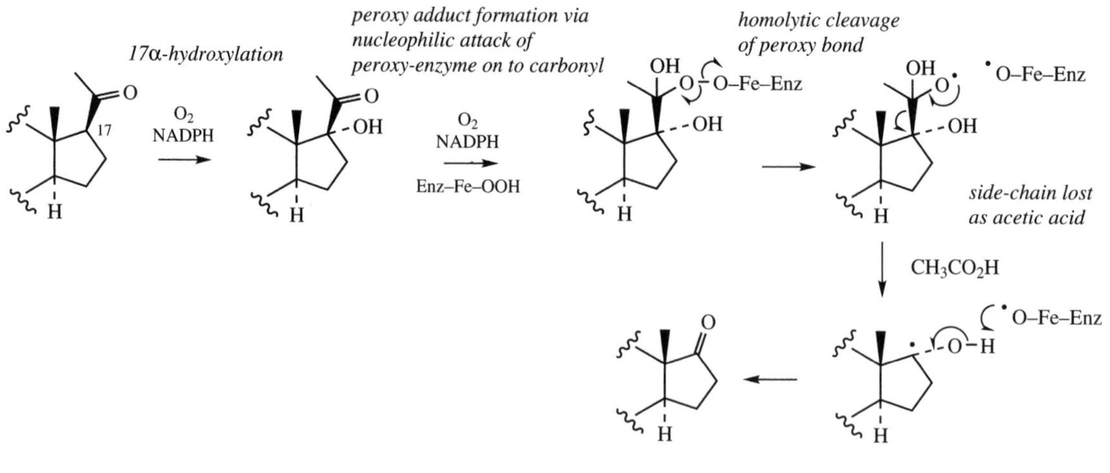

Figure 5.134

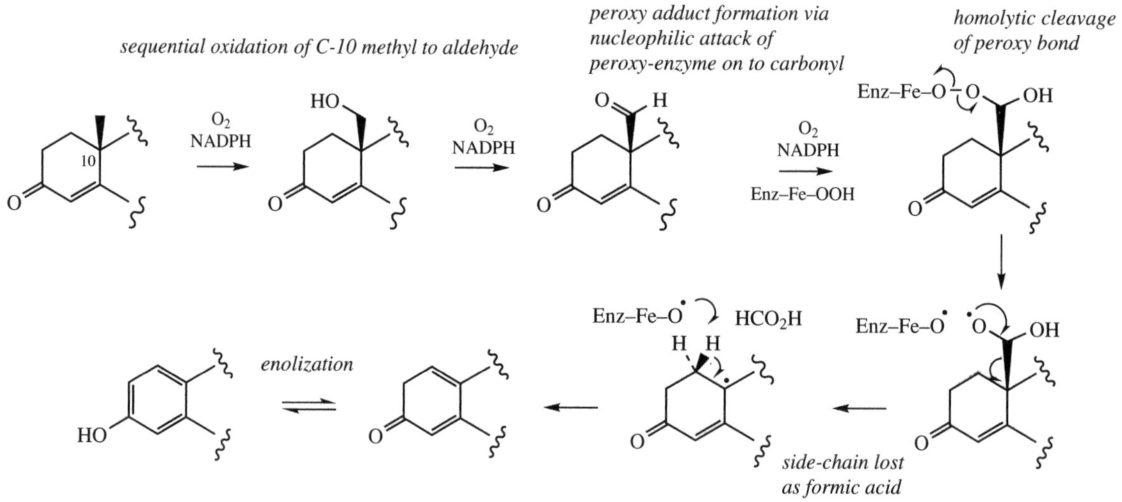

Figure 5.135

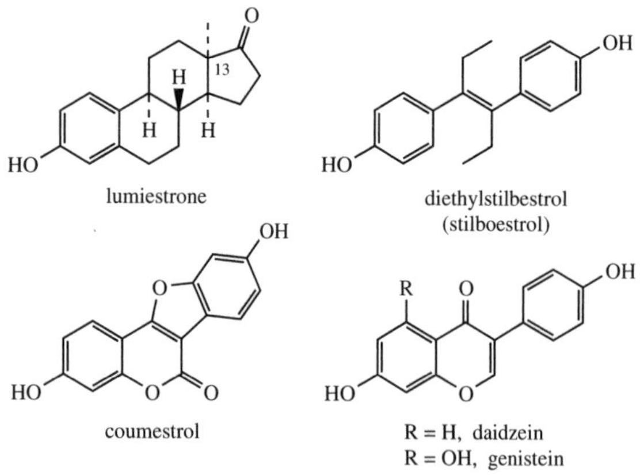

Figure 5.136

Thus, exposure of estrone to UV light leads to inversion of configuration at C-13 adjacent to the carbonyl function, and consequently formation of a *cis*-fused C/D ring system. The product, **lumiestrone** (Figure 5.136) is no longer biologically active. Some planar non-steroidal structures can also demonstrate oestrogenic activity as a result of a similar shape and relative spacing of oxygen functions. Thus, the synthetic **diethylstilbestrol (stilboestrol)** (Figure 5.136) has been widely used as an oestrogenic drug, and **coumestrol**, **daidzein**, and **genistein** (Figure 5.136) are naturally occurring isoflavonoids with oestrogenic properties from lucerne, clovers, and soya beans, and are termed

Oestrogen Drugs

Oestrogens suppress ovulation and with progestogens form the basis of combined oral contraceptives and hormone replacement therapy (HRT). They are also used to supplement natural oestrogen levels where these are insufficient as in some menstrual disorders, and to suppress androgen formation and thus tumour growth of cancers dependent on androgens, e.g. prostate cancers. Oestrogens appear to offer a number of beneficial effects to women, including protection against osteoporosis, heart attacks, and possibly Alzheimer's disease. However, some cancers, e.g. breast and uterine cancers, are dependent on a supply of oestrogen for growth, especially during the early stages, so high oestrogen levels are detrimental. Steroidal oestrogens for drug use were originally obtained by processing pregnancy urines, but the dramatic increase in demand due to the introduction of oral contraceptives required development of semi-synthetic procedures. Androstenolone formed via the Marker degradation of diosgenin (Figure 5.129) may be transformed to the dione by catalytic reduction of the Δ^5 double bond and oxidation of the 3-hydroxyl (Figure 5.137). This then allows production of androstadienedione by dibromination and base-catalysed elimination of HBr. Alternatively, it is now possible to achieve the synthesis of androstadienedione in a single step by a microbiological fermentation of either sitosterol obtained from soya beans (see page 256), or of cholesterol obtained in large quantities from the woolfat of sheep, or from the spinal cord of cattle (see page 236). These materials lack unsaturation in the side-chain and were not amenable to simple chemical oxidation processes, as for example with stigmasterol (see page 266). Their exploitation required the development of suitable biotransformations, and use of *Mycobacterium phlei* has now achieved this objective (Figure 5.137). The aromatization step to estrone can be carried out in low yields by vapour-phase free-radical-initiated thermolysis, or more recently with considerably better yields using a dissolving-metal reductive thermolysis. In both processes, the methyl at C-10 is lost. This sequence gives estrone, from which **estradiol (oestradiol)** may be obtained by reduction of the 17-carbonyl. However, by far the most commonly used medicinal estrogen is **ethinylestradiol (ethinyloestradiol)** (Figure 5.137), which is 12 times as effective as estradiol when administered orally. This analogue can be synthesized from estrone by treatment with potassium acetylide in liquid ammonia, which attacks from the less-hindered α-face (see page 273). The ethynyl substituent prevents oxidation at C-17, as in the metabolism of estradiol to the less active estrone. The phenol group allows synthesis of other derivatives, e.g. the 3-methyl ether **mestranol** (Figure 5.137), which acts as a pro-drug, being oxidized in the liver to ethinylestradiol. To retain oestrogenic activity, structural modifications appear effectively limited to the addition of the 17α-ethynyl group, and to substituents on the 3-hydroxyl. The ester **estradiol valerate (oestradiol valerate)** facilitates prolonged action through slower absorption and metabolism.

The lower activity metabolites **estriol (oestriol)** (about 2% activity of estradiol) and **estrone (oestrone)** (about 33% activity) (Figure 5.132) are sometimes used in **hormone replacement therapy** (HRT). Oestrogen and progesterone levels decline naturally at menopause when the menstrual cycle ceases. The sudden reduction in oestrogen levels can lead to a number of unpleasant symptoms, including tiredness, hot flushes, vaginal dryness, and mood changes. HRT reduces these symptoms, and delays other long-term consequences of reduced oestrogen levels, including osteoporosis and atherosclerosis. HRT currently provides the best therapy for preventing osteoporosis, a common disease in post-menopausal women.

(Continues)

(Continued)

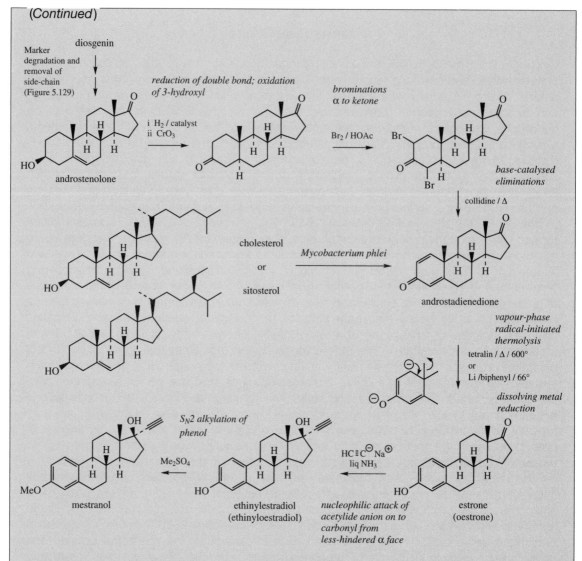

Figure 5.137

Osteoporosis is characterized by a generalized loss of bone mass leading to increased risk of fracture, and a sharp reduction in endogenous oestrogen levels is recognized as a critical factor. Oestrogen and progestogen combinations are used in HRT unless the woman has had a hysterectomy, in which case oestrogen alone is prescribed. Natural oestrogen structures are preferred to synthetic structures such as ethinylestradiol or mestranol. Before the availability of plant-derived semi-synthetic oestrogens, extraction of urine from pregnant women and pregnant horses allowed production of oestrogen mixtures for drug use. **Conjugated equine oestrogens** are still widely prescribed for HRT and are obtained by extraction from the urine of pregnant mares and subsequent purification, predominantly in Canada and the USA. Mares are about 120 days pregnant when urine collection begins, and collection continues for another 150–160 days. During this period, a mare will produce about 400–500 litres of urine.

(Continues)

(Continued)

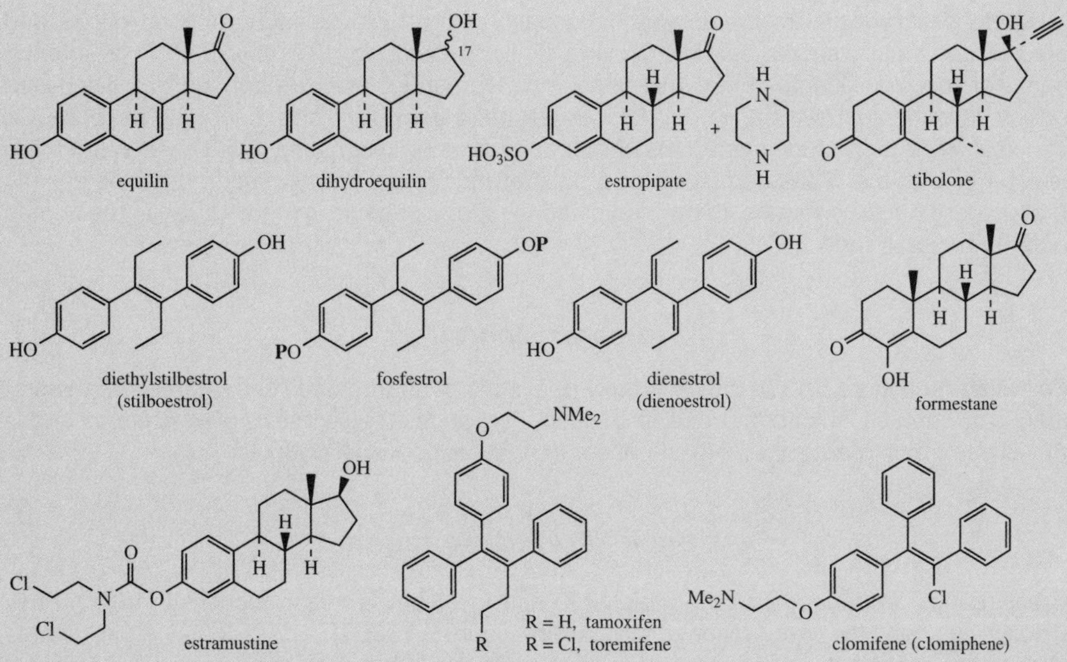

Figure 5.138

Horses are maintained in an almost continuous state of pregnancy, and need to be kept in confined stalls and fitted with a suitable urine collection device, though the use of catheters to collect urine has been discontinued. Animal welfare groups urge women to reject these drug preparations in favour of plant-derived alternatives. Conjugated equine oestrogens provide a profile of natural oestrogens based principally on estrone and equilin (Figure 5.138). It consists of a mixture of oestrogens in the form of sodium salts of their sulphate esters, comprising mainly estrone (50–60%) and equilin (20–30%), with smaller amounts of 17α-dihydroequilin, 17α-estradiol, and 17β-dihydroequilin. The semi-synthetic **estropipate** is also a conjugated oestrogen, the piperazine salt of estrone sulphate.

The structure of **tibolone** (Figure 5.138) probably resembles that of a progestogen more than it does an oestrogen. Although it does not contain an aromatic A-ring, the 5(10)-double bond ensures a degree of planarity. This agent combines both oestrogenic and progestogenic activity, and also has weak androgenic activity, and has been introduced for treatment of vasomotor symptoms of menopause.

Diethylstilbestrol (**stilboestrol**) and **dienestrol** (**dienoestrol**) (Figure 5.138) are the principal non-steroidal oestrogen drugs, used topically via the vagina. **Fosfestrol** (Figure 5.138) has value in the treatment of prostate cancer, being hydrolysed by the enzyme phosphatase to produce diethylstilbestrol as the active agent.

In **estramustine** (Figure 5.138), estradiol is combined with a cytotoxic alkylating agent of the nitrogen mustard class via a carbamate linkage. This drug has a dual function, a hormonal effect by suppressing androgen (testosterone) formation, and an antimitotic effect from the mustine residue. It is of value in treating prostate cancers.

(Continues)

(Continued)

Phyto-oestrogens are predominantly isoflavonoid derivatives found in food plants and are used as dietary supplements to provide similar benefits to HRT, especially in countering some of the side-effects of the menopause in women. These compounds are discussed under isoflavonoids (see page 156). **Dioscorea (wild yam)** root or extract (see page 239) is also marketed to treat the symptoms of menopause as an alternative to HRT. Although there is a belief that this increases levels of progesterone, which is then used as a biosynthetic precursor of other hormones, there is no evidence that diosgenin is metabolized in the human body to progesterone.

Aromatase Inhibitors

Formestane (Figure 5.138), the 4-hydroxy derivative of androstenedione, represents the first steroid aromatase inhibitor to be used clinically. It reduces the synthesis of oestrogens and is of value in treating advanced breast cancer in post-menopausal patients.

Oestrogen Receptor Antagonists

Breast cancer is dependent on a supply of oestrogen, and a major success in treating this disease has been the introduction of **tamoxifen** (Figure 5.138). This drug contains the stilbene skeleton seen in diethylstilbestrol and related oestrogens, but acts as an oestrogen-receptor antagonist rather than as an agonist in breast tissue, and deprives the cells of oestrogen. However, it is an agonist in bone and uterine tissue. The chlorinated analogue **toremifene** is also available, but is used primarily in post-menopausal women. Oestrogen antagonists can also be used as fertility drugs, occupying oestrogen receptors and interfering with feedback mechanisms and leading to ova release. **Clomifene (clomiphene)** (Figure 5.138), and to a lesser extent tamoxifen, are used in this way, but can lead to multiple pregnancies.

phyto-oestrogens (see page 156). Dietary natural isoflavonoids are believed to give some protection against breast cancers, and are also recommended to alleviate the symptoms of menopause.

Androgens

The primary male sex hormone, or **androgen**, is **testosterone** (Figure 5.139). This is secreted by the testes and is responsible for development and maintenance of the male sex characteristics. Androgens* also have a secondary physiological effect, an anabolic activity which stimulates growth of bone and muscle, and promotes storage of protein. The biosynthetic pathway to testosterone is outlined in Figure 5.133, where it can feature as an intermediate in the pathway to oestrogens. Low levels of testosterone are also synthesized in females in the ovary. Testosterone

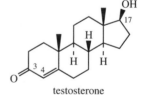

testosterone

Characteristic features of androgens:
- C_{19} androstane skeleton
- no side-chain
- Δ^4-3-keto
- 17β-hydroxyl

Figure 5.139

lacks any side-chain and has a 17β-hydroxyl as in estradiol, but still contains the methyl C-19 and the Δ^4-3-one system in ring A. This C_{19} skeleton is designated androstane.

It is particularly worthy of note that the routes to corticosteroids, progestogens, oestrogens, and androgens involve common precursors or partial pathways. This means that these processes need to be under very tight control if a person's normal physiological functions and characteristics are to be maintained. This balanced production is regulated primarily by gonadotrophic and hypothalamic proteins from the pituitary (see page 411).

Androgen Drugs

Testosterone can be produced from androstenolone (Figure 5.129) by chemical routes requiring reduction of the 17-carbonyl and oxidation of the 3-hydroxyl, and necessitating appropriate protecting groups. A simple high-yielding process (Figure 5.140) exploits yeast, in which fermentation firstly under aerobic conditions oxidizes the 3-hydroxyl, and then in the absence of air reduces the 17-keto group. Testosterone is not active orally since it is easily metabolized in the liver, and it has to be implanted, or injected in the form of esters. Transdermal administration from impregnated patches has also proved successful, and is now the method of choice for treating male sexual impotence caused by low levels of sex hormones (hypogonadism). Testosterone may also be prescribed for menopausal women as an adjunct to hormone replacement therapy (see page 279) to improve sex drive, and occasionally in the treatment of oestrogen-dependent breast cancer. The ester testosterone undecanoate is orally active, as is **mesterolone** (Figure 5.141), which features introduction of a 1α-methyl group and reduction of the Δ^4 double bond. Oral activity may also be attained by adding a 17α-alkyl group to reduce metabolism, e.g. 17α-methyltestosterone (Figure 5.140), such groups being introduced using appropriate Grignard reagents. However, these types of 17α-alkyl derivative are being replaced since they can sometimes cause jaundice as a side-effect.

The ratio of androgenic to anabolic activity can vary in different molecules. There have been attempts to produce steroids with low androgenic but high anabolic activity to use for various

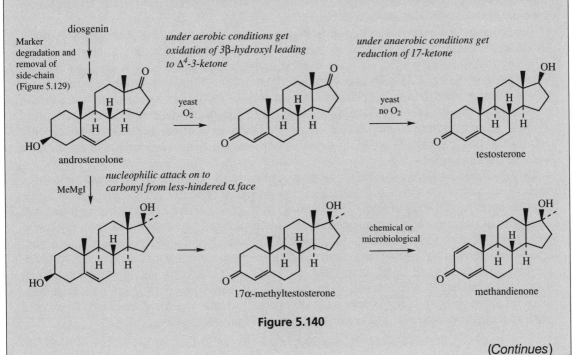

Figure 5.140

(Continues)

(Continued)

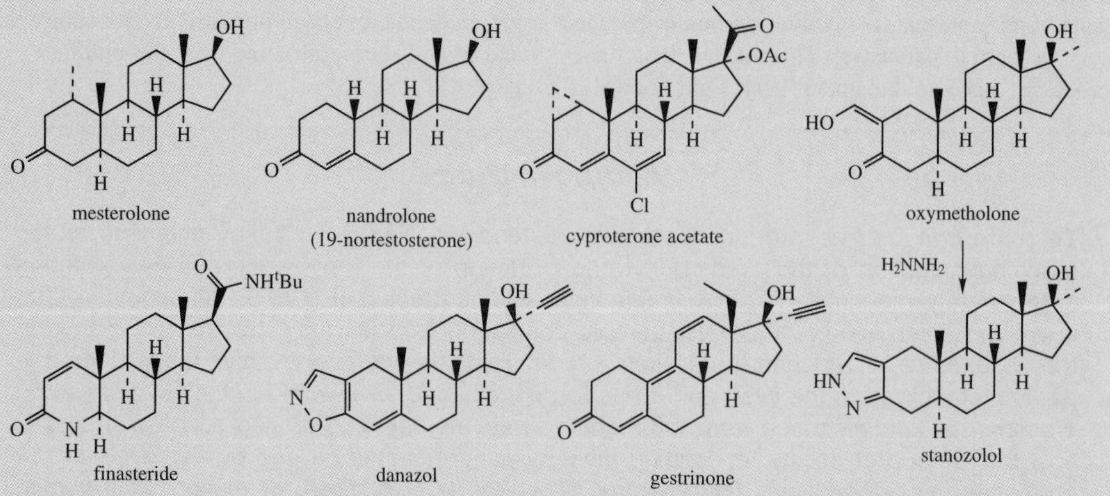

Figure 5.141

metabolic and endocrine disorders. **Methandrostenolone (methandienone)** (Figure 5.140), **oxymetholone**, and **nandrolone** (19-nortestosterone) (Figure 5.141) are modifications used in this way. Treatment of oxymetholone with hydrazine leads to formation of a pyrazole ring fused to a saturated ring A as seen in **stanozolol** (Figure 5.141). However, it is difficult to completely remove the androgenic activity from anabolic steroids. These materials are frequently abused by athletes wishing to promote muscle development and strength. Androgenic activity can affect the sexual characteristics of women, making them more masculine, whilst prolonged use of these drugs can lower fertility in either sex and endanger long term health by increasing the risks of heart and liver disease or cancer.

The progestogen **cyproterone acetate** (Figure 5.141) is a competitive androgen antagonist or anti-androgen, that reduces male libido and fertility, and finds use in the treatment of severe hypersexuality and sexual deviation in the male, as well as in prostate cancer. **Finasteride** (Figure 5.141) is another anti-androgen, which is of value in prostate conditions. It is a 4-aza-steroid and a specific inhibitor of the 5α-reductase involved in testosterone metabolism. This enzyme converts testosterone into dihydrotestosterone, which is actually a more potent androgen, so that its inhibition helps to reduce prostate tissue growth. Finasteride has also been noted to prevent hair loss in men, and is marketed to treat male-pattern baldness.

Dehydroepiandrosterone (DHEA) (Figure 5.133) is a precursor of androgens and oestrogens, and is the most abundant steroid in the blood of young adult humans, with levels peaking at about 20 years of age, then declining as the person ages. Whilst this hormone has a number of demonstrated biological activities, its precise physiological functions remain to be clarified. This material has become popular in the hope that it will maintain youthful vigour and health, countering the normal symptoms of age. These claims are as yet unsubstantiated, but taking large amounts of this androgen and oestrogen precursor can lead to side-effects associated with high levels of these hormones, e.g. increased risk of prostate cancer in men, or of breast cancer in women, who may also develop acne and facial hair. DHEA is not a precursor of glucocorticoids, mineralocorticoids, or of progestogens.

(Continues)

(Continued)

Danazol and **gestrinone** (Figure 5.141) are inhibitors of pituitary gonadotrophin release, combining weak androgenic activity with antioestrogenic and antiprogestogenic activity. These highly modified structures bear one or more of the features we have already noted in discussions of androgens, progestogens and oestrogens, possibly accounting for their complex activity. These compounds are used particularly to treat endometriosis, where endometrial tissue grows outside the uterus.

FURTHER READING

Biosynthesis, General

Bochar DA, Friesen JA, Stauffacher CV and Rodwell VW (1999) Biosynthesis of mevalonic acid from acetyl-CoA. *Comprehensive Natural Products Chemistry*, Vol 2. Elsevier, Amsterdam, pp 15–44.

Chappell J (1995) Biochemistry and molecular biology of the isoprenoid biosynthetic pathway in plants. *Annu Rev Plant Physiol Plant Mol Biol* **46**, 521–547.

Dewick PM (1999) The biosynthesis of C_5–C_{25} terpenoid compounds. *Nat Prod Rep* **16**, 97–130. *Earlier reviews*: 1997, **14**, 111–141; 1995, **12**, 507–534.

Eisenreich W, Schwarz M, Cartayrade A, Arigoni D, Zenk MH and Bacher A (1998) The deoxyxylulose phosphate pathway of terpenoid biosynthesis in plants and microorganisms. *Chem Biol* **5**, R221–R233.

Lichtenhaler HK (1999) The 1-deoxy-D-xylulose-5-phosphate pathway of isoprenoid biosynthesis in plants. *Annu Rev Plant Physiol Plant Mol Biol* **50**, 47–65.

McGarvey DJ and Croteau R (1995) Terpenoid metabolism. *Plant Cell* **7**, 1015–1026.

Rohmer M (1998) Isoprenoid biosynthesis via the mevalonate-independent route, a novel target for antibacterial drugs? *Prog Drug Res* **50**, 135–154.

Rohmer M (1999) The discovery of the mevalonate-independent pathway for isoprenoid biosynthesis in bacteria, algae and higher plants. *Nat Prod Rep* **16**, 565–574.

Rohmer M (1999) A mevalonate-independent route to isopentenyl diphosphate. *Comprehensive Natural Products Chemistry*, Vol 2. Elsevier, Amsterdam, pp 45–67.

Monoterpenoids

Croteau R (1991) Metabolism of monoterpenes in mint (*Mentha*) species. *Planta Med* **57**, S10-S14.

Grayson DH (2000) Monoterpenoids. *Nat Prod Rep* **17**, 385–419. *Earlier reviews*: 1998, **15**, 439–475; 1997, **14**, 477–522.

Wise ML and Croteau R (1999) Monoterpene biosynthesis. *Comprehensive Natural Products Chemistry*, Vol 2. Elsevier, Amsterdam, pp 97–153.

Volatile Oils

Mookherjee BD and Wilson RA (1996) Oils, essential. *Kirk–Othmer Encyclopedia of Chemical Technology*, 4th edn, Vol 17. Wiley, New York, pp 603–674.

Pyrethrins

Metcalf RL (1995) Insect control technology. *Kirk–Othmer Encyclopedia of Chemical Technology*, 4th edn, Vol 14. Wiley, New York, pp 524–602.

Ramos Tombo GM and Bellus D (1991) Chirality and crop protection. *Angew Chem Int Edn Engl* **30**, 1193–1215.

Iridoids

Houghton P (1994) Herbal products: valerian. *Pharm J* **253**, 95–96.

Junior P (1990) Recent developments in the isolation and structure elucidation of naturally occurring iridoid compounds. *Planta Med* **56**, 1–13.

Sesquiterpenoids

Berry M (1994) Herbal products: feverfew. *Pharm J* **253**, 806–808.

Berry M (1995) Herbal products: the chamomiles. *Pharm J* **254**, 191–193.

Cane DE (1999) Sesquiterpene biosynthesis: cyclization mechanisms. *Comprehensive Natural Products Chemistry*, Vol 2. Elsevier, Amsterdam, pp 155–200.

Christensen SB, Andersen A and Smitt UW (1997) Sesquiterpenoids from *Thapsia* species and medicinal chemistry of the thapsigargins. *Prog Chem Org Nat Prod* **71**, 129–167.

Fraga BM (2000) Natural sesquiterpenoids. *Nat Prod Rep* **17**, 483–504. *Earlier reviews*: 1999, **16**, 711–730; 1999, **16**, 21–38.

Groenewegen WA, Knight DA and Heptinstall S (1992) Progress in the medicinal chemistry of the herb feverfew. *Prog Med Chem* **29**, 217–238.

Knight DW (1995) Feverfew: chemistry and biological activity. *Nat Prod Rep* **12**, 271–276.

Robles M, Aregullin M, West J and Rodriguez E (1995) Recent studies on the zoopharmacognosy, pharmacology and neurotoxicology of sesquiterpene lactones. *Planta Med* **61**, 199–203.

Artemisinin

Bhattacharya AK and Sharma RP (1999) Recent developments on the chemistry and biological activity of artemisinin and related antimalarials – an update. *Heterocycles* **51**, 1681–1745.

Casteel DA (1999) Peroxy natural products. *Nat Prod Rep* **16**, 55–73.

Haynes RK and Vonwiller SC (1997) From qinghao, marvelous herb of antiquity, to the antimalarial trioxane qinghaosu – and some remarkable new chemistry. *Account Chem Res* **30**, 73–79.

Klayman DL (1992) Antiparasitic agents (antiprotozoals). *Kirk–Othmer Encyclopedia of Chemical Technology*, 4th edn, Vol 3. Wiley, New York, pp 489–526.

Robert A and Meunier B (1998) Is alkylation the main mechanism of action of the antimalarial drug artemisinin? *Chem Soc Rev* **27**, 273–279.

van Geldre E, Vergauwe A and van den Eeckhout E (1997) State of the art of the production of the antimalarial compound artemisinin in plants. *Plant Mol Biol* **33**, 199–209.

Ziffer H, Highet RJ and Klayman DL (1997) Artemisinin: an endoperoxidic antimalarial from *Artemisia annua* L. *Prog Chem Org Nat Prod* **72**, 121–214.

Gossypol

Jaroszewski JW, Strom-Hansen T, Hansen SH, Thastrup O and Kofod H (1992) On the botanical distribution of chiral forms of gossypol. *Planta Med* **58**, 454–458.

Trichothecenes

Grove JF (1993) Macrocyclic trichothecenes. *Nat Prod Rep* **10**, 429–448.

Grove JF (1996) Non-macrocyclic trichothecenes, part 2. *Prog Chem Org Nat Prod* **69**, 1–70.

Diterpenoids

Hanson JR (2001) Diterpenoids. *Nat Prod Rep* **18**, 88–94. *Earlier reviews*: 2000, **17**, 165–174; 1999, **16**, 209–219.

MacMillan J and Beale MH (1999) Diterpene biosynthesis. *Comprehensive Natural Products Chemistry*, Vol 2. Elsevier, Amsterdam, pp 217–243.

San Feliciano A, Gordaliza M, Salinero MA and del Corral JMM (1993) Abietane acids: sources, biological activities, and therapeutic uses. *Planta Med* **59**, 485–490.

Taxol

Amos LA and Löwe J (1999) How Taxol® stabilizes microtubule structure. *Chem Biol* **6**, R65–R69.

Appendino G (1995) The phytochemistry of the yew tree. *Nat Prod Rep* **12**, 349–360.

Appendino G (1996) Taxine. *Alkaloids, Chemical and Biological Perspectives* (ed Pelletier SW) Vol 11. Elsevier, Amsterdam, pp 237–268.

Baloglu E and Kingston DGI (1999) The taxane diterpenoids. *J Nat Prod* **62**, 1448–1472.

Guénard D, Guéritte-Voegelein F and Potier P (1993) Taxol and taxotere: discovery, chemistry, and structure-activity relationships. *Account Chem Res* **26**, 160–167.

Heinstein PF and Chang C-j (1994) Taxol. *Annu Rev Plant Physiol Plant Mol Biol* **45**, 663–674.

Jaziri M, Zhiri A, Guo Y-W, Dupont J-P, Shimomura K, Hamada H, Vanhaelen M and Homes J (1996) *Taxus* sp. cell, tissue and organ cultures as alternative sources for taxoids production: a literature survey. *Plant Cell Tissue Organ Culture* **44**, 59–75.

Jenkins P (1996) Taxol branches out. *Chem Brit* **32** (11), 43–46.

Kapoor VK and Mahindroo (1997) Recent advances in structure modifications of taxol (paclitaxel). *Ind J Chem* **36B**, 639–652.

Kingston DGI (2000) Recent advances in the chemistry of taxol. *J Nat Prod* **63**, 726–734.

Nicolaou KC, Dai W-M and Guy RK (1994) Chemistry and biology of taxol. *Angew Chem Int Edn Engl* **33**, 15–44.

van Rozendaal ELM, Leylveld GP and van Beek TA (2000) Screening of the needles of different yew species and cultivars for paclitaxel and related taxoids. *Phytochemistry* **53**, 383–389.

Vyas DM and Kadow JF (1995) Paclitaxel: a unique tubulin interacting anticancer agent. *Prog Med Chem* **32**, 289–337.

Ginkgo

Braquet P, Esanu A, Buisine E, Hosford D, Broquet C and Koltai M (1991) Recent progress in ginkgolide research. *Med Res Rev* **11**, 295–355.

Braquet P and Hosford D (1991) Ethnopharmacology and the development of natural PAF antagonists as therapeutic agents. *J Ethnopharmacol* **32**, 135–139.

Houghton P (1994) Herbal products: Ginkgo. *Pharm J* **253**, 122–123.

Sticher O (1993) Quality of *Ginkgo* preparations. *Planta Med* **59**, 2–11.

Forskolin

Bhat SV (1993) Forskolin and congeners. *Prog Chem Org Nat Prod* **62**, 1–74.

Hurley JH (1999) Structure, mechanism, and regulation of mammalian adenylyl cyclase. *J Biol Chem* **274**, 7599–7602.

Laurenza A, Sutkowski EM and Seamon KB (1989) Forskolin: a specific stimulator of adenylyl cyclase or a diterpene with multiple sites of action? *Trends Pharmacol Sci* **10**, 442–447.

Stevioside

Hanson JR and de Oliveira BH (1993) Stevioside and related sweet diterpenoid glycosides. *Nat Prod Rep* **10**, 301–309.

Triterpenoids

Abe I and Prestwich GD (1999) Squalene epoxidase and oxidosqualene:lanosterol cyclase–key enzymes in cholesterol biosynthesis. *Comprehensive Natural Products Chemistry*, Vol 2. Elsevier, Amsterdam, pp 267–298.

Abe I, Rohmer M and Prestwich GD (1993) Enzymatic cyclization of squalene and oxidosqualene to sterols and triterpenes. *Chem Rev* **93**, 2189–2206.

Abe I, Tomesch JC, Wattanasin S and Prestwich GD (1994) Inhibitors of squalene biosynthesis and metabolism. *Nat Prod Rep* **11**, 279–302.

Brown GD (1998) Biosynthesis of steroids and triterpenoids. *Nat Prod Rep* **15**, 653–696.

Connolly JD and Hill RA (2000) Triterpenoids. *Nat Prod Rep* **17**, 463–482. *Earlier reviews*: 1999, **16**, 221–240; 1997, **14**, 661–679.

Poralla K (1999) Cycloartenol and other triterpene cyclases. *Comprehensive Natural Products Chemistry*, Vol 2. Elsevier, Amsterdam, pp 299–319.

Shechter I, Guan G and Boettcher BR (1999) Squalene synthase. *Comprehensive Natural Products Chemistry*, Vol 2. Elsevier, Amsterdam, pp 245–266.

Wendt KU, Schultz GE, Corey EJ and Liu DR (2000) Enzyme mechanisms for polycyclic triterpene formation. *Angew Chem Int Edn* **39**, 2812–2833.

Triterpenoid Saponins

Mahato SB and Garai S (1998) Triterpenoid saponins. *Prog Chem Org Nat Prod* **74**, 1–196.

Yoshiki Y, Kodon S and Okubo K (1998) Relationship between chemical structures and biological activities of triterpenoid saponins from soybean. *Biosci Biotechnol Biochem* **62**, 2291–2199.

Liquorice

Lewis DA (1992) Antiulcer drugs from plants. *Chem Brit* 141–144.

Lewis DA and Hanson PJ (1991) Antiulcer drugs of plant origin. *Prog Med Chem* **28**, 201–231.

Nomura T and Fukai T (1998) Phenolic constituents of licorice (*Glycyrrhiza* species). *Prog Chem Org Nat Prod* **73**, 1–140 (note: includes triterpenoid saponins).

Ginseng

Chuang W-C, Wu H-K, Sheu S-J, Chiou S-H, Chang H-C and Chen Y-P (1995) A comparative study on commercial samples of Ginseng Radix. *Planta Med* **61**, 459–465.

Raman A and Houghton P (1995) Herbal products: ginseng. *Pharm J* **254**, 150–152.

Limonoids

Champagne DE, Koul O, Isman MB, Scudder GGE and Towers GHN (1992) Biological activity of limonoids from the Rutales. *Phytochemistry* **31**, 377–394

Ley SV, Denholm AA and Wood A (1993) The chemistry of azadirachtin. *Nat Prod Rep* **10**, 109–157.

Sesterterpenoids

Hanson JR (1996) The sesterterpenoids. *Nat Prod Rep* **13**, 529–535. *Earlier review*: 1992, **9**, 481–489.

Carotenoids

Bartley GE and Scolnik PA (1995) Plant carotenoids: pigments for photoprotection, visual attraction, and human health. *Plant Cell* **7**, 1027–1038.

Bramley PM (2000) Molecules of interest: is lycopene beneficial to human health? *Phytochemistry* **54**, 233–236.

Gordon MH (1996) Dietary antioxidants in disease prevention. *Nat Prod Rep* **13**, 265–273.

Krinsky NI (1994) The biological properties of carotenoids. *Pure Appl Chem* **66**, 1003–1010.

Sandmann G (1994) Carotenoid biosynthesis in microorganisms and plants. *Eur J Biochem* **223**, 7–24.

Stahl W and Sies H (1996) Lycopene: a biologically important carotenoid for humans? *Arch Biochem Biophys* **336**, 1–9.

Vitamin A

Bollag W (1994) Retinoids in oncology: experimental and clinical aspects. *Pure Appl Chem* **66**, 995–1002.

Gale JB (1993) Recent advances in the chemistry and biology of retinoids. *Prog Med Chem* **30**, 1–55.

Nakanishi K (2000) Recent bioorganic studies on rhodopsin and visual transduction. *Chem Pharm Bull* **48**, 1399–1409.

Van Arnum SD (1998) Vitamins (vitamin A). *Kirk–Othmer Encyclopedia of Chemical Technology*, 4th edn, Vol 25. Wiley, New York, pp 172–192.

Williams CM (1993) Diet and cancer prevention. *Chem Ind* 280–283.

Steroids, General

Brown GD (1998) Biosynthesis of steroids and triterpenoids. *Nat Prod Rep* **15**, 653–696.

Hanson JR (2000) Steroids: reactions and partial synthesis. *Nat Prod Rep* **17**, 423–434. *Earlier reviews*: 1999, **16**, 607–617; 1998, **15**, 261–273.

Morgan BP and Moynihan MS (1997) Steroids. *Kirk–Othmer Encyclopedia of Chemical Technology*, 4th edn, Vol 22. Wiley, New York, pp 851–921.

Various authors (1992) A history of steroid chemistry, part I. *Steroids* **57**, 354–418; part II. *Steroids* **57**, 578–657.

Watson NS and Procopiou PA (1996) Squalene synthase inhibitors: their potential as hypocholesterolaemic agents. *Prog Med Chem* **33**, 331–378.

Cardioactive Glycosides

Cervoni P, Crandall DL and Chan PS (1993) Cardiovascular agents – agents used in the treatment of congestive heart failure. *Kirk–Othmer Encyclopedia of Chemical Technology*, 4th edn, Vol 5. Wiley, New York, pp 251–257.

Deepak D, Srivastava S, Khare NK and Khare A (1996) Cardiac glycosides. *Prog Chem Org Nat Prod* **69**, 71–155.

Gupta SP (2000) Quantitative structure–activity relationships of cardiotonic agents. *Prog Drug Res* **55**, 235–282.

Krenn L and Kopp B (1998) Bufadienolides from animal and plant tissue. *Phytochemistry* **48**, 1–29.

Paul DB and Dobberstein RH (1994) Expectorants, antitussives, and related agents. *Kirk–Othmer Encyclopedia of Chemical Technology*, 4th edn, Vol 9. Wiley, New York, pp 1061–1081.

Repke KRH, Megges R, Weiland J and Schön R (1995) Digitalis research in Berlin-Buch – retrospective and perspective views. *Angew Chem Int Edn Engl* **34**, 282–294.

Rüegg UT (1992) Ouabain – a link in the genesis of high blood pressure? *Experientia* **48**, 1102–1106.

Schoner W (1993) Endogenous digitalis-like factors. *Prog Drug Res* **41**, 249–291.

Steyn PS and van Heerden FR (1998) Bufadienolides of plant and animal origin. *Nat Prod Rep* **15**, 397–413.

Digitalis

Ikeda Y, Fujii Y, Nakaya I and Yamazaki M (1995) Quantitative HPLC analysis of cardiac glycosides in *Digitalis purpurea* leaves. *J Nat Prod* **58**, 897–901.

Ikeda Y, Fujii Y and Yamazaki M (1992) Determination of lanatoside C and digoxin in *Digitalis lanata* by HPLC and its application to analysis of the fermented leaf powder. *J Nat Prod* **55**, 748–752.

Kreis W, Hensel A and Stuhlemmer U (1998) Cardenolide biosynthesis in foxglove. *Planta Med* **64**, 491–499.

Sterols

Dyas L and Goad LJ (1993) Steryl fatty acyl esters in plants. *Phytochemistry* **34**, 17–29.

Gopinath L (1996) Cholesterol drug dilemma. *Chem Brit* 38–41.

Krieger M (1998) The 'best' of cholesterols, the 'worst' of cholesterols: a tale of two receptors. *Proc Natl Acad Sci USA* **95**, 4077–4080.

Mercer EI (1991) Sterol biosynthesis inhibitors: their current status and modes of action. *Lipids* **26**, 584–597.

Parks LW and Casey WM (1995) Physiological implications of sterol biosynthesis in yeast. *Annu Rev Microbiol* **49**, 95–116.

Fusidic Acid

Verbist L (1990) The antimicrobial activity of fusidic acid. *J Antimicrob Chemother* **25** (Suppl B), 1–5.

Vitamin D

Beckman MJ and DeLuca HF (1998) Modern view of vitamin D_3 and its medicinal uses. *Prog Med Chem* **35**, 1–56.

Hirsch AL (1998) Vitamins (vitamin D). *Kirk–Othmer Encyclopedia of Chemical Technology*, 4th edn, Vol 25. Wiley, New York, pp 217–256.

Bile Acids

Russell DW and Setchell KDR (1992) Bile acid biosynthesis. *Biochemistry* **31**, 4737–4749.

Corticosteroids

Bernstein PR (1992) Antiasthmatic agents. *Kirk–Othmer Encyclopedia of Chemical Technology*, 4th edn, Vol 2. Wiley, New York, pp 830–854.

Cervoni P and Chan PS (1993) Diuretic agents. *Kirk–Othmer Encyclopedia of Chemical Technology*, 4th edn, Vol 8. Wiley, New York, pp 398–432.

Hirschmann R (1991) Medicinal chemistry in the golden age of biology: lessons from steroid and peptide research. *Angew Chem Int Edn Engl* **30**, 1278–1301.

Wainwright M (1998) The secret of success. *Chem Brit* **34** (1) 46–47 (prednisone and prednisolone).

Progestogens

Gunnet JW and Dixon LA (1995) Hormones (sex hormones). *Kirk–Othmer Encyclopedia of Chemical Technology*, 4th edn, Vol 13. Wiley, New York, pp 433–480.

Spitz IM and Agranat I (1995) Antiprogestins: modulators in reproduction. *Chem Ind* 89–92.

Oestrogens

Banting L (1996) Inhibition of aromatase. *Prog Med Chem* **33**, 147–184.

Guo JZ, Hahn DW and Wachter MP (1995) Hormones (estrogens and antiestrogens). *Kirk–Othmer Encyclopedia of Chemical Technology*, 4th edn, Vol 13. Wiley, New York, pp 481–512.

Hahn DW, McGuire JL and Bialy G (1993) Contraceptives. *Kirk–Othmer Encyclopedia of Chemical Technology*, 4th edn, Vol 7. Wiley, New York, pp 219–250.

Androgens

Jarman M, Smith HJ, Nicholls PJ and Simons C (1998) Inhibitors of enzymes of androgen biosynthesis: cytochrome $P450_{17\alpha}$ and 5α-steroid reductase. *Nat Prod Rep* **15**, 495–512.

6

ALKALOIDS

Alkaloids are classified according to the amino acid that provides both the nitrogen atom and the fundamental portion of the alkaloid skeleton, and these are discussed in turn. Ornithine gives rise to pyrrolidine and tropane alkaloids, lysine to piperidine, quinolizidine, and indolizidine alkaloids, and nicotinic acid to pyridine alkaloids. Tyrosine produces phenylethylamines and simple tetrahydroisoquinoline alkaloids, but also many others in which phenolic oxidative coupling plays an important role, such as modified benzyltetrahydroisoquinoline, phenethylisoquinoline, terpenoid tetrahydroisoquinoline, and Amaryllidaceae alkaloids. Alkaloids derived from tryptophan are subdivided into simple indole, simple β-carboline, terpenoid indole, quinoline, pyrroloindole, and ergot alkaloids. Anthranilic acid acts as a precursor to quinazoline, quinoline and acridine alkaloids, whilst histidine gives imidazole derivatives. However, many alkaloids are not derived from an amino acid core, but arise by amination of another type of substrate, which may be acetate derived, phenylalanine derived, a terpene or a steroid, and examples are discussed. Purine alkaloids are constructed by pathways that resemble those for purines in nucleic acids. Monograph topics giving more detailed information on medicinal agents include belladonna, stramonium, hyoscyamus, duboisia and allied drugs, hyoscyamine, hyoscine and atropine, coca, lobelia, vitamin B_3, tobacco, areca, catecholamines, lophophora, curare, opium, colchicum, ipecacuanha, galanthamine, serotonin, psilocybe, rauwolfia, catharanthus, iboga, nux-vomica, ellipticine, cinchona, camptothecin, physostigma, ergot, morning glories, pilocarpus, *Conium maculatum*, ephedra, khat, aconite, *Solanum* alkaloids, caffeine, theobromine and theophylline, coffee, tea, cola, cocoa, mate tea, guarana, saxitoxin, and tetrodotoxin.

The alkaloids are organic nitrogenous bases found mainly in plants, but also to a lesser extent in microorganisms and animals. One or more nitrogen atoms are present, typically as primary, secondary, or tertiary amines, and this usually confers basicity to the alkaloid, facilitating their isolation and purification since water-soluble salts can be formed in the presence of mineral acids. The name alkaloid is in fact derived from alkali. However, the degree of basicity varies greatly, depending on the structure of the alkaloid molecule, and the presence and location of other functional groups. Indeed, some alkaloids are essentially neutral. Alkaloids containing quaternary amines are also found in nature. The biological activity of many alkaloids is often dependent on the amine function being transformed into a quaternary system by protonation at physiological pHs.

Alkaloids are often classified according to the nature of the nitrogen-containing structure, e.g. pyrrolidine, piperidine, quinoline, isoquinoline, indole, etc, though the structural complexity of some examples rapidly expands the number of subdivisions. The nitrogen atoms in alkaloids originate from an amino acid, and, in general, the carbon skeleton of the particular amino acid precursor is also largely retained intact in the alkaloid structure, though the carboxylic acid carbon is often lost through decarboxylation. Accordingly, subdivision of alkaloids into groups based on amino acid precursors forms a rational and often illuminating approach to classification. Relatively few amino acid precursors are actually involved in alkaloid biosynthesis, the principal ones being ornithine, lysine, nicotinic acid, tyrosine, tryptophan, anthranilic acid, and histidine. Building blocks from the acetate, shikimate, or deoxyxylulose phosphate pathways are also frequently incorporated into the alkaloid structures. However, a large group of alkaloids are found to acquire their nitrogen atoms via transamination reactions, incorporating only the nitrogen from an amino acid, whilst the rest of the molecule may be derived from acetate or shikimate, or may be terpenoid or steroid in origin. The term 'pseudoalkaloid' is sometimes used to distinguish this group.

ALKALOIDS DERIVED FROM ORNITHINE

L-**Ornithine** (Figure 6.1) is a non-protein amino acid forming part of the urea cycle in animals, where it is produced from L-arginine in a reaction catalysed by the enzyme arginase. In plants it is formed mainly from L-glutamate (Figure 6.2). Ornithine contains both δ- and α-amino groups, and it is the nitrogen from the former group which is incorporated into alkaloid structures along with the carbon chain, except for the carboxyl group.

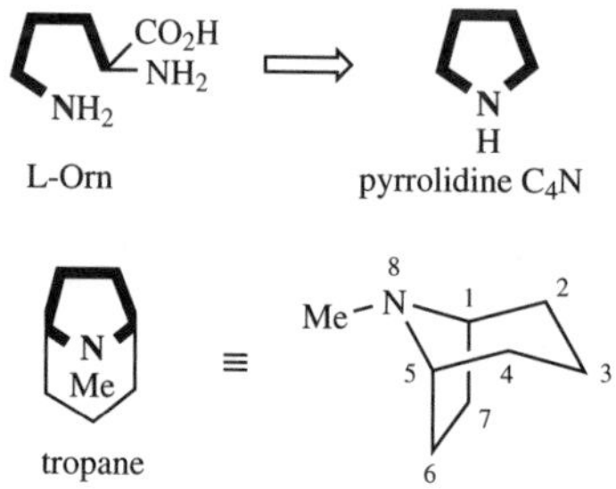

Figure 6.1

Thus ornithine supplies a C_4N building block to the alkaloid, principally as a pyrrolidine ring system, but also as part of the tropane alkaloids (Figure 6.1). Most of the other amino acid alkaloid precursors typically supply nitrogen from their solitary α-amino group. However, the reactions of ornithine are almost exactly paralleled by those of L-lysine, which incorporates a C_5N unit containing its ε-amino group (see page 307).

Pyrrolidine and Tropane Alkaloids

Simple pyrrolidine-containing alkaloid structures are exemplified by hygrine and cuscohygrine, found in those plants of the Solanaceae that accumulate medicinally valuable tropane alkaloids such as hyoscyamine or cocaine (see Figure 6.3). The pyrrolidine ring system is formed initially as a Δ^1-pyrrolinium cation (Figure 6.2). PLP-dependent decarboxylation (see page 20) of ornithine gives **putrescine**, which is then methylated to *N*-methylputrescine. Oxidative deamination of *N*-methylputrescine by the action of a diamine oxidase (see page 28) gives the aldehyde, and Schiff base (imine) formation produces the *N*-methyl-Δ^1-pyrrolinium cation. Indeed, the aminoaldehyde in aqueous solution is known to exist as an equilibrium mixture with the

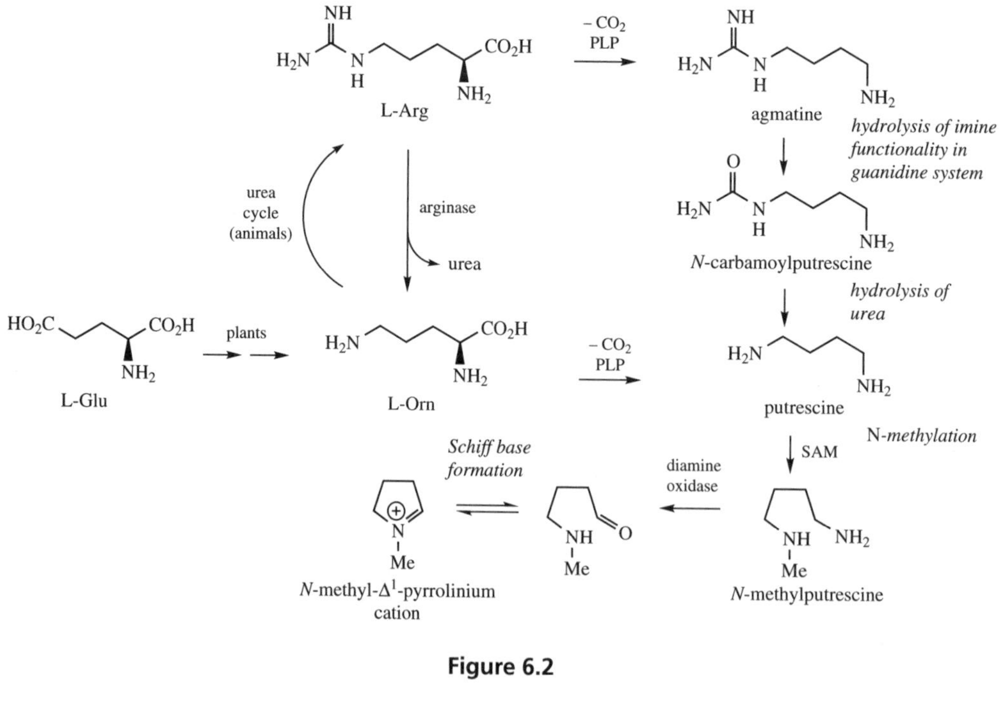

Figure 6.2

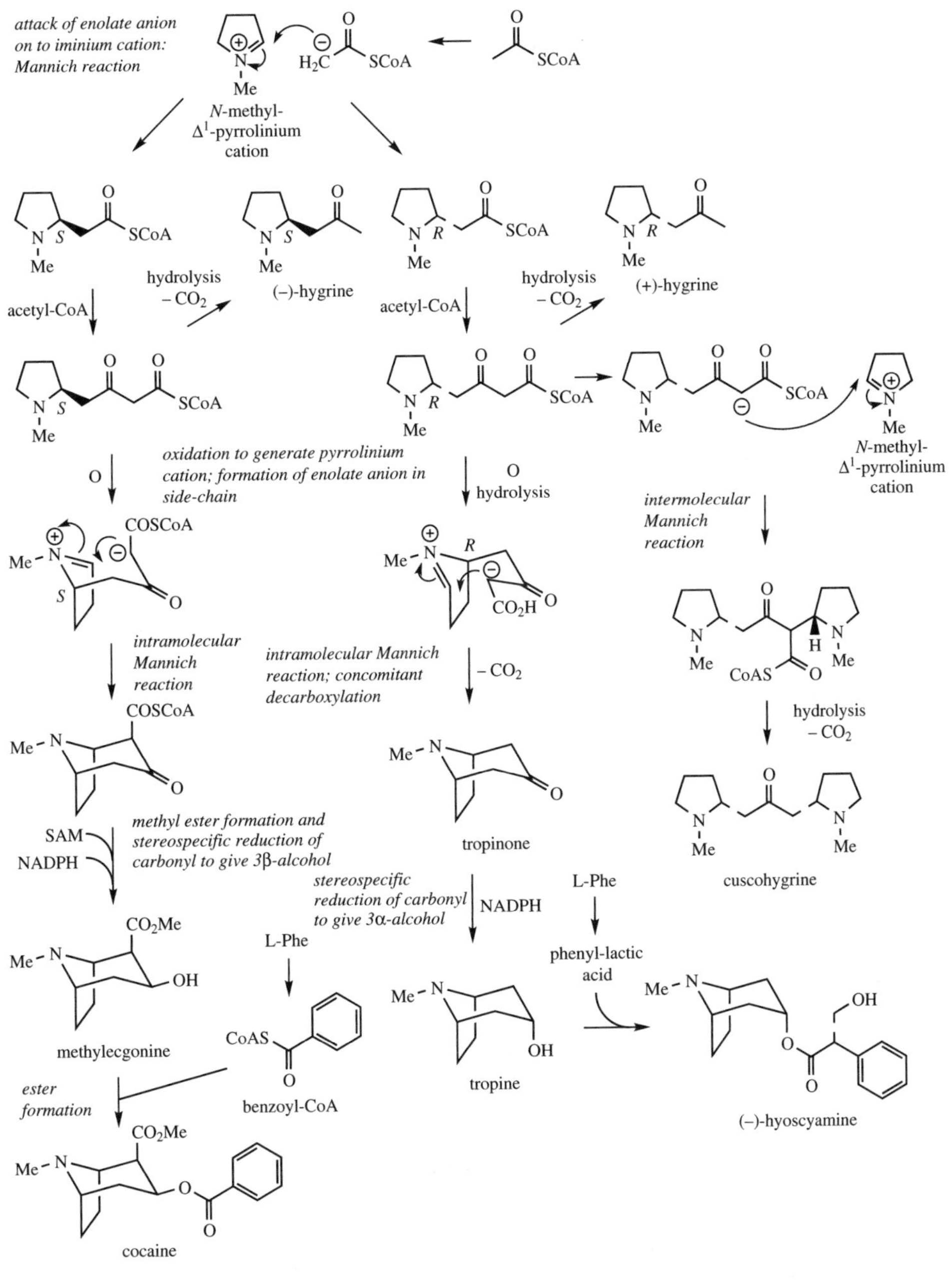

Figure 6.3

Schiff base. An alternative sequence to putrescine starting from arginine also operates concurrently as indicated in Figure 6.2. The arginine pathway also involves decarboxylation, but requires additional hydrolysis reactions to cleave the guanidine portion.

The extra carbon atoms required for hygrine formation are derived from acetate via acetyl-CoA, and the sequence appears to involve stepwise addition of two acetyl-CoA units (Figure 6.3). In the first step, the enolate anion from acetyl-CoA acts as nucleophile towards the pyrrolinium ion in a Mannich-like reaction, which could yield products with either R or S stereochemistry. The second addition is then a Claisen condensation extending the side-chain, and the product is the

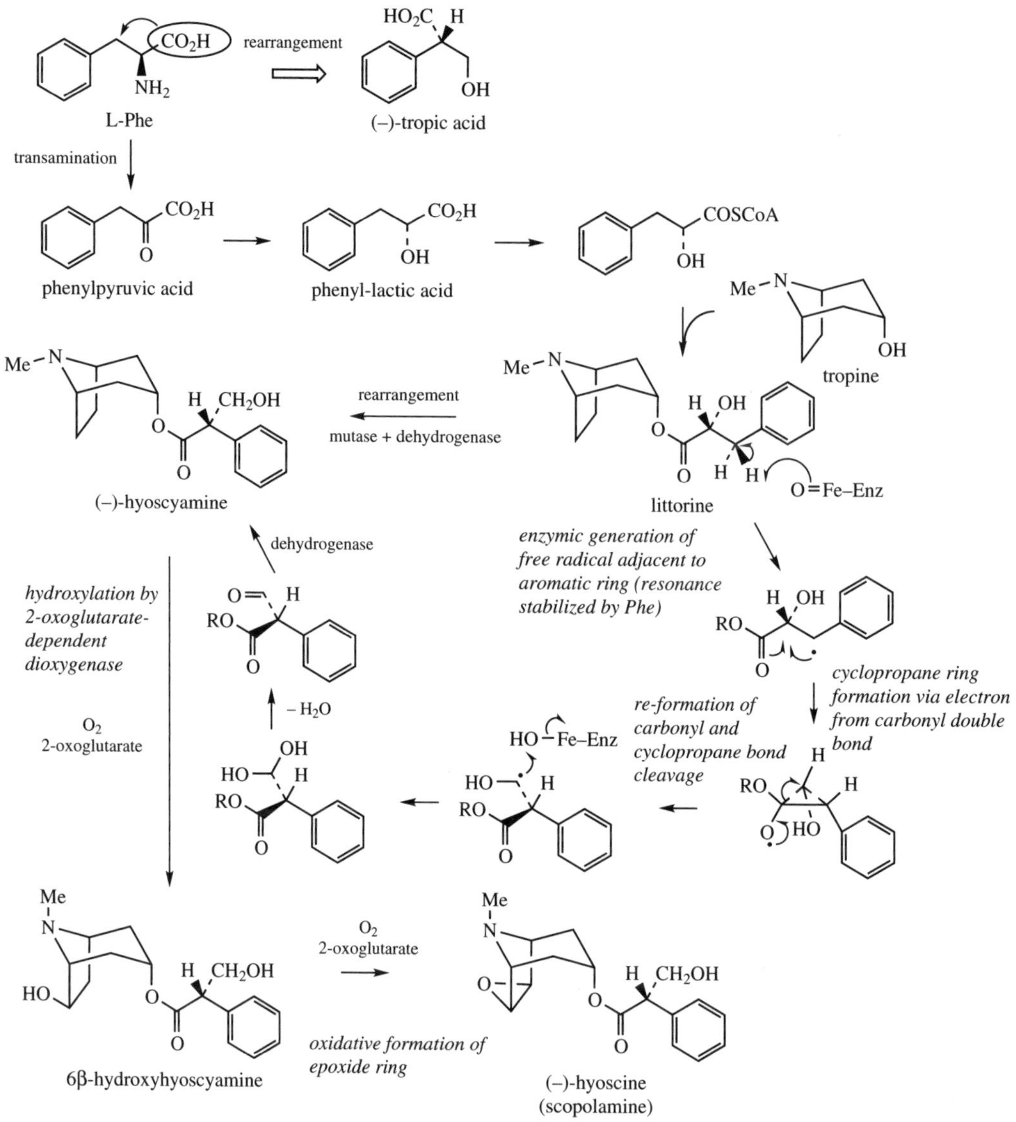

Figure 6.4

2-substituted pyrrolidine, retaining the thioester group of the second acetyl-CoA. **Hygrine** and most of the natural tropane alkaloids lack this particular carbon atom, which is lost by suitable hydrolysis/decarboxylation reactions. The bicyclic structure of the tropane skeleton in **hyoscyamine** and **cocaine** is achieved by a repeat of the Mannich-like reaction just observed. This requires an oxidation step to generate a new Δ^1-pyrrolinium cation, and removal of a proton α to the carbonyl. The *intramolecular* Mannich reaction on the R enantiomer accompanied by decarboxylation

generates **tropinone**, and stereospecific reduction of the carbonyl yields **tropine** with a 3α-hydroxyl. Hyoscyamine is the ester of tropine with (S)-**tropic acid** (Figure 6.4), which is derived from L-phenylalanine. A novel rearrangement process occurs in the phenylalanine → tropic acid transformation, in which the carboxyl group apparently migrates to the adjacent carbon (Figure 6.4). Phenylpyruvic acid and phenyl-lactic acid have been shown to be involved and tropine becomes esterified with phenyl-lactic acid (as the coenzyme-A ester) to form **littorine** before the rearrangement occurs. The mechanism of this rearrangement has yet to be proven, though a free radical process (Figure 6.4) with an intermediate cyclopropane-containing radical would accommodate the available data. Further modifications to the tropane skeleton then occur on the ester, and not on the free alcohol. These include hydroxylation to 6β-hydroxyhyoscyamine and additional oxidation allowing formation of an epoxide grouping as in **hyoscine (scopolamine)**. Both of these reactions are catalysed by a single 2-oxoglutarate-dependent dioxygenase (see page 27). Other esterifying acids may be encountered in tropane alkaloid structures, e.g. tiglic acid in **meteloidine** (Figure 6.5) from *Datura meteloides* and phenyl-lactic acid in littorine, above which is a major alkaloid in *Anthocercis littorea*. Tiglic acid is known to be derived from the amino acid L-isoleucine (see page 197).

The structure of **cuscohygrine** arises by an *intermolecular* Mannich reaction involving a second N-methyl-Δ^1-pyrrolinium cation (Figure 6.3).

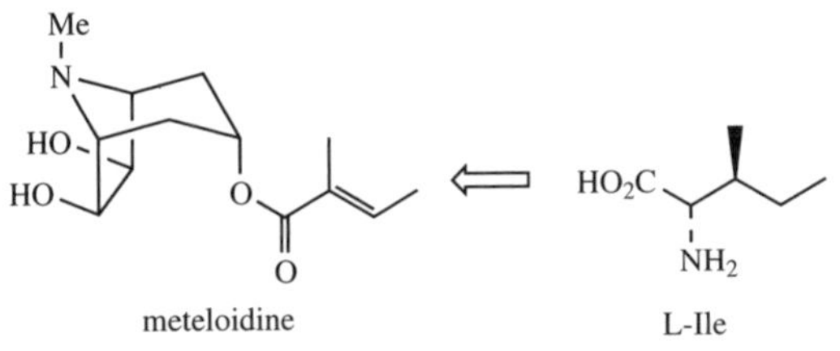

Should the carboxyl carbon from the acetoacetyl side-chain not be lost as it was in the formation of tropine, the subsequent intramolecular Mannich reaction will generate a tropane skeleton with an additional carboxyl substituent (Figure 6.3). However, this event is rare, and is only exemplified by the formation of ecgonine derivatives such as **cocaine** in *Erythroxylum coca* (Erythroxylaceae). The pathway is in most aspects analogous to that already described for hyoscyamine, but must proceed through the S-enantiomer of the N-methylpyrrolidineacetoacetyl-CoA. The ester function is then modified from a coenzyme A thioester to a simple methyl oxygen ester, and **methylecgonine** is subsequently obtained from the methoxycarbonyltropinone by stereospecific reduction of the carbonyl. Note that in this case reduction of the carbonyl occurs from the opposite face to that noted with the tropinone → tropine conversion and thus yields the 3β configuration in ecgonine. **Cocaine** is a diester of ecgonine, the benzoyl moiety arising from phenylalanine via cinnamic acid and benzoyl-CoA (see page 141).

The tropane alkaloids (−)-hyoscyamine* and (−)-hyoscine* are among the most important of the natural alkaloids used in medicine. They are found in a variety of solanaceous plants, including *Atropa belladonna** (deadly nightshade), *Datura stramonium** (thornapple) and other *Datura* species, *Hyoscyamus niger** (henbane), and *Duboisia** species. These alkaloids

meteloidine   L-Ile

Figure 6.5

Belladonna

The deadly nightshade *Atropa belladonna* (Solanaceae) has a long history as a highly poisonous plant. The generic name is derived from *Atropos*, in Greek mythology the Fate who cut the thread of life. The berries are particularly dangerous, but all parts of the plant contain toxic alkaloids, and even handling of the plant can lead to toxic effects since the alkaloids are readily absorbed through the skin. Although humans are sensitive to the toxins, some animals, including sheep, pigs, goats, and rabbits, are less susceptible. Cases are known where the consumption of rabbits or birds that have ingested belladonna has led to human poisoning. The plant is a tall perennial herb producing dull-purple bell-shaped flowers followed by conspicuous shiny black fruits, the size of a small cherry. *Atropa belladonna* is indigenous to Central and Southern Europe, though it is not especially common. It is cultivated for drug use in Europe and the United States. The tops of the plant are harvested two or three times per year and dried to give **belladonna** herb. Roots from plants some 3–4 years old are less commonly employed as a source of alkaloids.

(Continues)

(*Continued*)

Belladonna herb typically contains 0.3–0.6% of alkaloids, mainly (−)-hyoscyamine (Figure 6.4). Belladonna root has only slightly higher alkaloid content at 0.4–0.8%, again mainly (−)-hyoscyamine. Minor alkaloids including (−)-hyoscine (Figure 6.4) and cuscohygrine (Figure 6.3) are also found in the root, though these are not usually significant in the leaf. The mixed alkaloid extract from belladonna herb is still used as a gastrointestinal sedative, usually in combination with antacids. Root preparations can be used for external pain relief, e.g. in belladonna plasters.

Stramonium

Datura stramonium (Solanaceae) is commonly referred to as thornapple on account of its spikey fruit. It is a tall bushy annual plant widely distributed in Europe and North America, and because of its alkaloid content is potentially very toxic. Indeed, a further common name, Jimson or Jamestown weed, originates from the poisoning of early settlers near Jamestown, Virginia. At subtoxic levels, the alkaloids can provide mild sedative action and a feeling of well-being. In the Middle Ages, stramonium was employed to drug victims prior to robbing them. During this event, the victim appeared normal and was cooperative, though afterwards could usually not remember what had happened. For drug use, the plant is cultivated in Europe and South America. The leaves and tops are harvested when the plant is in flower. **Stramonium** leaf usually contains 0.2–0.45% of alkaloids, principally (−)-hyoscyamine and (−)-hyoscine in a ratio of about 2:1. In young plants, (−)-hyoscine can predominate.

The generic name *Datura* is derived from dhat, an Indian poison used by the Thugs. The narcotic properties of *Datura* species, especially *D. metel*, have been known and valued in India for centuries. The plant material was usually absorbed by smoking. Most species of *Datura* contain similar tropane alkaloids and are potential sources of medicinal alkaloids. In particular, *Datura sanguinea*, a perennial of treelike stature with blood-red flowers, is cultivated in Ecuador, and yields leaf material with a high (0.8%) alkaloid content in which the principal component is (−)-hyoscine. The plants can be harvested several times a year. *Datura sanguinea*, and several other species of the tree-daturas (now classified as a separate genus *Brugmansia*) are widely cultivated as ornamentals, especially for conservatories, because of their attractive large tubular flowers. The toxic potential of these plants is not always recognized.

Hyosycamus

Hyoscyamus niger (Solanaceae), or henbane, is a European native with a long history as a medicinal plant. Its inclusion in mediaeval concoctions and its power to induce hallucinations with visions of flight may well have contributed to our imaginary view of witches on broomsticks. The plant has both annual and biennial forms, and is cultivated in Europe and North America for drug use, the tops being collected when the plant is in flower, and then dried rapidly. The alkaloid content of **hyoscyamus** is relatively low at 0.045–0.14%, but this can be composed of similar proportions of (−)-hyoscine and (−)-hyoscyamine. Egyptian henbane, *Hyoscyamus muticus*, has a much higher alkaloid content than *H. niger*, and although it has mainly been collected from the wild, especially from Egypt, it functions as a major commercial source for alkaloid production. Some commercial cultivation occurs in California. The alkaloid content of the leaf is from 0.35% to 1.4%, of which about 90% is (−)-hyoscyamine.

(*Continues*)

(Continued)

Duboisia

Duboisia is a small genus of trees, containing only three species, found in Australia, again from the family Solanaceae. Two of these, *Duboisia myoporoides* and *D. leichhardtii* are grown commercially in Australia for tropane alkaloid production. The small trees are kept as bushes to allow frequent harvesting, with up to 70–80% of the leaves being removed every 7–8 months. The alkaloid content of the leaf is high, up to 3% has been recorded, and it includes (−)-hyoscyamine, (−)-hyoscine, and a number of related structures. The proportion of hyoscyamine to hyoscine varies according to the species used, and the area in which the trees are grown. The hyoscine content is frequently much higher than that of hyoscyamine. Indeed, interest in *Duboisia* was very much stimulated by the demand for hyoscine as a treatment for motion sickness in military personnel in the Second World War. Even higher levels of alkaloids, and higher proportions of hyoscine, can be obtained from selected *D. myoporoides* × *D. leichhardtii* hybrids, which are currently cultivated. The hybrid is superior to either parent, and can yield 1–2.5% hyoscine and 0–1% hyoscyamine. *Duboisia* leaf is an important commercial source of medicinal tropane alkaloids.

The third species of *Duboisia*, *D. hopwoodi*, contains little tropane alkaloid content, but produces mainly nicotine and related alkaloids, e.g. nornicotine (see page 313). Leaves of this plant were chewed by aborigines for their stimulating effects.

Allied Drugs

Tropane alkaloids, principally hyoscyamine and hyoscine, are also found in two other medicinal plants, scopolia and mandrake, but these plants find little current use. Scopolia (*Scopolia carniolica*; Solanaceae) resembles belladonna in appearance, though it is considerably smaller. Both root and leaf materials have been employed medicinally. The European mandrake (*Mandragora officinarum*; Solanaceae) has a complex history as a hypnotic, a general panacea, and an aphrodisiac. Its collection has been surrounded by much folklore and superstition, in that pulling it from the ground was said to drive its collector mad due to the unearthly shrieks emitted. The roots are frequently forked and are loosely likened to a man or woman. Despite the Doctrine of Signatures, which teaches that the appearance of an object indicates its special properties, from a pharmacological point of view, this plant would be much more efficient as a pain-reliever than as an aphrodisiac.

Hyoscyamine, Hyoscine and Atropine

All the above solanaceous plants contain as main alkaloidal constituents the tropane esters (−)-**hyoscyamine** and (−)-**hyoscine**, together with other minor tropane alkaloids. The piperidine ring in the bicyclic tropane system has a chairlike conformation, and there is a ready inversion of configuration at the nitrogen atom so that the *N*-methyl group can equilibrate between equatorial and axial positions (Figure 6.6). An equatorial methyl is strongly favoured provided there are no substituents on the two-carbon bridge, in which case the axial form may predominate. (−)-Hyoscyamine is the ester of tropine (Figure 6.4) with (−)-(*S*)-tropic acid, whilst (−)-hyoscine contains scopine (Figure 6.8) esterified with (−)-(*S*)-tropic acid. The optical activity of both hyoscyamine and hyoscine stems from the chiral centre

(Continues)

(Continued)

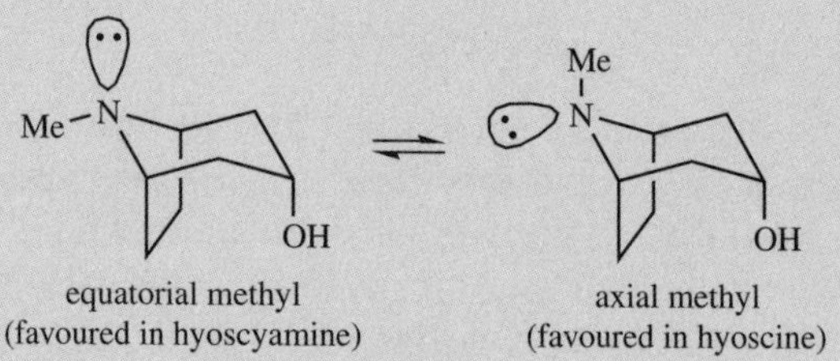

equatorial methyl
(favoured in hyoscyamine)

axial methyl
(favoured in hyoscine)

Figure 6.6

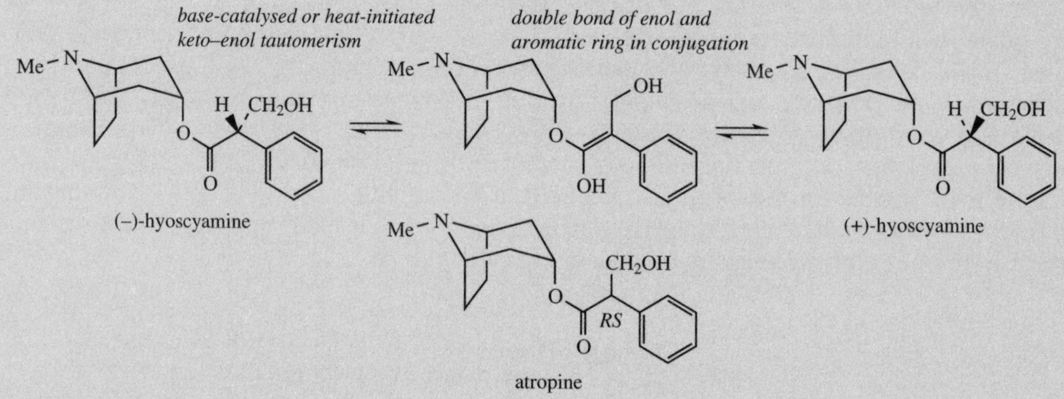

base-catalysed or heat-initiated
keto–enol tautomerism

double bond of enol and
aromatic ring in conjugation

(–)-hyoscyamine

(+)-hyoscyamine

atropine

Figure 6.7

in the acid portion, (S)-tropic acid. Tropine itself, although containing chiral centres, is a symmetrical molecule and is optically inactive; it can be regarded as a *meso* structure. The chiral centre in the tropic acid portion is adjacent to a carbonyl and the aromatic ring, and racemization can be achieved under mild conditions by heating or treating with base. This will involve an intermediate enol (or enolate) which is additionally favoured by conjugation with the aromatic ring (Figure 6.7). Indeed, normal base assisted fractionation of plant extracts to isolate the alkaloids can sometimes result in production of significant amounts of racemic alkaloids. The plant material itself generally contains only the enantiomerically pure alkaloids. Hyoscyamine appears to be much more easily racemized than hyoscine. Hydrolysis of the esters using acid or base usually gives racemic tropic acid. Note that littorine (Figure 6.4), in which the chiral centre is not adjacent to the phenyl ring, is not readily racemized, and base hydrolysis gives optically pure phenyl-lactic acid. The racemic form of hyoscyamine is called **atropine** (Figure 6.7), whilst that of hyoscine is called atroscine. In each case, the biological activity of the (+)-enantiomer is some 20–30 times less than that of the natural (–)-form. Chemical hydrolysis of hyoscine in an attempt to obtain the alcohol scopine is not feasible. Instead, the alcohol oscine is generated because of the proximity of the 3α-hydroxyl group to the reactive epoxide function (Figure 6.8).

Probably for traditional reasons, salts of both (–)-hyoscyamine and (±)-hyoscyamine (atropine) are used medicinally, whereas usage of hyoscine is restricted to the

(Continues)

(Continued)

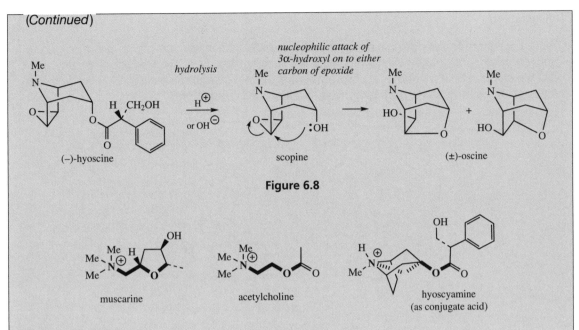

Figure 6.8

Figure 6.9

natural laevorotatory form. These alkaloids compete with acetylcholine for the muscarinic site of the parasympathetic nervous system, thus preventing the passage of nerve impulses, and are classified as anticholinergics. Acetylcholine binds to two types of receptor site, described as muscarinic or nicotinic, from the specific triggering of a response by the *Amanita muscaria* alkaloid muscarine or the tobacco alkaloid nicotine (see page 314) respectively. The structural similarity between acetylcholine and muscarine (Figure 6.9) can readily be appreciated, and hyoscyamine is able to occupy the same receptor site by virtue of the spatial relationship between the nitrogen atom and the ester linkage (Figure 6.9). The side-chain also plays a role in the binding, explaining the difference in activities between the two enantiomeric forms. The agonist properties of hyoscyamine and hyoscine give rise to a number of useful effects, including antispasmodic action on the gastrointestinal tract, antisecretory effect controlling salivary secretions during surgical operations, and as mydriatics to dilate the pupil of the eye. Hyoscine has a depressant action on the central nervous system and finds particular use as a sedative to control motion sickness. One of the side-effects from oral administration of tropane alkaloids is dry mouth (the antisecretory effect) but this can be much reduced by transdermal administration. In motion sickness treatment, hyoscine can be supplied via an impregnated patch worn behind the ear. Hyoscine under its synonym scopolamine is also well known, especially in fiction, as a 'truth drug'. This combination of sedation, lack of will, and amnesia was first employed in child-birth, giving what was termed 'twilight sleep', and may be compared with the mediaeval use of stramonium. The mydriatic use also has a very long history. Indeed, the specific name *belladonna* for deadly nightshade means 'beautiful lady' and refers to the practice of ladies at court who applied the juice of the fruit to the eyes, giving widely dilated pupils and a striking appearance, though at the expense of blurred vision through an inability to focus. Atropine also has useful antidote action in cases of poisoning caused by cholinesterase inhibitors, e.g. physostigmine and neostigmine (see page 366) and organophosphate insecticides.

(Continues)

(Continued)

It is valuable to reiterate here that the tropane alkaloid-producing plants are all regarded as very toxic, and that since the alkaloids are rapidly absorbed into the blood stream, even via the skin, first aid must be very prompt. Initial toxicity symptoms include skin flushing with raised body temperature, mouth dryness, dilated pupils, and blurred vision.

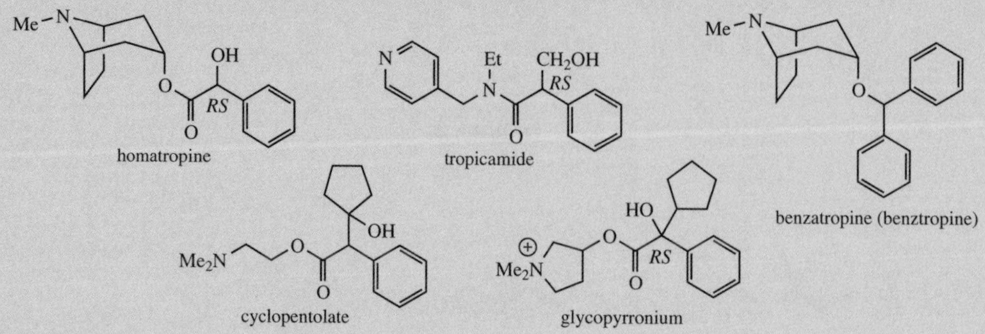

Figure 6.10

Homatropine (Figure 6.10) is a semi-synthetic ester of tropine with racemic mandelic (2-hydroxyphenylacetic) acid and is used as a mydriatic, as are **tropicamide** and **cyclopentolate** (Figure 6.10). Tropicamide is an amide of tropic acid, though a pyridine nitrogen is used to mimic that of the tropane. Cyclopentolate is an ester of a tropic acid-like system, but uses a non-quaternized amino alcohol resembling choline. **Glycopyrronium** (Figure 6.10) has a quaternized nitrogen in a pyrrolidine ring, with an acid moiety similar to that of cyclopentolate. This drug is an antimuscarinic used as a premedicant to dry bronchial and salivary secretions. **Hyoscine butylbromide** (Figure 6.11) is a gastro-intestinal antispasmodic synthesized from (−)-hyoscine by quaternization of the amine function with butyl bromide. The quaternization of tropane alkaloids by *N*-alkylation proceeds such that the incoming alkyl group always

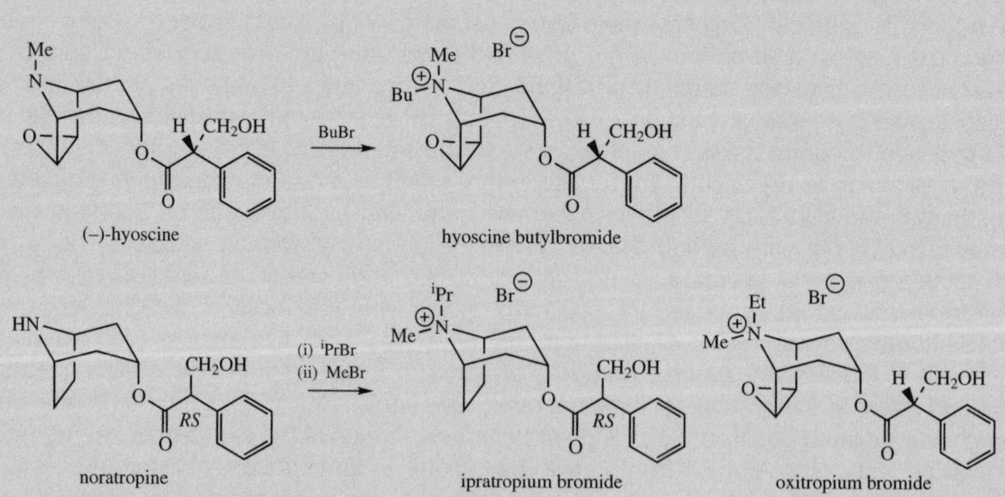

Figure 6.11

(Continues)

(Continued)

approaches from the equatorial position. The potent bronchodilator **ipratropium bromide** (Figure 6.11) is thus synthesized from noratropine by successive isopropyl and methyl alkylations whilst **oxitropium bromide** is produced from norhyoscine by *N*-ethylation and then *N*-methylation. Both drugs are used in inhalers for the treatment of chronic bronchitis.

Benzatropine (benztropine) (Figure 6.10) is an ether of tropine used as an antimuscarinic drug in the treatment of Parkinson's disease. It is able to inhibit dopamine reuptake, helping to correct the deficiency which is characteristic of Parkinsonism.

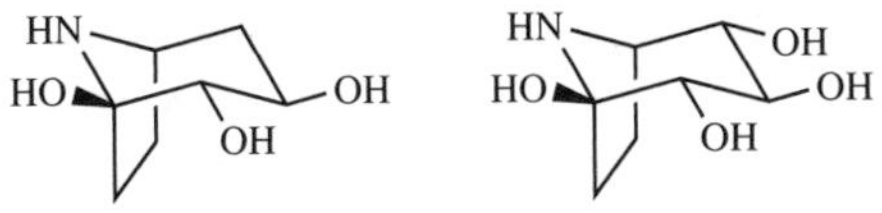

calystegine A₃   calystegine B₂

Figure 6.12

are also responsible for the pronounced toxic properties of these plants.

The **calystegines** (Figure 6.12) are a group of recently discovered, water-soluble, polyhydroxy nortropane derivatives that are found in the leaves and roots of many of the solanaceous plants, including *Atropa*, *Datura*, *Duboisia*, *Hyoscyamus*, *Mandragora*, *Scopolia*, and *Solanum*. They were first isolated from *Calystegia sepium* (Convolvulaceae). These compounds, e.g. calystegin A₃ and calystegin B₂ (Figure 6.12), are currently of great interest as glycosidase inhibitors and have similar potential as the polyhydroxyindolizidines such as castanospermine (see page 310) and the aminosugars such as deoxynojirimycin (see

page 477) in the development of drugs with activity against the AIDS virus HIV. It is likely that these alkaloids are produced by a similar pathway to that which yields tropine, but the stereochemistry of reduction of tropinone (or nortropinone) yields the 3β-alcohol, and further hydroxylation steps are necessary. Examples of tri-, tetra-, and penta-hydroxy calystegines are currently known.

Cocaine (Figure 6.3) is a rare alkaloid restricted to some species of *Erythroxylum* (Erythroxylaceae). *Erythroxylum coca* (coca)* is the most prominent as a source of cocaine, used medicinally as a local anaesthetic, and as an illicit drug for its euphoric properties. Coca also contains significant amounts of **cinnamoylcocaine (cinnamylcocaine)** (Figure 6.13), where cinnamic acid rather than benzoic acid is the esterifying acid, together with some typical tropine derivatives without the extra carboxyl, e.g. **tropacocaine** (Figure 6.13). Tropacocaine still retains the 3β-configuration, showing that the stereospecific carbonyl reduction is the same as with the cocaine route, and not as with the hyoscyamine pathway.

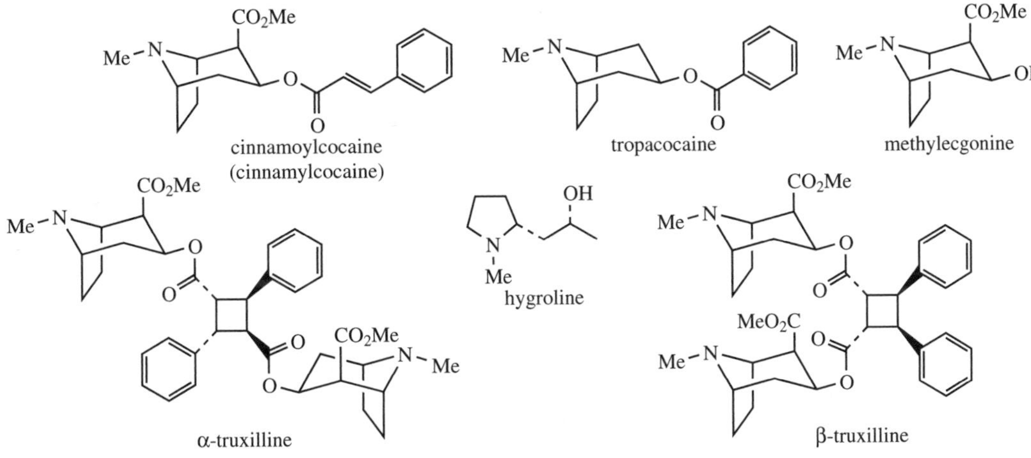

cinnamoylcocaine (cinnamylcocaine)   tropacocaine   methylecgonine

α-truxilline   hygroline   β-truxilline

Figure 6.13

Coca

Coca leaves are obtained from species of *Erythroxylum* (Erythroxylaceae), small shrubs native to the Andes region of South America, namely Colombia, Ecuador, Peru, and Bolivia. Plants are cultivated there, and in Indonesia. Two main species provide drug materials, *Erythroxylum coca* (Bolivian or Huanaco coca) and *E. truxillense* (Peruvian or Truxillo coca). Cultivated plants are kept small by pruning and a quantity of leaves is harvested from each plant three or more times per year.

Coca-leaf chewing has been practised by South American Indians for many years and is an integral part of the native culture pattern. Leaf is mixed with lime, this liberating the principal alkaloid cocaine as the free base, and the combination is then chewed. Cocaine acts as a potent antifatigue agent, and this allows labourers to ignore hunger, fatigue, and cold, enhancing physical activity and endurance. Originally the practice was limited to the Inca high priests and favoured individuals, but became widespread after the Spanish conquest of South America. It is estimated that 25% of the harvest is consumed in this way by the local workers, who may each use about 50 g of leaf per day ($\equiv$ 350 mg cocaine). Only a tiny amount (1–2%) of the coca produced is exported for drug manufacture. The rest contributes to illicit trade and the world's drug problems. Efforts to stem the supply of illicit coca and cocaine have been relatively unsuccessful.

Coca leaf contains 0.7–2.5% of alkaloids, the chief component (typically 40–50%) of which is (−)-cocaine (Figure 6.3), a diester of (−)-ecgonine. Note that although tropine is an optically inactive *meso* structure, ecgonine contains four chiral centres, is no longer symmetrical, and is therefore optically active. Cinnamoylcocaine (cinnamylcocaine), α-truxilline, β-truxilline, and methylecgonine (Figure 6.13) are minor constituents also based on ecgonine. The truxillines contain dibasic acid moieties, α-truxillic and β-truxinic acids, which are cycloaddition products from two cinnamic acid units (Figure 6.14). Other alkaloids present include structures based on φ-tropine (the 3β-isomer of tropine) such as tropacocaine (Figure 6.13) and on hygrine, e.g. hygrine, hygroline (Figure 6.13) and cuscohygrine (Figure 6.3). Cuscohygrine typically accounts for 20–30% of the alkaloid content.

Illegal production of cocaine is fairly unsophisticated, but can result in material of high quality. The alkaloids are extracted from crushed leaf using alkali (lime) and petrol. The petrol extract is then re-extracted with aqueous acid, and this alkaloid fraction is basified and allowed to stand, yielding the free alkaloid as a paste. Alternatively, the hydrochloride or sulphate salts may be prepared. The coca alkaloids are often diluted with carrier to give a

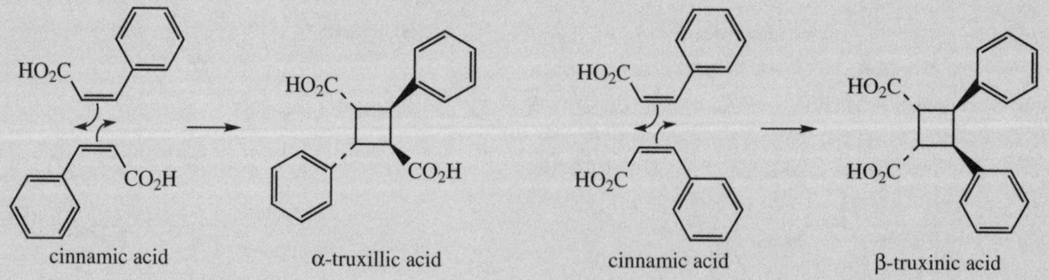

cinnamic acid   α-truxillic acid   cinnamic acid   β-truxinic acid

Figure 6.14

(*Continues*)

(Continued)

preparation with 10–12% of cocaine. The illicit use of cocaine and cocaine hydrochloride is a major problem worldwide. The powder is usually sniffed into the nostrils, where it is rapidly absorbed by the mucosa, giving stimulation and short-lived euphoria through inhibiting reuptake of neurotransmitters dopamine, noradrenaline, and serotonin, so prolonging and augmenting their effects. Regular usage induces depression, dependence, and damage to the nasal membranes. The drug may also be injected intravenously, or the vapour inhaled. For inhalation, the free base or 'crack' is employed to increase volatility. The vaporized cocaine is absorbed extremely rapidly and carried to the brain within seconds, speeding up and enhancing the euphoric lift. Taken in this form, cocaine has proved highly addictive and dangerous. Cocaine abuse is currently a greater problem than heroin addiction, and, despite intensive efforts, there is no useful antagonist drug available to treat cocaine craving and addiction.

In the 1800s, coca drinks were fashionable, and one in particular, Coca-Cola, became very popular. This was originally based on extracts of coca (providing cocaine) and cola (supplying caffeine) (see page 395), but although the coca content was omitted from 1906 onwards, the name and popularity continue.

Medicinally, **cocaine** is of value as a local anaesthetic for topical application. It is rapidly absorbed by mucous membranes and paralyses peripheral ends of sensory nerves. This is achieved by blocking ion channels in neural membranes. It was widely used in dentistry, but has been replaced by safer drugs, though it still has applications in ophthalmic and ear, nose, and throat surgery. As a constituent of Brompton's cocktail (cocaine and heroin in sweetened alcohol) it is available to control pain in terminal cancer patients. It increases the overall analgesic effect, and its additional CNS stimulant properties counteract the sedation normally associated with heroin (see page 332).

The essential functionalities of cocaine required for activity were eventually assessed to be the aromatic carboxylic acid ester and the basic amino group, separated by a lipophilic hydrocarbon chain. Synthetic drugs developed from the cocaine structure have been introduced to provide safer, less toxic local anaesthetics (Figure 6.15). **Procaine**, though little used now, was the first major analogue employed. **Benzocaine** is used topically, but has a short duration of action. **Tetracaine (amethocaine)**, **oxybuprocaine**, and **proxymetacaine** are valuable local anaesthetics employed principally in ophthalmic work. The ester function can be replaced by an amide, and this gives better stability towards hydrolysis in aqueous solution or by esterases. **Lidocaine (lignocaine)** is an example of an amino amide analogue and is perhaps the most widely used local anaesthetic, having rapid action, effective absorption, good stability, and may be used by injection or topically. Other amino amide local anaesthetic structures include **prilocaine**, with similar properties to lidocaine and very low toxicity, and **bupivacaine**, which has a long duration of action. **Ropivacaine**, **mepivacaine**, and **articaine (carticaine)** are some recently introduced amide-type local anaesthetics, the latter two being used predominantly in dentistry. **Cinchocaine** is often incorporated into preparations to soothe haemorrhoids.

Lidocaine, although introduced as a local anaesthetic, was subsequently found to be a potent antiarrhythmic agent, and it now finds further use as an antiarrhythmic drug, for treatment of ventricular arrhythmias especially after myocardial infarction. Other cocaine-related structures also find application in the same way, including **tocainide**, **procainamide**, and **flecainide** (Figure 6.15). Tocainide is a primary amine analogue of lidocaine, whilst procainamide is an amide analogue of procaine. In **mexiletene**, a congener of lidocaine, the amide group has been replaced by a simple ether linkage.

(Continues)

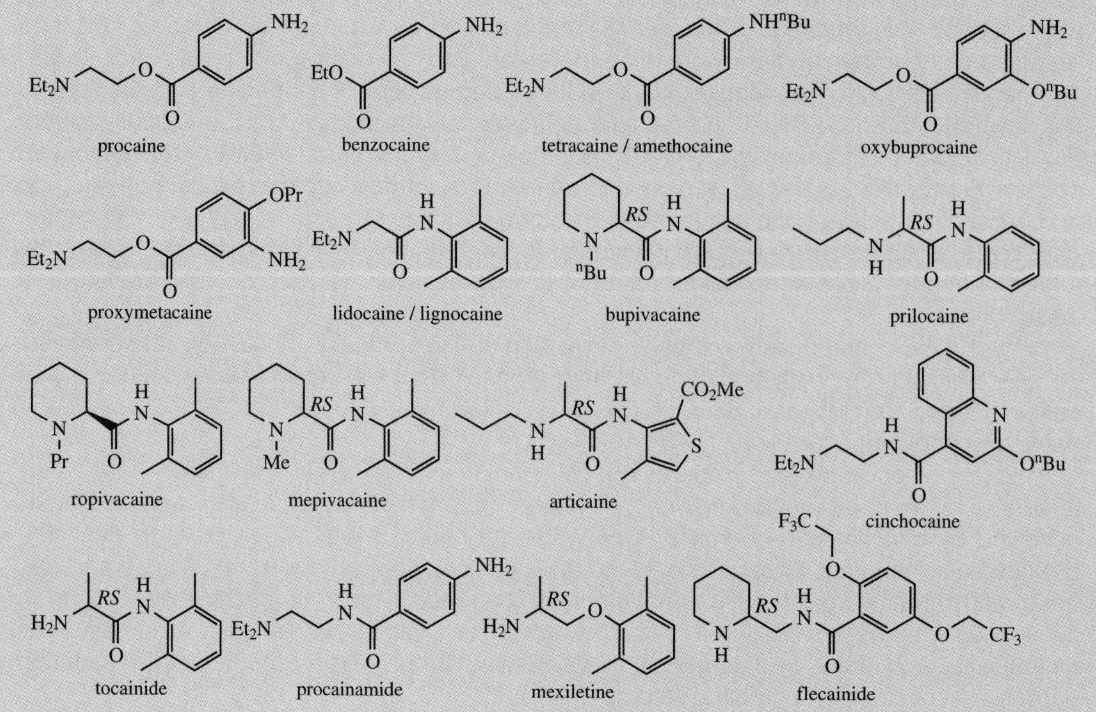

Figure 6.15

Anatoxin-*a* (Figure 6.16) is a toxic tropane-related alkaloid produced by a number of cyanobacteria, e.g. *Anabaena flos-aquae* and *Aphanizomenon flos-aquae*, species which proliferate in lakes and reservoirs during periods of hot, calm weather. A number of animal deaths have been traced back to consumption of water containing the cyanobacteria, and ingestion of the highly potent neurotoxin anatoxin-*a*, which has been termed Very Fast Death Factor. Anatoxin-*a* is a powerful agonist at nicotinic acetylcholine receptors, and has become a useful pharmacological probe. The ring system may be regarded as a homotropane, and it has been suggested that the pyrrolidine ring originates from ornithine via putrescine and Δ^1-pyrroline, in a way similar to the tropane alkaloids (Figure 6.16). The remaining carbons may originate from acetate precursors. A remarkable compound with a nortropane ring system has been isolated in tiny amounts from

the highly coloured skin of the Ecuadorian poison frog *Epipedobates tricolor*. This compound, called **epibatidine** (Figure 6.17), is exciting considerable interest as a lead compound for analgesic drugs. It is 200–500 times more potent than morphine (see page 331), and does not act by the normal opioid mechanism, but is a specific agonist at nicotinic acetylcholine receptors. Whether or not this structure is ornithine derived remains to be established.

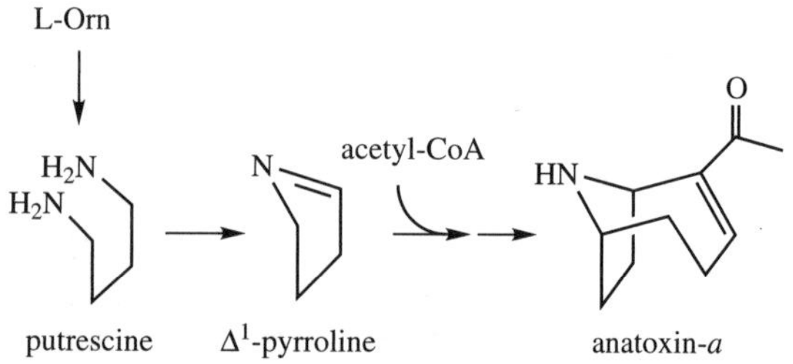

Figure 6.16

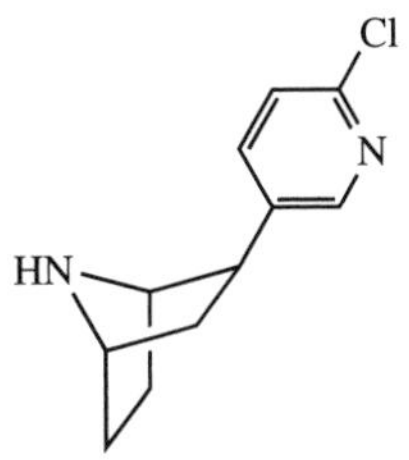

epibatidine

Figure 6.17

Pyrrolizidine Alkaloids

Two molecules of ornithine are utilized in formation of the bicyclic pyrrolizidine skeleton, the pathway (Figure 6.18) proceeding via the intermediate **putrescine**. Because plants synthesizing pyrrolizidine alkaloids appear to lack the decarboxylase enzyme transforming ornithine into putrescine, ornithine is actually incorporated by way of arginine (Figure 6.2). Two molecules of putrescine are condensed in an NAD^+-dependent oxidative deamination reaction to give the imine, which is then converted into **homospermidine** by NADH reduction. On paper, one might predict that one molecule of putrescine is converted by oxidative deamination into the aldehyde, which

condenses with a second putrescine molecule to give the imine, but this mechanism has been proven to be incorrect. The pyrrolizidine skeleton is formed from homospermidine by a sequence of oxidative deamination, imine formation, and an intramolecular Mannich reaction, which exploits the enolate anion generated from the aldehyde. This latter reaction is analogous to that proposed in formation of the tropane ring system (see page 293). A typical simple natural pyrrolizidine structure is that of **retronecine** (Figure 6.18), which can be derived from the pyrrolizidine aldehyde by modest oxidative and reductive steps. The pyrrolizidine skeleton thus incorporates a C_4N unit from ornithine, plus a further four carbons from the same amino acid precursor.

Pyrrolizidine alkaloids have a wide distribution, but are characteristic of certain genera of the Boraginaceae (e.g. *Heliotropium*, *Cynoglossum*, and *Symphytum*), the Compositae/Asteraceae (e.g. *Senecio*) and the Leguminosae/Fabaceae (e.g. *Crotalaria*). The pyrrolizidine bases rarely occur in the free form, but are generally found as esters with rare mono- or di-basic acids, the necic acids. Thus, **senecionine** (Figure 6.18) from *Senecio* species is a diester of retronecine with senecic acid. Inspection of the ten-carbon skeleton

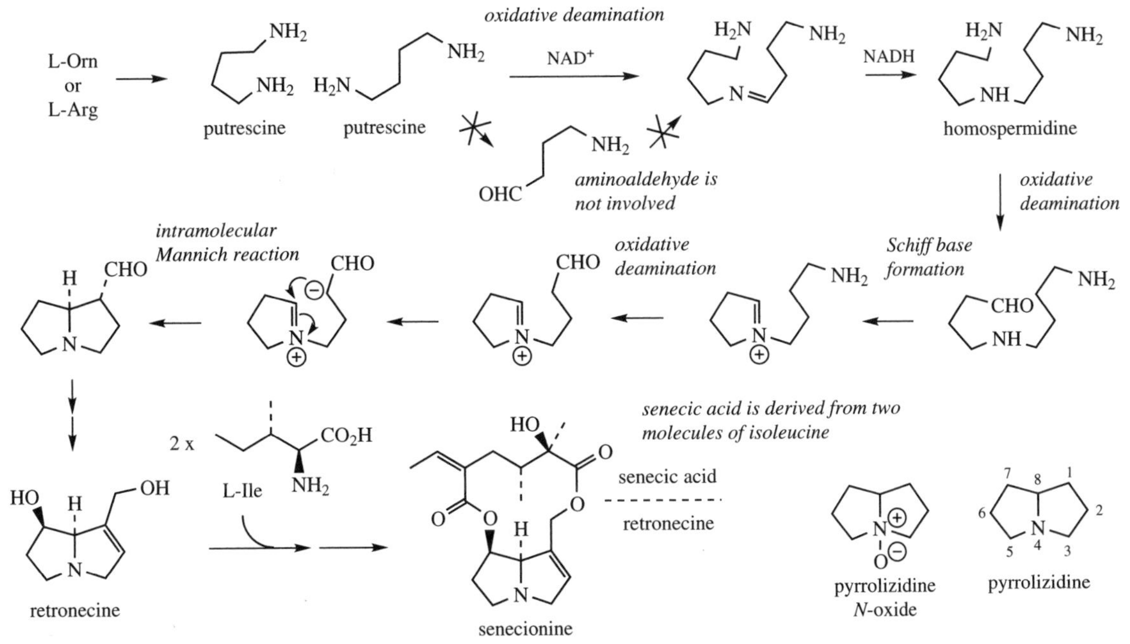

Figure 6.18

of senecic acid suggests it is potentially derivable from two isoprene units, but experimental evidence has demonstrated that it is in fact obtained by incorporation of two molecules of the amino acid L-isoleucine. Loss of the carboxyl from isoleucine supplies a carbon fragment analogous to isoprene units (compare tiglic acid in the tropane alkaloid meteloidine, Figure 6.5). Other necic acid structures may incorporate fragments from valine, threonine, leucine, or acetate. It is also worthy of note that, in general, the pyrrolizidine alkaloids accumulate in the plant as polar *N*-oxides, facilitating their transport, and above all, maintaining them in a non-toxic form. The *N*-oxides are easily changed back to the tertiary amines by mild reduction, as will occur in the gut of a herbivore.

Many pyrrolizidine alkaloids are known to produce pronounced hepatic toxicity and there are many recorded cases of livestock poisoning. Potentially toxic structures have 1,2-unsaturation in the pyrrolizidine ring and an ester function on the side-chain. Although themselves non-toxic, these alkaloids are transformed by mammalian liver oxidases into reactive pyrrole structures, which are potent alkylating agents and react with suitable cell nucleophiles, e.g. nucleic acids and proteins (Figure 6.19). *N*-oxides are not transformed by these oxidases, only the free bases. The presence of pyrrolizidine alkaloids, e.g. **acetyl-intermedine** and **acetyl-lycopsamine** (Figure 6.20) in medicinal comfrey (*Symphytum officinale*; Boraginaceae) has emphasized potential dangers of using this traditional herbal drug as a remedy for inflammatory, rheumatic, and gastrointestinal disorders. Prolonged usage may lead to liver damage. Caterpillars of the cinnabar moth *Tyria jacobaeae* feed on species of *Senecio* (e.g. ragwort, *S. jacobaea*, and groundsel, *S. vulgaris*) with impunity, building up levels of pyrrolizidine alkaloids in their bodies (in the form of non-toxic *N*-oxides) making them distasteful to predators, and potentially toxic should the predator convert the alkaloids into the free bases. **Indicine-*N*-oxide** (Figure 6.20) from *Heliotropium indicum* (Boraginaceae) demonstrated significant antileukaemic activity in clinical trials but undesirable hepatotoxicity prevented any further development.

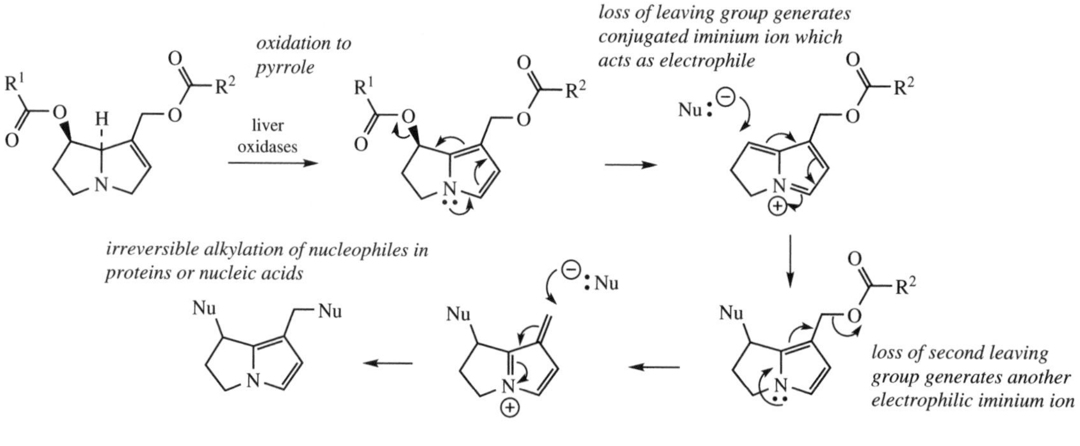

Figure 6.19

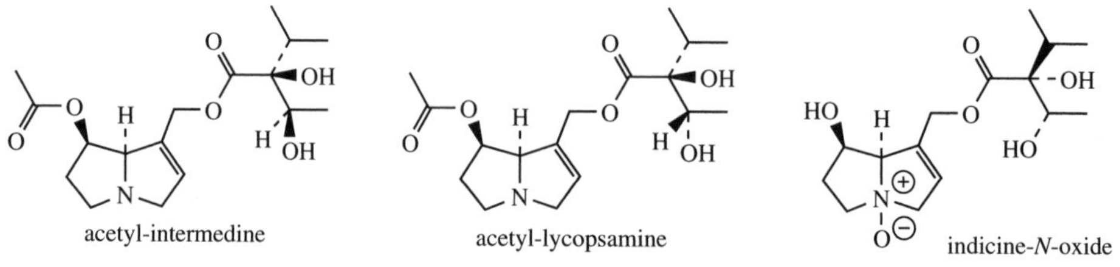

acetyl-intermedine   acetyl-lycopsamine   indicine-*N*-oxide

Figure 6.20

The tobacco alkaloids, especially nicotine, are derived from nicotinic acid (see page 311) but also contain a pyrrolidine ring system derived from ornithine as a portion of their structure.

ALKALOIDS DERIVED FROM LYSINE

L-Lysine is the homologue of L-ornithine, and it too functions as an alkaloid precursor, using pathways analogous to those noted for ornithine. The extra methylene group in lysine means this amino acid participates in forming six-membered piperidine rings, just as ornithine provided five-membered pyrrolidine rings. As with ornithine, the carboxyl group is lost, the ε-amino nitrogen rather than the α-amino nitrogen is retained, and lysine thus supplies a C_5N building block (Figure 6.21).

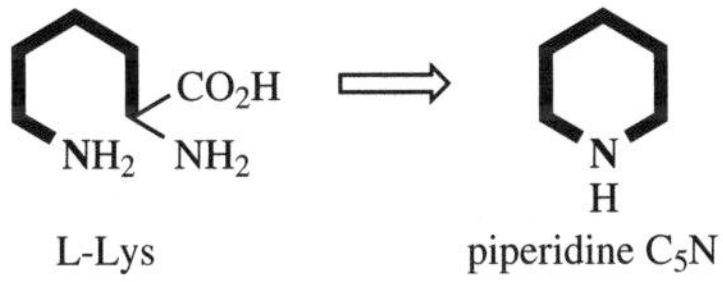

Figure 6.21

Piperidine Alkaloids

N-Methylpelletierine (Figure 6.22) is an alkaloidal constituent of the bark of pomegranate (*Punica granatum*; Punicacae), where it cooccurs with **pelletierine** and **pseudopelletierine** (Figure 6.22), the mixture of alkaloids having activity against intestinal tapeworms. *N*-Methylpelletierine and pseudopelletierine are homologues of hygrine and tropinone respectively, and a pathway similar to Figure 6.3 using the diamine **cadaverine** (Figure 6.22) may be proposed. (The rather distinctive names cadaverine and putrescine reflect early isolations of these compounds from decomposing animal flesh.) In practice, the Mannich reaction involving the Δ^1-piperidinium salt utilizes the more nucleophilic acetoacetyl-CoA rather than acetyl-CoA, and the carboxyl carbon from acetoacetate appears to be lost during the reaction by suitable hydrolysis/decarboxylation reactions (Figure 6.22). **Anaferine** (Figure 6.22) is an analogue of cuscohygrine in which a further piperidine ring is added via an intermolecular Mannich reaction.

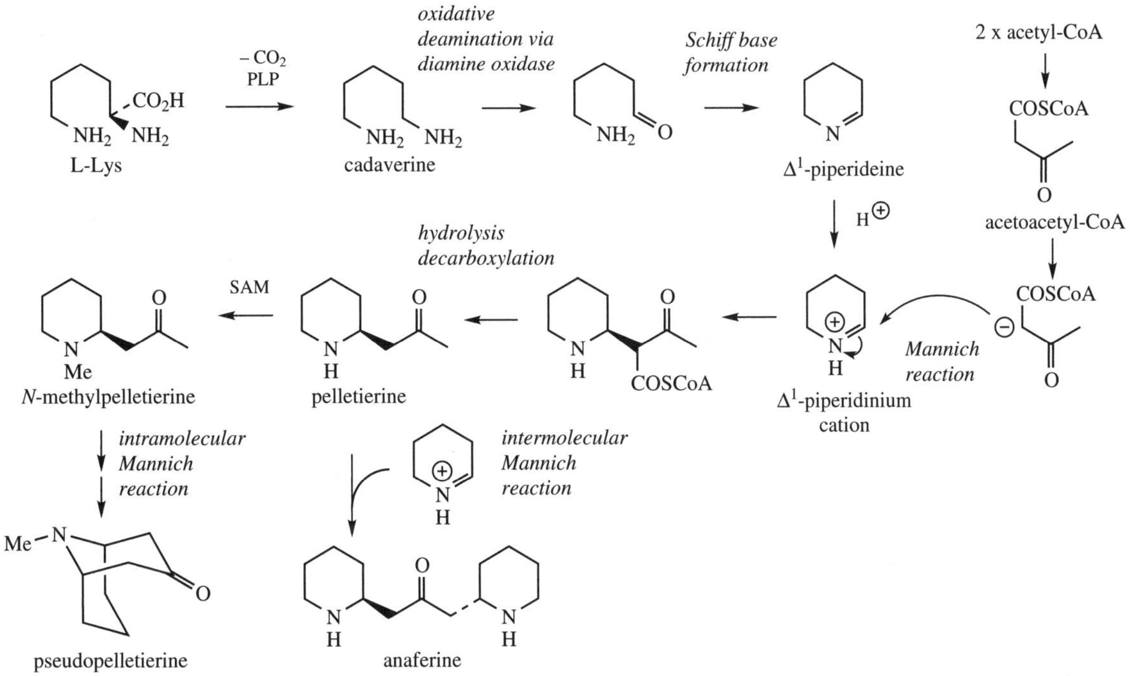

Figure 6.22

The alkaloids found in the antiasthmatic plant *Lobelia inflata** (Campanulaceae) contain piperidine rings with alternative C_6C_2 side-chains derived from phenylalanine via cinnamic acid. These alkaloids are produced as in Figure 6.23, in which benzoylacetyl-CoA, an intermediate in the β-oxidation of cinnamic acid (see page 141) provides the nucleophile for the Mannich reaction. Oxidation in the piperidine ring gives a new iminium species, and this can react further with a second molecule of benzoylacetyl-CoA, again via a Mannich reaction. Naturally, because of the nature of the side-chain, the second intramolecular Mannich reaction, as involved in pseudopelletierine biosynthesis, is not feasible. Alkaloids such as **lobeline** and **lobelanine** from *Lobelia inflata*, or **sedamine** from *Sedum acre* (Crassulaceae), are products from further *N*-methylation and/or carbonyl reduction reactions (Figure 6.23).

The simple piperidine alkaloid coniine from poison hemlock is not derived from lysine, but originates by an amination process and is discussed on page 381.

The pungency of the fruits of black pepper (*Piper nigrum*; Piperaceae), a widely used condiment, is mainly due to the piperidine alkaloid **piperine** (Figure 6.24). In this structure, the piperidine ring forms part of a tertiary amide structure, and is incorporated via piperidine itself, the reduction product of Δ^1-piperideine (Figure 6.22).

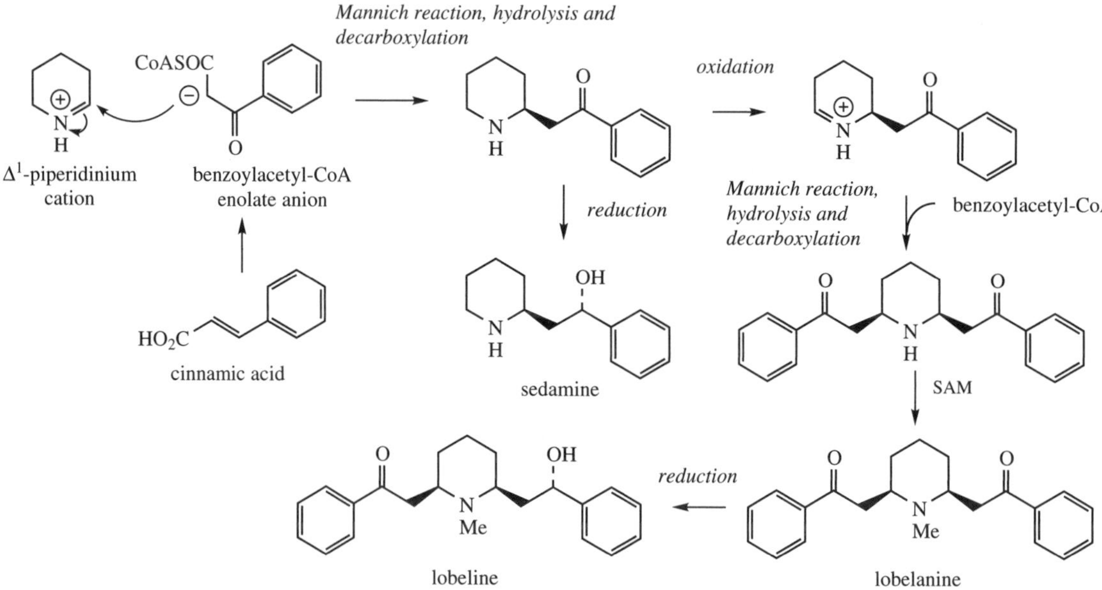

Figure 6.23

Lobelia

Lobelia or Indian tobacco consists of the dried leaves and tops of *Lobelia inflata* (Campanulaceae), an annual herb from the USA and Canada. Lobelia contains about 0.2–0.4% of alkaloids, of which the piperidine derivative lobeline (Figure 6.23) is the chief constituent. Minor alkaloids identified include closely related structures, e.g. lobelanine (Figure 6.23). The North American Indians employed lobelia as an alternative or substitute for tobacco (*Nicotiana tabacum*; Solanaceae), and it is found that lobeline stimulates nicotinic receptor sites in a similar way to nicotine, but with a weaker effect. **Lobeline** has been employed in preparations intended as smoking deterrents. The crude plant drug has also long been used to relieve asthma and bronchitis, though in large doses it can be quite toxic.

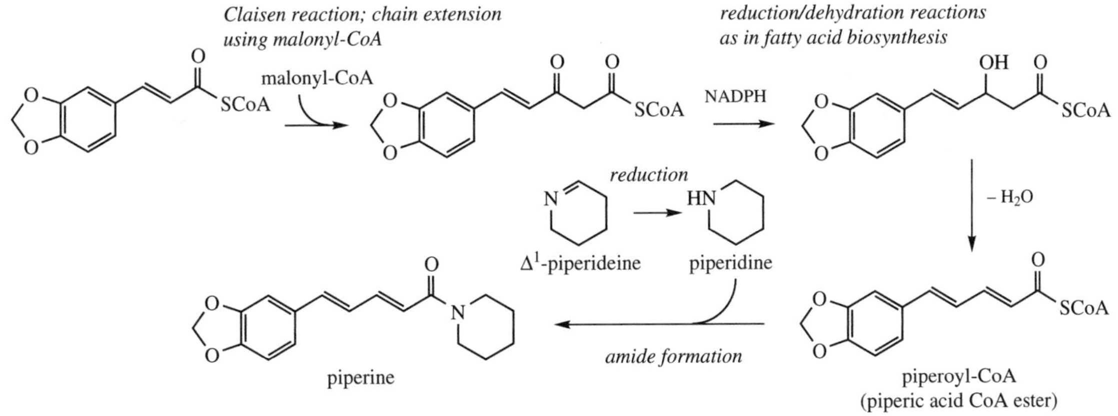

Claisen reaction; chain extension using malonyl-CoA

reduction/dehydration reactions as in fatty acid biosynthesis

malonyl-CoA

NADPH

reduction

Δ¹-piperideine   piperidine

− H₂O

amide formation

piperine

piperoyl-CoA
(piperic acid CoA ester)

Figure 6.24

The piperic acid portion is derived from a cinnamoyl-CoA precursor, with chain extension using acetate/malonate (compare flavonoids, page 149), and combines as its CoA ester with piperidine.

Quinolizidine Alkaloids

The lupin alkaloids, found in species of *Lupinus* (Leguminosae/Fabaceae), and responsible for the toxic properties associated with lupins, are characterized by a quinolizidine skeleton (Figure 6.25). This bicyclic ring system is closely related to the ornithine-derived pyrrolizidine system, but is formed from two molecules of lysine. **Lupinine** from *Lupinus luteus* is a relatively simple structure very comparable to the basic ring system of the pyrrolizidine alkaloid retronecine (see page 305), but other lupin alkaloids, e.g. **lupanine** and **sparteine** (Figure 6.25) contain a tetracyclic bis-quinolizidine ring system, and are formed by incorporation of a third lysine molecule. Sparteine is also the major alkaloid in broom (*Cytisus scoparius*; Leguminosae/Fabaceae). The alkaloid **cytisine**, a toxic component of *Laburnum* species (Leguminosae/Fabaceae) contains a modified tricyclic ring system, and comparison with the structures of lupanine or sparteine shows its likely relationship by loss of carbon atoms from the tetracyclic system (Figure 6.25). However, the structural similarity of lupinine and retronecine is not fully reflected in the biosynthetic pathways. Experimental evidence shows lysine to be

incorporated into lupinine via **cadaverine**, but the intermediate corresponding to homospermidine is excluded. Δ¹-Piperideine seems to be an important intermediate after cadaverine and the pathway proposed (Figure 6.25) invokes coupling of two such molecules. The two tautomers of Δ¹-piperideine, as *N*-analogues of corresponding carbonyl compounds, are able to couple by an aldol-type mechanism (see page 19). Indeed, this coupling occurs in solution at physiological pHs, though stereospecific coupling to the product shown in Figure 6.25 would require appropriate enzymic participation. Following the coupling, it is suggested that the imine system is hydrolysed, the primary amine group then oxidized, and formation of the quinolizidine ring is achieved by Schiff base formation. **Lupinine** is then synthesized by two reductive steps.

The pathway to **sparteine** and **lupanine** undoubtedly requires participation of another molecule of cadaverine or Δ¹-piperideine. Experimental data are not clear-cut and Figure 6.25 merely indicates how incorporation of a further piperidine ring might be envisaged. Loss of one or other of the outermost rings and oxidation to a pyridone system offers a potential route to **cytisine**.

Quinolizidine alkaloids are mainly found in plants of the Leguminosae/Fabaceae family. They deter or repel feeding of herbivores, and are toxic to them by a variety of mechanisms. A number of plants (*Laburnum*, *Cytisus*, *Lupinus*) containing significant quantities of these alkaloids must be regarded as potentially toxic to humans, and are known to be responsible for human poisoning.

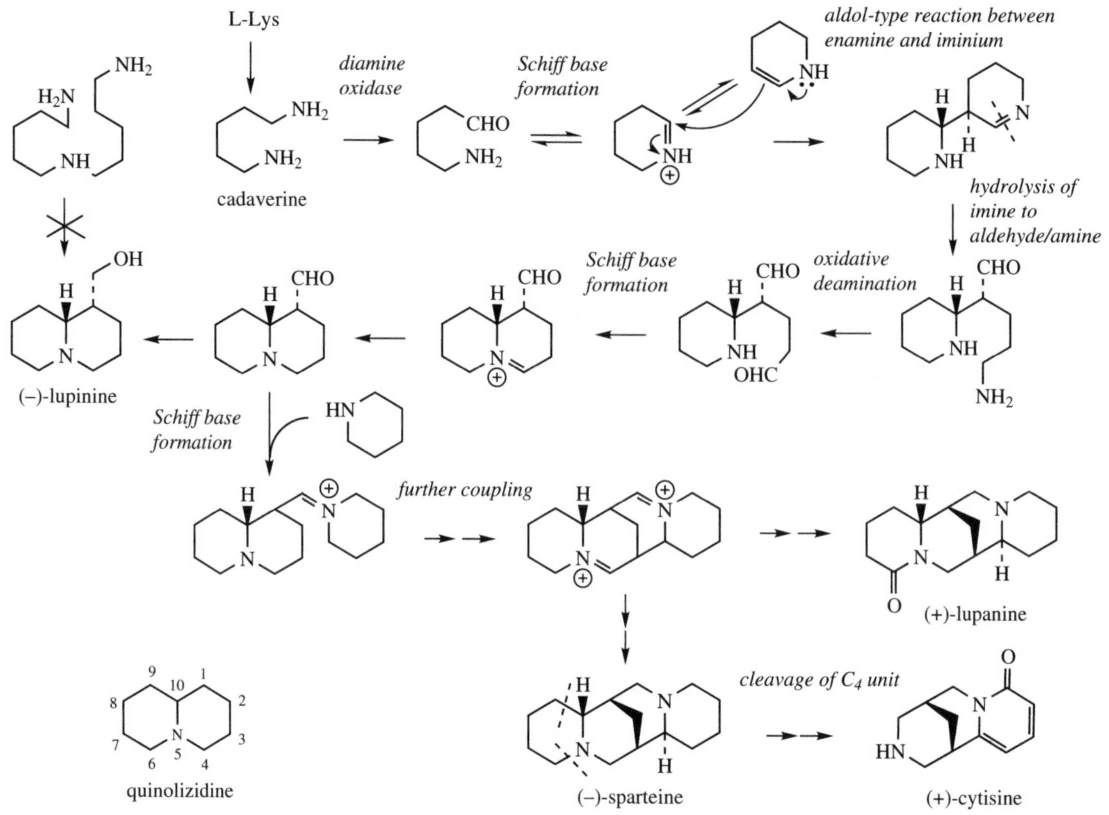

Figure 6.25

The widely planted and ornamental laburnum trees offer a particular risk, since all parts, including the pealike seeds, contain dangerously high amounts of alkaloids. So-called 'sweet lupins' are selected strains with an acceptably low alkaloid content (typically about a quarter of the total alkaloids of 'bitter' strains), which are grown as a high protein crop.

Indolizidine Alkaloids

Indolizidine alkaloids (Figure 6.26) are characterized by fused six- and five-membered rings, with a nitrogen atom at the ring fusion, e.g. **swainsonine** from *Swainsona canescens* (Leguminosae/Fabaceae) and **castanospermine** from the Moreton Bay chestnut *Castanospermum australe* (Leguminosae/Fabaceae). In this respect, they appear to be a hybrid between the pyrrolizidine and quinolizidine alkaloids described above. Although

they are derived from lysine, their origin deviates from the more common lysine-derived structures in that L-**pipecolic acid** is an intermediate in the pathway. Two routes to pipecolic acid are known in nature as indicated in Figure 6.26, and these differ with respect to whether the nitrogen atom originates from the α- or the ε-amino group of lysine. For indolizidine alkaloid biosynthesis, pipecolic acid is formed via the aldehyde and Schiff base with retention of the α-amino group nitrogen. The indolizidinone may then be produced by incorporating a C_2 acetate unit by simple reactions, though no details are known. This compound leads to castanospermine by a sequence of hydroxylations, but is also a branch-point compound to alkaloids such as swainsonine, which have the opposite configuration at the ring fusion. Involvement of a planar iminium ion could account for the change in stereochemistry. Polyhydroxyindolizidines such as swainsonine and castanospermine have demonstrated activity against

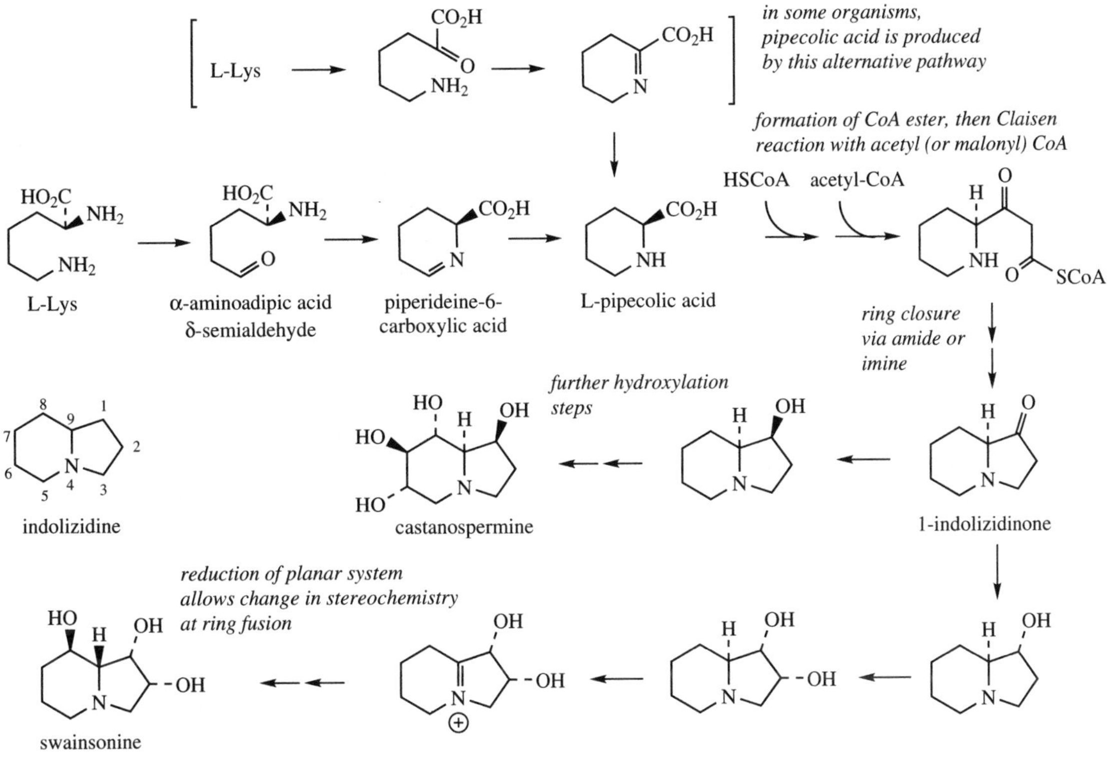

Figure 6.26

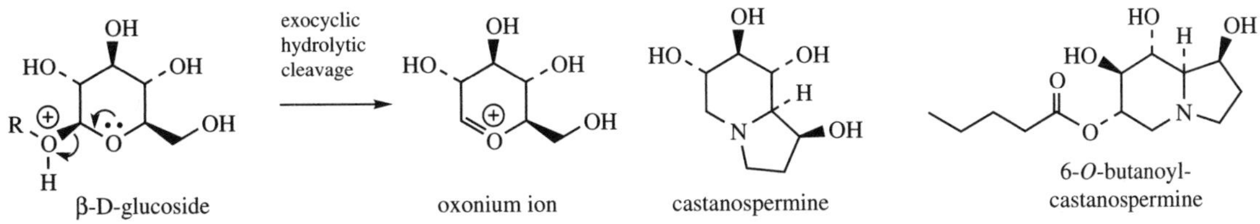

Figure 6.27

the AIDS virus HIV, by their ability to inhibit glycosidase enzymes involved in glycoprotein biosynthesis. The glycoprotein coating is essential for the proliferation of the AIDS virus. This has stimulated considerable research on related structures and their mode of action. The ester 6-*O*-butanoyl-castanospermine (Figure 6.27) is currently in clinical trials as an anti-AIDS agent. There is a strong similarity between castanospermine and the oxonium ion formed by hydrolytic cleavage of a glucoside (Figure 6.27) (see page 30), but there appears to be little stereochemical relationship with some other sugars, whose hydrolytic enzymes are also strongly inhibited. These alkaloids are also toxic to

animals, causing severe gastro-intestinal upset and malnutrition by severely affecting intestinal hydrolases. Indolizidine alkaloids are found in many plants in the Leguminosae/Fabaceae (e.g. *Swainsona*, *Astragalus*, *Oxytropis*) and also in some fungi (e.g. *Rhizoctonia leguminicola*).

ALKALOIDS DERIVED FROM NICOTINIC ACID

Pyridine Alkaloids

The alkaloids found in tobacco* (*Nicotiana tabacum*; Solanaceae) include **nicotine** and

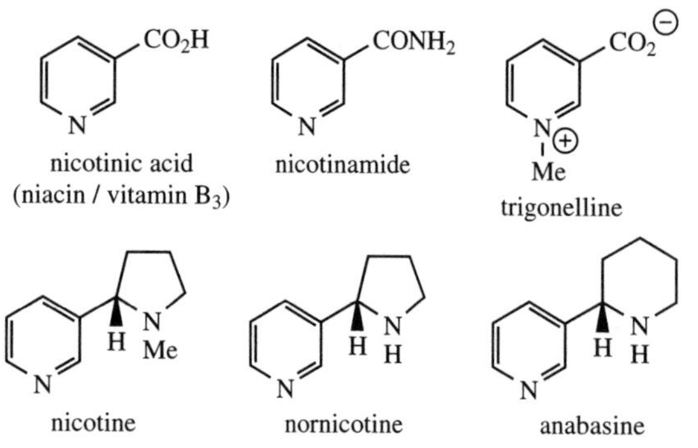

nicotinic acid
(niacin / vitamin B$_3$)

nicotinamide

trigonelline

nicotine

nornicotine

anabasine

Figure 6.28

anabasine (Figure 6.28). The structures contain a pyridine ring together with a pyrrolidine ring (in nicotine) or a piperidine unit (in anabasine),

the latter rings arising from ornithine and lysine respectively. The pyridine unit has its origins in **nicotinic acid** (**vitamin B$_3$**)* (Figure 6.28), the vitamin sometimes called **niacin**, which forms an essential component of coenzymes such as NAD$^+$ and NADP$^+$ (see page 24). The nicotinic acid component of nicotinamide is synthesized in animals by degradation of L-tryptophan through the **kynurenine** pathway and **3-hydroxyanthranilic acid** (Figure 6.29) (see also dactinomycin, page 432), the pyridine ring being formed by oxidative cleavage of the benzene ring and subsequent inclusion of the amine nitrogen (Figure 6.29). However, plants such as *Nicotiana* use a different pathway employing glyceraldehyde 3-phosphate and L-aspartic acid precursors (Figure 6.30). The dibasic

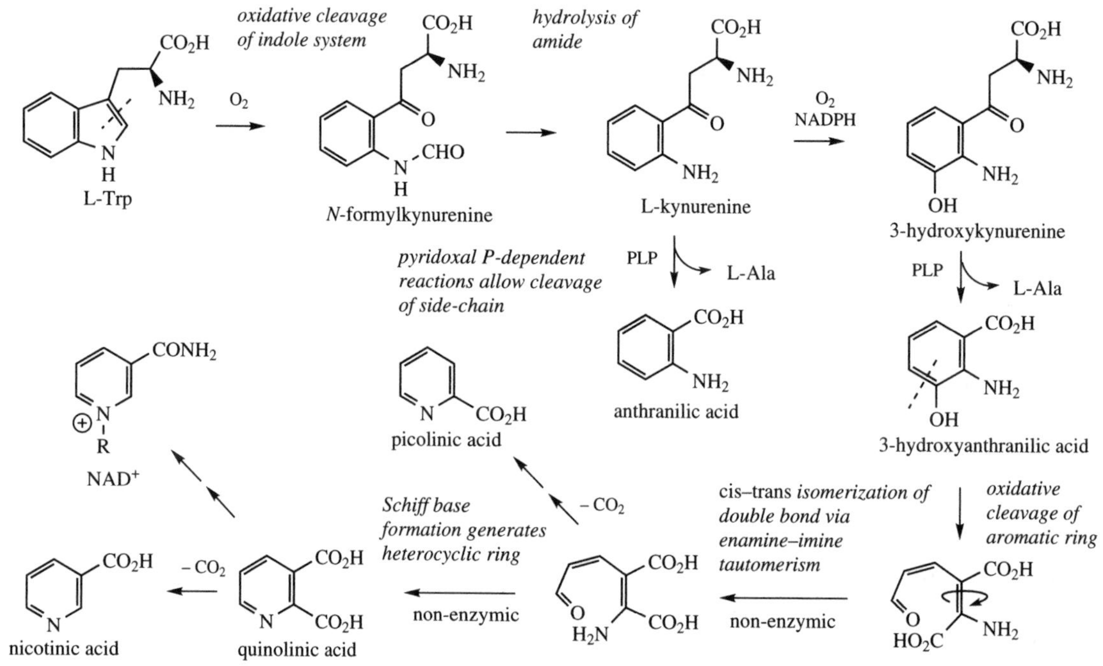

L-Trp

*oxidative cleavage
of indole system*

N-formylkynurenine

*hydrolysis of
amide*

L-kynurenine

3-hydroxykynurenine

*pyridoxal P-dependent
reactions allow cleavage
of side-chain*

anthranilic acid

3-hydroxyanthranilic acid

NAD$^+$

picolinic acid

*Schiff base
formation generates
heterocyclic ring*

*cis–trans isomerization of
double bond via
enamine–imine
tautomerism*

*oxidative
cleavage
of aromatic ring*

nicotinic acid

quinolinic acid

non-enzymic

non-enzymic

Figure 6.29

3-phospho-
glyceraldehyde

L-Asp

*dehydration and
dehydrogenation*

quinolinic acid

nicotinic acid

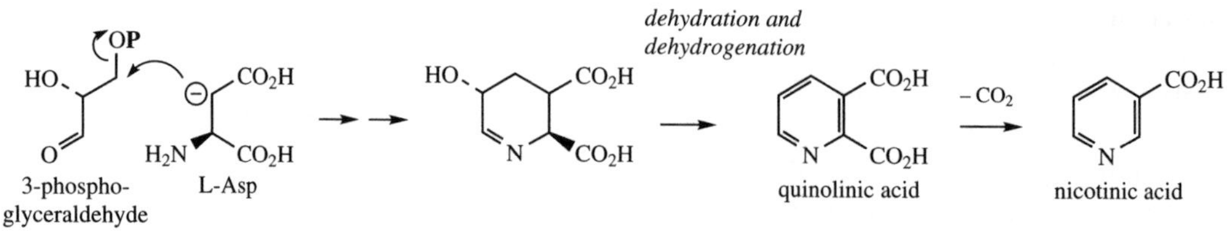

Figure 6.30

Vitamin B₃

Vitamin B₃ (nicotinic acid, niacin) (Figure 6.28) is a stable water-soluble vitamin widely distributed in foodstuffs, especially meat, liver, fish, wheat germ, and yeast. However, in some foods, e.g. maize, it may be present in a bound form, and is not readily available. Diets based principally on maize may lead to deficiencies. The amino acid tryptophan can be converted in the body into nicotinic acid (Figure 6.29), and may provide a large proportion of the requirements. Nicotinic acid is also produced during the roasting of coffee from the decomposition of the *N*-methyl derivative trigonelline (Figure 6.28). Nicotinic acid is converted into nicotinamide (Figure 6.28), though this compound also occurs naturally in many foods. The term vitamin B₃ is often used for the combined nicotinamide–nicotinic acid complement. In the form of the coenzymes NAD⁺ and NADP⁺, nicotinamide plays a vital role in oxidation–reduction reactions (see page 24), and is the most important electron carrier in primary metabolism. Deficiency in nicotinamide leads to pellagra, which manifests itself in diarrhoea, dermatitis, and dementia. Oral lesions and a red tongue may be more noticeable than the other symptoms. **Nicotinamide** is usually preferred over **nicotinic acid** for dietary supplements since there is less risk of gastric irritation. Both are produced synthetically. It is common practice to enrich many foods, including bread, flour, corn, and rice products.

Nicotinic acid in large doses can lower both cholesterol and triglyceride concentrations by inhibiting their synthesis.

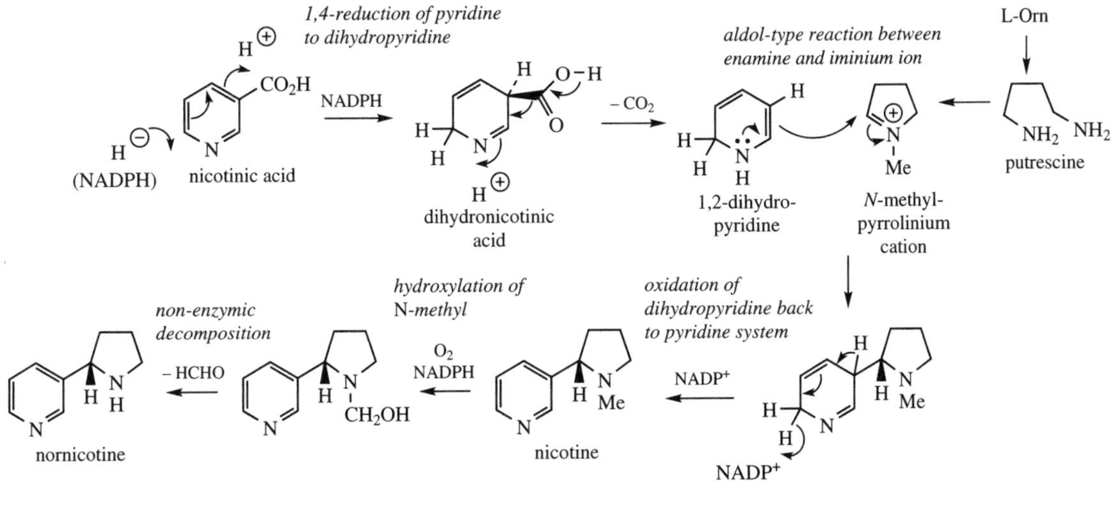

Figure 6.31

acid **quinolinic acid** features in both pathways, decarboxylation yielding nicotinic acid.

In the formation of **nicotine**, a pyrrolidine ring derived from ornithine, most likely as the *N*-methyl-Δ^1-pyrrolinium cation (see Figure 6.2) is attached to the pyridine ring of nicotinic acid, displacing the carboxyl during the sequence (Figure 6.31). A dihydronicotinic acid intermediate is likely to be involved allowing decarboxylation to the enamine 1,2-dihydropyridine.

This allows an aldol-type reaction with the *N*-methylpyrrolinium cation, and finally dehydrogenation of the dihydropyridine ring back to a pyridine gives nicotine. **Nornicotine** is derived by oxidative demethylation of nicotine. **Anabasine** is produced from nicotinic acid and lysine via the Δ^1-piperidinium cation in an essentially analogous manner (Figure 6.32). A subtle anomaly has been exposed in that a further *Nicotiana* alkaloid **anatabine** appears to be derived by combination of two nicotinic acid units, and the Δ^3-piperideine ring is *not* supplied by lysine (Figure 6.33).

Nicotinic acid undoubtedly provides the basic skeleton for some other alkaloids. **Ricinine** (Figure 6.35) is a 2-pyridone structure and contains a nitrile grouping, probably formed by dehydration of a nicotinamide derivative. This alkaloid is a toxic constituent of castor oil seeds (*Ricinus communis*; Euphorbiaceae), though the toxicity of the seeds results mainly from the polypeptide ricin (see page 434). **Arecoline** (Figure 6.36) is found in Betel nuts (*Areca catechu*: Palmae/Arecaceae)* and is a tetrahydronicotinic acid derivative. Betel nuts are chewed in India and Asia for the stimulant effect of arecoline.

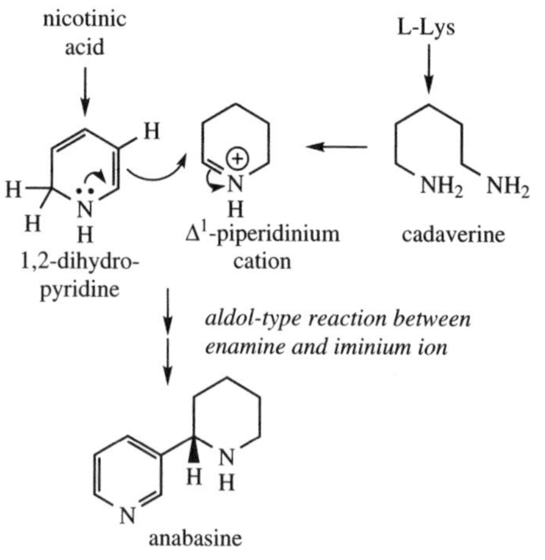

Figure 6.32

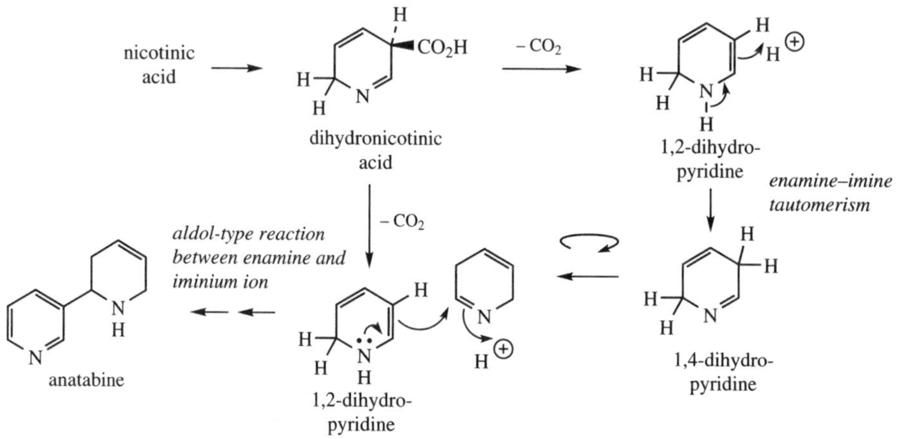

Figure 6.33

Tobacco

Tobacco is the cured and dried leaves of *Nicotiana tabacum* (Solanaceae), an annual herb indigenous to tropical America, but cultivated widely for smoking. Tobacco leaves may contain from 0.6–9% of (−)-nicotine (Figure 6.28), an oily, volatile liquid alkaloid, together with smaller amounts of structurally related alkaloids, e.g. anabasine and nornicotine (Figure 6.28). In the leaf, the alkaloids are typically present as salts with malic and citric acids. Nicotine in small doses can act as a respiratory stimulant, though in larger doses it causes respiratory depression. Despite the vast array of evidence linking tobacco smoking and cancer, the smoking habit continues throughout the world, and tobacco remains a major crop plant. Tobacco smoke contains a number of highly carcinogenic chemicals formed by incomplete combustion, including benzpyrene, 2-naphthylamine, and 4-aminobiphenyl. Metabolism by the body's P-450 system leads to further reactive intermediates, which can combine with DNA and cause mutations. Tobacco smoking also contributes to atherosclerosis, chronic

(Continues)

(Continued)

bronchitis, and emphysema, and is regarded as the single most preventable cause of death in modern society. **Nicotine** is being used by former smokers who wish to stop the habit. It is available in the form of chewing gum or nasal sprays, or can be absorbed transdermally from nicotine-impregnated patches.

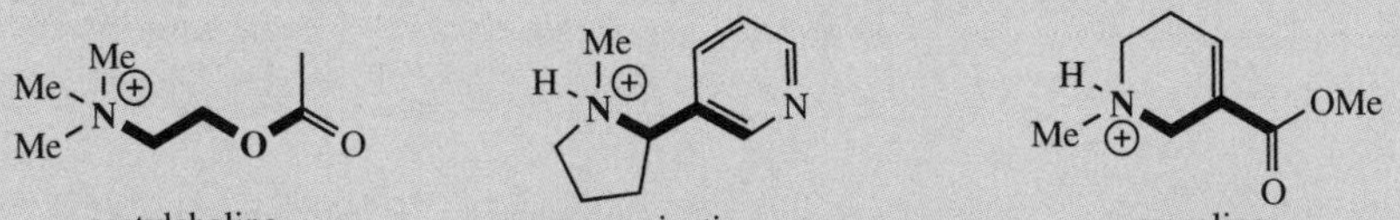

acetylcholine

nicotine
(as conjugate acid)

arecoline
(as conjugate acid)

Figure 6.34

Powdered tobacco leaves have long been used as an insecticide, and nicotine from *Nicotiana tabacum* or *N. rustica* has been formulated for agricultural and horticultural use. The free base is considerably more toxic than salts, and soaps were included in the formulations to ensure a basic pH and to provide a surfactant. Other *Nicotiana* alkaloids, e.g. anabasine and nornicotine, share this insecticidal activity. Although an effective insecticide, nicotine has been replaced by other agents considered to be safer. Nicotine is toxic to man due to its effect on the nervous system, interacting with the nicotinic acetylcholine receptors, though the tight binding observed is only partially accounted for by the structural similarity between acetylcholine and nicotine (Figure 6.34). Recent studies suggest that nicotine can improve memory by stimulating the transmission of nerve impulses, and this finding may account for the lower incidence of Alzheimer's disease in smokers. Any health benefits conferred by smoking are more than outweighed by the increased risk of heart, lung, and respiratory diseases.

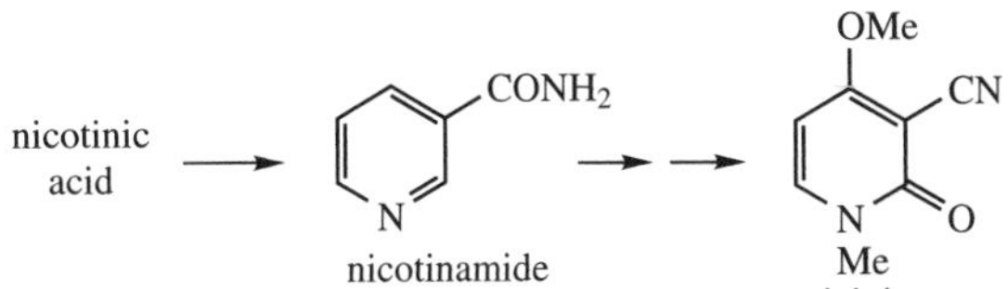

nicotinic acid

nicotinamide

ricinine

Figure 6.35

ALKALOIDS DERIVED FROM TYROSINE

Phenylethylamines and Simple Tetrahydroisoquinoline Alkaloids

PLP-dependent decarboxylation of L-**tyrosine** gives the simple phenylethylamine derivative

Areca

Areca nuts (betel nuts) are the seeds of *Areca catechu* (Palmae/Arecaceae), a tall palm cultivated in the Indian and Asian continents. These nuts are mixed with lime, wrapped in leaves of the betel pepper (*Piper betle*) and then chewed for their stimulant effect, and subsequent feeling of well-being and mild intoxication. The teeth and saliva of chewers stain bright red. The major stimulant alkaloid is arecoline (up to 0.2%) (Figure 6.36), the remainder of the alkaloid content (total about 0.45%) being composed of related reduced pyridine

(Continues)

structures, e.g. arecaidine, guvacine (tetrahydronicotinic acid), and guvacoline (Figure 6.36). Arecoline is an agonist for muscarinic acetylcholine receptors (see Figure 6.34), although it possesses a reversed ester profile compared with acetylcholine. **Arecoline** has been employed in veterinary practice as a vermicide to eradicate worms.

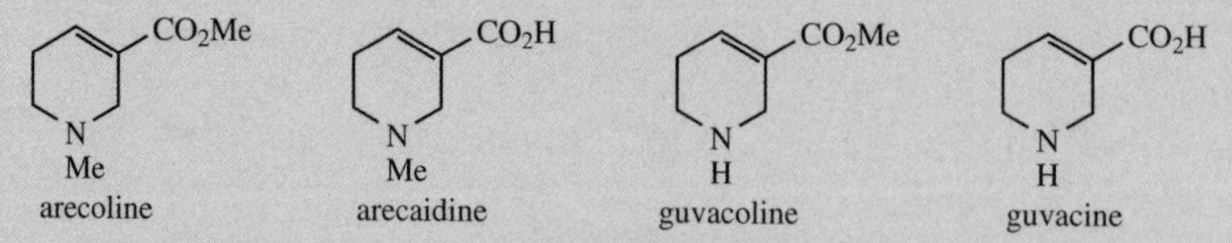

Figure 6.36

tyramine, which on di-*N*-methylation yields **hordenine**, a germination inhibitory alkaloid from barley (*Hordeum vulgare*; Graminae/Poaceae) (Figure 6.37). More commonly, phenylethylamine derivatives possess 3,4-di- or 3,4,5-tri-hydroxylation patterns, and are derived via **dopamine** (Figure 6.37), the decarboxylation product from L-**DOPA** (L-dihydroxyphenylalanine) (see page 129). Pre-eminent amongst these are the catecholamines* **noradrenaline** (**norepinephrine**), a mammalian neurotransmitter, and **adrenaline** (**epinephrine**), the 'fight or flight' hormone released in animals from the adrenal gland as a result of stress. These compounds are synthesized by successive β-hydroxylation and *N*-methylation reactions on dopamine (Figure 6.37). Aromatic hydroxylation and *O*-methylation reactions in the cactus *Lophophora williamsii** (Cactaceae) convert dopamine into **mescaline** (Figure 6.37), an alkaloid with pyschoactive and hallucinogenic properties. Note that the sequence of hydroxylations and methylations exactly parallel those described for the cinnamic acids (see page 131).

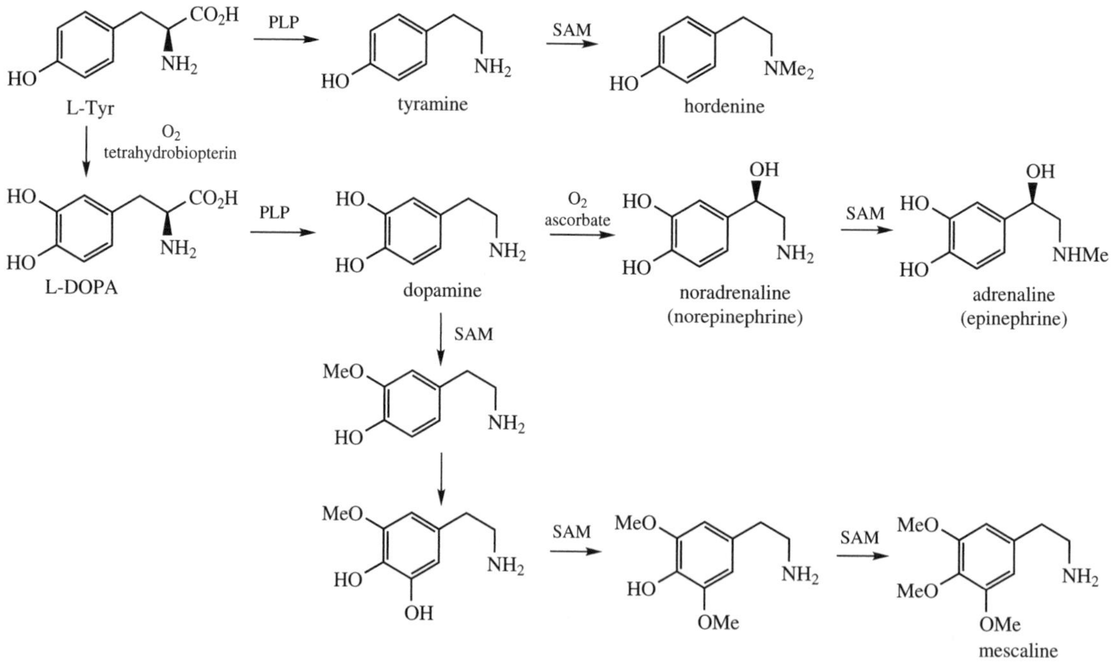

Figure 6.37

Catecholamines

The catecholamines dopamine, noradrenaline (norepinephrine), and adrenaline (epinephrine) are produced in the adrenal glands and nervous tissue and act as neurotransmitters in mammals. Several adrenergic receptors have been identified. α-Receptors are usually excitatory and produce a constricting effect on vascular, uterine, and intestinal muscles. β-Receptors are usually inhibitory on smooth muscle, but stimulatory on heart muscles. **Dopamine** (Figure 6.37) can act on both vascular α_1 and cardiac β_1 receptors, but also has its own receptors in several other structures. In Parkinson's disease, there is a deficiency of dopamine due to neural degeneration, affecting the balance between excitatory and inhibitory transmitters. Treatment with L-**DOPA** (**levodopa**) (Figure 6.37) helps to increase the dopamine levels in the brain. Unlike dopamine, DOPA can cross the blood–brain barrier, but needs to be administered with a DOPA-decarboxylase inhibitor, e.g. **carbidopa** (Figure 6.38), to prevent rapid decarboxylation in the bloodstream. Injections of dopamine or **dobutamine** (Figure 6.38) are valuable as cardiac stimulants in cases of cardiogenic shock. These agents act on β_1 receptors; **dopexamine** (Figure 6.38) is also used for chronic heart failure but acts on β_2 receptors in cardiac muscle.

Noradrenaline (norepinephrine) (Figure 6.37) is a powerful peripheral vasoconstrictor predominantly acting on α-adrenergic receptors, and is useful in restoring blood pressure in cases of acute hypotension. The structurally related alkaloid **ephedrine** (see page 384) may be used in the same way, and synthetic analogues of noradrenaline, e.g. **phenylephrine**, **methoxamine**, and **metaraminol** (Figure 6.38), have also been developed. **Methyldopa** is used to treat hypertension; it is a centrally acting agent that becomes decarboxylated and hydroxylated to form the false transmitter α-methylnoradrenaline, which competes with noradrenaline.

Adrenaline (epinephrine) (Figure 6.37) is released from the adrenal glands when an animal is confronted with an emergency situation, markedly stimulating glycogen breakdown in muscle, increasing respiration, and triggering catabolic processes that result in energy release. Adrenaline interacts with both α- and β- receptors, an α-response being vasoconstriction of smooth muscle in the skin. β-Responses include mediation of cardiac muscle contractions and the relaxation of smooth muscle in the bronchioles of the lung. Injection of adrenaline is thus of value in cases of cardiac arrest, or in allergic emergencies such as bronchospasm or severe allergy (anaphylactic shock). It is not effective orally. A wide range of cardioactive β-adrenoceptor blocking agents (**beta-blockers**) has been developed to selectively bind

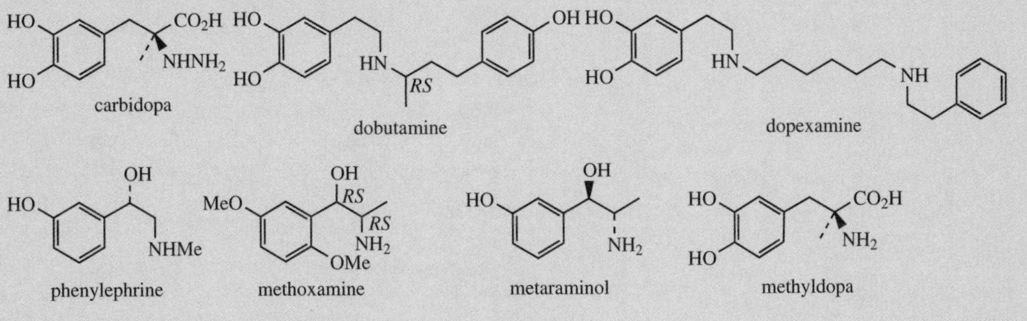

Figure 6.38

(Continues)

(*Continued*)

to β-receptors to control the rate and force of cardiac contractions in the management of hypertension and other heart conditions. The prototype of the beta-blocker drugs is **propranolol** (Figure 6.39), in which the catechol ring system has been modified to a

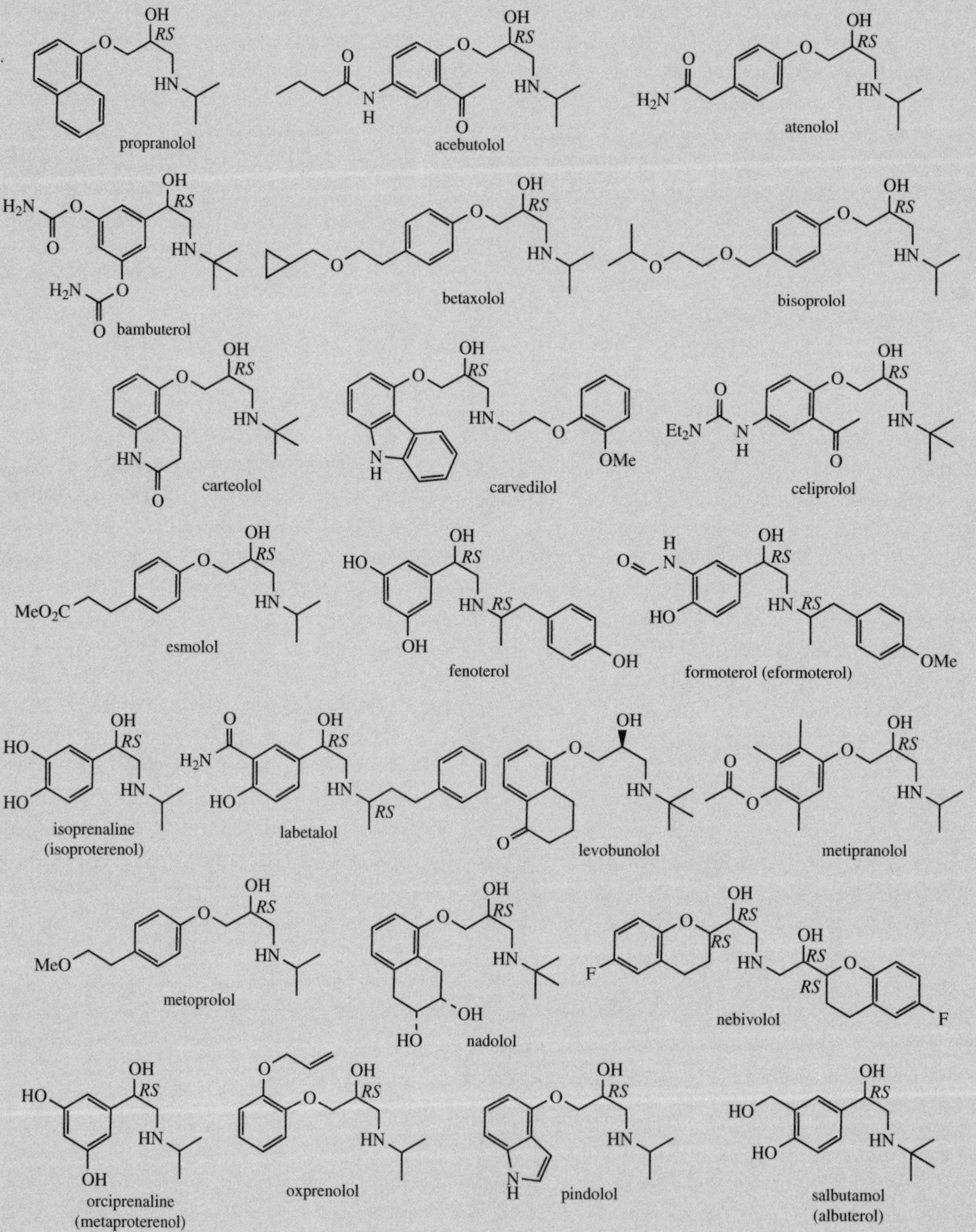

propranolol

acebutolol

atenolol

bambuterol

betaxolol

bisoprolol

carteolol

carvedilol

celiprolol

esmolol

fenoterol

formoterol (eformoterol)

isoprenaline (isoproterenol)

labetalol

levobunolol

metipranolol

metoprolol

nadolol

nebivolol

orciprenaline (metaproterenol)

oxprenolol

pindolol

salbutamol (albuterol)

Figure 6.39   (*continues*)

(*Continues*)

(Continued)

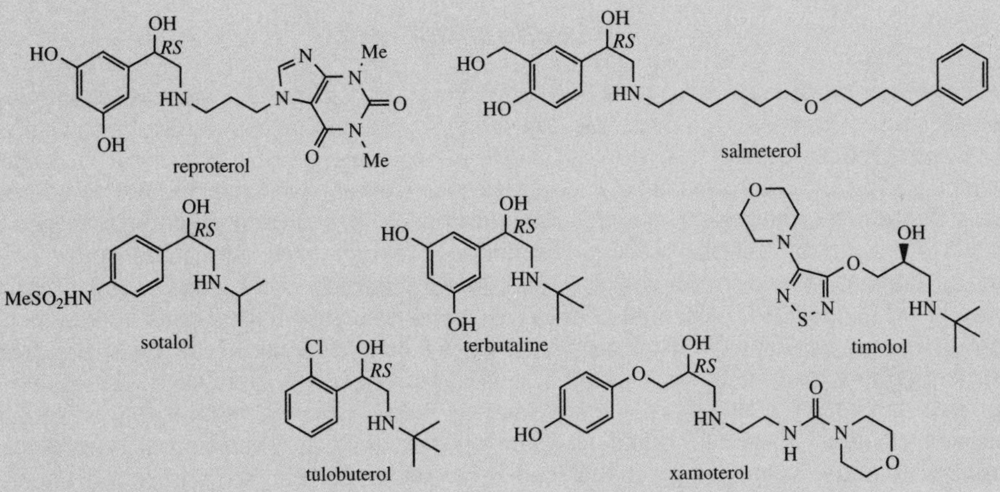

Figure 6.39 (*continued*)

naphthalene ether, and a bulky *N*-alkyl substituent has been incorporated. Many structural variants have been produced and there is now a huge, perhaps bewildering, variety of beta-blockers in regular use, with subtle differences in properties and action affecting the choice of drug for a particular condition or individual patient. These are shown in Figure 6.39. **Atenolol, betaxolol, bisoprolol, metoprolol, nebivolol**, and to a lesser extent **acebutolol**, have less effect on the β_2 bronchial receptors and are thus relatively cardioselective. Most other agents are non-cardioselective, and could also provoke breathing difficulties. **Esmolol** and **sotalol** are used only in the management of arrhythmias.

Other β-agonists are valuable as antiasthmatic drugs. Important examples include **salbutamol (albuterol)** and **terbutaline**, which are very widely prescribed, principally for administration by inhalation at the onset of an asthma attack, but, as with cardioactive beta-blockers, a wide range of agents is in current use (Figure 6.39). These agents are mainly selective towards the β_2-receptors, and supersede the earlier less selective bronchodilator drugs such as **isoprenaline (isoproterenol)** and **orciprenaline (metaproterenol)** (Figure 6.39). Topical application of a beta-blocker to the eye reduces intraocular pressure by reducing the rate of production of aqueous humour. Some drugs in this class, namely **betaxolol, carteolol, levobunolol, metipranolol**, and **timolol**, are thus useful in treating glaucoma. **Propranolol, metoprolol, nadolol**, and **timolol** also have additional application in the prophylaxis of migraine.

Catecholamine neurotransmitters are subsequently inactivated by enzymic methylation of the 3-hydroxyl (via catechol-*O*-methyltransferase) or by oxidative removal of the amine group via monoamine oxidase. Monoamine oxidase inhibitors are sometimes used to treat depression, and these drugs cause an accumulation of amine neurotransmitters. Under such drug treatment, simple amines such as tyramine in cheese, beans, fish, and yeast extracts are also not metabolized and can cause dangerous potentiation of neurotransmitter activity.

(Continues)

(Continued)

Lophophora

Lophophora or peyote consists of the dried sliced tops of *Lophophora williamsii* (Cactaceae), a small cactus from Mexico and the SW United States. The plant has been used by the Aztecs and since by the Mexican Indians for many years, especially in religious ceremonies to produce hallucinations and establish contact with the gods. The so-called mescal buttons were ingested and this caused unusual and bizarre coloured images. The plant is still used by people seeking drug-induced experiences. The most active of the range of alkaloids found in lophophora (total 8–9% alkaloids in the dried mescal buttons) is mescaline (Figure 6.37), a simple phenylethylamine derivative. Other constituents include anhalamine, anhalonidine, and anhalonine (Figure 6.40). **Mescaline** has been used as a hallucinogen in experimental psychiatry. The dosage required is quite large (300–500 mg), but the alkaloid can readily be obtained by total synthesis, which is relatively uncomplicated. Mescaline is also found in other species of cactus, e.g. *Trichocereus pachanoi*, a substantially larger columnar plant that can grow up to 20 feet tall, and found mainly in the Andes.

Closely-related alkaloids cooccurring with mescaline are **anhalamine**, **anhalonine**, and **anhalonidine** (Figure 6.40), which are representatives of simple tetrahydroisoquinoline derivatives.

The additional carbon atoms, two in the case of anhalonidine and anhalonine, and one for anhalamine, are supplied by pyruvate and glyoxylate respectively. In each case, a carboxyl group is

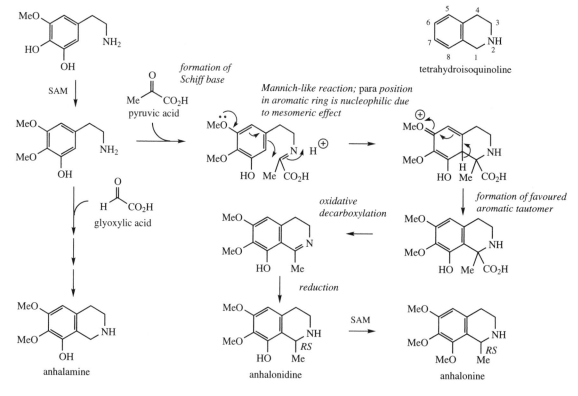

Figure 6.40

lost from this additional precursor. The keto acid pyruvate reacts with a suitable phenylethylamine, in this case the dimethoxy-hydroxy derivative, giving a Schiff base (Figure 6.40). In a Mannich-like mechanism, cyclization occurs to generate the isoquinoline system, the mesomeric effect of an oxygen substituent providing the nucleophilic site on the aromatic ring. Restoration of aromaticity via proton loss gives the tetrahydroisoquinoline, overall a biosynthetic equivalent of the Pictet–Spengler synthesis. The carboxyl group is then removed, not by a simple decarboxylation, but via an unusual oxidative decarboxylation first generating the intermediate imine, reduction finally leading to **anhalonidine** with further methylation giving **anhalonine**. **Anhalamine** is derived from the same phenylethylamine precursor utilizing glyoxylic acid (Figure 6.40).

The chemical synthesis of tetrahydroisoquinolines by the Pictet–Spengler reaction does not usually employ keto acids like pyruvate or aldehyde acids like glyoxylate. Instead, simple aldehydes, e.g. acetaldehyde or formaldehyde, could be used (Figure 6.41, a), giving the same product directly without the need for a decarboxylation step to convert the intermediate tetrahydroisoquinolinecarboxylic acid (Figure 6.41, b). In nature, both routes are in fact found to operate, depending on the complexity of the R group. Thus, the keto acid (route b) is used for relatively simple substrates (R = H, Me) whilst more complex precursors (R = ArCH_2, ArCH_2CH_2, etc) are incorporated via the corresponding aldehydes (route a). The stereochemistry in the product is thus controlled by the condensation/Mannich reactions (route a), or by the final reduction reaction (route b). Occasionally, both types of transformation have been demonstrated in the production of a single compound, an example being the *Lophophora schotti* alkaloid **lophocerine** (Figure 6.42). This requires

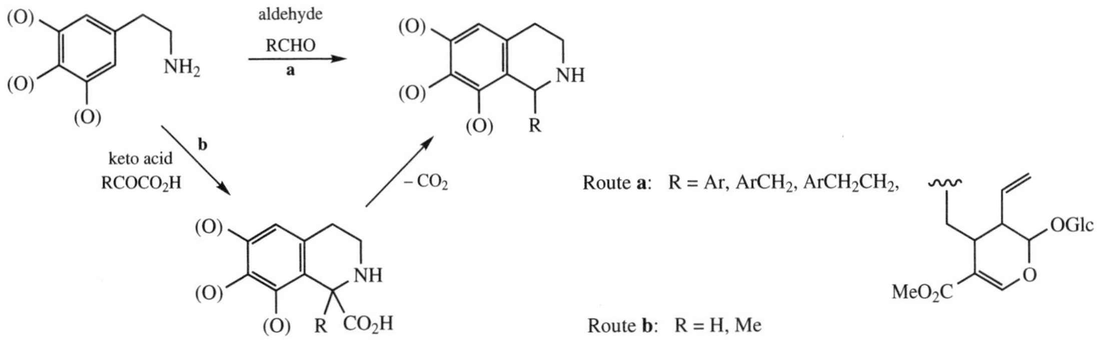

Route a: R = Ar, ArCH_2, ArCH_2CH_2,

Route b: R = H, Me

Figure 6.41

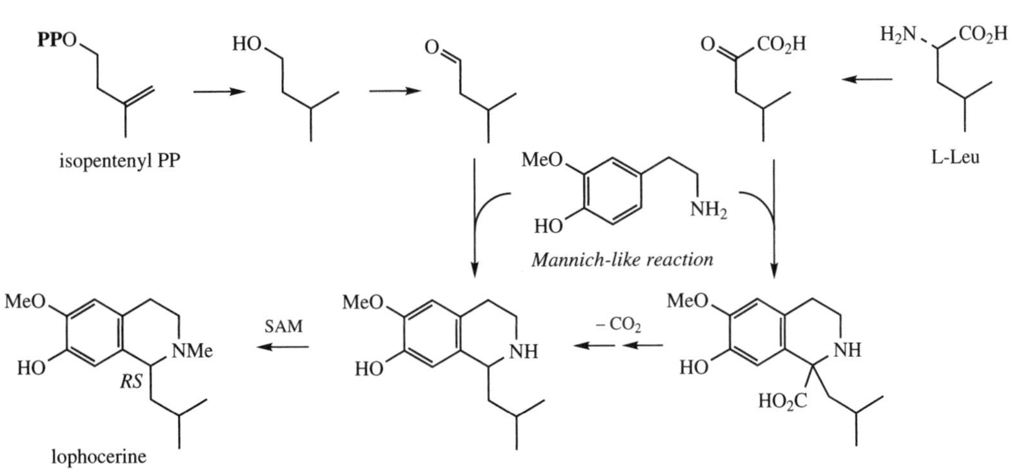

Figure 6.42

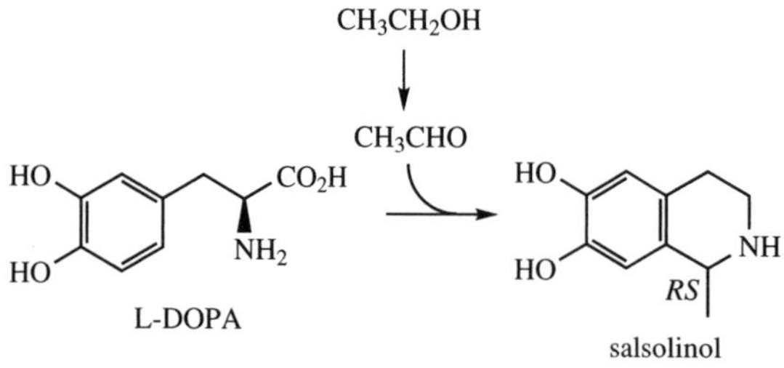

Figure 6.43

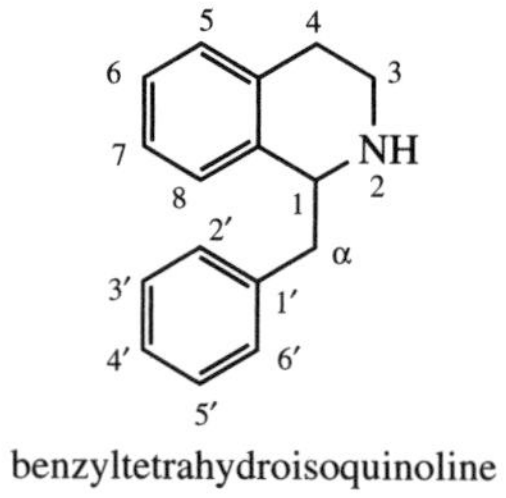

Figure 6.44

utilization of a C_5 isoprene unit, incorporated via an aldehyde. However, a second route using the keto acid derived from the amino acid L-leucine by transamination has also been demonstrated. The

alkaloid **salsolinol** (Figure 6.43) is found in plants, e.g. *Corydalis* spp. (Papaveraceae), but can also be detected in the urine of humans as a product from dopamine and acetaldehyde combining via

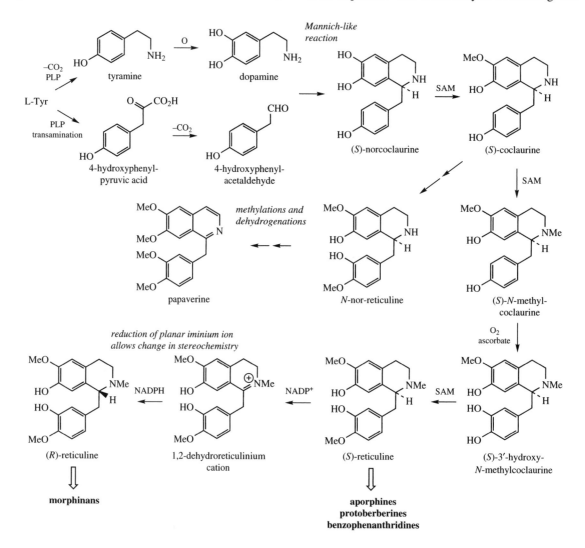

Figure 6.45

a Pictet–Spengler reaction. Acetaldehyde is typically formed after ingestion of ethanol.

Incorporation of a phenylethyl unit into the phenylethylamine gives rise to a benzyltetrahydroisoquinoline skeleton (Figure 6.44), which can undergo further modifications to produce a wide range of plant alkaloids, many of which feature as important drug materials. Fundamental changes to the basic skeleton increase the diversity of structural types as described under 'modified benzyltetrahydroisoqinolines'. Most examples of benzyltetrahydroisoquinoline alkaloids and modified structures contain *ortho* di-oxygenation in each aromatic ring, which pattern is potentially derivable from the utilization of two DOPA molecules. Although two tyrosine molecules are used in the biosynthetic pathway, only the phenylethylamine fragment of the tetrahydroisoquinoline ring system is formed via DOPA, the remaining carbons coming from tyrosine via 4-hydroxyphenylpyruvic acid and 4-hydroxyphenylacetaldehyde (Figure 6.45). The product from the Mannich-like reaction is thus the trihydroxy alkaloid **norcoclaurine**, formed stereospecifically as the (*S*)-enantiomer. The tetrahydroxy substitution pattern is built up by further hydroxylation in the benzyl ring, though *O*-methylation [giving (*S*)-**coclaurine**] and *N*-methylation steps precede this. Eventually,

(*S*)-**reticuline**, a pivotal intermediate to other alkaloids, is attained by *N*-methylation. Surprisingly, some alkaloids, such as the opium alkaloids morphine, codeine, and thebaine (see page 327) are elaborated from (*R*)-reticuline rather than the first-formed (*S*)-isomer. The change in configuration is known to be achieved by an oxidation–reduction process and the intermediate 1,2-dehydroreticulinium ion, as shown in Figure 6.45. **Papaverine**, a benzylisoquinoline alkaloid found in opium (see page 331), is formed from *N*-nor-reticuline by successive *O*-methylations and oxidation in the heterocyclic ring (Figure 6.45).

Structures in which two (or more) benzyltetrahydroisoquinoline units are linked together are readily explained by a phenolic oxidative coupling mechanism (see page 28). Thus, **tetrandrine** (Figure 6.46), a bis-benzyltetrahydroisoquinoline alkaloid isolated from *Stephania tetrandra* (Menispermaceae) is easily recognized as a coupling product from two molecules of (*S*)-*N*-methylcoclaurine (Figure 6.46). The two diradicals, formed by one-electron oxidations of a free phenol group in each ring, couple to give ether bridges, and the product is then methylated to tetrandrine. The pathway is much more likely to follow a stepwise coupling process requiring two oxidative enzymes rather

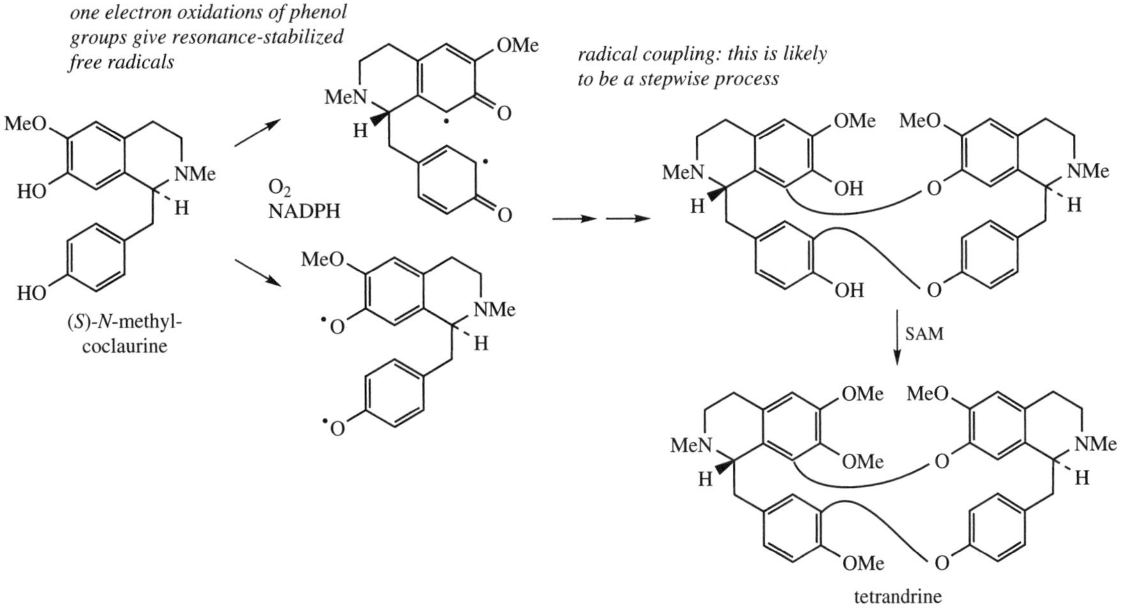

Figure 6.46

than the combined one suggested in Figure 6.46. Tetrandrine is currently of interest for its ability to block calcium channels, and may have applications in the treatment of cardiovascular disorders. By a similar mechanism, **tubocurarine** (Figure 6.47),

the principal active component in the arrow poison curare* from *Chondrodendron tomentosum* (Menispermaceae), can be elaborated by a different coupling of one molecule each of (S)- and (R)-N-methylcoclaurine (Figure 6.47).

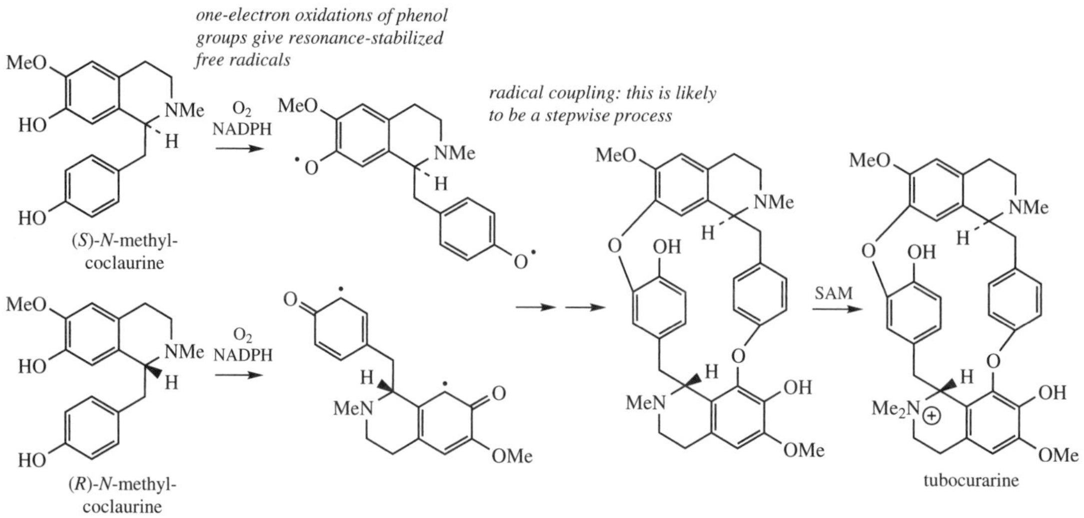

Figure 6.47

Curare

Curare is the arrow poison of the South American Indians, and it may contain as many as 30 different plant ingredients, which may vary widely from tribe to tribe according to local custom. Curare is prepared in the rain forests of the Amazon and Orinoco, and represents the crude dried extract from the bark and stems of various plants. The young bark is scraped off, pounded, and the fibrous mass percolated with water in a leaf funnel. The liquor so obtained is then concentrated by evaporation over a fire. Further vegetable material may be added to make the preparation more glutinous so that it will stick to the arrows or darts. The product is dark brown or black, and tarlike.

In the 1880s, it was found that the traditional container used for curare was fairly indicative of the main ingredients that had gone into its preparation. Three main types were distinguished. Tube curare was packed in hollow bamboo canes, and its principal ingredient was the climbing plant *Chondrodendron tomentosum* (Menispermaceae). Calabash curare was packed in gourds, and was derived from *Strychnos toxifera* (Loganiaceae). Pot curare was almost always derived from a mixture of loganiaceous and menispermaceous plants, and was packed in small earthenware pots. Current supplies of curare are mainly of the menispermaceous type, i.e. derived from *Chondrodendron*.

The potency of curare as an arrow poison is variable and consequently needs testing. A frequently quoted description of this testing is as follows: 'If a monkey hit by a dart is only able to get from one tree to the next before it falls dead, this is "one-tree curare", the superior grade. "Two-tree curare" is less satisfactory, and "three-tree curare" is so weak that it can be used to bring down live animals that the Indians wish to keep in captivity.' Thus, the poison

(Continues)

(Continued)

does not necessarily cause death; it depends on the potency. Curare is only effective if it enters the bloodstream, and small amounts taken orally give no ill effects provided there are no open sores in the mouth or throat.

Curare kills by producing paralysis, a limp relaxation of voluntary muscles. It achieves this by competing with acetylcholine at nicotinic receptor sites (see page 299), thus blocking nerve impulses at the neuromuscular junction. Death occurs because the muscles of respiration cease to operate, and artificial respiration is an effective treatment prior to the effects gradually wearing off through normal metabolism of the drug. Anticholinesterase drugs such as physostigmine and neostigmine are specific antidotes for moderate curare poisoning. Curare thus found medicinal use as a muscle relaxant, especially in surgical operations such as abdominal surgery, tonsillectomy, etc, where tense muscles needed to be relaxed. Curare was also found to be of value in certain neurological conditions, e.g. multiple sclerosis, tetanus, and Parkinson's disease, to temporarily relax rigid muscles and control convulsions, but was not a curative. However, the potency of curare varied markedly, and supplies were sometimes limited.

The alkaloid content of curare is from 4% to 7%. The most important constituent in menispermaceous curare is the bis-benzyltetrahydroisoquinoline alkaloid (+)-tubocurarine (Figure 6.48). This is a monoquaternary ammonium salt, and is water soluble. Other main alkaloids include non-quaternary dimeric structures, e.g. isochondrodendrine and curine (bebeerine) (Figure 6.48), which appear to be derived from two molecules of (R)-N-methylcoclaurine, with the former also displaying a different coupling mode. The constituents in loganiaceous curare (from calabash curare, i.e. *Strychnos toxifera*) are even more complex, and a series of 12 quaternary dimeric strychnine-like alkaloids has been identified, e.g. C-toxiferine (toxiferine-1) (see page 359).

Tubocurarine (Figure 6.48) is still extracted from menispermaceous curare and injected as a muscle relaxant in surgical operations, reducing the need for deep anaesthesia. Artificial respiration is required until the drug has been inactivated (about 30 minutes) or antagonized (e.g. with neostigmine). The limited availability of tubocurarine has led to the development of a series of synthetic analogues, some of which have improved characteristics and are now

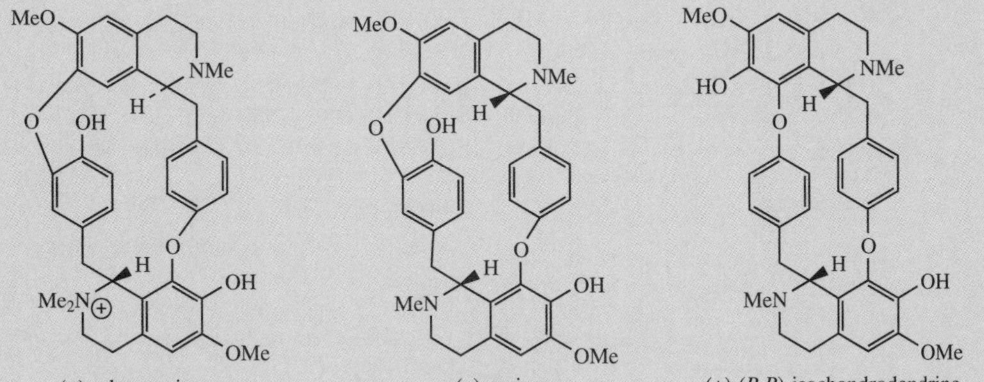

(+)-tubocurarine

(−)-curine
[(−)-(R,R)-bebeerine]

(+)-(R,R)-isochondrodendrine

Figure 6.48

(Continues)

326   ALKALOIDS

(Continued)

preferred over the natural product. Interestingly, the structure of tubocurarine was originally formulated incorrectly as a diquaternary salt, rather than the monoquaternary salt, and analogues were based on the pretext that curare-like effects might be obtained from compounds containing two quaternary nitrogens separated by a polymethylene chain. This was borne out in practice, and the separation was found to be optimal at about ten carbons.

Decamethonium (Figure 6.49) was the first synthetic curare-like muscle relaxant, but has since been superseded. In tubocurarine, the two nitrogens are also separated by ten atoms, and at physiological pHs it is likely that both centres will be positively charged. Obviously, the interatomic distance (1.4 nm in tubocurarine) is very dependent on the structure and stereochemistry rather than just the number of atoms separating the centres, but an extended conformation of decamethonium approximates to this distance. **Suxamethonium**

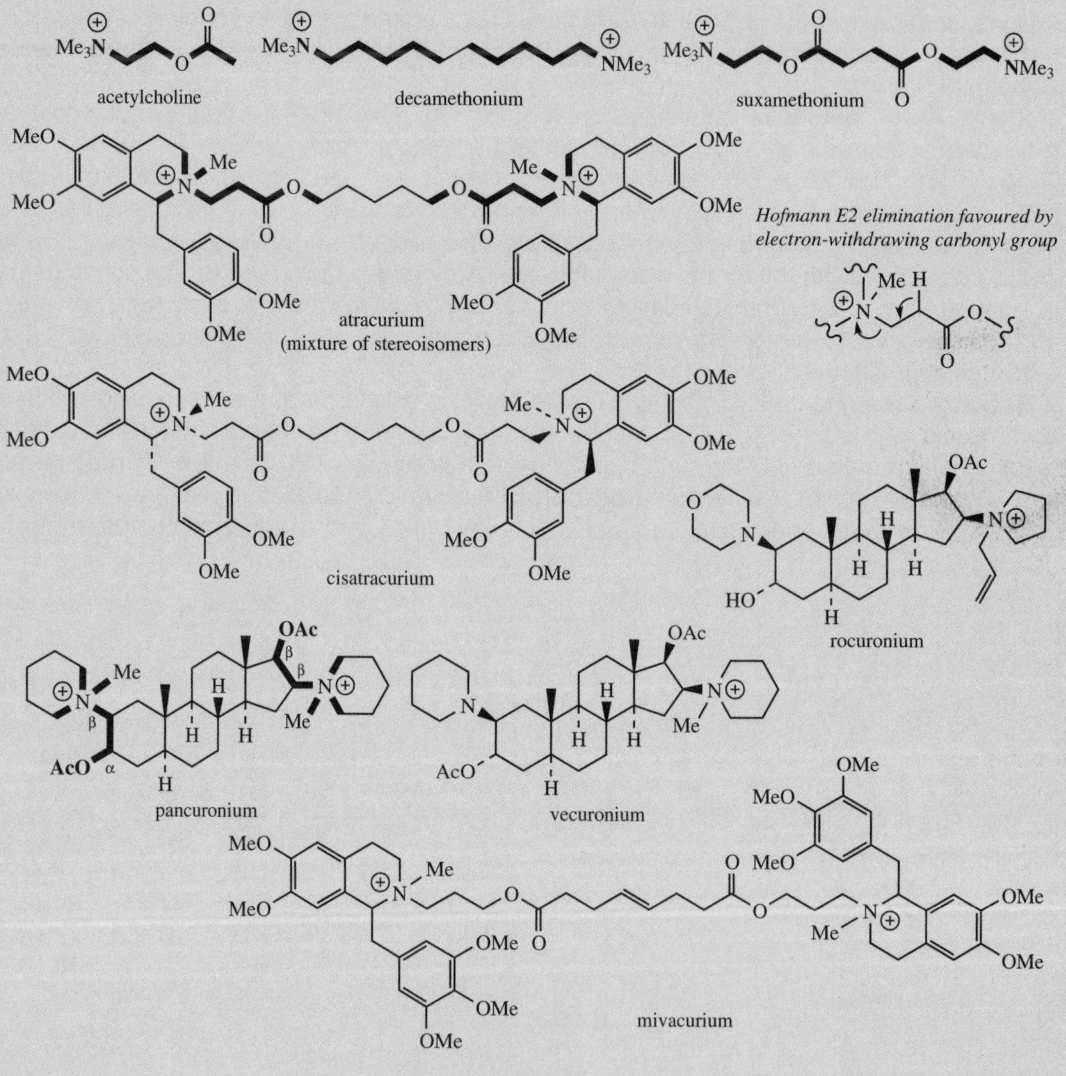

Hofmann E2 elimination favoured by electron-withdrawing carbonyl group

Figure 6.49

(Continues)

(Continued)

(Figure 6.49) is an effective agent with a very short duration of action, due to the two ester functions, which are rapidly metabolized by an esterase (a pseudocholinesterase) in the body, and this means the period during which artificial respiration is required is considerably reduced. It also has ten-atom separation between the quaternary nitrogens. **Atracurium** (Figure 6.49) is a recent development, containing two quaternary nitrogens in benzyltetrahydroisoquinoline structures separated by 13 atoms. In addition to enzymic ester hydrolysis, atrocurium is also degraded in the body by non-enzymic E2 Hofmann elimination (Figure 6.49), which is independent of liver or kidney function. Normally, this elimination would require strongly alkaline conditions and a high temperature, but the presence of the carbonyl group increases the acidity and thus facilitates loss of the proton, and the elimination can proceed readily under physiological conditions, giving atracurium a half life of about 20 minutes. This is particularly valuable where patients have low or atypical pseudocholinesterase enzymes. Atracurium contains four chiral centres (including the quaternary nitrogens) and is supplied as a mixture of stereoisomers; the single isomer **cisatracurium** has now been introduced. This isomer is more potent than the mixture, has a slightly longer duration of action, and produces fewer cardiovascular side-effects. **Mivacurium** (Figure 6.49) has similar benzyltetrahydroisoquinoline structures to provide the quaternary centres, but the separation has now increased to 16 atoms. In **pancuronium**, separation of the two quaternary centres is achieved by a steroidal skeleton. This agent is about five times as potent as tubocurarine. **Vecuronium** is the equivalent monoquaternary structure, and has the fewest side-effects. **Rocuronium** is also based on a steroidal skeleton, and provides rapid action with no cardiovascular effects.

The toxiferines (see Figure 6.85, page 359) also share the diquaternary character. **Alcuronium** is a semi-synthetic skeletal muscle relaxant containing the dimeric strychnine-like structure and is produced chemically from C-toxiferine (see page 360).

These neuromuscular blocking agents act by occupying nicotinic acetylcholine (Figure 6.49) receptor sites. All the structures have two acetylcholine-like portions, which can interact with the receptor. Where these are built into a rigid framework, e.g. tubocurarine and pancuronium, the molecule probably spans and blocks several receptor sites. Tubocurarine and the heterocyclic analogues are termed non-depolarizing or competitive muscle relaxants. The straight chain structures, e.g. decamethonium and suxamethonium, initially mimic the action of acetylcholine but then persist at the receptor, and are termed depolarizing blocking agents. Thus they trigger a response, a brief contraction of the muscle, which is then followed by a prolonged period of muscular paralysis until the compound is metabolized.

Modified Benzyltetrahydroisoquinoline Alkaloids

The concept of phenolic oxidative coupling is a crucial theme in modifying the basic benzyltetrahydroisoquinoline skeleton to many other types of alkaloid. Tetrandrine (Figure 6.46) and tubocurarine (Figure 6.47) represent coupling of two benzyltetrahydroisoquinoline molecules by ether bridges, but this form of coupling is perhaps less frequent than that involving carbon–carbon bonding between aromatic rings. The principal

opium* alkaloids **morphine**, **codeine**, and **thebaine** (Figure 6.50) are derived by this type of coupling, though the subsequent reduction of one aromatic ring to some extent disguises their benzyltetrahydroisoquinoline origins. (R)-**Reticuline** (Figure 6.45) is firmly established as the precursor of these morphinan alkaloids.

(R)-Reticuline, rewritten as in Figure 6.50, is the substrate for one-electron oxidations via the phenol group in each ring, giving the diradical. Coupling *ortho* to the phenol group in the tetrahydroisoquinoline, and *para* to the phenol in the benzyl substituent, then yields the dienone

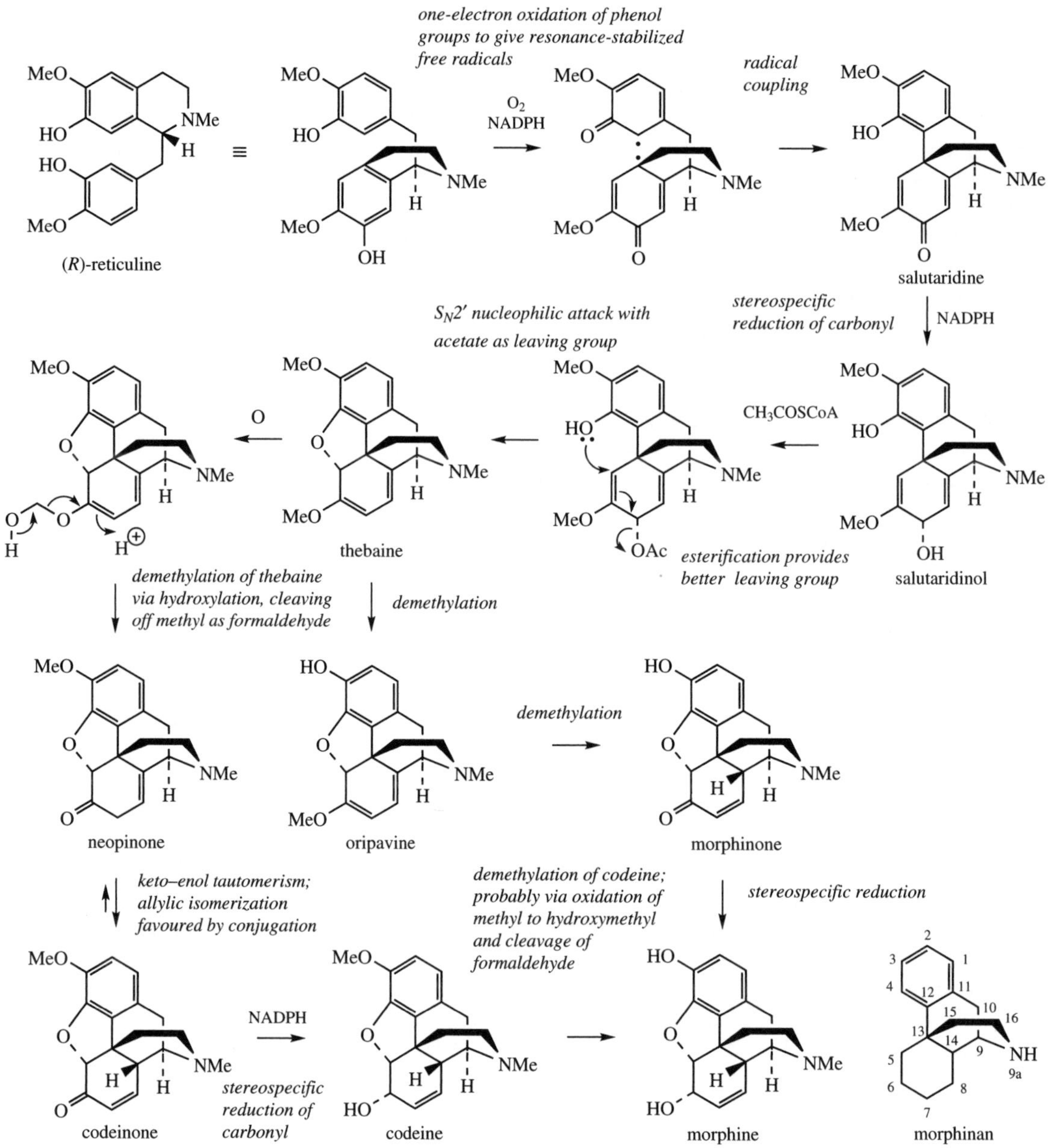

one-electron oxidation of phenol
groups to give resonance-stabilized
free radicals

radical
coupling

O_2
NADPH

(R)-reticuline

salutaridine

S_N2' nucleophilic attack with
acetate as leaving group

stereospecific
reduction of carbonyl

NADPH

$CH_3COSCoA$

thebaine

esterification provides
better leaving group

salutaridinol

demethylation of thebaine
via hydroxylation, cleaving
off methyl as formaldehyde

demethylation

demethylation

neopinone

oripavine

morphinone

keto–enol tautomerism;
allylic isomerization
favoured by conjugation

demethylation of codeine;
probably via oxidation of
methyl to hydroxymethyl
and cleavage of
formaldehyde

stereospecific reduction

NADPH

stereospecific
reduction of
carbonyl

codeinone

codeine

morphine

morphinan

Figure 6.50

salutaridine, found as a minor alkaloid constituent in the opium poppy *Papaver somniferum* (Papaveraceae). Only the original benzyl aromatic ring can be restored to aromaticity, since the tetrahydroisoquinoline fragment is coupled *para* to the phenol function, a position which is already substituted. The alkaloid **thebaine** is obtained by way of **salutaridinol**, formed from salutaridine

by stereospecific reduction of the carbonyl group. Ring closure to form the ether linkage in thebaine would be the result of nucleophilic attack of the phenol group on to the dienol system and subsequent displacement of the hydroxyl. This cyclization step can be demonstrated chemically by treatment of salutaridinol with acid. *In vivo*, however, an additional reaction is used to improve the nature of the leaving group, and this is achieved by acetylation with acetyl-CoA. The cyclization then occurs readily, and without any enzyme participation. Subsequent reactions involve conversion of thebaine into **morphine** by way of **codeine**, a process which modifies the oxidation state of the diene ring, but most significantly removes two O-methyl groups. One is present as an enol ether, removal generating **neopinone**, which gives **codeinone** and then codeine by allylic isomerization and reduction respectively. The last step, demethylation of the phenol ether codeine to the phenol morphine, is the type of reaction only achievable in the laboratory by the use of powerful and reactive demethylating agents, e.g. HBr or BBr$_3$. Because of the other functional groups present, chemical conversion of codeine into morphine is not usually a satisfactory process. However, the enzyme-mediated conversion in *P. somniferum* proceeds smoothly and efficiently. The enzymic demethylations of both the enol ether and the phenol ether probably involve initial hydroxylation followed by loss of the methyl groups as formaldehyde (Figure 6.50).

The involvement of these O-demethylation reactions is rather unusual; secondary metabolic pathways tend to increase the complexity of the product by adding methyls rather than removing them. In this pathway, it is convenient to view the methyl groups in reticuline as protecting groups, which reduce the possible coupling modes available during the oxidative coupling process, and these groups are then removed towards the end of the synthetic sequence. There is also some evidence that the later stages of the pathway in Figure 6.50 are modified in some strains of opium poppy. In such strains, thebaine is converted by way of **oripavine** and **morphinone**, this pathway removing the phenolic O-methyl before that of the enol ether, i.e. carrying out the same steps but in a different order. The enzymic transformation of thebaine into morphine, and the conversion of (R)-reticuline into salutaridinol have also been observed in mammalian tissues, giving strong evidence that the trace amounts of morphine and related alkaloids which can sometimes be found in mammals are actually of endogenous origin rather than dietary.

Opium

Opium is the air-dried milky exudate, or latex, obtained by incising the unripe capsules of the opium poppy *Papaver somniferum* (Papaveraceae). The plant is an annual herb with large solitary flowers, of white, pink, or dull red-purple colour. For opium production, the ripening capsules, which are just changing colour from blue-green to yellow, are carefully incised with a knife to open the latex tubes, but not to cut through to the interior of the capsule. These latex tubes open into one another, so it is not necessary to incise them all. Cuts are made transversely or longitudinally according to custom. The initially white milky latex quickly oozes out, but rapidly turns brown and coagulates. This material, the raw opium, is then removed early the following morning, being scraped off and moulded into balls or blocks. Typically, these are wrapped in poppy leaves and shade-dried. The blocks may be dusted with various plant materials to prevent cohering. Fresh opium is pale to dark brown and plastic, but it becomes hard and brittle when stored.

Opium has been known and used for 4000 years or more. In recent times, attempts have been made at governmental and international levels to control the cultivation of the opium poppy, but with only limited success. In endeavours to reduce drug problems involving opium-derived materials, especially heroin, where extremely large profits can be made from smuggling relatively small amounts of opium, much pharmaceutical production has been replaced by the processing of the bulkier 'poppy straw'. The entire plant tops are harvested and dried, then extracted for their alkaloid content in the pharmaceutical industry. Poppy straw now accounts for most of the medicinal opium alkaloid production, but there is still

(Continues)

(Continued)

considerable trade in illicit opium. In addition to opium, the opium poppy yields seeds, which are used in baking and are also pressed to give poppy seed oil. The remaining seed cake is used as cattle feed, and it is held that these poppy seed products cover all the growing expenses, with opium providing the profit. Poppy seeds do not contain any significant amounts of alkaloids.

The main producer of medicinal opium is India, whilst poppy straw is cultivated in Turkey, Russia, and Australia. Opium destined for the black market originates from the Golden Triangle (Burma, Laos, and Thailand), the Golden Crescent (Iran, Pakistan, and Afghanistan), and Mexico.

Crude opium has been used since antiquity as an analgesic, sleep-inducer (narcotic), and for the treatment of coughs. It has been formulated in a number of simple preparations for general use, though these are now uncommon. Laudanum, or opium tincture, was once a standard analgesic and narcotic mixture. Paregoric, or camphorated opium tincture, was used in the treatment of severe diarrhoea and dysentery, but is still an ingredient in the cough and cold preparation Gee's linctus. In Dover's powder, powdered opium was combined with powdered ipecacuanha (see page 344) to give a popular sedative and diaphoretic (promotes perspiration) to take at the onset of colds and influenza. Opium has traditionally been smoked for pleasure, but habitual use develops a craving for the drug followed by addiction. An unpleasant abstinence syndrome is experienced if the drug is withdrawn.

In modern medicine, only the purified opium alkaloids and their derivatives are commonly employed. Indeed, the analgesic preparation '**papaveretum**' (see below), which once contained the hydrochlorides of total opium alkaloids, is now formulated from selected purified alkaloids, in the proportions likely to be found in opium. Although the ripe poppy capsule can contain up to 0.5% total alkaloids, opium represents a much concentrated form and up to 25% of its mass is composed of alkaloids. Of the many (>40) alkaloids identified, some six represent almost all of the total alkaloid content. Actual amounts vary widely, as shown by the following figures: morphine (Figure 6.50) (4–21%); codeine (Figure 6.50) (0.8–2.5%); thebaine (Figure 6.50) (0.5–2.0%); papaverine (Figure 6.45) (0.5–2.5%); noscapine (narcotine) (Figure 6.51) (4–8%); narceine (Figure 6.51; see also Figure 6.63, page 340) (0.1–2%). A typical commercial sample of opium would probably have a morphine content of about 12%. **Powdered opium** is standardized to contain 10% of anhydrous morphine, usually by dilution with an approved diluent, e.g. lactose or cocoa husk powder. The alkaloids are largely combined in salt form with meconic acid (Figure 6.51), opium containing some 3–5% of this material. Meconic acid is invariably found in opium, but, apart from its presence in other *Papaver* species, has not been detected elsewhere. It gives a deep red-coloured complex with ferric chloride, and this has thus been used as a rapid and reasonably specific test for opium. In the

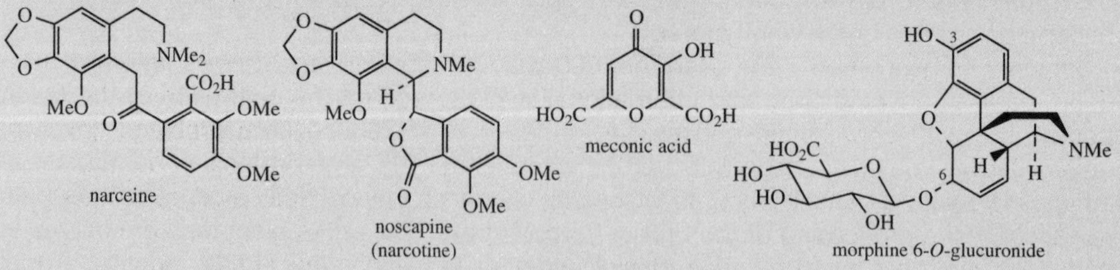

Figure 6.51

(Continues)

(Continued)

past, the urine of suspected opium smokers could also be tested in this way. Of the main opium alkaloids, only morphine and narceine display acidic properties, as well as the basic properties due to the tertiary amine. Narceine has a carboxylic acid function, whilst morphine is acidic due to its phenolic hydroxyl. This acidity can be exploited for the preferential extraction of these alkaloids (principally morphine) from an organic solvent by partitioning with aqueous base.

Morphine (Figure 6.50) is a powerful analgesic and narcotic, and remains one of the most valuable analgesics for relief of severe pain. It also induces a state of euphoria and mental detachment, together with nausea, vomiting, constipation, tolerance, and addiction. Regular users experience withdrawal symptoms, including agitation, severe abdominal cramps, diarrhoea, nausea, and vomiting, which may last for 10–14 days unless a further dose of morphine is taken. This leads to physical dependence, which is difficult to overcome, so that the major current use of morphine is thus in the relief of terminal pain. Although orally active, it is usually injected to obtain rapid relief of acute pain. The side-effect of constipation is utilized in some anti-diarrhoea preparations, e.g. kaolin and morphine. Morphine is metabolized in the body to glucuronides, which are readily excreted. Whilst morphine 3-O-glucuronide is antagonistic to the analgesic effects of morphine, morphine 6-O-glucuronide (Figure 6.51) is actually a more effective and longer lasting analgesic than morphine, with fewer side-effects.

Codeine (Figure 6.50) is the 3-O-methyl ether of morphine, and is the most widely used of the opium alkaloids. Because of the relatively small amounts found in opium, most of the material prescribed is manufactured by semi-synthesis from morphine. Its action is dependent on partial demethylation in the liver to produce morphine, so it produces morphine-like analgesic effects, but little if any euphoria. As an analgesic, codeine has about one-tenth the potency of morphine. Codeine is almost always taken orally and is a component of many compound analgesic preparations. Codeine is a relatively safe non-addictive medium analgesic, but is still too constipating for long-term use. Codeine also has valuable antitussive action, helping to relieve and prevent coughing. It effectively depresses the cough centre, raising the threshold for sensory cough impulses.

Thebaine (Figure 6.50) differs structurally from morphine/codeine mainly by its possession of a conjugated diene ring system. It is almost devoid of analgesic activity, but may be used as a morphine antagonist. Its main value is as substrate for the semi-synthesis of other drugs (see below).

Papaverine (Figure 6.45) is a benzylisoquinoline alkaloid, and is structurally very different from the morphine, codeine, thebaine group of alkaloids (morphinans). It has little or no analgesic or hypnotic properties put possesses spasmolytic and vasodilator activity. It has been used in some expectorant preparations, and in the treatment of gastrointestinal spasms, but its efficacy was not substantiated. It is sometimes used as an effective treatment for male impotence, being administered by direct injection to achieve erection of the penis.

Noscapine (Figure 6.51) is a member of the phthalideisoquinoline alkaloids (see page 339) and provides a further structural variant in the opium alkaloids. Noscapine has good antitussive and cough suppressant activity comparable to that of codeine, but no analgesic or narcotic action. Its original name 'narcotine' was changed to reflect this lack of narcotic action. Despite many years of use as a cough suppressant, the finding that noscapine may have teratogenic properties (i.e. may deform a fetus) has resulted in noscapine preparations being deleted. In recent studies, antitumour activity has been noted from noscapine, which binds to tubulin as do podophyllotoxin and colchicine (see pages 136 and 343), thus arresting cells at mitosis. The chemotherapeutic potential of this orally effective agent merits further evaluation.

(Continues)

(Continued)

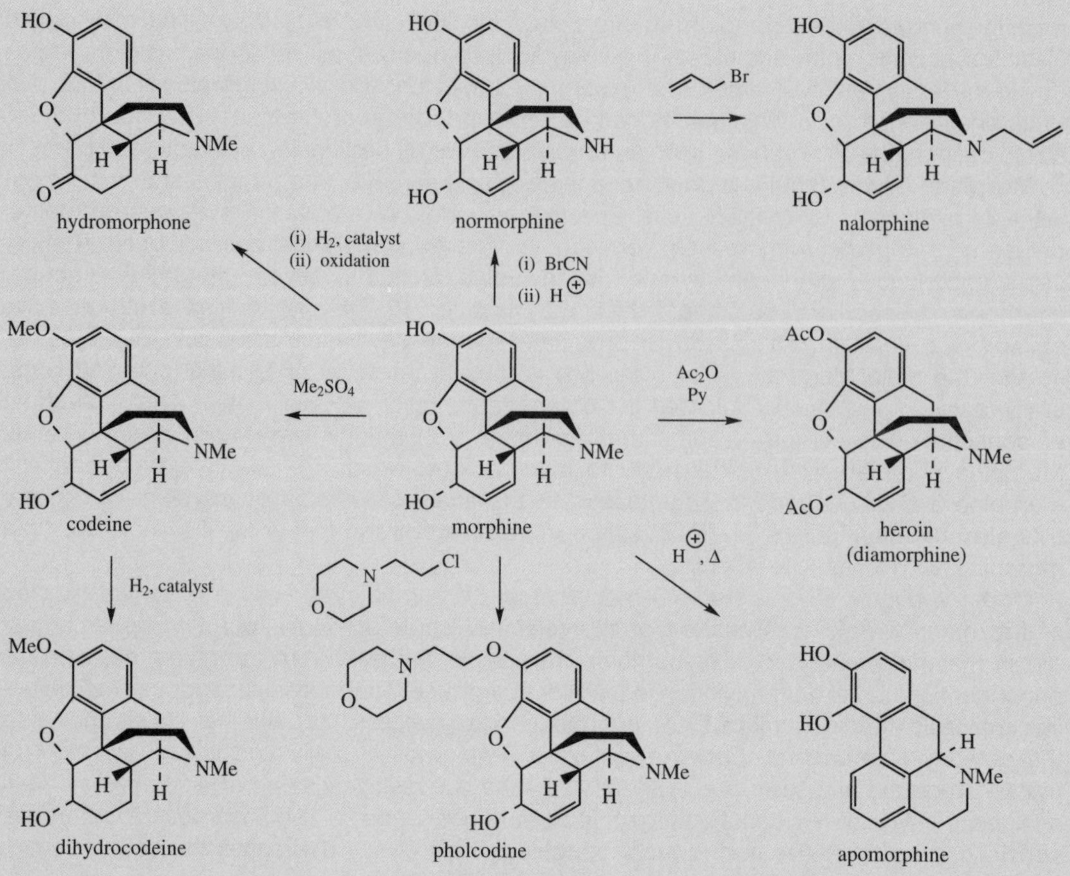

Figure 6.52

Papaveretum is a mixture of purified opium alkaloids, as their hydrochlorides, and is now formulated to contain only morphine (85.5%), codeine (7.8%), and papaverine (6.7%). It is used for pain relief during operations.

A vast range of semi-synthetic or totally synthetic morphine-like derivatives have been produced. These are collectively referred to as 'opioids'. Many have similar narcotic and pain-relieving properties as morphine, but are less habit forming. Others possess the cough-relieving activity of codeine, but without the analgesic effect. More than 90% of the morphine extracted from opium (or poppy straw) is currently processed to give other derivatives (Figure 6.52). Most of the codeine is obtained by semi-synthesis from morphine, mono-O-methylation occurring at the acidic phenolic hydroxyl. Similarly, **pholcodine** (Figure 6.52), an effective and reliable antitussive, can be obtained by alkylation with N-(chloroethyl)morpholine. **Dihydrocodeine** (Figure 6.52) is a reduced form of codeine with similar analgesic properties, the double bond not being essential for activity. In **hydromorphone**, the double bond of morphine has been reduced, and in addition the 6-hydroxyl has been oxidized to a ketone. This increases the analgesic effects, but also the side-effects. **Diamorphine**, or **heroin** (Figure 6.52) is merely the diacetate of morphine, and is a highly addictive analgesic and hypnotic. The increased lipophilic

(Continued)

character results in better transport and absorption, though the active agent is probably the 6-acetate, the 3-acetate group being hydrolysed by esterases in the brain. Heroin was synthesized originally as a cough suppressant, and though most effective in this role has unpleasant addictive properties, with users developing a psychological craving for the drug. It is widely used for terminal care, e.g. cancer sufferers, both as an analgesic and cough suppressant. The euphoria induced by injection of heroin has resulted in much abuse of the drug, and creation of a world-wide major drug problem.

The *N*-methyl group of morphine can be removed by treatment with cyanogen bromide, then hydrolysis. A variety of *N*-alkyl derivatives, e.g. *N*-allyl-normorphine (**nalorphine**) (Figure 6.52) may be produced by use of appropriate alkyl bromides. Nalorphine has some analgesic activity, but was also found to counter the effects of morphine, and is thus a mixed agonist–antagonist. It is sometimes used as a narcotic antagonist, but is principally regarded as the forerunner of pure opiate antagonists such as naloxone (see below). Treatment of morphine with hot acid induces a rearrangement process, resulting in a highly modified structural skeleton, a representative of the aporphine group of alkaloids (see page 337). The product **apomorphine** (Figure 6.52) has no analgesic properties, but morphine's side-effects of nausea and vomiting are highly emphasized. Apomorphine is a powerful emetic, and can be injected for emergency treatment of poisoning. This is now regarded as dangerous, but apomorphine is also valuable to control the symptoms of Parkinson's disease, being a stimulator of D_1 and D_2 dopamine receptors. Apomorphine's structure contains a dihydroxyphenylethylamine (dopamine) fragment, conferring potent dopamine agonist properties to this agent.

It has been found that a common structural feature required for centrally acting analgesic activity in the opioids is the combination of an aromatic ring, and a piperidine ring which maintain the stereochemistry at the chiral centre as shown in Figure 6.53. The three-dimensional disposition of the nitrogen function to the aromatic ring allows morphine and other analgesics to bind to a pain-reducing receptor in the brain. Several different receptors and groups of receptors are known. The natural agonists include peptides called enkephalins, Met-enkephalin and Leu-enkephalin (Figure 6.53), produced from a larger peptide endorphin (see page 419). The terminal tyrosine residue in the enkephalins is mimicked by portions of

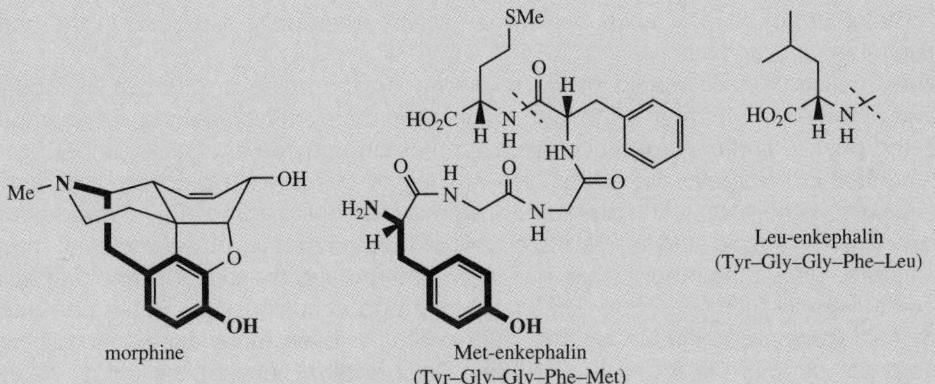

morphine

Met-enkephalin
(Tyr–Gly–Gly–Phe–Met)

Leu-enkephalin
(Tyr–Gly–Gly–Phe–Leu)

Figure 6.53

(Continues)

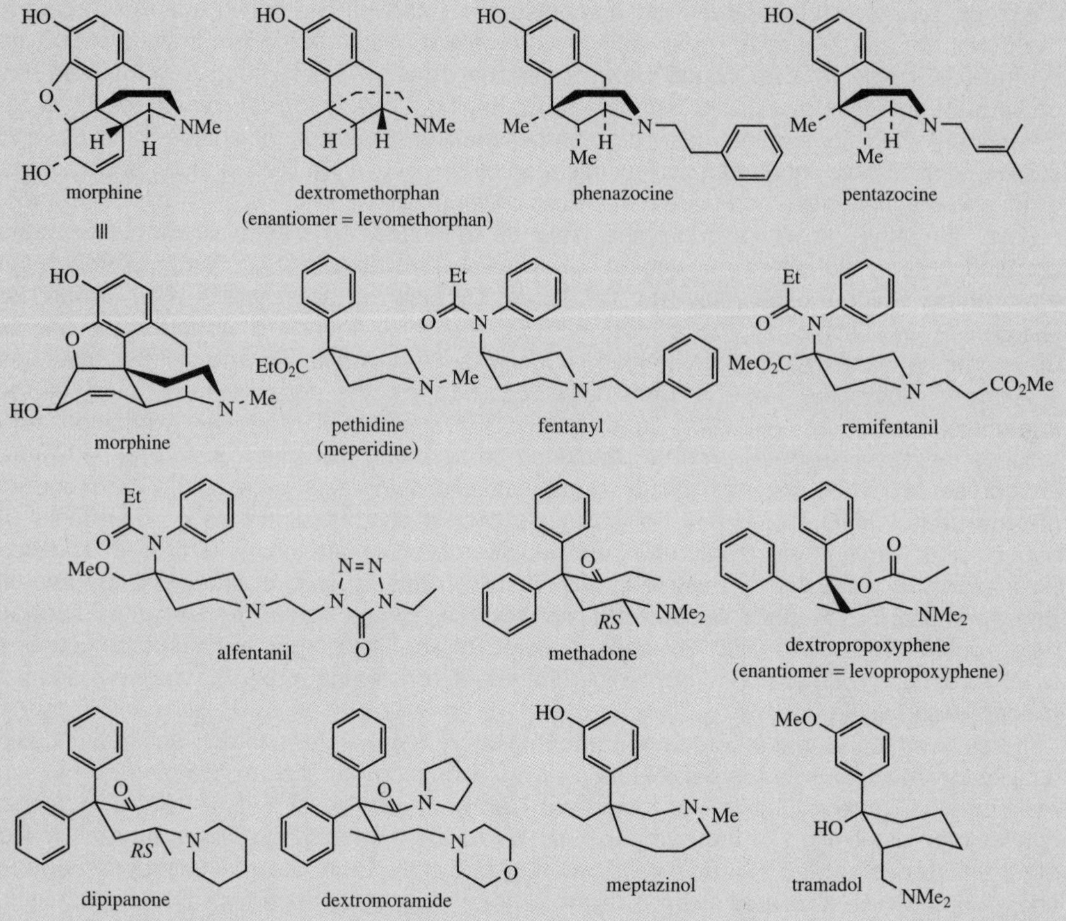

Figure 6.54

the morphine structure. The enkephalins themselves are rapidly degraded in the body and are unsuitable for drug use.

Some totally synthetic opioid drugs modelled on morphine are shown in Figure 6.54. Removal of the ether bridge and the functionalities in the cyclohexene ring are exemplified in levomethorphan and **dextromethorphan**. Levomethorphan has analgesic properties, whilst both enantiomers possess the antitussive activity of codeine. In practice, the 'unnatural' isomer dextromethorphan is the preferred drug material, being completely non-addictive and possessing no analgesic activity. **Pentazocine** and **phenazocine** are examples of morphine-like structures where the ether bridge has been omitted and the cyclohexene ring has been replaced by simple methyl groups. These drugs are good analgesics and are non-addictive, though pentazocine can induce withdrawal symptoms. Even more drastic simplification of the morphine structure is found in **pethidine** (**meperidine**), one of the most widely used synthetic opiates. Only the aromatic ring and the piperidine systems are retained. Pethidine is less potent than morphine, but produces prompt, short-acting analgesia, and is also less

(Continued)

constipating than morphine. It can be addictive. **Fentanyl** has a 4-anilino- rather than a 4-phenyl-piperidine structure, and is 50–100 times more active than morphine due to its high lipophilicity and excellent transport properties. **Alfentanil** and **remifentanil** are further variants on the fentanyl structure; all three drugs are rapid-acting and used during operative procedures. The piperidine ring system is no longer present in **methadone**, though this diphenylpropylamine derivative can be drawn in such a way as to mimic the piperidine ring conformation. Methadone is orally active, has similar activity to morphine, but is less euphorigenic and has a longer duration of action. Although it is as potentially addictive as morphine, the withdrawal symptoms are different and much less severe than with other drugs such as heroin, and methadone is widely used for the treatment and rehabilitation of heroin addicts. However, it only replaces one addiction with another, albeit a less dangerous one. **Dextropropoxyphene** mimics the piperidine ring in a rather similar manner, but this agent has only low analgesic activity, about half that of codeine, and finds application in combination formulations with aspirin or paracetamol. The enantiomeric **levopropoxyphene** has antitussive activity, but no analgesic properties. **Dipipanone** and **dextromoramide** are structural variants on methadone, and are used for moderate to severe pain; dipipanone is usually administered in combination with an anti-emetic. **Meptazinol** is structurally unlike the other opiate analgesics in that it contains a seven-membered nitrogen heterocycle. It is an effective analgesic, and produces relatively few side-effects with a low incidence of respiratory depression. **Tramadol** is a recent drug claimed to produce analgesia by an opioid mechanism and by enhancement of serotoninergic and adrenergic pathways, with few typical opioid side-effects.

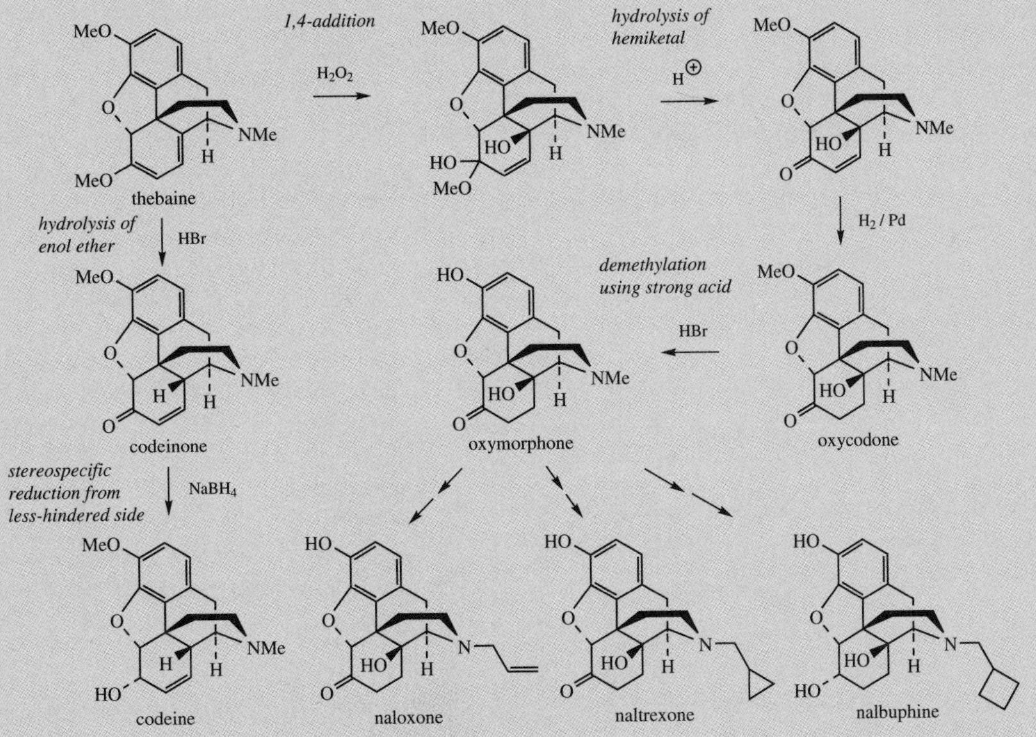

Figure 6.55

(Continues)

(Continued)

Thebaine, for many years regarded as an unwanted by-product from opium, is now utilized for the semi-synthesis of useful new drugs. On treatment with hydrogen peroxide, the conjugated diene undergoes 1,4-addition, and hydrolysis results in formation of a 4-hydroxy cyclohexenone system (Figure 6.55). Reduction and demethylation lead respectively to **oxycodone** and **oxymorphone**, which are potent analgesics. The conjugated diene system can also be exploited in a Diels–Alder reaction, building on another ring system (Figure 6.56). Some of these adducts have quite remarkable levels of analgesic activity, but are too powerful for human use. Some, e.g. **etorphine** (Figure 6.57), are used in veterinary practice to sedate large animals (elephants, rhinos) by means of tranquillizer darts. Etorphine is some 5000–10 000 times more potent than morphine. **Buprenorphine** (Figure 6.57) is an etorphine analogue with an *N*-cyclopropylmethyl substituent and *tert*-butyl instead of *N*-propyl in the side-chain. This material has both opioid agonist and antagonist properties. Mixed agonist–antagonist properties offer scope for producing analgesia whilst negating the effects of other opioids to which a patient may be addicted. Buprenorphine has a long duration of action, and only low dependence potential, but may precipitate withdrawal symptoms in patients dependent on other opioids. It is now being used as an alternative to methadone in the treatment of opioid dependence. **Nalbuphine** (Figure 6.55), produced semi-sythetically from thebaine, also displays mixed agonist–antagonist properties, and has similar agonist activity as morphine, but produces less side-effects and has less abuse potential. **Naloxone** (Figure 6.55) shows hardly any agonist activity but is a potent antagonist, and is used to treat opiate poisoning, including that in children born to heroin addicts. **Naltrexone** (Figure 6.55)

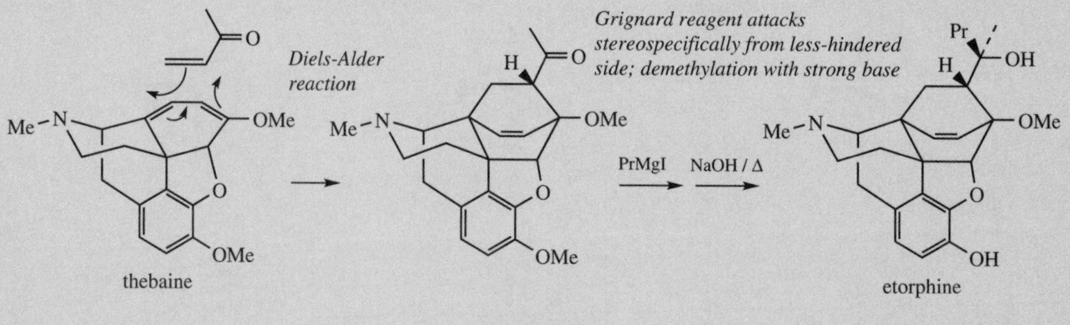

Figure 6.56

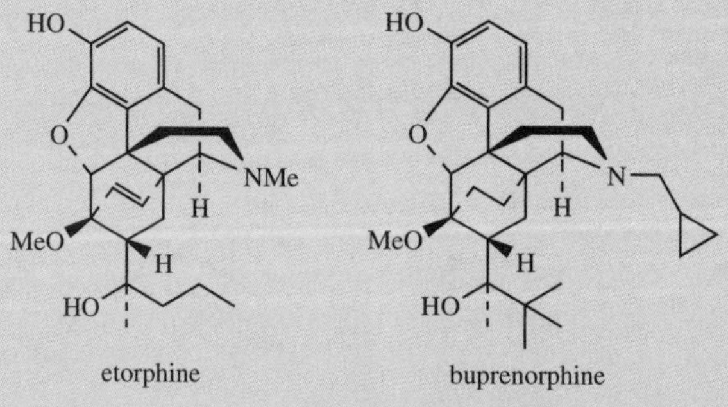

Figure 6.57

(Continues)

(Continued)

also has antagonist activity similar to naloxone. These agents are *N*-alkyl derivatives related to oxymorphone/oxycodone.

Thebaine may also be transformed very efficiently into codeine in about 75% yield (Figure 6.55). The two-stage synthesis involves acid-catalysed hydrolysis of the enol ether function to give codeinone (this being the more favoured tautomer of the first-formed β, γ-unsaturated ketone) followed by selective borohydride reduction of the carbonyl. This opens up possibilities for producing codeine (the most widely used of the opium alkaloids) without using morphine. At present, most of the codeine is synthesized by methylation of morphine. The advantage of using thebaine is that the raw material for the pharmaceutical industry could be shifted away from morphine and opium. This might then help in the battle to eliminate illicit morphine production and its subsequent conversion into heroin. Conversion of thebaine into morphine and heroin is much more difficult and low yielding. Thus, there is interest in cultivating *Papaver bracteatum* rather than *P. somniferum*. This plant produces mainly thebaine, no morphine, and only faint traces of codeine. Experiments have shown it has the enzymic activity to convert codeinone into codeine, but it appears to lack enzymes which carry out the late demethylation steps in Figure 6.50. The capsules can produce up to 3% thebaine, but, regrettably, there have been difficulties in making this a commercially viable project, and this idea has not materialized. Other species of *Papaver* seem to lack the enzyme that reduces salutaridine to salutaridinol (Figure 6.50) and they thus do not synthesize morphine-like alkaloids. Remarkably, there is now considerable evidence that various animals, including humans and other mammals, are also able to synthesize morphine and related alkaloids in small amounts. These compounds have been detected in various tissues, including brain, liver, spleen, adrenal glands and skin, and endogenous morphine may thus play a role in pain relief, combining its effects with those provided by the enkephalin peptides.

A minor constituent of *P. somniferum* is the aporphine alkaloid **isoboldine** (Figure 6.58). Other species of poppy, e.g. *Papaver orientale* and *P. pseudoorientale*, are known to synthesize aporphine alkaloids as principal constituents rather than morphinan structures. (*S*)-Isoboldine is readily appreciated to be the product of oxidative coupling of (*S*)-**reticuline**, coupling *ortho* to the phenol group in the tetrahydroisoquinoline, and *para* to the phenol of the benzyl substituent

(Figure 6.58). Some structures, e.g. **isothebaine** (Figure 6.59) from *P. orientale*, are not as easily rationalized. (*S*)-**Orientaline** is a precursor of isothebaine (Figure 6.59). This benzyltetrahydroisoquinoline, with a different methylation pattern to reticuline, is able to participate in oxidative coupling, but inspection of the structures indicates a phenol group is lost in the transformation. The pathway (Figure 6.59) involves an unexpected rearrangement process, however. Thus, oxidative

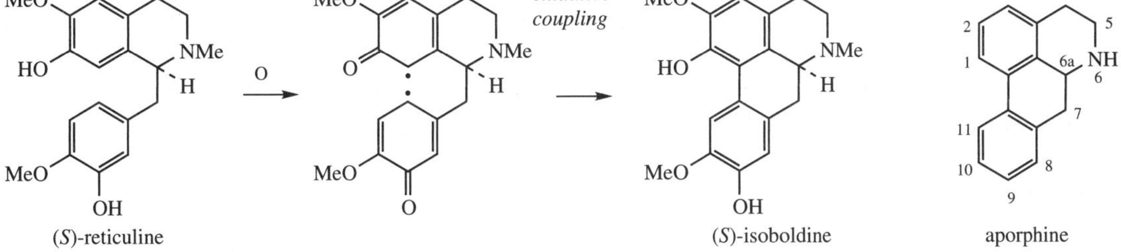

Figure 6.58

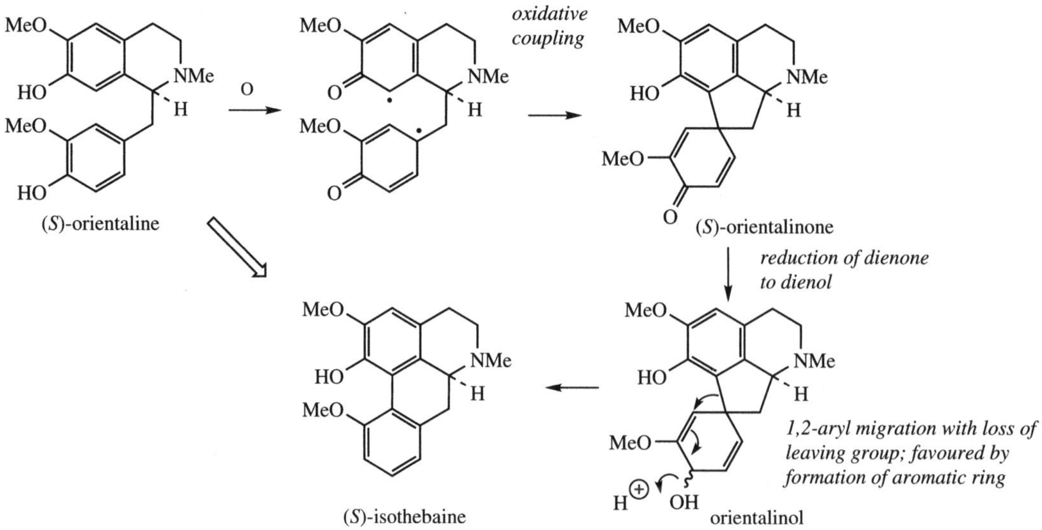

Figure 6.59

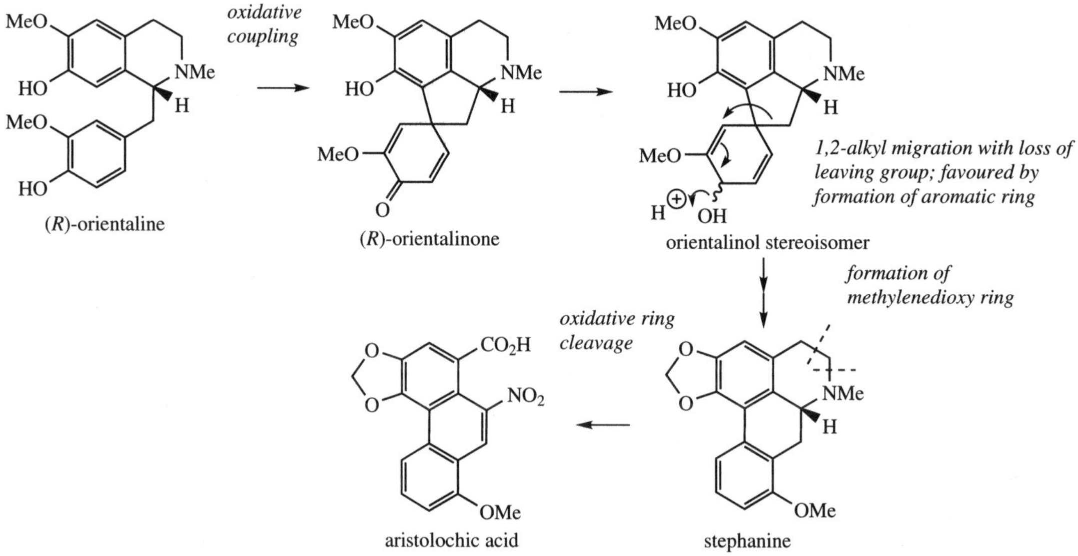

Figure 6.60

coupling *ortho–para* to the phenol groups gives a dienone **orientalinone** (compare the structure of salutaridine (Figure 6.50)). After reduction of the carbonyl group, a rearrangement occurs, restoring aromaticity and expelling the hydroxyl (originally a phenol group) to produce **isothebaine**. This type of rearrangement, for which good chemical analogies are available, is a feature of many other alkaloid biosynthetic pathways, and occurs

because normal keto–enol tautomerism is not possible for rearomatization when coupling involves positions already substituted. The process is fully borne out by experimental evidence, including the subsequent isolation of orientalinone and orientalinol from *P. orientale*.

Stephanine (Figure 6.60) from *Stephania* species (Menispermaceae) is analogous to isothebaine and shares a similar pathway, though from

(*R*)-**orientaline**. The different substitution pattern in stephanine compared to isothebaine is a consequence of the intermediate dienol suffering migration of the alkyl rather than aryl group (Figure 6.60). **Aristolochic acid** is a novel modified aporphine containing a nitro group and is produced from stephanine by oxidative reactions leading to ring cleavage (Figure 6.60). Aristolochic acid is present in many species of *Aristolochia* (Aristolochiaceae) used in traditional medicine, e.g. snake-root *A. serpentina*. However, because aristolochic acid is now known to be nephrotoxic and to cause acute kidney failure, the use of *Aristolochia* species in herbal medicines, especially Chinese remedies, has been banned in several countries.

The alkaloid **berberine** (Figure 6.61) is found in many members of the Berberidaceae (e.g. *Berberis*, *Mahonia*), the Ranunculaceae (e.g. *Hydrastis*), and other families. Berberine has antiamoebic, antibacterial, and anti-inflammatory properties and plants containing berberine have long been used in traditional medicine. Its tetracyclic skeleton is derived from a benzyltetrahydroisoquinoline system with the incorporation of an extra carbon atom, supplied from *S*-adenosylmethionine via an *N*-methyl group (Figure 6.61). This extra skeletal carbon is known as a 'berberine bridge'. Formation of the berberine bridge is readily rationalized as an oxidative process in which the *N*-methyl group is oxidized to an iminium ion, and a cyclization to the aromatic ring occurs by virtue of the phenolic group (Figure 6.62).

The oxidative cyclization process is analogous to the formation of a methylenedioxy group (see page 27), whilst the mechanism of cyclization is exactly the same as that invoked in formation of a tetrahydroisoquinoline ring, i.e. a Mannich-like

reaction (see page 320). The product from the enzymic transformation of (*S*)-**reticuline** is the protoberberine alkaloid (*S*)-**scoulerine**, the berberine bridge enzyme requiring molecular oxygen as oxidant and releasing H_2O_2 as by-product (Figure 6.62). Its role in the cyclization reaction completed, the phenol group in scoulerine is then methylated, and **tetrahydrocolumbamine** is oxidized further to give the quaternary isoquinoline system in **columbamine**. This appears to involve two separate oxidation steps, both requiring molecular oxygen, though H_2O_2 and H_2O are produced in the successive processes. The mechanistic sequence through an iminium ion has been suggested to account for these observations. Finally, **berberine** is produced by transformation of the *ortho*-methoxyphenol to a methylenedioxy group, via the O_2-, NADPH-, and cytochrome P-450-dependent enzyme.

The protoberberine skeleton of scoulerine may be subjected to further modifications, some of which are given in Figure 6.63. Cleavage of the heterocyclic ring systems adjacent to the nitrogen atom as shown give rise to new skeletal types: protopine, e.g. **protopine** from *Chelidonium majus* (Papaveraceae), phthalideisoquinoline, e.g. **hydrastine** from *Hydrastis canadensis* (Ranunculaceae), and benzophenanthridine, e.g. **chelidonine**, also from *Chelidonium majus*. The non-heterocyclic system seen in the opium alkaloid **narceine** from *Papaver somniferum* can be visualized as the result of cleavage of two of these bonds. Some alkaloids of the phthalide type are medicinally important. **Noscapine** (Figure 6.64) is one of the opium alkaloids and although it lacks any analgesic activity it is an effective cough suppressant (see page 331). **Hydrastine** is beneficial as a traditional remedy in the control of uterine bleeding. *Hydrastis* also contains berberine, indicating the

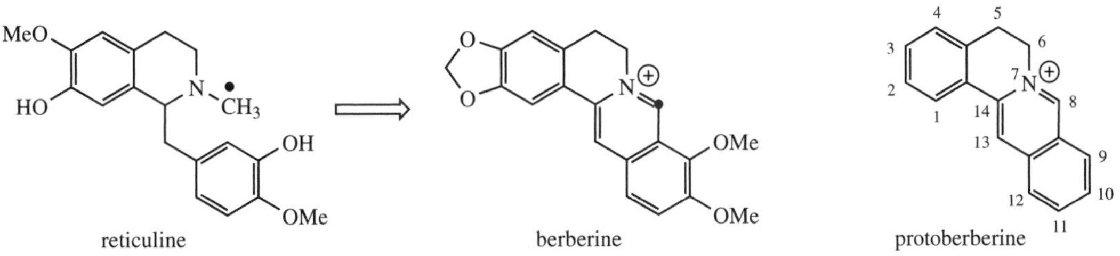

reticuline

berberine

protoberberine

Figure 6.61

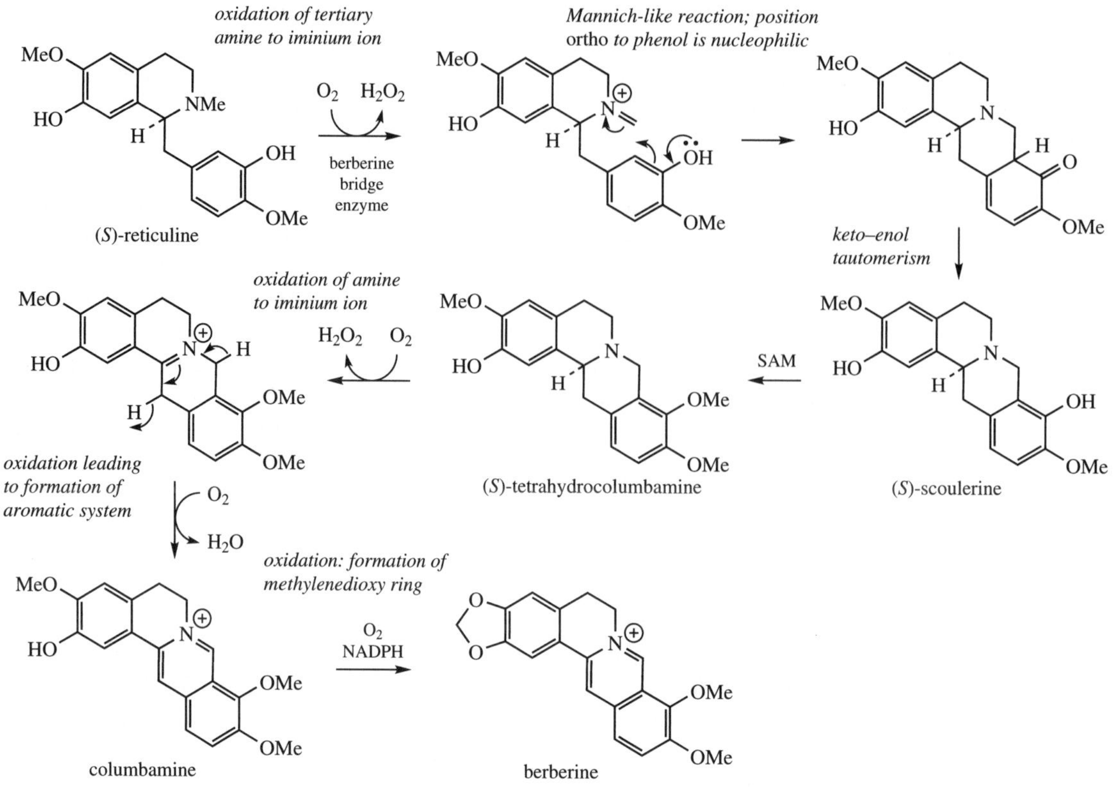
oxidation of tertiary amine to iminium ion
O_2 H_2O_2
berberine bridge enzyme
(S)-reticuline
Mannich-like reaction; position ortho to phenol is nucleophilic
keto–enol tautomerism
(S)-scoulerine
SAM
(S)-tetrahydrocolumbamine
oxidation of amine to iminium ion
H_2O_2 O_2
oxidation leading to formation of aromatic system
O_2
H_2O
oxidation: formation of methylenedioxy ring
O_2
NADPH
columbamine
berberine

Figure 6.62

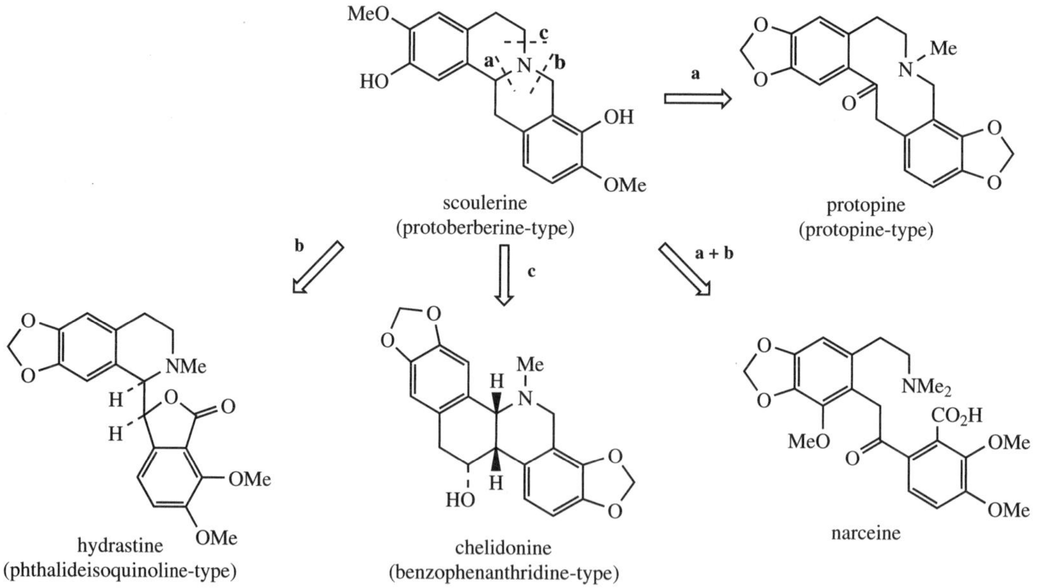
scoulerine
(protoberberine-type)
protopine
(protopine-type)
hydrastine
(phthalideisoquinoline-type)
chelidonine
(benzophenanthridine-type)
narceine

Figure 6.63

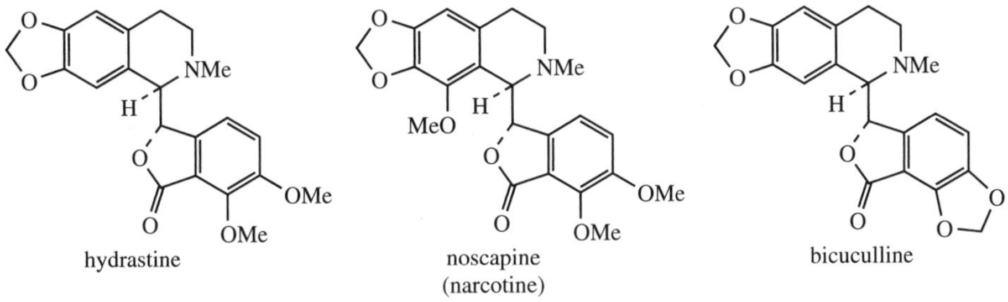

Figure 6.64

close biosynthetic relationship of the two types of alkaloid. **Bicuculline** (Figure 6.64) from species of *Corydalis* and *Dicentra* (Fumariaceae) and its quaternary methiodide have been identified as potent GABA (γ-aminobutyric acid) antagonists and have found widespread application as pharmacological probes for convulsants acting at GABA neuroreceptors.

Phenethylisoquinoline Alkaloids

Several genera in the lily family (Liliaceae) are found to synthesize analogues of the benzyltetrahydroisoquinoline alkaloids, e.g. **autumnaline**

(Figure 6.65), which contain an extra carbon between the tetrahydroisoquinoline and the pendant aromatic rings. This skeleton is formed in a similar way to that in the benzyltetrahydroisoquinolines from a phenylethylamine and an aldehyde (Figure 6.65), but a whole C_6C_3 unit rather than a C_6C_2 fragment functions as the reacting aldehyde. Typically, dopamine (from tyrosine) and 4-hydroxydihydrocinnamaldehyde (from phenylalanine) are involved in the initial condensation, and further hydroxylation and methylation steps then build up the substitution pattern to that of autumnaline. Phenolic oxidative coupling accounts for the occurrence of homoaporphine alkaloids such as **floramultine** and **kreysigine** in

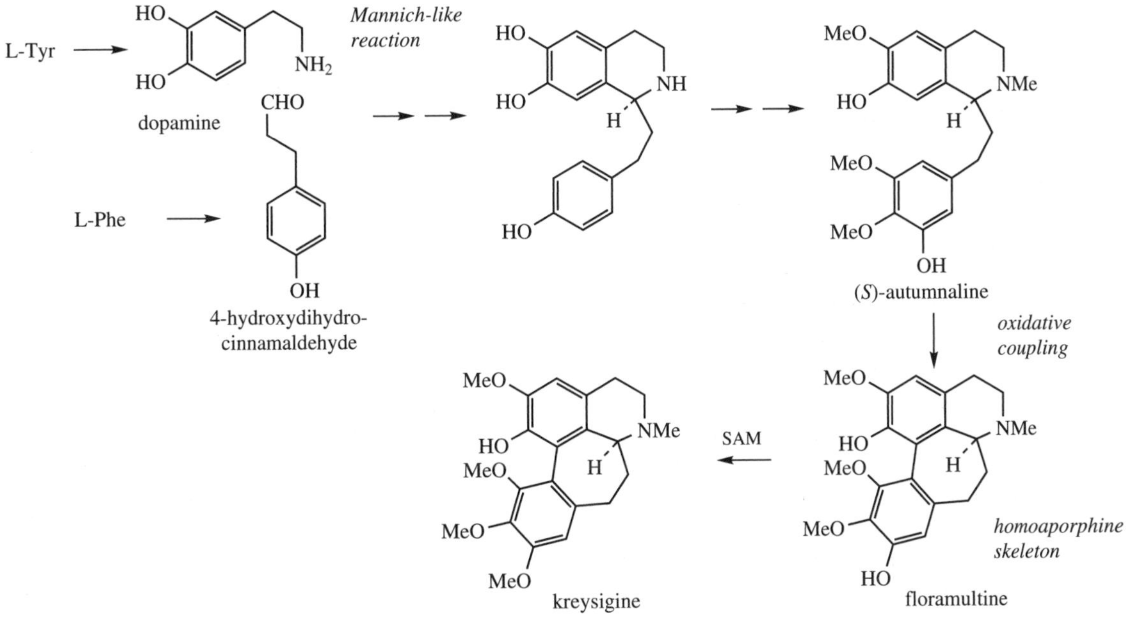

Figure 6.65

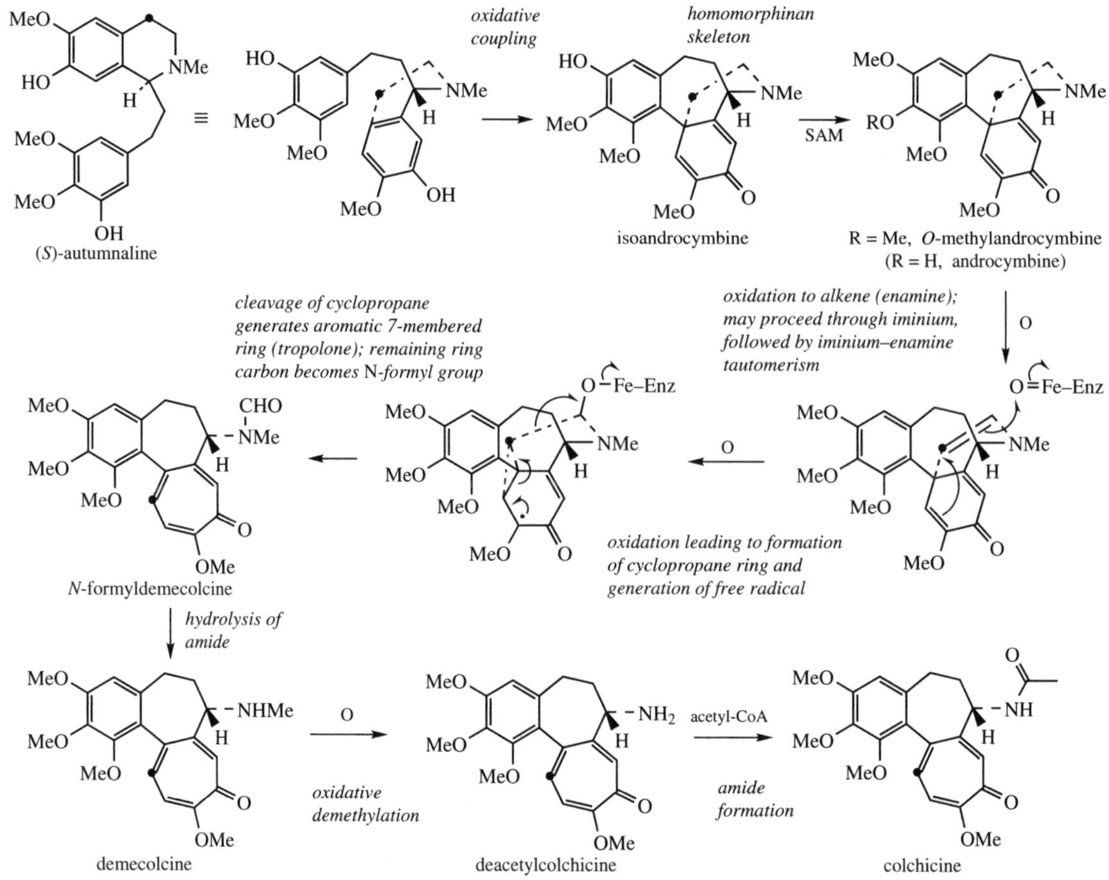

Figure 6.66

Kreysigia multiflora (Liliaceae/Convallariaceae). (*S*)-**Autumnaline** has also been found to act as a precursor for **colchicine** (Figure 6.66), an alkaloid containing an unusual tropolone ring. Colchicine is found in species of *Colchicum**, e.g. *Colchicum autumnale* (Liliaceae/Colchicaceae), as well as many other plants in the Liliaceae. Colchicine no longer has its nitrogen atom in a ring system, and extensive reorganization of the autumnaline structure is thus necessary. The seven-membered tropolone ring was shown by labelling experiments to originate by ring expansion of the tyrosine-derived aromatic ring taking in the adjacent benzylic carbon (Figure 6.66). Prior to these remarkable rearrangements, oxidative coupling of autumnaline in the *para–para* sense features in the pathway giving the dienone **isoandrocymbine**, which has a homomorphinan skeleton (compare salutaridine, Figure 6.50). The isomer **androcymbine** (Figure 6.66) had been isolated from *Androcymbium melanthioides* (Liliaceae/Colchicaceae), thus giving a clue to the biosynthetic pathway. Methylation follows giving *O*-methylandrocymbine, and it is then proposed that enzymic oxidation to an enamine yields the substrate for ring modification. Experimental labelling studies are then best explained by formation of a cyclopropane ring followed by ring opening to generate the 6π electron aromatic tropolone system, incorporating the original tyrosine benzylic carbon into the seven-membered ring, and also breaking the original phenylethylamine side-chain between the carbons. One carbon is left on the nitrogen as a formyl group, and this can be lost by hydrolysis. **Colchicine** is produced by exchanging the *N*-methyl group for an *N*-acetyl group, by way of an oxidative demethylation followed by acetylation using acetyl-CoA. **Demecolcine** and **deacetylcolchicine** are intermediates in the process.

Colchicum

Colchicum seed and corm are obtained from *Colchicum autumnale* (Liliaceae/Colchicaceae), the autumn crocus or meadow saffron. The plant, though not a crocus, produces crocus-like flowers in the autumn, the leaves not emerging until the spring. It is a native of Europe, is widely cultivated as an ornamental garden plant, and is grown for drug use, mainly in Europe and North Africa. The principal alkaloid is colchicine (Figure 6.66), which occurs to the level of about 0.8% in the seed, and 0.6% in the corm. As an *N*-acetyl derivative, colchicine does not display any significant basicity, and does not form well-defined salts. Demecolcine (*N*-deacetyl-*N*-methylcolchicine) (Figure 6.66) is a minor constituent in both corm and seeds.

Extracts of *Colchicum autumnale*, and later **colchicine** itself, have been used in the treatment of gout, a painful condition in which impaired purine metabolism leads to a build-up of uric acid crystals in the joints. Colchicine is an effective treatment for acute attacks, but it is very toxic, and this restricts its general use. It appears to act primarily as an anti-inflammatory agent, and does not itself affect uric acid metabolism, which needs to be treated with other agents, e.g. a xanthine oxidase inhibitor such as allopurinol. The cytotoxic properties of colchicine and related alkaloid structures from *C. autumnale* led to their being tested as potential anticancer agents, though they still proved too toxic for medicinal use. Colchicine binds to tubulin in the mitotic spindle, preventing polymerization and assembly into microtubules as do podophyllotoxin (see page 136) and vincristine (see page 356), and is a useful biochemical probe. However, the ability of colchicine to act as a mitotic poison is exploited in plant breeding, since the interference with mitosis results in multiplication of chromosomes in the cell nucleus without the process of cell division. Cell division recommences on cessation of treatment. This allows the generation of mutations (polyploids) and possible new varieties of plant. Colchicine is also found in other species of *Colchicum*, as well as many other plants in the Liliaceae (e.g. *Bulbocodium*, *Gloriosa*, *Merendera*, and *Sandersonia*), a group of plants now classified as the family Colchicaceae.

Terpenoid Tetrahydroisoquinoline Alkaloids

The alkaloids found in ipecacuanha*, the dried rhizome and roots of *Cepahaelis ipecacuanha* (Rubiaceae), have a long history of use in the treatment of amoebic dysentery, and provide unusual examples of tetrahydroisoquinoline structures. The principal alkaloids, e.g. **emetine** and **cephaeline** (Figure 6.67), possess a skeleton with two tetrahydroisoquinoline ring systems, plus a further fragment that has its origin in a terpenoid-derived molecule. This terpenoid substrate is the secoiridoid **secologanin** (see page 189), a compound that also features in the biosynthesis of many complex indole alkaloids (see page 350). Secologanin is an aldehyde and can condense with dopamine to give the tetrahydroisoquinoline alkaloid *N*-deacetylisoipecoside (Figure 6.67). **Ipecoside** itself is found in ipecacuanha, though it has

the opposite stereochemistry at C-1 to this biosynthetic intermediate. The secologanin fragment contains an acetal function, which can be restored to its component aldehyde and alcohol fragments by hydrolysis of the glucosidic bond. The newly liberated aldehyde can then bond with the secondary amine to give the quaternary Schiff base. This intermediate is converted into an aldehyde by a sequence of reactions: reduction of iminium, reduction of alkene, plus hydrolysis of ester and subsequent decarboxylation, though not necessarily in that order. The decarboxylation step is facilitated by the β-aldehyde function, shown as an enol in Figure 6.67. Most of the reactions taking place in the secologanin-derived part of the structure are also met in discussions of terpenoid indole alkaloids (see page 350). The resultant aldehyde is now able to participate in formation of a second tetrahydroisoquinoline ring system, by reaction with a second dopamine molecule. Methylation gives **cephaeline** and **emetine**.

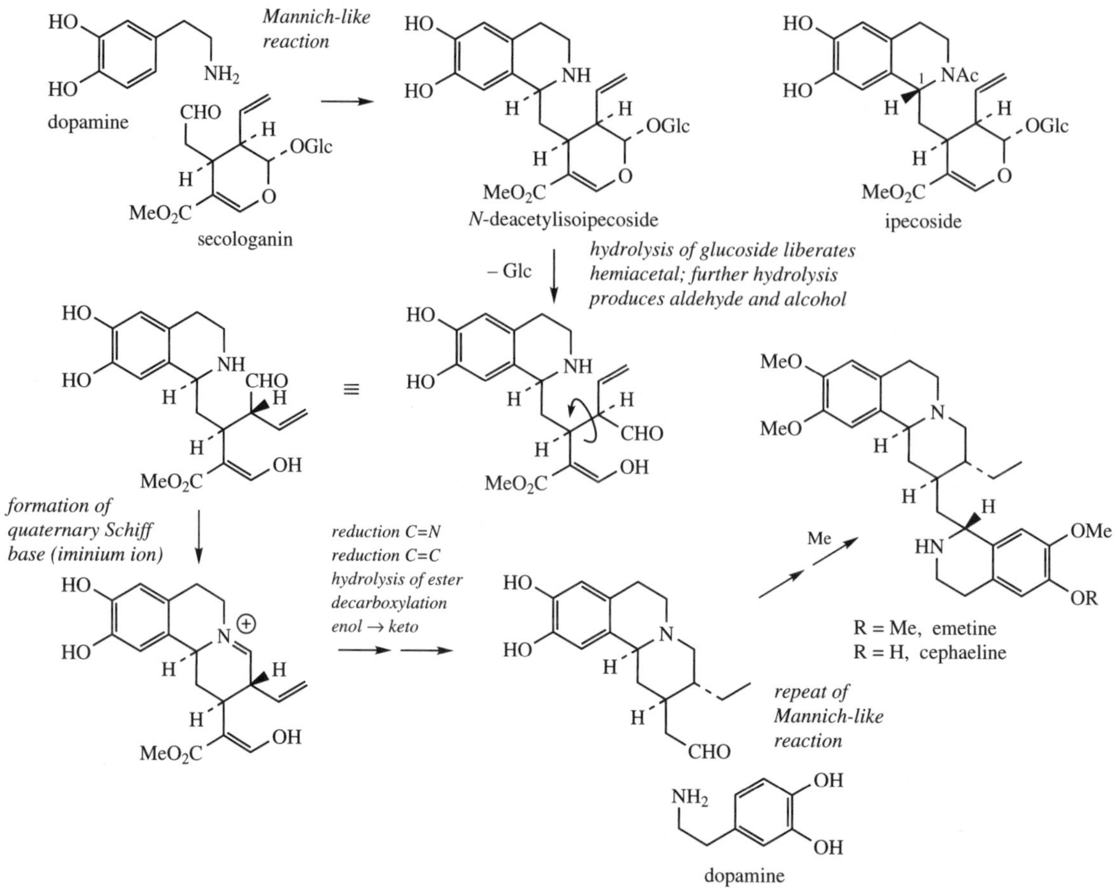

Figure 6.67

Ipecacuanha

Ipecacuanha or **ipecac** is derived from the dried rhizome and roots of *Cephaelis ipecacuanha* or *C. acuminata* (Rubiaceae). These are low straggling shrubs having horizontal rhizomes with prominently ridged roots. *Cephaelis ipecacuanha* yields what is termed Rio or Brazilian ipecac, and is cultivated mainly in Brazil, whilst *C. acuminata* gives Cartagena, Nicaragua, or Panama ipecac, and comes principally from Colombia and Nicaragua. Most of the commercial ipecac now derives from *C. acuminata*. Ipecac is an age-old remedy of the South American Indians, who used it for the treatment of dysentery. More recently it was mixed with powdered opium to give Dover's powder (see page 330), where the ipecac content functioned as a diaphoretic.

Ipecac contains 2–2.5% of alkaloids, the principal ones being emetine and cephaeline (Figure 6.67). Typically, in *C. ipecacuanha* the emetine to cephaeline ratio might be about 2:1, whereas in *C. acuminata* the ratio ranges from about 1:2 to 1:1. Minor alkaloids characterized include psychotrine and *O*-methylpsychotrine (Figure 6.68), which are dehydro variants of cephaeline and emetine respectively.

Both **emetine** and the synthetic **2,3-dehydroemetine** (Figure 6.68) have been useful as anti-amoebics, particularly in the treatment of amoebic dysentery. However, they also cause

(Continues)

(*Continued*)

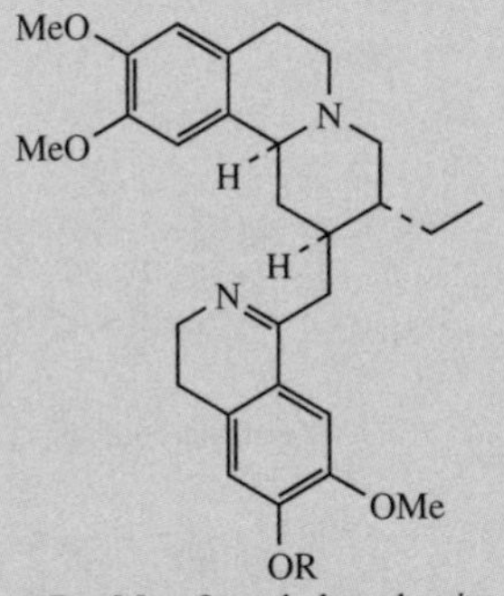

R = Me, *O*-methylpsychotrine
R = H, psychotrine

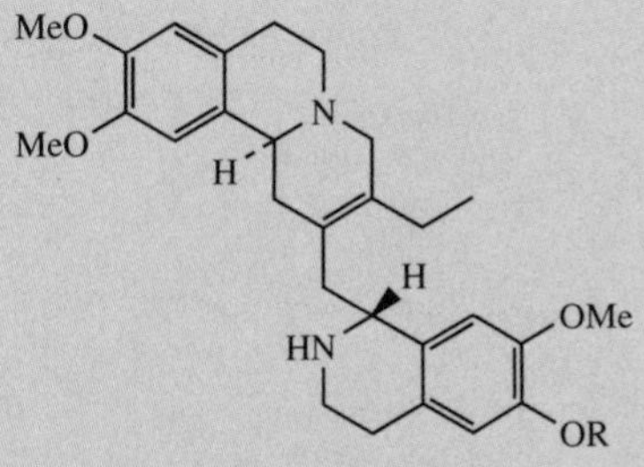

2,3-dehydroemetine

Figure 6.68

nausea, and this has now made other drugs preferable. The emetic action of the alkaloids is particularly valuable though, and the crude drug extract in the form of **ipecacuanha emetic mixture** is an important preparation used for drug overdose or poisoning. The emetic mixture is often a standard component in poison antidote kits. Ipecacuanha also has expectorant activity and extracts are still components of a number of compound expectorant preparations. Emetine has more expectorant and less emetic action than cephaeline, and thus the Brazilian drug is preferred for such mixtures. If required, emetine may be obtained in larger amounts by also methylating the cephaeline component of the plant material.

Emetine and cephaeline are both potent inhibitors of protein synthesis, inhibiting at the translocation stage. They display antitumour and antiviral as well as antiamoebic activity, but are too toxic for therapeutic use. In recent studies, *O*-methylpsychotrine has displayed fairly low effects on protein synthesis, but a quite potent ability to curb viral replication through inhibition of HIV-reverse transcriptase. This may give it potential in the treatment of AIDS.

Amaryllidaceae Alkaloids

Various types of alkaloid structure are encountered in the daffodil family, the Amaryllidaceae, and they can be rationalized better through biosynthesis than by structural type. The alkaloids arise by alternative modes of oxidative coupling of precursors related to **norbelladine** (Figure 6.69), which is formed through combination of 3,4-dihydroxybenzaldehyde with tyramine, these two preursors arising from phenylalanine and tyrosine respectively. Three structural types of alkaloid can be related to **4′-*O*-methylnorbelladine** by different alignments of the phenol rings, allowing coupling *para–ortho* (A), *para–para* (B), or *ortho–para* (C) as shown in Figure 6.69. For **galanthamine**, the dienone formed via oxidative coupling (C) undergoes nucleophilic addition from the phenol group, forming an ether linkage

(compare opium alkaloids, page 328), and the sequence is completed by reduction and methylation reactions. For **lycorine** and **crinine**, although details are not given in Figure 6.69, it is apparent that the nitrogen atom acts as a nucleophile towards the dienone system in a similar manner, generating the new heterocyclic ring systems. Alkaloids such as lycorine, crinine, and galanthamine can undergo further modifications, which include ring cleavage reactions, generating many more variations than can be considered here. The Amaryllidaceae family includes *Amaryllis*, *Narcissus*, and *Galanthus*, and the alkaloid content of bulbs from most members makes these toxic. Lycorine was first isolated from *Lycorus radiata*, but is common and found throughout the family. **Galanthamine** from snowdrops (*Galanthus* species) is currently an important drug material of value in treating Alzheimer's disease.

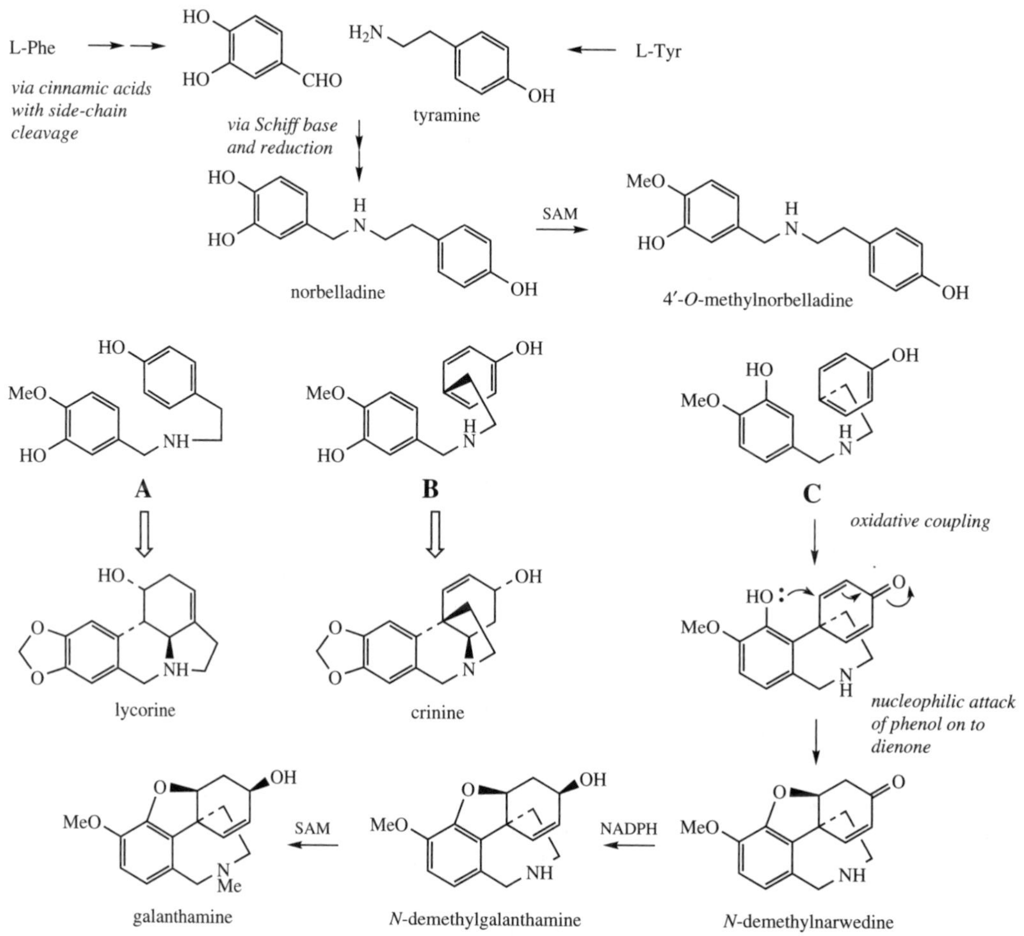

Figure 6.69

Galanthamine

Galantamine (galanthamine) can be isolated from a number of species of the Amaryllidaceae, including snowdrops (*Galanthus* species), daffodils (*Narcissus pseudonarcissus*), and snowflakes (*Leucojum* species), where typical content varies from about 0.05 to 0.2% in the bulbs. It is currently isolated for drug use from the bulbs of wild *Leucojum aestivum* and *Galanthus* species, since commercial synthesis is not economic. Galantamine acts as a competitive cholinesterase inhibitor, and enhances cognitive function in the treatment of Alzheimer's disease by raising acetylcholine levels in brain areas lacking cholinergic neurones. In common with other treatments for Alzheimer's disease, it does not cure the condition, but merely slows the rate of cognitive decline.

ALKALOIDS DERIVED FROM TRYPTOPHAN

L-**Tryptophan** is an aromatic amino acid containing an indole ring system, having its origins in the shikimate pathway (Chapter 4) via anthranilic acid. It acts as a precursor of a wide range of indole alkaloids, but there is also definite proof that major rearrangement reactions can convert the indole ring system into a quinoline ring, thus increasing further the ability of this amino acid to act as an alkaloid precursor (see page 359).

Simple Indole Alkaloids

Tryptamine and its *N*-methyl and *N,N*-dimethyl derivatives (Figure 6.70) are widely distributed in plants, as are simple hydroxylated derivatives such as **5-hydroxytryptamine (serotonin)**. These are formed (Figure 6.70) by a series of decarboxylation, methylation, and hydroxylation reactions, though the sequences of these reactions are found to vary according to final product and/or

organism involved. 5-Hydroxytryptamine is also found in mammalian tissue, where it acts as a neurotransmitter in the central nervous system. It is formed from tryptophan by hydroxylation and then decarboxylation, paralleling the tyrosine → dopamine pathway (see page 316). In the formation of **psilocin** (Figure 6.70), decarboxylation precedes *N*-methylation, and hydroxylation occurs last. Phosphorylation of the hydroxyl in psilocin gives **psilocybin**. These two compounds are responsible for the hallucinogenic properties of so-called magic mushrooms, which include species of *Psilocybe**, *Panaeolus*, etc.

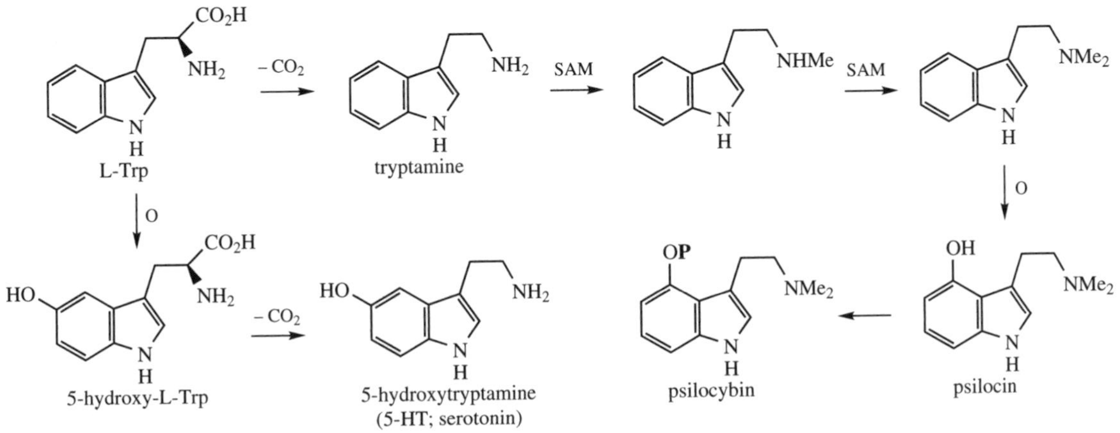

Figure 6.70

5-Hydroxytryptamine (Serotonin)

5-Hydroxytryptamine (5-HT, serotonin) is a monoamine neurotransmitter found in cardiovascular tissue, the peripheral nervous system, blood cells, and the central nervous system. It mediates many central and peripheral physiological functions, including contraction of smooth muscle, vasoconstriction, food intake, sleep, pain perception, and memory, a consequence of it acting on several distinct receptor types. Although 5-HT may be metabolized by monoamine oxidase, platelets and neurons possess a high affinity 5-HT reuptake mechanism. This mechanism may be inhibited by widely-prescribed antidepressant drugs termed selective serotonin re-uptake inhibitors (SSRIS), e.g. fluoxetine (Prozac®), thereby increasing levels of 5-HT in the CNS.

Migraine headaches that do not respond to analgesics may be relieved by the use of an agonist of the 5-HT_1 receptor, since these receptors are known to mediate vasoconstriction. Though the causes of migraine are not clear, they are characterized by dilation of cerebral blood vessels. 5-HT_1 agonists based on the 5-HT structure in current use include the sulphonamide derivative **sumatriptan**, and the more recent agents **naratriptan, rizatriptan,**

(Continues)

(Continued)

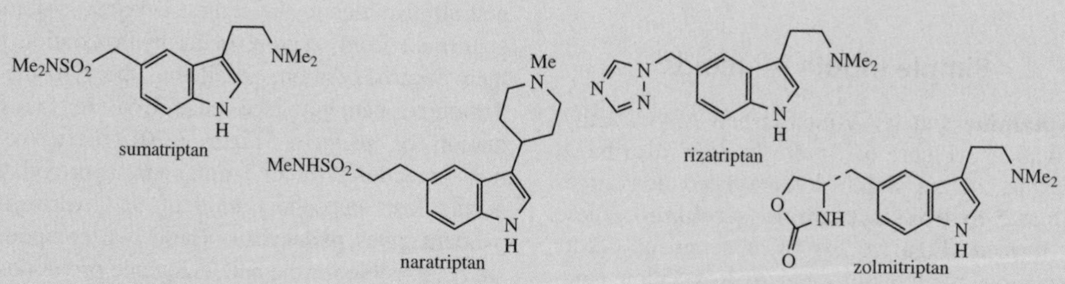

Figure 6.71

and **zolmitriptan** (Figure 6.71). These are of considerable value in treating acute attacks. Several of the ergot alkaloids (page 371) also interact with 5-HT receptors.

Psilocybe

The genus *Psilocybe* constitutes a group of small mushrooms with worldwide distribution. It has achieved notoriety on account of hallucinogenic experiences produced following ingestion of several species, particularly those from Mexico, and has led to the description 'magic mushrooms'. Over 80 species of *Psilocybe* have been found to be psychoactive, whereas over 50 species are inactive. More than 30 of the hallucinogenic species have been identified in Mexico, but active species may be found in all areas of the world. *Psilocybe mexicana* has been used by the Mexican Indians in ancient ceremonies for many years, and its history can be traced back to the Aztecs. In temperate regions, *Psilocybe semilanceata*, the liberty cap, is a common species with similar activity. All the psychoactive members of the genus are said to stain blue when the fresh tissue, particularly that near the base of the stalk, is damaged, though the converse is not true. Ingestion of the fungus causes visual hallucinations with rapidly changing shapes and colours, and different perceptions of space and time, the effects gradually wearing off and causing no lasting damage or addiction.

The active hallucinogens, present at about 0.3%, are the tryptamine derivatives psilocybin and psilocin (Figure 6.70), which are structurally related to the neurotransmitter 5-HT, thus explaining their neurological effects. Psilocybin is probably the main active ingredient, and to produce hallucinations a dose of some 6–20 mg is required. In addition to species of *Psilocybe*, these compounds may be found in some fungi from other genera, including *Conocybe*, *Panaeolus*, and *Stropharia*. Misidentification of fungi can lead to the consumers experiencing possible unwanted toxic effects, especially gastro-intestinal upsets, instead of the desired psychedelic visions.

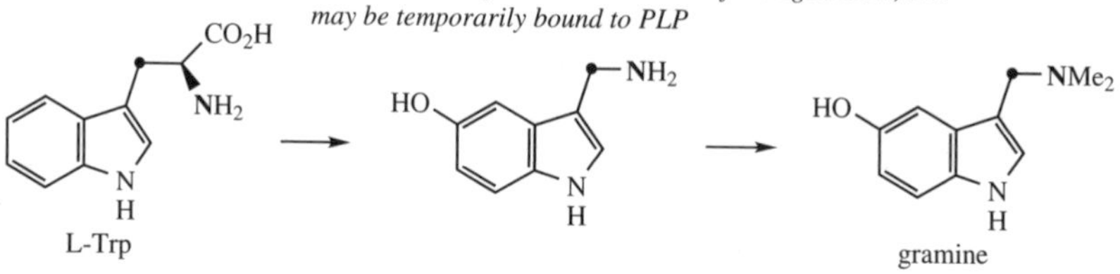

chain shortening involves retention of nitrogen atom; this may be temporarily bound to PLP

Figure 6.72

Gramine (Figure 6.72) is a simple amine found in barley (*Hordeum vulgare*; Graminae/Poaceae) and is derived from tryptophan by a biosynthetic pathway which cleaves off two carbon atoms, yet surprisingly retains the tryptophan nitrogen atom. Presumably, the nitrogen reacts with a cofactor, e.g. pyridoxal phosphate, and is subsequently transferred back to the indolemethyl group after the chain shortening.

Simple β-Carboline Alkaloids

Alkaloids based on a β-carboline system (Figure 6.73) exemplify the formation of a new six-membered heterocyclic ring using the ethylamine side-chain of tryptamine in a process analogous to generation of tetrahydroisoquinoline alkaloids (see page 320). Position 2 of the indole system is nucleophilic due to the adjacent nitrogen, and can participate in a Mannich/Pictet–Spengler type reaction,

attacking a Schiff base generated from tryptamine and an aldehyde (or keto acid) (Figure 6.73). Aromaticity is restored by subsequent loss of the C-2 proton. (It should be noted that the analogous chemical reaction actually involves nucleophilic attack from C-3, and then a subsequent rearrangement occurs to give bonding at C-2; there is no evidence yet for this type of process in biosynthetic pathways.) Extra carbons are supplied by aldehyes or keto acids, according to the complexity of the substrate (compare tetrahydroisoquinoline alkaloids, page 321). Thus, complex β-carbolines, e.g. the terpenoid indole alkaloid ajmalicine (see page 351), are produced by a pathway using an aldehyde such as secologanin. Simpler structures employ keto acids, e.g. **harmine** (Figure 6.74) incorporates two extra carbons from pyruvate. In such a case, an acid is an intermediate, and oxidative decarboxylation gives the dihydro-β-carboline, from which reduced

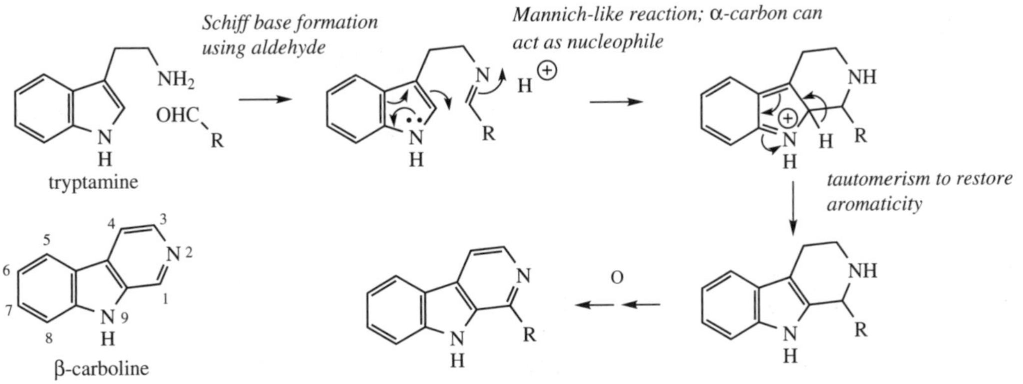

Figure 6.73

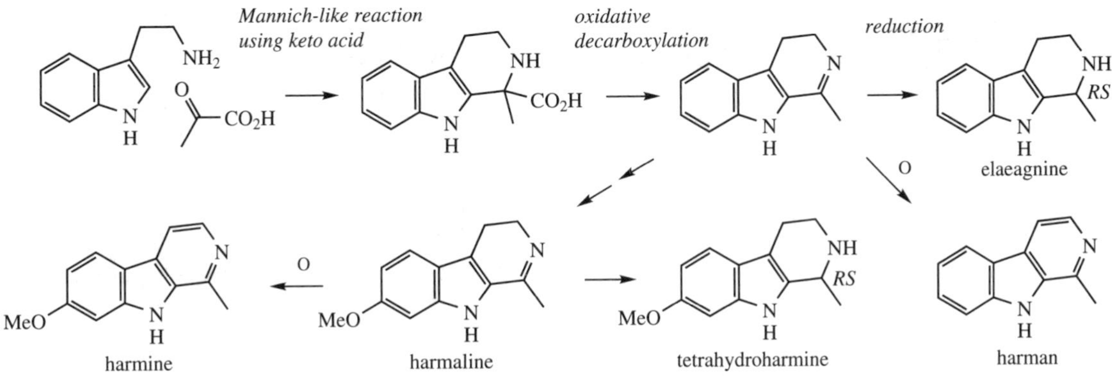

Figure 6.74

tetrahydro-β-carboline structures, e.g. **elaeagnine** from *Elaeagnus angustifolia* (Elaeagnaceae), or fully aromatic β-carboline structures, e.g. **harman** and **harmine** from *Peganum harmala* (Zygophyllaceae) are derived (Figure 6.74). The methoxy substitution in the indole system of harmine is introduced at some stage in the pathway by successive hydroxylation and methylation reactions. A sequence from 6-hydroxytryptamine is also feasible. The reported psychoactive properties of the plants *Peganum harmala* and *Banisteriopsis caapi* (Malpighiaceae) is due to the β-carboline alkaloids such as harmine, harmaline, and tetrahydroharmine (Figure 6.74).

Terpenoid Indole Alkaloids

More than 3000 terpenoid indole alkaloids are recognized, making this one of the major groups of alkaloids in plants. They are found mainly in

eight plant families, of which the Apocynaceae, the Loganiaceae, and the Rubiaceae provide the best sources. In terms of structural complexity, many of these alkaloids are quite outstanding, and it is a tribute to the painstaking experimental studies of various groups of workers that we are able to rationalize these structures in terms of their biochemical origins. In virtually all structures, a tryptamine portion can be recognized. The remaining fragment is usually a C_9 or C_{10} residue, and three main structural types are discernable. These are termed the *Corynanthe* type, as in **ajmalicine** and **akuammicine**, the *Aspidosperma* type, as in **tabersonine**, and the *Iboga* type, exemplified by **catharanthine** (Figure 6.75). The C_9 or C_{10} fragment was shown to be of terpenoid origin, and the secoiridoid **secologanin** (see page 189) was identified as the terpenoid derivative, which initially combined with the tryptamine portion of the molecule. Furthermore, the *Corynanthe*, *Aspidosperma*, and *Iboga* groups of alkaloids could

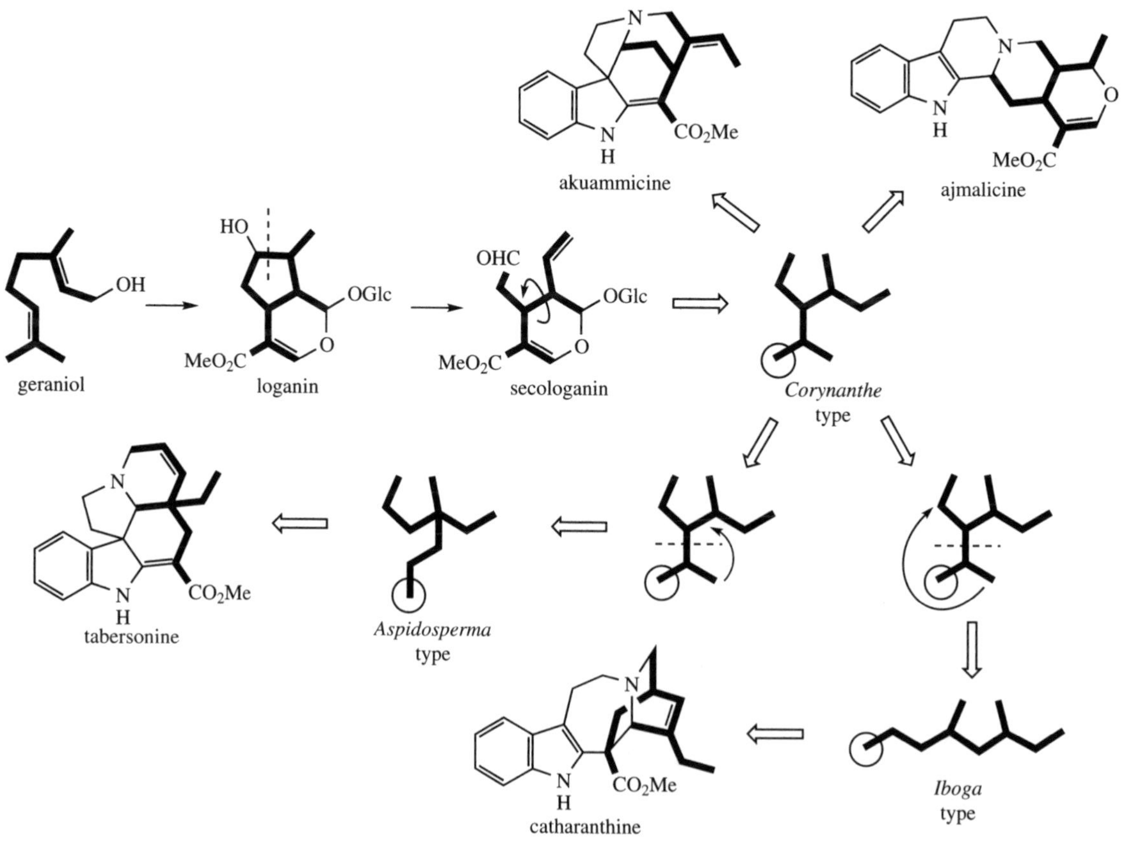

Figure 6.75

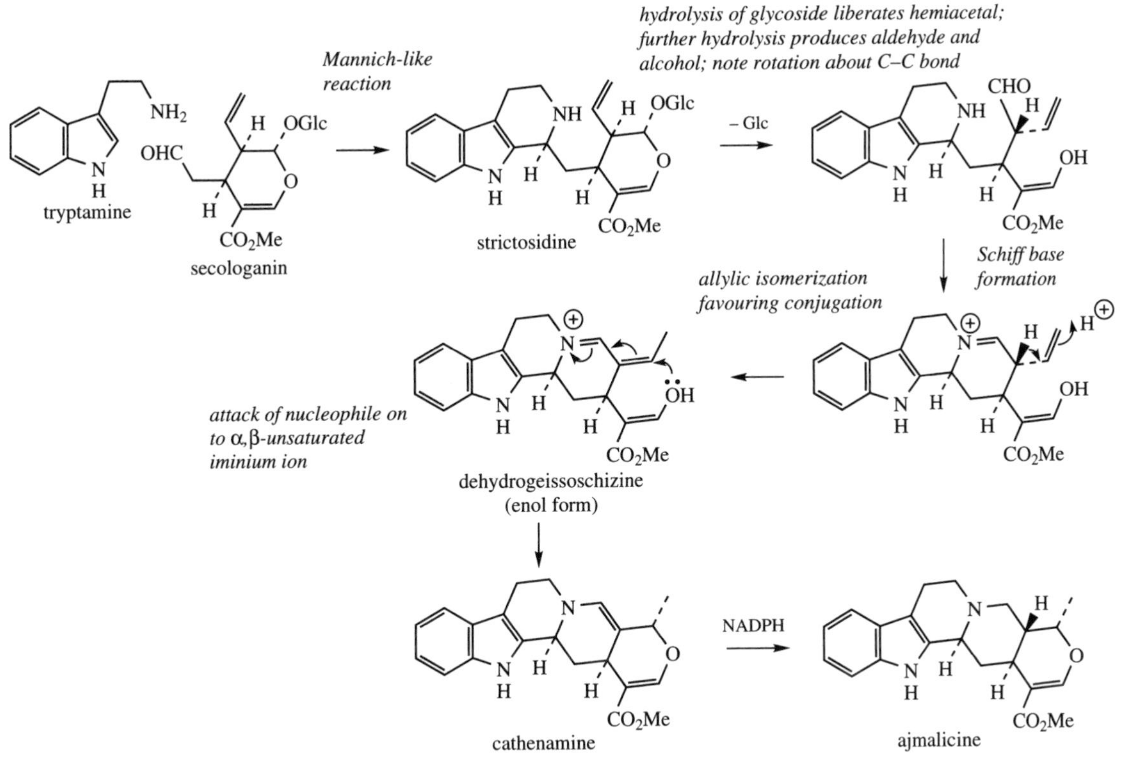

Figure 6.76

be related and rationalized in terms of rearrangements occurring in the terpenoid part of the structures (Figure 6.75). Secologanin itself contains the ten-carbon framework typical of the *Corynanthe* group. The *Aspidosperma* and *Iboga* groups could then arise by rearrangement of the *Corynanthe* skeleton as shown. This is represented by detachment of a three-carbon unit, which is then rejoined to the remaining C_7 fragment in one of two different ways. Where C_9 terpenoid units are observed, the alkaloids normally appear to have lost the carbon atom indicated in the circle. This corresponds to the carboxylate function of secologanin and its loss by hydrolysis/decarboxylation is now understandable.

The origins of loganin and secologanin have already been discussed in Chapter 5 (see page 189). Condensation of secologanin with tryptamine in a Mannich-like reaction generates the tetrahydro-β-carboline system and produces **strictosidine** (Figure 6.76). Hydrolysis of the glycoside function allows opening of the hemiacetal, and exposure of an aldehyde group, which can react with

the secondary amine function giving a quaternary Schiff base. These reactions are also seen in the pathway to ipecac alkaloids (see page 343). Allylic isomerization, moving the vinyl double bond into conjugation with the iminium generates **dehydrogeissoschizine**, and cyclization to **cathenamine** follows. Cathenamine is reduced to **ajmalicine** in the presence of NADPH.

Carbocyclic variants related to ajmalicine such as **yohimbine** are likely to arise from dehydrogeissoschizine by the mechanism indicated in Figure 6.77. Yohimbine is found in Yohimbe bark (*Pausinystalia yohimbe*; Rubiaceae) and Aspidosperma bark (*Aspidosperma* species; Apocynaceae) and has been used in folk medicine as an aphrodisiac. It does have some pharmacological activity and is known to dilate blood vessels. More important examples containing the same carbocyclic ring system are the alkaloids found in species of *Rauwolfia**, especially *R. serpentina* (Apocynaceae). **Reserpine** and **deserpidine** (Figure 6.78) are trimethoxybenzoyl esters of yohimbine-like alkaloids, whilst **rescinnamine** is

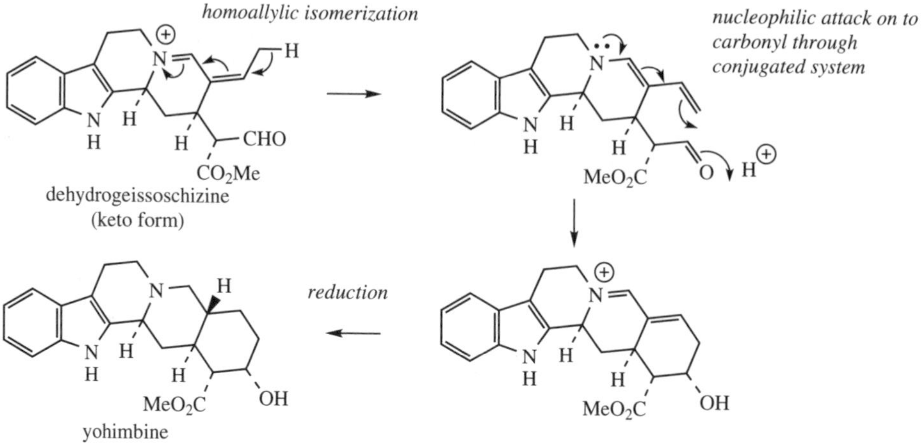

homoallylic isomerization

nucleophilic attack on to carbonyl through conjugated system

dehydrogeissoschizine
(keto form)

reduction

yohimbine

Figure 6.77

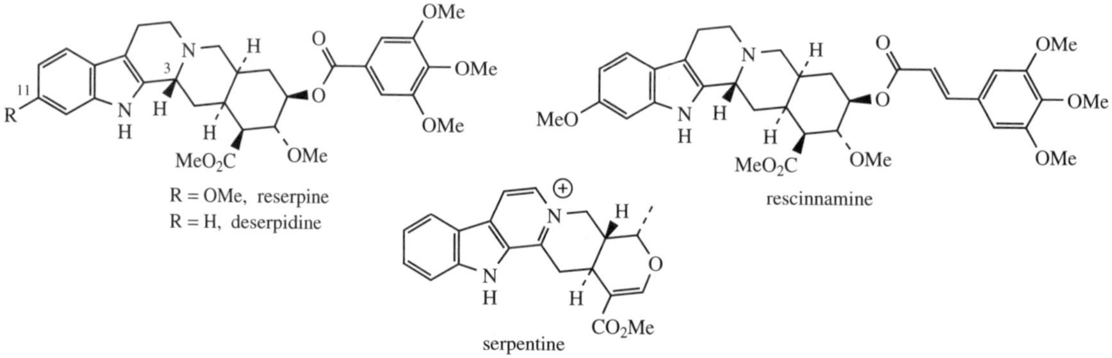

R = OMe, reserpine
R = H, deserpidine

rescinnamine

serpentine

Figure 6.78

Rauwolfia

Rauwolfia has been used in Africa for hundreds of years, and in India for at least 3000 years. It was used as an antidote to snake-bite, to remove white spots in the eyes, against stomach pains, fever, vomiting, and headache, and to treat insanity. It appeared to be a universal panacea, and was not considered seriously by Western scientists until the late 1940s/early 1950s. Clinical tests showed the drug to have excellent antihypertensive and sedative activity. It was then rapidly and extensively employed in treating high blood pressure and to help mental conditions, relieving anxiety and restlessness, and thus initiating the tranquillizer era. The 'cure for insanity' was thus partially justified, and rauwolfia was instrumental in showing that mental disturbance has a chemical basis and may be helped by the administration of drugs.

Rauwolfia is the dried rhizome and roots of *Rauwolfia* (sometimes *Rauvolfia*) *serpentina* (Apocynaceae) or snakeroot, a small shrub from India, Pakistan, Burma, and Thailand. Other species used in commerce include *R. vomitoria* from tropical Africa, a small tree whose leaves after ingestion cause violent vomiting, and *R. canescens* (= *R. tetraphylla*) from India and the

(Continues)

(Continued)

Caribbean. Most of the drug material has been collected from the wild. *Rauwolfia serpentina* contains a wide range of indole alkaloids, totalling 0.7–2.4%, though only 0.15–0.2% consists of desirable therapeutically active compounds, principally reserpine, rescinnamine, and deserpidine (Figure 6.78). Other alkaloids of note are serpentine (Figure 6.78), ajmalicine (Figure 6.76), and ajmaline (see Figure 6.82). Reserpine and deserpidine are major alkaloids in *R. canescens*, and *R. vomitoria* contains large amounts of rescinnamine and reserpine.

Reserpine and **deserpidine** (Figure 6.78) have been widely used as antihypertensives and mild tranquillizers. They act by interfering with catecholamine storage, depleting levels of available neurotransmitters. Prolonged use of the pure alkaloids, reserpine in particular, has been shown to lead to severe depression in some patients, a feature not so prevalent when the powdered root was employed. The complex nature of the alkaloidal mixture means the medicinal action is somewhat different from that of reserpine alone. Accordingly, crude powdered rauwolfia remained an important drug for many years, and selected alkaloid fractions from the crude extract have also been widely used. The alkaloids can be fractionated according to basicity. Thus, serpentine and similar structures are strongly basic, whilst reserpine, rescinnamine, deserpidine and ajmalicine are weak bases. Ajmaline and related compounds have intermediate basicity.

The rauwolfia alkaloids are now hardly ever prescribed in the UK, either as antihypertensives or as tranquillizers. Over a period of a few years, they have been rapidly superseded by synthetic alternatives. Reserpine has also been suggested to play a role in the promotion of breast cancers. Both **ajmalicine** (= raubasine) (Figure 6.76) and **ajmaline** (Figure 6.82) are used clinically in Europe, though not in the UK. Ajmalicine is employed as an antihypertensive, whilst ajmaline is of value in the treatment of cardiac arrhythmias. Ajmalicine is also extracted commercially from *Catharanthus roseus* (see page 357).

a trimethoxycinnamoyl ester. Both reserpine and rescinnamine contain an additional methoxyl substituent on the indole system at position 11, the result of hydroxylation and methylation at a late stage in the pathway. A feature of these alkaloids is that they have the opposite stereochemistry at position 3 to yohimbine and strictosidine. *Rauwolfia serpentina* also contains significant amounts of ajmalicine (Figure 6.76), emphasizing the structural and biosynthetic relationships between the two types of alkaloid.

The structural changes involved in converting the *Corynanthe* type skeleton into those of the *Aspidosperma* and *Iboga* groups are quite complex, and are summarized in Figure 6.79. Early intermediates are alkaloids such as **preakuammicine**, which, although clearly of the *Corynanthe* type, is sometimes designated as *Strychnos* type (compare strychnine, page 358). This is because the *Corynanthe* terpenoid unit, originally attached to the indole α-carbon, is now bonded to the β-carbon, and a new bonding between the rearrangeable C_3 unit and C-α is in place. **Stemmadenine** arises through fission

of the bond to C-β, and then further fission yields a hypothetical intermediate, the importance of which is that the rearrangeable C_3 unit has been cleaved from the rest of the terpenoid carbons. Alkaloids of the *Aspidosperma* type, e.g. **tabersonine** and **vindoline**, and *Iboga* type, e.g. **catharanthine**, then arise from this intermediate by different bonding modes (Figure 6.79).

Many of the experimental studies that have led to an understanding of terpenoid indole alkaloid biosynthesis have been carried out using plants of the Madagascar periwinkle (*Catharanthus roseus**, formerly *Vinca rosea*; Apocynaceae). Representatives of all the main classes of these alkaloids are produced, including **ajmalicine** (*Corynanthe*), **catharanthine** (*Iboga*), and **vindoline** (*Aspidosperma*). The sequence of alkaloid formation has been established initially by noting which alkaloids become labelled as a feeding experiment progresses, and more recently by appropriate enzymic studies. However, the extensive investigations of the *Catharanthus roseus* alkaloids have also been prompted by the anticancer activity

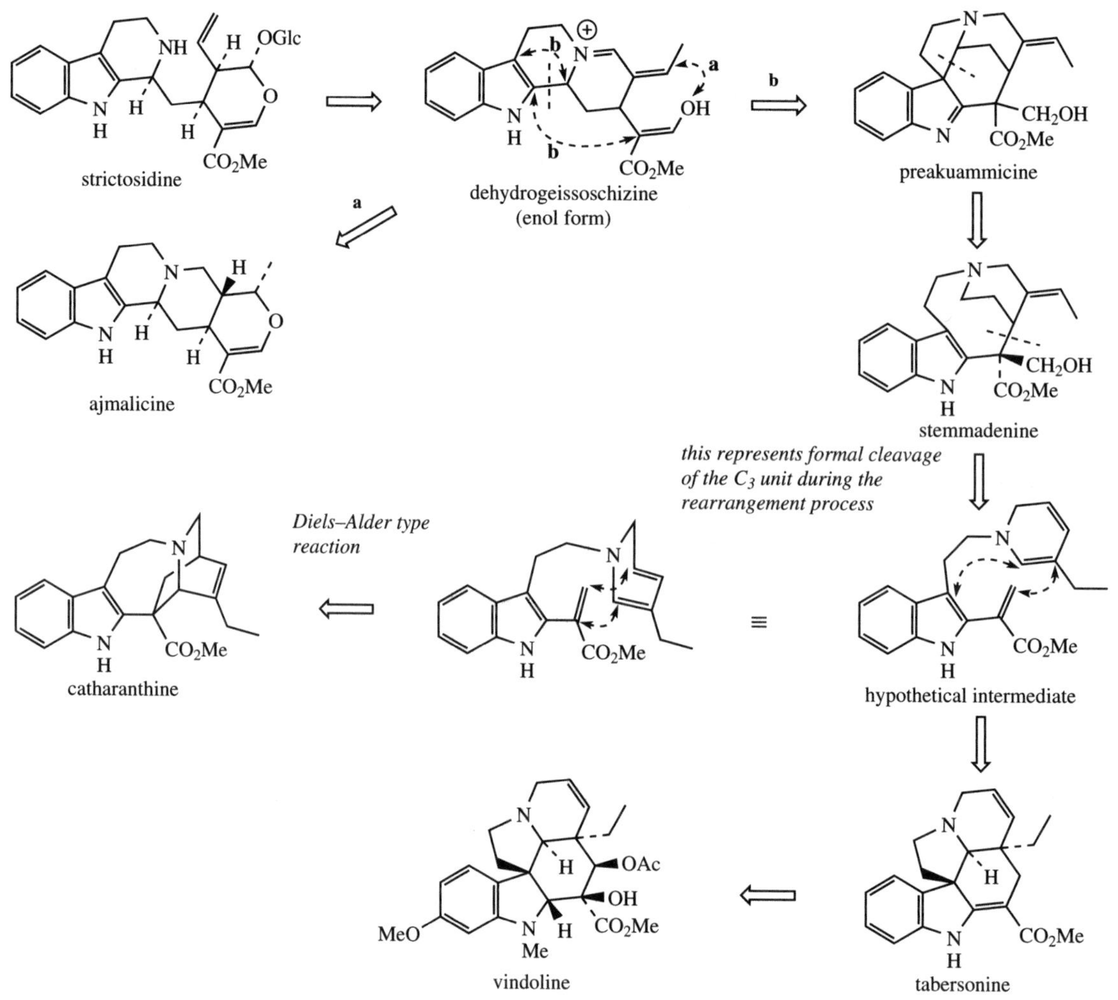

Figure 6.79

detected in a group of bisindole alkaloids. Two of these, **vinblastine** and **vincristine** (Figure 6.80), have been introduced into cancer chemotherapy and feature as some of the most effective anti-cancer agents available. These structures are seen to contain the elements of catharanthine and vindoline, and, indeed, they are derived by coupling of these two alkaloids. The pathway is believed to involve firstly an oxidative reaction on **catharanthine**, catalysed by a peroxidase, generating a peroxide which loses the peroxide as a leaving group, breaking a carbon–carbon bond as shown (Figure 6.81). This intermediate electrophilic ion is attacked by the nucleophilic vindoline, C-5 of the indole nucleus being suitably activated by the OMe at C-6 and also by the indole nitrogen. The adduct

is then reduced in the dihydropyridinium ring by NADH-dependent 1,4-addition, giving the substrate for hydroxylation. Finally, reduction yields **vinblastine**. **Vincristine**, with its N-formyl group rather than N-methyl on the vindoline fragment, may be an oxidized product from vinblastine.

Further variants on the terpenoid indole alkaloid skeleton (Figure 6.82) are found in **ibogaine** from *Tabernanthe iboga**, **vincamine** from *Vinca minor*, and **ajmaline** from *Rauwolfia serpentina*. Ibogaine is simply a C_9 *Iboga* type alkaloid, but is of interest as an experimental drug to treat heroin addiction. In a number of European countries, vincamine is used clinically as a vasodilator to increase cerebral blood flow in cases of senility, and ajmaline for cardiac arrhythmias. Ajmaline

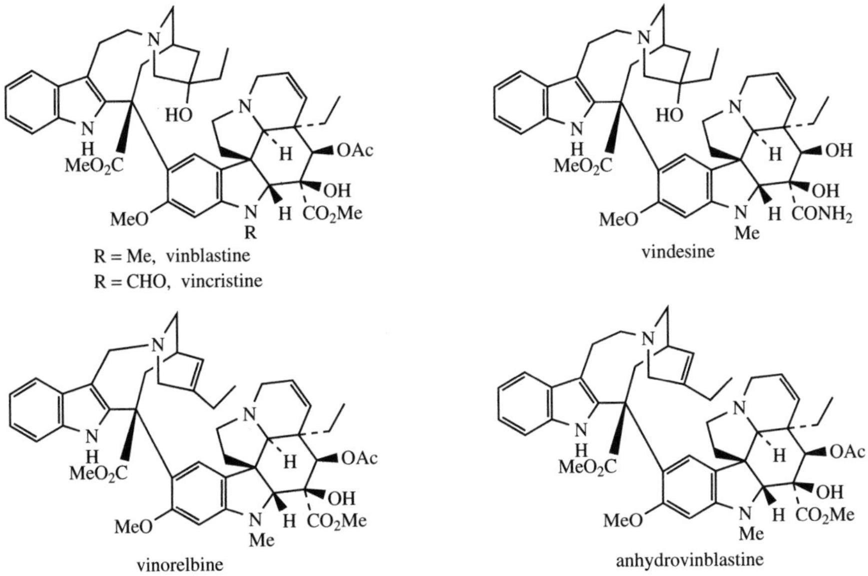

R = Me, vinblastine
R = CHO, vincristine

vindesine

vinorelbine

anhydrovinblastine

Figure 6.80

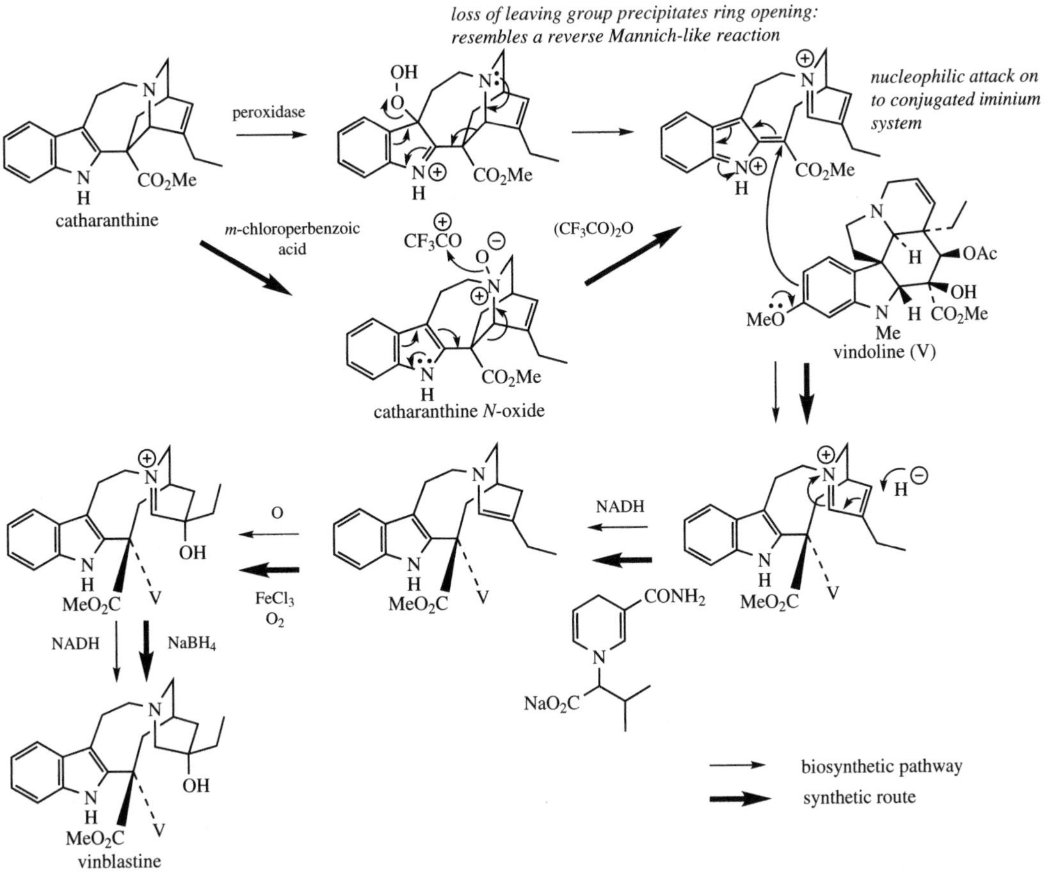

loss of leaving group precipitates ring opening:
resembles a reverse Mannich-like reaction

nucleophilic attack on
to conjugated iminium
system

peroxidase

catharanthine

m-chloroperbenzoic
acid

(CF₃CO)₂O

catharanthine *N*-oxide

vindoline (V)

NADH

O

FeCl₃
O₂

NADH   NaBH₄

vinblastine

⟶ biosynthetic pathway

⟹ synthetic route

Figure 6.81

Catharanthus

The Madagascar periwinkle *Catharanthus roseus* (= *Vinca rosea*) (Apocynaceae) is a small herb or shrub originating in Madagascar, but now common in the tropics and widely cultivated as an ornamental for its shiny dark green leaves and pleasant five-lobed flowers. Drug material is now cultivated in many parts of the world, including the USA, Europe, India, Australia, and South America.

Because of its folklore usage as a tea for diabetics, the plant was originally investigated for potential hypoglycaemic activity. Although plant extracts had no effects on blood sugar levels in rabbits, the test animals succumbed to bacterial infection due to depleted white blood cell levels (leukopenia). The selective action suggested anticancer potential for the plant, and an exhaustive study of the constituents was initiated. The activity was found in the alkaloid fraction, and more than 150 alkaloids have been characterized in the plant. These are principally terpenoid indole alkaloids, many of which are known in other plants, especially from the same family. Useful antitumour activity was demonstrated in a number of dimeric indole alkaloid structures (more correctly bis-indole alkaloids), including vincaleukoblastine, leurosine, leurosidine, and leurocristine. These compounds became known as vinblastine, vinleurosine, vinrosidine, and vincristine respectively, the vin- prefix being a consequence of the earlier botanical nomenclature *Vinca rosea*, which was commonly used at that time. The alkaloids vinblastine and vincristine (Figure 6.80) were introduced into cancer chemotherapy and have proved to be extremely valuable drugs.

Despite the minor difference in structure between vinblastine and vincristine, a significant difference exists in the spectrum of human cancers that respond to the drugs. **Vinblastine** (Figure 6.80) is used mainly in the treatment of Hodgkin's disease, a cancer affecting the lymph glands, spleen, and liver. **Vincristine** (Figure 6.80) has superior antitumour activity compared to vinblastine but is more neurotoxic. It is clinically more important than vinblastine, and is especially useful in the treatment of childhood leukaemia, giving a high rate of remission. Some other cancer conditions, including lymphomas, small cell lung cancer, and cervical and breast cancers, also respond favourably. The alkaloids need to be injected, and both generally form part of a combination regimen with other anticancer drugs. **Vindesine** (Figure 6.80) is a semi-synthetic derivative of vinblastine, which has been introduced for the treatment of acute lymphoid leukaemia in children. **Vinorelbine** (Figure 6.80), an anhydro derivative of 8′-norvinblastine, is a newer semi-synthetic modification obtained from anhydrovinblastine (Figure 6.80), where the indole.C_2N bridge in the catharanthine-derived unit has been shortened by one carbon. It is orally active and has a broader anticancer activity yet with lower neurotoxic side-effects than either vinblastine or vincristine. These compounds all inhibit cell mitosis, acting by binding to the protein tubulin in the mitotic spindle, preventing polymerization into microtubules, a mode of action shared with other natural agents, e.g. colchicine (see page 343) and podophyllotoxin (see page 136).

A major problem associated with the clinical use of vinblastine and vincristine is that only very small amounts of these desirable alkaloids are present in the plant. Although the total alkaloid content of the leaf can reach 1% or more, over 500 kg of catharanthus is needed to yield 1 g of vincristine. This yield (0.0002%) is the lowest of any medicinally important alkaloid isolated on a commercial basis. Extraction is both costly and tedious, requiring large quantities

(Continues)

(Continued)

of raw material and extensive use of chromatographic fractionations. The growing importance of vincristine relative to vinblastine as drugs is not reflected in the plant, which produces a much higher proportion of vinblastine. Fortunately, it is possible to convert vinblastine into vincristine by controlled chromic acid oxidation or via a microbiological *N*-demethylation using *Streptomyces albogriseolus*. Considerable effort has been expended on the semi-synthesis of the 'dimeric' alkaloids from 'monomers' such as catharanthine and vindoline, which are produced in *C. roseus* in much larger amounts. Efficient, stereospecific coupling has eventually been achieved, and it is now possible to convert catharanthine and vindoline into vinblastine in about 40% yield. The process used is a biomimetic one, virtually identical to the suggested biosynthetic process, and is also included in Figure 6.81. Catharanthine-*N*-oxide is employed instead of the peroxidase-generated peroxide, and this couples readily in trifluoroacetic anhydride with vindoline in almost quantitative yield. Subsequent reduction, oxidation, and reduction steps then give vinblastine via the same 'biosynthetic' intermediates. It is particularly interesting that the most effective reducing agents for the transformation of the dihydropyridinium compound into the tetrahydropyridine were *N*-substituted 1,4-dihydronicotinamides, simpler analogues of NADH, the natural reducing agent. Excellent yields of anhydrovinblastine (the starting material for vinorelbine production) (Figure 6.80) can also be obtained by electrochemical oxidation of catharanthine/vindoline. These syntheses should improve the supply of these alkaloids and derivatives, and also allow more detailed studies of structure–activity relationships to be undertaken. This group of compounds is still of very high interest, and development programmes for analogues continue.

Ajmalicine (see rauwolfia, page 353) is present in the roots of *Catharanthus roseus* at a level of about 0.4%, and this plant is used as a commercial source in addition to *Rauwolfia serpentina*.

Iboga

The *Iboga* group of terpenoid indole alkaloids takes its name from *Tabernanthe iboga* (Apocynaceae), a shrub from the Congo and other parts of equatorial Africa. Extracts from the root bark of this plant have long been used by indigenous people in rituals, to combat fatigue, and as an aphrodisiac. The root bark contains up to 6% indole alkaloids, the principal component of which is ibogaine (Figure 6.82). Ibogaine is a CNS stimulant, and is also psychoactive. In large doses, it can cause paralysis and respiratory arrest. Ibogaine is of interest as a potential drug for relieving heroin craving in drug addicts. Those who use the drug experience hallucinations from the ibogaine, but it is claimed they emerge from this state with a significantly reduced opiate craving. A number of deaths resulting from the unsupervised use of ibogaine has led to its being banned in some countries.

---

contains a C_9 *Corynanthe* type unit and its relationship to **dehydrogeissoschizine** is indicated in Figure 6.82. Vincamine still retains a C_{10} *Aspidosperma* unit, and it originates from **tabersonine** by a series of reactions that involve cleavage of bonds to both α and β positions of the indole (Figure 6.82).

Alkaloids like **preakuammicine** (Figure 6.79) and **akuammicine** (Figure 6.75) contain the C_{10} and C_9 *Corynanthe* type terpenoid units respectively. They are, however, representatives of a subgroup of *Corynanthe* alkaloids termed the *Strychnos* type because of their structural similarity to many of the alkaloids found in *Strychnos*

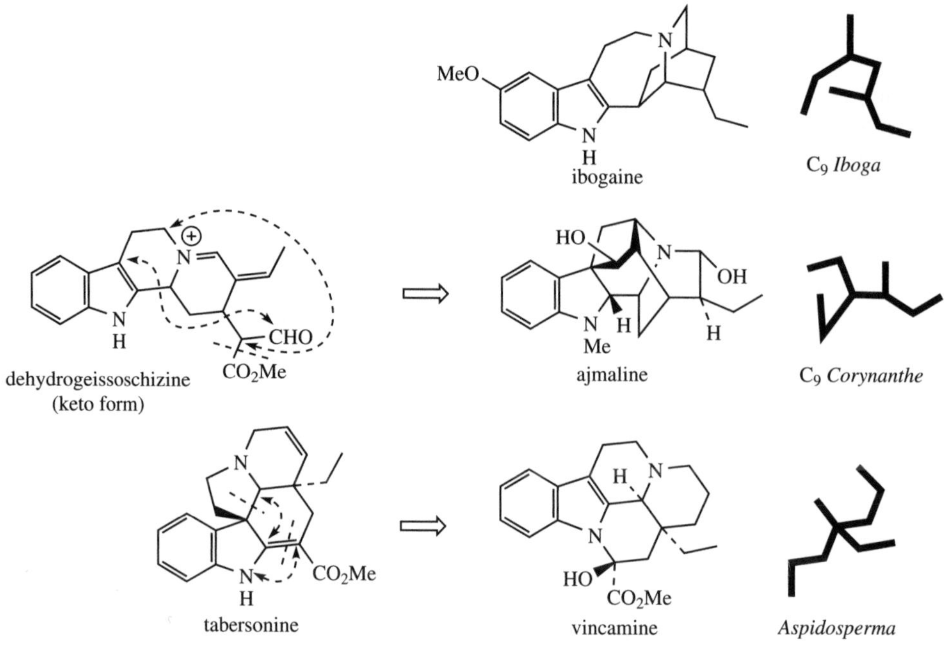

Figure 6.82

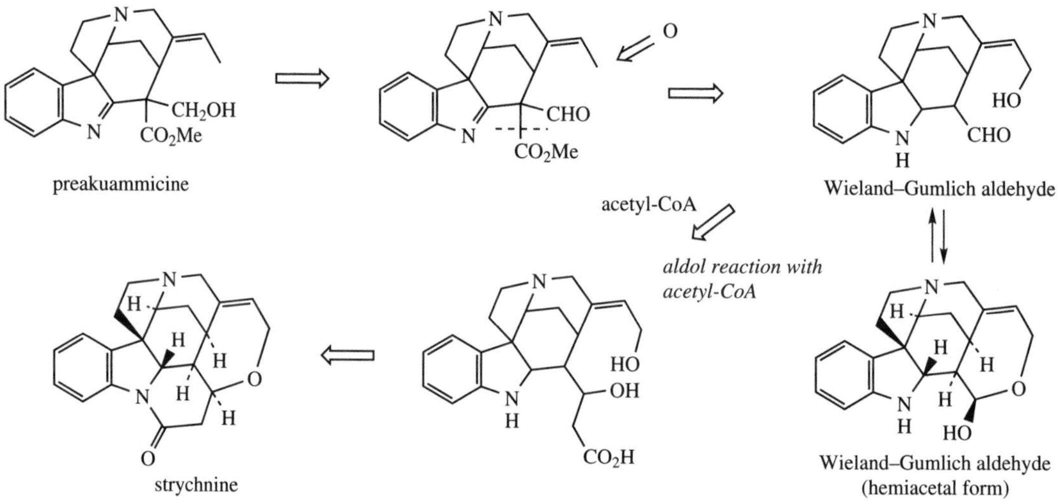

Figure 6.83

species (Loganiaceae), e.g. *S. nux-vomica**, note-worthy examples being the extremely poisonous **strychnine** (Figure 6.83) and its dimethoxy ana-logue **brucine** (Figure 6.84). The non-tryptamine portion of these compounds contains 11 carbons, and is constructed from an iridoid-derived C_9 unit, plus two further carbons supplied from acetate. The pathway to **strychnine** in Figure 6.83 involves loss of one carbon from a preakuammicine-like struc-ture via hydrolysis/decarboxylation and then addi-tion of the extra two carbons by aldol condensation with the formyl group, complexed as a hemiacetal in the so-called Wieland–Gumlich aldehyde. The subsequent formation of strychnine from this hemi-acetal is merely construction of ether and amide linkages.

Nux-vomica

Nux-vomica consists of the dried ripe seeds of *Strychnos nux-vomica* (Loganiaceae), a small tree found in a wide area of East Asia extending from India to Northern Australia. The fruit is a large berry with a hard coat and a pulpy interior containing three to five flattish grey seeds. These seeds contain 1.5–5% of alkaloids, chiefly strychnine (about 1.2%) and brucine (about 1.6%) (Figure 6.82). **Strychnine** is very toxic, affecting the CNS and causing convulsions. This is a result of binding to receptor sites in the spinal cord that normally accommodate glycine. Fatal poisoning (consumption of about 100 mg by an adult) would lead to asphyxia following contraction of the diaphragm. It has found use as a vermin killer, especially for moles. Its only medicinal use is in very small doses as an appetite stimulant and general tonic, sometimes with iron salts if the patient is anaemic. Brucine is considerably less toxic. Both compounds have been regularly used in synthetic chemistry as optically active bases to achieve optical resolution of racemic acids. Seeds of the related *Strychnos ignatii* have also served as a commercial source of strychnine and brucine.

Of biochemical interest is the presence of quite significant amounts (up to 5%) of the iridoid glycoside loganin (see page 188) in the fruit pulp of *Strychnos nux-vomica*. This compound is, of course, an intermediate in the biosynthesis of strychnine and other terpenoid indole alkaloids.

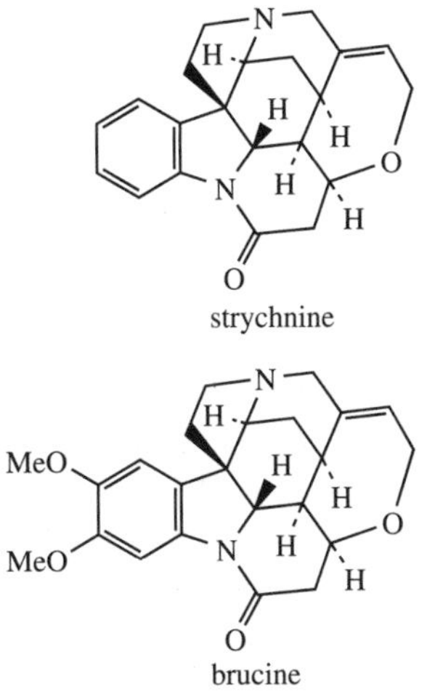

strychnine

brucine

Figure 6.84

The arrow poison curare, when produced from *Chondrodendron* species (Menispermaceae), contains principally the bis-benzyltetrahydroisoquinoline alkaloid tubocurarine (see page 324). Species of *Strychnos*, especially *S. toxifera*, are employed in making loganiaceous curare, and biologically active alkaloids isolated from such preparations have been identified as a series of toxiferines, e.g. **C-toxiferine** (Figure 6.85). The structures appear remarkably complex, but may be envisaged as a combination of two Wieland–Gumlich aldehyde-like molecules (Figure 6.85). The presence of two quaternary nitrogens, separated by an appropriate distance, is responsible for the curare-like activity (compare tubocurarine and synthetic analogues, page 326). **Alcuronium** (Figure 6.85) is a semi-synthetic skeletal muscle relaxant produced from C-toxiferine (see curare, page 327).

Ellipticine* (Figure 6.86) contains a pyrido-carbazole skeleton, which is also likely to be formed from a tryptamine–terpenoid precursor. Although little evidence is available, it is suggested that a precursor like **stemmadenine** may undergo transformations that effectively remove the two-carbon bridge originally linking the indole and the nitrogen in tryptamine (Figure 6.86). The remaining C_9 terpenoid fragment now containing the tryptamine nitrogen can then be used to generate the rest of the skeleton. Ellipticine is found in *Ochrosia elliptica* (Apocynaceae) and related species and has useful anticancer properties.

Quinoline Alkaloids

Some of the most remarkable examples of terpenoid indole alkaloid modifications are to be found in the genus *Cinchona** (Rubiaceae), in the alkaloids **quinine**, **quinidine**, **cinchonidine**,

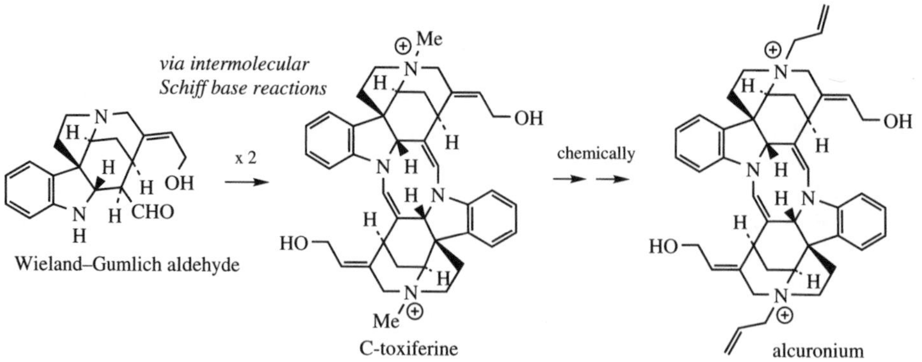

via intermolecular
Schiff base reactions

Wieland–Gumlich aldehyde

x 2

chemically

C-toxiferine

alcuronium

Figure 6.85

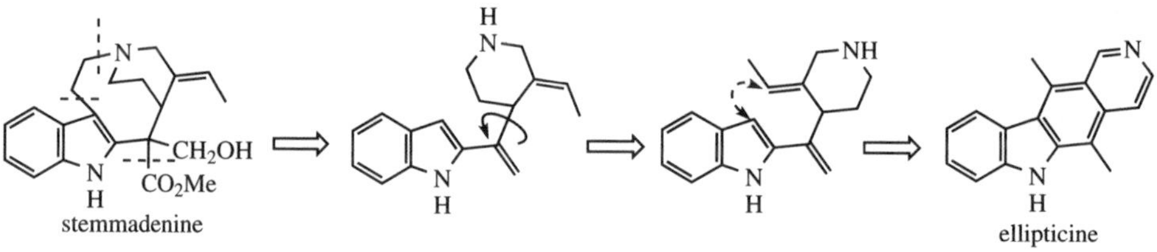

stemmadenine

ellipticine

Figure 6.86

Ellipticine

Ellipticine (Figure 6.86) and related alkaloids, e.g. 9-methoxyellipticine (Figure 6.87), are found in the bark of *Ochrosia elliptica* (Apocynaceae) and other *Ochrosia* species. Clinical trials with these alkaloids and a number of synthetic analogues showed them to be potent inhibitors of several cancerous disorders, but pre-clinical toxicology indicated a number of side-effects, including haemolysis and cardiovascular effects. Ellipticines are planar molecules that intercalate between the base pairs of DNA and cause a partial unwinding of the helical array. There is a some correlation between the degree of unwinding and the biological properties, those showing the largest unwinding inhibiting the greatest number of cancerous cells. Recent research suggests there may be more than one mechanism of action, however. Ellipticine is oxidized *in vivo* mainly to 9-hydroxyellipticine, which has an increased activity, and it is believed that this may in fact be the active agent. Poor water-solubility of ellipticine and derivatives gave problems in formulation for clinical use, but quaternization of 9-hydroxyellipticine to give the water-soluble 9-hydroxy-2-*N*-methylellipticinium acetate (**elliptinium acetate**) (Figure 6.87) has produced a highly active material, of value in some forms of breast cancer, and perhaps also in renal cell cancer. A variety of such quaternized derivatives is being tested, and some water-soluble *N*-glycosides also show high activity.

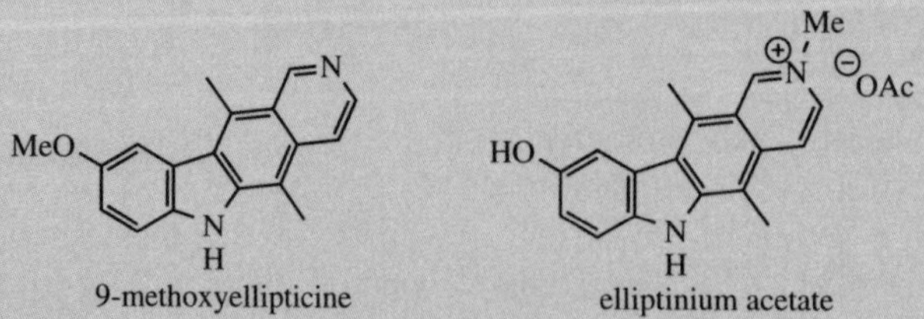

9-methoxyellipticine

elliptinium acetate

Figure 6.87

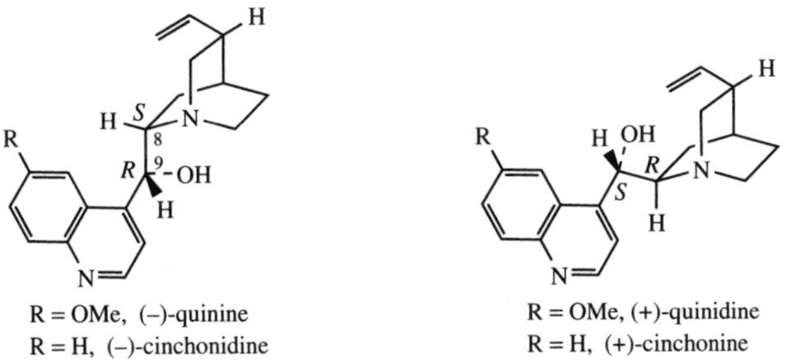

Figure 6.88

R = OMe, (–)-quinine
R = H, (–)-cinchonidine

R = OMe, (+)-quinidine
R = H, (+)-cinchonine

and **cinchonine** (Figure 6.88), long prized for their antimalarial properties. These structures are remarkable in that the indole nucleus is no longer present, having been rearranged into a quinoline system (Figure 6.89). The relationship was suspected quite early on, however, since the indole derivative **cinchonamine** (Figure 6.90) was known

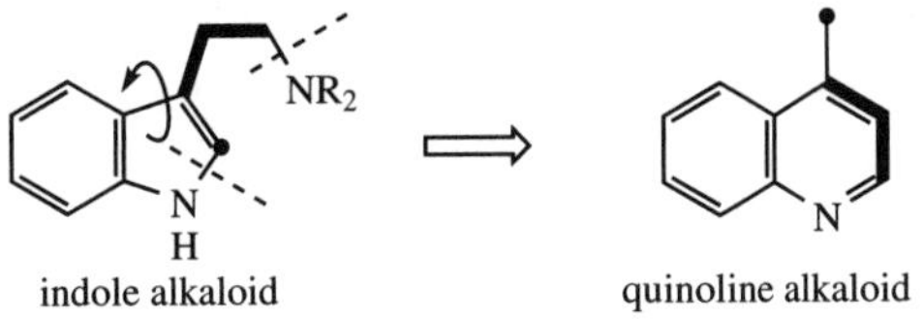

indole alkaloid

quinoline alkaloid

Figure 6.89

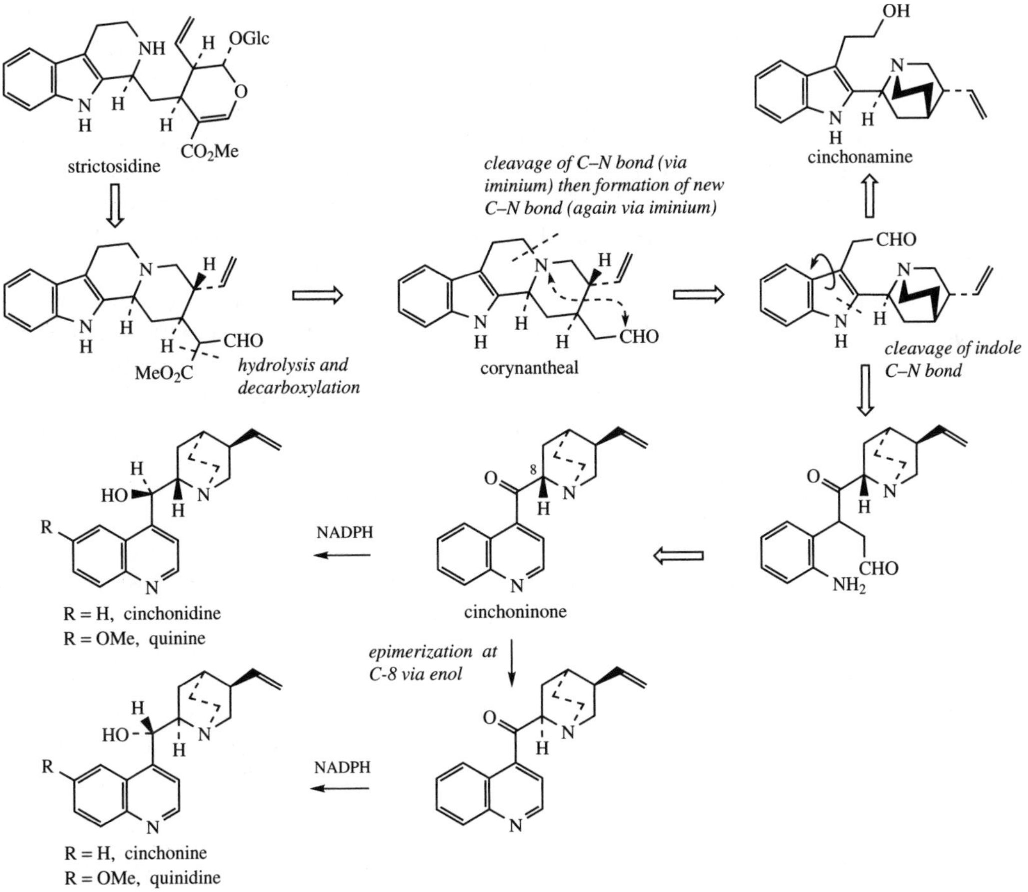

strictosidine

cleavage of C–N bond (via iminium) then formation of new C–N bond (again via iminium)

cinchonamine

hydrolysis and decarboxylation

corynantheal

cleavage of indole C–N bond

R = H, cinchonidine
R = OMe, quinine

cinchoninone

R = H, cinchonine
R = OMe, quinidine

epimerization at C-8 via enol

NADPH

NADPH

Figure 6.90

to co-occur with these quinoline alkaloids. An outline of the pathway from the *Corynanthe*-type indole alkaloids to cinchonidine is shown in Figure 6.90. The conversion is dependent on the reversible processes by which amines plus aldehydes or ketones, imines (Schiff bases), and their reduction products are related in nature. Suitable modification of strictosidine leads to an aldehyde (compare the early reactions in the ajmalicine pathway (Figure 6.76)). Hydrolysis/decarboxylation would initially remove one carbon from the iridoid portion and produce **corynantheal**. An intermediate of the cinchonamine type would then result if the tryptamine side-chain were cleaved adjacent to the nitrogen, and if this nitrogen were then bonded to the acetaldehyde function. Ring opening in the indole heterocyclic ring

could generate new amine and keto functions. The new heterocycle would then be formed by combining this amine with the aldehyde produced in the tryptamine side-chain cleavage. Finally, reduction of the ketone gives **cinchonidine** or **cinchonine**. Hydroxylation and methylation at some stage allows biosynthesis of **quinine** and **quinidine**. Quinine and quinidine, or cinchonidine and cinchonine, are pairs of diastereoisomers, which have opposite chiralities at two centres (Figure 6.88). Stereospecific reduction of the carbonyl in cinchoninone can control the stereochemistry adjacent to the quinoline ring (C-9). The stereochemistry at the second centre (C-8) is also determined during the reduction step, presumably via the enol form of cinchoninone (Figure 6.90).

Cinchona

Cinchona bark is the dried bark from the stem and root of species of *Cinchona* (Rubiaceae), which are large trees indigenous to South America. Trees are cultivated in many parts of the world, including Bolivia, Guatemala, India, Indonesia, Zaire, Tanzania, and Kenya. About a dozen different *Cinchona* species have been used as commercial sources, but the great variation in alkaloid content, and the range of alkaloids present, has favoured cultivation of three main species, together with varieties, hybrids, and grafts. *Cinchona succirubra* provides what is called 'red' bark (alkaloid content 5–7%), *C. ledgeriana* gives 'brown' bark (alkaloid content 5–14%), and *C. calisaya* 'yellow' bark with an alkaloid content of 4–7%. Selected hybrids can yield up to 17% total alkaloids. Bark is stripped from trees which are about 8–12 years old, the trees being totally uprooted by tractor for the process.

A considerable number of alkaloids have been characterized in cinchona bark, four of which account for some 30–60% of the alkaloid content. These are quinine, quinidine, cinchonidine, and cinchonine, quinoline-containing structures representing two pairs of diastereoisomers (Figure 6.88). Quinine and quinidine have opposite configurations at two centres. Cinchonidine and cinchonine are demethoxy analogues, but unfortunately use of the -*id*- syllable in the nomenclature does not reflect a particular stereochemistry. Quinine is usually the major component (half to two-thirds total alkaloid content) but the proportions of the four alkaloids vary according to species or hybrid. The alkaloids are often present in the bark in salt combination with quinic acid (see page 122) or a tannin material called cinchotannic acid. Cinchotannic acid decomposes due to enzymic oxidation during processing of the bark to yield a red pigment, which is particularly prominent in the 'red' bark.

Cinchona and its alkaloids, particularly **quinine**, have been used for many years in the treatment of malaria, a disease caused by protozoa, of which the most troublesome is *Plasmodium falciparum*. The beneficial effects of cinchona bark were first discovered in South America in the 1630s, and the bark was then brought to Europe by Jesuit missionaries. Religious intolerance initially restricted its universal acceptance, despite the widespread occurrence of malaria in Europe and elsewhere. The name cinchona is a mis-spelling derived from Chinchon. In an often quoted tale, now historically disproved, the Spanish Countess of Chinchon, wife of the viceroy of Peru, was reputedly cured of malaria by the bark. For

(Continues)

(Continued)

many years, the bark was obtained from South America, but cultivation was eventually established by the English in India, and by the Dutch in Java, until just before the Second World War, when almost all the world's supply came from Java. When this source was cut off by Japan in the Second World War, a range of synthetic antimalarial drugs was hastily produced as an alternative to quinine. Many of these compounds were based on the quinine structure. Of the wide range of compounds produced, **chloroquine, primaquine**, and **mefloquine** (Figure 6.91) are important antimalarials. Primaquine is exceptional in having an 8-aminoquinoline structure, whereas chloroquine and mefloquine retain the 4-substituted quinoline as in quinine. The acridine derivative **mepacrine** (Figure 6.91), though not now used for malaria treatment, is of value in other protozoal infections. **Halofantrine** (Figure 6.91) dispenses with the heterocyclic ring system completely, and is based on phenanthrene. At one time, synthetic antimalarials had almost entirely superseded natural quinine, but the emergence of *Plasmodium falciparum* strains resistant to the synthetic drugs, especially the widely used prophylactic chloroquine, has resulted in reintroduction of quinine. Mefloquine is currently active against chloroquine-resistant strains, but, whilst ten times as active as quinine, does produce gastrointestinal upsets and dizziness, and can trigger psychological problems such as depression, panic, or psychosis in some patients. The ability of *P. falciparum* to develop resistance to modern drugs means malaria still remains a huge health problem, and is probably the major single cause of deaths in the modern world. **Chloroquine** and its derivative **hydroxychloroquine** (Figure 6.91), although antimalarials, are also used to suppress the disease process in rheumatoid arthritis.

   Quinine (Figure 6.88), administered as free base or salts, continues to be used for treatment of multidrug-resistant malaria, though it is not suitable for prophylaxis. The specific mechanism of action is not thoroughly understood, though it is believed to prevent polymerization of toxic haemoglobin breakdown products formed by the parasite (see artemisinin, page 200). Vastly larger amounts of the alkaloid are consumed in beverages, including vermouth and tonic water. It is amusing to realize that gin was originally added to quinine to make the bitter antimalarial more palatable. Typically, the quinine dosage was up to 600 mg three times a

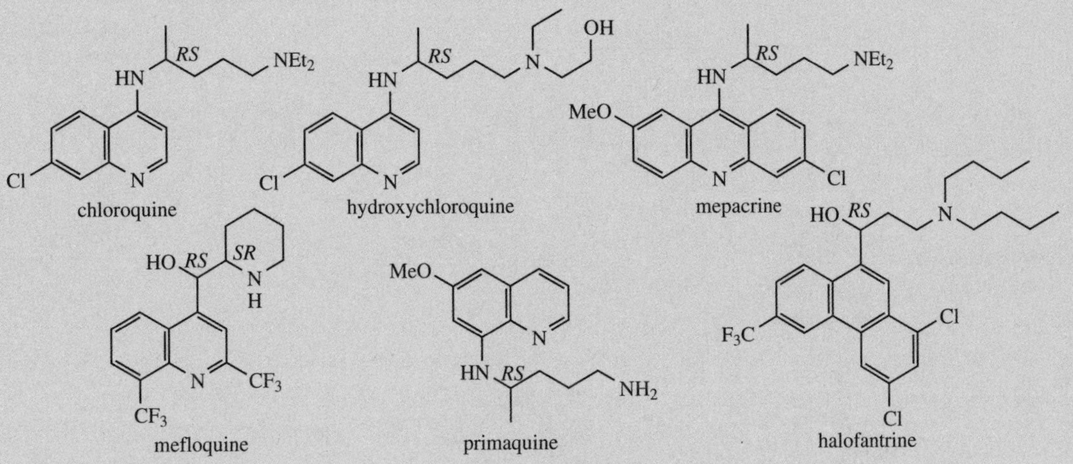

Figure 6.91

(Continues)

(Continued)

day. Quinine in tonic water is now the mixer added to gin, though the amounts of quinine used (about 80 mg l⁻¹) are well below that providing antimalarial protection. Quinine also has a skeletal muscle relaxant effect with a mild curare-like action. It thus finds use in the prevention and treatment of nocturnal leg cramp, a painful condition affecting many individuals, especially the elderly.

Quinidine (Figure 6.88) is the principal cinchona alkaloid used therapeutically, and is administered to treat cardiac arrhymias. It inhibits fibrillation, the uncoordinated contraction of muscle fibres in the heart. It is rapidly absorbed by the gastrointestinal tract and overdose can be hazardous, leading to diastolic arrest.

Quinidine, cinchonine, and cinchonidine also have antimalarial properties, but these alkaloids are not as effective as quinine. The cardiac effect makes quinidine unsuitable as an antimalarial. However, mixtures of total *Cinchona* alkaloids, even though low in quinine content, are acceptable antimalarial agents. This mixture, termed totaquine, has served as a substitute for quinine during shortages. Quinine-related alkaloids, especially quinidine, are also found in the bark of *Remija pendunculata* (Rubiaceae).

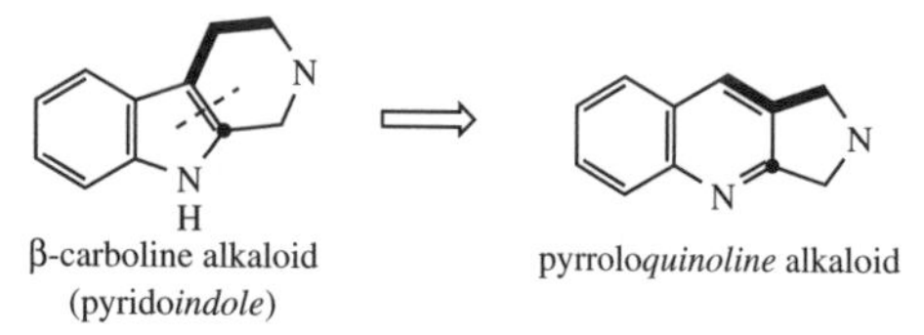

β-carboline alkaloid
(pyrido*indole*)

pyrrolo*quinoline* alkaloid

Figure 6.92

Camptothecin* (Figure 6.93) from *Camptotheca acuminata* (Nyssaceae) is a further example of a quinoline-containing structure that is actually derived by modification of an indole system. The main rearrangement process is that the original β-carboline 6–5–6 ring system becomes a 6–6–5 pyrroloquinoline by ring expansion of the indole

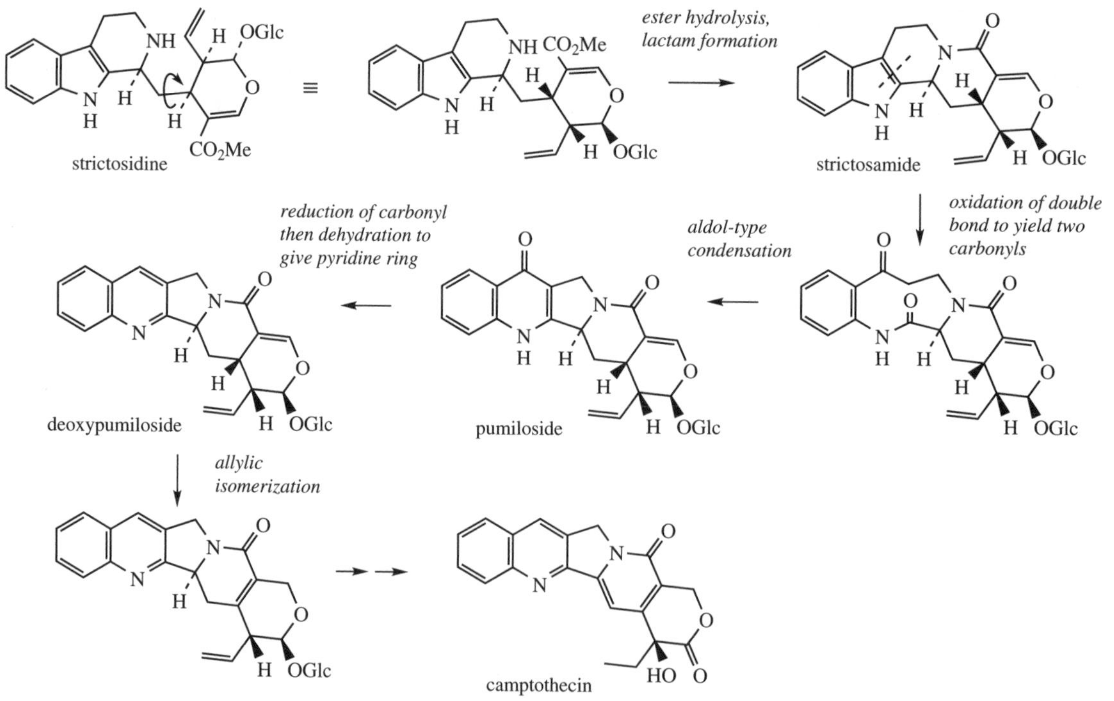

strictosidine

strictosamide

ester hydrolysis, lactam formation

deoxypumiloside

pumiloside

reduction of carbonyl then dehydration to give pyridine ring

aldol-type condensation

oxidation of double bond to yield two carbonyls

allylic isomerization

camptothecin

Figure 6.93

heterocycle (Figure 6.92). In camptothecin, the iridoid portion from strictosidine is effectively still intact, the original ester function being utilized in forming an amide linkage to the secondary amine. This occurs relatively early, in that **strictosamide** is an intermediate. **Pumiloside** (also isolated from *C. acuminata*) and **deoxypumiloside** are potential intermediates. Steps beyond are not yet defined, but involve relatively straightforward oxidation and reduction processes (Figure 6.93).

Pyrroloindole Alkaloids

Both C-2 and C-3 of the indole ring can be regarded as nucleophilic, but reactions involving C-2 appear to be the most common in alkaloid biosynthesis. There are examples where the nucleophilic character of C-3 is exploited, however, and the rare pyrroloindole skeleton typified by **physostigmine** (**eserine**) (Figure 6.95) is a likely case. A suggested pathway to physostigmine is by C-3 methylation of tryptamine, followed by ring

---

Camptothecin

Camptothecin (Figure 6.93) and derivatives are obtained from the Chinese tree *Camptotheca acuminata* (Nyssaceae). Seeds yield about 0.3% camptothecin, bark about 0.2%, and leaves up to 0.4%. *Camptotheca acuminata* is found only in Tibet and West China, but other sources of camptothecin such as *Nothapodytes foetida* (formerly *Mappia foetida*) (Icacinaceae), *Merilliodendron megacarpum* (Icacinaceae), *Pyrenacantha klaineana* (Icacinaceae), *Ophiorrhiza mungos* (Rubiaceae), and *Ervatmia heyneana* (Apocynaceae) have been discovered. In limited clinical trials camptothecin showed broad-spectrum anticancer activity, but toxicity and poor solubility were problems. The natural 10-hydroxycamptothecin (about 0.05% in the bark of *C. acuminata*) is more active than camptothecin, and is used in China against cancers of the neck and head. Synthetic analogues 9-aminocamptothecin (Figure 6.94) and the water-soluble derivatives **topotecan** and **irinotecan** (Figure 6.94) showed good responses in a number of cancers; topotecan and irinotecan are now available for the treatment of ovarian cancer and colorectal cancer, respectively. Irinotecan is a carbamate pro-drug of 10-hydroxy-7-ethylcamptothecin, and is converted into the active drug by liver enzymes. These agents act by inhibition of the enzyme topoisomerase I, which is involved in DNA replication and reassembly, by binding to and stabilizing a covalent DNA–topoisomerase complex (see page 137). Camptothecin has also been shown to have potentially useful activity against pathogenic protozoa such as *Trypanosoma brucei* and *Leishmania donovani*, which cause sleeping sickness and leishmaniasis respectively. Again, this is due to topoisomerase I inhibition.

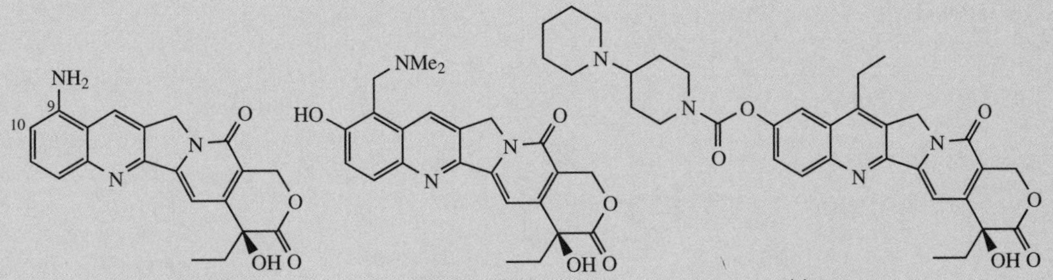

9-aminocamptothecin   topotecan   irinotecan

Figure 6.94

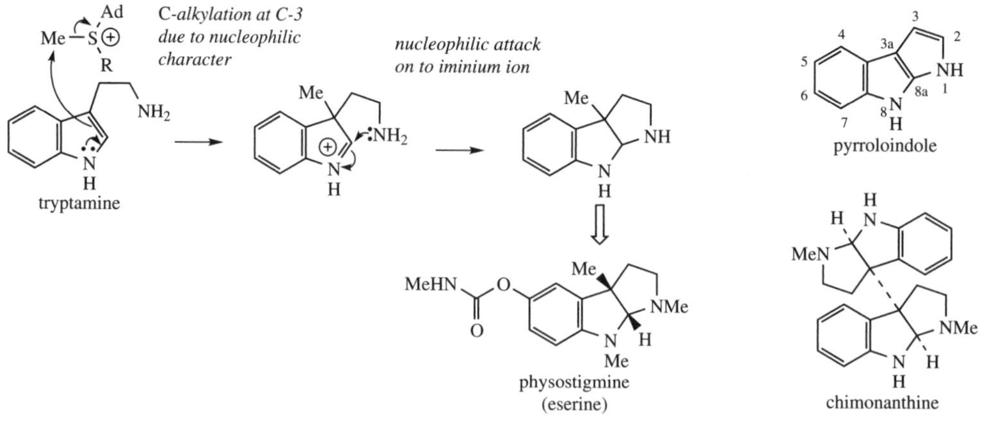

Figure 6.95

formation involving attack of the primary amine function on to the iminium ion (Figure 6.95). Further substitution is then necessary. Dimers with this ring system are also known, e.g. **chimonanthine**

(Figure 6.95) from *Chimonanthus fragrans* (Calycanthaceae), the point of coupling being C-3 of the indole, and an analogous radical reaction may be proposed. Physostigmine is found in seeds of

Physostigma

Physostigma venenosum (Leguminosae/Fabaceae) is a perennial woody climbing plant found on the banks of streams in West Africa. The seeds are known as Calabar beans (from Calabar, now part of Nigeria) and have an interesting history in the native culture as an ordeal poison. The accused was forced to swallow a potion of the ground seeds, and if the mixture was subsequently vomited, he/she was judged innocent and set free. If the poison took effect, the prisoner suffered progressive paralysis and died from cardiac and respiratory failure. It is said that slow consumption allows the poison to take effect, whilst emesis is induced by a rapid ingestion of the dose.

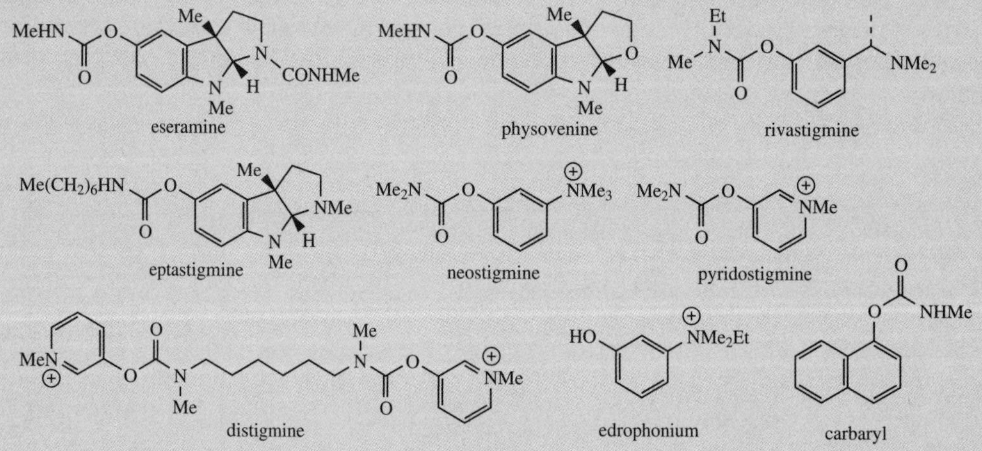

Figure 6.96

(Continues)

(Continued)

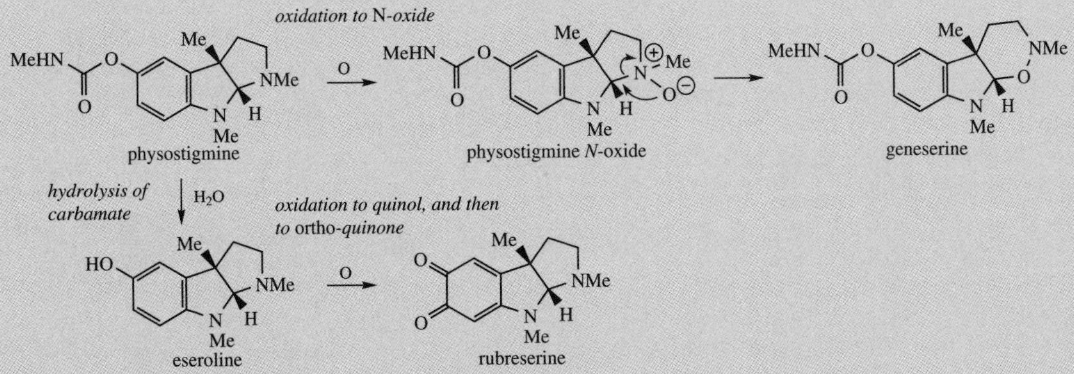

Figure 6.97

The seeds contain several alkaloids (alkaloid content about 1.5%), the major one (up to 0.3%) being physostigmine (eserine) (Figure 6.95). The unusual pyrroloindole ring system is also present in some of the minor alkaloids, e.g. eseramine (Figure 6.96), whilst physovenine (Figure 6.96) contains an undoubtedly related furanoindole system. Another alkaloid, geneserine (Figure 6.97), is an artefact produced by oxidation of physostigmine, incorporating oxygen into the ring system, probably by formation of an N-oxide and ring expansion. Solutions of physostigmine are not particularly stable in the presence of air and light, especially under alkaline conditions, oxidizing to a red quinone, rubreserine (Figure 6.97).

Physostigmine (**eserine**) is a reversible inhibitor of cholinesterase, preventing normal destruction of acetylcholine and thus enhancing cholinergic activity. Its major use is as a miotic, to contract the pupil of the eye, often to combat the effect of mydriatics such as atropine (see page 297). It also reduces intraocular pressure in the eye by increasing outflow of the aqueous humour, and is a valuable treatment for glaucoma, often in combination with pilocarpine (see page 380). Because it prolongs the effect of endogenous acetylcholine, physostigmine can be used as an antidote to anticholinergic poisons such as hyoscyamine/atropine (see page 297), and it also reverses the effects of competitive muscle relaxants such as curare, tubocurarine, atracurium, etc (see page 324). Anticholinesterase drugs are also of value in the treatment of Alzheimer's disease, which is characterized by a dramatic decrease in functionality of the central cholinergic system. Use of acetylcholinesterase inhibitors can result in significant memory enhancement in patients, and analogues of physostigmine are presently in use (e.g. **rivastigmine**) or in advanced clinical trials (e.g. eptastigmine (Figure 6.96)). These analogues have a longer duration of action and less toxicity than physostigmine.

The biological activity of physostigmine resides primarily in the carbamate portion, which is transferred to the hydroxyl group of an active site serine in cholinesterase (Figure 6.98). The enzyme is only slowly regenerated by hydrolysis of this group, since resonance contributions reduce the reactivity of the carbonyl in the amide relative to the ester. Accordingly, cholinesterase becomes temporarily inactivated. Synthetic analogues of physostigmine which have been developed retain the carbamate residue, an aromatic ring to achieve binding and to provide a good leaving group, whilst ensuring water-solubility through possession of a quaternary ammonium system. **Neostigmine**, **pyridostigmine**, and **distigmine** (Figure 6.96) are examples of synthetic anticholinesterase drugs used primarily for enhancing neuromuscular transmission in the rare autoimmune condition myasthenia

(Continued)

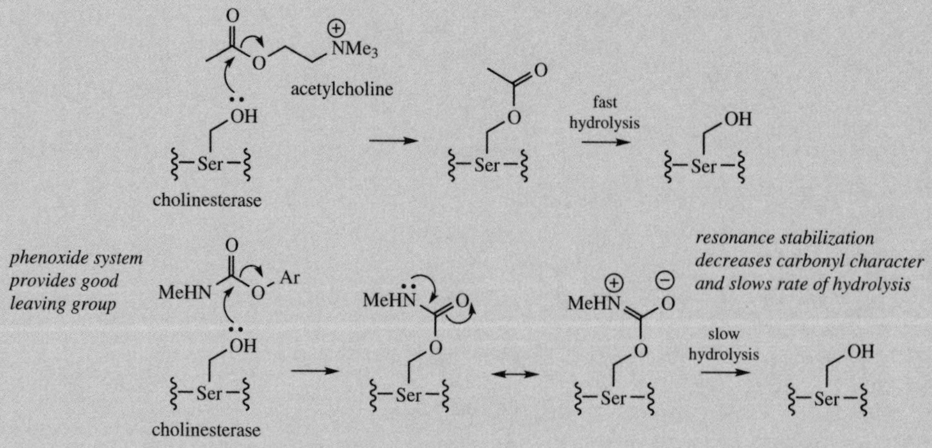

Figure 6.98

gravis, in which muscle weakness is caused by faulty transmission of nerve impulses. **Edrophonium** is a short-acting competitive blocker of the acetylcholinesterase active site, which is used to help diagnose myasthenia gravis. A number of carbamate insecticides, e.g. carbaryl (Figure 6.96), also depend on inhibition of cholinesterase for their action, insect acetylcholinesterase being more susceptible to such agents than the mammalian enzyme. Physostigmine displays little insecticidal action because of its poor lipid solubility.

*Physostigma venenosum** (Leguminosae/Fabaceae) and has played an important role in pharmacology because of its anticholinesterase activity. The inherent activity is in fact derived from the carbamate side-chain rather than the heterocyclic ring system, and this has led to a range of synthetic materials being developed.

Ergot Alkaloids

Ergot* is a fungal disease commonly found on many wild and cultivated grasses, and is caused by species of *Claviceps*. The disease is eventually characterized by the formation of hard, seedlike 'ergots' instead of normal seeds, these structures,

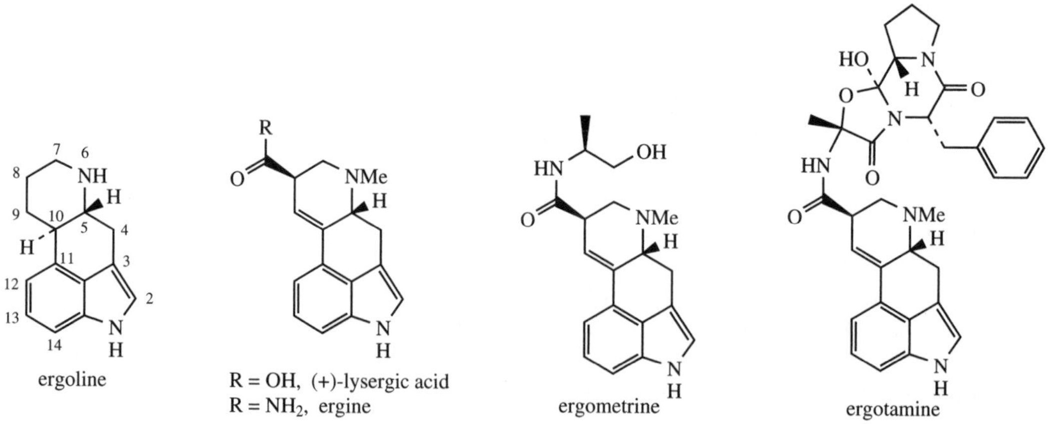

ergoline

R = OH, (+)-lysergic acid
R = NH$_2$, ergine

ergometrine

ergotamine

Figure 6.99

called sclerotia, forming the resting stage of the fungus. The poisonous properties of ergots in grain, especially rye, for human or animal consumption have long been recognized, and the causative agents are known to be a group of indole alkaloids, referred to collectively as the ergot alkaloids or ergolines (Figure 6.99). Under natural conditions the alkaloids are elaborated by a combination of fungal and plant metabolism, but they can

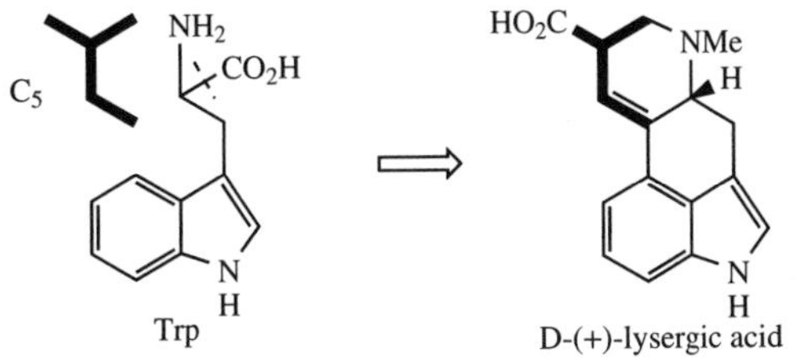

Figure 6.100

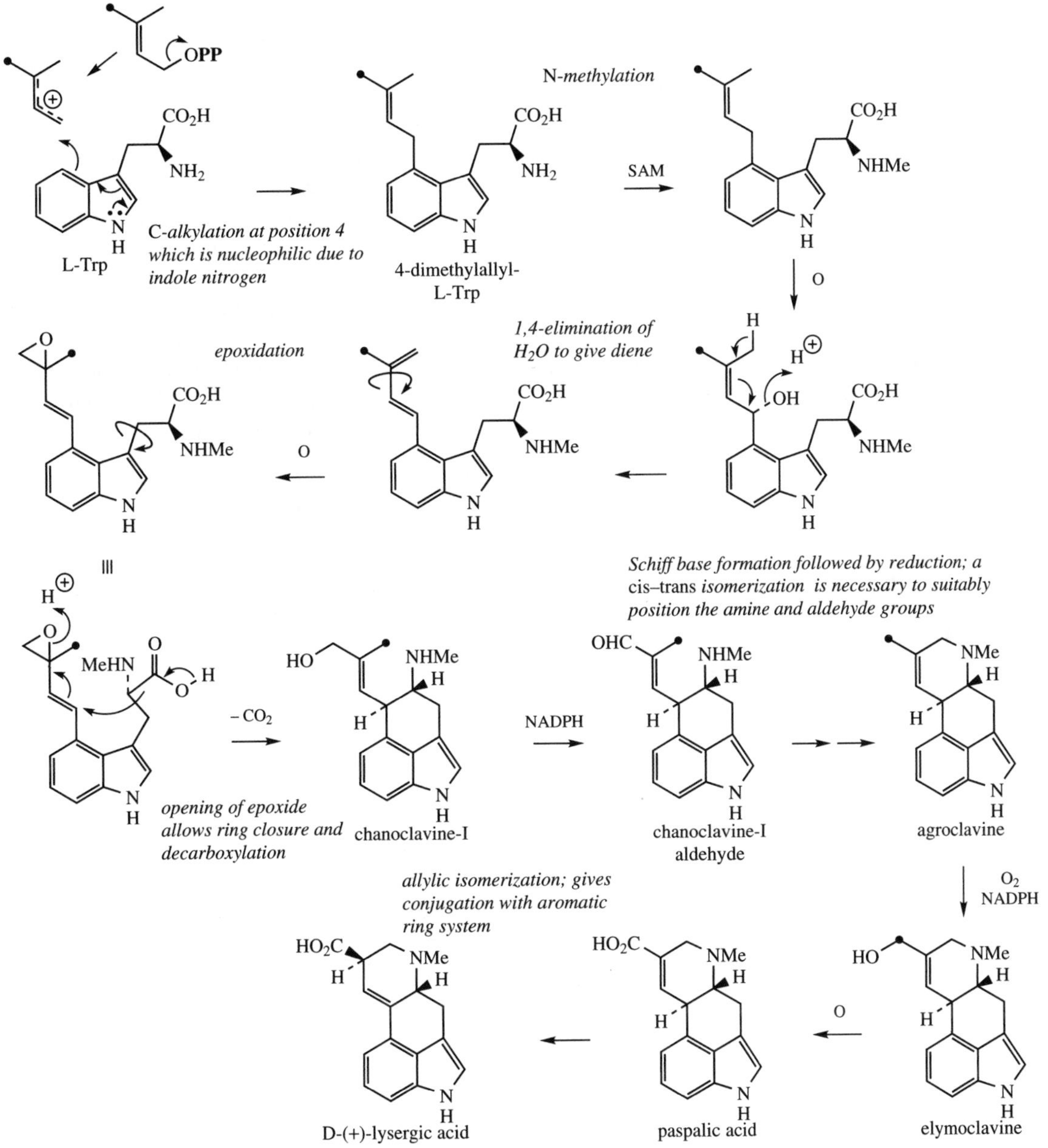

Figure 6.101

be synthesized in cultures of suitable *Claviceps* species. Ergoline alkaloids have also been found in fungi belonging to genera *Aspergillus*, *Rhizopus*, and *Penicillium*, as well as *Claviceps*, and simple examples are also found in some plants of the Convolvulaceae such as *Ipomoea* and *Rivea* (morning glories)*. Despite their toxicity, some of these alkaloids have valuable pharmacological activities and are used clinically on a routine basis.

Medicinally useful alkaloids are derivatives of (+)-**lysergic acid** (see Figure 6.99), which is typically bound as an amide with an amino alcohol as in **ergometrine**, or with a small polypeptide structure as in **ergotamine**. The building blocks for lysergic acid are tryptophan (less the carboxyl group) and an isoprene unit (Figure 6.100). Alkylation of tryptophan with dimethylallyl diphosphate gives 4-dimethylallyl-L-tryptophan, which then undergoes *N*-methylation (Figure 6.101). Formation of the tetracyclic ring system of lysergic acid is known to proceed through **chanoclavine-I** and **agroclavine**, though the mechanistic details are far from clear. Labelling studies have established that the

double bond in the dimethylallyl substituent must become a single bond on two separate occasions, allowing rotation to occur as new rings are established. This gives the appearance of *cis–trans* isomerizations as 4-dimethylallyl-L-tryptophan is transformed into chanoclavine-I, and as chanoclavine-I aldehyde cyclizes to agroclavine (Figure 6.101). A suggested sequence to account for the first of these is shown. In the later stages, agroclavine is hydroxylated to **elymoclavine**, further oxidation of the primary alcohol occurs giving **paspalic acid**, and **lysergic acid** then results from a spontaneous allylic isomerization.

Simple derivatives of lysergic acid require the formation of amides; for example, **ergine** (Figure 6.99) in *Rivea* and *Ipomoea* species is lysergic acid amide, whilst **ergometrine** from *Claviceps purpurea* is the amide with 2-aminopropanol. The more complex structures containing peptide fragments, e.g. **ergotamine** (Figure 6.99), are formed by sequentially adding amino acid residues to thioester-bound lysergic acid, giving a linear lysergyl–tripeptide covalently attached to the enzyme complex (Figure 6.102). Peptide formation

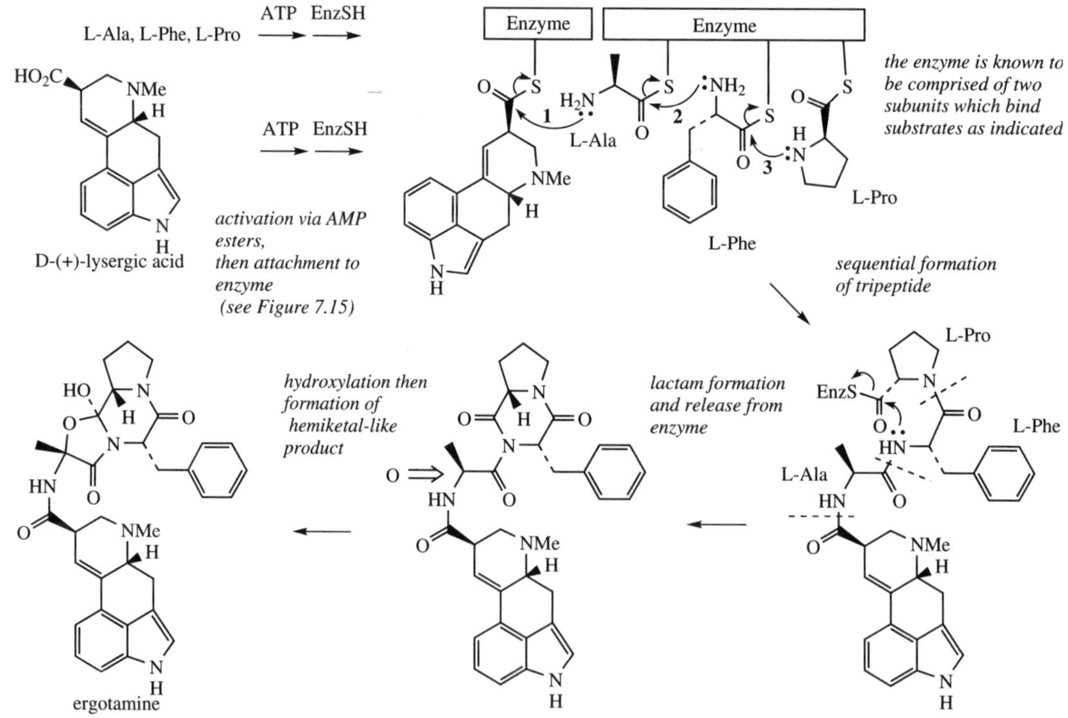

Figure 6.102

involves the same processes seen in the non-ribosomal biosynthesis of peptides, and involves ATP-mediated activation of the amino acids prior to attachment to the enzyme complex through thioester linkages (see page 421). A phosphopantatheine arm is used to enable the growing chain to reach the various active sites (compare fatty acid biosynthesis, page 36). The cyclized tripeptide residue is readily rationalized by the formation of a lactam (amide) which releases the product from the enzyme, followed by generation of a hemiketal-like linkage as shown (Figure 6.102).

Ergot

Medicinal ergot is the dried sclerotium of the fungus *Claviceps purpurea* (Clavicipitaceae) developed on the ovary of rye, *Secale cereale* (Graminae/Poaceae). Ergot is a fungal disease of wild and cultivated grasses, and initially affects the flowers. In due course, a dark sclerotium, the resting stage of the fungus, is developed instead of the normal seed. This protrudes from the seed head, the name ergot being derived from the French word argot – a spur. The sclerotia fall to the ground, germinating in the spring and reinfecting grasses or grain crops by means of spores. Two types of spore are recognized: ascospores, which are formed in the early stages and are dispersed by the wind, whilst later on conidiospores are produced, which are insect distributed. The flowers are only susceptible to infection before pollination. Ergots may subsequently be harvested with the grain and contaminate flour or animal feed. The consumption of ergot-infected rye has resulted in the disease ergotism, which has a long, well documented history.

There are three broad clinical features of ergot poisoning, which are due to the alkaloids present and the relative proportions of each component:

- Alimentary upsets, e.g. diarrhoea, abdominal pains, and vomiting.
- Circulatory changes, e.g. coldness of hands and feet due to a vasoconstrictor effect, a decrease in the diameter of blood vessels, especially those supplying the extremeties.
- Neurological symptoms, e.g. headache, vertigo, convulsions, psychotic disturbances, and hallucinations.

These effects usually disappear on removal of the source of poisoning, but much more serious problems develop with continued ingestion, or with heavy doses of ergot-contaminated food. The vasoconstrictor effect leads to restricted blood flow in small terminal arteries, death of the tissue, the development of gangrene, and even the shedding of hands, feet, or limbs. Gangrenous ergotism was known as St Anthony's Fire, the Order of St Anthony traditionally caring for sufferers in the Middle Ages. The neurological effects were usually manifested by severe and painful convulsions. Outbreaks of the disease in both humans and animals were relatively frequent in Europe in the Middle Ages, but once the cause had been established it became relatively simple to avoid contamination. Separation of the ergots from grain, or the use of fungicides during cultivation of the crop, have removed most of the risks, though infection of crops is still common.

The ergot sclerotia contain from 0.15–0.5% alkaloids, and more than 50 have been characterized. The medicinally useful compounds are derivatives of (+)-lysergic acid (Figure 6.103), and can be separated into two groups, the water-soluble amino alcohol derivatives (up to about 20% of the total alkaloids), and water-insoluble peptide derivatives (up to 80% of the total alkaloids). Ergometrine (Figure 6.103) (also known as ergonovine in the USA and ergobasine in Switzerland) is an amide of lysergic acid and 2-aminopropanol, and is the only significant member of the first group. The peptide derivatives contain a

(Continues)

(Continued)

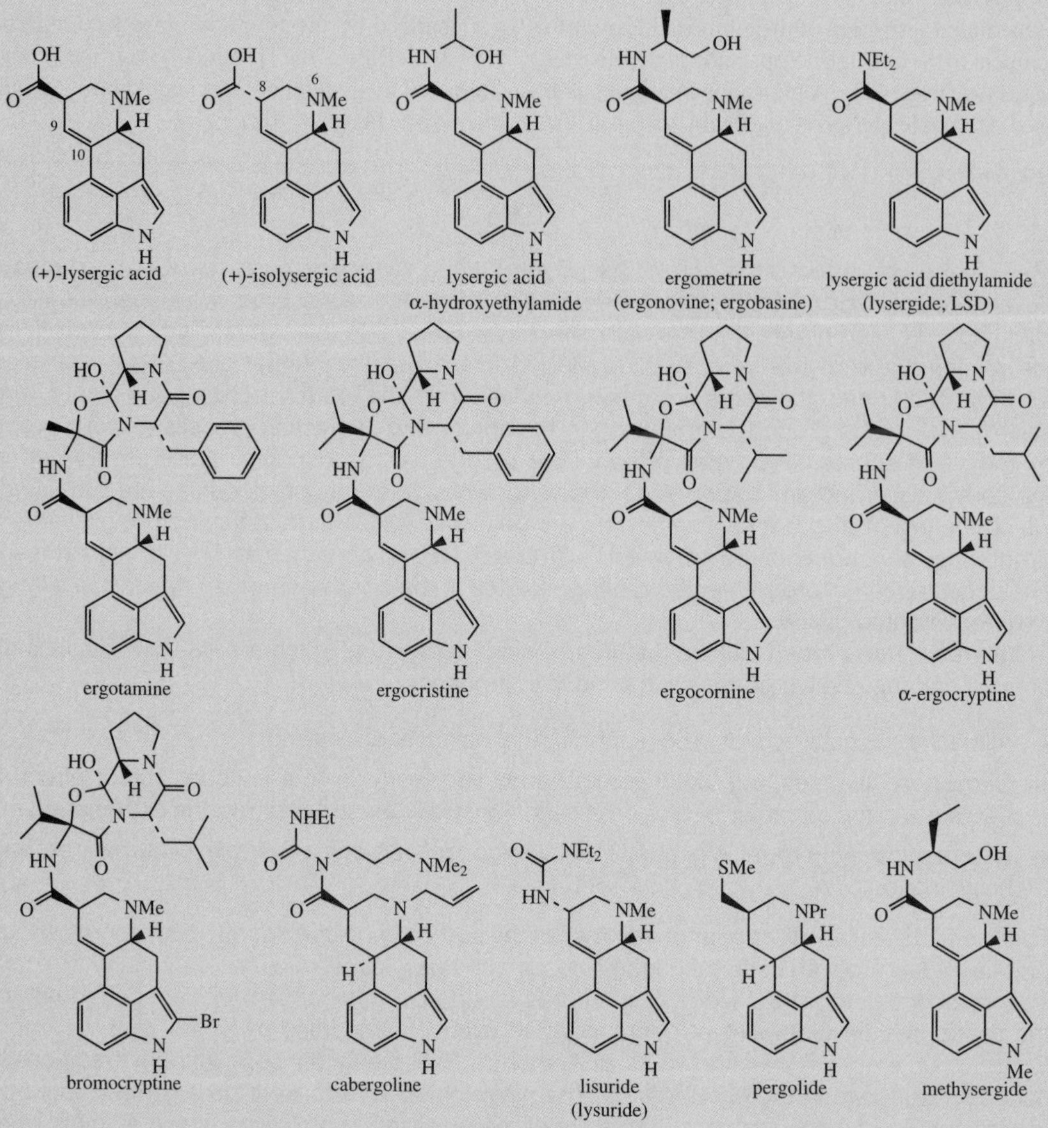

Figure 6.103

cyclized tripeptide fragment bonded to lysergic acid via an amide linkage. Based on the nature of the three amino acids, these structures can be subdivided into three groups, the ergotamine group, the ergoxine group, and the ergotoxine group (Table 6.1). The amino acids involved are alanine, valine, leucine, isoleucine, phenylalanine, proline, and α-aminobutyric acid, in various combinations (see Figure 6.104). All contain proline in the tripeptide, and one of the amino acids is effectively incorporated into the final structure in the form of an α-hydroxy-α-amino acid. Thus, ergotamine incorporates alanine, phenylalanine, and proline residues in its peptide portion. Hydrolysis gives (+)-lysergic acid, proline, and phenylalanine,

(Continues)

(Continued)

Table 6.1 Peptide alkaloids in ergot

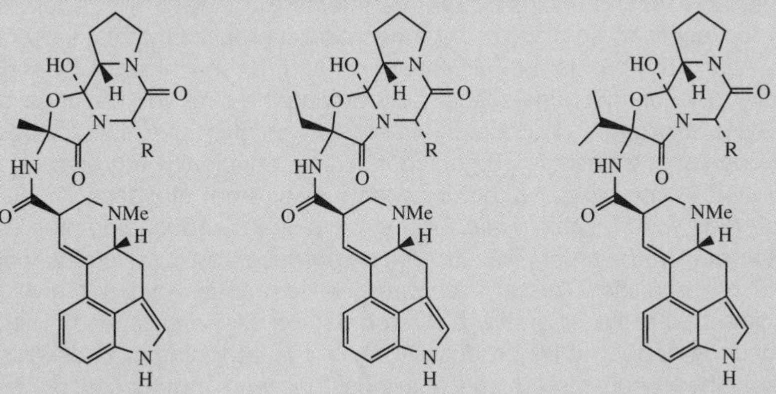

	ergotamine group	ergoxine group	ergotoxine group
R = CH₂Ph	ergotamine	ergostine	ergocristine
R = CH₂CHMe₂	ergosine	ergoptine	α-ergocryptine
S R = CH(Me)Et	[β-ergosine]	[β-ergoptine]	β-ergocryptine
R = CHMe₂	ergovaline	ergonine	ergocornine
R = Et	ergobine	ergobutine	ergobutyrine

[ ] not yet known in nature

together with pyruvic acid and ammonia, the latter hydrolysis products a consequence of the additional hydroxylation involving alanine (Figure 6.104). Hydrolysis of the ergotoxine group of alkaloids results in the proximal valine unit being liberated as dimethylpyruvic acid (not systematic nomenclature) and ammonia, and the ergoxine group similarly yields α-oxobutyric acid from the α-aminobutyric acid fragment. The alkaloid 'ergotoxine' was originally thought to be a single compound, but was subsequently shown to be a mixture of alkaloids. The proposed structures β-ergosine and β-ergoptine, which complete the combinations shown in Table 6.1, have not yet been isolated as natural products.

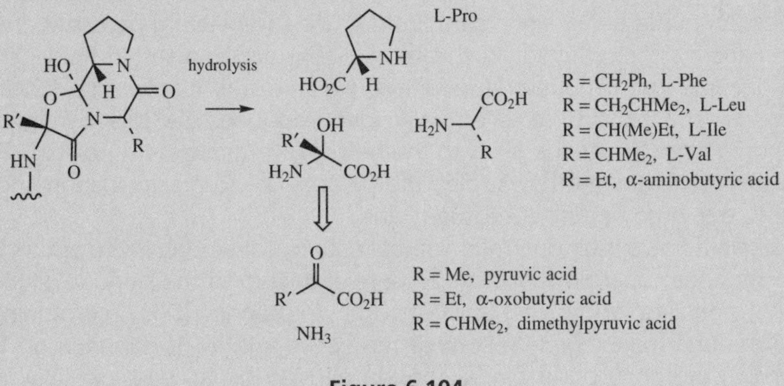

Figure 6.104

(Continues)

(Continued)

Medicinal ergot is cultivated in the Czech Republic, Germany, Hungary, Switzerland, Austria, and Poland. Fields of rye are infected artificially with spore cultures of *Claviceps purpurea*, either by spraying or by a mechanical process that uses needles dipped in a spore suspension. The ergots are harvested by hand, by machine, or by separation from the ripe grain by flotation in a brine solution. By varying the strain of the fungal cultures, it is possible to maximize alkaloid production (0.4–1.2%), or give alkaloid mixtures in which particular components predominate. Ergots containing principally ergotamine in concentrations of about 0.35% can be cultivated. In recent times, ergot of wheat (*Triticum aestivum*), and the wheat–rye hybrid triticale (*Triticosecale*) have been produced commercially.

Alternatively, the ergot alkaloids can be produced by culturing the fungus. Initially, cultures of the rye parasite *Claviceps purpurea* in fermentors did not give the typical alkaloids associated with the sclerotia, e.g. ergometrine and ergotamine. These medicinally useful compounds appear to be produced only in the later stages of development of the fungus. Instead, the cultures produced alkaloids that were not based on lysergic acid, and are now recognized as intermediates in the biosynthesis of lysergic acid, e.g. chanoclavine-I, agroclavine, and elymoclavine (Figure 6.101). Ergot alkaloids that do not yield lysergic acid on hydrolysis have been termed clavine alkaloids. Useful derivatives based on lysergic acid can be obtained by fermentation growth of another fungal species, namely *Claviceps paspali*. Although some strains are available that produce peptide alkaloids in culture, other strains produce high yields of simple lysergic acid derivatives. These include lysergic acid α-hydroxyethylamide (Figure 6.103), lysergic acid amide (ergine) (Figure 6.99), which is also an acid-catalysed decomposition product from lysergic acid α-hydroxyethylamide, and the $\Delta^{8,9}$-isomer of lysergic acid, paspalic acid (Figure 6.101). Lysergic acid is obtained from the first two by hydrolysis, or from paspalic acid by allylic isomerization. Other alkaloids, e.g. ergometrine and ergotamine, can then be produced semi-synthetically. High yielding fermentation methods have also been developed for direct production of ergotamine and the ergotoxine group of peptide alkaloids.

The pharmacologically active ergot alkaloids are based on (+)-lysergic acid (Figure 6.103), but since one of the chiral centres in this compound (and its amide derivatives) is adjacent to a carbonyl, the configuration at this centre can be changed as a result of enolization brought about by heat or base (compare tropane alkaloids, page 298; again note that enolization is favoured by conjugation with the aromatic ring). The new diastereomeric form of (+)-lysergic acid is (+)-isolysergic acid (Figure 6.103), and alkaloids based on this compound are effectively pharmacologically inactive. They are frequently found along with the (+)-lysergic acid derivatives, amounts being significant if old ergot samples are processed, or unsuitable isolation techniques are employed. In the biologically active lysergic acid derivatives, the amide group occupies an 8-equatorial position, whilst in the inactive iso-forms, this group is axial. However, the axial form is actually favoured because this configuration allows hydrogen-bonding from the amide N–H to the hetrocyclic nitrogen at position 6. Derivatives of (+)-isolysergic acid are named by adding the syllable -*in*- to the corresponding (+)-lysergic acid compound, e.g. ergomet*in*ine, ergotam*in*ine.

The ergot alkaloids owe their pharmacological activity to their ability to act at α-adrenergic, dopaminergic and serotonergic receptors. The relationship of the general alkaloid structure to those of noradrenaline, dopamine, and 5-hydroxytryptamine (5-HT, serotonin) is shown in Figure 6.105. The pharmacological response may be complex. It depends on the preferred receptor to which the compound binds, though all may be at least partially involved, and whether the alkaloid is an agonist or antagonist.

(Continues)

(*Continued*)

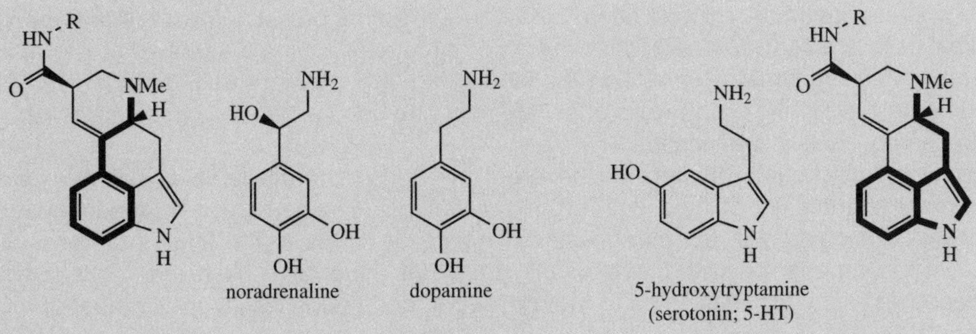

noradrenaline   dopamine   5-hydroxytryptamine
                               (serotonin; 5-HT)

Figure 6.105

Despite the unpleasant effects of ergot as manifested by St Anthony's Fire, whole ergot preparations have been used since the 16th century to induce uterine contractions during childbirth, and to reduce haemorrhage following the birth. This oxytocic effect (oxytocin is the pituitary hormone that stimulates uterine muscle, see page 415) is still medicinally valuable, but is now achieved through use of the isolated alkaloid ergometrine. The deliberate use of ergot to achieve abortions is dangerous and has led to fatalities.

Ergometrine (ergonovine) (Figure 6.103) is used as an oxytocic, and is injected during the final stages of labour and immediately following childbirth, especially if haemorrhage occurs. Bleeding is reduced because of its vasoconstrictor effects, and it is valuable after Caesarian operations. It is sometimes administered in combination with oxytocin itself (see page 415). Ergometrine is also orally active. It produces faster stimulation of uterine muscle than do the other ergot alkaloids, and probably exerts its effect by acting on α-adrenergic receptors, though it may also stimulate 5-HT receptors.

Ergotamine (Figure 6.103) is a partial agonist of α-adrenoceptors and 5-HT receptors. It is not suitable for obstetric use because it also produces a pronounced peripheral vasoconstrictor action. This property is exploited in the treatment of acute attacks of migraine, where it reverses the dilatation of cranial blood vessels. Ergotamine is effective orally, or by inhalation in aerosol form, and may be combined with caffeine, which is believed to enhance its action. The semi-synthetic **dihydroergotamine** is produced by hydrogenation of the lysergic acid $\Delta^{9,10}$ double bond (giving C-10 stereochemistry as in ergoline) and is claimed to produce fewer side-effects, especially digestive upsets.

The 'ergotoxine' alkaloid mixture also has oxytocic and vasoconstrictor activity but is only employed medicinally as the 9,10-dihydro derivatives **dihydroergotoxine (co-dergocrine)**, a mixture in equal proportions of **dihydroergocornine**, **dihydroergocristine**, and the **dihydroergocryptines** (α- and β- in the ratio 2:1). In the case of these alkaloids, reduction of the double bond appears to reverse the vasoconstrictor effect, and dihydroergotoxine has a cerebral vasodilator activity. The increased blood flow is of benefit in some cases of senility and mild dementia, and helps to improve both mental function and physical performance.

A number of semi-synthetic lysergic acid derivatives act by stimulation of dopamine receptors in the brain, and are of value in the treatment of neurological disorders such as Parkinson's disease. **Bromocriptine** (2-bromo-α-ergocryptine), **cabergoline**, **lisuride (lysuride)**, and **pergolide** (Figure 6.103) are all used in this way. Bromocriptine and cabergoline find wider use in that they also inhibit release of prolactin by the pituitary (see page 414), and can

(*Continues*)

(Continued)

thus suppress lactation and be used in the treatment of breast tumours. **Methysergide** (Figure 6.103) is a semi-synthetic analogue of ergometrine, having a modified amino alcohol side-chain and an *N*-methyl group on the indole ring. It is a potent 5-HT antagonist and as such is employed in the prophylaxis of severe migraine headaches, though its administration has to be very closely supervised.

Prolonged treatment with any of the ergot alkaloids is undesirable and it is vital that the clinical features associated with ergot poisoning are recognized. Treatment must be withdrawn immediately if any numbness or tingling develops in the fingers or toes. Side-effects will disappear on withdrawal of the drug, but there have been many cases where misdiagnosis has unfortunately led to foot or toe rot, and the necessity for amputation of the dead tissue.

Undoubtedly the most notorious of the lysergic acid derivatives is lysergide (lysergic acid diethylamide or LSD) (Figure 6.103). This widely abused hallucinogen, known as 'acid', is probably the most active and specific psychotomimetic known, and is a mixed agonist–anatagonist at 5-HT receptors, interfering with the normal processes. An effective oral dose is from 30 to 50 µg. It was synthesized from lysergic acid, and even the trace amounts absorbed during its handling were sufficient to give its creator quite dramatic hallucinations. LSD intensifies perceptions and distorts them. How the mind is affected depends on how the user is feeling at the time, and no two 'trips' are alike. Experiences can vary from beautiful visions to living nightmares, sometimes lasting for days. Although the drug is not addictive, it can lead to schizophrenia and there is danger of serious physical accidents occurring whilst the user is under the influence of the drug.

Morning Glories

Lysergic acid derivatives have also been characterized in the seeds of morning glory (*Ipomoea violacea*), *Rivea corymbosa*, and other members of the Convolvulaceae. Such seeds formed the ancient hallucinogenic drug ololiuqui still used by the Mexican Indians in religious and other ceremonies. Extracts from the ground seeds are swallowed and the narcotic and hallucinogenic effects are said to provide contact with the gods. The active constituent has been identified to be principally ergine (lysergic acid amide) (Figure 6.99), and this has an activity about one-20th that of LSD, but is more narcotic than hallucinogenic. The alkaloid content of the seeds is usually low, at about 0.05%, but higher levels (0.5–1.3%) have been recorded. Minor ergot-related constituents include ergometrine (Figure 6.103), lysergic acid α-hydroxyethylamide (Figure 6.103), the inactive isolysergic acid amide (erginine), and some clavine alkaloids.

Since morning glories are widely cultivated ornamentals and seeds are readily available, deliberate ingestion by thrill-seekers has been considerable. Although the biological activity is well below that of LSD, the practice is potentially dangerous.

ALKALOIDS DERIVED FROM ANTHRANILIC ACID

Anthranilic acid (Figure 6.106) is a key intermediate in the biosynthesis of L-tryptophan (see page 126) and so contributes to the elaboration of indole alkaloids. During this conversion, the anthranilic acid residue is decarboxylated, so that only the C_6N skeleton is utilized. However, there are also many examples of where anthranilic acid itself functions as an alkaloid precursor, using processes which retain the full skeleton and exploit the carboxyl (Figure 6.106). It should also be

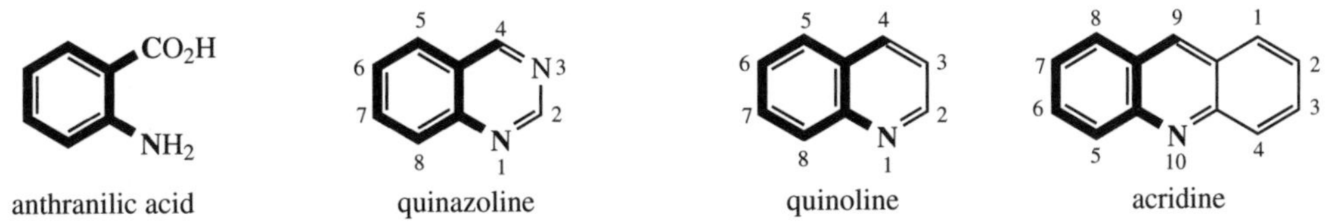

Figure 6.106

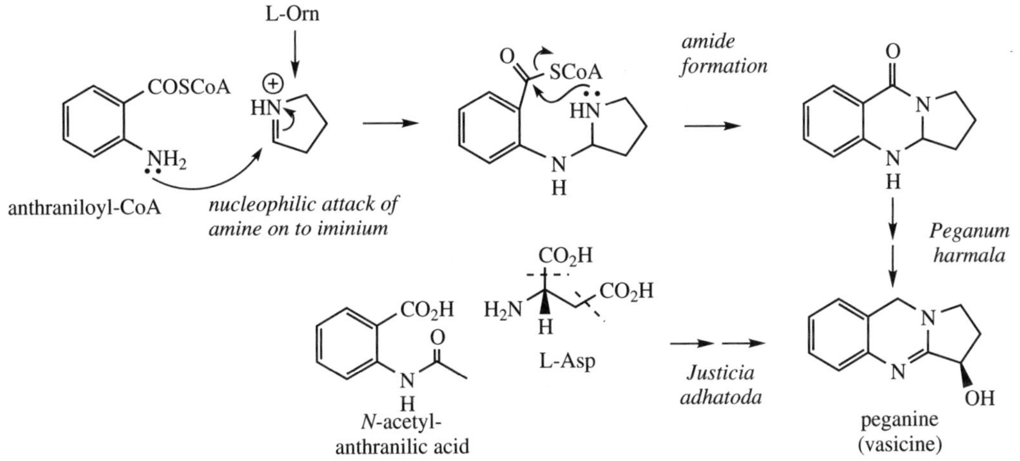

Figure 6.107

appreciated that, in mammals, L-tryptophan can be degraded back to anthranilic acid (see page 127), but this is not a route of importance in plants.

Quinazoline Alkaloids

Peganine (Figure 6.107) is a quinazoline alkaloid found in *Peganum harmala* (Zygophyllaceae), where it co-occurs with the β-carboline alkaloid harmine (see page 349). It is also responsible for the bronchodilator activity of *Justicia adhatoda* (*Adhatoda vasica*) (Acanthaceae), a plant used in the treatment of respiratory ailments. As a result, the alternative name **vasicine** is also sometimes used for peganine. Studies in *Peganum harmala* have clearly demonstrated peganine to be derived from anthranilic acid, the remaining part of the structure being a pyrrolidine ring supplied by ornithine (compare Figure 6.1, page 292). The peganine skeleton is readily rationalized as a result of nucleophilic attack from the anthranilate nitrogen on to the pyrrolinium cation, followed by amide formation (Figure 6.107). Remarkably, this

pathway is not operative in *Justicia adhatoda*, and a much less predictable sequence from *N*-acetylanthranilic acid and aspartic acid is observed (Figure 6.107).

Quinoline and Acridine Alkaloids

Alkaloids derived from anthranilic acid undoubtedly occur in greatest abundance in plants from the family the Rutaceae. Particularly well represented are alkaloids based on quinoline and acridine skeletons (Figure 6.106). Some quinoline alkaloids such as quinine and camptothecin have been established to arise by fundamental rearrangement of indole systems and have their origins in tryptophan (see page 359). A more direct route to the quinoline ring system is by the combination of anthranilic acid and acetate/malonate, and an extension of this process also accounts for the origins of the acridine ring system (see Figure 3.47, page 81). Thus, anthraniloyl-CoA (Figure 6.108) can act as a starter unit for chain extension via one molecule of malonyl-CoA, and amide

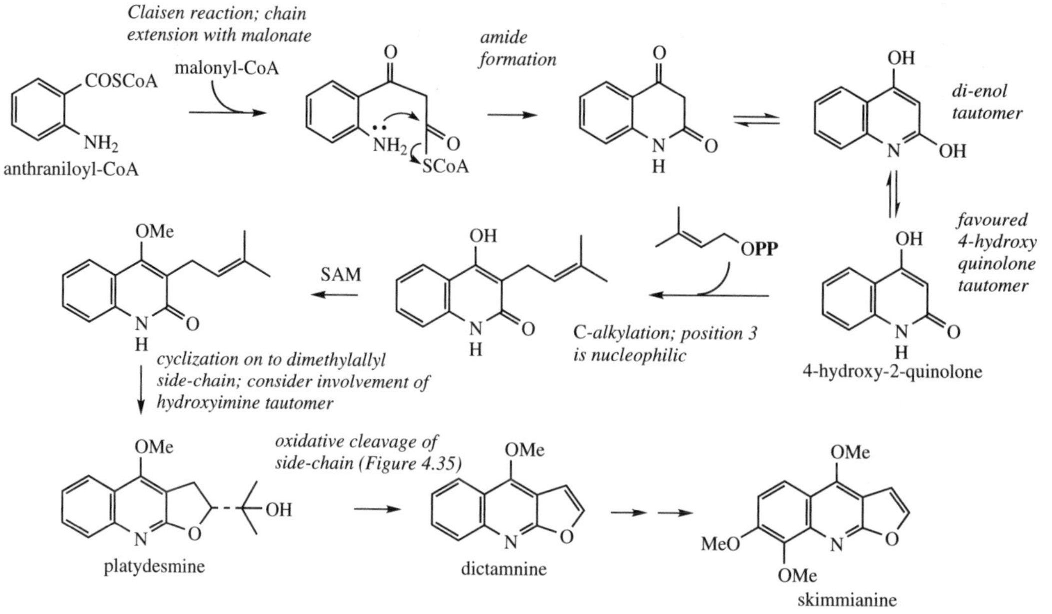

Figure 6.108

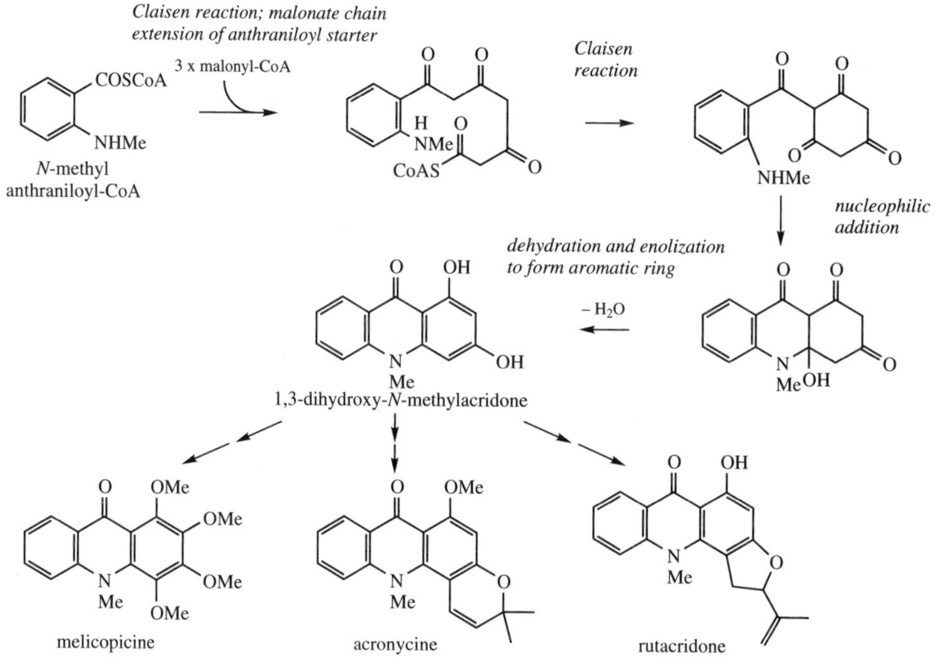

Figure 6.109

formation generates the heterocyclic system, which will adopt the more stable 4-hydroxy-2-quinolone form (Figure 6.108). Position 3 is highly nucleophilic and susceptible to alkylation, especially via dimethylallyl diphosphate in the case of these alkaloids. This allows formation of additional six- and five-membered oxygen heterocyclic rings, as seen with other systems, e.g. coumarins, isoflavonoids, etc (see pages 145, 155). By an analogous series of reactions, the dimethylallyl derivative can act as a precursor of furoquino-line alkaloids such as **dictamnine** and **skimmianine** (Figure 6.108). These alkaloids are found in both *Dictamnus albus* and *Skimmia japonica* (Rutaceae). To simplify the mechanistic interpretation of these reactions, it is more convenient to consider the di-enol form of the quinolone system.

Should chain extension of anthranilyl-CoA (as the *N*-methyl derivative) incorporate three acetate/malonate units, a polyketide would result (Figure 6.109). The acridine skeleton is then produced by sequential Claisen reaction and C–N linkage by an addition reaction, dehydration, and enolization leading to the stable aromatic tautomer. Again, the acetate-derived ring, with its alternate oxygenation, is susceptible to electrophilic attack, and this can lead to alkylation (with dimethylallyl diphosphate) or further hydroxylation. Alkaloids **melicopicine** from *Melicope fareana*, **acronycine** from *Acronychia baueri*, and **rutacridone** from *Ruta graveolens* (Rutaceae) typify some of the structural variety that may then ensue (Figure 6.109).

ALKALOIDS DERIVED FROM HISTIDINE

Imidazole Alkaloids

The amino acid L-**histidine** (Figure 6.110) contains an imidazole ring, and is thus the likely

Figure 6.110

presursor of alkaloids containing this ring system. There are relatively few examples, however, and definite evidence linking them to histidine is often lacking.

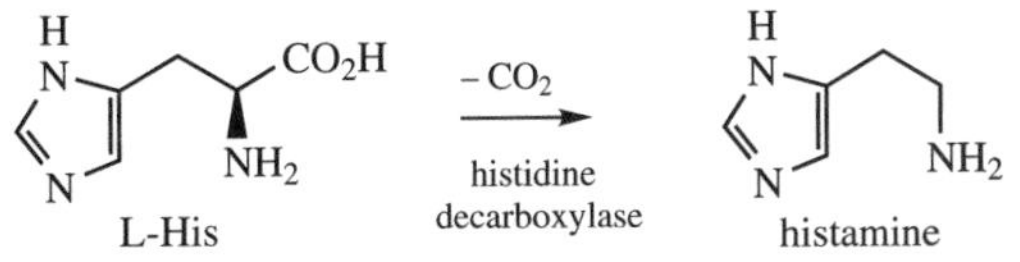

Histamine (Figure 6.110) is the decarboxylation product from histidine and is often involved in human allergic responses, e.g. to insect bites or pollens. Stress stimulates the action of the enzyme histidine decarboxylase and histamine is released from mast cells (Figure 6.110). Topical antihistamine creams are valuable for pain relief, and oral antihistamines are widely prescribed for nasal allergies such as hay-fever. Major effects of histamine include dilation of blood vessels, inflammation and swelling of tissues, and narrowing of airways. In serious cases, life-threatening anaphylactic shock may occur, caused by a dramatic fall in in blood pressure.

Histidine is a proven precursor of **dolichotheline** (Figure 6.111) in *Dolichothele sphaerica* (Cactaceae), the remaining carbon atoms originating from leucine via isovaleric acid (see page 197). The imidazole alkaloids found in

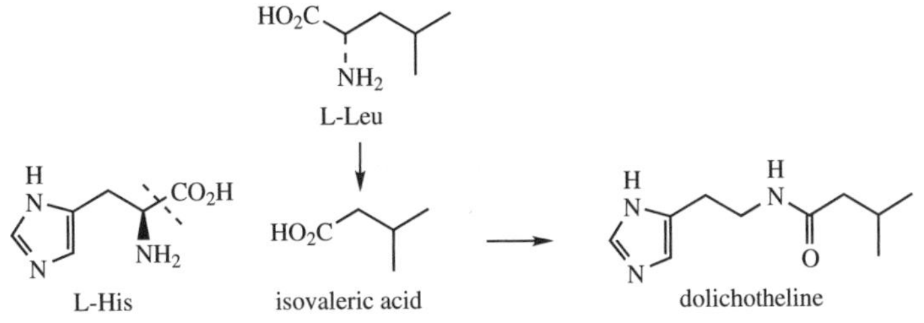

Figure 6.111

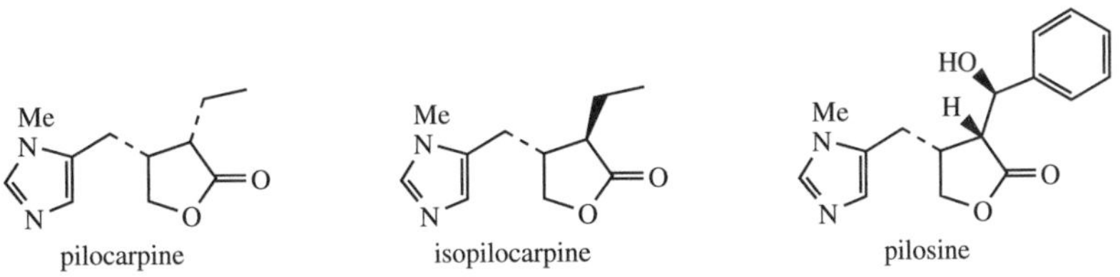

Figure 6.112

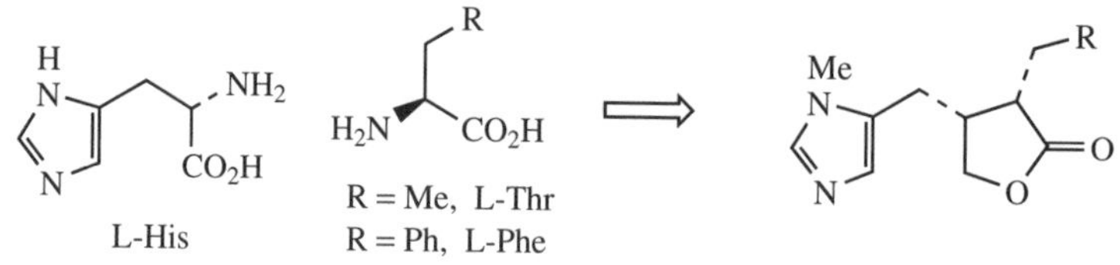

Figure 6.113

Jaborandi leaves (*Pilocarpus microphyllus* and *P. jaborandi*; Rutaceae)* are also probably derived from histidine, but experimental data are lacking. Jaborandi leaves contain primarily **pilocarpine** and **pilosine** (Figure 6.112). Pilocarpine is valuable in ophthalmic work as a miotic and as a treatment for glaucoma. Additional carbon atoms may originate from acetate or perhaps the amino acid threonine in the case of pilocarpine, whilst pilosine incorporates a phenylpropane C_6C_3 unit (Figure 6.113).

Pilocarpus

Pilocarpus or jaborandi consists of the dried leaflets of *Pilocarpus jaborandi*, *P. microphyllus*, or *P. pennatifolius* (Rutaceae), small shrubs from Brazil and Paraguay. *Pilocarpus microphyllus* is currently the main source. The alkaloid content (0.5–1.0%) consists principally of the imidazole alkaloid pilocarpine (Figure 6.112), together with small amounts of pilosine (Figure 6.112) and related structures. Isomers such as isopilocarpine (Figure 6.112) and isopilosine are readily formed if base or heat is applied during extraction of the alkaloids. This is a result of enolization in the lactone ring, followed by adoption of the more favourable *trans* configuration rather than the natural *cis*. However, the *iso* alkaloids lack biological activity. The alkaloid content of the leaf rapidly deteriorates on storage.

Pilocarpine salts are valuable in ophthalmic practice and are used in eyedrops as miotics and for the treatment of glaucoma. Pilocarpine is a cholinergic agent and stimulates the muscarinic receptors in the eye, causing constriction of the pupil and enhancement of outflow of aqueous humour. The structural resemblance to muscarine and acetylcholine is shown in Figure 6.114. Pilocarpine gives relief for both narrow angle and wide angle glaucoma. However, the ocular bioavailability of pilocarpine is low, and it is rapidly eliminated, thus resulting in a rather short duration of action. The effects are similar to those of physostigmine (see page 366) and the two agents are sometimes combined. Pilocarpine is antagonistic to atropine. It has been found that pilocarpine gives relief for dryness of the mouth that results in patients undergoing radiotherapy for mouth and throat cancers. As muscarinic agonists, pilocarpine and analogues are also being investigated for potential treatment of Alzheimer's disease.

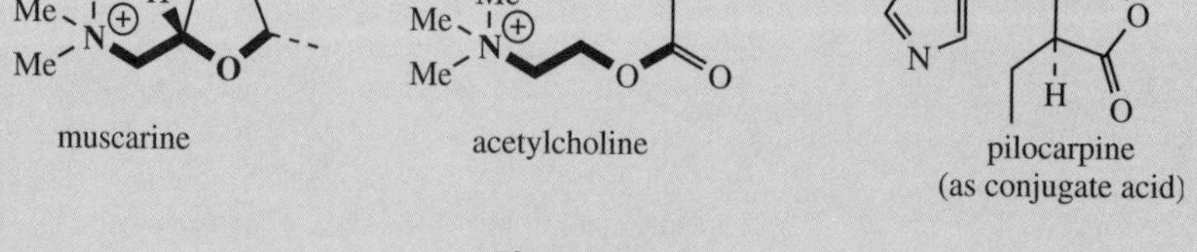

muscarine

acetylcholine

pilocarpine
(as conjugate acid)

Figure 6.114

ALKALOIDS DERIVED BY AMINATION REACTIONS

The majority of alkaloids are derived from amino acid precursors by processes that incorporate into the final structure the nitrogen atom together with the amino acid carbon skeleton or a large proportion of it. Many alkaloids do not conform with this description, however, and are synthesized primarily from non-amino acid precursors, with the nitrogen atom being inserted into the structure at a relatively late stage. Such structures are frequently based on terpenoid or steroidal skeletons, though some relatively simple alkaloids also appear to be derived by similar late amination processes. In most of the examples studied, the nitrogen atom is donated from an amino acid source through a transamination reaction with a suitable aldehyde or ketone (see page 20).

Acetate-derived Alkaloids

The poison hemlock (*Conium maculatum**; Umbelliferae/Apiaceae) accumulates a range of simple piperidine alkaloids, e.g. **coniine** and **γ-coniceine** (Figure 6.115). These alkaloids would appear to be related to simple lysine-derived compounds such as pelletierine (see page 307), but, surprisingly, a study of their biosynthetic origins excluded lysine as a precursor, and demonstrated the sequence shown in Figure 6.115 to be operative. A fatty acid precursor, capric (octanoic) acid, is utilized, and this is transformed into the ketoaldehyde by successive oxidation and reduction steps. This ketoaldehyde is then the substrate for a transamination reaction, the amino group originating from L-alanine. Subsequent transformations are imine formation giving the heterocyclic ring of γ-coniceine, and then reduction to **coniine**. **Pinidine** (Figure 6.116) from *Pinus* species is found

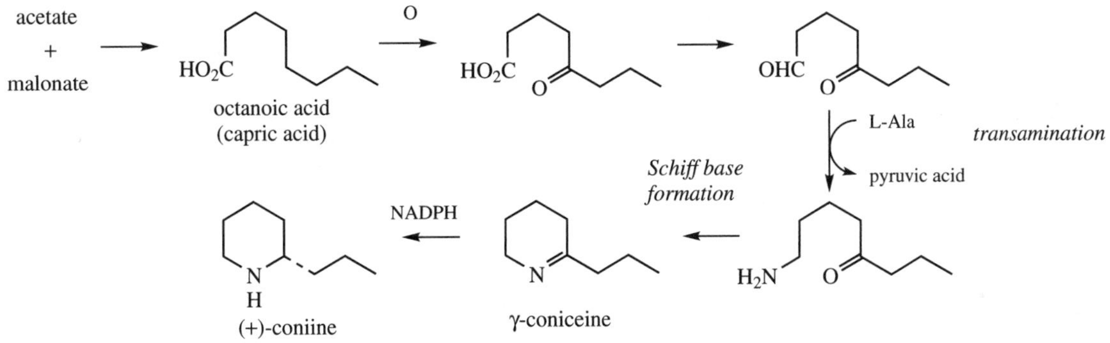

Figure 6.115

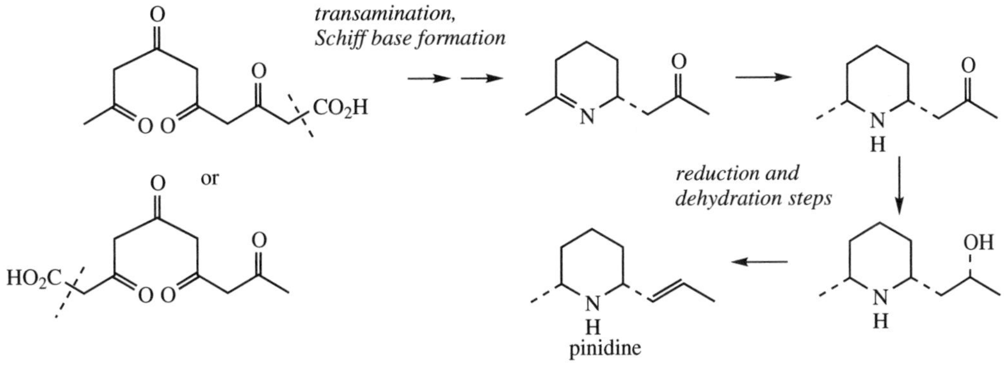

Figure 6.116

Conium maculatum

Conium maculatum (Umbelliferae/Apiaceae) or poison hemlock is a large biennial herb indigenous to Europe and naturalized in North and South America. As a common poisonous plant, recognition is important, and this plant can be differentiated from most other members of the Umbelliferae/Apiaceae by its smooth purple-spotted stem. The dried unripe fruits were formerly used as a pain reliever and sedative, but have no medicinal use now. The ancient Greeks are said to have executed condemned prisoners, including Socrates, using poison hemlock. The poison causes gradual muscular paralysis followed by convulsions and death from respiratory paralysis. All parts of the plant are poisonous due to the alkaloid content, though the highest concentration of alkaloids is found in the green fruit (up to 1.6%). The major alkaloid (about 90%) is the volatile liquid coniine (Figure 6.115), with smaller amounts of structurally related piperidine alkaloids, including *N*-methylconiine and γ-coniceine (Figure 6.115).

In North America, the name hemlock refers to species of *Tsuga* (Pinaceae), a group of coniferous trees, which should not be confused with the poison hemlock.

to have a rather similar origin in acetate, and most likely a poly-β-keto acid. During the sequence outlined in Figure 6.116, the carboxyl group is lost. Note that an alternative folding of the poly-β-keto acid and loss of carboxyl might be formulated.

Phenylalanine-derived Alkaloids

Whilst the aromatic amino acid L-tyrosine is a common and extremely important precursor

of alkaloids (see page 315), L-**phenylalanine** is less frequently utilized, and usually it contributes only carbon atoms, e.g. C_6C_3, C_6C_2, or C_6C_1 units, without providing a nitrogen atom from its amino group (see colchicine, page 342, lobeline, page 308, etc). **Ephedrine** (Figure 6.117), the main alkaloid in species of *Ephedra** (Ephedraceae) and a valuable nasal decongestant and bronchial dilator, is a prime example. Whilst ephedrine contains the same carbon and nitrogen skeleton as seen in phenylalanine, and L-phenylalanine is a

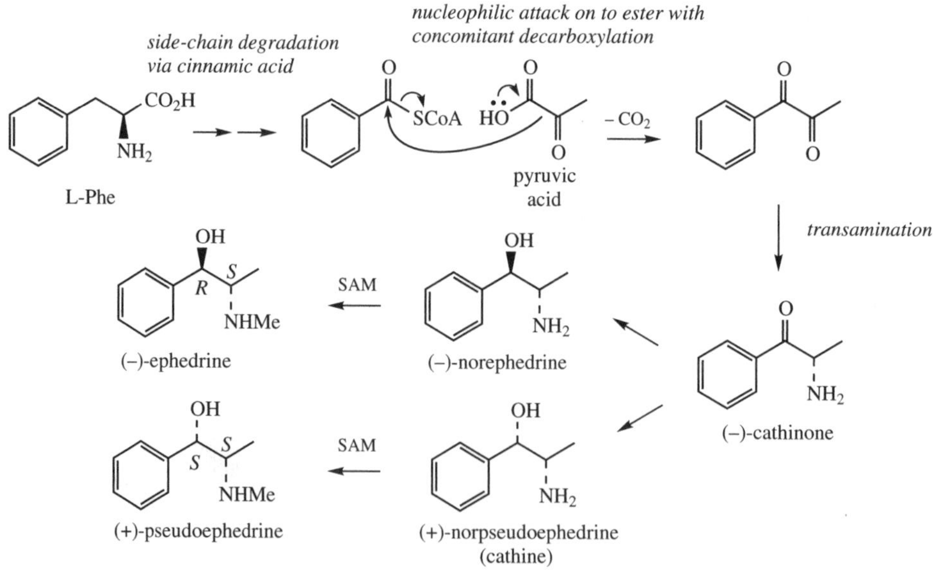

Figure 6.117

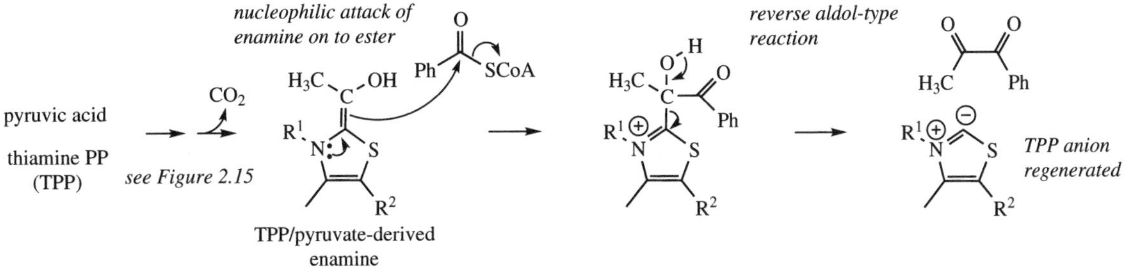

Figure 6.118

precursor, only seven carbons, a C_6C_1 fragment, are actually incorporated. It is found that phenylalanine is metabolized, probably through cinnamic acid to benzoic acid (see page 141), and this, perhaps as its coenzyme A ester, is acylated with pyruvate, decarboxylation occurring during the addition (Figure 6.117). The use of pyruvate as a nucleophilic reagent in this way is unusual in secondary metabolism, but occurs in primary metabolism during isoleucine and valine biosynthesis. A thiamine PP-mediated mechanism is suggested (Figure 6.118; compare decarboxylation of pyruvate, page 21, and formation of deoxyxylulose phosphate, page 170). This process yields the diketone, and a transamination reaction would then give **cathinone** (Figure 6.115). Reduction of the carbonyl group from either face provides the diastereomeric **norephedrine** or **norpseudoephedrine** (**cathine**). Finally, *N*-methylation would provide **ephedrine** or **pseudoephedrine** (Figure 6.117). Typically, all four of the latter compounds can be found in *Ephedra* species, the proportions varying according to species. Norpseudoephedrine is also a major constituent of the leaves of khat* (*Catha edulis*; Celastraceae), chewed in African and Arab countries as a stimulant. Most of the CNS stimulant action comes

from the more active cathinone, the corresponding carbonyl derivative. These natural compounds are structurally similar to the synthetic amfetamine/dexamfetamine; (amphetamine/dexamphetamine) (Figure 6.119) and have similar properties.

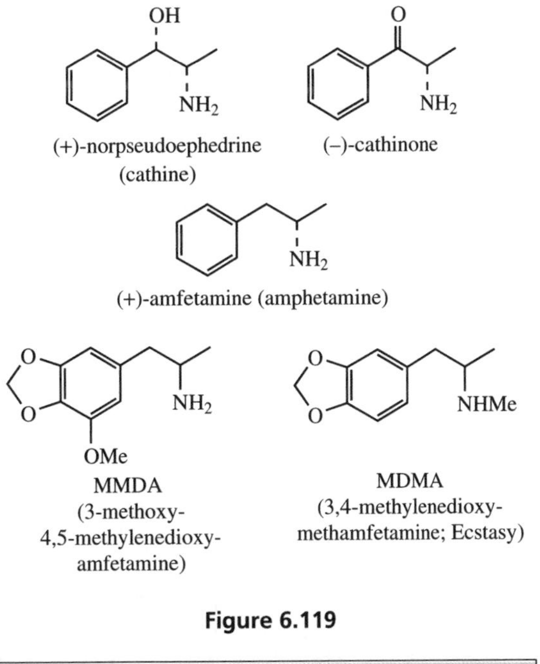

(+)-norpseudoephedrine
(cathine)

(−)-cathinone

(+)-amfetamine (amphetamine)

MMDA
(3-methoxy-
4,5-methylenedioxy-
amfetamine)

MDMA
(3,4-methylenedioxy-
methamfetamine; Ecstasy)

Figure 6.119

Ephedra

Ephedra or Ma Huang is one of the oldest known drugs, having being used by the Chinese for at least 5000 years. It consists of the entire plant or tops of various *Ephedra* species (Ephedraceae), including *E. sinica* and *E. equisetina* from China, and *E. geriardiana*, *E. intermedia* and *E. major* from India and Pakistan. The plants are small bushes with slender aerial stems and minute leaves, giving the appearance of being effectively leafless. The plants typically contain 0.5–2.0% of alkaloids, according to species, and from 30–90%

(Continues)

(Continued)

of the total alkaloids is (−)-ephedrine (Figure 6.117). Related structures, including the diastereoisomeric (+)-pseudoephedrine and the demethyl analogues (−)-norephedrine and (+)-norpseudoephedrine (Figure 6.117) are also present. In *E. intermedia*, the proportion of pseudoephedrine exceeds that of ephedrine.

Ephedrine is an indirectly acting sympathomimetic amine with effects similar to noradrenaline (see page 317). Lacking the phenolic groups of the catecholamines, it has only weak action on adrenoreceptors, but it is able to displace noradrenaline from storage vesicles in the nerve terminals, which can then act on receptors. It is orally active and has a longer duration of action than noradrenaline. It also has bronchodilator activity, giving relief in asthma, plus a vasoconstrictor action on mucous membranes, making it an effective nasal decongestant. **Pseudoephedrine** is also widely used in compound cough and cold preparations and as a decongestant. The ephedrine and pseudoephedrine used medicinally are usually synthetic. One commercial synthesis of ephedrine involves a fermentation reaction on benzaldehyde using brewer's yeast (*Saccharomyces* sp.), giving initially an alcohol, then reductive condensation with methylamine yields (−)-ephedrine with very high enantioselectivity (Figure 6.120). The fermentation reaction is similar to that shown in Figure 6.118, in that an activated acetaldehyde bound to TPP is produced by the yeast by decarboxylation of pyruvate, and this unit is added stereospecifically to benzaldehyde in an aldol-like reaction.

The herbal drug ephedra/Ma Huang is currently being traded as 'herbal ecstasy'. Consumption gives CNS stimulation, but in high amounts can lead to hallucinations, paranoia, and psychosis.

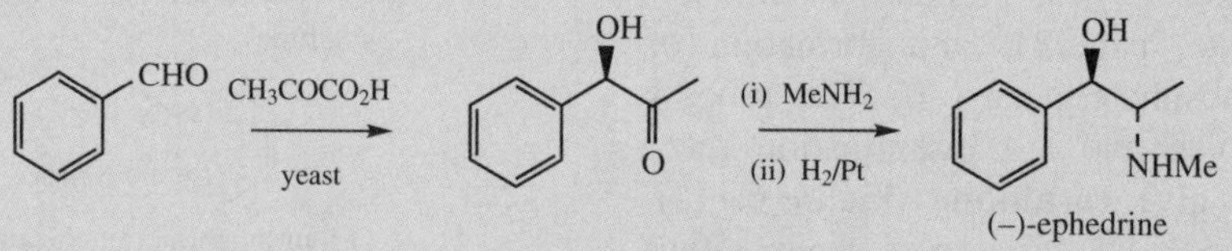

Figure 6.120

Khat

Khat, or Abyssinian tea, consists of the fresh leaves of *Catha edulis* (Celastraceae), a small tree cultivated in Ethiopia, East and South Africa, and the Yemen. The leaves are widely employed in African and Arabian countries, where they are chewed for a stimulant effect. This traditional use alleviates hunger and fatigue, but also gives a sensation of general well-being (compare coca, page 302). Users become cheerful and talkative, and khat has become a social drug. Prolonged usage can lead to hypertension, insomnia, or even mania. Khat consumption may lead to psychological dependence, but not normally physical dependence. There is presently little usage outside of Africa and Arabia, although this is increasing due to immigration from these areas. However, for maximum effects, the leaves must be fresh, and this somewhat restricts international trade. Dried leaves contain up to 1% cathine ((+)-norpseudoephedrine (Figure 6.119)), but young fresh leaves contain (−)-cathinone (Figure 6.119) as the principal CNS stimulant. Cathinone has similar pharmacological properties as the synthetic CNS stimulant (+)-amfetamine/dexamfetamine (amphetamine/dexamphetamine) (Figure 6.119), with a similar potency. Both compounds act by inducing release of catecholamines.

(Continues)

(*Continued*)

Medicinal use of amfetamine has declined markedly as drug dependence and the severe depression generated on withdrawal have been appreciated. Nevertheless, amfetamine abuse is significant. Amfetamines are taken orally, sniffed, or injected to give a long period of CNS stimulation (hours to days). Users often then take a depressant drug (alcohol, barbiturates, opioids) to terminate the effects. Users rapidly become dependent and develop tolerance. The consumption of khat is not yet restricted in the UK, even though both cathine and cathinone are now controlled drugs. It remains to be seen whether khat will be reclassified and its use restricted in any way. Other amfetamine-like derivatives of note are methoxymethylenedioxyamfetamine (MMDA) and methylenedioxymethamfetamine (MDMA) (Figure 6.119). MMDA is thought to be formed in the body after ingestion of nutmeg (*Myristica fragrans*; Myristicaceae), by an amination process on myristicin (see page 138), and it may be the agent responsible for the euphoric and hallucinogenic effects of nutmeg. MDMA is the illicit drug Ecstasy, a synthetic amfetamine-like stimulant popular among young people. The use of Ecstasy has resulted in a number of deaths, brought about by subsequent heatstroke and dehydration.

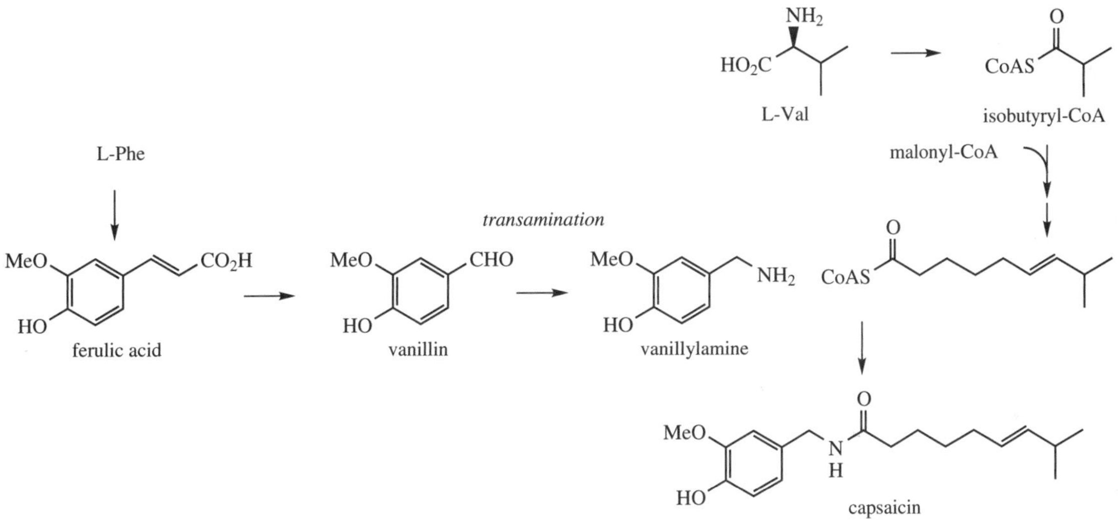

Figure 6.121

The amide **capsaicin** (Figure 6.121) constitutes the powerfully pungent principle in chilli peppers (*Capsicum annuum*; Solanaceae). Apart from its culinary importance, it is also used medicinally in creams to counter neuralgia caused by herpes infections and in other topical pain-relieving preparations. The initial burning effect of capsaicin is found to affect the pain receptors, making them less sensitive. The aromatic portion of capsaicin is derived from phenylalanine through ferulic acid and vanillin (Figure 6.121, compare page 141), this aldehyde being the substrate for

transamination to give vanillylamine. The acid portion of the amide structure is of polyketide origin, with a branched-chain fatty acyl-CoA being produced by chain extension of isobutyryl-CoA. This starter unit is valine derived (see page 197).

Terpenoid Alkaloids

A variety of alkaloids based on mono-, sesqui-, di-, and tri-terpenoid skeletons have been characterized, but information about their formation in nature is still somewhat sparse. Monoterpene alkaloids are in the main structurally related to iridoid materials (see page 187), the oxygen heterocycle being replaced by a nitrogen-containing ring. **β-Skytanthine** from *Skytanthus acutus* (Apocynaceae) and **actinidine** from *Actinidia polygama* (Actinidiaceae) serve as examples (Figure 6.122). The iridoid loganin, so important in the biosynthesis of terpenoid indole alkaloids (see page 350) and the ipecac alkaloids (see page 343), is not a precursor of these structures, and a modified series of reactions starting from geraniol is proposed (Figure 6.122). The formation of the dialdehyde follows closely elaboration of its stereoisomer in loganin biosynthesis (see page 189). This could then act as a substrate for amination via an amino acid, followed by ring formation as

seen with coniine (see page 381). Reduction and methylation would yield **β-skytanthine**, whereas further oxidation could provide the pyridine ring of **actinidine**. Valerian root (*Valeriana officinalis*; Valerianaceae) (see page 190) is known to contain alkaloids such as that shown in Figure 6.122, as well as iridoid structures. Whilst this alkaloid may be the result of *N*-alkylation on actinidine, an alternative pathway in which tyramine is condensed with the dialdehyde could be proposed. In the latter case, this alkaloid could be regarded as an alkaloid derived from tyrosine, rather than in this group of terpenoid alkaloids.

Gentianine (Figure 6.123) is probably the most common of the monoterpene alkaloids, but it is

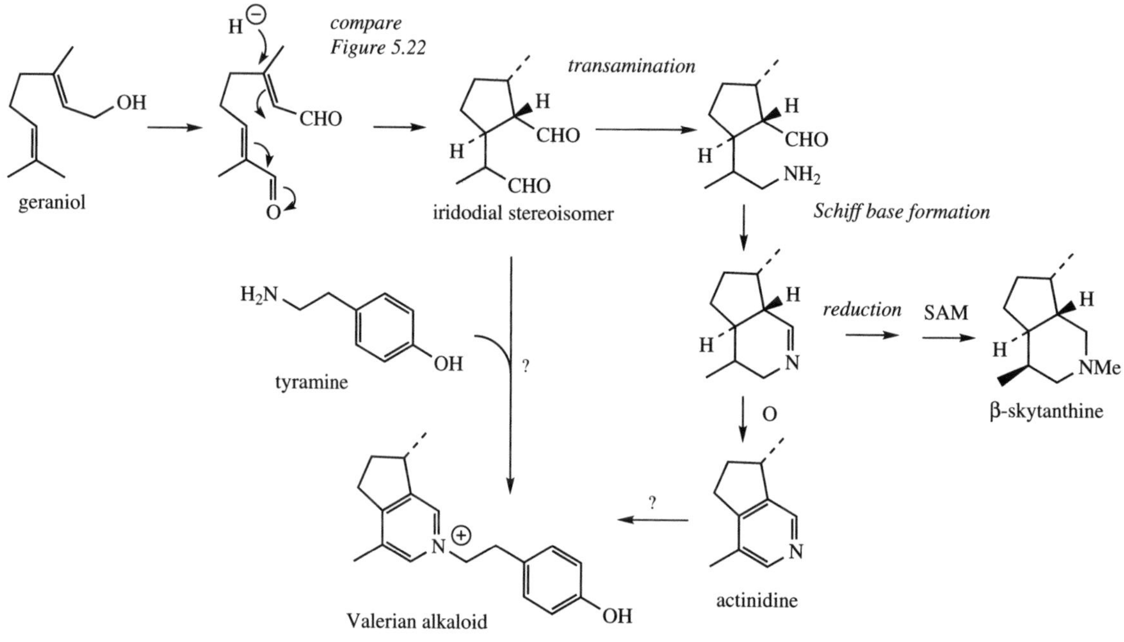

Figure 6.122

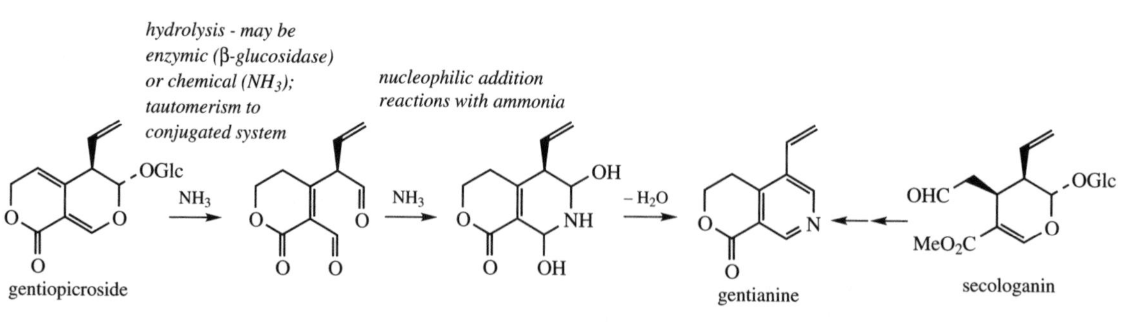

Figure 6.123

sometimes formed as an artefact when a suitable plant extract is treated with ammonia, the base commonly used during isolation of alkaloids. Thus, **gentiopicroside** (see page 190) from *Gentiana lutea* (Gentianaceae) is known to react with ammonia to give gentianine (Figure 6.123). Many other iridoid structures are known to react with ammonia to produce alkaloid artefacts. In some plants, gentianine can be found when no ammonia treatment has been involved, and one may speculate that loganin and secologanin may be precursors.

Perhaps the most important examples of terpenoid alkaloids from a pharmacological point of view are those found in aconite* or wolfsbane (*Aconitum* species; Ranunculaceae) and species of *Delphinium* (Ranunculaceae). Whilst *Aconitum*

napellus has had some medicinal uses, plants of both genera owe their highly toxic nature to diterpenoid alkaloids. Aconite in particular is regarded as extremely toxic, due to the presence of **aconitine** (Figure 6.124) and related C_{19} norditerpenoid alkaloids. Species of *Delphinium* accumulate diterpenoid alkaloids such as **atisine** (Figure 6.124), which tend to be less toxic than aconitine. An appreciation of their structural relationship to diterpenes, e.g. *ent*-**kaurene** (see page 208), is given in Figure 6.125, though little experimental evidence is available. It appears feasible that a pre-*ent*-kaurene carbocation undergoes Wagner–Meerwein rearrangements, and that the atisine skeleton is then produced by incorporating an N–CH₂CH₂–O fragment (e.g.

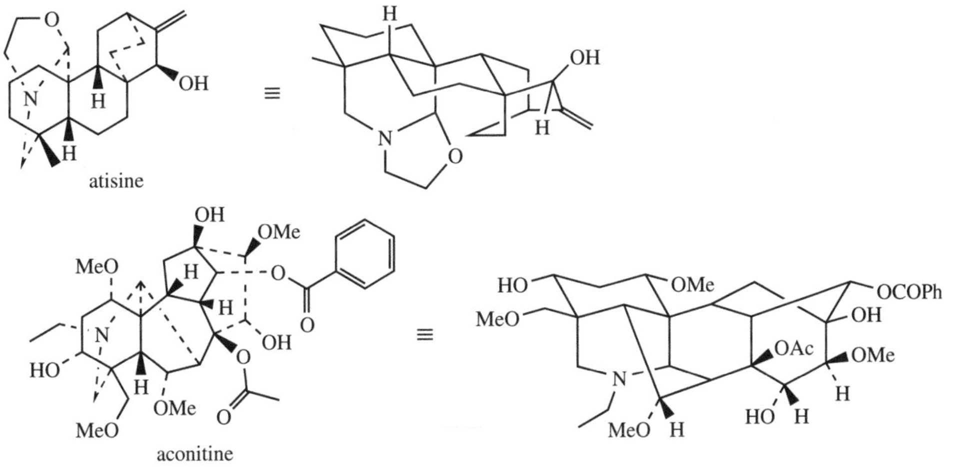

Figure 6.124

Aconite

Aconites, commonly called wolfsbane or monkshood, are species of *Aconitum* (Ranunculaceae), valued ornamental herbaceous plants, grown for their showy blue or purple flowers, which are shaped like a monk's cowl. Their alkaloid content, mainly in the roots, makes them some of the most toxic plants commonly encountered. The dried roots of *Aconitum napellus* were once used, mainly externally for relief of pain, e.g. in rheumatism. The toxic alkaloids (0.3–1.5%) are complex diterpene-derived esters. Aconitine (Figure 6.124) is the principal component (about 30%) and is a diester of aconine with acetic and benzoic acids. Hydrolysis products benzoylaconine and aconine are also present in dried plant material. These alkaloids appear to behave as neurotoxins by acting on sodium channels. All species of *Aconitum* and *Delphinium* are potentially toxic to man and animals and must be treated with caution.

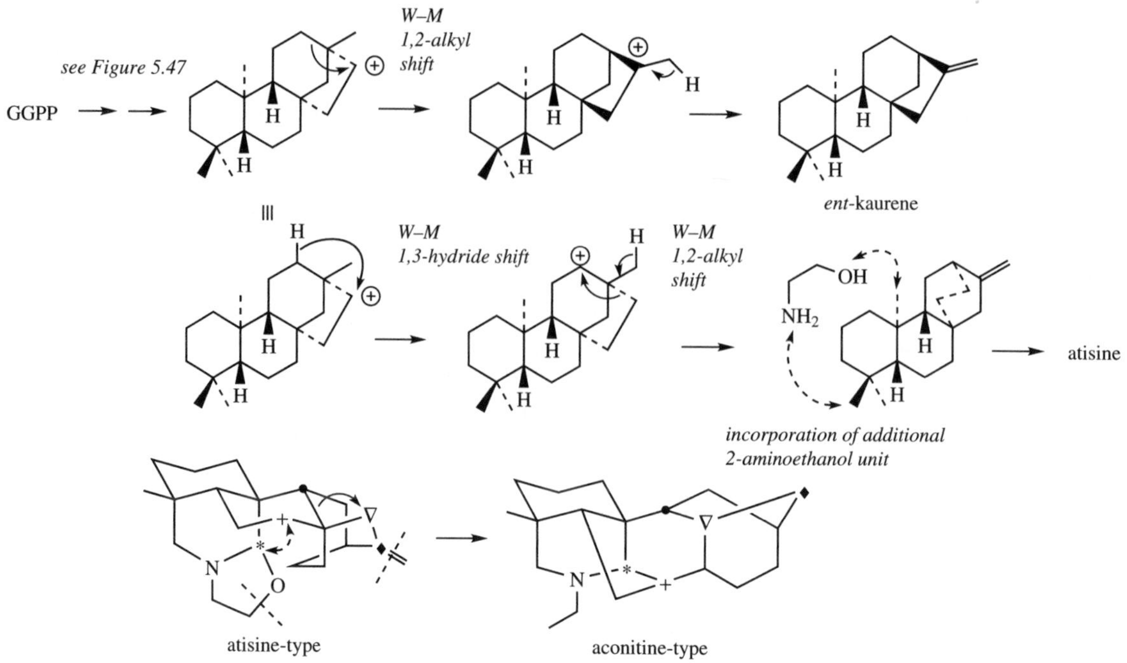

Figure 6.125

from 2-aminoethanol) to form the heterocyclic rings. The aconitine akeleton is probably formed from the atisine skeleton by further modifications as indicated. Note that a rearrangement process converts two fused six-membered rings into a (7 + 5)-membered bicyclic system, and that one carbon, that from the exocyclic double bond, is lost.

Steroidal Alkaloids

Many plants in the Solanaceae accumulate steroidal alkaloids based on a C_{27} cholestane skeleton, e.g. **solasodine** and **tomatidine** (Figure 6.126). These are essentially nitrogen analogues of steroidal saponins (see page 240) and have already been briefly considered along with these compounds. In contrast to the oxygen analogues, these compounds all have the same stereochemistry at C-25 (methyl always equatorial), but C-22 isomers do exist, as solasodine and tomatidine exemplify. They are usually present as glycosides which have surface activity and haemolytic properties as do the saponins, but these compounds are also toxic if ingested. **Solasonine** from *Solanum* species and **tomatine** (Figure 6.126) from tomato

(*Lycopersicon esculente*) are typical examples of such glycosides.

As with the sapogenins, this group of steroidal alkaloids is derived from cholesterol, with appropriate side-chain modifications during the sequence (Figure 6.127). Amination appears to employ L-arginine as the nitrogen source, probably via a substitution process on 26-hydroxycholesterol. A second substitution allows 26-amino-22-hydroxycholesterol to cyclize, generating a piperidine ring. After 16β-hydroxylation, the secondary amine is oxidized to an imine, and the spiro-system can be envisaged as the result of a nucleophilic addition of the 16β-hydroxyl on to the imine (or iminium via protonation). Whether the 22*R* (as in **solasodine**) or 22*S* (as in **tomatidine**) configuration is established may depend on this reaction.

A variant on the way the cholesterol side-chain is cyclized can be found in **solanidine** (Figure 6.126), which contains a condensed ring system with nitrogen at the bridgehead. Solanidine is found in potatoes (*Solanum tuberosum*), typically as the glycoside **α-solanine** (Figure 6.126). This condensed ring system appears to be produced by a branch from the main pathway to solasodine/

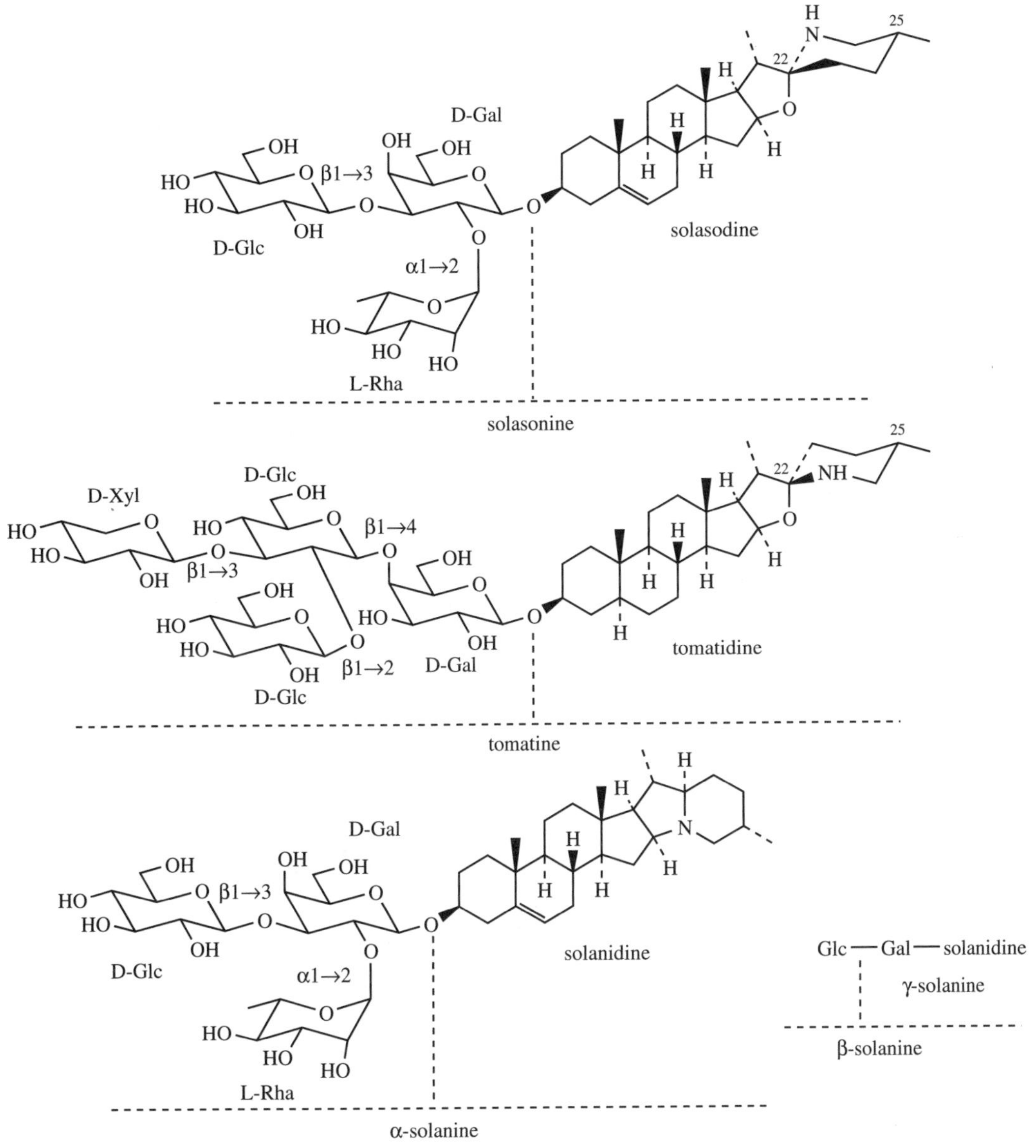

Figure 6.126

tomatidine structures. Thus, a substitution process will allow generation of the new ring system (Figure 6.128).

Since the production of medicinal steroids from steroidal saponins (see page 266) requires preliminary degradation to remove the ring systems containing the original cholesterol side-chain, it is immaterial whether these rings contain oxygen or nitrogen. Thus, plants rich in **solasodine** or **tomatidine** could also be employed for commercial steroid production. Similarly, other *Solanum* alkaloids* such as **solanidine** with nitrogen in a condensed ring system might also be exploited.

Several plants in the Liliaceae, notably the genus *Veratrum* (Liliaceae/Melanthiaceae), contain a remarkable group of steroidal alkaloids in which a fundamental change to the basic steroid nucleus has taken place. This change expands ring D by one carbon at the expense of ring C, which consequently becomes five-membered.

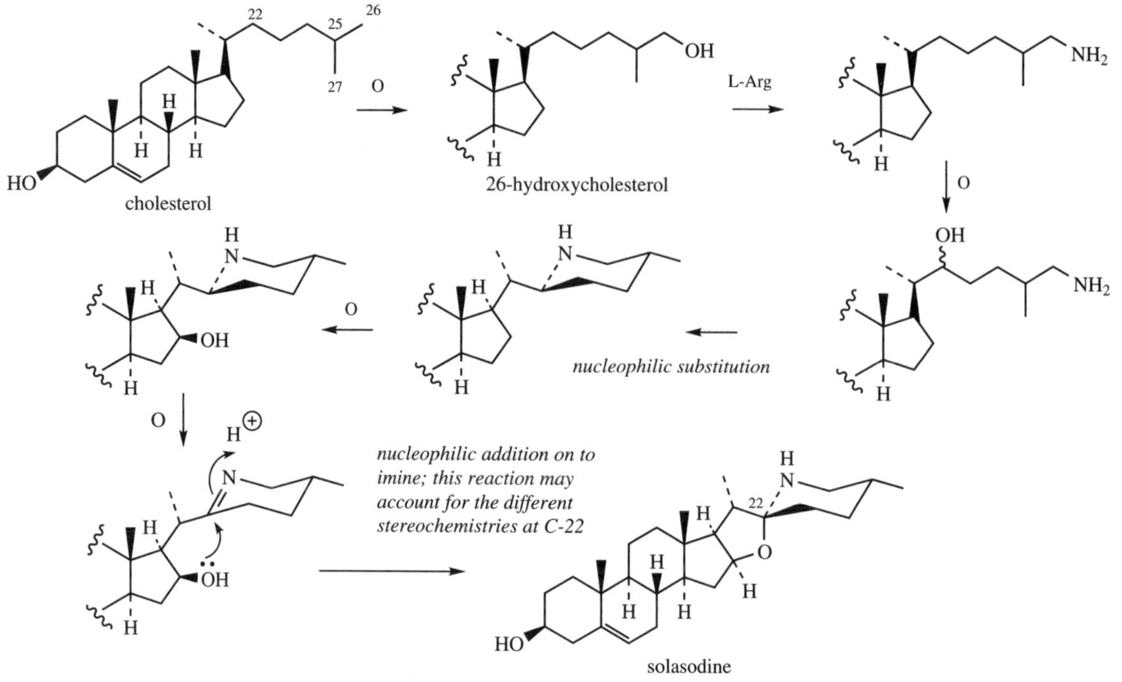

Figure 6.127

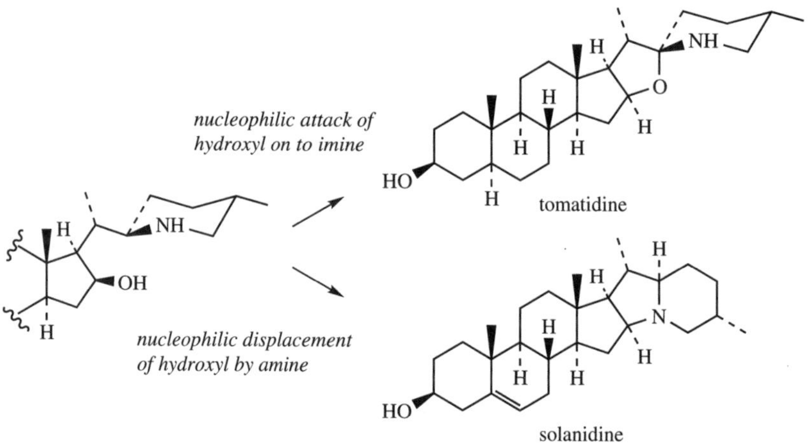

Figure 6.128

The resulting skeleton is termed a C-nor-D-homosteroid in keeping with these alterations in ring size. Cholesterol is a precursor of this group of alkaloids, and a mechanism accounting for the ring modifications is shown in Figure 6.129, where the changes are initiated by loss of a suitable leaving group from C-12. Typical representatives of C-nor-D-homosteroids are **jervine** and **cyclopamine**

(Figure 6.130) from *Veratrum californicum*, toxic components in this plant that are responsible for severe teratogenic effects. Animals grazing on *V. californicum* and some other species of *Veratrum* frequently give birth to young with cyclopia, a malformation characterized by a single eye in the centre of the forehead. The teratogenic effects of jervine, cyclopamine, and cyclopamine

Solanum Alkaloids

Solasodine (Figure 6.126) is a major component in many *Solanum* species, where it is present as glycosides in the leaves, and especially in the unripe fruits. Trial cultivations of a number of species, including *Solanum laciniatum* and *S. aviculare* (indigenous to New Zealand), *S. khasianum* (from India), and *S. marginatum* (from Ecuador), have been conducted in various countries. Alkaloid levels of 1–2% have been obtained, and these plants are especially suitable for long term cultivation if the fruits provide suitable quantities, being significantly easier to cultivate than disogenin-producing *Dioscorea* species. Solasodine may be converted into progesterone by means of the Marker degradation shown in Figure 5.119 (see page 266).

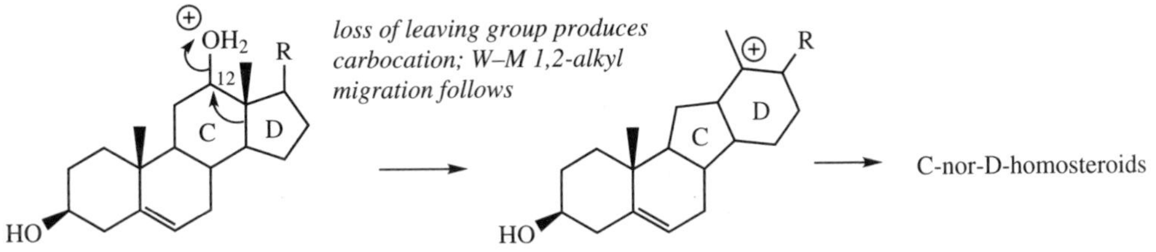

loss of leaving group produces carbocation; W–M 1,2-alkyl migration follows

C-nor-D-homosteroids

Figure 6.129

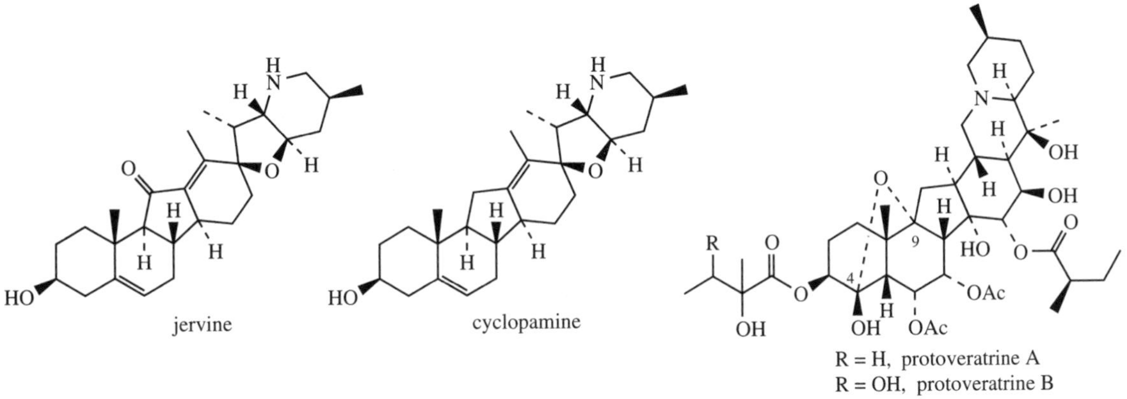

jervine

cyclopamine

R = H, protoveratrine A
R = OH, protoveratrine B

Figure 6.130

glucoside (cycloposine) on the developing fetus have now been well established. Other *Veratrum* alkaloids, especially those found in *V. album* and *V. viride*, have been employed medicinally as hypotensive agents, and used in the same way as *Rauwolfia* alkaloids (see page 352), often in combination with *Rauwolfia*. These medicinal alkaloids, e.g. **protoveratrine A** and **protoveratrine B** (Figure 6.130), which are esters of protoverine, are characterized by fusion of

two more six-membered rings on to the C-nor-D-homosteroid skeleton. This hexacyclic system is extensively oxygenated, and a novel hemiketal linkage bridges C-9 with C-4. Both the jervine and protoverine skeletons are readily rationalized through additional cyclization reactions involving a piperidine ring, probably formed by processes analogous to those seen with the *Solanum* alkaloids (Figure 6.127). These are outlined in Figure 6.131, which suggests the participation of the piperidine

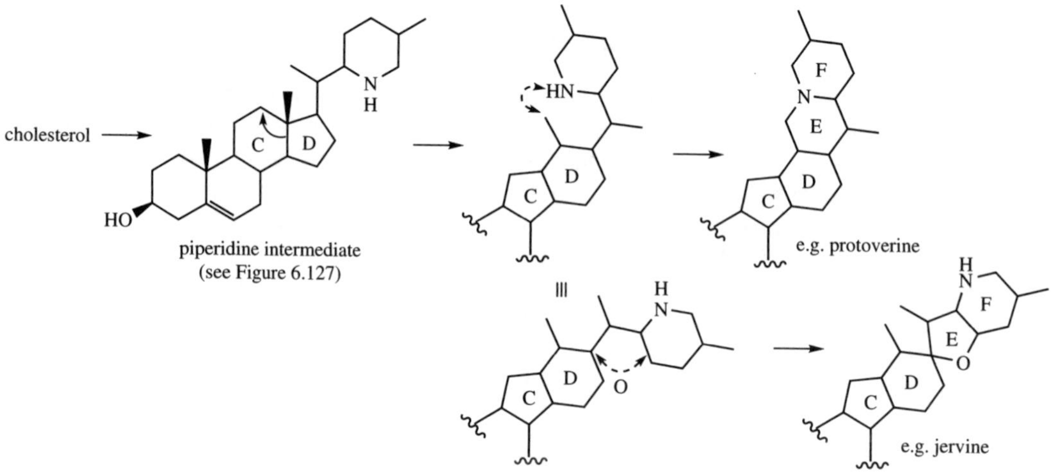

Figure 6.131

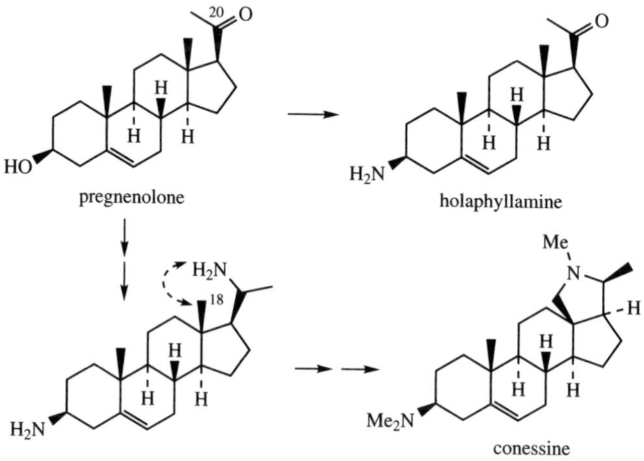

Figure 6.132

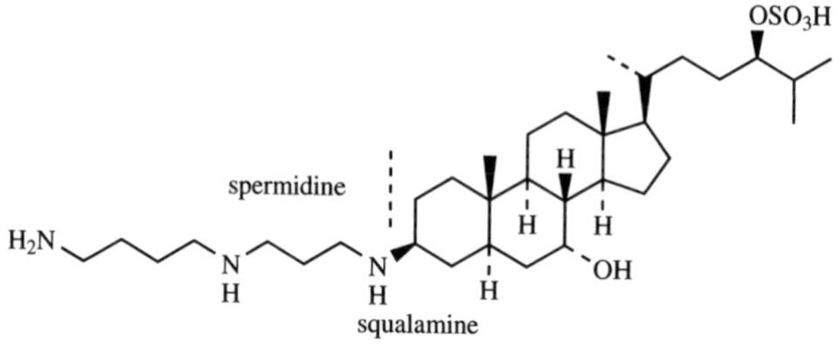

Figure 6.133

intermediate from Figure 6.127. Typically, both types of alkaloid are found co-occurring in *Veratrum* species.

Many steroidal derivatives are formed by truncation of the original C_8 side-chain, and C_{21}

pregnane derivatives are important animal hormones (see page 273) or intermediates on the way to other natural steroidal derivatives, e.g. cardioactive glycosides (see page 241). Alkaloids based on a pregnane skeleton are found in plants, particularly in the Apocynaceae and Buxaceae, and **pregnenolone** (Figure 6.132) is usually involved in their production. **Holaphyllamine** from *Holarrhena floribunda* (Apocynaceae) is obtained from pregnenolone by replacement of the 3-hydroxyl with an amino group (Figure 6.132). **Conessine** (Figure 6.132) from *Holarrhena antidysenterica* is also derived from pregnenolone, and requires two amination reactions, one at C-3 as for holaphyllamine, plus a further one, originally at C-20, probably via the C-20 alcohol. The new ring system in conessine is then the result of attack of the C-20 amine on to the C-18 methyl, suitably activated, of course. The bark of *H. antidysenterica* has long been used, especially in India, as a treatment for amoebic dysentery.

The novel steroidal polyamine **squalamine** (Figure 6.133) has been isolated in very small amounts (about 0.001%) from the liver of the dogfish shark (*Squalus acanthias*), and is

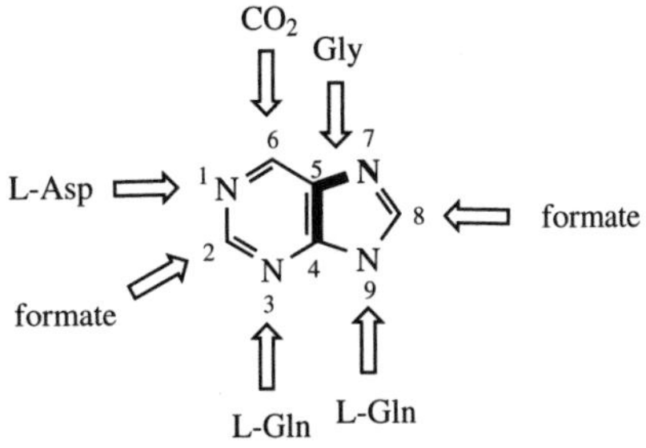

Figure 6.134

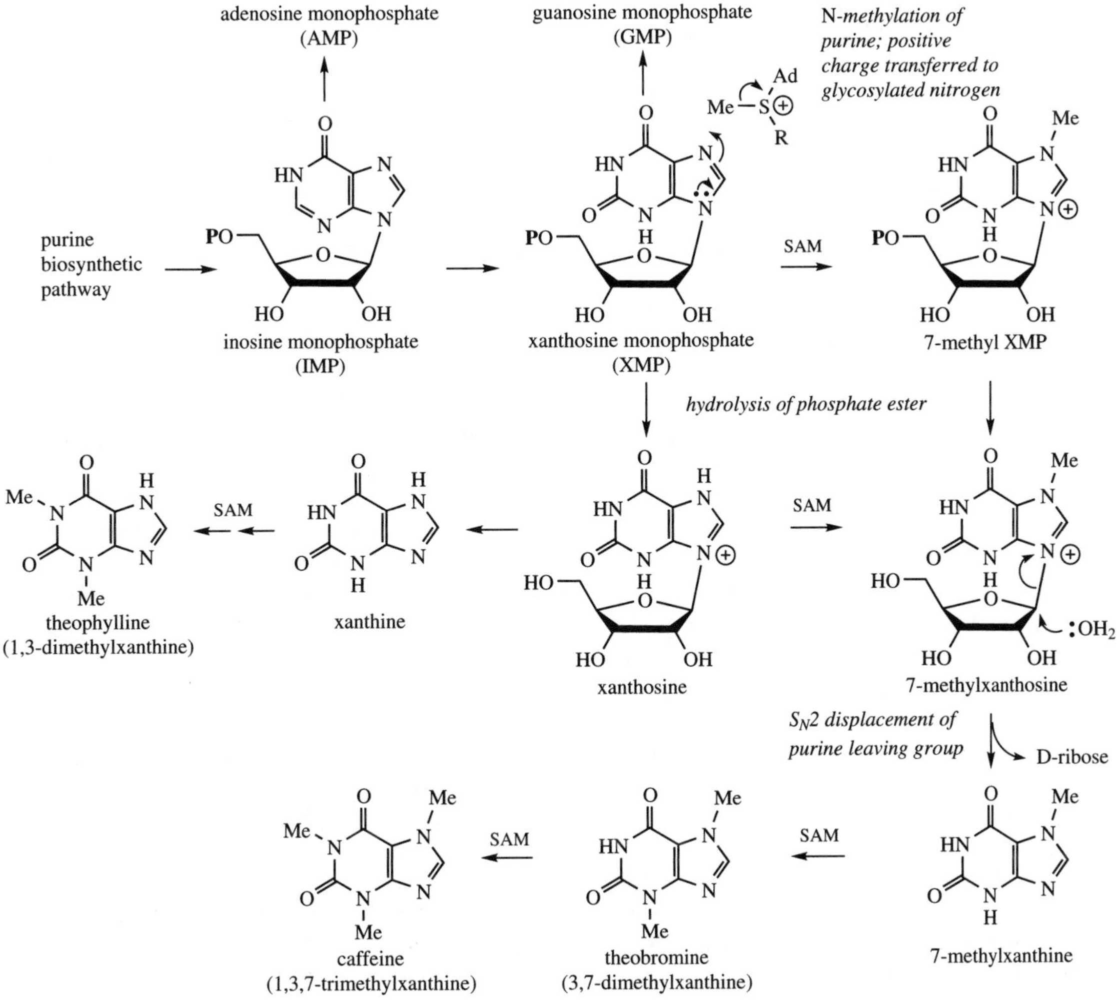

Figure 6.135

attracting attention because of its remarkable antimicrobial activity. This compound is a broad-spectrum agent effective at very low concentrations against Gram-positive and Gram-negative bacteria, and also fungi, protozoa, and viruses including HIV. The sulphated side-chain helps to make squalamine water soluble. The polyamine portion is spermidine, a compound widely distributed in both animals and plants. Related aminosterol derivatives with similar high antimicrobial activity have also been isolated from the liver extracts.

PURINE ALKALOIDS

The purine derivatives **caffeine***, **theobromine***, and **theophylline*** (Figure 6.135) are usually referred to as purine alkaloids. As alkaloids they have a limited distribution, but their origins are very closely linked with those of the purine bases adenine and guanine, fundamental components of nucleosides, nucleotides, and the nucleic acids. Caffeine, in the form of beverages such as tea*, coffee*, and cola* is one of the most widely consumed and socially accepted natural stimulants. It is also used medicinally, but theophylline is much more important as a drug compound because of its muscle relaxant properties, utilized in the relief of bronchial asthma. Theobromine is a major constituent of cocoa*, and related chocolate products.

The purine ring is gradually elaborated by piecing together small components from primary metabolism (Figure 6.134). The largest component incorporated is glycine, which provides a C_2N unit, whilst the remaining carbon atoms come from formate (by way of N^{10}-formyl-tetrahydrofolate (see page 126)) and bicarbonate. Two of the four nitrogen atoms are supplied by glutamine, and a third by aspartic acid. Synthesis of the nucleotides adenosine 5'-monophosphate (AMP) and guanosine 5'-monophosphate (GMP) is by way of inosine 5'-monophosphate (IMP) and xanthosine 5'-monophosphate (XMP) (Figure 6.135), and the purine alkaloids then branch away through XMP. Methylation, then loss of phosphate, generates the nucleoside **7-methylxanthosine**, which is then released from the sugar. Successive methylations on the nitrogens give **caffeine** by way of **theobromine**, whilst a different methylation sequence can account for the formation of **theophylline**.

Caffeine, Theobromine, and Theophylline

The purine alkaloids caffeine, theobromine, and theophylline (Figure 6.135) are all methyl derivatives of xanthine and they commonly co-occur in a particular plant. The major sources of these compounds are the beverage materials such as tea, coffee, cocoa, and cola, which owe their stimulant properties to these water-soluble alkaloids. They competitively inhibit phosphodiesterase, resulting in an increase in cyclic AMP and subsequent release of adrenaline. This leads to a stimulation of the CNS, a relaxation of bronchial smooth muscle, and induction of diuresis, as major effects. These effects vary in the three compounds. **Caffeine** is the best CNS stimulant, and has weak diuretic action. **Theobromine** has little stimulant action, but has more diuretic activity and also muscle relaxant properties. **Theophylline** also has low stimulant action and is an effective diuretic, but it relaxes smooth muscle better than caffeine or theobromine.

Caffeine is used medicinally as a CNS stimulant, usually combined with another therapeutic agent, as in compound analgesic preparations. **Theobromine** is of value as a diuretic and smooth muscle relaxant, but is not now routinely used. **Theophylline** is an important smooth muscle relaxant for relief of bronchospasm, and is frequently dispensed in slow-release formulations to reduce side-effects. It is also available as **aminophylline** (a more soluble preparation containing theophylline with ethylenediamine) and **choline theophyllinate** (theophylline and choline). The alkaloids may be isolated from natural sources, or obtained by total or partial synthesis.

(Continues)

(Continued)

It has been estimated that beverage consumption may provide the following amounts of caffeine per cup or average measure: coffee, 30–150 mg (average 60–80 mg); instant coffee, 20–100 mg (average 40–60 mg); decaffeinated coffee, 2–4 mg; tea, 10–100 mg (average 40 mg); cocoa, 2–50 mg (average 5 mg); cola drink, 25–60 mg. The maximal daily intake should not exceed about 1 g to avoid unpleasant side-effects, e.g. headaches, restlessness. An acute lethal dose is about 5–10 g. The biological effects produced from the caffeine ingested via the different drinks can vary, since its bioavailability is known to be modified by the other constituents present, especially the amount and nature of polyphenolic tannins.

Coffee

Coffee consists of the dried ripe seed of *Coffea arabica*, *C. canephora*, *C. liberica*, or other *Coffea* species (Rubiaceae). The plants are small evergreen trees, widely cultivated in various parts of the world, e.g. Brazil and other South American countries, and Kenya. The fruit is deprived of its seed coat, then dried and roasted to develop its characteristic colour, odour, and taste. Coffee seeds contain 1–2% of caffeine and traces of theophylline and theobromine. These are mainly combined in the green seed with chlorogenic acid (see page 132) (5–7%), and roasting releases them and also causes some decomposition of chlorogenic acid to quinic acid and caffeic acid. The nicotinic acid derivative trigonelline is present in green seeds to the extent of about 0.25–1%; during roasting, this is extensively converted into nicotinic acid (vitamin B_3, see page 313). Volatile oils and tannins provide odour and flavour. A proportion of the caffeine may sublime off during the roasting process, providing some commercial caffeine. Decaffeinated coffee, containing up to 0.08% caffeine, is obtained by removing caffeine, usually by aqueous percolation prior to roasting. This process provides another source of caffeine.

Tea

Tea is the prepared leaves and leaf buds of *Camellia sinensis* (*Thea sinensis*) (Theaceae), an evergreen shrub cultivated in China, India, Japan, and Sri Lanka. For black tea, the leaves are allowed to ferment, allowing enzymic oxidation of the polyphenols, whilst green tea is produced by steaming and drying the leaves to prevent oxidation. During oxidation, colourless catechins (up to 40% in dried leaf) (see page 150) are converted into intensely coloured theaflavins and thearubigins. Oolong tea is semi-fermented. Tea contains 1–4% caffeine, and small amounts (up to 0.05%) of both theophylline and theobromine. Astringency and flavour come from tannins and volatile oils, the latter containing monoterpene alcohols (geraniol, linalool) and aromatic alcohols (benzyl alcohol, 2-phenylethanol). Theaflavins (see page 151) are believed to act as radical scavengers/antioxidants, and to provide beneficial effects against cardiovascular disease, cancers, and the ageing process generally. Tea leaf dust and waste is a major source of caffeine.

Cola

Cola, or kola, is the dried cotyledon from seeds of various species of *Cola* (Sterculiaceae), e.g. *C. nitida* and *C. acuminata*, trees cultivated principally in West Africa and the West Indies. Seeds are prepared by splitting them open and drying. Cola seeds contain up to 3% caffeine

(Continues)

(Continued)

and about 0.1% theobromine, partly bound to tannin materials. Drying allows some oxidation of polyphenols, formation of a red pigment, and liberation of free caffeine. Fresh cola seeds are chewed in tropical countries as a stimulant, and vast quantities of dried seeds are processed for the preparation of cola drinks, e.g. Coca-Cola and Pepsi-Cola.

Cocoa

Although cocoa as a drink is now rather unfashionable, it provides the raw material for the manufacture of chocolate and is commercially very important. Cocoa (or cacao) is derived from the roasted seeds of *Theobroma cacao* (Sterculiaceae), a tree widely cultivated in South America and West Africa. The fruits develop on the trunk of the tree, and the seeds from them are separated, allowed to ferment, and are then roasted to develop the characteristic chocolate flavour. The kernels are then separated from the husks, ground up, and processed in various ways to give chocolate, cocoa, and cocoa butter.

Cocoa seeds contain 35–50% of oil (cocoa butter or theobroma oil), 1–4% theobromine and 0.2–0.5% caffeine, plus tannins and volatile oils. During fermentation and roasting, most of the theobromine from the kernel passes into the husk, which thus provides a convenient source of the alkaloid. Theobroma oil or cocoa butter is obtained by hot expression from the ground seeds as a whitish solid with a mild chocolate taste. It is a valuable formulation aid in pharmacy where it is used as a suppository base. It contains glycerides of oleic (35%), stearic (35%), palmitic (26%), and linoleic (3%) acids (see page 44).

Maté Tea

Maté tea is consumed in South America as a stimulant drink. Maté or Paraguay tea consists of the leaves of *Ilex paraguensis* (Aquifoliaceae), South American shrubs of the holly genus. The dried leaf contains 0.8–1.7% caffeine and smaller amounts of theobromine (0.3–0.9%) with little or no theophylline. Considerable amounts (10–16%) of chlorogenic acid (see page 132) are also present.

Guarana

The seeds of the Brazilian plant *Paullinia cupana* (Sapindaceae) are used to make a stimulant drink. Crushed seeds are mixed with water to a paste, which is then sun dried. Portions of this are then boiled with hot water to provide a refreshing drink. The principal constituent, previously called guaranine, has been shown to be identical to caffeine, and the seeds may contain 3–5%. Small amounts of theophylline (0–0.25%) and theobromine (0.02–0.06%) are also present. Guarana is widely available as tablets and capsules, or as extracts, in health food shops where it is promoted to relieve mental and physical fatigue. Labels on such products frequently show the active constituent to be guaranine, but may not indicate that this is actually caffeine.

Saxitoxin and Tetrodotoxin

The structure of **saxitoxin*** (Figure 6.136) contains a reduced purine ring system, but it is not biosynthetically related to the purine alkaloids described above. Not all features of its biosynthetic origin have been established, but the amino acid supplying most of the ring system is known to be L-arginine (Figure 6.136). Acetate and a C_1 unit from methionine are

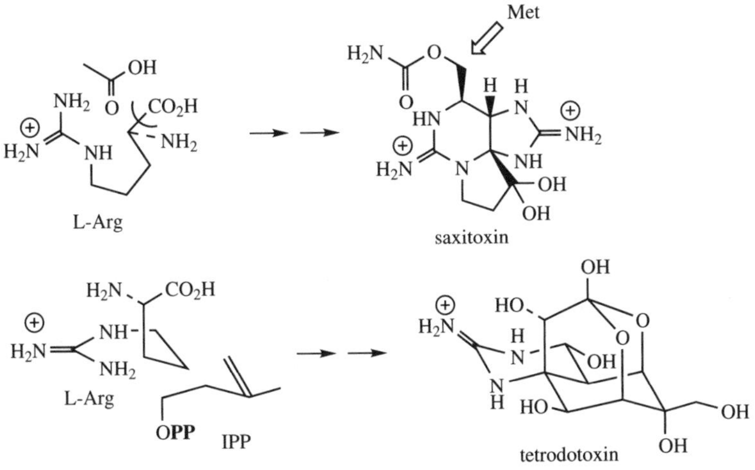

Figure 6.136

also utilized. Saxitoxin contains two highly polar guanidino functions, one of which is provided by arginine, and is a fast acting neurotoxin inhibiting nerve conduction by blocking sodium channels. It is one of a group of marine toxins referred to as paralytic shellfish poisons (PSP) found in a range of shellfish, but ultimately derived from toxic strains of dinoflagellates consumed by the shellfish. Arginine is also a precursor for **tetrodotoxin*** (Figure 6.136), another

marine neurotoxin containing a polar guanidino group. It has been established that the remainder of the carbon skeleton in tetrodotoxin is a C_5 isoprene unit, probably supplied as isopentenyl diphosphate (Figure 6.136). Tetrodotoxin is well known as the toxic principle in the puffer fish (*Tetraodon* species), regarded as delicacy in Japanese cuisine. As potent sodium channel blockers, both saxitoxin and tetrodotoxin are valuable pharmacological tools.

Saxitoxin

Saxitoxin (Figure 6.136) was first isolated from the Alaskan butter clam (*Saxidomus giganteus*) but has since been found in many species of shellfish, especially bivalves such as mussels, scallops, and oysters. These filter feeders consume dinoflagellates (plankton) and can accumulate toxins synthesized by these organisms, particularly during outbreaks known as red tides, when conditions favour formation of huge blooms of the dinoflagellates (see also brevetoxin A, page 109). Species of the dinoflagellate *Gonyaulax* in marine locations, or the cyanobacterium *Aphanizomenon* in freshwater, have been identified among the causative organisms, and the problem is encountered widely in temperate and tropical areas (including Europe, North America, and Japan). Commercial production of shellfish is routinely monitored for toxicity, which will slowly diminish as conditions change and the causative organism disappears from the water. About a dozen natural saxitoxin-related structures have been characterized, and mixtures in various proportions are typically synthesized by a producer, with the possibility that the shellfish may also structurally modify the toxins further. Acute and often fatal poisonings caused by the consumption of contaminated shellfish are termed paralytic shellfish poisoning (PSP), which involves paralysis of the neuromuscular system, death resulting from respiratory failure. Saxitoxin is a cationic molecule, which binds to sodium channels, blocking the influx of sodium ions through excitable nerve membranes,

(Continues)

(Continued)

and is a valuable pharmacological tool for the study of this process. Saxitoxin and tetrodotoxin (below) are some of the most potent non-protein neurotoxins known, and are active at very low concentrations (μg kg^{-1}).

Tetrodotoxin

Tetrodotoxin (Figure 6.136) is traditionally associated with the puffer fish, *Tetraodon* species, a fish known as fugu, a highly prized delicacy eaten in Japan. Preparation of fugu is a skilled operation in which organs containing the highest levels of toxin, e.g. liver, ovaries, and testes, are carefully separated from the flesh. Even so, deaths from fugu poisoning are not uncommon, and the element of risk presumably heightens culinary appreciation of the fish. As with saxitoxin, tetrodotoxin appears to be produced by microorganisms, and symbiotic marine bacteria, e.g. *Vibrio* species, have been implicated as the synthesizers. In addition to fugu, several other species of fish, newts, and frogs have been found to accumulate tetrodotoxin or related structures. The mode of action of tetrodotoxin is exactly the same as that of saxitoxin above, though there are some subtle differences in the mechanism of binding.

FURTHER READING

Biosynthesis, General

Dalton DR (1991) Alkaloids. *Kirk–Othmer Encyclopedia of Chemical Technology*, 4th edn, Vol 1. Wiley, New York, pp 1039–1087.

Herbert RB (2001) The biosynthesis of plant alkaloids and nitrogenous microbial metabolites. *Nat Prod Rep* **18**, 50–65. Earlier reviews: 1999, **16**, 199–208; 1997, **14**, 359–372.

Kutchan TM (1995) Alkaloid biosynthesis – the basis for metabolic engineering of medicinal plants. *Plant Cell* **7**, 1059–1070.

Kutchan TM (1998) Molecular genetics of plant alkaloid biosynthesis. *The Alkaloids, Chemistry and Pharmacology* (ed Cordell GA) Vol 50. Academic, San Diego, pp 257–316.

Misra N, Luthra R, Singh KL and Kumar S (1999) Recent advances in biosynthesis of alkaloids. *Comprehensive Natural Products Chemistry*, Vol 4. Elsevier, Amsterdam, pp 25–59.

Tropane Alkaloids

Bernstein PR (1992) Antiasthmatic agents. *Kirk–Othmer Encyclopedia of Chemical Technology*, 4th edn, Vol 2. Wiley, New York, pp 830–854.

Evans WC (1990) *Datura*, a commercial source of hyoscine. *Pharm J* **244**, 651–653.

Griffin WJ and Lin GD (2000) Chemotaxonomy and geographical distribution of tropane alkaloids. *Phytochemistry* **53**, 623–637.

Gritsanapan W and Griffin WJ (1992) Alkaloids and metabolism in a *Duboisia* hybrid. *Phytochemistry* **31**, 471–477.

Lounasmaa M and Tamminen T (1993) The tropane alkaloids. *The Alkaloids, Chemistry and Pharmacology* (ed Cordell GA) Vol 44. Academic, San Diego, 1–114.

Molyneux RJ, Nash RJ and Asano N (1996) Chemistry and biological activity of the calystegines and related nortropane alkaloids. *Alkaloids, Chemical and Biological Perspectives* (ed Pelletier SW) Vol 11. Elsevier, Amsterdam, pp 303–343.

O'Hagan D (2000) Pyrrole, pyrrolidine, pyridine, piperidine and tropane alkaloids. *Nat Prod Rep* **17**, 435–446. Earlier review: 1997, **14**, 637–651.

O'Hagan D and Robins RJ (1998) Tropic acid ester biosynthesis in *Datura stramonium* and related species. *Chem Soc Rev* **27**, 207–212.

Cocaine

Johnson EL and Emche SD (1994) Variation in alkaloid content in *Erythroxylum coca* leaves from leaf bud to leaf drop. *Ann Bot* **73**, 645–650.

Lenz GR, Schoepke HG and Spaulding TC (1992) Anesthetics. *Kirk–Othmer Encyclopedia of Chemical Technology*, 4th edn, Vol 2. Wiley, New York, 778–800.

Singh S (2000) Chemistry, design, and structure–activity relationship of cocaine antagonists. *Chem Rev* **100**, 925–1024.

Woolverton WL and Johnson KM (1992) Neurobiology of cocaine abuse. *Trends in Pharmacological Sciences* **13**, 193–200.

Epibatidine

Daly JW, Garaffo HM, Spande TF, Decker MW, Sullivan JP and Williams M (2000) Alkaloids from frog skin: the discovery of epibatidine and the potential for developing novel non-opioid analgesics. *Nat Prod Rep* **17**, 131–135.

Szantay C, Kardos-Balogh Z and Szantay C Jr (1995) Epibatidine. *The Alkaloids, Chemistry and Pharmacology* (ed Cordell GA) Vol 46. Academic, San Diego, pp 95–125.

Pyrrolizidine Alkaloids

Hartmann T (1999) Chemical ecology of pyrrolizidine alkaloids. *Planta* **207**, 483–495.

Hartmann T and Witte L (1995) Chemistry, biology and chemoecology of the pyrrolizidine alkaloids. *Alkaloids, Chemical and Biological Perspectives* (ed Pelletier SW) Vol 9. Wiley, New York, 155–233.

Liddell JR (2000) Pyrrolizidine alkaloids. *Nat Prod Rep* **17**, 455–462. Earlier reviews: 1999, **16**, 499–507; 1998, **15**, 363–370.

Robins DJ (1995) Biosynthesis of pyrrolizidine and quinolizidine alkaloids. *The Alkaloids, Chemistry and Pharmacology* (ed Cordell GA) Vol 46. Academic, San Diego, pp 1–61.

Piperidine Alkaloids

O'Hagan D (2000) Pyrrole, pyrrolidine, pyridine, piperidine and tropane alkaloids. *Nat Prod Rep* **17**, 435–446. Earlier reviews: 1997, **14**, 637–651; Plunkett AO (1994) **11**, 581–590.

Schneider MJ (1996) Pyridine and piperidine alkaloids: an update. *Alkaloids, Chemical and Biological Perspectives* (ed Pelletier SW) Vol 10. Elsevier, Amsterdam, pp 155–299.

Indolizidine and Quinolizidine Alkaloids

Michael JP (2000) Indolizidine and quinolizidine alkaloids. *Nat Prod Rep* **17**, 579–602. Earlier reviews: 1999, **16**, 675–696; 1998, **15**, 571–594.

Takahata H and Momose T (1993) Simple indolizidine alkaloids. *The Alkaloids, Chemistry and Pharmacology* (ed Cordell GA) Vol 44. Academic, San Diego, pp 189–256.

Ohmiya S, Saito K and Murakoshi I (1995) Lupine alkaloids. *The Alkaloids, Chemistry and Pharmacology* (ed Cordell GA) Vol 47. Academic, San Diego, pp 1–114.

Robins DJ (1995) Biosynthesis of pyrrolizidine and quinolizidine alkaloids. *The Alkaloids, Chemistry and Pharmacology* (ed Cordell GA) Vol 46. Academic, San Diego, pp 1–61.

Castanospermine

Burgess K and Henderson I (1992) Synthetic approaches to stereoisomers and analogues of castanospermine. *Tetrahedron* **48**, 4045–4066.

Nash RJ, Watson AA and Asano N (1996) Polyhydroxylated alkaloids that inhibit glycosidases. *Alkaloids, Chemical and Biological Perspectives* (ed Pelletier SW) Vol 11. Elsevier, Amsterdam, pp 345–376.

Piperidine Alkaloids

O'Hagan D (2000) Pyrrole, pyrrolidine, pyridine, piperidine and tropane alkaloids. *Nat Prod Rep* **17**, 435–446. Earlier reviews: 1997, **14**, 637–651; Plunkett AO (1994) **11**, 581–590.

Schneider MJ (1996) Pyridine and piperidine alkaloids: an update. *Alkaloids, Chemical and Biological Perspectives* (ed Pelletier SW) Vol 10. Elsevier, Amsterdam, pp 155–299.

Vitamin B$_3$

Van Arnum SD (1998) Vitamins (niacin, nicotinamide, and nicotinic acid). *Kirk–Othmer Encyclopedia of Chemical Technology*, 4th edn, Vol 25. Wiley, New York, pp 83–99.

Tobacco

Baker R and Proctor C (2001) Where there's smoke. *Chem Brit* **37**(1), 38–41.

Botting NP (1995) Chemistry and biochemistry of the kynurenine pathway of tryptophan metabolism. *Chem Soc Rev* **24**, 401–412.

Hucko F, Tsetlin VI and Machold (1996) The emerging three-dimensional structure of a receptor. The nicotinic acetylcholine receptor. *Eur J Biochem* **239**, 539–557.

Nathan A (1998) Smoking cessation products. *Pharm J* **260**, 340–343.

Wills S (1994) Drugs and substance misuse: tobacco and alcohol. *Pharm J* **253**, 158–160.

Catecholamines

Bernstein PR (1992) Antiasthmatic agents. *Kirk–Othmer Encyclopedia of Chemical Technology*, 4th edn, Vol 2. Wiley, New York, 830–854.

Cervoni P, Crandall DL and Chan PS (1993) Cardiovascular agents – antiarrhythmic agents. *Kirk–Othmer Encyclopedia of Chemical Technology*, 4th edn, Vol 5. Wiley, New York, pp 207–238.

Cervoni P, Crandall DL and Chan PS (1993) Cardiovascular agents – antihypertensive agents. *Kirk–Othmer Encyclopedia of Chemical Technology*, 4th edn, Vol 5. Wiley, New York, pp 262–289.

Pugsley TA (1994) Epinephrine and norepinephrine. *Kirk–Othmer Encyclopedia of Chemical Technology*, 4th edn, Vol 9. Wiley, New York, pp 715–730.

Isoquinoline Alkaloids

Bentley KW (2000) β-Phenylethylamines and the isoquinoline alkaloids. *Nat Prod Rep* **17**, 247–268. Earlier reviews: 1999, **16**, 367–388; 1998, **15**, 341–362.

Curare

Bisset NG (1992) War and hunting poisons of the new World. Part 1. Notes on the early history of curare. *J Ethnopharmacol* **36**, 1–26.

Bisset NG (1992) Curare. Alkaloids, *Chemical and Biological Perspectives* (ed Pelletier SW) Vol 8. Wiley, New York, pp 1–150.

Menachery MD (1996) The alkaloids of South American Menispermaceae. *Alkaloids, Chemical and Biological Perspectives* (ed Pelletier SW) Vol 11. Elsevier, Amsterdam, pp 269–302.

Opium Alkaloids

Brownstein MJ (1993) A brief history of opiates, opioid peptides, and opiate receptors. *Proc Natl Acad Sci USA* **90**, 5391–5393.

Chiara GD and North RA (1992) Neurobiology of opiate abuse. *Trends Pharmacol Sci* **13**, 185–193.

Herbert RB, Venter H and Pos S (2000) Do mammals make their own morphine? *Nat Prod Rep* **17**, 317–322.

Lenz GR, Schoepke HG and Spaulding TC (1992) Anesthetics. *Kirk–Othmer Encyclopedia of Chemical Technology*, 4th edn, Vol 2. Wiley, New York, pp 778–800.

Nugent RA and Hall CM (1992) Analgesic, antipyretics, and antiinflammatory agents. *Kirk–Othmer Encyclopedia of Chemical Technology*, 4th edn, Vol 2. Wiley, New York, pp 729–748.

Schmidhammer H (1998) Opioid receptor antagonists. *Prog Med Chem* **35**, 83–132.

Szantay C, Dornyei G and Blasko G (1994) The morphine alkaloids. *The Alkaloids, Chemistry and Pharmacology* (ed Cordell GA) Vol 45. Academic, San Diego, pp 127–232.

Colchicine

Boye O and Brossi A (1992) Tropolonic *Colchicum* alkaloids and allo congeners. *The Alkaloids, Chemistry and Pharmacology* (ed Brossi A and Cordell GA) Vol 41. Academic, San Diego, pp 125–176.

Ipecacuanha

Fujii T and Ohba M (1998) The ipecac alkaloids and related bases. *The Alkaloids, Chemistry and Biology* (ed Cordell GA) Vol 51. Academic, San Diego, pp 271–321.

Klayman DL (1992) Antiparasitic agents (antiprotozoals). *Kirk–Othmer Encyclopedia of Chemical Technology*, 4th edn, Vol 3. Wiley, New York, pp 489–526.

Paul DB and Dobberstein RH (1994) Expectorants, antitussives, and related agents. *Kirk–Othmer Encyclopedia of Chemical Technology*, 4th edn, Vol 9. Wiley, New York, pp 1061–1081.

Amaryllidaceae Alkaloids

Hoshino O (1998) The Amaryllidaceae alkaloids. *The Alkaloids, Chemistry and Biology* (ed Cordell GA) Vol 51. Academic, San Diego, pp 324–424.

Lewis JR (2001) Amaryllidaceae, *Sceletium*, imidazole, oxazole, thiazole, peptide and miscellaneous alkaloids. *Nat Prod Rep* **18**, 95–128. Earlier reviews: 2000, **17**, 57–84; 1999, **16**, 389–416.

Scott LJ and Goa KL (2000) Galantamine: a review of its use in Alzheimer's disease. *Drugs* **60**, 1095–1122.

Indole Alkaloids

Ihara M and Fukumoto K (1996) Recent progress in the chemistry of non-monoterpenoid indole alkaloids. *Nat Prod Rep* **13**, 241–261. Earlier review: 1995, **12**, 277–301.

Psilocybin

Antkowiak R and Antkowiak WZ (1991) Alkaloids from mushrooms. *The Alkaloids, Chemistry and Pharmacology* (ed Brossi A) Vol 40. Academic, San Diego, pp 189–340.

Wills S (1993) Drugs and substance misuse: plants. *Pharm J* **251**, 227–229.

Terpenoid Indole Alkaloids

Hibino S and Choshi T (2001) Simple indole alkaloids and those with a nonrearranged monoterpenoid unit. *Nat Prod Rep.* **18**, 66–87. Earlier review: Lounasmaa M and Tolvanen A (2000) **17**, 175–191.

Kutchan TM (1993) Strictosidine: from alkaloid to enzyme to gene. *Phytochemistry* **32**, 493–506.

Leonard J (1999) Recent progress in the chemistry of monoterpenoid alkaloids derived from secologanin. *Nat Prod Rep* **16**, 319–338.

Stöckigt J and Ruppert M (1999) Strictosidine – the biosynthetic key to monoterpenoid indole alkaloids. *Comprehensive Natural Products Chemistry*, Vol 4. Elsevier, Amsterdam, pp 109–138.

Toyota M and Ihara M (1998) Recent progress in the chemistry of non-monoterpenoid indole alkaloids. *Nat Prod Rep* **15**, 327–340. Earlier reviews: Ihara M and Fukumoto K (1997) **14**, 413–429; 1996, **13**, 241–261.

Verpoorte R, van der Heiden R and Memelink J (1998) Plant biotechnology and the production of alkaloids: prospects of metabolic engineering. *The Alkaloids, Chemistry and Pharmacology* (ed Cordell GA) Vol 50. Academic, San Diego, pp 453–508.

Rauwolfia

Curzon G (1990) How reserpine and chlorpromazine act: the impact of key discoveries on the history of psychopharmacology. *Trends Pharmacol Sci* **11**, 61–63.

Stöckigt J (1995) Biosynthesis in *Rauwolfia serpentina*: modern aspects of an old medicinal plant. *The Alkaloids, Chemistry and Pharmacology* (ed Cordell GA) Vol 47. Academic, San Diego, pp 115–172.

Stöckigt J, Obitz P, Falkenhagen H, Lutterbach R and Endreß S (1995) Natural products and enzymes from plant cell cultures. *Plant Cell Tissue Organ Culture* **43**, 97–109.

Catharanthus Alkaloids

Kutney JP (1990) Biosynthesis and synthesis of indole and bisindole alkaloids in plant cell cultures: a personal overview. *Nat Prod Rep* **7**, 85–103.

Kutney JP (1993) Plant cell culture combined with chemistry: a powerful route to complex natural products. *Account Chem Res* **26**, 559–566.

Noble RL (1990) The discovery of the Vinca alkaloids – chemotherapeutic agents against cancer. *Biochem Cell Biol* **68**, 1344–1351.

Verpoorte R, van der Heijden R and Moreno PRH (1997) Biosynthesis of terpenoid indole alkaloids in *Catharanthus roseus* cells. *The Alkaloids, Chemistry and Pharmacology* (ed Cordell GA) Vol 49. Academic, San Diego, pp 221–299.

Ibogaine

Popik P and Skolnick P (1999) Pharmacology of ibogaine and ibogaine-related alkaloids. *The Alkaloids, Chemistry and Biology* (ed Cordell GA) Vol 52. Academic, San Diego, pp 197–231.

Strychnine

Bosch J, Bonjoch J and Amat M (1996) The *Strychnos* alkaloids. *The Alkaloids, Chemistry and Pharmacology* (ed Cordell GA) Vol 48. Academic, San Diego, pp 75–189.

Quetin-Leclercq J, Angenot L and Bisset NG (1990) South American *Strychnos* species. Ethnobotany (except curare) and alkaloid screening. *J Ethnopharmacol* **28**, 1–52.

Ellipticine

Gribble GW (1990) Synthesis and antitumour activity of ellipticine alkaloids and related compounds. *The Alkaloids, Chemistry and Pharmacology* (ed Brossi A) Vol 39. Academic, San Diego, pp 239–352.

Cinchona

Cervoni P, Crandall DL and Chan PS (1993) Cardiovascular agents – antiarrhythmic agents. *Kirk–Othmer*

Encyclopedia of Chemical Technology, 4th edn, Vol 5. Wiley, New York, 207–238.

Goodyer L (2000) Travel medicine (4): malaria. *Pharm J* **264**, 405–410.

Klayman DL (1992) Antiparasitic agents (antiprotozoals). *Kirk–Othmer Encyclopedia of Chemical Technology*, 4th edn, Vol 3. Wiley, New York, pp 489–526.

Lee M (1996) Malaria: in search of solutions. *Chem Brit* **32** (8), 28–30.

Camptothecin

Cordell GA (1993) The discovery of plant anticancer agents. *Chem Ind* 841–844.

Kawato Y and Terasawa H (1997) Recent advances in the medicinal chemistry and pharmacology of camptothecin. *Prog Med Chem* **34**, 69–109.

Wall ME and Wani MC (1998) History and future prospects of camptothecin and taxol. *The Alkaloids, Chemistry and Pharmacology* (ed Cordell GA) Vol 50. Academic, San Diego, pp 509–536.

Physostigmine

Brossi A, Pei X-F and Greig NH (1996) Phenserine, a novel anticholinesterase related to physostigmine: total synthesis and biological properties. *Aust J Chem* **49**, 171–181.

Takano S and Ogasawara K (1989) Alkaloids of the calabar bean. *The Alkaloids, Chemistry and Pharmacology* (ed Brossi A) Vol 36. Academic, San Diego, pp 225–251.

Ergot

Groger D and Floss H (1998) Biochemistry of ergot alkaloids: achievements and challenges. *The Alkaloids, Chemistry and Pharmacology* (ed Cordell GA) Vol 50. Academic, San Diego, pp 171–218.

Kren V (1991) Bioconversions of ergot alkaloids. *Adv Biochem Eng Biotech* **44**, 123–144.

Kren V, Harazim P and Malinka Z (1994) *Claviceps purpurea* (ergot): culture and bioproduction of ergot alkaloids. *Biotechnology in Agriculture and Forestry, Vol 58 Medicinal and Aromatic Plants VII* (ed Bajaj YPS). Springer, Heidelberg, pp 139–156.

Quinoline and Acridine Alkaloids

Michael JP (2000) Quinoline, quinazoline and acridone alkaloids. *Nat Prod Rep* **17**, 603–620. Earlier reviews: 1999, **16**, 697–709; 1998, **15**, 595–606.

Tillequin F, Michel S and Skaltsounis A-L (1998) Acronycine-type alkaloids: chemistry and biology. *Alkaloids, Chemical and Biological Perspectives* (ed Pelletier SW) Vol 12. Elsevier, Amsterdam, pp 1–102.

Pilocarpine

Maat L and Beyerman HC (1983) The imidazole alkaloids. *The Alkaloids, Chemistry and Pharmacology* (ed Brossi A) Vol 22. Academic, San Diego, pp 282–331.

Ephedra, Khat

Bernstein PR (1992) Antiasthmatic agents. *Kirk–Othmer Encyclopedia of Chemical Technology*, 4th edn, Vol 2. Wiley, New York, pp 830–854.

Crombie L, Crombie WML and Whiting DA (1990) Alkaloids of khat (*Catha edulis*). *The Alkaloids, Chemistry and Pharmacology* (ed Brossi A) Vol 39. Academic, San Diego, pp 139–164.

Kalix P (1991) The pharmacology of psychoactive alkaloids from *Ephedra* and *Catha*. *J Ethnopharmacol* **32**, 201–208.

Wills S (1993) Drugs and substance misuse: plants. *Pharm J* **251**, 227–229.

Capsaicin

Fusco BM and Giacovazzo M (1997) Peppers and pain: the promise of capsaicin. *Drugs* **53**, 909–914.

Terpenoid and Steroidal Alkaloids

Amiya T and Bando H (1988) *Aconitum* alkaloids. *The Alkaloids, Chemistry and Pharmacology* (ed Brossi A) Vol 34. Academic, San Diego, pp 95–179.

Atta-ur-Rahman and Choudhary MI (1998) Chemistry and biology of steroidal alkaloids. *The Alkaloids, Chemistry and Pharmacology* (ed Cordell GA) Vol 50. Academic, San Diego, pp 61–108.

Atta-ur-Rahman and Choudhary MI (1999) Diterpenoid and steroidal alkaloids. *Nat Prod Rep* **16**, 619–635. Earlier reviews: 1997, **14**, 191–203; 1995, **12**, 361–379.

Cordell GA (1999) The monoterpene alkaloids. *The Alkaloids, Chemistry and Biology* (ed Cordell GA) Vol 52. Academic, San Diego, pp 261–376.

Joshi BS and Pelletier SW (1999) Recent developments in the chemistry of norditerpenoid and diterpenoid alkaloids. *Alkaloids, Chemical and Biological Perspectives* (ed Pelletier SW) Vol 13. Elsevier, Amsterdam, pp 289–370.

Ripperger H (1998) *Solanum* steroid alkaloids: an update. *Alkaloids, Chemical and Biological Perspectives* (ed Pelletier SW) Vol 12. Elsevier, Amsterdam, pp 103–185.

Wang FP and Liang XT (1992) Chemistry of the diterpenoid alkaloids. *The Alkaloids, Chemistry and Pharmacology* (ed Cordell GA) Vol 42. Academic, San Diego, 151–247.

Purine Alkaloids

Atta-ur-Rahman and Choudhary MI (1990) Purine alkaloids. *The Alkaloids, Chemistry and* Pharmacology (ed Brossi A) Vol 38. Academic, San Diego, pp 225–323.

Baumann TW, Schulthess BH and Hänni K (1995) Guarana (*Paullinia cupana*) rewards seed dispersers without intoxicating them by caffeine. *Phytochemistry* **39**, 1063–1070.

Bernstein PR (1992) Antiasthmatic agents. *Kirk–Othmer Encyclopedia of Chemical Technology*, 4th edn, Vol 2. Wiley, New York, pp 830–854.

Houghton P (1995) Herbal products: guarana. *Pharm J* **254**, 435–436.

Wills S (1994) Drugs and substance misuse: caffeine. *Pharm J* **252**, 822–824.

Saxitoxin, Tetrodotoxin

Shimizu Y (1993) Microalgal metabolites. *Chem Rev* **93**, 1685–1698.

Yasumoto T and Murata M (1993) Marine toxins. *Chem Rev* **93**, 1897–1909.

7

PEPTIDES, PROTEINS, AND OTHER AMINO ACID DERIVATIVES

The fundamental structures of peptides and proteins are considered, and their biosynthesis by ribosomal and nonribosomal (multi-enzyme) processes are discussed. Ribosomal peptide biosynthesis leads to peptide hormones, and major groups of these are described, including thyroid, hypothalamic, anterior pituitary, posterior pituitary, and pancreatic hormones, as well as interferons, opioid peptides, and enzymes. Nonribosomal peptide biosynthesis is responsible for the formation of peptide antibiotics, peptide toxins, and modified peptides such as the penicillins, cephalosporins, and other β-lactams. Also covered in this chapter are some other amino acid derivatives, such as cyanogenic glycosides, glucosinolates, and the cysteine sulphoxides characteristic of garlic. Monograph topics giving more detailed information on medicinal agents include thyroxine, calcitonin, thyrotrophin-releasing hormone, luteinizing hormone-releasing hormone, growth hormone-releasing hormone, somatostatin, corticotropin, growth hormone, prolactin, gonadotrophins, oxytocin, vasopressin, insulin, glucagon, interferons, pharmaceutically important enzymes, cycloserine, polymyxins, bacitracins, tyrothricin and gramicidins, capreomycin, vancomycin and teicoplanin, bleomycin, cyclosporins, streptogramins, dactinomycin, death cap, ricin, botulinum toxin, microcystins, snake venoms, penicillins, cephalosporins, cephamycins, carbacephems, clavulanic acid, carbapenems, monobactams, and garlic.

Although the participation of amino acids in the biosynthesis of some shikimate metabolites and particularly in the pathways leading to alkaloids has already been explored in Chapters 4 and 6, amino acids are also the building blocks for other important classes of natural products. The elaboration of shikimate metabolites and alkaloids utilized only a limited range of amino acid precursors. Peptides, proteins, and the other compounds considered in this chapter are synthesized from a very much wider range of amino acids. Peptides and proteins represent another grey area between primary metabolism and secondary metabolism, in that some materials are widely distributed in nature and found, with subtle variations, in many different organisms, whilst others are of very restricted occurrence.

PEPTIDES AND PROTEINS

Peptides and proteins are both polyamides composed of α-amino acids linked through their carboxyl and α-amino functions (Figure 7.1). In biochemistry, the amide linkage is traditionally referred to as a peptide bond. Whether the resultant polymer is classified as a peptide or a protein is not clearly defined; generally a chain length of more than 40 residues confers protein status, whilst the term polypeptide can be used to cover all chain lengths. Although superficially similar, peptides and proteins display a wide variety of biological functions and many have marked physiological properties. For example, they function as structural molecules in tissues, as enzymes, as antibodies, as

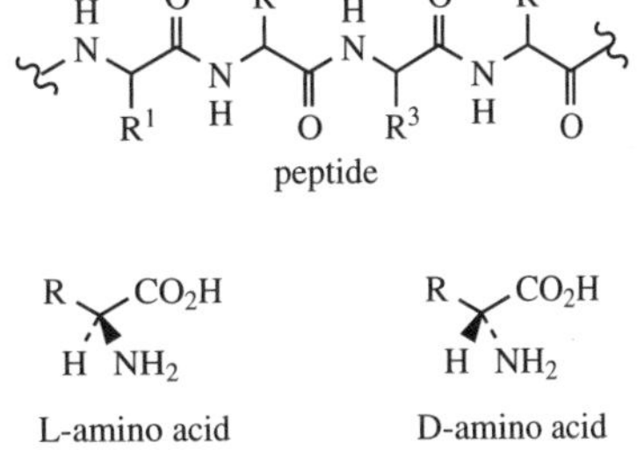

peptide

L-amino acid   D-amino acid

Figure 7.1

neurotransmitters, and acting as hormones can control many physiological processes, ranging from gastric acid secretion and carbohydrate metabolism to growth itself. The toxic components of snake and spider venoms are usually peptide in nature, as are some plant toxins. These different activities arise as a consequence of the sequence of amino acids in the peptide or protein (the primary structure), the three-dimensional structure which the molecule then adopts as a result of this sequence

Table 7.1   Amino acids: structures and standard abbreviations

Amino acids encoded by DNA

Alanine		Ala	A	Leucine		Leu	L
Arginine		Arg	R	Lysine		Lys	K
Asparagine		Asn	N	Methionine		Met	M
Aspartic acid		Asp	D	Phenylalanine		Phe	F
Cysteine		Cys	C	Proline		Pro	P
Glutamic acid		Glu	E	Serine		Ser	S
Glutamine		Gln	Q	Threonine		Thr	T
Glycine		Gly	G	Tryptophan		Trp	W
Histidine		His	H	Tyrosine		Tyr	Y
Isoleucine		Ile	I	Valine		Val	V

Some common amino acids not encoded by DNA

| Pyroglutamic acid (5-oxoproline) | | Glp oxoPro <Glu | Ornithine | | Orn |
| Hydroxyproline | | HPro | Sarcosine (N-methylglycine) | | Sar |

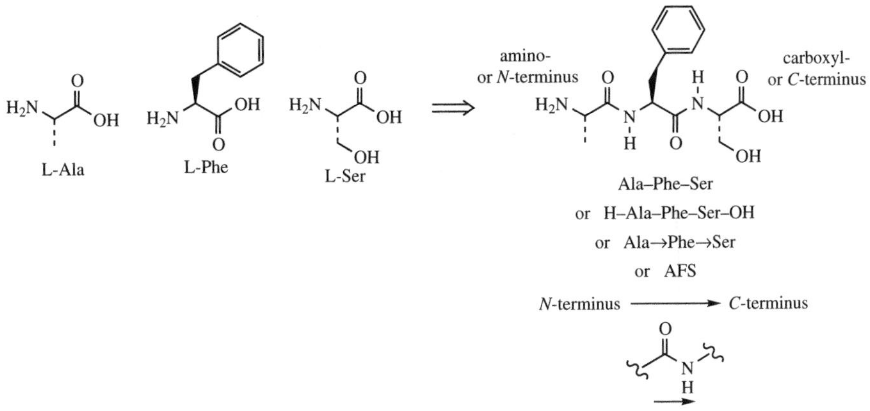

Figure 7.2

(the secondary and tertiary structures), and the specific nature of individual side-chains in the molecule. Many structures have additional modifications to the basic polyamide system shown in Figure 7.1, and these features may also contribute significantly to their biological activity.

The tripeptide formed from L-alanine, L-phenylalanine, and L-serine in Figure 7.2 by two condensation reactions is alanyl–phenylalanyl–serine, commonly represented as Ala–Phe–Ser, using the standard three-letter abbreviations for amino acids as shown in Table 7.1, which gives the structures of the 20 L-amino acids which are encoded by DNA. By convention, the left hand amino acid in this sequence is the one with a free amino group, the *N*-terminus, whilst the right hand amino acid has the free carboxyl, the *C*-terminus. Sometimes, the termini are emphasized by showing H– and –OH (Figure 7.2). In cyclic peptides, this convention can have no significance, so arrows are incorporated into the sequence to indicate peptide bonds in the direction CO→NH. As sequences become longer, one letter abbreviations for amino acids are commonly used instead of the three-letter abbreviations, thus Ala–Phe–Ser becomes AFS. The amino acid components of peptides and proteins predominantly have the L-configuration, but many peptides contain one or more D-amino acids in their structures. Abbreviations thus assume the L-configuration applies, and D-amino acids must be specifically noted, e.g. Ala–D-Phe–Ser. Some amino acids that are not encoded by DNA, but which are frequently encountered in peptides, are shown in Table 7.1, and these also have their

appropriate abbreviations. Modified amino acids may be represented by an appropriate variation of the normal abbreviation, e.g. *N*-methyltyrosine as Tyr(Me). A frequently encountered modification is the conversion of the *C*-terminal carboxyl into an amide, and this is represented as Phe–NH₂ for example, which must not be interpreted as an indication of the *N*-terminus.

RIBOSOMAL PEPTIDE BIOSYNTHESIS

Protein biosynthesis takes place on the ribosomes, and a simplified representation of the process as characterized in *Escherichia coli* is shown in Figure 7.3. The messenger RNA (mRNA) contains a transcription of one of the genes of DNA, and carries the information necessary to direct the biosynthesis of a specific protein. The message is stored as a series of three-base sequences (codons) in its nucleotides, and is read (translated) in the 5′ to 3′ direction along the mRNA molecule. The mRNA is bound to the smaller 30S subunit of the bacterial ribosome. Initially, the amino acid is activated by an ATP-dependent process and it then binds via an ester linkage to an amino acid-specific transfer RNA (tRNA) molecule through a terminal adenosine group, giving an aminoacyl-tRNA (Figure 7.4). The aminoacyl-tRNA contains in its nucleotide sequence a combination of three bases (the anticodon) which allows binding via hydrogen bonding to the appropriate codon on mRNA. In prokaryotes, the first amino acid encoded in the sequence is *N*-formylmethionine, and the corresponding aminoacyl-tRNA is thus bound and

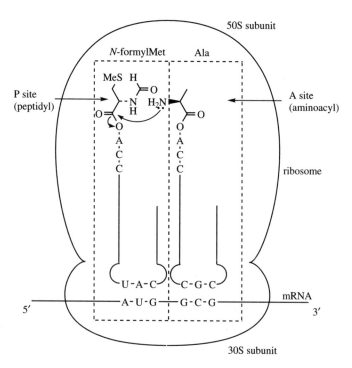

Figure 7.3

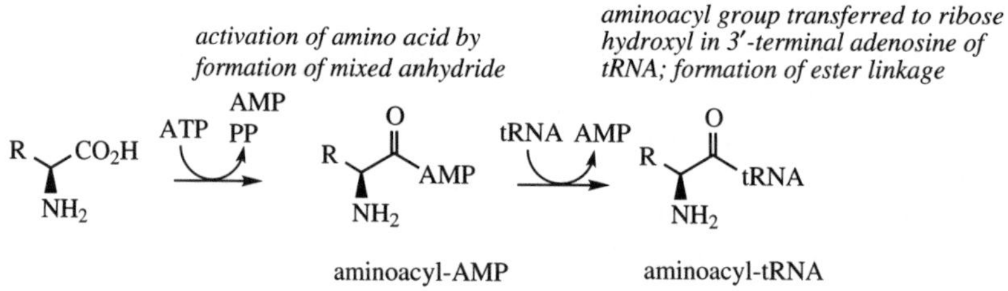

*activation of amino acid by
formation of mixed anhydride*

*aminoacyl group transferred to ribose
hydroxyl in 3′-terminal adenosine of
tRNA; formation of ester linkage*

Figure 7.4

positioned at the P (for peptidyl) site on the ribosome (Figure 7.3). The next aminoacyl-tRNA (Figure 7.3 shows a tRNA specific for alanine) is also bound via a codon–anticodon interaction and is positioned at an adjacent A (for aminoacyl) site on the ribosome. This allows peptide bond formation to occur, the amino group of the amino acid in the A site attacking the activated ester in the P site. The peptide chain is thus initiated and has become attached to the tRNA located in the A site. The tRNA at the P site is no longer required and is released from the ribosome. Then the peptidyl-tRNA at the A site is translocated to the P site by the ribosome moving along the mRNA a codon at a time, exposing the A site for

a new aminoacyl-tRNA appropriate for the particular codon, and a repeat of the elongation process occurs. The cycles of elongation and translocation continue until a termination codon is reached, and the peptide or protein is then hydrolysed and released from the ribosome. The individual steps of protein biosynthesis all seem susceptible to disruption by specific agents. Many of the antibiotics used clinically are active by their ability to inhibit protein biosynthesis in bacteria. They may interfere with the binding of the aminoacyl-tRNA to the A site (e.g. tetracyclines, see page 91), the formation of the peptide bond (e.g. chloramphenicol, see page 130), or the translocation step (e.g. erythromycin, see page 99).

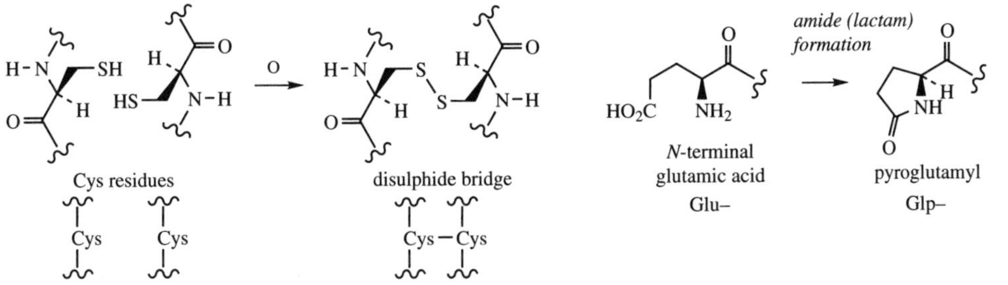

Figure 7.5

Glp–Gly–Pro–Trp–Leu–Glu–Glu–Glu–Glu–Glu–Ala–Tyr–Gly–Trp–Gly–Trp–Met–Asp–**Phe–NH₂**

human gastrin

Figure 7.6

Many of the peptides and proteins synthesized on the ribosome then undergo enzymic post-translational modifications. In prokaryotes, the *N*-formyl group on the *N*-terminal methionine is removed, and in both prokaryotes and eukaryotes a number of amino acids from the *N*-terminus may be hydrolysed off, thus shortening the original chain. This type of chain shortening is a means by which a protein or peptide can be stored in an inactive form, and then transformed into an active form when required, e.g. the production of insulin from proinsulin (see page 416). **Glycoproteins** are produced by adding sugar residues via *O*-glycoside linkages to the hydroxyls of serine and threonine residues or via *N*-glycoside linkages to the amino of asparagine. **Phosphoproteins** have the hydroxyl groups of serine or threonine phosphorylated. Importantly, post-translational modifications allow those amino acids that are not encoded by DNA, yet are found in peptides and proteins, to be formed by the transformation of encoded ones. These include the hydroxylation of proline to hydroxyproline and of lysine to hydroxylysine, *N*-methylation of histidine, and the oxidation of the thiol groups of two cysteine residues to form a **disulphide bridge**, allowing cross-linking of polypeptide chains. This latter process (Figure 7.5) will form loops in a single polypeptide chain, or may join separate chains together as in insulin (see page 416). Many peptides contain a **pyroglutamic acid** residue (Glp) at the *N*-terminus, a consequence of intramolecular cyclization between

the γ-carboxylic acid and the α-amino of an *N*-terminal glutamic acid (Figure 7.5). The *C*-terminal carboxylic acid may also frequently be converted into an amide. Both modifications are exemplified in the structure of **gastrin** (Figure 7.6), a peptide hormone that stimulates secretion of HCl in the stomach. Such terminal modifications comprise a means of protecting a peptide from degradation by exopeptidases, which remove amino acids from the ends of peptides, thus increasing its period of action.

With the rapid advances made in genetic engineering, it is now feasible to produce relatively large amounts of ribosome-constructed polypeptides by isolating or constructing DNA sequences encoding the particular product, and inserting these into suitable organisms, commonly *Escherichia coli*. Of course, such procedures will not duplicate any post-translational modifications, and these will have to be carried out on the initial polypeptide by chemical means, or by suitable enzymes if these are available. Not all of the transformations can be carried out both efficiently and selectively, thus restricting access to important polypeptides by this means. Small peptides for drug use are generally synthesized chemically, though larger peptides and proteins may be extracted from human and animal tissues or bacterial cultures. In the design of semi-synthetic or synthetic analogues for potential drug use, enzymic degradation can be reduced by the use of *N*-terminal pyroglutamic acid and *C*-terminal amide residues, the inclusion of D-amino acids, and the removal of specific residues. These ploys to change recognition by specific degradative

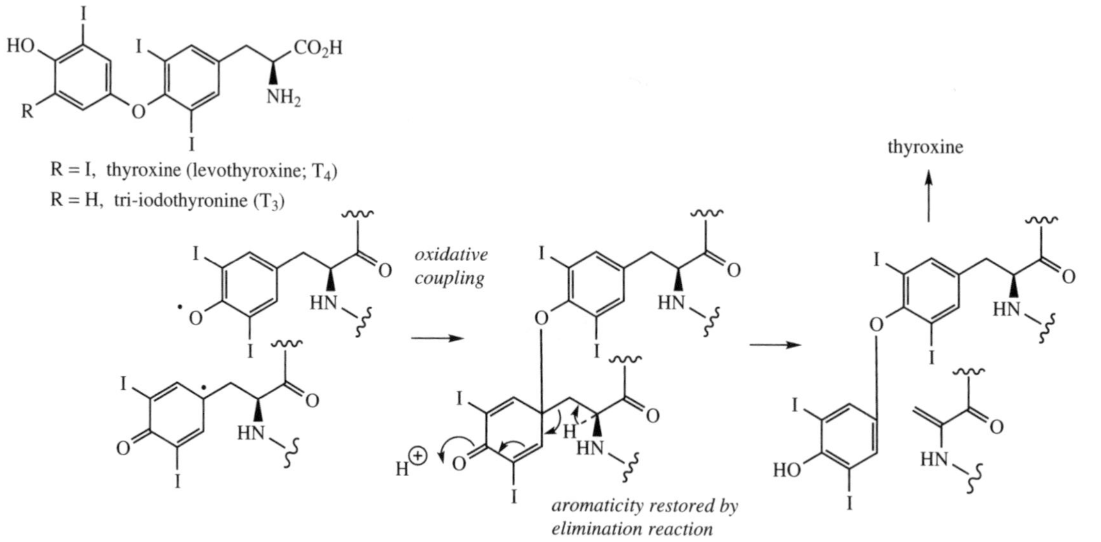

R = I, thyroxine (levothyroxine; T$_4$)
R = H, tri-iodothyronine (T$_3$)

oxidative coupling

aromaticity restored by elimination reaction

thyroxine

Figure 7.7

peptidases can increase the activity and lifetime of the peptide. Very few peptide drugs may be administered orally since they are rapidly inactivated or degraded by gastrointestinal enzymes, and they must therefore be given by injection.

PEPTIDE HORMONES

Hormones are mammalian metabolites released into the blood stream to elicit specific responses on a target tissue or organ. They are chemical messengers, and may be simple amino acid derivatives, e.g. adrenaline (see page 317), or polypeptides, e.g. insulin (see page 417), or they may be steroidal in nature, e.g. progesterone (see page 273). They may exert their effects in a variety of ways, e.g. by influencing the rate of synthesis of enzymes or proteins, by affecting the catalytic activity of an enzyme, or by altering the permeability of cell membranes. Hormones are not enzymes; they act by regulating existing processes. Frequently, this action depends on the involvement of a second messenger, such as cyclic AMP (cAMP).

Thyroid Hormones

Thyroid hormones are necessary for the development and function of cells throughout the body. The thyroid hormones **thyroxine*** and **tri-iodothyronine*** (Figure 7.7) are not peptides, but are actually simple derivatives of tyrosine. However, they are believed to be derived by degradation of a larger protein molecule. One

---

Thyroxine

The thyroid hormones thyroxine (T$_4$) and tri-iodothyronine (T$_3$) (Figure 7.7) are derivatives of tyrosine and are necessary for development and function of cells throughout the body. They increase protein synthesis in almost all types of body tissue and increase oxygen consumption dependent upon Na$^+$/K$^+$ ATPase (the Na pump). Excess thyroxine causes hyperthyroidism, with increased heart rate, blood pressure, overactivity, muscular weakness, and loss of weight. Too little thyroxine may lead to cretinism in children, with poor growth and mental deficiency, or myxoedema in adults, resulting in a slowing down of all body processes. Tri-iodothyronine is also

(Continues)

(Continued)

produced from thyroxine by mono-deiodination in tissues outside the thyroid, and is actually three to five times more active than thyroxine. **Levothyroxine (thyroxine)** and **liothyronine (tri-iodothyronine)** are both used to supplement thyroid hormone levels, and may be administered orally. These materials are readily produced by chemical synthesis.

Cys–**Gly**–Asn–Leu–Ser–Thr–Cys–**Met**–Leu–Gly–
Thr–**Tyr**–**Thr**–Gln–**Asp**–**Phe**–**Asn**–**Lys**–**Phe**–**His**–
Thr–**Phe**–Pro–**Gln**–Thr–**Ala**–**Ile**–Gly–**Tyr**–Gly–
Ala–Pro–NH$_2$

human calcitonin

Cys–**Ser**–Asn–Leu–Ser–Thr–Cys–**Val**–Leu–Gly–
Lys–**Leu**–**Ser**–Gln–**Glu**–**Leu**–**His**–Lys–**Leu**–**Gln**–
Thr–**Tyr**–Pro–**Arg**–Thr–**Asn**–**Thr**–Gly–Ser–Gly–
Thr–Pro–NH$_2$

calcitonin (salmon), salcatonin

Figure 7.8

Calcitonin

Also produced in the thyroid gland, calcitonin is involved along with parathyroid hormone and 1α,25-dihydroxyvitamin D$_3$ (see page 259) in the regulation of bone turnover and maintenance of calcium balance. Under the influence of high blood calcium levels, calcitonin is released to lower these levels by inhibiting uptake of calcium from the gastrointestinal tract, and by promoting its storage in bone. It also suppresses loss of calcium from bone when levels are low. **Calcitonin** is used to treat weakening of bone tissue and hypercalcaemia. Human calcitonin (Figure 7.8) is a 32-residue peptide with a disulphide bridge, and synthetic material may be used. However, synthetic salmon calcitonin (**calcitonin (salmon)**; **salcatonin**) (Figure 7.8) is found to have greater potency and longer duration than human calcitonin. Calcitonins from different sources have quite significant differences in their amino acid sequences; the salmon peptide shows 16 changes from the human peptide. Porcine calcitonin is also available.

---

hypothesis for their formation invokes tyrosine residues in the protein thyroglobulin, which are iodinated to mono- and di-iodotyrosine, with suitably placed residues reacting together by a phenolic oxidative coupling process (Figure 7.7). Re-aromatization results in cleavage of the side-chain of one residue, and thyroxine (or triiodothyronine) is released from the protein by proteolytic cleavage. Alternative mechanisms have been proposed, however. **Calcitonin*** (Figure 7.8) is also produced in the thyroid gland, and is involved in the regulation of bone turnover and maintenance of calcium balance. This is a relatively simple peptide structure containing a disulphide bridge.

Hypothalamic Hormones

Hypothalamic hormones (Figure 7.9) can modulate a wide variety of actions throughout the body, via the regulation of anterior pituitary hormone secretion. **Thyrotrophin-releasing hormone (TRH)*** is a tripeptide with an *N*-terminal pyroglutamyl residue, and a *C*-terminal pro-lineamide, whilst **luteinizing hormone-releasing hormone (LH-RH)*** is a straight-chain decapeptide that also has both *N*- and *C*-termini blocked, through pyroglutamyl and glycineamide respectively. **Somatostatin*** (**growth hormone-release inhibiting factor; GHRIH**) merely displays a disulphide bridge in its 14-amino acid chain. One of the largest of the hypothalamic hormones is **corticotropin-releasing hormone (CRH)** which contains 41 amino acids, with only the *C*-terminal blocked as an amide. This hormone controls release from the anterior pituitary of **corticotropin (ACTH)** (see page 414), which in turn is responsible for production of corticosteroids.

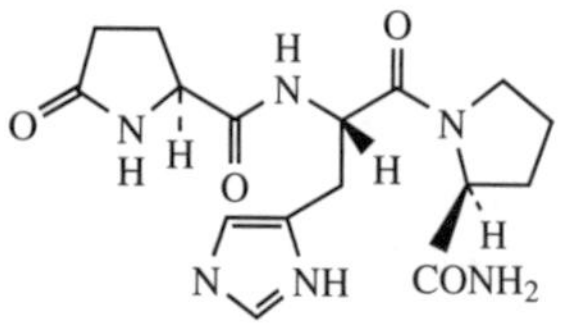

Glp–His–Pro–NH$_2$

thyrotrophin-releasing hormone (TRH)
(protirelin)

Tyr–Ala–Asp–Ala–Ile–Phe–Thr–Asn–Ser–Tyr–
Arg–Lys–Val–Leu–Gly–Gln–Leu–Ser–Ala–Arg–
Lys–Leu–Leu–Gln–Asp–Ile–Met–Ser–Arg–NH$_2$

sermorelin

Ala–Gly–Cys–Lys–Asn–Phe–**Phe**–**Trp**
         | ↓
Cys–Ser–Thr–Phe–**Thr**–**Lys**

growth hormone-release inhibiting factor (GHRIH)
(somatostatin)

Ser–Gln–Glu–Pro–Pro–Ile–Ser–Leu–Asp–Leu–
Thr–Phe–His–Leu–Leu–Arg–Glu–Val–Leu–Glu–
Met–Thr–Lys–Ala–Asp–Gln–Leu–Ala–Gln–Gln–
Ala–His–Ser–Asn–Arg–Lys–Leu–Leu–Asp–Ile–
Ala–NH$_2$

corticotropin-releasing hormone (CRH)

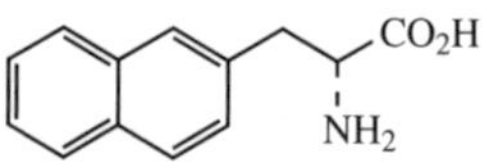

Glp–His–Trp–Ser–Tyr–**Gly**–Leu–Arg–Pro–**Gly**–NH$_2$

luteinizing hormone-releasing factor (LH-RH)
(gonadorelin)

Glp–His–Trp–Ser–Tyr–**D-Trp**–Leu–Arg–Pro–Gly–NH$_2$

triptorelin

Glp–His–Trp–Ser–Tyr–**D-Ser(^{t}Bu)**–Leu–Arg–Pro–NHEt

buserelin

Glp–His–Trp–Ser–Tyr–**D-Ser(^{t}Bu)**–Leu–Arg–Pro–NH–NH–CONH$_2$

goserelin

Glp–His–Trp–Ser–Tyr–**D-Leu**–Leu–Arg–Pro–NHEt

leuprorelin

Glp–His–Trp–Ser–Tyr–**3-(2-naphthyl)-D-Ala**–Leu–Arg–Pro–Gly–NH$_2$

nafarelin

D-Phe–Cys–**Phe**–**D-Trp**
      | ↓
Thr(ol)–Cys–**Thr**–**Lys**

octreotide

3-(2-naphthyl)-D-Ala–Cys–**Tyr**–**D-Trp**
              | ↓
H$_2$N–Thr–Cys–**Val**–**Lys**

lanreotide

3-(2-naphthyl)-D-Ala

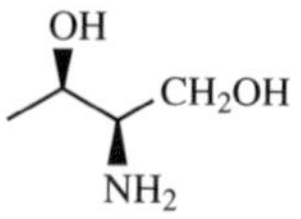

Thr(ol)

Figure 7.9

Thyrotrophin-Releasing Hormone

Thyrotrophin-releasing hormone (TRH) (Figure 7.9) acts directly on the anterior pituitary to stimulate release of thyroid-stimulating hormone (TSH), prolactin, and growth hormone. Many other hormones may be released by direct or indirect effects. Synthetic material (known as **protirelin**) is used to assess thyroid function and TSH reserves.

Luteinizing Hormone-Releasing Hormone

Luteinizing hormone-releasing hormone (LH-RH) (also gonadotrophin-releasing hormone, GnRH) (Figure 7.9) is the mediator of gonadotrophin secretion from the anterior pituitary, stimulating both luteinizing hormone (LH) and follicle-stimulating hormone (FSH) release (see page 415). These are both involved in controlling male and female reproduction, inducing the production of oestrogens and progestogens in the female, and of androgens in the male. LH is essential for causing ovulation, and for the development and maintenance of the corpus luteum in the ovary, whilst FSH is required for maturation of both ovarian follicles in women, and of the testes in men.

(Continues)

(*Continued*)

Gonadorelin is synthetic LH-RH and is used for the assessment of pituitary function, and also for the treatment of infertility, particularly in women. Analogues of LH-RH, e.g. **triptorelin**, **buserelin**, **goserelin**, **leuprorelin**, and **nafarelin** (Figure 7.9), have been developed and find use to inhibit ovarian steroid secretion, and to deprive cancers such as prostate and breast cancers of essential steroid hormones. All of these analogues include a D-amino acid residue at position 6, and modifications to the terminal residue 10, including omission of this residue in some cases. Such changes increase activity and half-life compared with gonadorelin, e.g. leuprorelin is 50 times more potent, and half-life is increased from 4 minutes to 3–4 hours.

Growth Hormone-Releasing Hormone/Factor

Growth hormone-releasing hormone/factor (**GHRH/GHRF**) contains 40–44 amino acid residues and stimulates secretion of growth hormone (see page 414) from the anterior pituitary. Synthetic material containing the first 29-amino acid sequence has the full activity and potency of the natural material. This peptide (**sermorelin**) (Figure 7.9) is used as a diagnostic aid to test the secretion of growth hormone.

Somatostatin

Somatostatin (**growth hormone-release inhibiting factor**; **GHRIH**) (Figure 7.9) is a 14-amino acid peptide containing a disulphide bridge, but is derived from a larger precursor protein. It is found in the pancreas and gastrointestinal tract as well as in the hypothalamus. It inhibits the release of growth hormone (see page 414) and thyrotrophin from the anterior pituitary, and also secretions of hormones from other endocrine glands, e.g. insulin and glucagon. Receptors for somatostatin are also found in most carcinoid tumours. Somatostatin has a relatively short duration of action (half-life 2–3 minutes), and the synthetic analogue **octreotide** (Figure 7.9), which is much longer acting (half-life 60–90 minutes), is currently in drug use for the treatment of various endocrine and malignant disorders, especially treatment of neuroendocrine tumours and the pituitary-related growth condition acromegaly.

Octreotide contains only eight residues, but retains a crucial Phe–Trp–Lys–Thr sequence, even though the Trp now has the D-configuration. This prevents proteolysis between Trp and Lys, a major mechanism in somatostatin degradation. The *C*-terminus is no longer an amino acid, but an amino alcohol related to threonine. Radiolabelled octreotide has considerable potential for the visualization of neuroendocrine tumours. **Lanreotide** (Figure 7.9) is a recently introduced somatostatin analogue used in a similar way to octreotide. Lanreotide retains only a few of the molecular characteristics of octreotide, with the Phe–D-Trp–Lys–Thr sequence now modified to Tyr–D-Trp–Lys–Val and the *N*-terminal amino acid is the synthetic analogue 3-(2-napthyl)-D-alanine, also seen in the LH-RH analogue nafarelin.

Anterior Pituitary Hormones

The main hormones of the anterior pituitary are **corticotropin*** and **growth hormone***, which each consist of a single long polypeptide chain, and the **gonadotrophins***, which are glycoproteins containing two polypeptide chains. These hormones regulate release of glucocorticoids, human growth, and sexual development respectively. A further hormone, **prolactin***, which controls milk production in females, is structurally related to growth hormone.

Posterior Pituitary Hormones

The two main hormones of the posterior pituitary are **oxytocin***, which contracts the smooth muscle of the uterus, and **vasopressin***,

Corticotropin

Corticotropin (corticotrophin; adrenocorticotrophin; ACTH) is a straight-chain polypeptide with 39 amino acid residues, and its function is to control the activity of the adrenal cortex, particularly the production of corticosteroids. Secretion of the hormone is controlled by corticotropin-releasing hormone (CRH) from the hypothalamus. ACTH was formerly used as an alternative to corticosteroid therapy in rheumatoid arthritis, but its value was limited by variable therapeutic response. ACTH may be used to test adrenocortical function. It has mainly been replaced for this purpose by the synthetic analogue **tetracosactide (tetracosactrin)** (Figure 7.10), which contains the first 24 amino acid residues of ACTH, and is preferred because of its shorter duration of action and lower allergenicity.

Ser–Tyr–Ser–Met–Glu–His–Phe–Arg–Trp–Gly–Lys–Pro–Val–Gly–Lys–Lys–Arg–Arg–Pro–Val–Lys–Val–Tyr–Pro

tetracosactide (tetracosactrin)

Figure 7.10

Growth Hormone

Growth Hormone (GH) (human growth hormone (HGH) or **somatotrophin)** is necessary for normal growth characteristics, especially the lengthening of bones during development. A lack of HGH in children results in dwarfism, whilst continued release can lead to gigantism, or acromegaly, in which only the bones of the hands, feet, and face continue to grow. Growth hormones from animal sources are very species specific, so it has not been possible to use animal hormones for drug use. HGH contains 191 amino acids with two disulphide bridges, one of which creates a large loop (bridging residues 53 and 165) and the other a very small loop (bridging residues 182 and 189) near the C-terminus. It is synthesized via a prohormone containing 26 extra amino acids. Production of material with the natural amino acid sequence has become possible as a result of recombinant DNA technology; this is termed **somatropin**. This drug is used to improve linear growth in patients whose short stature is known to be caused by a lack of pituitary growth hormone. There is also some abuse of the drug by athletes wishing to enhance performance. Although HGH increases skeletal mass and strength, its use can result in some abnormal bone growth patterns.

Prolactin

Prolactin has structural similarities to growth hormone, in that it is a single-chain polypeptide (198 amino acid residues) with three loops created by disulphide bonds. It is synthesized via a prohormone containing 29 extra amino acids. Growth hormone itself can bind to the prolactin receptor, but this is not significant under normal physiological conditions. Prolactin release is controlled by dopamine produced by the hypothalamus, and its main function is to control milk production. Prolactin has a synergistic action with oestrogen to promote mammary tissue proliferation during pregnancy, then at parturition, when oestrogen levels fall, prolactin levels rise and lactation is initiated. New nursing mothers have high levels of prolactin and this also inhibits gonadotrophin release and/or the response of the ovaries to these hormones. As a

(Continues)

(Continued)

result, ovulation does not usually occur during breast feeding, preventing further conception. No prolactin derivatives are currently used in medicine, but the ergot alkaloid derivatives bromocriptine and cabergoline (see page 375) are dopamine agonists employed to inhibit prolactin release by pituitary tumours. They are not now recommended for routine suppression of lactation.

Gonadotrophins

Follicle-stimulating hormone (FSH) and **luteinizing hormone (LH)** are involved in controlling both male and female reproduction (see LH-RH, page 412). These are glycoproteins both composed of two polypeptide chains of 89 and 115 amino acid residues. The shorter chains are essentially identical, and differences in activity are caused by differences in the longer chain. Each chain has asparagine-linked oligosaccharide residues: the short chains each have one, the long chains two in the case of FSH or one for LH. FSH and LH together (**human menopausal gonadotrophins; menotrophin**) are purified from the urine of post-menopausal women and used in the treatment of female infertility. FSH alone is also available for this purpose, either natural material from urine termed **urofollitropin (urofollitrophin)**, or the recombinant proteins **follitropin alfa** and **follitropin beta. Chorionic gonadotrophin (human chorionic gonadotrophin; HCG)** is a gonad-stimulating glycoprotein hormone which is obtained for drug use from the urine of pregnant women. It may be used to stimulate testosterone production in males with delayed puberty.

the antidiuretic hormone. These nonapeptides containing a disulphide bridge are structurally very similar, differing in only two amino acid residues (Figure 7.11). Structurally related peptides are classified as belonging to the vasopressin family when the amino acid residue at position 8 is basic, e.g. Arg or Lys, or to the oxytocin family when this amino acid is neutral.

Cys–Tyr–**Ile**–Gln–Asn–Cys–Pro–**Leu**–Gly–NH$_2$

oxytocin

(desamino)Cys–Tyr–**Phe**–Gln–Asn–Cys–Pro–**D-Arg**–Gly–NH$_2$

desmopressin (1-desamino-8-D-Arg-VP)

Cys–Tyr–**Phe**–Gln–Asn–Cys–Pro–**Arg**–Gly–NH$_2$

argipressin (arginine vasopressin) (human/bovine)

Gly–Gly–Gly–Cys–Tyr–**Phe**–Gln–Asn–Cys–Pro–**Lys**–Gly–NH$_2$

terlipressin (Gly-Gly-Gly-8-L-Lys-VP)

Cys–Tyr–**Phe**–Gln–Asn–Cys–Pro–**Lys**–Gly–NH$_2$

lypressin (lysine vasopressin) (porcine)

Figure 7.11

Oxytocin

Oxytocin (Figure 7.11) stimulates the pregnant uterus, causing contractions, and also brings about ejection of milk from the breasts. It thus plays a major role in the normal onset of labour at the end of pregnancy. Oxytocin for drug use is produced by synthesis, and is employed to induce or augment labour, as well as to minimize subsequent blood loss.

(Continues)

(Continued)

Vasopressin

Vasopressin (antidiuretic hormone, ADH) is a hormone that has an antidiuretic action on the kidney, regulating the reabsorption of water. A deficiency in this hormone leads to diabetes insipidus, where the patient suffers increased urine output and intense thirst, typically consuming enormous quantities of fluid. Vasopressin is used to treat this condition. At high dosage, vasopressin promotes contraction of arterioles and capillaries, and brings about an increase in blood pressure. The structure of human and bovine vasopressin (arginine vasopressin; **argipressin**) (Figure 7.11) differs from that of oxytocin only in two amino acid residues. Lysine vasopressin (**lypressin**) (Figure 7.11) from pigs differs from arginine vasopressin in the second amino acid from the *C*-terminus. Both bovine and porcine peptides have been used medicinally, but these have been replaced by synthetic materials. The 1-desamino-8-D-Arg-vasopressin analogue **desmopressin** (Figure 7.11) has a longer duration of action than vasopressin and may also be administered orally. In contrast to the natural hormone, desmopressin has no vasoconstrictor effect. **Terlipressin** (Figure 7.11) is a lypressin pro-drug in which the polypeptide chain has been extended by three glycine residues. Enzymic hydrolysis liberates lypressin. It is mainly used for control of oesophageal bleeding.

Pancreatic Hormones

The hormone **insulin*** (Figure 7.12) plays a key role in the regulation of carbohydrate, fat, and protein metabolism. In particular, it has a hypoglycaemic effect, lowering the levels of glucose in the blood. A deficiency in insulin synthesis leads to the condition diabetes, treatment of which requires daily injections of insulin. Insulin is composed of two straight chain polypeptides joined by disulphide bridges. This structure is known to arise from a straight chain polypeptide preproinsulin containing 100 amino acid residues. This loses a 16-residue portion of its chain, and forms proinsulin with disulphide bridges connecting the terminal portions of the chain in a loop (Figure 7.13). A central portion of the loop (the C chain) is then cleaved out, leaving the A chain (21 residues) bonded to the B chain (30 residues) by two disulphide bridges. This is the resultant insulin. Mammalian insulins (Figure 7.12) from different sources are very similar, showing variations in the sequence

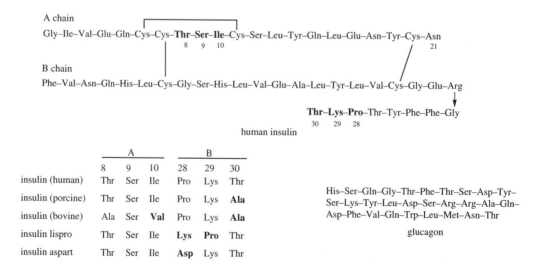

human insulin

	A			B		
	8	9	10	28	29	30
insulin (human)	Thr	Ser	Ile	Pro	Lys	Thr
insulin (porcine)	Thr	Ser	Ile	Pro	Lys	**Ala**
insulin (bovine)	Ala	Ser	**Val**	Pro	Lys	**Ala**
insulin lispro	Thr	Ser	Ile	**Lys**	**Pro**	Thr
insulin aspart	Thr	Ser	Ile	**Asp**	Lys	Thr

His–Ser–Gln–Gly–Thr–Phe–Thr–Ser–Asp–Tyr–
Ser–Lys–Tyr–Leu–Asp–Ser–Arg–Arg–Ala–Gln–
Asp–Phe–Val–Gln–Trp–Leu–Met–Asn–Thr

glucagon

Figure 7.12

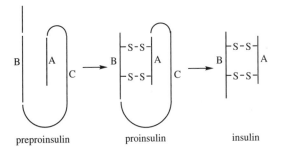

preproinsulin   proinsulin   insulin

Figure 7.13

Interferons

The **interferons*** are a family of proteins secreted by animal cells in response to viral and parasitic infections, and are part of the host's defence mechanism. They display multiple activities, affecting the functioning of the immune system, cell proliferation, and cell differentiation, primarily by inducing the synthesis of other proteins. Accordingly, they have potential as antiviral, antiprotozoal, immunomodulatory, and cell growth regulatory agents.

of amino acid residues 8–10 in chain A, and at amino acid 30 in chain B. **Glucagon*** (Figure 7.12) is a straight-chain polypeptide hormone containing 29 amino acids that is secreted by the pancreas when blood sugar levels are low, thus stimulating breakdown of glycogen in the liver. Unlike insulin, its structure is identical in all animals.

Opioid Peptides

Although the pain-killing properties of morphine and related compounds have been known for a considerable time (see page 329), the existence of endogenous peptide ligands for the receptors to which these compounds bind is a more recent discovery. It is now appreciated that the body

Insulin

Insulin is a hormone produced by the pancreas that plays a key role in the regulation of carbohydrate, fat, and protein metabolism. In particular, it has a hypoglycaemic effect, lowering the levels of glucose in the blood. If a malfunctioning pancreas results in a deficiency in insulin synthesis or secretion, the condition known as diabetes mellitus ensues. This results in increased amounts of glucose in the blood and urine, diuresis, depletion of carbohydrate stores, and subsequent breakdown of fat and protein. Incomplete breakdown of fat leads to the accumulation of ketones in the blood, severe acidosis, coma, and death. Where the pancreas is still functioning, albeit less efficiently, the condition is known as type 2 diabetes (non-insulin-dependent diabetes, NIDDM) and can be controlled by a controlled diet or oral antidiabetic drugs. In type 1 diabetes (insulin-dependent diabetes, IDDM) pancreatic cells no longer function, and injections of insulin are necessary, one to four times daily, depending on the severity of the condition. These need to be combined with a controlled diet and regular monitoring of glucose levels, but do not cure the disease, so treatment is lifelong. Mammalian insulins from different sources are very similar and may be used to treat diabetes. **Porcine insulin** and **bovine insulin** (Figure 7.12) are extracted from the pancreas of pigs and cattle respectively. **Human insulin** (Figure 7.12) is produced by the use of recombinant DNA technology in *Escherichia coli* to obtain the two polypeptide chains, and linking these chemically to form the disulphide bridges (such material is coded 'crb'), or by modification of proinsulin produced in genetically modified *E. coli* (coded 'prb'). Human insulin may also be obtained from porcine insulin by semi-synthesis, replacing the terminal alanine in chain B with threonine by enzymic methods (coded 'emp'). Human insulin does not appear to be less immunogenic than animal insulin, but genetic engineering offers significant advantages over animal sources for obtaining highly purified material. Insulin may be provided in a rapid-acting soluble form, as suspensions of the zinc complex which have longer duration, or as suspensions with protamine

(Continues)

(Continued)

(as **isophane insulin**). **Protamine** is a basic protein from the testes of fish of the salmon family, e.g. *Salmo* and *Onchorhynchus* species, which complexes with insulin, thereby reducing absorption and providing a longer period of action. Recently introduced recombinant human insulin analogues **insulin lispro** and **insulin aspart** have a faster onset and a shorter duration of action than soluble insulin. Insulin lispro has the reverse sequence for the 28 and 29 amino acids in the B chain, i.e. B28-Lys–B29-Pro, whilst insulin aspart has a single substitution of aspartic acid for proline at position 28 in the B chain (Figure 7.12). These changes in the primary structure affect the tendency of the molecule to associate into dimers and larger oligomers, thus increasing the availability of absorbable monomers. Insulin aspart is produced by expression in *Saccharomyces cerevisiae* (baker's yeast). **Insulin glargine** is a new ultra-long-acting analogue that differs from human insulin by replacing the terminal asparagine at position 21 in chain A with glycine, and also adding two arginines to the end of the B chain, i.e. positions B31 and B32. These changes result in enhanced basicity, causing precipitation at neutral pH post-injection, and consequently a delayed, very gradual and prolonged activity profile (up to 24–48 hour duration of action) and allowing once-daily dosing.

Glucagon

Glucagon (Figure 7.12) is a straight-chain polypeptide hormone containing 29 amino acids that is secreted by the pancreas when blood sugar levels are low, thus stimulating breakdown of glycogen in the liver. It may be isolated from animal pancreas, or be produced by recombinant DNA processes using *Saccharomyces cerevisiae*. It may be administered for the emergency treatment of diabetes patients suffering from hypoglycaemia as a result of building up a dangerously high insulin level. Normally, a patient would counter this by eating some glucose or sucrose, but hypoglycaemia can rapidly cause unconsciousness, requiring very prompt action.

produces a family of endogenous opioid peptides, which bind to a series of receptors in different locations. These peptides include **enkephalins**, **endorphins**, and **dynorphins**, and are produced primarily, but not exclusively, in the pituitary gland. The pentapeptides **Met-enkephalin** and **Leu-enkephalin** (Figure 7.14) were the first to be characterized. The largest peptide is **β-endorphin** (*'end*ogenous *m*orphine') (Figure 7.14), which is several times more potent than morphine in relieving pain. Although β-endorphin contains the sequence for Met-enkephalin, the latter peptide and Leu-enkephalin are derived from a larger peptide proenkephalin A, whilst β-endorphin itself is formed by cleavage of pro-opiomelanocortin. The proenkephalin A structure contains four

Interferons

Interferons were originally discovered as proteins that interfered with virus replication. When mice were injected with antibodies to interferons, they became markedly susceptible to virus-mediated disease, including virus-related tumour induction. Interferons can be detected at low levels in most human tissues, but amounts increase upon infection with viruses, bacteria, protozoa, and exposure to certain growth factors. Interferons were initially classified according to the cellular source, but recent nomenclature is based primarily on sequencing data. Thus leukocyte interferon (a mixture of proteins) is now known as interferon alfa, fibroblast interferon as interferon beta, and immune interferon as interferon gamma.

(Continues)

(Continued)

These typically range in size from 165 to 172 amino acid residues. The interferon system can often impair several steps in viral replication, and can modulate the immune system, and affect cell growth and differentiation. They have potential in treating many human diseases, including leukaemias and solid tumours, and viral conditions such as chronic infection with hepatitis B and C. Much of their current drug use is still research based. **Interferon alfa** is a protein containing 166 amino acid residues with two disulphide bridges and is produced by recombinant DNA techniques using *Escherichia coli*, or by stimulation of specific human cell lines. Variants with minor differences in sequence may be obtained according to the gene used, and are designated as alfa-2a, alfa-2b, etc. Interferon alfa is employed as an antitumour drug against certain lymphomas and solid tumours, and in the management of chronic hepatitis. **Interferon beta** is of value in the treatment of multiple sclerosis patients, though not all patients respond. Variants are designated beta-1a (a glycoprotein with 165 amino acid residues and one disulphide bridge) or beta-1b (a non-glycosylated protein with 164 amino acid residues and one disulphide bridge). **Interferon gamma** produced by recombinant DNA methods contains an unbridged polypeptide chain, and is designated gamma-1a (146 residues) or gamma-1b (140 residues) according to sequence. The immune interferon interferon gamma-1b is used to reduce the incidence of serious infection in patients with chronic granulomatous disease.

Met-enkephalin sequences and one of Leu-enkephalin. The dynorphins, e.g. **dynorphin A** (Figure 7.14) are also produced by cleavage of a larger precursor, namely proenkephalin B (prodynorphin), and all contain the Leu-enkephalin sequence. Some 20 opioid ligands have now been characterized. When released, these endogenous opioids act upon specific receptors, inducing analgesia, and depressing respiratory function and several other processes. The individual peptides have relatively high specificity towards different receptors. It is known that morphine, β-endorphin, and Met-enkephalin are agonists for the same site. The opioid peptides are implicated in analgesia brought about by acupuncture, since opiate antagonists can reverse the effects. The hope of exploiting similar peptides as ideal, non-addictive analgesics has yet to be attained; repeated doses of endorphin or enkephalin produce addiction and withdrawal symptoms.

Enzymes

Enzymes are proteins that act as biological catalysts. They facilitate chemical modification of substrate molecules by virtue of their specific binding properties, which arise from particular combinations of functional groups in the constituent amino acids at the so-called active site. In many cases, an essential cofactor, e.g. NAD^+, PLP, or TPP, may also be bound to participate in the transformation. The involvement of enzymes in biochemical reactions has been a major theme throughout this book. The ability of enzymes to carry out quite complex chemical reactions, rapidly, at room temperature, and under essentially neutral conditions is viewed with envy by synthetic chemists, who are making rapid progress in harnessing this ability for their own uses. Several enzymes are currently of importance commercially, or for medical use, and

Tyr–Gly–Gly–Phe–**Met**–Thr–Ser–Glu–Lys–Ser–
Gln–Thr–Pro–Leu–Val–Thr–Leu–Phe–Lys–Asn–
Ala–Ile–Val–Lys–Asn–Ala–His–Lys–Lys–Gly–
Gln

        β-endorphin

Tyr–Gly–Gly–Phe–Met   Tyr–Gly–Gly–Phe–Leu
  Met-enkephalin   Leu-enkephalin

Tyr–Gly–Gly–Phe–Leu–Arg–Arg–Ile–Arg–Pro–Lys–Leu–Lys–Trp–Asp–Asn–Gln

                dynorphin A

Figure 7.14

Table 7.2 Pharmaceutically important enzymes

Enzyme	Action	Source	Use
Hydrolytic enzymes			
Pancreatin	Hydrolysis of starch (amylase), fat (lipase), and protein (protease)	Porcine pancreas	Digestive aid
Papain	Hydrolysis of proteins	Papaya fruit (*Carica papaya*; Caricaceae)	Meat tenderizer; cleaning of contact lenses
Chymotrypsin	Hydrolysis of proteins	Bovine pancreas	Zonal lysis in cataract removal
Hyaluronidase	Hydrolysis of mucopolysaccharides	Mammalian testes	Renders tissues more permeable for subcutaneous or intramuscular injections
Pepsin	Hydrolysis of proteins	Porcine stomach	Digestive aid
Trypsin	Hydrolysis of proteins	Bovine pancreas	Wound and ulcer cleansing
Fibrinolytic enzymes			
Streptokinase	No enzymic activity, until it complexes with and activates plasminogen in blood plasma to produce the proteolytic enzyme plasmin, which hydrolyses fibrin clots	*Streptococcus haemolyticus*	Treatment of venous thrombosis and pulmonary embolism
Urokinase	A protease which activates plasminogen in blood plasma to form plasmin, which hydrolyses fibrin clots	Human urine, or human kidney tissue cultures	Treatment of venous thrombosis and pulmonary embolism; thrombolysis in the eye
Alteplase (recombinant tissue-type plasminogen activator; rt-PA)	A protease which binds to fibrin converting it to a potent plasminogen activator; only active at the surface of the blood clot	Recombinant genetic engineering: human gene expressed in Chinese hamster ovary cells	Treatment of acute myocardial infarction
Anistreplase (acylated plasminogen-streptokinase activator complex; APSAC)	An inactive acylated form of the plasminogen–strepto-kinase activator complex; the acyl group (*p*-anisoyl) is slowly hydrolysed in the blood to give the active agent	Semi-synthesis from urokinase	Treatment of acute myocardial infarction
Reteplase	A fibrinolytic protease; a genetically-engineered human tissue-type plasminogen activator differing from alteplase at four amino acid residues	Recombinant genetic engineering	Treatment of acute myocardial infarction

(Continues)

Table 7.2 (*Continued*)

Enzyme	Action	Source	Use
Others Asparaginase (crisantaspase)	Degradation of L-asparagine	*Erwinia chrysanthemi*	Treatment of acute lymphoblastic leukaemia; results in death of those tumour cells which require increased levels of exogenous L-asparagine; side-effects nausea and vomiting, allergic reactions and anaphylaxis
Streptodornase	Depolymerization of polymerized deoxyribonucleo-proteins	*Streptococcus haemolyticus*	In combination with streptokinase as desloughing agent to cleanse ulcers and promote healing

these are described in Table 7.2. Enzymes are typically larger than most of the polypeptides discussed above, and are thus extracted from natural sources. Recombinant DNA procedures are likely to make a very significant contribution in the future.

NONRIBOSOMAL PEPTIDE BIOSYNTHESIS

In marked contrast to the ribosomal biosynthesis of peptides and proteins, where a biological production line interprets the genetic code, many natural peptides are known to be synthesized by a more individualistic sequence of enzyme-controlled processes, in which each amino acid is added as a result of the specificity of the enzyme involved. The many stages of the whole process appear to be carried out by a multi-functional enzyme (nonribosomal peptide synthase, NRPS) with a modular arrangement comparable to that seen with type I polyketide synthases (see page 114). The linear sequence of modules in the enzyme usually corresponds to the generated amino acid sequence in the peptide product. The amino acids are first activated by conversion into AMP esters, which then bind to the enzyme through thioester linkages. The residues are held so as to allow a sequential series of peptide bond formations (Figure 7.15

gives a simplified representation), until the peptide is finally released from the enzyme. A typical module consists of an adenylation (A) domain, a peptidyl carrier protein (PCP) domain, and a condensation (C) or elongation domain. The A domain activates a specific amino acid as an aminoacyl adenylate, which is then transferred to the PCP domain forming an aminoacyl thioester. Pantothenic acid (vitamin B_5, see page 31) bound to the enzyme as pantatheine is used to carry the growing peptide chain through its thiol group (Figure 7.15). The significance of this is that the long 'pantatheinyl arm' allows different active sites on the enzyme to be reached in the chain assembly process (compare biosynthesis of fatty acids, page 36, and polyketides, page 62). Nucleophilic attack by the amino group of the neighbouring aminoacyl thioester is catalysed by the C domain and results in amide (peptide) bond formation. Enzyme-controlled biosynthesis in this manner is a feature of many microbial peptides, especially those containing unusual amino acids not encoded by DNA and where post-translational modification is unlikely, and also for the cyclic structures which are frequently encountered. As well as activating the amino acids and catalysing formation of the peptide linkages, the enzyme may possess other domains that are responsible for epimerizing L-amino acids to D-amino acids, probably through

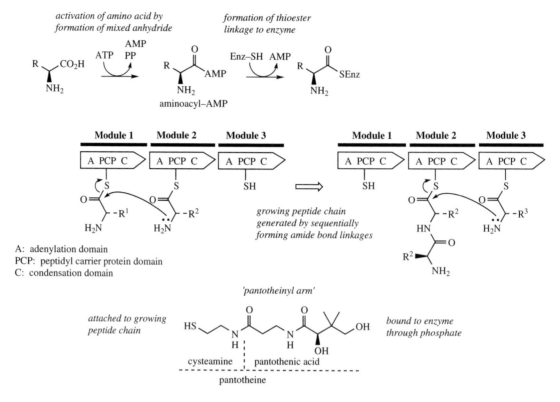

A: adenylation domain
PCP: peptidyl carrier protein domain
C: condensation domain

Figure 7.15

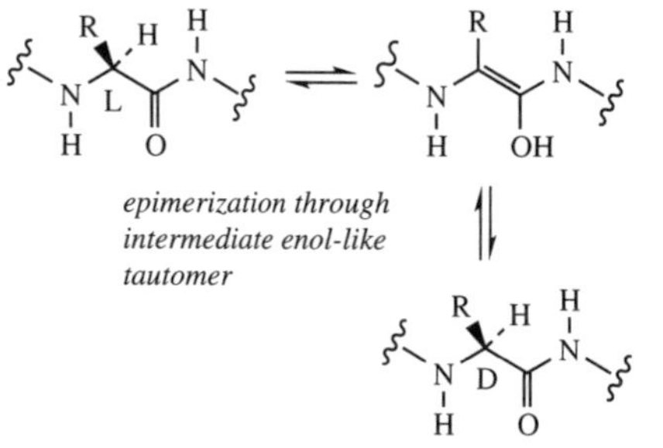

epimerization through
intermediate enol-like
tautomer

Figure 7.16

enol-like tautomers in the peptide (Figure 7.16). A terminal thioesterase domain is responsible for

terminating chain extension and releasing the peptide from the enzyme. Many of the medicinally useful peptides have cyclic structures. Cyclization may result if the amino acids at the two termini of a linear peptide link up to form another peptide bond, but very often, ring formation can be the result of ester or amide linkages, which utilize side-chain functionalities in the constituent amino acids. As with polyketide synthases, genetic manipulation of nonribosomal peptide synthases allows production of peptide derivatives in which rational modifications may be programmed according to the genes encoded.

PEPTIDE ANTIBIOTICS

Cycloserine

D-**Cycloserine** (Figure 7.17) is produced by cultures of *Streptomyces orchidaceus*, or may be prepared synthetically, and is probably the simplest substance with useful antibiotic

(Continues)

activity. Cycloserine is water soluble and has a broad spectrum of antibacterial activity, but it is primarily employed for its activity against *Mycobacterium tuberculosis*. It behaves as a structural analogue of D-alanine, and inhibits the incorporation of D-alanine into bacterial cell walls. Since it can produce neurotoxicity in patients, it is reserved for infections resistant to first-line drugs.

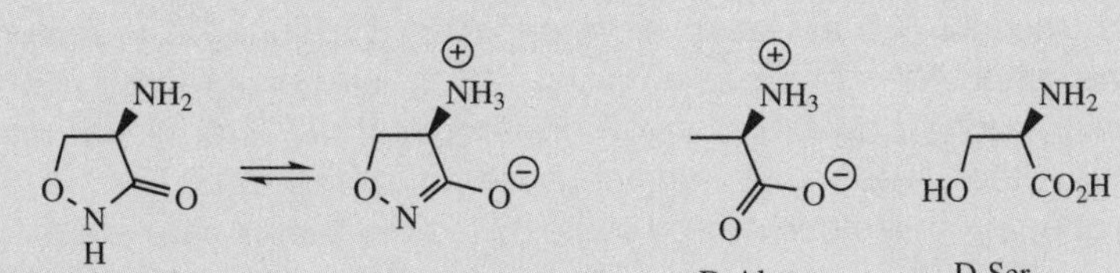

D-cycloserine   D-Ala   D-Ser

Figure 7.17

Polymyxins

The polymyxins are a group of cyclic polypeptide antibiotics produced by species of *Bacillus*. Polymyxins A–E were isolated from *Bacillus polymyxa*, though polymyxin B and polymyxin E were both subsequently shown to be mixtures of two components. A polypeptide mixture called colistin isolated from *Bacillus colistinus* was then found to be identical to polymyxin E. **Polymyxin B** and **colistin (polymyxin E)** are both used clinically. These antibiotic mixtures respectively contain principally polymyxin B_1 with small amounts of polymyxin B_2, or predominantly polymyxin E_1 ($\equiv$ colistin A) with small amounts of polymyxin E_2 ($\equiv$ colistin B) (Figure 7.18). These molecules contain ten amino acids, six of which are L-α,γ-diaminobutyric acid (L-Dab), with a fatty acid (6-methyloctanoic acid or 6-methylheptanoic acid) bonded to the *N*-terminus, and a cyclic peptide portion constructed via an amide bond between the carboxyl terminus and the γ-amino of one of the Dab residues. The γ-amino groups of the remaining Dab residues confer a strongly basic character to the antibiotics. This results in detergent-like properties and allows them to bind to and damage bacterial membranes. These peptides have been used for the treatment of infections with Gram-negative bacteria such as *Pseudomonas aeruginosa*, but are seldom used now because of

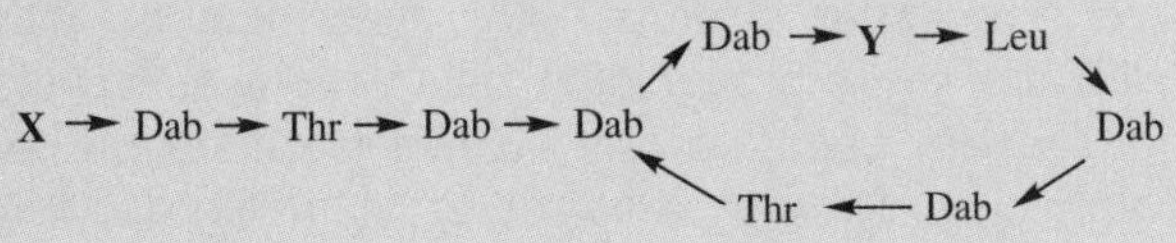

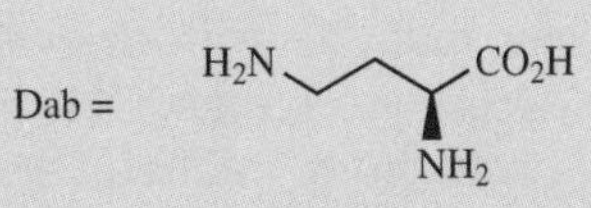

	X	Y
polymyxin B_1	6-methyloctanoic acid	D-Phe
polymyxin B_2	6-methylheptanoic acid	D-Phe
polymyxin E_1 (colistin A)	6-methyloctanoic acid	D-Leu
polymyxin E_2 (colistin B)	6-methylheptanoic acid	D-Leu

Figure 7.18

(Continues)

(Continued)

neurotoxic and nephrotoxic effects. However, they are included in some topical preparations such as ointments, eye drops, and ear drops, frequently in combination with other antibiotics.

Bacitracins

Bacitracin is a mixture of at least nine peptides produced by cultures of *Bacillus subtilis*, with the principal component being bacitracin A (Figure 7.19). The structure contains a cyclic peptide portion, involving the carboxyl terminus and the ε-amino of lysine, and at the *N*-terminus an unusual thiazolinecarboxylic acid, which is a condensation product from isoleucine and cysteine residues (compare epothilones, page 105, and bleomycin, page 429). Bacitracin is active against a wide range of Gram-positive bacteria, and appears to affect biosynthesis of the bacterial cell wall by binding to and sequestering a polyprenyl diphosphate carrier of intermediates; this binding also requires a divalent metal ion, with zinc being especially active. It is rarely used systemically because some bacitracin components are nephrotoxic, but as zinc bacitracin, it is a component of ointment formulations for topical application. The vast majority of bacitracin manufactured is used at subtherapeutic doses as an animal feed additive, to increase feed efficiency, and at therapeutic dosage to control a variety of disorders in poultry and animals.

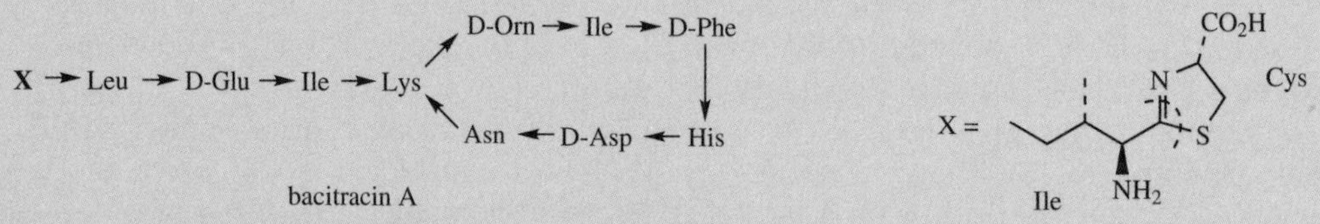

Figure 7.19

Tyrothricin and Gramicidins

Tyrothricin is a mixture of polypeptide antibiotics produced by cultures of *Bacillus brevis*. The mixture contains about 20–30% linear polypeptides called gramicidins (Figure 7.20), and 70–80% of cyclic structures called tyrocidines (Figure 7.20). Tyrothricin is active against many Gram-positive bacteria, with the linear gramicidins being more active than the cyclic tyrocidines. The two groups are readily separated by solvent fractionation, and the **gramicidin** fraction, also termed gramicidin D, a mixture of at least eight closely related compounds, is used principally in ophthalmic preparations. The gramicidins are neutral polypeptides having the *N*-terminal amino group formylated, and the carboxy group linked to ethanolamine. Most of the gramicidin mixture is composed of valine-gramicidin A (about 80%) (Figure 7.20). Apart from the glycine residue, these compounds have a sequence of alternating D- and L-amino acids. Gramicidins act by producing ion channels in bacterial membranes. The tyrocidines are too toxic for therapeutic use on their own, but the tyrothricin mixture is incorporated into lozenges for relief of throat infections.

    Gramicidin S is a mixture of cyclic peptides obtained from another strain of *Bacillus brevis*. Its main component is gramicidin S_1 (Figure 7.20), a symmetrical

(Continues)

(Continued)

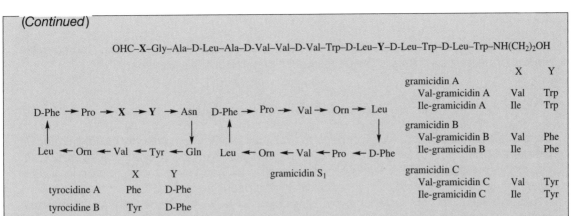

OHC–**X**–Gly–Ala–D-Leu–Ala–D-Val–Val–D-Val–Trp–D-Leu–**Y**–D-Leu–Trp–D-Leu–Trp–NH(CH$_2$)$_2$OH

	X	Y
gramicidin A		
Val-gramicidin A	Val	Trp
Ile-gramicidin A	Ile	Trp
gramicidin B		
Val-gramicidin B	Val	Phe
Ile-gramicidin B	Ile	Phe
gramicidin C		
Val-gramicidin C	Val	Tyr
Ile-gramicidin C	Ile	Tyr

D-Phe → Pro → **X** → **Y** → Asn

Leu ← Orn ← Val ← Tyr ← Gln

	X	Y
tyrocidine A	Phe	D-Phe
tyrocidine B	Tyr	D-Phe
tyrocidine C	Tyr	D-Tyr

D-Phe → Pro → Val → Orn → Leu

Leu ← Orn ← Val ← Pro ← D-Phe

gramicidin S$_1$

Figure 7.20

decapeptide known to be formed by the joining together of two separate chains. Gramicidin S is fairly toxic, and its use is restricted to topical preparations. Gramicidin S also acts on bacterial membranes, increasing permeability and loss of barrier function.

Capreomycin

Capreomycin is a mixture of cyclic polypeptides obtained from cultures of *Streptomyces capreolus*, and contains about 90% of capreomycins I, principally capreomycin IB (Figure 7.21). The capreomycin structures incorporate an L-capreomycidine moiety derived by cyclization of L-arginine, and three molecules of 2,3-diaminopropionic acid (Dap), which originate from serine via dehydroalanine. This antibiotic is given intramuscularly to treat tuberculosis patients who do not respond to first-line drugs, or where patients are sensitive to

Figure 7.21

(Continues)

(Continued)

streptomycin. It can cause irreversible hearing loss and impair kidney function. Capreomycin inhibits protein biosynthesis at the translocation step in sensitive bacteria.

Vancomycin and Teicoplanin

Vancomycin (Figure 7.22) is a glycopeptide antibiotic produced in cultures of *Amycolatopsis orientalis* (formerly *Streptomyces orientalis*), and has activity against Gram-positive bacteria, especially resistant strains of staphylococci, streptococci, and enterococci. It is an important agent in the control of methicillin-resistant *Staphylococcus aureus* (MRSA), with some strains now being sensitive only to vancomycin or teicoplanin (below). Vancomycin is not absorbed orally, and must be administered by intravenous injection. However, it can be given orally in the treatment of pseudomembranous colitis caused by *Clostridium difficile*, which may occur after administration of other antibiotics. Vancomycin acts by its ability to form a complex with terminal *N*-acyl–D-Ala–D-Ala residues of growing peptidoglycan chains (see Figure 7.36 and page 444), preventing their cross-linking to adjacent strands and thus inhibiting bacterial cell wall biosynthesis. The –D-Ala–D-Ala residues are accommodated in a 'carboxylate-binding pocket' in the vancomycin structure. By preventing peptidoglycan polymerization and cross-linking, it weakens the bacterial cell wall and ultimately causes cell lysis.

The novel feature of vancomycin, and several other related antibiotics, is the tricyclic structure generated by three phenolic oxidative coupling reactions. The β-hydroxychlorotyrosine and 4-hydroxyphenylglycine residues in vancomycin originate from L-tyrosine, but the 3,5-dihydroxyphenylglycine ring is actually acetate derived. These modified aromatic rings are presumably present in the heptapeptide before coupling occurs (Figure 7.22).

The teicoplanins (Figure 7.23) possess the same basic structure as vancomycin, but the *N*-terminal (4-hydroxyphenylglycine) and third (3,5-dihydroxyphenylglycine) amino acids are also aromatic, and this allows further phenolic oxidative coupling and generation of yet another ring system. **Teicoplanin** for drug use is a mixture of five teicoplanins produced by cultures of *Actinoplanes teichomyceticus*, which differ only in the nature and length of the fatty acid chain attached to the sugar residue. Teicoplanin has similar antibacterial activity to vancomycin, but has a longer duration of action, and may be administered by intramuscular as well as by intravenous injection. It is also used against Gram-positive pathogens resistant to established antibiotics.

Vancomycin, teicoplanin, and structurally related glycopeptides are often referred to as dalbaheptides (from *D-al*anyl–D-alanine-*binding* *hepta*pept*ide*), reflecting their mechanism of action and their chemical nature. Unfortunately, with increasing use of vancomycin and teicoplanin, there have even been reports of these agents becoming ineffective because resistant bacterial strains have emerged, particularly in enterococci. In resistant strains, the terminal –D-Ala–D-Ala residues, to which the antibiotic normally binds, have become replaced by –D-Ala–D-lactate. The incorporation of D-lactate into the peptide intermediates results in loss of crucial hydrogen-bonding interactions and a thousand-fold lowering in binding efficiency.

(Continues)

(Continued)

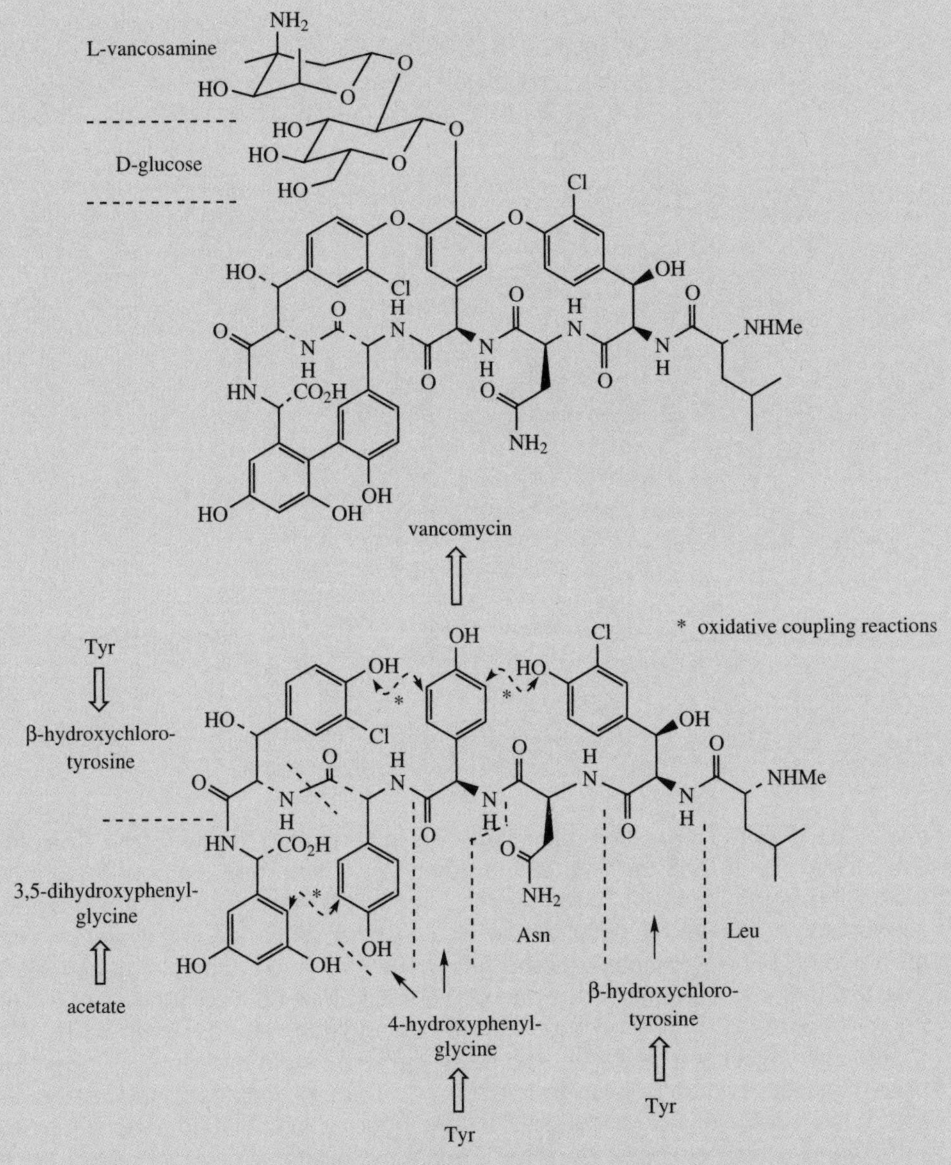

L-vancosamine

D-glucose

vancomycin

* oxidative coupling reactions

Tyr

β-hydroxychloro-
tyrosine

3,5-dihydroxyphenyl-
glycine

acetate

4-hydroxyphenyl-
glycine

Tyr

Asn

Leu

β-hydroxychloro-
tyrosine

Tyr

Figure 7.22

Bleomycin

Bleomycin is a mixture of glycopeptide antibiotics isolated from cultures of *Streptomyces verticillus*, used for its anticancer activity. The major component (55–70%) of the mixture

(Continues)

(Continued)

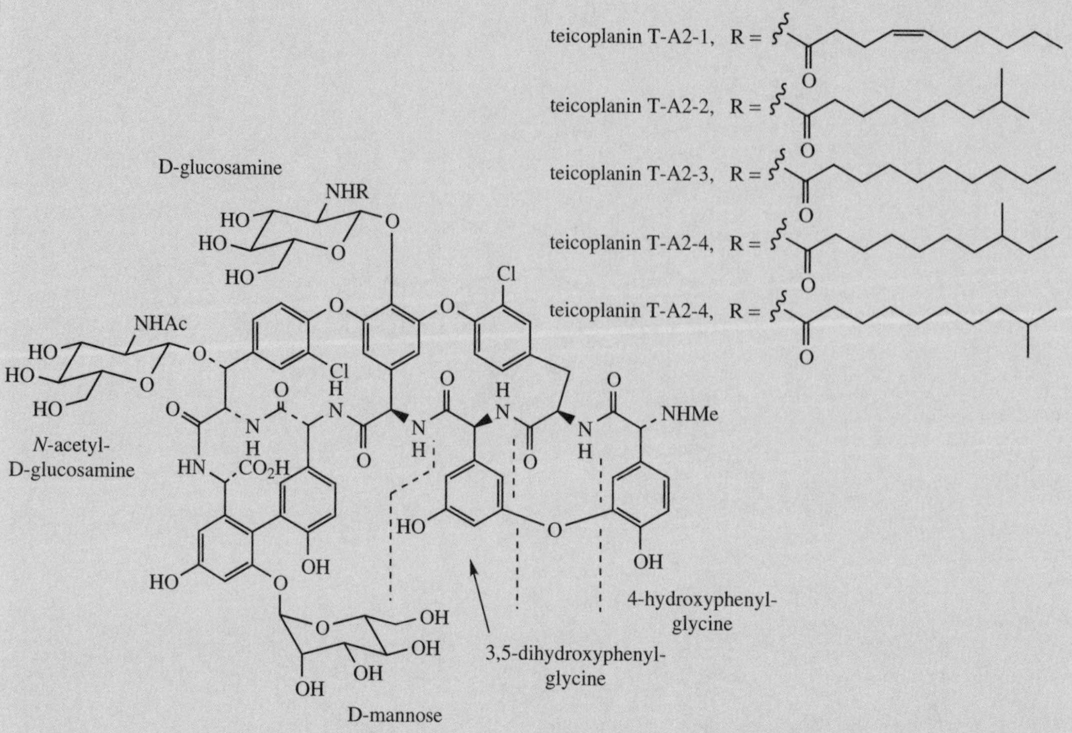

teicoplanin T-A2-1, R =

teicoplanin T-A2-2, R =

teicoplanin T-A2-3, R =

teicoplanin T-A2-4, R =

teicoplanin T-A2-4, R =

D-glucosamine

NHR

NHAc

HO

HO

HO

N-acetyl-
D-glucosamine

HN   CO₂H

Cl

Cl

NHMe

HO

HO

OH

OH

OH

OH

OH

OH

D-mannose

4-hydroxyphenyl-
glycine

3,5-dihydroxyphenyl-
glycine

Figure 7.23

is bleomycin A$_2$ (Figure 7.24), with bleomycin B$_2$ constituting about 30%. The various bleomycins differ only in their terminal amine functions, the parent compound bleomycinic acid (Figure 7.24) being inactive. The molecules contain several unusual amino acids and sugars, an asparagine-derived pyrimidine ring, and a planar dithiazole ring system which has its origins in two cysteines (compare epothilones, page 105, and bacitracin, page 424). A C$_2$ unit supplied by malonyl-CoA also forms part of the main chain as a component of the amino acid 4-amino-3-hydroxy-2-methylvaleric acid, which in addition features a methionine-derived methyl group. Bleomycin is a DNA-cleaving drug, causing single and double-strand breaks in DNA. The dithiazole system is involved in binding to DNA, probably by intercalation, whilst other parts of the molecule near the *N*-terminus are involved in chelating a metal ion, usually Fe^{2+}, and oxygen, which are necessary for the DNA degradation reaction. More recently, bleomycin A$_2$ has been shown to cleave RNA as well as DNA.

Bleomycin is used alone, or in combination with other anticancer drugs, to treat squamous cell carcinomas of various organs, lymphomas, and some solid tumours. It is unusual amongst antitumour antibiotics in producing very little bone-marrow suppression, making it particularly useful in combination therapies with other drugs that do cause this response. However, there is some lung toxicity associated with bleomycin treatment. Various bleomycin analogues have been made by adding different precursor amines to the culture medium, or by semi-synthesis from bleomycinic acid. **Peplomycin** (Figure 7.24) is an example with some promise, in that it is more resistant to enzymes that cause *in vivo* hydrolysis of bleomycin at the *N*-terminal β-aminoalanine group.

(Continued)

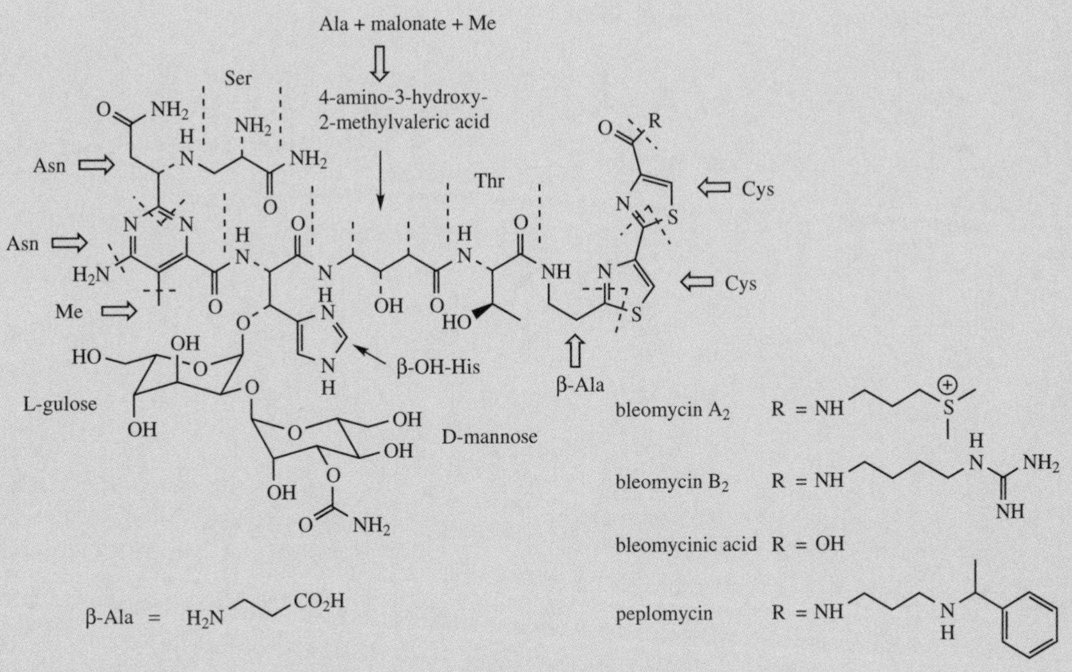

Figure 7.24

Cyclosporins

The cyclosporins are a group of cyclic peptides produced by fungi such as *Cylindrocarpon lucidum* and *Tolypocladium inflatum*. These agents showed a narrow range of antifungal activity, but high levels of immunosuppressive and anti-inflammatory activities. The main component from the culture extracts is cyclosporin A (**ciclosporin, cyclosporin**) (Figure 7.25), but some 25 naturally occurring cyclosporins have been characterized. Cyclosporin A contains several *N*-methylated amino acid residues, together with the less common L-α-aminobutyric acid and an *N*-methylated butenylmethylthreonine. This latter amino acid has been shown to originate via the acetate pathway, and it effectively comprises a C$_8$ polyketide chain, plus a methyl group from SAM. The sequence shown in Figure 7.26 is consistent with experimental data, and is analogous to other acetate-derived compounds (see Chapter 3). The assembly of the polypeptide chain is known to start from the D-alanine residue. Many of the other natural cyclosporin structures differ only with respect to a single amino acid (the α-aminobutyric acid residue) or the amount of *N*-methylation. Of all the natural analogues, and many synthetic ones produced, cyclosporin A is the most valuable for drug use. It is now widely exploited in organ and tissue transplant surgery, to prevent rejection following bone-marrow, kidney, liver, and heart transplants. It has revolutionized organ transplant surgery, substantially increasing survival rates in transplant patients. It may be administered orally or by intravenous injection, and the primary side-effect is nephrotoxicity, necessitating careful monitoring of kidney function. Cyclosporin A has the same mode of action as the macrolide FK-506 (tacrolimus) (see page 104) and is believed to inhibit T-cell activation in the immunosuppressive mechanism

(Continues)

(Continued)

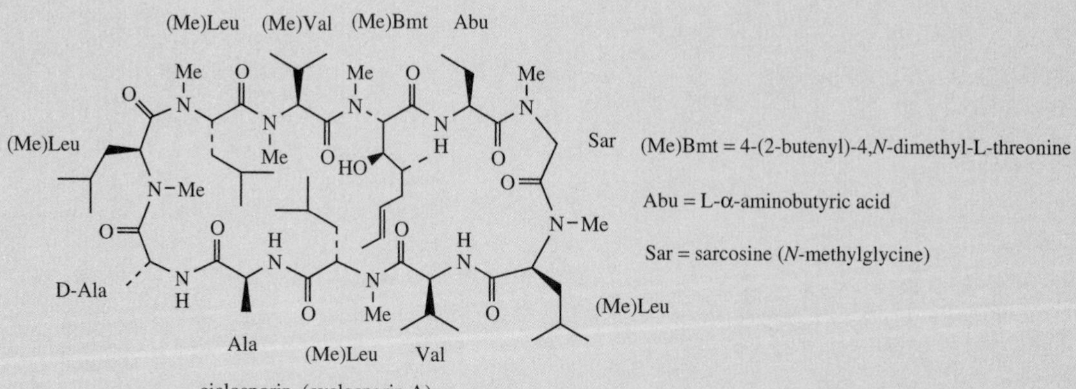

(Me)Leu   (Me)Val   (Me)Bmt   Abu

(Me)Leu

D-Ala

Ala   (Me)Leu   Val

(Me)Bmt = 4-(2-butenyl)-4,*N*-dimethyl-L-threonine

Abu = L-α-aminobutyric acid

Sar = sarcosine (*N*-methylglycine)

Sar

(Me)Leu

ciclosporin (cyclosporin A)

Figure 7.25

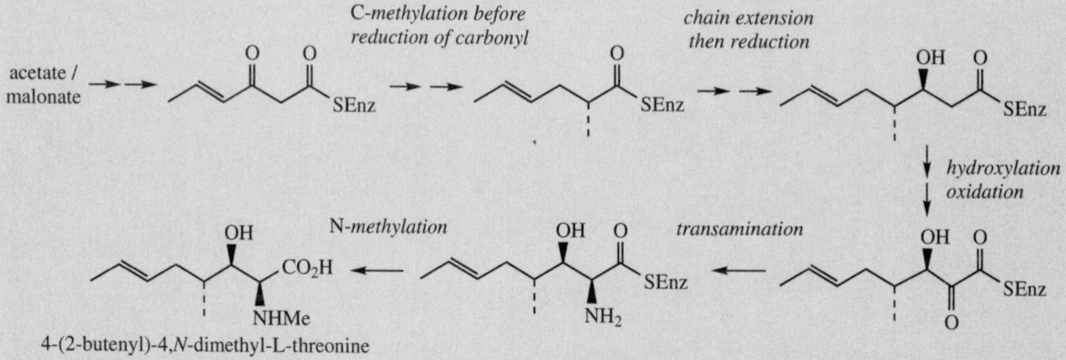

acetate /
malonate

*C-methylation before
reduction of carbonyl*

*chain extension
then reduction*

*hydroxylation
oxidation*

N-methylation

transamination

4-(2-butenyl)-4,*N*-dimethyl-L-threonine

Figure 7.26

by first binding to a receptor protein, giving a complex that then inhibits a phosphatase enzyme called calcineurin. The resultant aberrant phosphorylation reactions prevent appropriate gene transcription and subsequent T-cell activation. Cyclosporin A also finds use in the specialist treatment of severe resistant psoriasis.

Streptogramins

The names streptogramin and virginiamycin have been applied to antibiotic mixtures isolated from strains of *Streptomyces virginiae*, and individual components have thus acquired multiple synonyms; as a family these antibiotics have now been termed streptogramin antibiotics. These compounds fall into two distinct groups, group A, containing a 23-membered unsaturated ring with peptide and lactone bonds, and group B, which are depsipeptides (essentially peptides cyclized via a lactone). These structures contain many nonprotein amino acids (Figure 7.27). Until recently, most commercial production of these antibiotics was directed towards animal feed additives, but the growing emergence of antibiotic-resistant

(Continues)

(*Continued*)

bacterial strains has led to the drug use of some streptogramin antibiotics. Thus, **dalfopristin** and **quinupristin** (Figure 7.27) are water-soluble drugs that may be used in combination for treating infections caused by Gram-positive bacteria that have failed to respond to other antibiotics, including methicillin-resistant *Staphylococcus aureus* (MRSA), vancomycin-resistant enterococci and staphylococci, and drug-resistant *Streptococcus pneumoniae*. They may need to be combined also with other agents where mixed infections involve Gram-negative organisms. Dalfopristin is a semi-synthetic sulphonyl derivative of streptogramin A (also termed virginiamycin M1, mikamycin A, pristinamycin IIA, and other names), and quinupristin is a modified form of streptogramin B (also mikamycin B, pristinamycin IA, and other names). Members of the A group tend to be less powerful antibiotics than those of the B group, but together they act synergistically, providing greater activity than the combined activity expected from the separate components. The dalfopristin and quinupristin combination is supplied in a 70:30 ratio, which provides maximum synergy (a 100-fold increase in activity compared to the single agents), and also corresponds to the natural proportion of group A to group B antibiotics in the producer organism. The streptogramins bind to the peptidyl transferase domain of the 50S ribosomal subunit; the remarkable synergism arises because initial binding of the group A derivative causes a conformational change to the ribosome, increasing affinity for the group B derivative and formation of an extremely stable

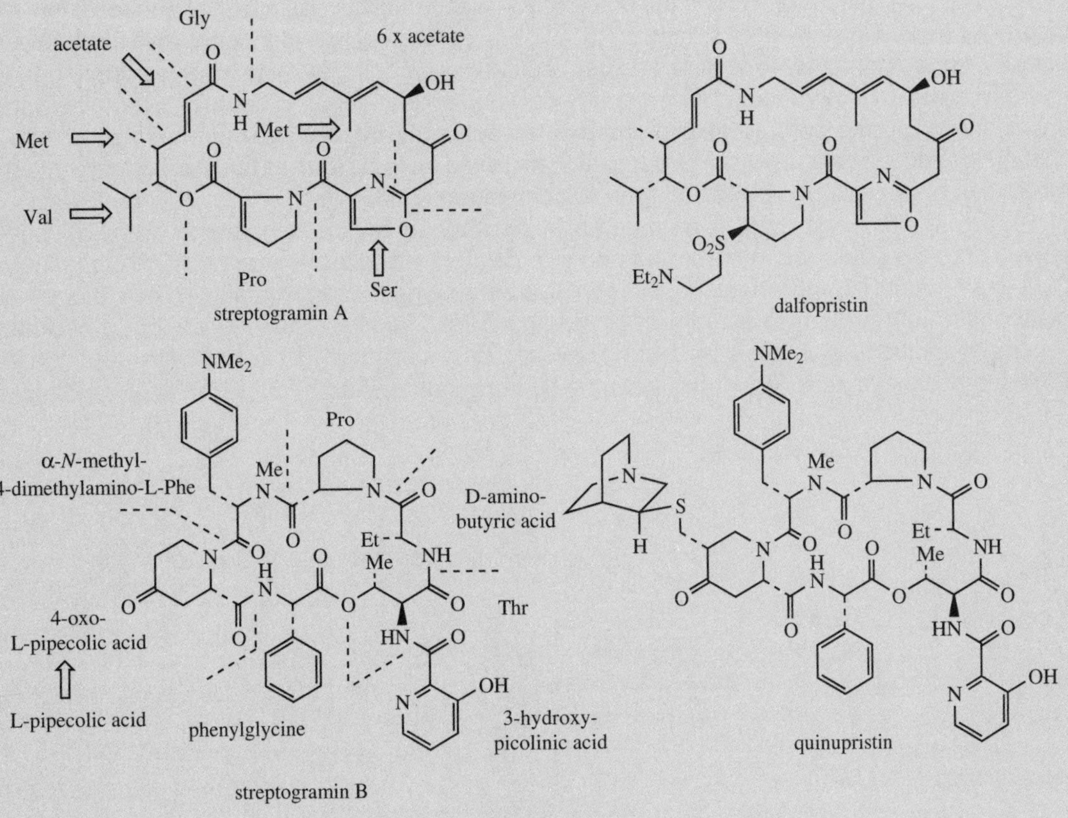

Figure 7.27

(*Continues*)

(Continued)

ternary complex. This makes the streptogramin combination bactericidal, whereas the single agents provide only bacteriostatic activity.

Streptogramin A is known to be biosynthesized from four amino acids, namely valine, glycine, serine, and proline, a polyketide-like chain containing six acetate units, a further isolated acetate unit, and two methionine-derived methyl groups. The oxazole ring is formed from serine by incorporating the carboxyl terminus of the polyketide chain (compare the thiazole rings in bleomycin, page 429). Streptogramin B is formed by a typical nonribosomal peptide synthase, utilizing several rare modified amino acid precursors. All except one are modified before assembly; oxidation of the pipecolic acid residue (see page 310) to 4-oxopipecolic acid is carried out post-cyclization. The starter unit is 3-hydroxypicolinic acid, which arises via picolinic acid in the kynurenine pathway (see page 312); *p*-aminophenylalanine has been met previously in chloramphenicol biosynthesis (page 129).

Dactinomycin

Dactinomycin (actinomycin D) (Figure 7.28) is an antibiotic produced by *Streptomyces parvullus* (formerly *S. antibioticus*), which has antibacterial and antifungal activity, but whose high toxicity limits its use to anticancer therapy. Several related natural actinomycins are known, but only dactinomycin is used medicinally. Dactinomycin has a planar phenoxazinone dicarboxylic acid system in its structure, to which are attached two identical cyclic pentapeptides via amide bonds to threonine. The peptides are cyclized by lactone linkages utilizing the hydroxyl group of this threonine. The peptide portions of dactinomycin contain *N*-methylvaline and *N*-methylglycine (sarcosine) residues. In other actinomycins, the two peptides are not necessarily identical. The phenoxazinone ring system is known to be formed by fusing together two molecules of 3-hydroxy-4-methylanthranilic acid (Figure 7.28), which arises by *C*-methylation of 3-hydroxyanthranilic acid, a metabolite of tryptophan by the kynurenine pathway (see page 312). This planar phenoxazinone ring intercalates with double-stranded DNA inhibiting DNA-dependent RNA polymerases, but can also cause single-strand breaks in DNA. It is principally used to treat paediatric cancers, including Wilms' tumour of the kidney, but produces several serious and painful side-effects. However, as a selective inhibitor of DNA-dependent RNA synthesis (transcription), it has become an important research tool in molecular biology.

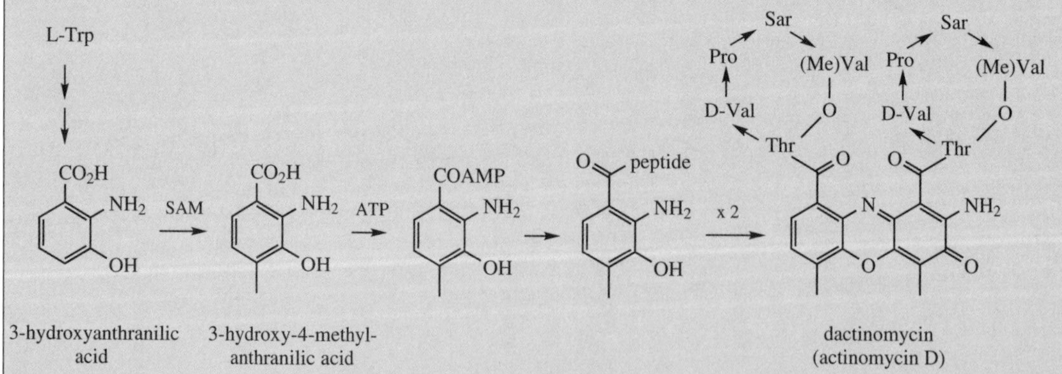

Figure 7.28

(Continues)

(Continued)

PEPTIDE TOXINS

Death Cap (Amanita phalloides)

The death cap, *Amanita phalloides*, is a highly poisonous European fungus with a mushroom-like fruiting body. The death cap has a whitish-green cap, and white gills. It has a superficial similarity to the common mushroom, *Agaricus campestris*, and may sometimes be collected in error. Some 90% of human fatalities due to mushroom poisoning are attributed to the death cap. Identification of the death cap as a member of the genus *Amanita*, which includes other less poisonous species, is easily achieved by the presence of a volva at the base of the stem. This cuplike membranous structure is the remains of the universal veil in which the immature fruiting body was enclosed. Ingestion of the death cap produces vomiting and diarrhoea during the first 24 hours, followed after 3–5 days by coma and death. Some recoveries do occur, but the fatality rate is probably from 30 to 60%. There is no guaranteed treatment for death cap poisoning, though removal of material from the gastrointestinal tract, replacement of lost fluids, blood dialysis, and blood transfusions may be undertaken. The antihepatotoxic agent silybin (see page 153) has been used successfully. The toxic principles are cyclic polypeptides, which bring about major degeneration of the liver and kidneys. At least ten toxins have been identified, which may be subdivided into two groups, the phallotoxins and the amatoxins. The most extensively studied compounds are phalloin, a phallotoxin, and α-amanitin, an amatoxin (Figure 7.29). The phallotoxins are much less toxic than the amatoxins since after ingestion they are not well absorbed into the blood stream. When injected, they can cause severe damage to the membranes of liver cells, releasing potassium ions and enzymes. The amatoxins are extremely toxic when ingested, with a lethal dose of 5–7 mg for an adult human, and an average specimen of the death cap containing about 7 mg. The amatoxins cause lesions to the stomach and

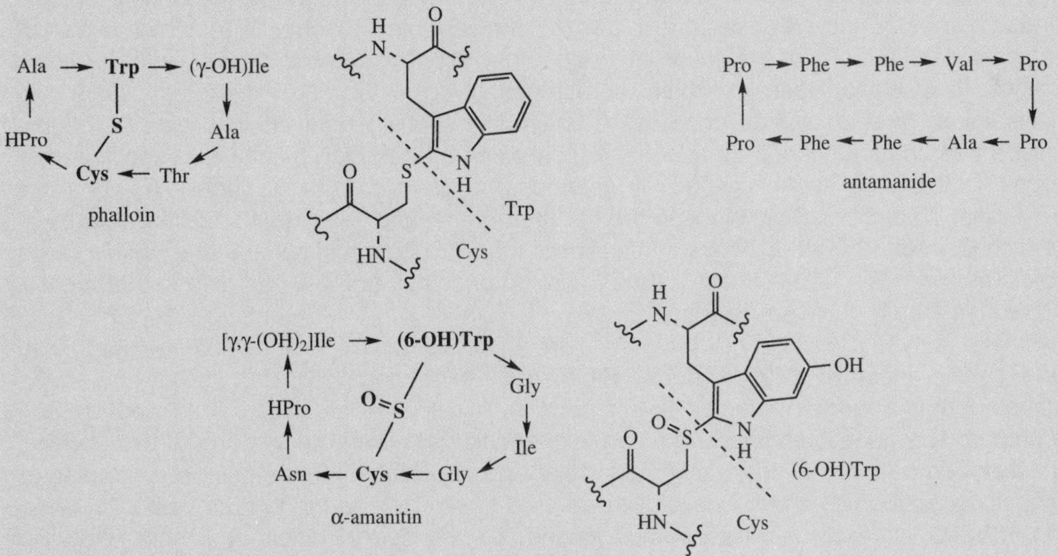

Figure 7.29

(Continues)

(Continued)

intestine, and then irreversible loss of liver and kidney function. α-Amanitin has been shown to be a powerful inhibitor of RNA polymerase, blocking the elongation step. Structurally, the two groups of toxins appear similar. They both contain a γ-hydroxylated amino acid essential for toxic action, and have a sulphur bridge between a cysteine residue and the 2-position of a tryptophan or 6-hydroxytryptophan residue (Figure 7.29). In the case of the phallotoxins, this is a simple sulphide bridge, but in the amatoxins it is in the form of a sulphoxide. Interestingly, a rather more simple cyclic peptide antamanide (anti-amanita peptide) has also been isolated from *Amanita phalloides*. When pre-administered to laboratory animals, antamanide (Figure 7.29) provided prophylactic protection from the lethal affects of the phallotoxins.

Ricin

The distinctive mottled-brown seeds of the castor oil plant *Ricinus communis* (Euphorbiaceae) are crushed to produce castor oil, which is predominantly composed of glycerides of ricinoleic acid (see page 47 and Table 3.2). The seed itself and the seed cake remaining after expression of the oil are highly toxic, due to the presence of polypeptide toxins, termed ricins. One or more forms of ricin are present according to the strain of plant. Seeds typically contain about 1mg g^{-1} ricin, representing about 5% of the protein content. The toxicity of ricins to mammals is so high that the ricin content of one seed (about 250 μg) is sufficient to kill an adult human, though, because of considerable variations in absorption and metabolism, consumption of a single seed might not be fatal, but would certainly lead to severe poisoning. The toxic symptoms include irritation and haemorrhage in the digestive tract, leading to vomiting and bloody diarrhoea, with subsequent convulsions and circulatory collapse. If death does not occur within 3–5 days, the patient may recover.

Ricin is probably the best studied of a group of polypeptide toxins, known as ribosome-inactivating proteins (the acronym RIP seems most appropriate!). These toxins are potent inhibitors of eukaryotic protein biosynthesis by virtue of their cleavage of an *N*-glycosidic bond at a specific nucleotide residue in the 28S molecule of ribosomal RNA, which is part of the larger (60S) subunit of the eukaryotic ribosome. RIPs fall into two types. Type I proteins comprise a single polypeptide chain, sometimes glycosylated. Type II proteins have two chains linked by a disulphide bond: an A chain is essentially equivalent to type I proteins, whilst the B chain functions as a lectin, which means it has high affinity for specific sugar groups, in this case sugars (especially galactose) in glycolipids or glycoproteins on a cell membrane. Thus the B chain binds to the cell membrane, and, in so doing, facilitates entry of the A chain into the cytosol, where it inactivates the 60S ribosomal subunit and rapidly stops protein biosynthesis. The A chain is thus the toxic principle, but it is nontoxic to intact cells, and requires the B chain for its action.

Ricin is a type II toxin. The A chain (ricin A) contains 267 amino acid residues, and the B chain (ricin B) 262 residues. Ricin A is exceptionally toxic, and it has been estimated that a single molecule is sufficient to kill an individual cell. This peptide can be prepared by genetic engineering using *Escherichia coli*. The potent action of this material on eukaryotic cells has been investigated in anticancer therapy. Ricin A has been coupled to monoclonal antibodies and successfully delivered specifically to the tumour cells. However, *in vitro* toxicity of ricin A-based immunotoxins is enhanced significantly if ricin B is also present.

(Continues)

(Continued)

Similar toxic RIPs are found in other plants. Examples are trichosanthin, a type I toxin from the root tubers of *Trichosanthes kirilowii* (Cucurbitaceae), abrin, a type II toxin from the small brightly coloured red and black jequirity seeds (*Abrus precatorius*; Leguminosae/Fabaceae), and viscumin, a type II toxin from the leaf and stems of mistletoe (*Viscum album*, Loranthaceae).

Botulinum Toxin

The Gram-positive bacterium *Clostridium botulinum* produces one of the most toxic materials known to man, botulinum toxin. Poisoning by the neurotoxins from this source, known as botulism, is not uncommon, and is a life-threatening form of food poisoning. It has been estimated that as many as 50 million people could be killed by one gram of the toxin. *Clostridium botulinum* is an anaerobic organism that is significantly heat resistant, though botulinum toxin is easily destroyed by heat. Food poisoning is almost always associated with foods such as canned meats and fish that have been incompletely sterilized, allowing growth of the bacterium, after which the food is then consumed without further cooking. Botulinum toxin is an extremely potent neurotoxin that acts by blocking calcium-dependent acetylcholine release at the peripheral neuromuscular junctions. Poisoning results in paralysis and death from respiratory failure. Damage to the nervous system is usually preceded by vomiting, diarrhoea, and severe abdominal pains.

Seven different neurotoxins, types A–G, have been characterized, though only four of these, types A, B, E, and F, are clearly associated with human poisoning. A particular strain of the bacterium usually produces only one type of toxin. Each of these proteins is a single chain polypeptide (mass about 150 kDa) containing two subunits, a 'light' subunit with a mass of about 50 kDa, and a 'heavy' subunit with a mass of about 100 kDa. These subunits are linked by at least one disulphide bridge. The heavy subunit is responsible for toxin binding, whilst the light subunit possesses zinc metalloprotease activity, cleaving one of the proteins involved in the docking and release of synaptic vesicles. There is considerable structural similarity between botulinum toxins and tetanus toxin. **Botulism antitoxin** is available for the treatment of botulism food poisoning. This is a mixture of globulins raised against types A, B, and E toxins.

Botulinum toxin A complexed with haemagglutinin is currently employed medicinally to counter involuntary facial muscle spasms, e.g. around the eye. Very small (nanogram) amounts are injected locally and result in the destruction of the acetylcholine release mechanism at the neuromuscular junction. Since new nerve junctions will gradually be formed over two months or so, the result is not permanent, and the treatment will need to be repeated. It has also been found useful in easing muscle spasticity in children with cerebral palsy.

Microcystins

The microcystins are a group of some65 cyclic heptapeptides, produced by certain fresh water blue–green algae (cyanobacteria), including *Microcystis aeruginosa*, *M. viridis*, *Anabaena flos-aquae*, and others. These organisms form blooms on the surface of lakes and reservoirs during periods of calm hot weather, and can pollute drinking water for animals and humans. The microcystins cause acute hepatotoxicity and liver haemorrhage, and have been

(Continues)

(Continued)

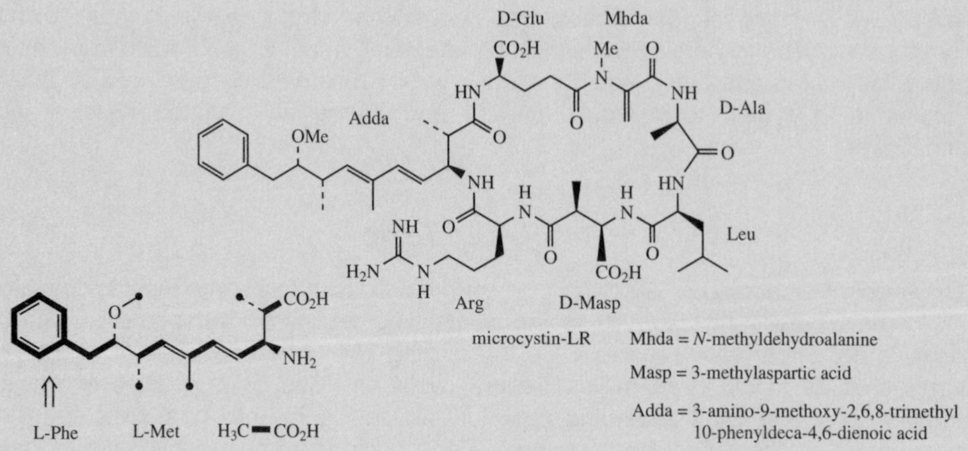

Figure 7.30

responsible for numerous animal deaths, though there is much less risk of human fatalities. They strongly inhibit protein phosphatases, and this activity may be associated with the hepatotoxicity and the promotion of liver tumours. The microcystins contain a mixture of D- and L-amino acids, and mainly differ in their combination of L-amino acids. The most frequently encountered compound, microcystin-LR (Figure 7.30) (LR = leucine–arginine) exemplifies the structures. Unusual amino acids are N-methyldehydroalanine (from N-methylserine), 3-methylaspartic acid (which originates from acetate and pyruvate), and the long chained aromatic-containing Adda. This amino acid is elaborated from phenylalanine, which supplies a C_6C_2 unit to act as a starter unit for a polyketide chain. Methyl substituents are provided by SAM (Figure 7.30).

Snake Venoms

It is estimated that some 1300 of the 3200 species of snake are venomous. Snake venoms are used to immobilize prey, and to facilitate its digestion. Most of the material is polypeptide in nature, and can include enzymes and polypeptide toxins. A number of enzymes have been identified in all venoms, and these include hyaluronidase (see Table 7.2), which facilitates the distribution of the other venom components through the tissues. Peptidases, phosphodiesterases, phospholipases, ribonuclease, and deoxyribonuclease are all hydrolytic enzymes designed to digest the tissue of the prey. Some enzymes induce direct toxic effects, for example L-amino acid oxidase liberates hydrogen peroxide, a powerful oxidizing agent. In some venoms, the enzyme acetylcholinesterase disturbs the normal physiological response of the prey by hydrolysing acetylcholine. Major groups of polypeptide toxins found in snake venoms may be classified as neurotoxins, cytotoxins (or cardiotoxins), dendrotoxins, proteinase inhibitors, or acetylcholinesterase inhibitors. α-Neurotoxins (curaremimetic neurotoxins) found in many mambas (*Dendroaspis*) and cobras (*Naja*) are capable of interacting with nicotinic acetylcholine receptors in the postsynaptic membranes of skeletal muscles, leading to paralysis, an action similar to that of curare (see page 324). κ-Neurotoxins, on the other hand, are selective for neuronal nicotinic acetylcholine receptors. Typically, the neurotoxins contain 60–74 amino acid residues with four or five

---

(Continued)

disulphide bridges. Cytotoxins are present in cobra (*Naja*) and ringhal (*Hemachatus*) venoms and produce a rapid effect on the heart and circulation though the mode of action is not well established. Most have 60–62 amino acid residues. Dendrotoxins from mambas are characterized by their ability to facilitate the release of acetylcholine from nerve endings, and also act as highly potent and selective blockers of potassium channels. These contain about 60 amino acid residues, with three disulphide bridges. Anticholinesterase toxins have also been found in mamba venoms, typically with about 60 amino acids and four disulphide bridges.

MODIFIED PEPTIDES: PENICILLINS, CEPHALOSPORINS, AND OTHER β-LACTAMS

Penicillins

The **penicillins*** are the oldest of the clinical antibiotics, and are still the most widely used. The first of the many penicillins to be employed on a significant scale was **penicillin G (benzylpenicillin)** (Figure 7.31), obtained from the fungus *Penicillium chrysogenum* by fermentation in a medium containing corn-steep liquor. Penicillins contain a β-lactam-thiazolidine structure, which has its biosynthetic origins in a tripeptide, the components of which are L-aminoadipic acid, formed in β-lactam-producing organisms from lysine via piperideine-6-carboxylic acid (see page 311), together with L-cysteine, and L-valine (Figure 7.31). Non-ribosomal peptide assembly then leads to the tripeptide known as ACV, but, during the condensation, the configuration of the

valine residue is also inverted to the D-form. (Caution: ACV is an acronym, and does not refer to the systematic abbreviation described on page 407; ACV is actually δ-(L-α-aminoadipyl)-L-cysteinyl-D-valine). ACV is then cyclized to **isopenicillin N**, with a single enzyme catalysing formation of the bicyclic ring system of the penicillins. The reaction is oxidative and requires molecular oxygen, and there is evidence that the four-membered β-lactam ring is formed first. The mechanism shown in Figure 7.31 is given to simplify what is quite a complex reaction. **Penicillin G** differs from isopenicillin N by the nature of the side-chain attached to the 6-amino group. The α-aminoadipyl side-chain of isopenicillin N is removed and replaced by another according to its availability from the fermentation medium. Phenylethylamine in the corn-steep liquor medium was transformed by the fungus into phenylacetic acid, which then reacted as its coenzyme A ester to produce the new amide penicillin G. Other penicillins are accessible by

---

Penicillins

Commercial production of **benzylpenicillin (penicillin G)** (Figure 7.31) is by fermentation of selected high-yielding strains of *Penicillium chrysogenum* in the presence of phenylacetic acid. Although benzylpenicillin was the earliest commercially available member of the penicillin group of antibiotics, it still remains an important and useful drug for the treatment of many Gram-positive bacteria, including streptococcal, pneumococcal, gonococcal, and meningococcal infections. Benzylpenicillin is destroyed by gastric acid, and is thus not suitable for oral administration, and is best given as intramuscular or intravenous injection of the water-soluble sodium salt. Decomposition under acidic conditions leads to formation of penicillic acid and/or penicillenic acid, depending on pH (Figure 7.32). The β-lactam ring is opened by a mechanism in which the side-chain carbonyl participates, resulting in formation of an oxazolidine ring. Penicillic acid arises as the result of nucleophilic attack of the thiazolidine nitrogen on to the iminium function, followed by expulsion of the carboxylate leaving group.

(Continues)

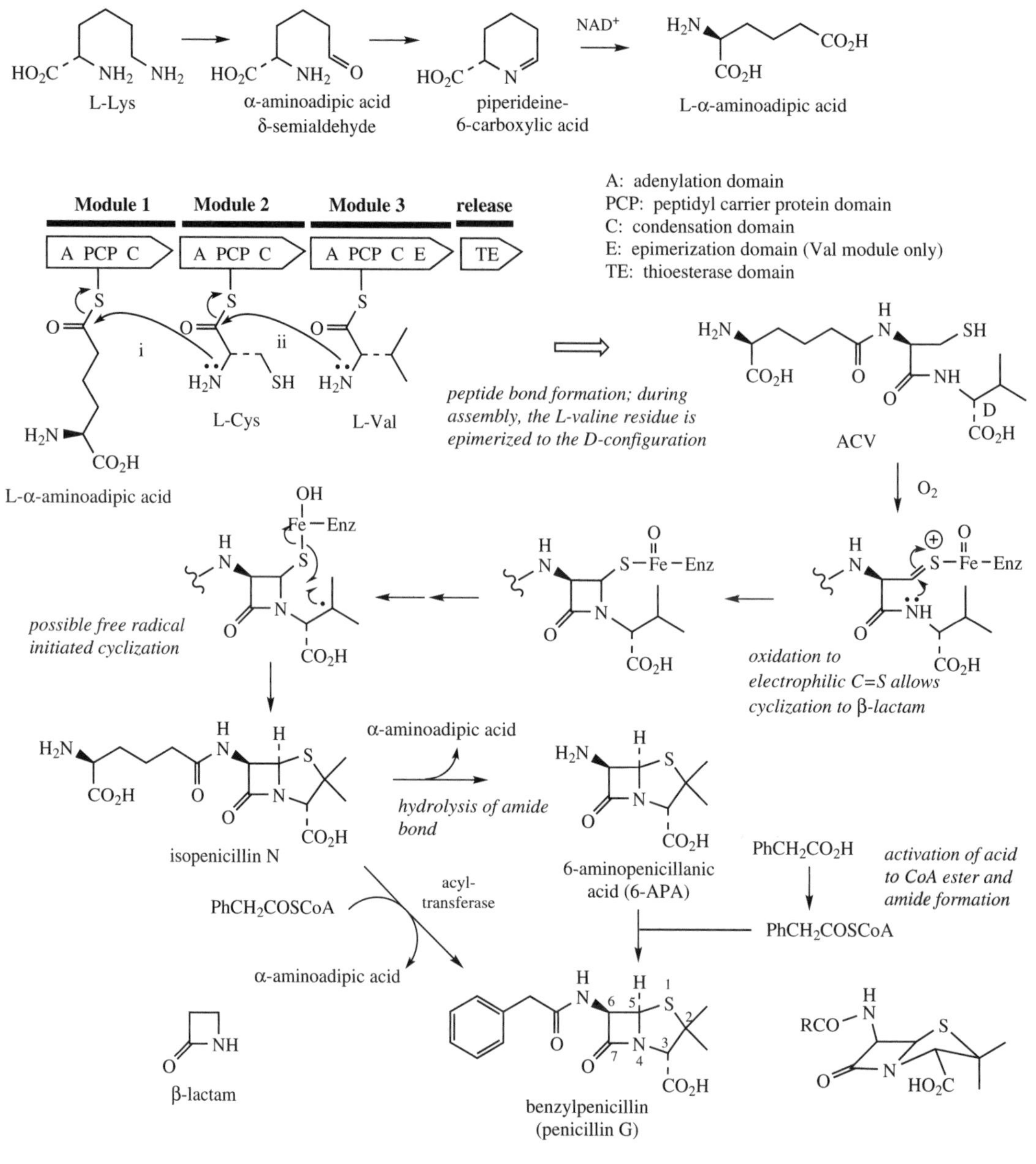

Figure 7.31

(Continued)

Alternatively, elimination of thiol accounts for formation of penicillenic acid. At higher pHs, benzylpenicillin suffers simple β-lactam ring opening and gives penicilloic acid (Figure 7.32). The strained β-lactam (cyclic amide) ring is more susceptible to hydrolysis than the unstrained side-chain amide function, since the normal stabilizing effect of the lone pair from the adjacent nitrogen is not possible due to the geometric restrictions (Figure 7.33).

(Continues)

(Continued)

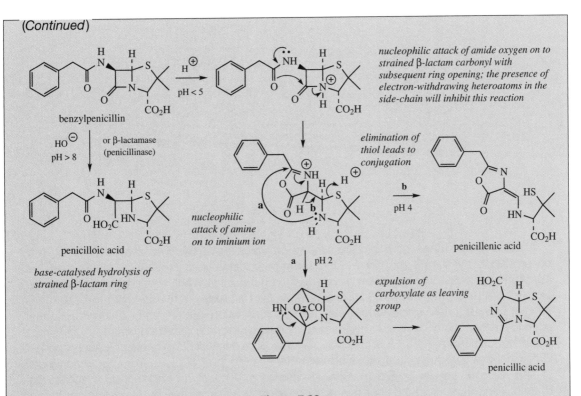

Figure 7.32

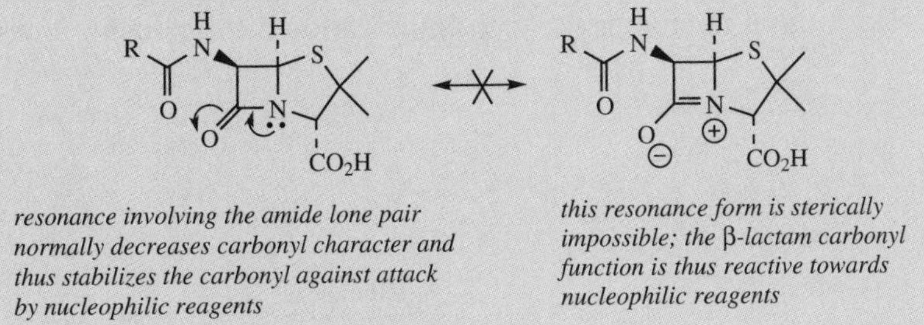

Figure 7.33

Supplementation of the fermentation medium with acids other than phenylacetic acid was used to provide structurally modified penicillins, though the scope was limited to series of monosubstituted acetic acids by the specificity of the fungal enzymes involved in the activation of the acids to their coenzyme A esters. The most important new penicillin produced was **phenoxymethylpenicillin (penicillin V)**, a result of adding phenoxyacetic acid to the culture (Figure 7.34). This new penicillin had the great advantage of being acid resistant, the introduction of an electron-withdrawing heteroatom into the side-chain inhibiting participation of the side-chain carbonyl in the reaction shown in Figure 7.32. Thus, penicillin V is suitable for oral administration, and still has particular value for respiratory tract infections and tonsillitis.

(Continues)

(*Continued*)

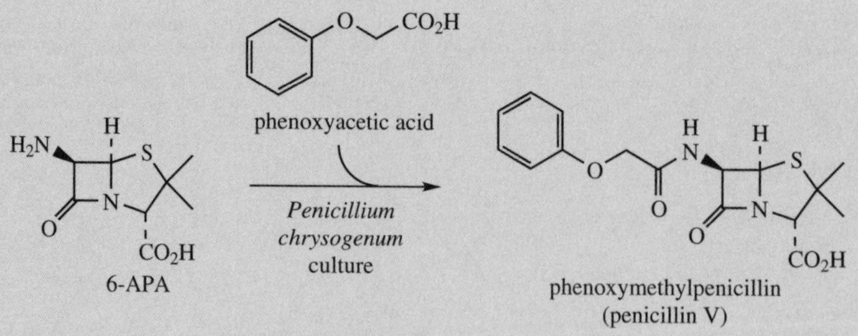

Figure 7.34

A much wider range of penicillins, many of which have become clinically useful, may be produced by semi-synthesis from 6-aminopenicillanic acid (6-APA). A multi-stage, but high-yielding, procedure has been developed to chemically hydrolyse a primary fermentation product like benzylpenicillin to 6-APA (Figure 7.35). This exploits the ability of the side-chain amide to adopt a resonance form, thus allowing conversion into an imidyl chloride, and then an imidyl ether, which is readily hydrolyzed. A new side-chain can then be added by simple esterification (Figure 7.35). Hydrolysis of penicillin G or penicillin V may also be accomplished enzymically in very high yield by using bacterial enzyme preparations from *Escherichia coli*, or species of *Fusarium* or *Erwinia*. Certain strains of *Penicillium chrysogenum* accumulate 6-aminopenicillanic acid, so that this compound may be produced by fermentation, though this is commercially less economic than the hydrolysis approach. Clinically useful penicillins produced by semi-synthesis or total synthesis are listed in Table 7.3. Penicillins with side-chains containing a basic amino group, e.g. **ampicillin** and **amoxicillin** (**amoxycillin**), are

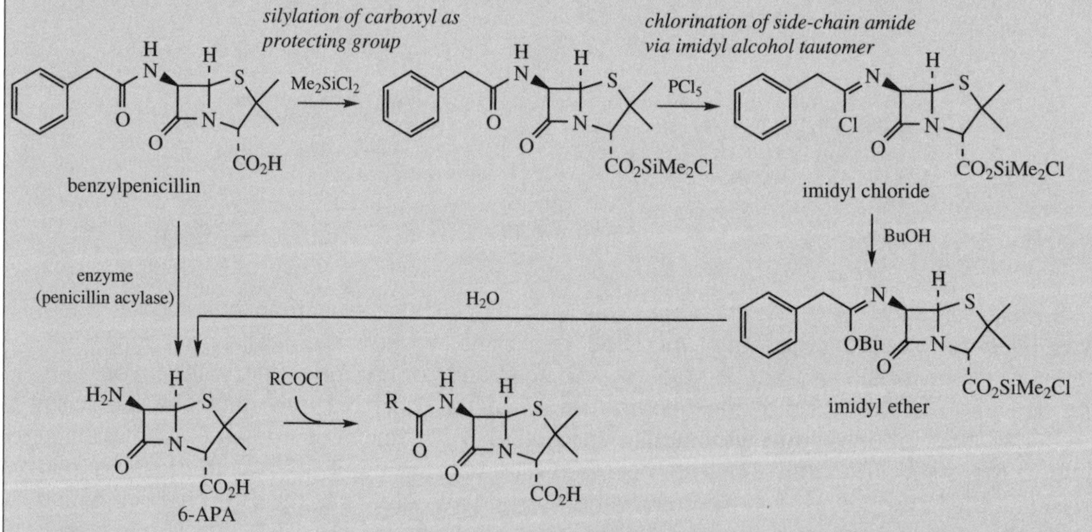

Figure 7.35

(*Continues*)

(Continued)

Table 7.3   Semi-synthetic and synthetic penicillins

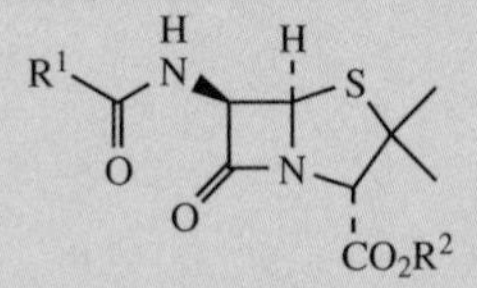

R¹	R²	Name	Notes
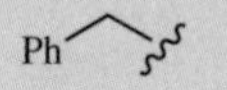 Ph (benzyl)	H	benzylpenicillin (penicillin G)	acid sensitive
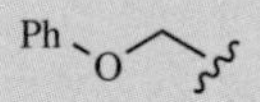 Ph—O—CH₂	H	phenoxymethyl-penicillin (penicillin V)	acid resistant
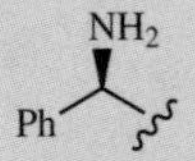 Ph—CH(NH₂)—	H	ampicillin	acid resistant, broad spectrum
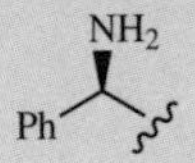 Ph—CH(NH₂)—	—CH₂OCOCMe₃	pivampicillin	pro-drug of ampicillin; better absorbed, and then hydrolysed by esterases
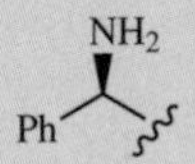 Ph—CH(NH₂)—	*RS* —CHMeOCO₂Et	bacampicillin	pro-drug of ampicillin; better absorbed, and then hydrolysed by esterases
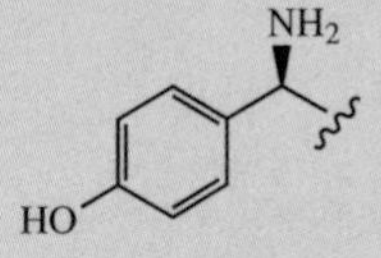 HO—C₆H₄—CH(NH₂)—	H	amoxicillin (amoxycillin)	acid resistant, broad spectrum; better absorption than ampicillin
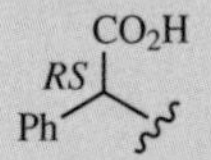 *RS* Ph—CH(CO₂H)—	H	carbenicillin	acid sensitive, broad spectrum; active against *Pseudomonas aeruginosa*
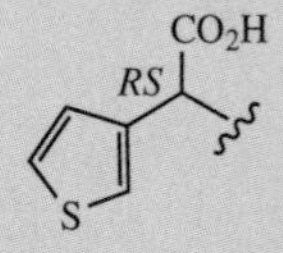 thienyl—CH(CO₂H)— *RS*	H	ticarcillin	acid sensitive, broad spectrum; 2 × more active than carbenicillin; used in combination with β-lactamase inhibitor clavulanic acid
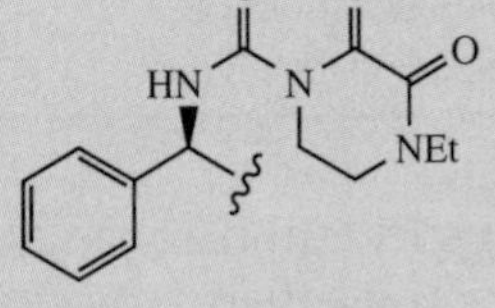 piperazinedione-Ph—CH—	H	piperacillin	broad spectrum, 8–10 × more active than carbenicillin against *P. aeruginosa*; used by injection for serious infections

(continued overleaf)

(Continued)

Table 7.3 (*continued*)

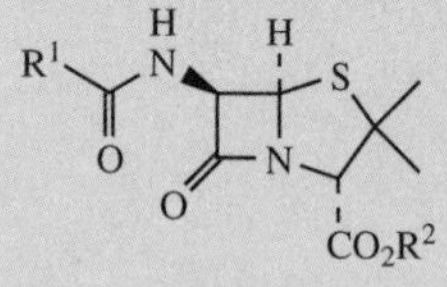

R^1	R^2	Name	Notes
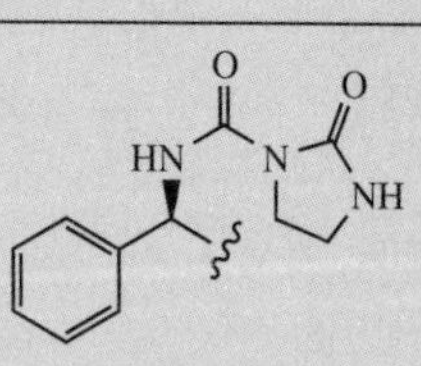	H	azlocillin	broad spectrum, more active than ticarcillin against *P. aeruginosa*; used by injection for serious infections
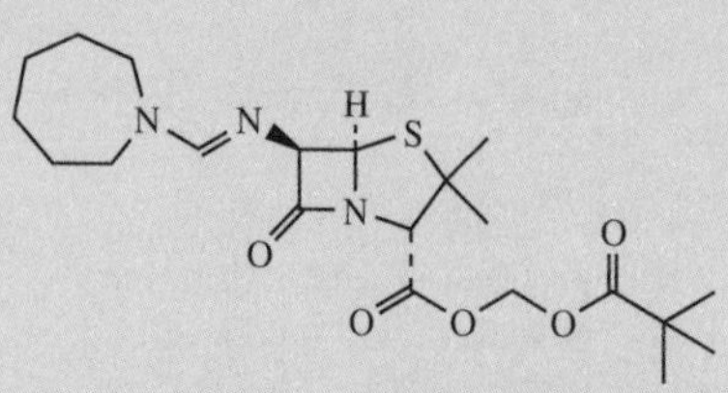		pivmecillinam	orally active pro-drug of mecillinam; β-lactamase sensitive, active against Gram-negative organisms (except *P. aeruginosa*)
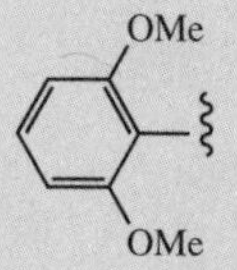	H	methicillin	β-lactamase resistant, acid sensitive
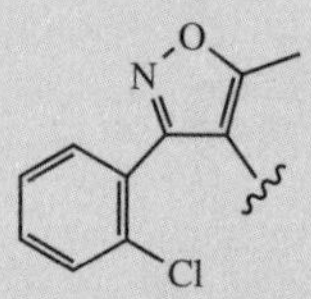	H	cloxacillin	β-lactamase and acid resistant
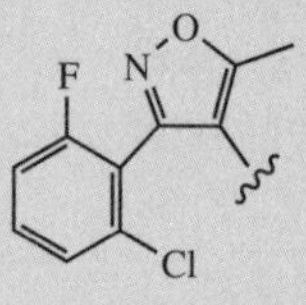	H	flucloxacillin	β-lactamase and acid resistant
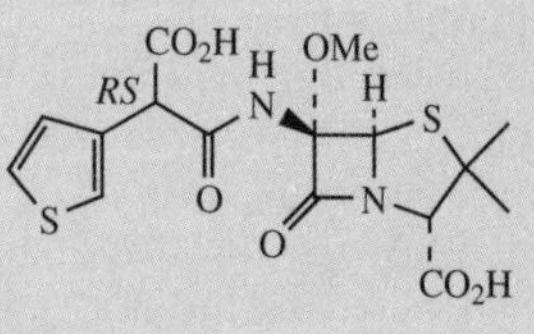		temocillin	β-lactamase resistant; active against Gram-negative (except *P. aeruginosa*) but not Gram-positive organisms

also acid resistant, since this nitrogen is protonated in preference to the lactam nitrogen. In addition, these agents were found to have a broader spectrum of activity than previous materials, in particular, activity against some Gram-negative bacteria which were not affected by penicillins G and V. The polar side-chain improves water-solubility and cell penetration into these microorganisms. Amoxicillin shows better oral absorption properties

(Continued)

than ampicillin, though this can also be achieved by using the pro-drugs **pivampicillin** and **bacampicillin**. These are acyloxymethyl esters through the thiazolidine carboxyl, and are hydrolysed to ampicillin by esterases in the gut. Broad spectrum activity is also found with penicillins containing a carboxyl group in the side-chain, e.g. **carbenicillin** and **ticarcillin**, and, although these compounds are still acid sensitive and are thus orally inactive, they demonstrate activity against pseudomonads, especially *Pseudomonas aeruginosa*. The activity of these agents is rather low, and it is necessary to inject very large doses. The acylureido penicillins **azlocillin** and **piperacillin** are much more active against *Pseudomonas aeruginosa*, and are also active against other Gram-negative bacteria such as *Klebsiella pneumoniae* and *Haemophilus influenzae*. **Pivmecillinam** is an acyloxymethyl ester pro-drug, and is hydrolysed to mecillinam after oral ingestion. It is unusual in being an amidino derivative instead of having an acyl side-chain; it has significant activity towards many Gram-negative bacteria.

Despite the dramatic successes achieved with the early use of penicillin antibiotics, it was soon realised that many bacteria previously susceptible to these agents were subsequently able to develop resistance. The principal mechanism of resistance lies in the ability of organisms to produce β-lactamase (penicillinase) enzymes capable of hydrolysing the β-lactam ring in the same manner as shown for the base-catalysed hydrolysis in Figure 7.32. Several distinct classes of bacterial β-lactamases are recognized, the main division being into serine enzymes and zinc enzymes. The former have an active site serine residue, which attacks the β-lactam carbonyl, forming an acyl–enzyme intermediate. On the basis of characteristic amino acid sequences they are then subdivided into three classes, A, C, and D. The zinc metallo-enzymes form class B, and appear to involve only non-covalently bound intermediates. Class A β-lactamases are the most common amongst pathogenic bacteria. Most staphylococci are now resistant to benzylpenicillin. The discovery of new penicillins that were not hydrolysed by bacterial β-lactamases was thus a major breakthrough. **Methicillin** (Table 7.3), though no longer used, was the first commercial β-lactamase-resistant penicillin, and the steric bulk of the side-chain appears to contribute to this valuable property, hindering the approach of β-lactamase enzymes. Methicillin is acid sensitive, since it lacks an electron-withdrawing side-chain, but other penicillins were developed that combined bulk and electron-withdrawing properties, and could thus be used orally. These include a group of isoxazole derivatives termed the oxacillins, of which **cloxacillin** and **flucloxacillin** are first-choice agents against penicillin-resistant *Staphylococcus aureus*. **Temocillin** also has excellent resistance to β-lactamases as well as high activity towards Gram-negative organisms. It differs from all the other penicillins described in possessing a 6α-methoxyl group (compare the cephamycins, page 450). Another way of overcoming the penicillin-degrading effects of β-lactamase is to combine a β-lactamase-sensitive agent, e.g. amoxicillin or ticarcillin, with clavulanic acid (see page 452), which is a specific inhibitor of β-lactamase. Other mechanisms of resistance which have been encountered include modification of the binding sites on penicillin-binding proteins (see below), thus reducing their affinity for the penicillin, and decreased cell permeability, leading to reduced uptake of the antibiotic. Strains of *Staphylococcus aureus* resistant to both methicillin and isoxazolylpenicillins, e.g. cloxacillin and flucloxacillin, are known to have modified and insensitive penicillin-binding proteins; such strains are termed methicillin-resistant *Staphylococcus aureus* (MRSA).

Penicillins and other β-lactam drugs exert their antibacterial effects by binding to proteins (penicillin-binding proteins) that are involved in the late stages of the biosynthesis of

(Continues)

(Continued)

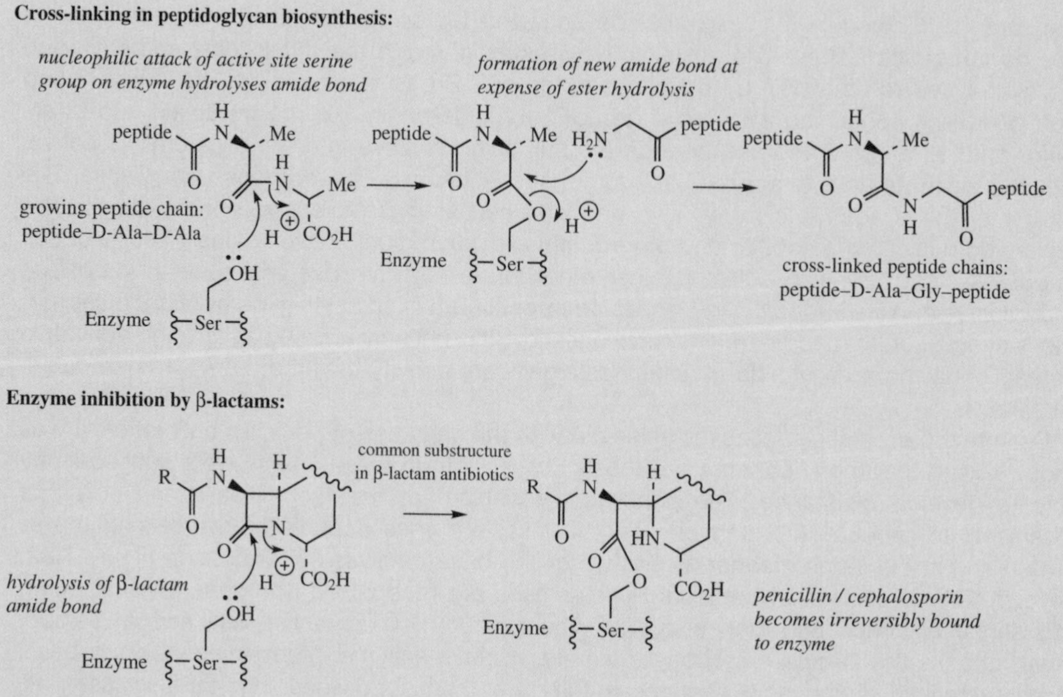

Cross-linking in peptidoglycan biosynthesis:

Figure 7.36

the bacterial cell wall. Cross-linking of the peptidoglycan chains which constitute the bacterial cell wall (see page 473) involves an acyl-D-Ala–D-Ala intermediate, which in its transition state conformation closely resembles the penicillin molecule (Figure 7.36). As a result, the penicillin occupies the active site of the enzyme, and becomes bound via an active site serine residue, this binding causing irreversible enzyme inhibition, and cessation of cell wall biosynthesis. Growing cells are then killed due to rupture of the cell membrane and loss of cellular contents. This binding reaction between penicillin-binding proteins and penicillins is chemically analogous to the action of β-lactamases, but in the latter case penicilloic acid is subsequently released from the β-lactamase, which can continue to function. The penicillins are very safe antibiotics for most individuals. The bacterial cell wall has no counterpart in mammalian cells, and the action is thus very specific. However, a significant proportion of patients can experience allergic responses ranging from a mild rash to fatal anaphylactic shock. Cleavage of the β-lactam ring through nucleophilic attack of an amino group in a protein is believed to lead to the formation of antigenic substances causing the allergic response.

supplying different acids. The new amide link may be achieved in two ways. Hydrolysis of isopenicillin N releases the amine **6-aminopenicillanic acid** (6-APA), which can then react with the coenzyme A ester. Alternatively, an acyltransferase enzyme converts isopenicillin N into penicillin G directly, without 6-aminopenicillanic acid actually being released from the enzyme.

Cephalosporins

The **cephalosporins***, e.g. **cephalosporin C** (Figure 7.37), are a penicillin-related group of antibiotics having a β-lactam-dihydrothiazine ring system, and are produced by species of *Cephalosporium*. The six-membered dihydrothiazine ring is produced from the five-membered

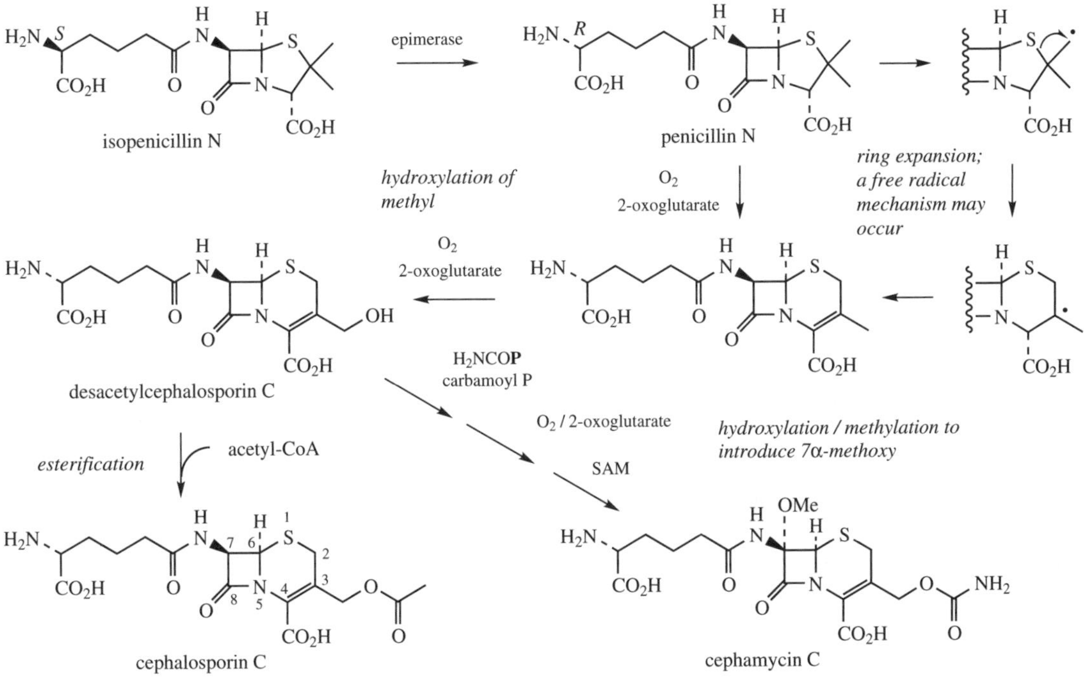

Figure 7.37

thiazolidine ring of the penicillin system by an oxidative process of ring expansion, incorporating one of the methyl groups. The pathway (Figure 7.37) diverges from that to penicillins at **isopenicillin N**, which is first epimerized in the α-aminoadipyl side-chain to give **penicillin N**. Ring expansion then occurs, incorporating one of the methyls into the heterocyclic ring, though the mechanism for this is not clearly defined. A free radical mechanism is suggested

in Figure 7.37 to rationalize the transformation. Hydroxylation of the remaining methyl gives desacetylcephalosporin C. **Cephalosporin C** is the acetyl ester of this, whilst a further group of antibiotics termed the **cephamycins**[*] are characterized by a 7α-methoxy group, and are produced by hydroxylation/methylation, and, in the case of **cephamycin C**, introduction of a carbamate group from carbamoyl phosphate on to the hydroxymethyl function.

Cephalosporins

Cephalosporin C (Figure 7.37) is produced commercially by fermentation using cultures of a high-yielding strain of *Acremonium chrysogenum* (formerly *Cephalosporium acremonium*). Initial studies of the antibiotic compounds synthesized by *C. acremonium* identified penicillin N (originally called cephalosporin N) as the major component, with small amounts of cephalosporin C. In contrast to the penicillins, cephalosporin C was stable under acidic conditions and also was not attacked by penicillinase (β-lactamase). Antibacterial activity was rather low, however, and the antibiotic was poorly absorbed after oral administration. However, the structure offered considerable scope for side-chain modifications, more so than with the penicillins since it has two side-chains, and this has led to a wide variety of cephalosporin drugs, many of which are currently in clinical use. As with the penicillins, removal of the amide

(Continues)

(Continued)

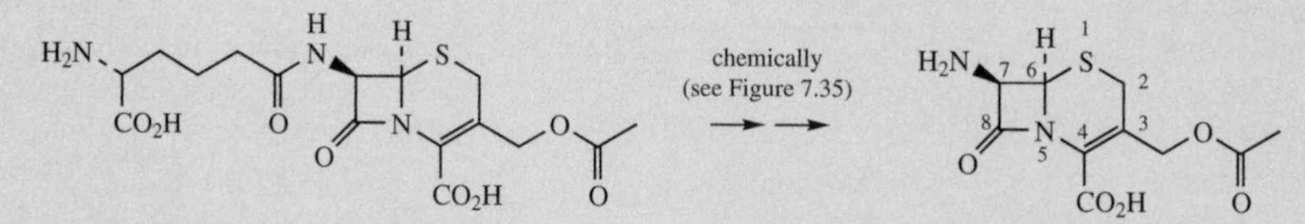

Figure 7.38

side-chain by the hydrolysis of cephalosporin C to 7-aminocephalosporanic acid (7-ACA) (Figure 7.38) was the key to semi-synthetic modifications, and this may be achieved chemically by the procedure used for the penicillins (compare Figure 7.35). Removal of this side-chain by suitable microorganisms or enzymes has proved elusive. The ester side-chain at C-3 may be hydrolysed enzymically by fermentation with a yeast, or, alternatively, the acetoxy group is easily displaced by nucleophilic reagents. It is also possible to convert readily available benzylpenicillin into the deacetoxy derivative of 7-ACA through a chemical ring expansion process and enzymic removal of the side-chain.

The semi-synthetic cephalosporins may be classified according to chemical structure, antibacterial spectrum, or β-lactamase resistance, but in practice, they tend to be classified by a more abitrary system, dividing them into 'generations' (Table 7.4). Note that all the cephalosporin antibiotics begin with the prefix *ceph-* or *cef-*, the latter spelling now being preferred, though both spellings are still encountered for some drugs. The classification into generations is based primarily on the antibacterial spectrum displayed by the drugs but it is also more or less related to the year of introduction. However, drugs in the second generation may have been introduced after the third generation of drugs had been established. There is no intention to suggest that third generation drugs automatically supersede second and first generation drugs, and, indeed, agents from all generations are still currently used. First generation cephalosporins, e.g. **cefalotin (cephalothin)**, **cefalexin (cephalexin)**, **cefradine (cephradine)**, **cefadroxil**, and **cefazolin (cephazolin)** have good activity against Gram-positive bacteria but low activity against Gram-negative organisms. They have comparable activity to ampicillin, and are effective against penicillinase-producing *Staphylococcus*. However, another β-lactamase (cephalosporinase) developed that inactivated these agents. Cefalotin, the first modified cephalosporin to be marketed, is poorly absorbed from the gut, and is thus not orally active. However, cefalexin, cefradine, and cefadroxil may be administered orally, a property that appears to be related to the 3-methyl side-chain. Second generation cephalosporins show a broader spectrum of activity, and are more active against aerobic Gram-negative bacteria like *Haemophilus influenzae* and *Neisseria gonorrhoeae*. This group of antibiotics includes **cefaclor**, **cefuroxime**, and **cefamandole (cephamandole)**, and in general displays better resistance to β-lactamases that inactivated first generation cephalosporins. The third generation of cephalosporin antibiotics, e.g. **cefotaxime**, **ceftazidime**, **ceftizoxime**, **cefodizime**, and **ceftriaxone**, have an extended Gram-negative spectrum, and are most active against enteric Gram-negative bacilli, but may be less active against some Gram-positive bacteria, especially *Staphylococcus aureus*. Many of the third generation cephalosporins are characterized by an aminothiazole ring on the amide side-chain, which appears to impart the high activity against Gram-negative bacteria. The *O*-substituted oxime group also improves potency, and confers resistance to β-lactamases. The oximes with *syn* stereochemistry as shown in Table 7.4 are considerably more active than the

(Continued)

Table 7.4   Cephalosporin antibiotics

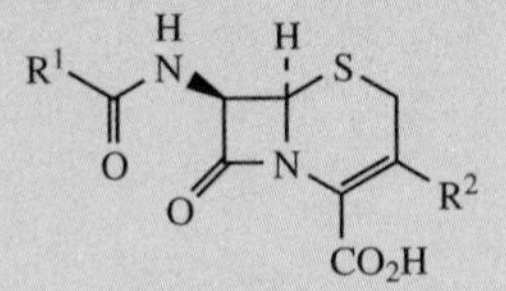

R^1	R^2	Name	Notes
First generation			
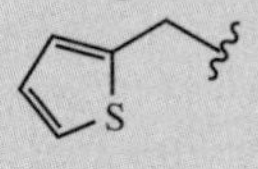 (thiophene-CH₂)	ξ—CH₂OCOMe	cefalotin (cephalothin)	R^2 group unstable to mammalian esterases; generally superseded
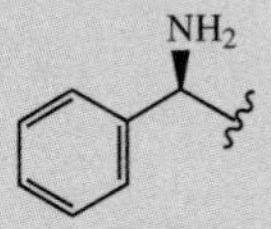 (phenyl, NH₂)	ξ—CH₃	cefalexin (cephalexin)	orally active
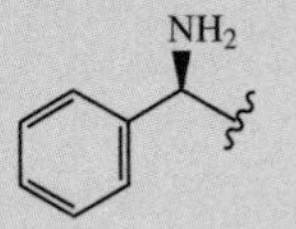 (cyclohexadienyl, NH₂)	ξ—CH₃	cefradine (cephradine)	orally active; generally superseded
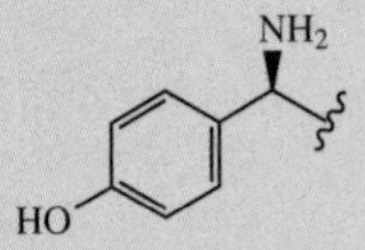 (HO-phenyl, NH₂)	ξ—CH₃	cefadroxil	orally active
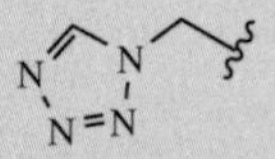 (tetrazolyl-CH₂)	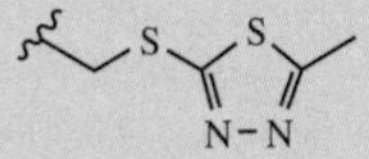 (—CH₂—S—thiadiazole-Me)	cefazolin (cephazolin)	generally superseded
Second generation			
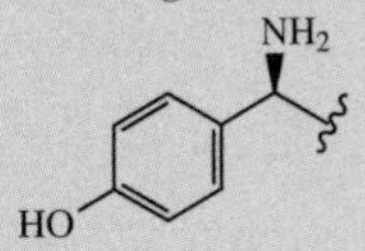 (HO-phenyl, NH₂)	ξ—Cl	cefaclor	orally active
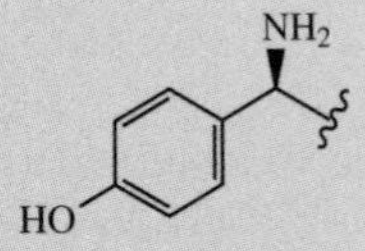 (HO-phenyl, NH₂)	ξ—CH=CH—CH₃	cefprozil	orally active
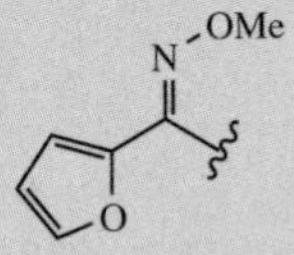 (furanyl, N—OMe)	ξ—CH₂OCONH₂	cefuroxime	high resistance to β-lactamases; resistant to mammalian esterases
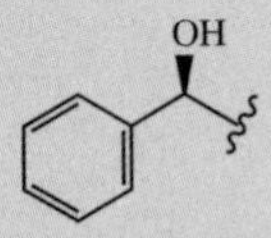 (phenyl, OH)	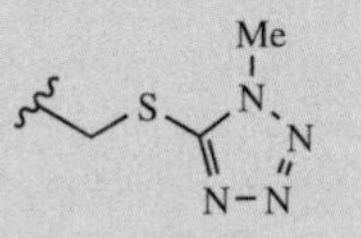 (—CH₂—S—tetrazole-Me)	cefamandole (cephamandole)	high resistance to β-lactamases

(continued overleaf)

(Continues)

(Continued)

Table 7.4 (*continued*)

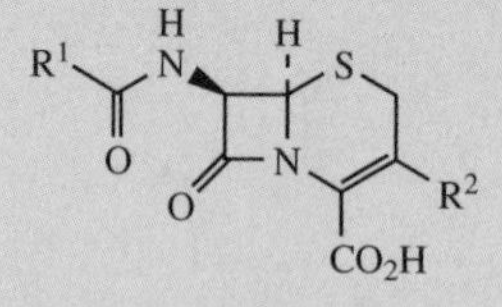

R^1	R^2	Name	Notes
Third generation			
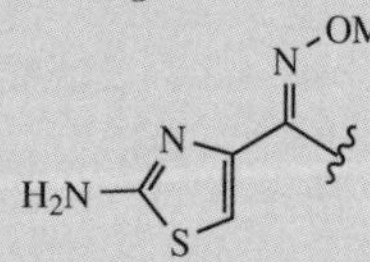	$-CH_2OCOMe$	cefotaxime	unstable to mammalian esterases, but desacetyl metabolite still has considerable antimicrobial activity
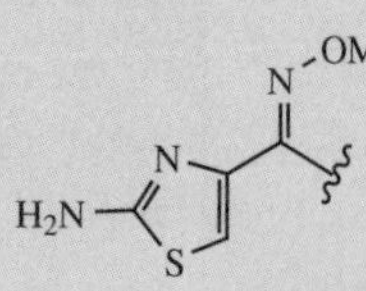	$-H$	ceftizoxime	broad-spectrum Gram-negative activity; generally superseded
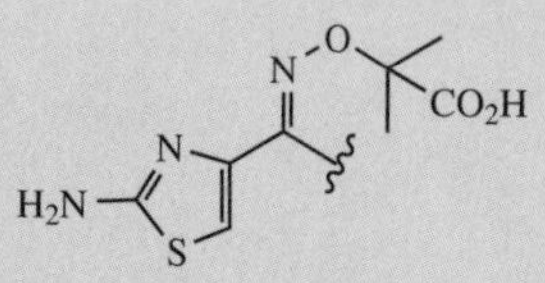	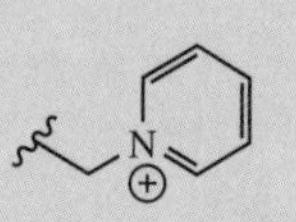	ceftazidime	broad-spectrum Gram-negative activity; good activity towards *Pseudomonas*
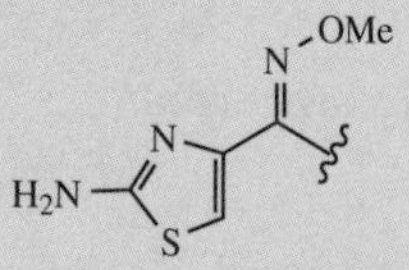	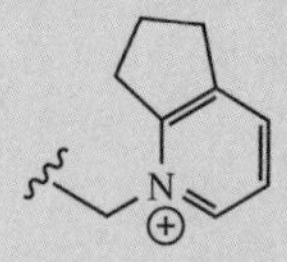	cefpirome	
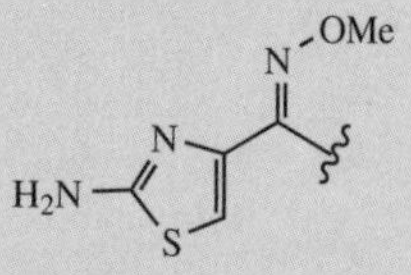	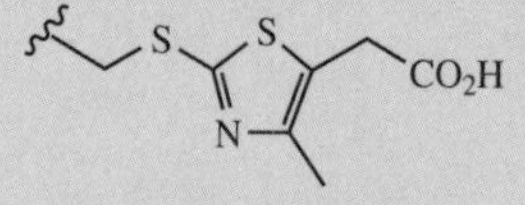	cefodizime	broad-spectrum Gram-negative activity
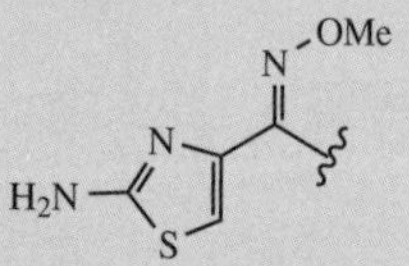	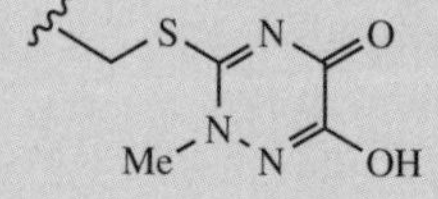	ceftriaxone	broad-spectrum Gram-negative activity; longer half-life than other cephalosporins
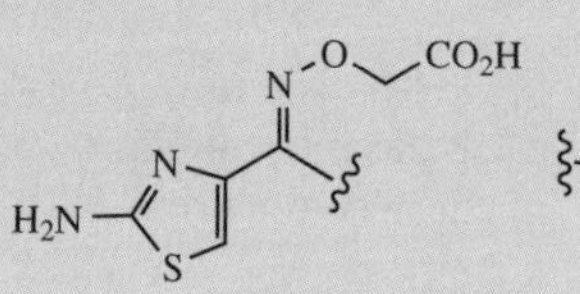	$-CH=CH_2$	cefixime	2nd/3rd generation, orally active; long duration of action
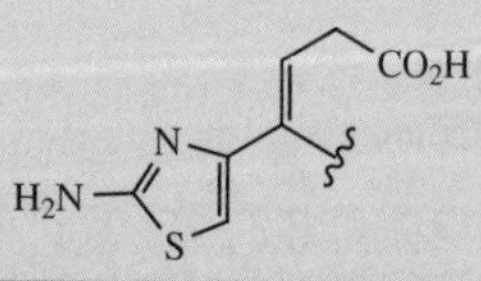	$-H$	ceftibuten	2nd/3rd generation, orally active; generally superseded

(*continued*)

(Continues)

(Continued)

Table 7.4 (continued)

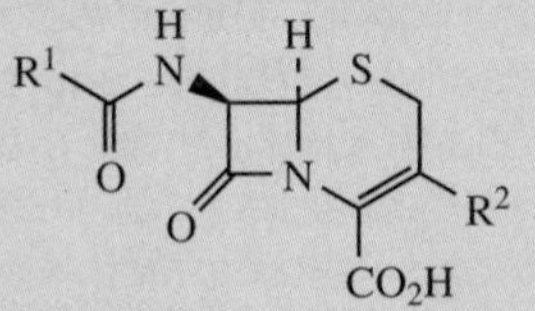

R¹	R²	Name	Notes
Pro-drugs			
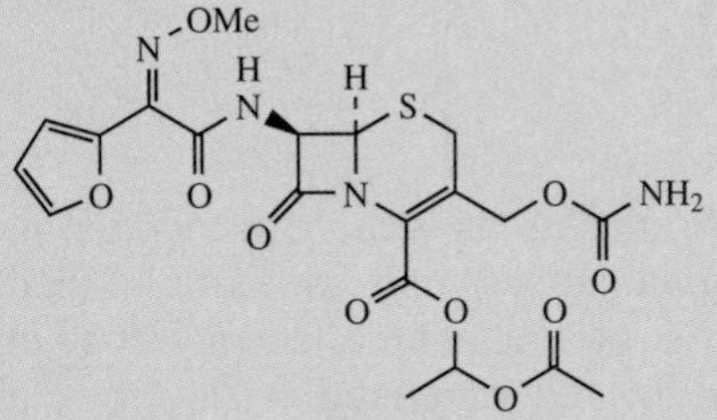		cefuroxime-axetil	2nd generation, orally active; hydrolysed by esterases to liberate cefuroxime
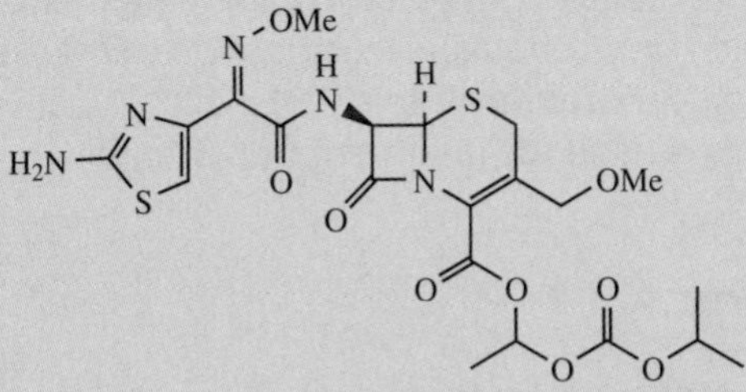		cefpodoxime-proxetil	2nd/3rd generation, orally active; hydrolysed by esterases to liberate cefpodoxime

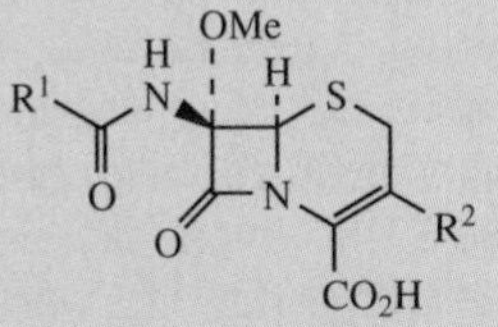

R¹	R²	Name	Notes
Cephamycins			
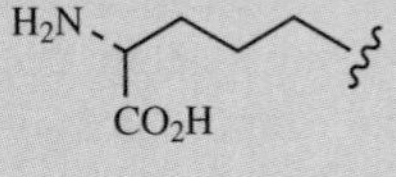	$-CH_2OCONH_2$	cephamycin C	
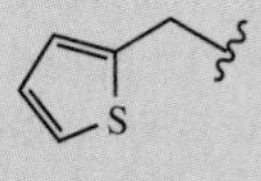	$-CH_2OCONH_2$	cefoxitin	stable to β-lactamases and mammalian esterases
Carbacephems			
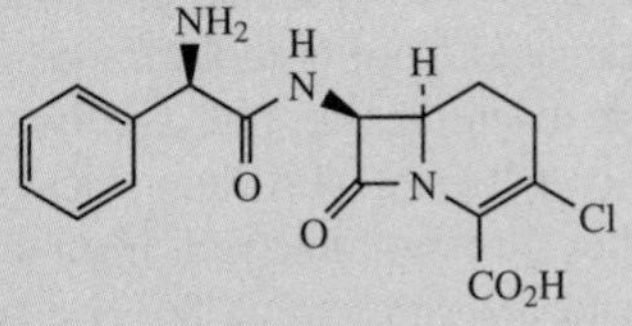		loracarbef	improved chemical stability, longer half-life, better oral bioavailability compared to cefaclor

(Continues)

(Continued)

anti isomers. A disadvantage of many of the current cephalosporin drugs is that they are not efficiently absorbed when administered orally. This is to some extent governed by the nature of the side-chain on C-3. The orally active **cefixime** and **ceftibuten** have spectra of activity between the second and third generations. Orally active pro-drugs such as **cefuroxime-axetil** and **cefpodoxime-proxetil** have been developed with an additional ester function on the C-4 carboxyl. These compounds are hydrolysed to the active agents by esterases.

Cephalosporin antibiotics are especially useful for treating infections in patients who are allergic to penicillins. Hypersensitivity to cephalosporins is much less common, and only about 5–10% of penicillin-sensitive patients will also be allergic to cephalosporins.

Cephamycins

The antibiotic cephamycin C (Figure 7.37) was isolated from *Streptomyces clavuligerus* and shown to have a 7α-methoxy group on the basic cephalosporin ring system. Although cephamycin C and other natural cephamycins have only weak antibacterial activity, they are resistant to β-lactamase hydrolysis, a property conferred by the increased steric crowding due to the additional methoxy group. Semi-synthetic analogues have been obtained either by modification of the side-chains of natural cephamycins, or by chemical introduction of the 7α-methoxyl. Currently, the only cephamycin in general use is **cefoxitin** (Table 7.4), which is active against bowel flora including *Bacterioides fragilis*, and is used for treatment of peritonitis.

Carbacephems

Although cephalosporin analogues where the sulphur heteroatom has been replaced with carbon are not known naturally (contrast carbapenems, natural penicillin analogues, page 451), synthetically produced carbacephems have shown good antibacterial activity with considerably improved chemical stability over cephalosporins. The first of these to be produced for drug use is **loracarbef** (Table 7.4), which has similar antibacterial activity to cefaclor, but considerably greater stability, a longer half-life, and better oral bioavailability.

Other β-Lactams

The fused β-lactam skeletons found in penicillins and cephalosporins are termed **penam** and **cephem** respectively (Figure 7.39). Other variants containing the basic β-lactam ring system are also found in nature. Of especial importance is the **clavam** (Figure 7.39) or oxapenam fused ring system typified by **clavulanic acid*** (Figure 7.40) from *Streptomyces clavuligerus*. The weak antibacterial activity of clavulanic acid is unimportant, for this compound is valuable as an efficient inhibitor of β-lactamases from both Gram-positive and Gram-negative bacteria. Despite the obvious structural similarity between clavulanic acid and the penicillins, they are not derived from common precursors, and there are some novel aspects associated with clavulanic acid biosynthesis. All

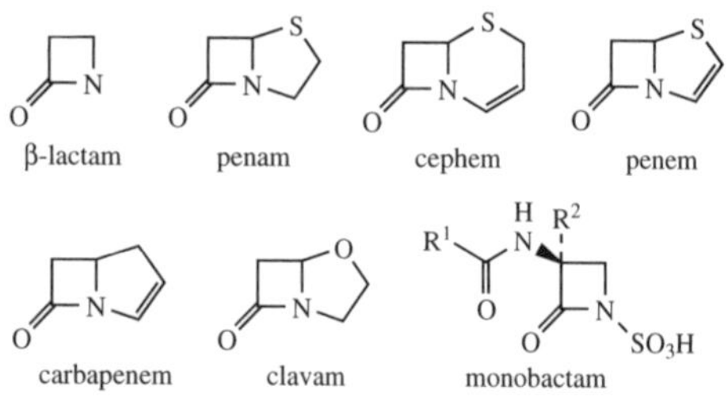

Figure 7.39

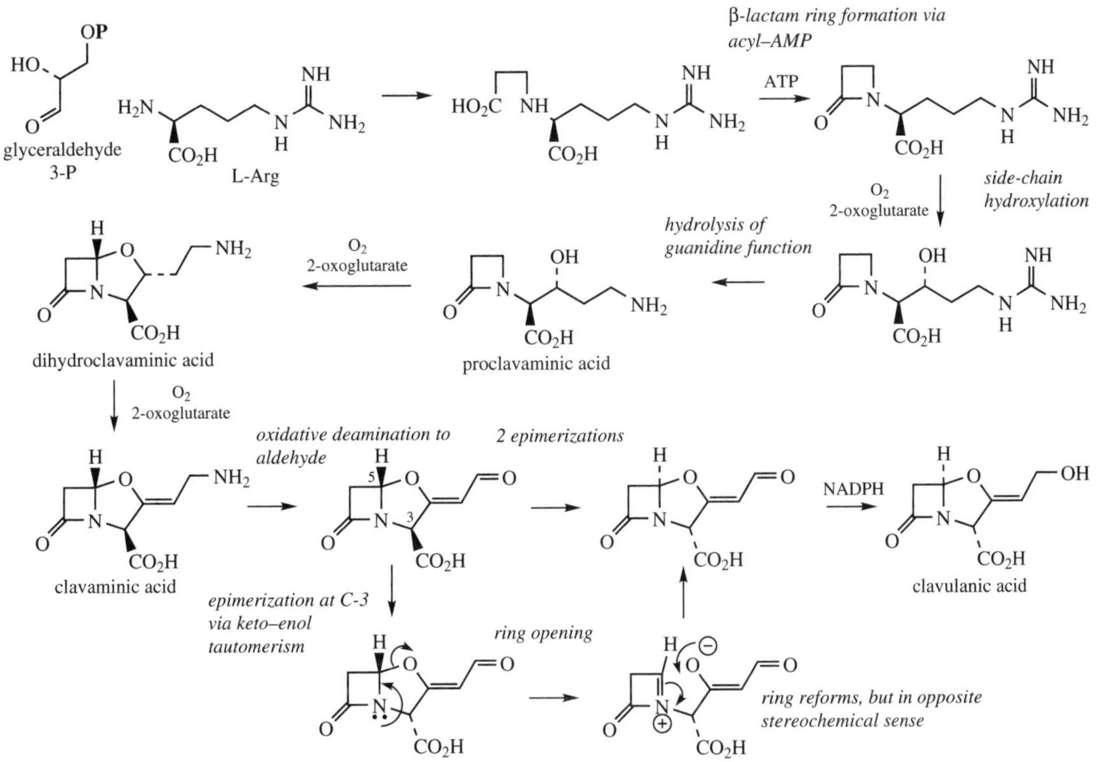

Figure 7.40

carbons are provided by two precursors, arginine and most likely glyceraldehyde 3-P. Glyceraldehyde 3-P supplies the β-lactam carbons whilst the α-amino group of arginine provides the β-lactam nitrogen. In contrast to the penicillins, the β-lactam ring is formed via an acyl-AMP activated intermediate. The sequence of reactions shown in Figure 7.40 leads to the monocyclic β-lactam **proclavaminic acid**, which is the substrate for oxidative cyclization to provide the oxazolidine ring then dehydrogenation to **clavaminic acid**. The three 2-oxoglutarate-dependent oxidations in this sequence are all catalysed by a single enzyme. The final transformation into **clavulanic acid** requires oxidative deamination of the terminal amine to an aldehyde and then reduction to an alcohol, but, more intriguingly, the two chiral centres C-3 and C-5 have to be epimerized. These reactions have yet to be fully clarified. The unsaturated aldehyde is easily susceptible to inversion of configuration at C-3 via keto–enol tautomerism, but a change

of chirality at C-5 (which actually retains the C-5 hydrogen) must invoke opening of the oxazolidine ring (Figure 7.40).

Another variant on the penicillin penam ring system is found in a group of compounds termed **carbapenems*** (Figure 7.39), where the sulphur heteroatom has been replaced by carbon. This is exemplified by **thienamycin** (Figure 7.43), an antibiotic isolated from cultures of *Streptomyces cattleya*. The sulphur-containing side-chain is a feature of many of the natural examples, and this may feature sulphide or sulphone moieties, e.g. the **olivanic acids** (Figure 7.44) from *S. olivaceus*. The olivanic acids are potent β-lactamase inhibitors, especially towards the cephalosporinases, which are poorly inhibited by clavulanic acid. Only the broad outlines of the pathways to carbapenems are established (Figure 7.43). The fundamental ring system is derived from glutamic acid and acetate. It is suggested that an activated form of glutamic acid, e.g. γ-glutamyl phosphate, is chain

Clavulanic Acid

Clavulanic acid (Figure 7.38) is produced by cultures of *Streptomyces clavuligerus*, the same actinomycete that produces cephamycin C. Although it has only weak antibacterial activity, it is capable of reacting with a wide variety of β-lactamase enzymes, opening the β-lactam ring, in a process initially analogous to that seen with penicillins (Figure 7.41). However, binding to the enzyme is irreversible, and the β-lactamase is inactivated. It seems likely that the side-chain with its double bond may contribute to this effect, causing further ring opening to give intermediates which react at the active site of the enzyme (Figure 7.41). Thus, dihydroclavulanic acid is no longer an enzyme inhibitor. Clavulanic acid is usually combined with a standard penicillin, e.g. with amoxicillin (as **co-amoxyclav**) or with ticarcillin, to act as a suicide substrate and provide these agents with protection against β-lactamases, thus extending their effectivity against a wider range of organisms. Similar success has been achieved using the semi-synthetic penicillanic acid sulphone, **sulbactam** (Figure 7.42), which is also a potent inhibitor of β-lactamase by a similar double ring opening mechanism (Figure 7.42). Sulbactam was used in combination with ampicillin, but has been superseded by the related sulphone **tazobactam** (Figure 7.42), which is a β-lactamase inhibitor used in combination with piperacillin.

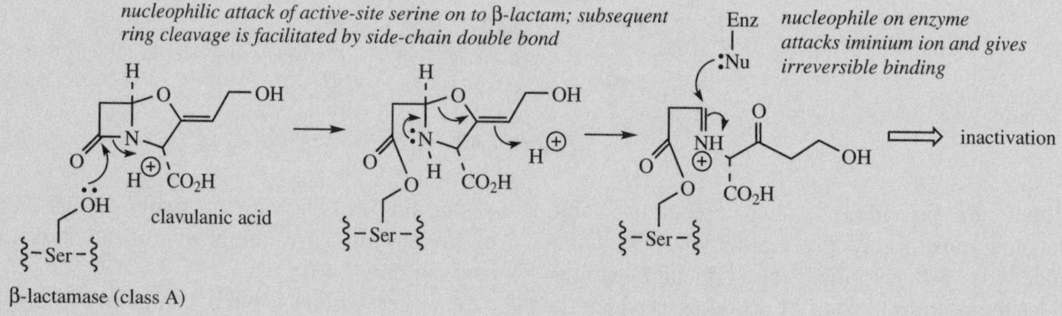

Figure 7.41

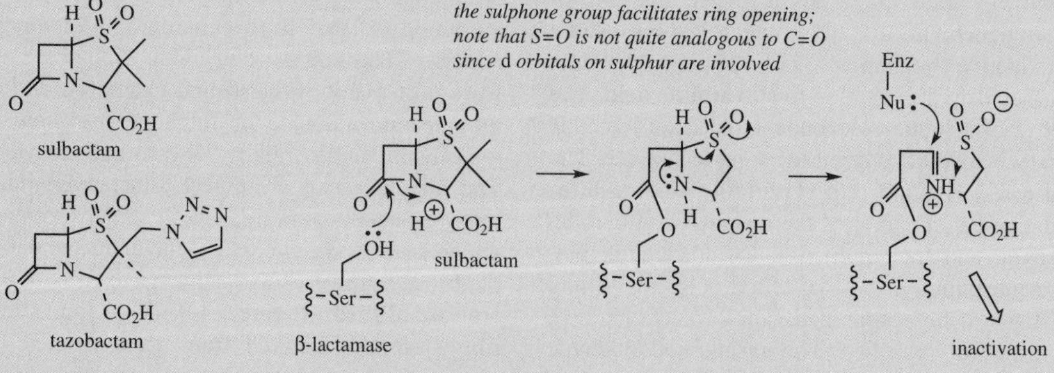

Figure 7.42

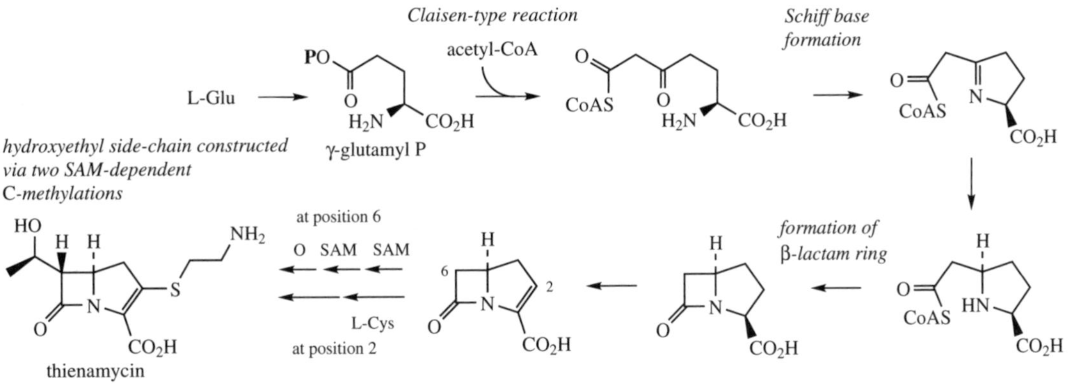

Figure 7.43

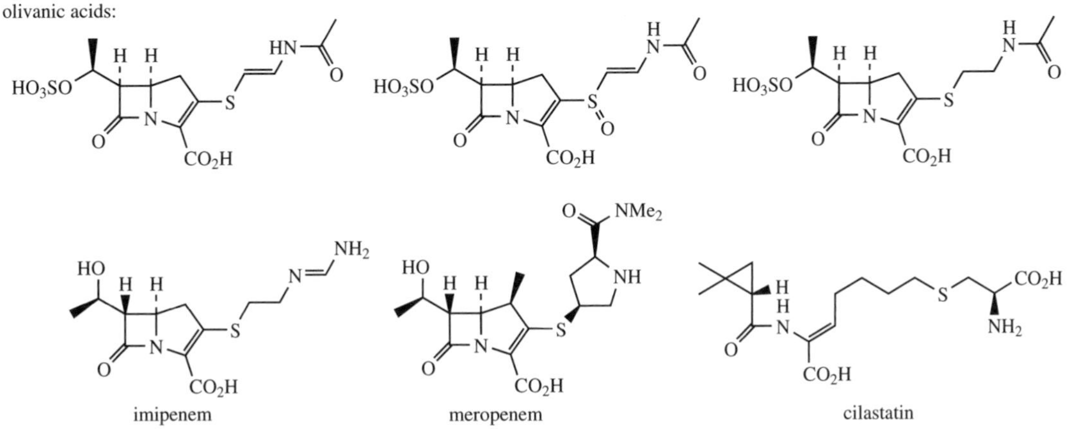

Figure 7.44

extended with acetyl-CoA, then cyclized to the imine (Figure 7.43). Reduction and β-lactam formation lead to the carbapenam, which yields the carbapenem by dehydrogenation. This material is modified to produce thienamycin by insertion of side-chains at C-2 and C-6; the detailed sequence has yet to be established. Both carbons of the two-carbon side-chain at C-2 are known to be derived from methionine, the result of a double methylation sequence reminiscent of the processes used for alkylating the side-chains of sterols (see page 254). The C-2 cysteaminyl side-chain is supplied by the amino acid cysteine.

The simple nonfused β-lactam ring is encountered in a number of natural structures, such as the **nocardicins**, e.g. nocardicin A (Figure 7.45), from *Nocardia uniformis*, and the so-called **monobactams*** (monocyclic β-lactams) based on the structure shown in Figure 7.39, which was

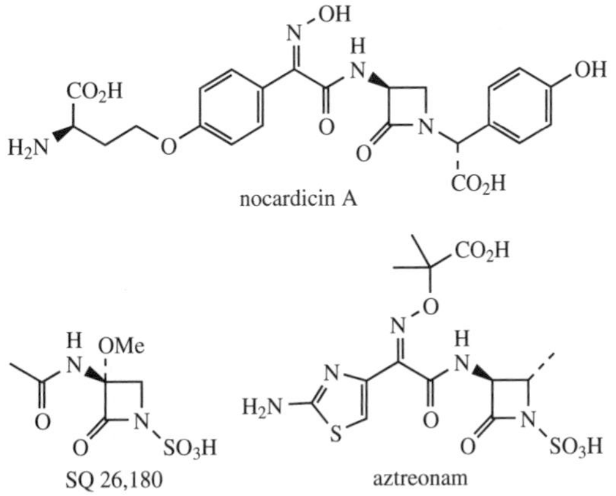

Figure 7.45

isolated from strains of bacteria. The simplest of these monobactams is the compound referred to by its research coding SQ 26,180 (Figure 7.45)

Carbapenems

Thienamycin (Figure 7.44) is produced by cultures of *Streptomyces cattleya*, but in insufficient amounts for commercial use. This compound is thus obtained by total synthesis. In addition, thienamycin is relatively unstable, its side-chain primary amino group reacting as a nucleophile with other species, including the β-lactam group in other molecules. For drug use, thienamycin is converted into its more stable *N*-formimidoyl derivative **imipenem** (Figure 7.44). Imipenem has a broad spectrum of activity which includes activity towards many aerobic and anaerobic Gram-positive and Gram-negative bacteria, it is resistant to hydrolysis by most classes of β-lactamases, and it also possesses β-lactamase inhibitory activity. However, it is partially inactivated by dehydropeptidase in the kidney, and is thus administered in combination with cilastatin (Figure 7.45), a specific inhibitor of this enzyme. The newer carbapenem **meropenem** (Figure 7.45) is stable to dihydropeptidase and can be administered as a single agent. It has good activity against all clinically significant aerobes and anaerobes, except MRSA (methicillin-resistant *S. aureus*) and *Enterococcus faecium*, and stability to serine-based β-lactamases.

discovered in *Chromobacterium violaceum*. Whilst many of the natural examples show a 3α-methoxyl (corresponding to the 7α-methoxyl in the cephamycins), the prominent feature in the monobactams is the *N*-sulphonic acid grouping. The fundamental precursor of the β-lactam ring in the monobactams is serine (Figure 7.46). In the simple structures, the β-lactam nitrogen presumably arises from ammonia, and cyclization occurs by displacement of the hydroxyl, in suitably activated form. The *N*-sulphonate function comes from sulphate. For the more complex structures such as the nocardicins, evidence points to a polypeptide origin with considerable parallels with penicillin and cephalosporin biosynthesis. The postulated D,L,D-tripeptide is formed from L-serine and two molecules of L-4-hydroxyphenylglycine, the latter having its origins in L-tyrosine (Figure 7.47). The inversion of configurations probably occurs during polypeptide formation (compare penicillins, page 438). The remaining portion in the carbon skeleton of nocardicin A is ether-linked to a 4-hydroxyphenylglycine unit, and actually derives from L-methionine. This probably involves an S_N2 displacement on SAM as with simple methylation reactions, though attack must be on the secondary rather than primary centre (Figure 7.47). Again, the configuration of the L-amino acid is inverted at a late stage, as with the isopenicillin N to penicillin N conversion in cephalosporin biosynthesis (see page 445).

Monobactams

The naturally occurring monobactams show relatively poor antibacterial activity, but alteration of the side-chain as with penicillins and cephalosporins has produced many potent new compounds. Unlike those structures, the nonfused β-lactam ring is readily accessible by synthesis, so all analogues are produced synthetically. The first of these to be used clinically is **aztreonam** (Figure 7.45), which combines the side-chain of the cephalosporin ceftazidime (Table 7.4) with the monobactam nucleus. Aztreonam is very active against Gram-negative bacteria, including *Pseudomonas aeruginosa*, *Haemophilus influenzae*, and *Neisseria meningitidis*, but has little activity against Gram-positive organisms. It also displays a high degree of resistance to enzymatic hydrolysis by most of the common β-lactamases. Oral absorption is poor, and this drug is administered by injection.

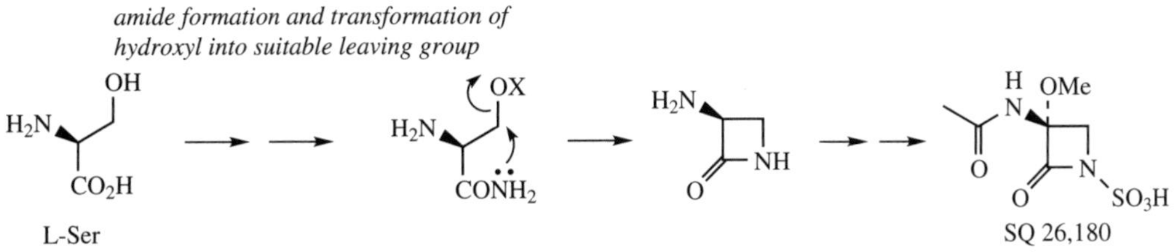

Figure 7.46

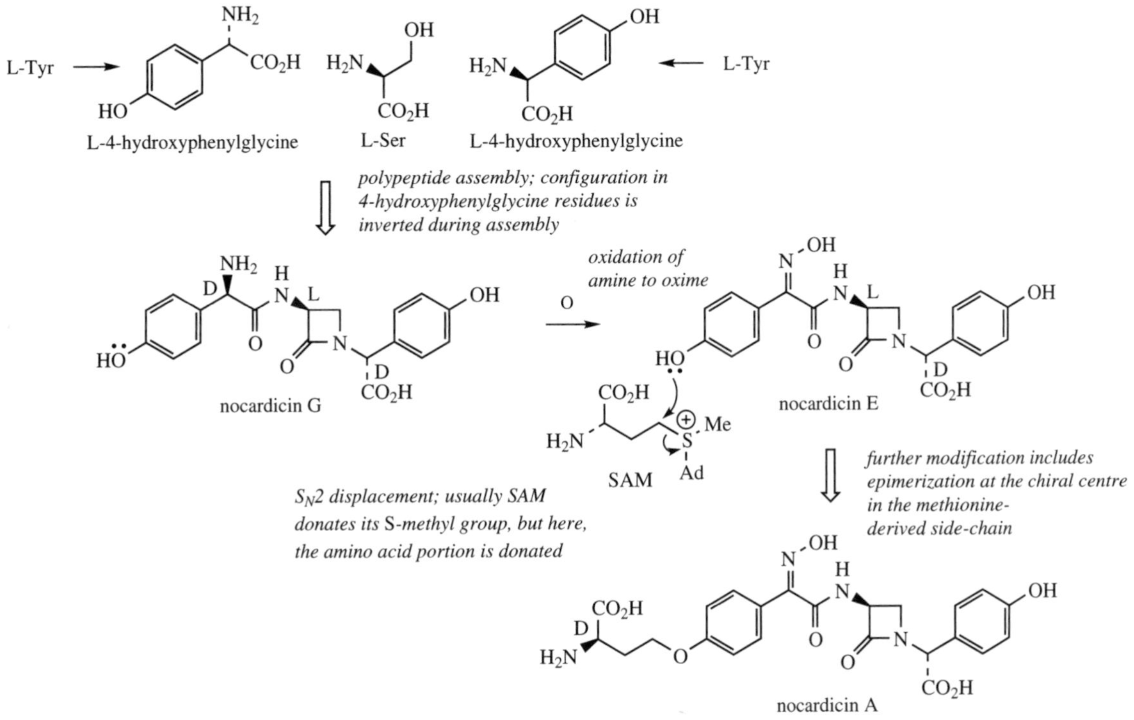

Figure 7.47

CYANOGENIC GLYCOSIDES

Cyanogenic glycosides are a group of mainly plant-derived materials, which liberate hydrocyanic acid (HCN) on hydrolysis, and are thus of concern as natural toxicants. The group is exemplified by **amygdalin** (Figure 7.48), a constituent in the kernels of bitter almonds (*Prunus amygdalus* var. *amara*; Rosaceae) and other *Prunus* species such as apricots, peaches, cherries, and plums. When plant tissue containing a cyanogenic glycoside is crushed, glycosidase enzymes also in the plant but usually located in different cells

are brought into contact with the glycoside and begin to hydrolyse it. Thus, amygdalin is hydrolysed sequentially by β-glucosidase-type enzymes to **prunasin** and then **mandelonitrile**, the latter compound actually being the cyanohydrin of benzaldehyde (Figure 7.48). Mandelonitrile is then hydrolysed to its component parts, benzaldehyde and toxic HCN, by the action of a further enzyme. The kernels of bitter almonds, which contain amygdalin and the hydrolytic enzymes, are thus potentially toxic if ingested, whilst kernels of sweet almonds (*Prunus amygdalus* var. *dulcis*) are not toxic, containing the enzymes

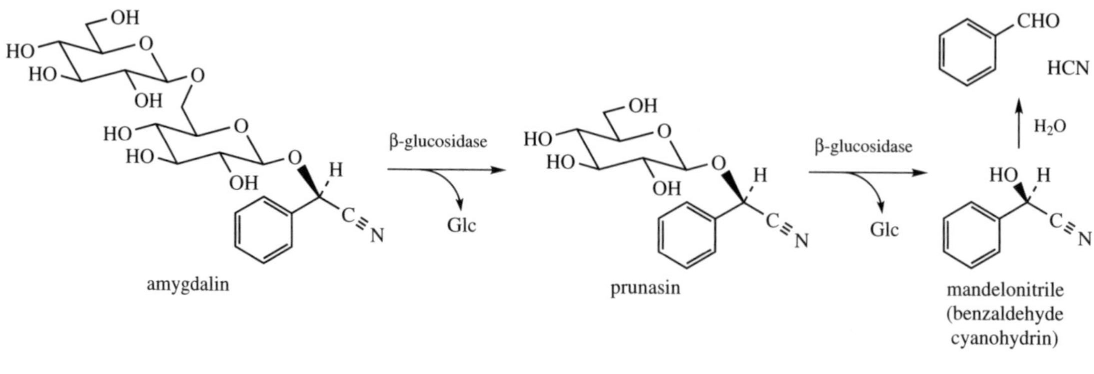

Figure 7.48

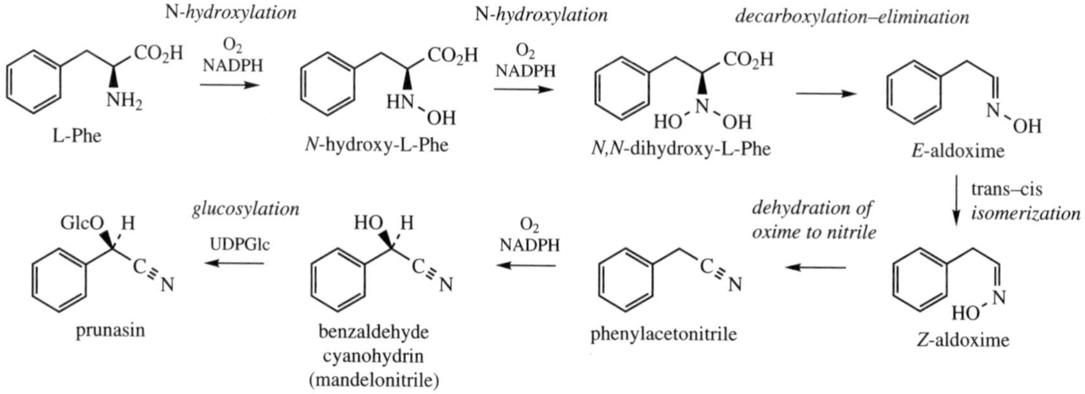

Figure 7.49

but no cyanogenic glycoside. Amygdalin itself is not especially toxic to animals; toxicity depends on the co-ingestion of the hydrolytic enzymes. Although formed by the hydrolysis of amygdalin, prunasin is also a natural cyanogenic glycoside and may be found in seeds of black cherry (*Prunus serotina*), and in the seeds and leaves of cherry laurel (*Prunus laurocerasus*). The food plant cassava (or tapioca) (*Manihot esculenta*; Euphorbiacae) also produces the cyanogenic glycosides **linamarin** and **lotaustralin** (see Figure 7.50), and preparation of the starchy tuberous roots involves prolonged hydrolysis and boiling to release and drive off the HCN before they are suitable for consumption.

Cyanogenic glycosides are produced from a range of amino acids by a common pathway (Figure 7.49). The amino acid precursor of prunasin is phenylalanine, which is

N-hydroxylated and then converted into the aldehyde oxime (aldoxime) by a sequence that involves further *N*-hydroxylation and subsequent decarboxylation–elimination, with all of these reactions catalysed by a single cytochrome P-450-dependent enzyme. The nitrile is formed by dehydration of the oxime, but this reaction actually proceeds on the *Z*-aldoxime produced by isomerization of the first formed *E*-aldoxime. (*R*)-Mandelonitrile is then the result of a stereoselective cytochrome P-450-dependent hydroxylation reaction. Finally, glycosylation occurs, the sugar unit usually being glucose as in the case of prunasin and amygdalin. The stereoselectivity of the nitrile hydroxylation step varies depending on the plant system, so that epimeric cyanohydrins are found in nature, though not in the same plant. Thus, the (*S*)-enantiomer of prunasin (called sambunigrin) is found in the leaves of elder (*Sambucus nigra*; Caprifoliaceae).

The main amino acids utilized in the biosynthesis of cyanogenic glycosides are phenylalanine (e.g. prunasin, sambunigrin, and amygdalin), tyrosine (e.g. **dhurrin** from sorghum (*Sorghum bicolor*; Graminae/Poaceae)), valine (e.g. linamarin from flax (*Linum usitatissimum*; Linaceae)), isoleucine (e.g. lotaustralin, also from flax), and leucine (e.g. **heterodendrin** from *Acacia* species (Leguminosae/Fabaceae)) (Figure 7.50). Although cyanogenic glycosides are widespread, they are particularly found in the families Rosaceae, Leguminosae/Fabaceae, Graminae/Poaceae, Araceae, Compositae/Asteraceae, Euphorbiaceae, and Passifloraceae. It is highly likely that plants synthesize these compounds as protecting agents against herbivores. Some insects also accumulate cyanogenic glycosides in their bodies, again as a protective device. Whilst many insects obtain these compounds by feeding on suitable plant sources, it is remarkable that others are known to synthesize cyanogenic glycosides themselves from amino acid precursors. There is hope that cyanogenesis may provide a means of destroying cancer cells. By targeting cancer cells with linamarase via a retrovirus

and then supplying linamarin, it has been possible to selectively generate toxic HCN in cancer cells.

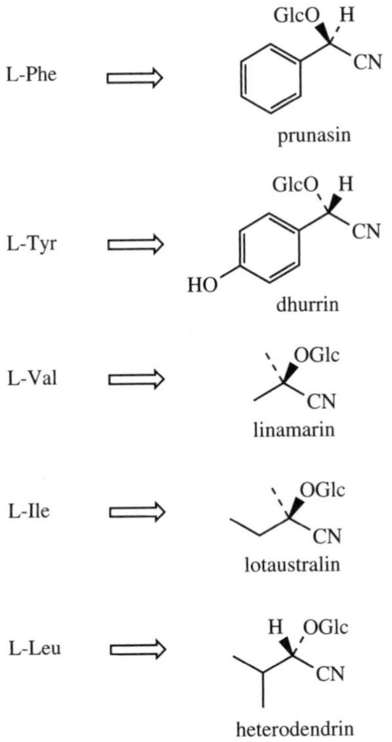

GLUCOSINOLATES

Glucosinolates have several features in common with the cyanogenic glycosides. They too are glycosides which are enzymically hydrolysed in damaged plant tissues giving rise to potentially toxic materials, and they share the early stages of the cyanogenic glycoside biosynthetic pathway for their formation in plants. A typical structure is **sinalbin** (Figure 7.51), found in seeds of white mustard (*Sinapis alba*; Cruciferae/Brassicaceae). Addition of water to the crushed or powdered seeds results in hydrolysis of the glucoside bond via the enzyme myrosinase (a thioglucosidase) to give a thiohydroximate sulphonate (Figure 7.51). This compound usually yields the isothiocyanate **acrinylisothiocyanate** by a Lössen-type rearrangement as shown, prompted by the sulphate leaving group. Under certain conditions, dependent on pH, or the presence of metal ions or other enzymes, related compounds such as thiocyanates (RSCN) or nitriles (RCN) may be formed from glucosinolates. Acrinylisothiocyanate is a pungent-tasting material (mustard oil) typical of many plants in the Cruciferae/Brassicaceae used as vegetables (e.g. cabbage, radish) and condiments (e.g. mustard, horseradish). Black mustard (*Brassica nigra*) contains **sinigrin** (Figure 7.51) in its seeds, which by a similar sequence is hydrolysed to **allylisothiocyanate**. Allylisothiocyanate is considerable more volatile than acrinylisothiocyanate, so that condiment mustard prepared from black mustard has a pungent aroma as well as taste.

The biosynthesis of **sinalbin** from tyrosine is indicated in Figure 7.52. The aldoxime is produced from the amino acid by the early part of the cyanogenic glycoside pathway shown in Figure 7.49. This aldoxime incorporates sulphur from methionine (or cysteine) to give the thiohydroximic acid, perhaps by attack of a thiolate ion on to the imine system, and this compound is then S-glucosylated using UDPglucose. In nature, sulphate groups are provided by PAPS (3′-phosphoadenosine-5′-phosphosulphate), and sulphation features as the last step in the pathway. Similarly, phenylalanine is the precursor of **benzylglucosinolate**

Figure 7.50

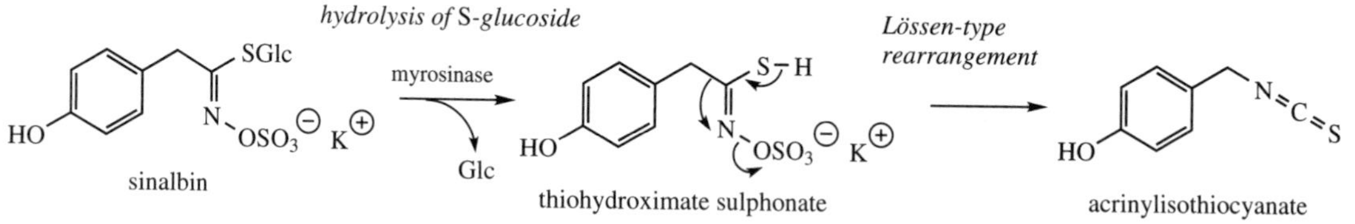

Figure 7.51

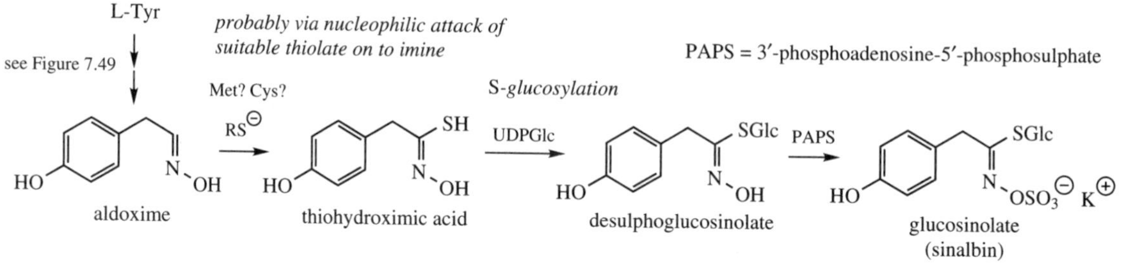

Figure 7.52

(Figure 7.53) in nasturtium (*Tropaeolum majus*; Tropaeolaceae), and tryptophan yields **glucobrassicin** in horseradish (*Armoracia rusticana*; Cruciferae/Brassicaceae). Interestingly, chain extension of methionine to **homomethionine** and of phenylalanine to **homophenylalanine** is involved in the formation of **sinigrin** in *Brassica nigra* and **gluconasturtiin** in rapeseed (*Brassica napus*) respectively (Figure 7.54). Two carbons derived from acetate are incorporated into the side-chain in each case, together with loss of the original carboxyl, and this allows biosynthesis of the appropriate glucosinolates. For elaboration of the allyl side-chain in the biosynthesis of sinigrin, loss of methanethiol occurs as a late step (Figure 7.54).

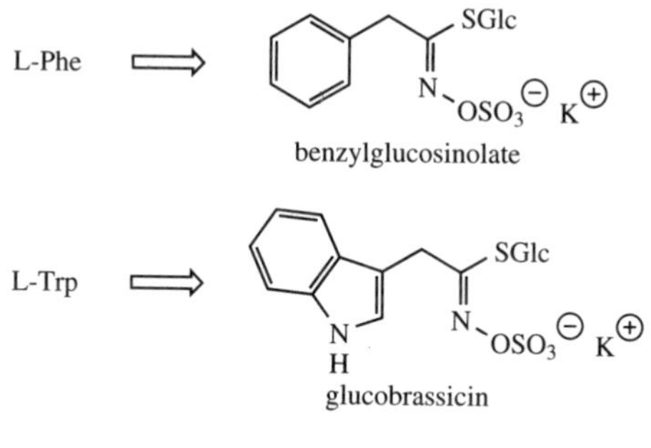

Glucosinolates are found in many plants of the Cruciferae/Brassicaceae, Capparidaceae, Euphorbiacae, Phytolaccaceae, Resedaceae, and Tropaeolaceae, contributing to the pungent properties of their crushed tissues. They are often at their highest concentrations in seeds rather than leaf tissue. These compounds and their degradation products presumably deter some predators, but may actually attract others, e.g. caterpillars on cabbages and

Figure 7.53

similar crops. There is evidence that consumption of the hydrolysis products from glucosinolates in food crops may induce goitre, an enlargement of the thyroid gland. Thus, **progoitrin** in oil seed rape (*Brassica napus*; Cruciferae/Brassicaceae) on hydrolysis yields the oxazolidine-2-thione **goitrin** (Figure 7.55), which is a potent goitrogen, inhibiting iodine incorporation and thyroxine formation (see page 410). The goitrogenic effects of glucosinolates cannot be alleviated merely by the administration of iodine. This severely limits economic utilization for animal foodstuffs of

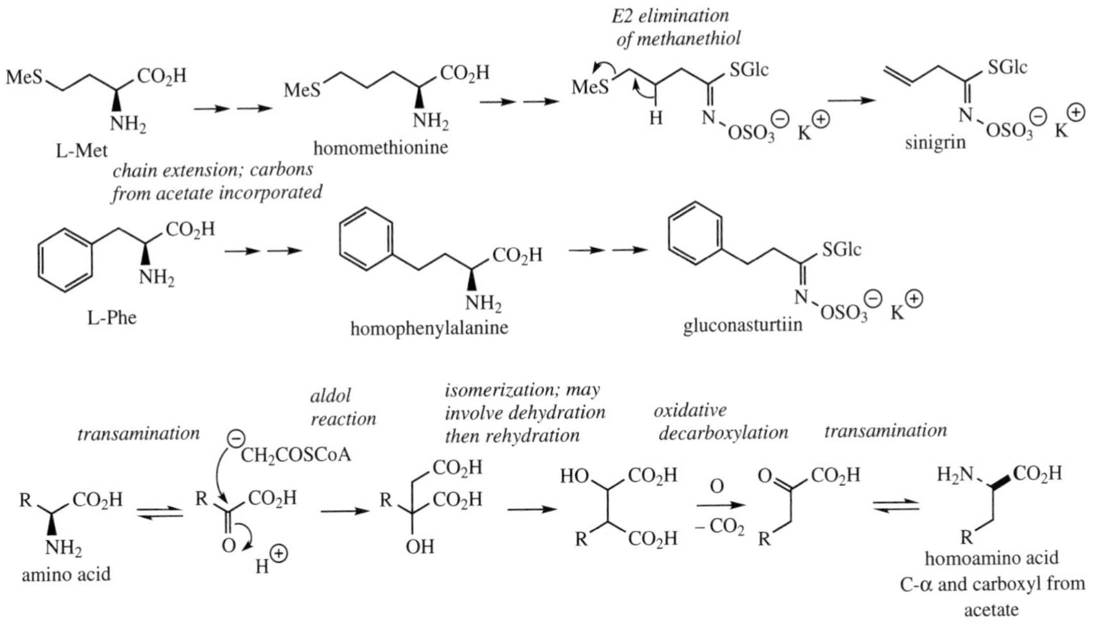
MeS
CO2H
NH2
L-Met
MeS
CO2H
NH2
homomethionine
chain extension; carbons
from acetate incorporated
E2 elimination
of methanethiol
MeS
H
SGlc
N
OSO3 K
SGlc
N
OSO3 K
sinigrin
CO2H
NH2
L-Phe
CO2H
NH2
homophenylalanine
SGlc
N
OSO3 K
gluconasturtiin
transamination
aldol
reaction
isomerization; may
involve dehydration
then rehydration
oxidative
decarboxylation
transamination
CH2COSCoA
R
CO2H
NH2
amino acid
R
CO2H
O
H
R
CO2H
CO2H
OH
HO
CO2H
R
CO2H
O
– CO2
O
CO2H
R
H2N
CO2H
R
homoamino acid
C-α and carboxyl from
acetate

Figure 7.54

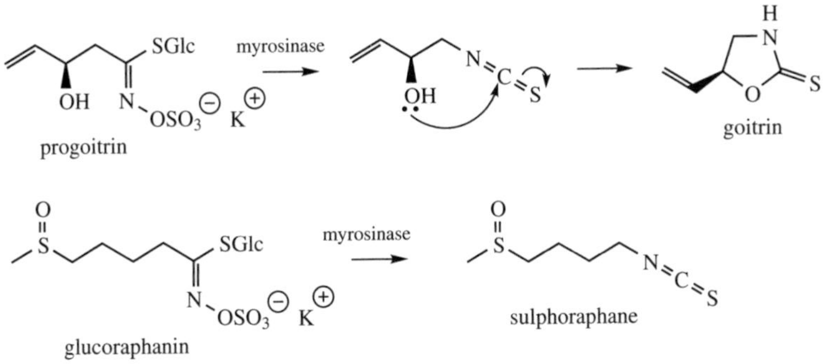
SGlc
OH
N
OSO3 K
progoitrin
myrosinase
N
C
S
OH
H
N
O
S
goitrin
O
S
SGlc
N
OSO3 K
glucoraphanin
myrosinase
O
S
N
C
S
sulphoraphane

Figure 7.55

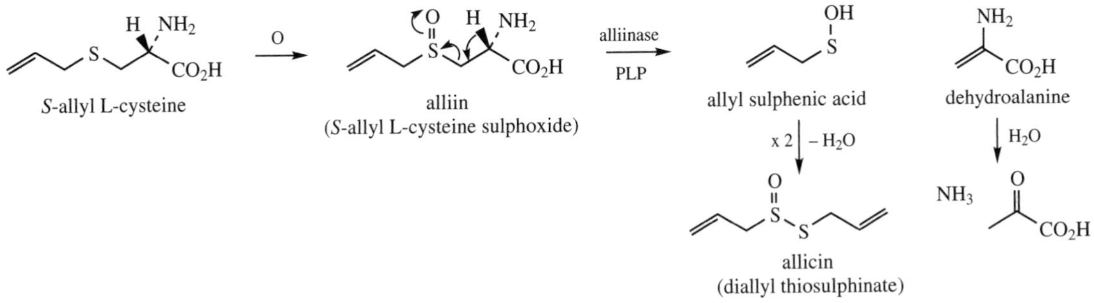
H NH2
S
CO2H
S-allyl L-cysteine
O
O
H NH2
S
CO2H
alliin
(S-allyl L-cysteine sulphoxide)
alliinase
PLP
OH
S
allyl sulphenic acid
x 2
– H2O
O
S
S
allicin
(diallyl thiosulphinate)
NH2
CO2H
dehydroalanine
H2O
NH3
O
CO2H

Figure 7.56

the rapeseed meal remaining after oil expression unless strains with very low levels of the glucosinolate are employed. On the other hand, **glucoraphanin**, the glucosinolate precursor of **sulphoraphane** (Figure 7.55) from broccoli (*Brassica oleracea italica*; Cruciferae/Brassicaceae), has been shown to have beneficial medicinal properties, in that it induces carcinogen-detoxifying enzyme systems and accelerates the removal of xenobiotics. Young sprouted seedlings contain some 10–100 times as much glucoraphanin as the mature plant, but, nevertheless, broccoli may be regarded as a valuable dietary vegetable.

CYSTEINE SULPHOXIDES

The major flavour component of garlic* (*Allium sativum*; Liliaceae/Alliaceae) is a thiosulphinate called **allicin** (Figure 7.56). This compound is formed when garlic tissue is damaged as a hydrolysis product of *S*-allyl cysteine sulphoxide (**alliin**) brought about by the pyridoxal phosphate-dependent enzyme alliinase (Figure 7.56). Under these conditions, alliin is cleaved by an elimination reaction, and two molecules of the sulphenic acid then form allicin. Pyruvic acid and ammonia are the other hydrolysis products. Allicin has considerable antibacterial and antifungal properties. There is widespread use of garlic, either fresh, dried, or as garlic oil, as a beneficial agent to reduce cholesterol levels and reduce the

S-methyl
L-cysteine sulphoxide

S-propyl
L-cysteine sulphoxide

S-1-propenyl
L-cysteine sulphoxide

Figure 7.57

1-propenyl
sulphenic acid

(Z)-propanethial S-oxide
(onion lachrymatory factor)

Figure 7.58

risk of heart attacks. *S*-Alkyl cysteine sulphoxides are characteristic components of the onion (*Allium*) genus. All *Allium* species contain *S*-methyl cysteine sulphoxide, though the *S*-propyl analogue (Figure 7.57) predominates in chives (*A. schoenoprasum*), the *S*-1-propenyl derivative in onions (*A. cepa*) and the *S*-allyl compound alliin in garlic. Propenyl sulphenic acid derived by hydrolysis of *S*-1-propenyl cysteine sulphoxide provides a common kitchen hazard, rearranging to the lachrymatory factor of onions (Figure 7.58).

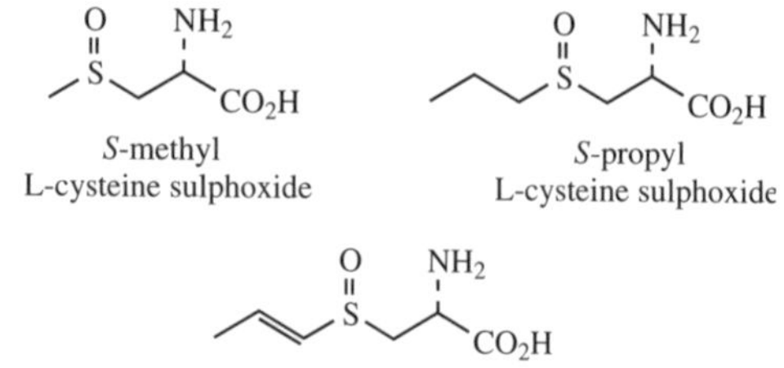

Garlic

Garlic (*Allium sativum*; Liliaceae/Alliaceae) has a long history of culinary and medicinal use. The compound bulb is composed of several smaller sections termed cloves. Allicin (Figure 7.56) is considered to be the most important of the biologically active components in the crushed bulb. It is not present in garlic, but is rapidly produced when the precursor alliin is cleaved by the action of the enzyme alliinase upon crushing the tissue. Both alliin and alliinase are stable when dry, and dried garlic still has the potential for releasing allicin when subsequently moistened. However, allicin itself is very unstable to heat or organic solvents degrading to many other compounds, including diallyl sulphides (mono-, di-, and oligo-sulphides), vinyldithiins, and ajoenes (Figure 7.59). Processed garlic preparations typically contain a range of different sulphur compounds. Garlic preparations used medicinally include steam-distilled oils, garlic macerated in vegetable oils (e.g. soybean oil), dried garlic powder, and gel-suspensions of garlic powder. Analyses indicate wide variations in the nature and amounts of constituents in the various preparations. Thus, freshly crushed garlic cloves typically contain allicin (about 0.4%) and other thiosulphinates (about 0.1%, chiefly allyl methyl thiosulphinate). The

(Continues)

(Continued)

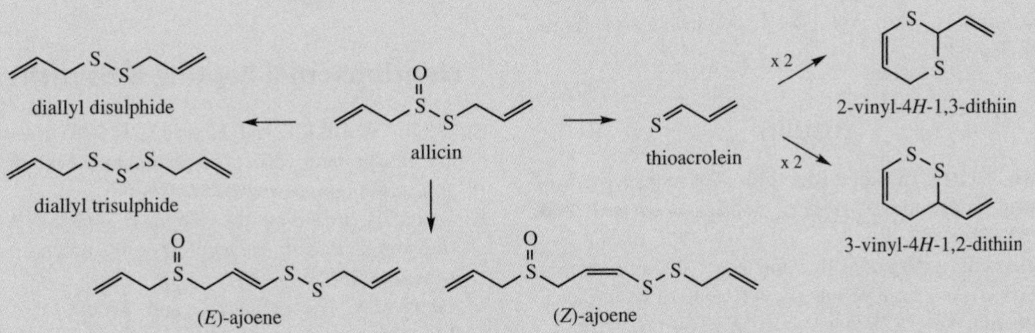

Figure 7.59

garlic powder usually produces less of these materials, though high quality samples may be similar to the fresh samples. Oil-macerated powders appear to lose up to 80% of their sulphur compounds, and vinyldithiins and ajoenes predominate. Steam-distilled garlic oil has dialk(en)yl sulphides (i.e. diallyl sulphide, allyl methyl sulphide, etc) as the major sulphur components (0.1–0.5%). Bad breath and perspiration odours which often follow the ingestion of garlic, either medicinally or culinary, are due to allyl methyl sulphide and disulphide, diallyl sulphide and disulphide, and 2-propenethiol.

Garlic is used for a variety of reasons, and some of the attributes associated with it, e.g. for cancer prevention, or to reduce heart attacks, may not be substantiated. Other properties such as antimicrobial activity, effects on lipid metabolism, and platelet aggregation inhibitory action have been demonstrated. Ajoene has been shown to be a potent antithrombotic agent through inhibition of platelet aggregation.

FURTHER READING

General

Lancini GC and Lorenzetti R (1993) *Biotechnology of Antibiotics and Other Microbial Metabolites*. Plenum, New York.

Lancini G, Parenti F and Gallo GG (1995) *Antibiotics – A Multidisciplinary Approach*. Plenum, New York.

Niccolai D, Tarsi L and Thomas RJ (1997) The renewed challenge of antibacterial chemotherapy. *Chem Commun* 2333–2342.

Russell AD (1998) Mechanisms of bacterial resistance to antibiotics and biocides. *Prog Med Chem* **35**, 133–197.

Peptide Hormones

Anon (1997) Thyroid and antithyroid preparations. *Kirk–Othmer Encyclopedia of Chemical Technology*, 4th edn, Vol 24. Wiley, New York, pp 89–104.

Becker GW, MacKellar WC, Riggin RM and Wroblewski VW (1995) Hormones (human growth hormone). *Kirk–Othmer Encyclopedia of Chemical Technology*, 4th edn, Vol 13. Wiley, New York, pp 406–418.

Engeland WC (1995) Hormones (survey). *Kirk–Othmer Encyclopedia of Chemical Technology*, 4th edn, Vol 13. Wiley, New York, pp 357–370.

Giannis A and Kolter T (1993) Peptidomimetics for receptor ligands – Discovery, development, and medicinal perspectives. *Angew Chem Int Edn Engl* **32**, 1244–1267.

Hirschmann R (1991) Medicinal chemistry in the golden age of biology: lessons from steroid and peptide research. *Angew Chem Int Edn Engl* **30**, 1278–1301.

Matteri RL (1995) Hormones (anterior pituitary hormones). *Kirk–Othmer Encyclopedia of Chemical Technology*, 4th edn, Vol 13. Wiley, New York, pp 370–380.

Matteri RL (1995) Hormones (anterior pituitary-like hormones). *Kirk–Othmer Encyclopedia of Chemical Technology*, 4th edn, Vol 13. Wiley, New York, pp 381–391.

Spatola AF (1995) Hormones (posterior pituitary hormones). *Kirk–Othmer Encyclopedia of Chemical Technology*, 4th edn, Vol 13. Wiley, New York, pp 391–406.

Insulin

Brown A (1998) Diabetes mellitus (2): management of insulin-dependent diabetes mellitus. *Pharm J* **260**, 753–756.

Shoelson SE (1995) Insulin and other diabetic agents. *Kirk–Othmer Encyclopedia of Chemical Technology*, 4th edn, Vol 14. Wiley, New York, pp 662–676.

Interferons

Sen GC and Lengyel P (1992) The interferon system: a bird's eye view of its biochemistry. *J Biol Chem* **267**, 5017–20.

Wong S (1995) Immunotherapeutic agents. *Kirk–Othmer Encyclopedia of Chemical Technology*, 4th edn, Vol 14. Wiley, New York, pp 64–86.

Opioid Peptides

Brownstein MJ (1993) A brief history of opiates, opioid peptides, and opiate receptors. *Proc Natl Acad Sci USA* **90**, 5391–5393.

Childers SR (1995) Hormones (brain oligopeptides). *Kirk–Othmer Encyclopedia of Chemical Technology*, 4th edn, Vol 13. Wiley, New York, pp 418–433.

Giannis A and Kolter T (1993) Peptidomimetics for receptor ligands – discovery, development, and medicinal perspectives. *Angew Chem Int Edn Engl* **32**, 1244–1267.

Nugent RA and Hall CM (1992) Analgesic, antipyretics, and antiinflammatory agents. *Kirk–Othmer Encyclopedia of Chemical Technology*, 4th edn, Vol 2. Wiley, New York, pp 729–748.

Schiller PW (1991) Development of receptor-specific opioid peptide analogues. *Prog Med Chem* **28**, 301–340.

Selley DE, Sim LJ and Childers SR (1995) Opioids, endogenous. *Kirk–Othmer Encyclopedia of Chemical Technology*, 4thedn, Vol 17. Wiley, New York, pp 858–882.

Enzymes

Cervoni P, Crandall DL and Chan PS (1993) Cardiovascular agents – thrombolytic agents. *Kirk–Othmer Encyclopedia of Chemical Technology*, 4th edn, Vol 5. Wiley, New York, pp 289–293.

Wiseman A (1994) Better by design: biocatalysts for the future. *Chem Brit* 571–573.

Nonribosomal Peptide Biosynthesis

Cane DE, Walsh CT and Khosla C (1998) Harnessing the biosynthetic code: combinations, permutations, and mutations. *Science* **262**, 63–68.

Kleinkauf H and von D "ohren H (1996) A nonribosomal system of peptide biosynthesis. *Eur J Biochem* **236**, 335–351.

Marahiel MA, Stachelhaus T and Mootz HD (1997) Modular peptide synthetases involved in nonribosomal peptide synthesis. *Chem Rev* **97**, 2651–2673.

Rajendran N and Marahiel MA (1999) Multifunctional peptide synthetases required for nonribosomal biosynthesis of peptide antibiotics. *Comprehensive Natural Products Chemistry*, Vol 4. Elsevier, Amsterdam, pp 195–200.

von D "ohrenH, Keller U, Vater J and Zocher R (1997) Multifunctional peptide synthetases. *Chem Rev* **97**, 2675–2705.

Peptide Antibiotics

Nicolau KC, Boddy CNC, Br "aseS and Winssinger N (1999) Chemistry, biology and medicine of the glycopeptide antibiotics. *Angew Chem Int Edn* **38**, 2096–2152.

Wise EM (1992) Antibiotics (peptides). *Kirk–Othmer Encyclopedia of Chemical Technology*, 4th edn, Vol 3. Wiley, New York, pp 266–306.

Vancomycin, Teichoplanin

Cavalleri B and Parenti F (1992) Antibiotics [glycopeptides (dalbaheptides)]. *Kirk–Othmer Encyclopedia of Chemical Technology*, 4th edn, Vol 2. Wiley, New York, pp 995–1018.

Williams DH (1996) The glycopeptide story – how to kill the deadly 'superbugs'. *Nat Prod Rep* **13**, 469–477.

Williams DH and Bardsley B (1999) The vancomycin group of antibiotics and the fight against resistant bacteria. *Angew Chem Int Edn* **38**, 1172–1193.

Cyclosporins

Clardy J (1995) The chemistry of signal transduction. *Proc Natl Acad Sci USA* **92**, 56–61.

Lawen A (1996) Biosynthesis and mechanism of action of cyclosporins. *Prog Med Chem* **33**, 53–97.

Rosen MK and Schreiber SL (1992) Natural products as probes of cellular function: studies of immunophilins. *Angew Chem Int Edn Engl* **31**, 384–400.

St "ahelin HF (1996) The history of cyclosporin A (Sandimmune®) revisited: another point of view. *Experientia* **52**, 5–13.

von D "ohren H and Kleinkauf H (1999) Cyclosporin: the biosynthetic path to a lipopeptide. *Comprehensive Natural Products Chemistry*, Vol 1. Elsevier, Amsterdam, pp 533–555.

Wong S (1995) Immunotherapeutic agents. *Kirk–Othmer Encyclopedia of Chemical Technology*, 4th edn, Vol 14. Wiley, New York, pp 64–86.

Bleomycin

Boger DL and Cai H (1999) Bleomycin: synthetic and mechanistic studies. *Angew Chem Int Edn* **38**, 448–476.

Hecht SM (2000) Bleomycin: new perspectives on the mechanism of action. *J Nat Prod* **63**, 158–168.

Shen B, Du L, Sanchez C, Chen M and Edwards DJ (1999) Bleomycin biosynthesis in Streptomyces verticillus ATCC15003: a model of hybrid peptide and polyketide biosynthesis. *Bioorg Chem* **27**, 155–171.

Streptogramins

Pechère J-C (1996) Streptogramins: a unique class of antibiotics. *Drugs* **51**(Suppl 1), 13–19.

Vannuffel P and Cocito C (1996) Mechanism of action of streptogramins and macrolides. *Drugs* **51** (Suppl 1), 20–30.

Peptide Toxins

Berressem P (1999) From bites and stings to medicines. *Chem Brit* **35** (4), 40–42.

Carmichael WW (1992) Cyanobacteria secondary metabolites – the cyanotoxins. *J Appl Bacteriol* **72**, 445–459.

Hucho F (1995) Toxins as tools in neurochemistry. *Angew Chem Int Edn Engl* **34**, 39–50.

Shimizu Y (1993) Microalgal metabolites. *Chem Rev* **93**, 1685–1698.

β-Lactam Antibiotics

Bentley R (1995) A secret arrangement. *Chem Brit* 793–796. (history of penicillin).

Brakhage A (1999) *Biosynthesis of ?-lactam compounds in microorganisms*. *Comprehensive Natural Products Chemistry*, Vol 4. Elsevier, Amsterdam, pp 159–193.

Bugg TDH (1999) Bacterial peptidoglycan biosynthesis and its inhibition. *Comprehensive Natural Products Chemistry*, Vol 3. Elsevier, Amsterdam, pp 241–294.

Bugg TDH and Walsh CT (1992) Intracellular steps of bacterial cell wall peptidoglycan biosynthesis: enzymology, antibiotics, and antibiotic resistance. *Nat Prod Rep* **9**, 199–215.

Coulton S and Francois I (1994) ?-Lactamases: targets for drug design. *Prog Med Chem* **31**, 297–349.

Massova I and Mobashery S (1997) Molecular bases for interactions between β-lactam antibiotics and β-lactamases. *Account Chem Res* **30**, 162–168.

Matagne A, Dubus A, Galleni M and Frère J-M (1999) The β-lactamase cycle: a tale of selective pressure and bacterial ingenuity. *Nat Prod Rep* **16**, 1–19.

Niccolai D, Tarsi L and Thomas RJ (1997) The renewed challenge of antibacterial chemotherapy. *Chem Commun* 2333–2342.

Page MI and Laws AP (1998) The mechanism of catalysis and the inhibition of ?-lactamases. *Chem Commun* 1609–1617.

Penicillins and Cephalosporins

Bryskier A (2000) Cephems: fifty years of continuous research. *J Antibiot* **53**, 1028–1037.

Luengo JM (1999) Enzymatic synthesis of penicillins. *Comprehensive Natural Products Chemistry*, Vol 4. Elsevier, Amsterdam, pp 239–274.

Nathwani D and Wood MJ (1993) Penicillins: a current review of their clinical pharmacology and therapeutic use. *Drugs* **45**, 866–894.

Ponsford RJ (1992) Antibiotics [β-lactams (penicillins and others)]. *Kirk–Othmer Encyclopedia of Chemical Technology*, 4th edn, Vol 3. Wiley, New York, pp 129–158.

Roberts J (1992) Antibiotics [β-lactams (cephalosporins)]. *Kirk–Othmer Encyclopedia of Chemical Technology*, 4th edn, Vol 3. Wiley, New York, pp 28–82.

Clavulanic Acid, Carbapenems, Monobactams

Baggaley KH, Brown A and Schofield CJ (1997) Chemistry and biosynthesis of clavulanic acid and other clavams. *Nat Prod Rep* **14**, 309–333.

Coulton S and Hunt E (1996) Recent advances in the chemistry and biology of carbapenem antibiotics. *Prog Med Chem* **33**, 99–145.

Lindner KR, Bonner DP and Koster WH (1992) Antibiotics [β-lactams (monobactams)]. *Kirk–Othmer*

Encyclopedia of Chemical Technology, 4th edn, Vol 3, John Wiley, New York, pp 107–129.

Southgate R and Osborne NF (1992) Antibiotics [?-lactams (carbapenems and penems)]. *Kirk–Othmer Encyclopedia of Chemical Technology*, 4th edn, Vol 3. Wiley, New York, pp 1–27.

Stam JG (1992) Antibiotics [?-lactams (?-lactamase inhibitors)]. *Kirk–Othmer Encyclopedia of Chemical Technology*, 4th edn, Vol 3. Wiley, New York, pp 83–107.

Cyanogenic Glycosides

Conn EE (1991) The metabolism of a natural product: lessons learned from cyanogenic glycosides. *Planta Med* **57**, S1–S9.

Hickel A, Hasslacher M and Griengl H (1996) Hydroxynitrile lyases: Functions and properties. *Physiol Plant* **98**, 891–898.

Hughes MA (1999) Biosynthesis and degradation of cyanogenic glycosides. *Comprehensive Natural Products Chemistry*, Vol 1. Elsevier, Amsterdam, pp 881–895.

Jones DA (1998) Why are so many food plants cyanogenic? *Phytochemistry* **47**, 155–162.

Poulton JE (1990) Cyanogenesis in plants. *Plant Physiol* **94**, 401–405.

Glucosinolates

Bones AM and Rossiter JT (1996) The myrosinase–glucosinolate system, its organisation and biochemistry. *Physiol Plant* **97**, 194–208.

Nestle M (1997) Broccoli sprouts as inducers of carcinogen-detoxifying enzyme systems: Clinical, dietary, and policy implications. *Proc Natl Acad Sci USA* **94**, 11 149–11 151.

Cysteine Sulphoxides, Garlic

Block E (1992) The organosulfur chemistry of the genus *Allium* – Implications for the organic chemistry of sulfur. *Angew Chem Int Edn Engl* **31**, 1135–1178.

Lawson LD, Wang ZJ and Hughes BG (1991) Identification and HPLC quantitation of the sulfides and dialk(en)yl thiosulfinates in commercial garlic products. *Planta Med* **57**, 363–370.

McElnay JC and Li Wan Po A (1991) Dietary supplements: garlic. *Pharm J* **246**, 324–326.

8

CARBOHYDRATES

Some fundamental modifications that occur in carbohydrate metabolism are outlined, then specific examples of monosaccharides, oligosaccharides, and polysaccharides are described. The formation of amino sugars follows, leading to a discussion of aminoglycoside antibiotics. Monograph topics giving more detailed information on medicinal agents include monosaccharides and disaccharides, vitamin C, polysaccharides, aminoglycoside antibiotics based on streptamine and 2-deoxystreptamine, acarbose, lincomycin, and clindamycin.

The main pathways of carbohydrate biosynthesis and degradation comprise an important component of primary metabolism that is essential for all organisms. Carbohydrates are among the most abundant constituents of plants, animals, and microorganisms. Polymeric carbohydrates function as important food reserves, and as structural components in cell walls. Animals and most microorganisms are dependent for their very existence on the carbohydrates produced by plants. Carbohydrates are the first products formed in photosynthesis, and are the products from which plants then synthesize their own food reserves as well as other chemical constituents. These materials then become the foodstuffs of other organisms. Secondary metabolites are also ultimately derived from carbohydrate metabolism, and the relationships of the acetate, shikimate, mevalonate, and deoxyxylulose phosphate pathways to primary metabolism have already been indicated. Many of the medicinally important secondary metabolites described in the earlier chapters have been seen to contain clearly recognizable carbohydrate portions in their structures, e.g. note the frequent occurrence of glycosides. In this chapter, some of the important natural materials which can be grouped together because they are composed entirely or predominantly of basic carbohydrate units are discussed. Because of their widespread use in medicinal preparations, some materials with no inherent biological activity, and which are clearly of primary metabolic status, e.g. sucrose and starch, are also included.

MONOSACCHARIDES

Six-carbon sugars (hexoses) and five-carbon sugars (pentoses) are the most frequently encountered carbohydrate units (monosaccharides) in nature. Photosynthesis produces initially the three-carbon sugar 3-phosphoglyceraldehyde, two molecules of which are used to synthesize glucose 6-phosphate by a sequence which effectively achieves the reverse of the glycolytic reactions (Figure 8.1). Alternatively, by the complex reactions of the Calvin cycle, 3-phosphoglyceraldehyde may be used in the construction of the pentoses ribose 5-phosphate, ribulose 5-phosphate, and xylulose 5-phosphate. These sequences incorporate some of the fundamental reactions which are used in the biochemical manipulation of monosaccharide structures:

- Mutation, where repositioning of a phosphate group in the monosaccharide phosphate molecule, e.g. the isomerization of glucose 6-phosphate and glucose 1-phosphate (Figure 8.2), is achieved via an intermediate diphosphate.

- Epimerization changes the stereochemistry at one of the chiral centres, e.g. the interconversion of ribulose 5-phosphate and xylulose 5-phosphate (Figure 8.3). This reaction involves epimerization adjacent to a carbonyl group and probably proceeds through a common enol tautomer, but some other epimerizations are

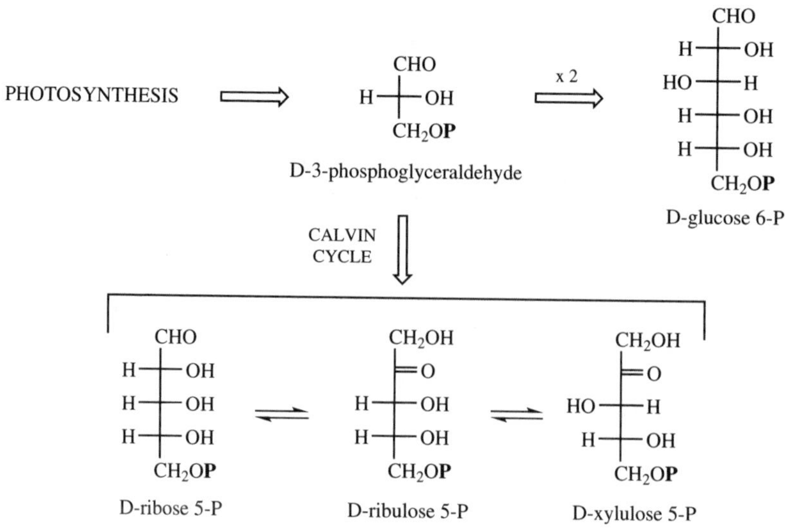

Figure 8.1

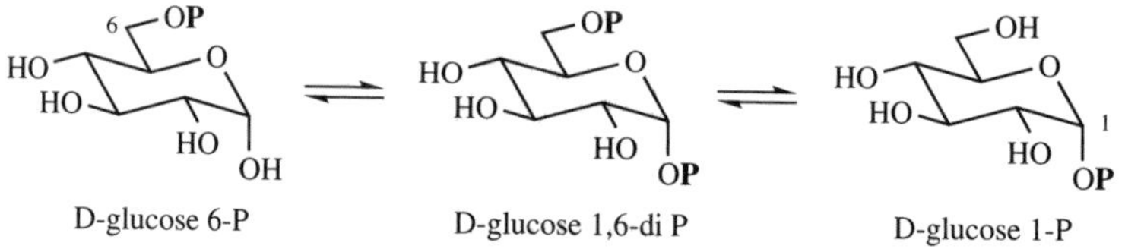

Figure 8.2

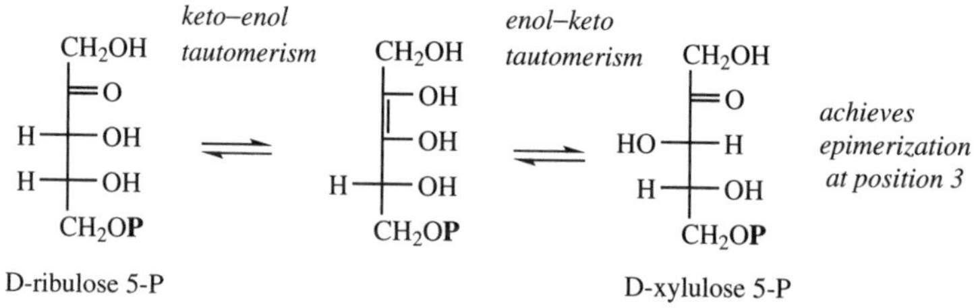

Figure 8.3

known proceed through oxidation to an inter-mediate carbonyl, followed by reduction to give the opposite configuration. The substrate for epimerization is often the UDPsugar rather than the monosaccharide phosphate.

- Aldose–ketose interconversions, e.g. glu-cose 6-phosphate to fructose 6-phosphate (Figure 8.4), also proceed through a common enol intermediate.

- Transfer of C_2 and C_3 units in reactions catal-ysed by transketolase and transaldolase respec-tively modify the chain length of the sugar.

Transketolase removes a two-carbon fragment from ketols such as fructose 6-phosphate (alter-natively xylulose 5-phosphate or sedoheptu-lose 7-phosphate) through the participation of thiamine diphosphate. Nucleophilic attack of the thiamine diphosphate anion on to the car-bonyl results in an addition product which then fragments by a reverse aldol reaction, gener-ating the chain-shortened aldose erythrose 4-phosphate, and the two-carbon carbanion unit attached to TPP (Figure 8.5) (compare the role of TPP in the decarboxylation of α-keto

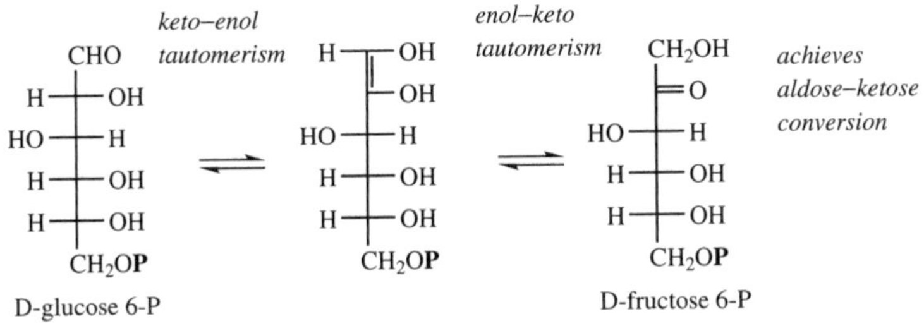

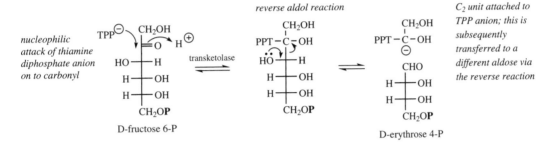

Figure 8.4

Figure 8.5

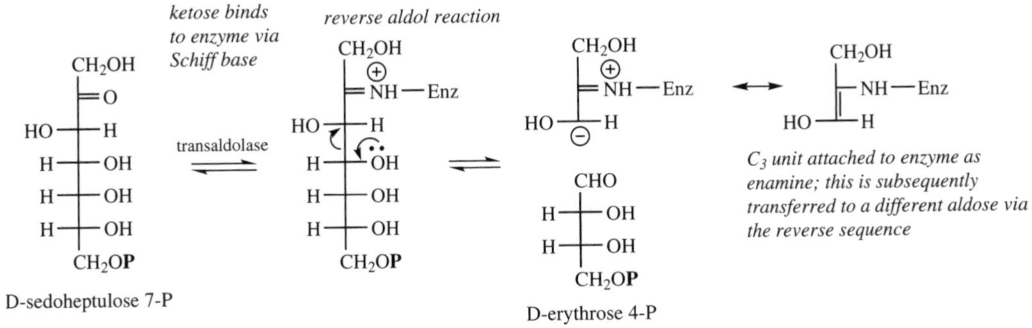

Figure 8.6

acids, page 21). Then, in what is formally the reverse of this reaction, this carbanion can attack another aldose such as ribose 5-phosphate (alternatively erythrose 4-phosphate or glyceraldehyde 3-phosphate), thus extending its chain length by two carbons. Transaldolase removes a three-carbon fragment from a ketose such as sedoheptulose 7-phosphate (alternatively fructose 6-phosphate) in a reverse aldol reaction, though this requires formation of a Schiff base between the carbonyl group and an active site lysine of the enzyme (Figure 8.6). Again, the reaction is completed by a reversal

of this process, but transferring the C_3 carbanion to another aldose such as glyceraldehyde 3-phosphate (alternatively erythrose 4-phosphate or ribose 5-phosphate) and thus increasing its length.

- Oxidation and reduction reactions, typically employing the NAD/NADP nucleotides, alter the oxidation state of the substrate. Oxidation at C-1 converts an aldose into an aldonic acid, e.g. glucose 6-phosphate gives gluconolactone 6-phosphate and then the open-chain gluconic acid 6-phosphate (Figure 8.7). Oxidation at C-6 yields the corresponding uronic acids,

oxidation at C-1   *hydrolysis of lactone*

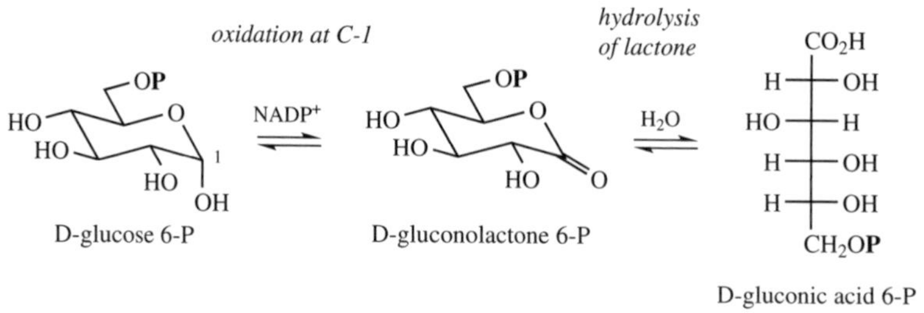

D-glucose 6-P   D-gluconolactone 6-P   D-gluconic acid 6-P

Figure 8.7

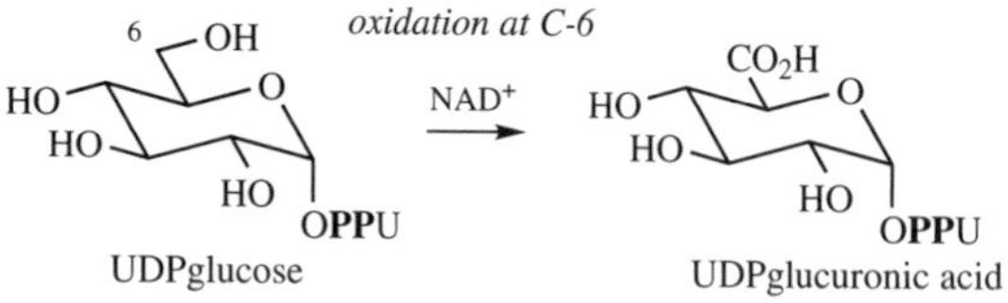

oxidation at C-6

UDPglucose   UDPglucuronic acid

Figure 8.8

but this takes place on UDPsugar derivatives, e.g. UDPglucose to UDPglucuronic acid (Figure 8.8). Reduction is exemplified by the conversion of both glucose and fructose into the sugar alcohol sorbitol (glucitol), and of mannose into mannitol (Figure 8.9).

- Transamination reactions on keto sugars allow the introduction of amino groups as seen in the amino sugars glucosamine and galactosamine (Figure 8.10). These compounds, as their *N*-acetyl derivatives, are part of the structures of several natural polysaccharides, and

other uncommon amino sugars are components of the aminoglycoside antibiotics (see page 478).

Monosaccharide structures may be depicted in open-chain forms showing their carbonyl character, or in cyclic hemiacetal or hemiketal forms. The compounds exist predominantly in the cyclic forms, which result from nucleophilic attack of an appropriate hydroxyl on to the carbonyl (Figure 8.11). Both six-membered pyranose and five-membered furanose structures are encountered, a particular ring size usually being characteristic for any one sugar. Since the carbonyl group may be attacked from either side, two epimeric structures (anomers) are possible in each case, and in solution, the two forms are frequently in equilibrium. In natural product structures, sugar units are most likely (but not always) to be encountered in just one of the epimeric forms. The two forms are designated α or β on the basis of the

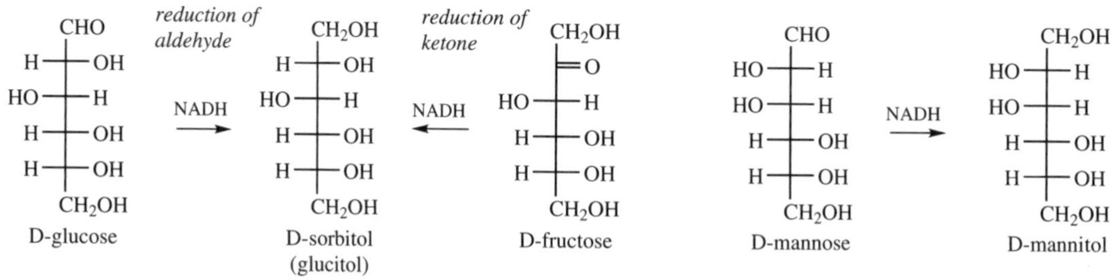

reduction of aldehyde   *reduction of ketone*

D-glucose   D-sorbitol (glucitol)   D-fructose   D-mannose   D-mannitol

Figure 8.9

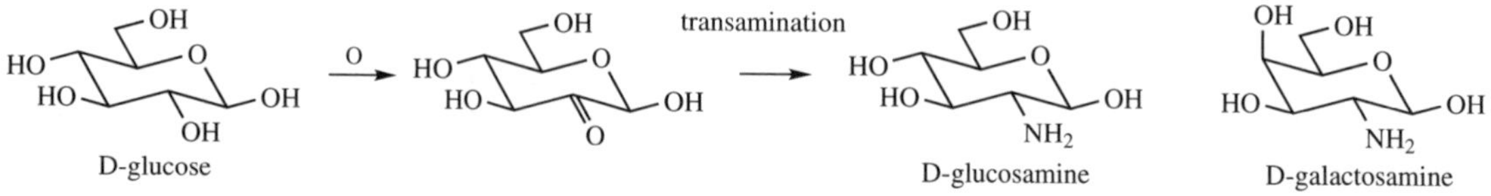

transamination

D-glucose   D-glucosamine   D-galactosamine

Figure 8.10

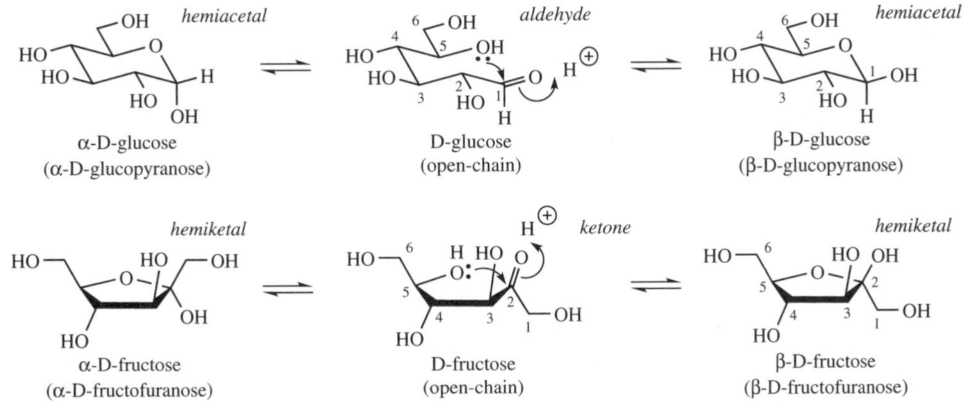

Figure 8.11

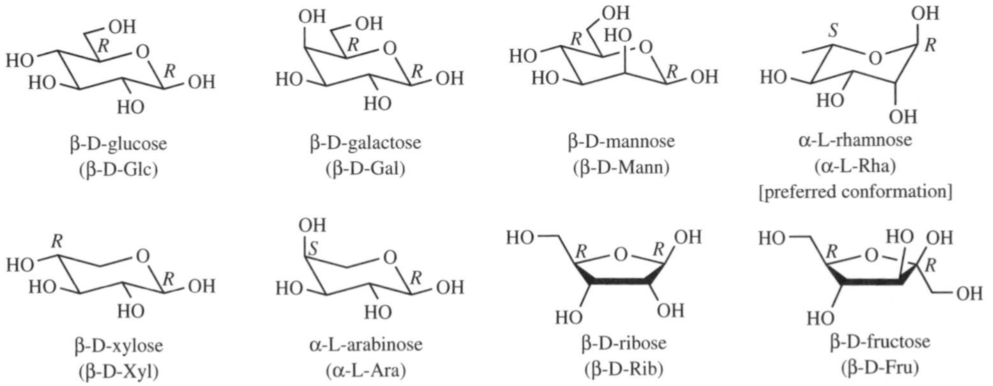

Figure 8.12

chiralities at the anomeric centre and at the highest numbered chiral centre. If these are the same (*RS* convention) the anomer is termed β, or α if they are different. The most commonly encountered monosaccharides, and their usual anomers are shown in Figure 8.12. Note that the D- and L- prefixes are assigned on the basis of the chirality (as depicted in Fischer projection) at the highest numbered chiral centre and its relationship to D-(*R*)-(+)-glyceraldehyde or L-(*S*)-(−)-glyceraldehyde (Figure 8.13).

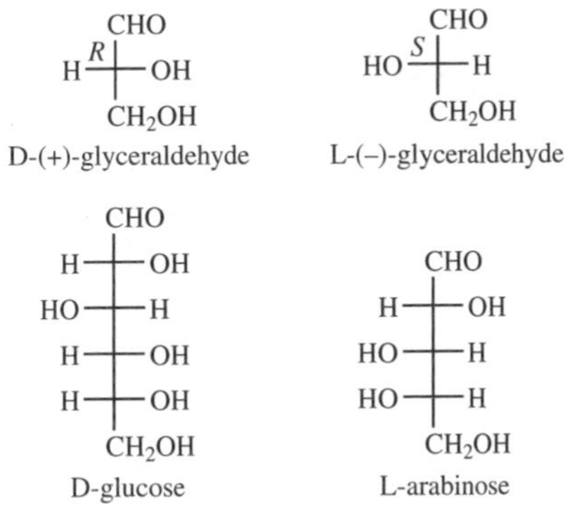

Figure 8.13

OLIGOSACCHARIDES

The formation of oligosaccharides (typically two to five monomers) and polysaccharides is dependent on the generation of an activated sugar bound to a nucleoside diphosphate. The nucleoside diphosphate most often employed is UDP, but ADP and GDP are sometimes involved. As outlined in Chapter 2 (see page 29), a UDPsugar is formed by the reaction of a sugar 1-phosphate with UTP, and then nucleophilic displacement of the UDP leaving group by a suitable nucleophile generates the new sugar derivative. This will be a glycoside if the nucleophile is a suitable aglycone molecule, or an oligosaccharide if the nucleophile is another sugar molecule (Figure 8.14). This reaction, if mechanistically of S_N2 type, should give an inversion of configuration at C-1 in the electrophile, generating a product with the β-configuration in the case of UDPglucose as shown. Many of the linkages formed between glucose monomers actually have the α-configuration, and it is believed that a double S_N2 mechanism operates, which also involves a nucleophilic group on the enzyme (Figure 8.14). Linkages are usually represented by a shorthand version, which indicates the atoms bonded and the configuration at the appropriate centre(s). Thus

maltose (Figure 8.15), a hydrolysis product from starch, contains two glucoses linked α1→4, whilst lactose, the main sugar component of cow's milk, has galactose linked β1→4 to glucose. In the systematic names, the ring size (pyranose or furanose) is also indicated. Sucrose ('sugar') (Figure 8.15) is composed of glucose and fructose, but these are both linked through their anomeric centres, so the shorthand representation becomes α1→β2. This means that both the hemiacetal structures are prevented from opening, and, in contrast to maltose and lactose, there can be no open-chain form in equilibrium with the cyclic form. Therefore sucrose does not display any of the properties usually associated with the masked carbonyl group, e.g. it is not a reducing sugar.

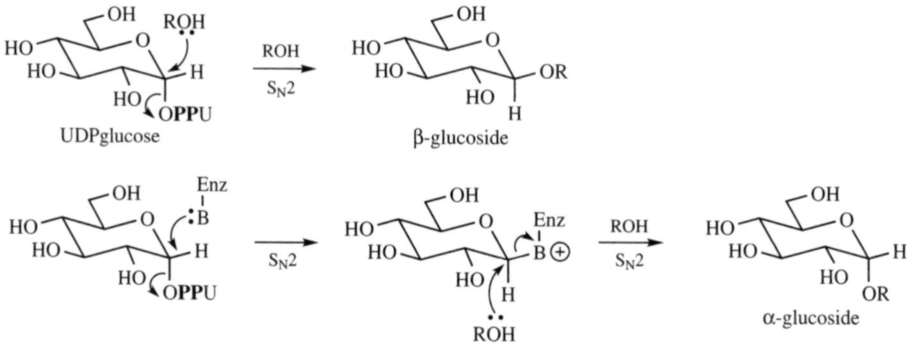

Figure 8.14

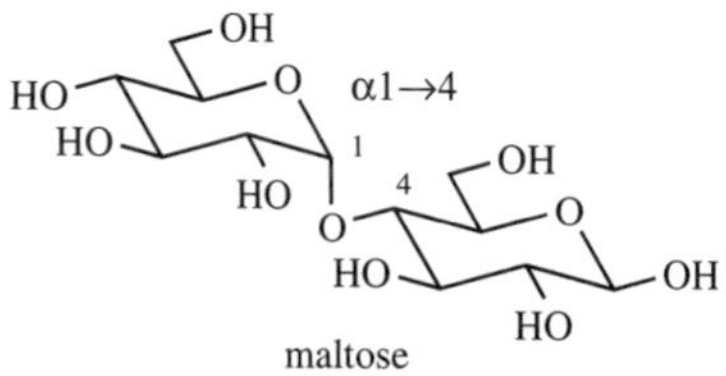

maltose
D-Glc(α1→4)D-Glc
4-O-(α-D-glucopyranosyl)-D-glucopyranose

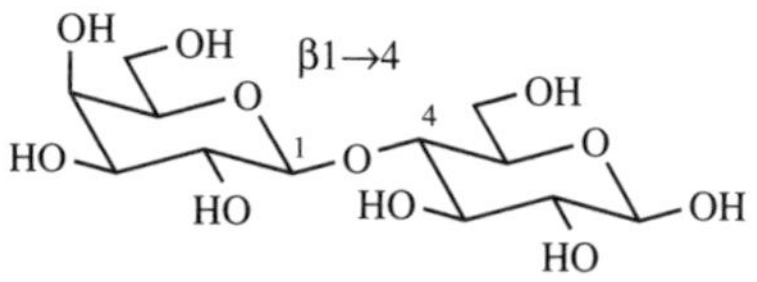

lactose
D-Gal(β1→4)D-Glc
4-O-(β-D-galactopyranosyl)-D-glucopyranose

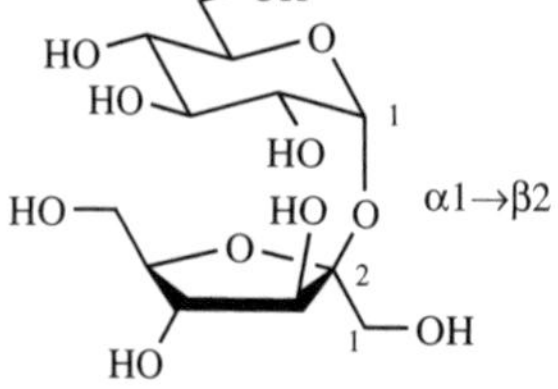

sucrose
D-Glc(α1→β2)D-Fru
α-D-glucopyranosyl-(1→2)-β-D-fructofuranoside

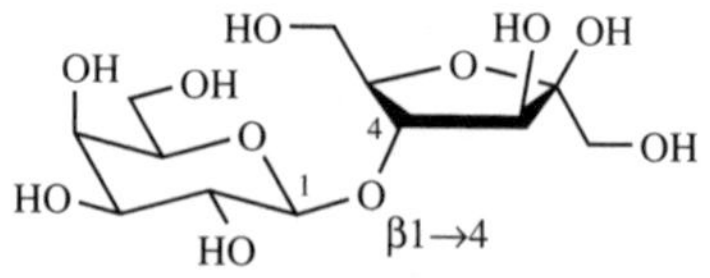

lactulose
D-Gal(β1→4)D-Fru
β-D-galactopyranosyl-(1→4)-β-D-fructofuranoside

Figure 8.15

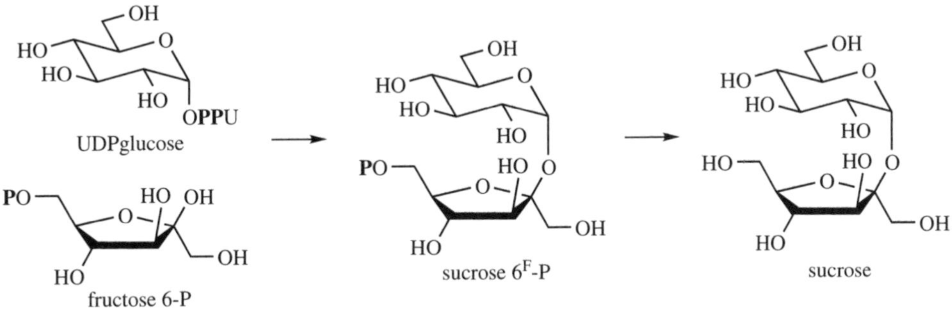

Figure 8.16

Sucrose is known to be formed predominantly by a slightly modified form of the sequence shown in Figure 8.16, in that UDPglucose is attacked by fructose 6-phosphate, and that the first formed product is sucrose 6^F-phosphate (F indicating the numbering refers to the fructose ring). Hydrolysis of the phosphate then generates sucrose (Figure 8.16).

Monosaccharides and Disaccharides

D-**Glucose** (**dextrose**) (Figure 8.9) occurs naturally in grapes and other fruits. It is usually obtained by enzymic hydrolysis of starch, and is used as a nutrient, particularly in the form of an intravenous infusion. Chemical oxidation of glucose produces gluconic acid. The soluble calcium salt **calcium gluconate** is used as an intravenous calcium supplement. D-**Fructose** (Figure 8.13) is usually obtained from invert sugar (see below) separating it from glucose, and is of benefit as a food and sweetener for patients who cannot tolerate glucose, e.g. diabetics. Fructose has the sweetness of sucrose, and about twice that of glucose. High fructose corn syrup for use as a food sweetener is a mixture of fructose and glucose containing up to 90% fructose and is produced by enzymic hydrolysis/isomerization of starch. The sugar alcohol D-**sorbitol** (Figure 8.9) is found naturally in the ripe berries of the mountain ash (*Sorbus aucuparia*; Rosaceae) but is prepared semi-synthetically from glucose. It is half as sweet as sucrose, is not absorbed orally, and is not readily metabolized in the body. It finds particular use as a sweetener for diabetic products. D-**Mannitol** (Figure 8.9) is also produced from glucose, but occurs naturally in manna, the exudate of the manna ash *Fraxinus ornus* (Oleaceae). This material has similar characteristics to sorbitol, but is used principally as a diuretic. It is injected intravenously, is eliminated rapidly into the urine, and removes fluid by an osmotic effect.

    Sucrose (Figure 8.15) is obtained from a variety of sources, including sugar cane (*Saccharum officinarum*; Graminae/Poaceae), sugar beet (*Beta vulgaris*; Chenopodiaceae), and sugar maple (*Acer saccharum*; Aceraceae). It is a standard sweetening agent for foods, syrups, and drug preparations. **Invert sugar** is an equimolar mixture of glucose and fructose, obtained from sucrose by hydrolysis with acid or the enzyme invertase. During this process, the optical activity changes from + to −, hence the reference to inversion. The high sweetness of fructose combined with that of glucose means invert sugar provides a cheaper, less calorific food sweetener than sucrose. Honey is also mainly composed of invert sugar. **Lactose** (Figure 8.15) can comprise up to 8% of mammalian milk, and is extracted from cow's milk, often as a by-product from cheese manufacture. It is only faintly sweet, and its principal use is as a diluent in tablet formulations. **Lactulose** (Figure 8.15)

(Continues)

(Continued)

is a semi-synthetic disaccharide prepared from lactose, and composed of galactose linked β1 → 4 to fructose. It is not absorbed from the gastrointestinal tract, and is predominantly excreted unchanged. It helps to retain fluid in the bowel by osmosis, and is thus used as a laxative.

Vitamin C

Vitamin C (ascorbic acid) (Figure 8.17) can be synthesized by most animals except humans, other primates, guinea pigs, bats and some birds, and for these it is obtained via the diet. Citrus fruits, peppers, guavas, rose hips, and blackcurrants are especially rich sources, but it is present in most fresh fruit and vegetables. Raw citrus fruits provide a good daily source. It is a water-soluble acidic compound (an enol; see Figure 8.18) and is rapidly degraded during cooking in the presence of air. Vitamin C deficiency leads to scurvy, characterized by muscular pain, skin lesions, fragile blood vessels, bleeding gums, and tooth loss. The vitamin is essential for the formation of collagen, the principal structural protein in skin, bone, tendons, and ligaments, being a cofactor in the hydroxylation of proline to 4-hydroxyproline, and of lysine to 5-hydroxylysine (see page 409), which account for up to 25% of the collagen structure. These reactions are catalysed by 2-oxoglutarate dioxygenases (see page 27), and the ascorbic acid requirement is to reduce an enzyme-bound iron–oxygen complex. Skin lesions characteristic of scurvy are a direct result of low levels of hydroxylation in the collagen structure synthesized in the absence of ascorbic acid. Ascorbic acid is also associated with the hydroxylation of tyrosine in the pathway to catecholamines (see page 316), and in the biosynthesis of homogentisic acid, the precursor of tocopherols and plastoquinones (see page 159). Ascorbic acid is usually prepared synthetically, and is used to treat or prevent deficiency. Natural ascorbic acid is extracted from rose hips, persimmons, and citrus fruits. Large doses have been given after surgery or burns to promote healing by increasing collagen synthesis. The benefits of consuming large doses of vitamin C to alleviate the common cold and other viral infections are not proven. Some sufferers believe it to be beneficial in the

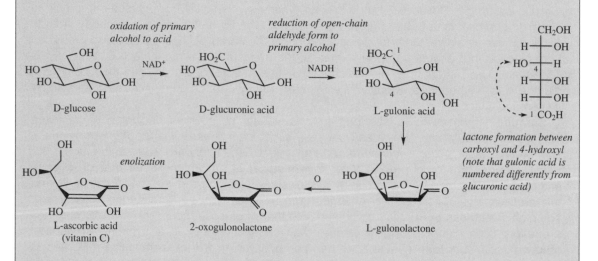

Figure 8.17

(Continued)

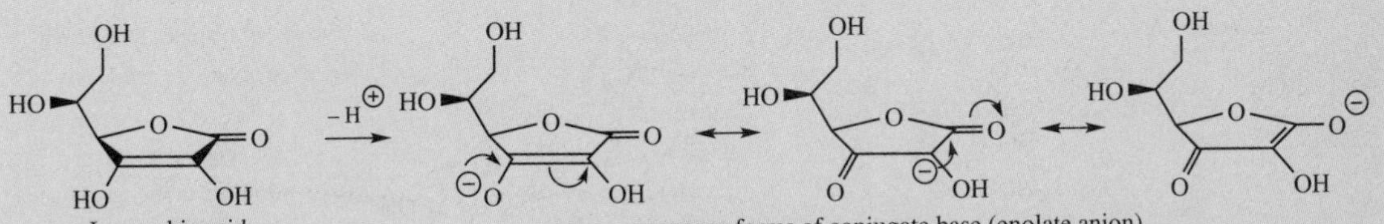

L-ascorbic acid

resonance forms of conjugate base (enolate anion)

Figure 8.18

prevention and therapy of cancer. Vitamin C does have valuable antioxidant properties, and these are exploited commercially in the food industries.

In animals, ascorbic acid is synthesized in the liver from glucose, by a pathway which initially involves oxidation to glucuronic acid. This is followed by reduction of the carbonyl function, lactone formation, oxidation of the secondary alcohol to a carbonyl, with subsequent enolization (Figure 8.17). In plants, glucose or galactose can be converted into ascorbic acid by an analogous pathway, though other sequences from glucose have also been observed to operate. Man and other primates appear to be deficient in the enzyme oxidizing gulonolactone to the ketolactone, and are thus dependent on a dietary source of vitamin C.

POLYSACCHARIDES

Polysaccharides fulfil two main functions in living organisms, as food reserves, and as structural elements. Plants accumulate starch as their main food reserve, a material that is composed entirely of glucopyranose units but in two types of molecule. **Amylose** (Figure 8.19) is a linear polymer containing some 1000–2000 glucopyranose units linked α1→4. **Amylopectin** (Figure 8.19) is a much larger molecule than amylose (the number of glucose residues varies widely but may be as high as 10^6), and it is a branched-chain molecule. In addition to α1→4 linkages, amylopectin has branches at about every 20 units through α1→6 linkages. These branches continue with α1→4 linkages, but then may have subsidiary branching, giving a treelike structure. The mammalian carbohydrate storage molecule is **glycogen**, which is analogous to amylopectin in structure, but is larger and contains more frequent branching, about every ten residues. The branching in amylopectin and glycogen is achieved by the enzymic removal of a portion of the α1→4 linked straight chain consisting of several glucose residues, then transferring this short chain to a suitable 6-hydroxyl group. A less common storage polysaccharide found in certain plants of the Compositae/Asteraceae and Campanulaceae is **inulin** (Figure 8.19), which is a relatively small polymer of fructofuranose, linked through β2→1 bonds.

Cellulose is reputedly the most abundant organic material on earth, being the main constituent in plant cell walls. It is composed of glucopyranose units linked β1→4 in a linear chain. Alternate residues are 'rotated' in the structure (Figure 8.19), allowing hydrogen bonding between adjacent molecules, and construction of the strong fibres characteristic of cellulose, as for example in cotton. The structure of **chitin** (Figure 8.19) is rather similar to cellulose, though it is composed of β1→4 linked N-acetylglucosamine residues. Chitin is a major constituent in the shells of crustaceans, e.g. crabs and lobsters, and insect skeletons, and its strength again depends on hydrogen bonding between adjacent molecules, producing rigid sheets. Chemical deacetylation of chitin provides chitosan, a valuable industrial material used for water purification because of its chelating properties, and in wound-healing preparations. Bacterial cell walls contain **peptidoglycan** structures in which carbohydrate chains composed of alternating β1→4 linked N-acetylglucosamine and N-acetylmuramic acid (O-lactyl-N-acetylglucosamine) residues are

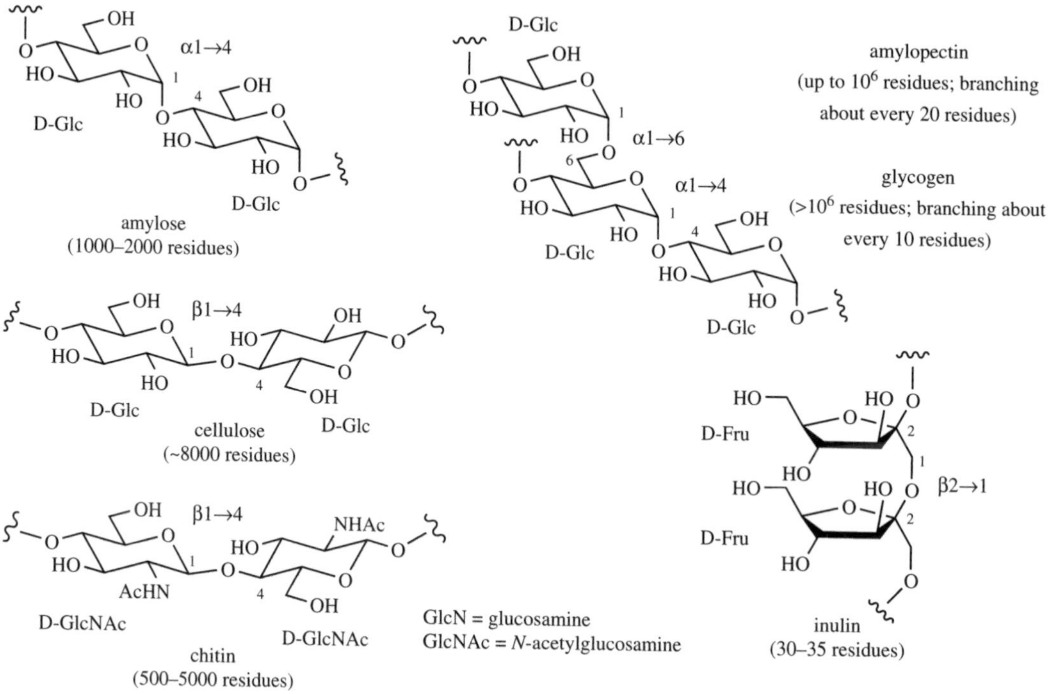

Figure 8.19

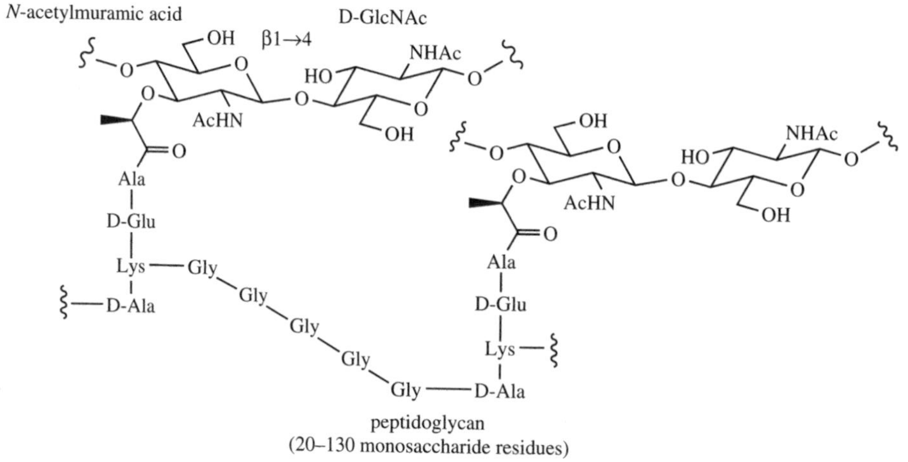

Figure 8.20

cross-linked via peptide structures. The pepti-doglycan of *Staphylococcus aureus* is illustrated in Figure 8.20, showing the involvement of the lactyl group of the *N*-acetylmuramic acid to link the peptide with the carbohydrate via an amide/peptide bond. During the cross-linking process, the peptide chains from the *N*-acetylmuramic acid residues have a terminal –Lys–D-Ala–D-Ala sequence, and the lysine from one chain is bonded to the penultimate D-alanine of another chain through five glycine residues, at the same time displacing the terminal D-alanine (see Figure 7.36, page 444). The biological activities of the β-lactam antibiotics, e.g. penicillins and cephalosporins (see page 444) and of the last-resort antibiotic vancomycin (see page 426)

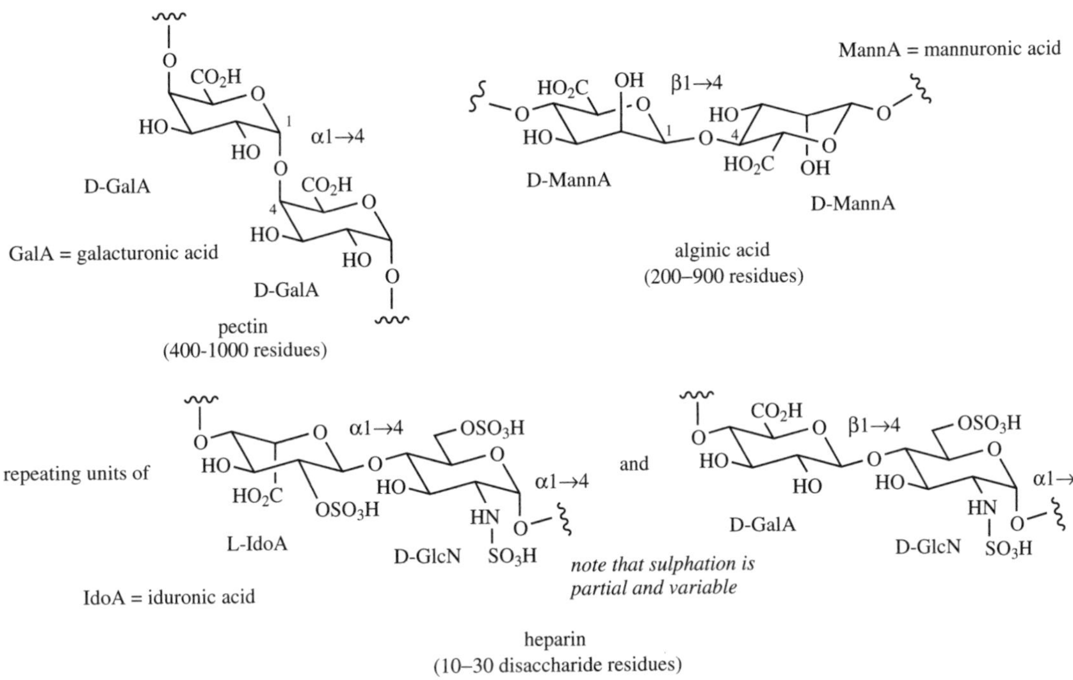

Figure 8.21

stem from an inhibition of the cross-linking mechanism during the biosynthesis of the bacterial cell wall, and relate to this terminal –D-Ala–D-Ala sequence during biosynthesis. The subdivision of bacteria into Gram-positive or Gram-negative reflects the ability of the peptidoglycan cell wall to take up Gram's dye stain. In Gram-negative organisms, an additional lipopolysaccharide cell membrane surrounding the peptidoglycan prevents attack of the dye.

Polymers of uronic acids are encountered in **pectins**, which are essentially chains of galacturonic acid residues linked α1→4 (Figure 8.21), though some of the carboxyl groups are present as methyl esters. These materials are present in the cell walls of fruit, and the property of aqueous solutions under acid conditions forming gels is the basis of jam making. **Alginic acid** (Figure 8.21) is formed by β1→4 linkage of mannuronic acid residues, and is the main cell wall constituent of brown algae (seaweeds). Salts of alginic acid are valuable thickening agents in the food industry, and the insoluble calcium salt is the basis of absorbable alginate surgical dressings. The mammalian blood anticoagulant **heparin** (Figure 8.21) is also a carbohydrate polymer containing uronic acid residues, but these alternate with glucosamine derivatives. Polymers of this kind are known as anionic mucopolysaccharides, or glycosaminoglycans. Heparin consists of two

Polysaccharides

Starch for medicinal and pharmaceutical use may be obtained from a variety of plant sources, including maize (*Zea mays*; Gramineae), wheat (*Triticum aestivum*; Gramineae), potato (*Solanum tuberosum*; Solanaceae), rice (*Oryza sativa*; Gramineae/Poaceae), and arrowroot (*Maranta arudinacea*; Marantaceae). Most contain about 25% amylose and 75% amylopectin (Figure 8.19), but these proportions can vary according to the plant tissue.

(Continues)

(Continued)

Starch is widely used in the food industry, and finds considerable applications in medicine. Its absorbent properties make it ideal for dusting powders, and its ability to swell in water makes it a valuable formulation aid, being the basis for tablet disintegrants. **Soluble starch** is obtained by partial acid hydrolysis, and is completely soluble in hot water.

Cellulose (Figure 8.19) may be extracted from wood pulp, and is usually partially hydrolysed with acid to give microcrystalline cellulose. These materials are used as tablet diluents. Semi-synthetic derivatives of cellulose, e.g. **methylcellulose**, **hydroxymethylcellulose**, and **carboxymethylcellulose**, are used as emulsifying and suspending agents. **Cellulose acetate phthalate** is cellulose with about half the hydroxyl groups acetylated, and the remainder esterified with phthalic acid. It is used as an acid-resistant enteric coating for tablets and capsules.

Alginic acid (Figure 8.21) is obtained by alkaline (Na_2CO_3) extraction of a range of brown seaweeds, chiefly species of *Laminaria* (Laminariaceae) and *Ascophyllum* (Phaeophyceae) in Europe, and species of *Macrocystis* (Lessoniaceae) on the Pacific coast of the USA. The carbohydrate material constitutes 20–40% of the dry weight of the algae. The acid is usually converted into its soluble sodium salt or insoluble calcium salt. Sodium alginate finds many applications as a stabilizing and thickening agent in a variety of industries, particularly food manufacture, and the pharmaceutical industry, where it is of value in the formulation of creams, ointments, and tablets. Calcium alginate is the basis of many absorbable haemostatic surgical dressings. Alginic acid or alginates are incorporated into many aluminium- and magnesium-containing antacid preparations to protect against gastro-oesophageal reflux. Alginic acid released by the action of gastric acid helps to form a barrier over the gastric contents.

Agar is a carbohydrate extracted using hot dilute acid from various species of red algae (seaweeds) including *Gelidium* (Gelidiaceae) and *Gracilaria* (Gracilariaceae) from Japan, Spain, Australasia, and the USA. Agar is a heterogeneous polymer, which may be fractionated into two main components, agarose and agaropectin. Agarose yields D- and L-galactose on hydrolysis, and contains alternating β1→3 linked D-galactose and α1→4 linked L-galactose, with the L-sugar in a 3,6-anhydro form. Agaropectin has a similar structure but some of the residues are methylated, sulphated, or in the form of a cyclic ketal with pyruvic acid. Agar's main application is in bacterial culture media, where its gelling properties are exploited. It is also used to some extent as a suspending agent and a bulk laxative. Agarose is now important as a support in affinity chromatography.

Tragacanth is a dried gummy exudate obtained from *Astragalus gummifer* (Leguminosae/Fabaceae) and other *Astragalus* species, small shrubs found in Iran, Syria, Greece, and Turkey. It is usually obtained by deliberate incision of the stems. This material swells in water to give a stiff mucilage with an extremely high viscosity, and provides a useful suspending and binding agent. It is chemically a complex material, and yields D-galacturonic acid, D-galactose, L-fucose, L-arabinose, and D-xylose on hydrolysis. Some of the uronic acid carboxyls are methylated.

Acacia (gum arabic) is a dried gum from the stems and branches of the tree *Acacia senegal* (Leguminosae/Fabaceae), abundant in the Sudan and Central and West Africa. Trees are tapped by removing a portion of the bark. The gum is used as a suspending agent, and adhesive and binder for tablets. The carbohydrate is a complex branched-chain material, which yields L-arabinose, D-galactose, D-glucuronic acid, and L-rhamnose on hydrolysis. Occluded enzymes (oxidases, peroxidases, and pectinases) can cause problems in some formulations, unless inactivated by heat.

(Continues)

(Continued)

Karaya or **Sterculia Gum** is a dried gum obtained from the trunks of the tree *Sterculia urens* (Sterculiaceae) or other *Sterculia* species found in India. It exudes naturally, or may be obtained by incising through the bark. It contains a branched polysaccharide comprising L-rhamnose, D-galactose, and a high proportion of D-galacturonic acid and D-glucuronic acid residues. The molecule is partially acetylated, and the gum typically has an odour of acetic acid. It is used as a bulk laxative, and as a suspending agent. It has proved particularly effective as an adhesive for stomal appliances, rings of the purified gum being used to provide a non-irritant seal between the stomal bag and the patient's skin.

Heparin is usually extracted from the mucosa of bovine lung or porcine intestines, where it is present in the mast cells. It is a blood anticoagulant and is used clinically to prevent or treat deep-vein thrombosis. It is administered by injection or intravenous infusion and provides rapid action. It is also active *in vitro*, and is used to prevent the clotting of blood in research preparations. Heparin acts by complexing with enzymes in the blood which are involved in the clotting process. Although not strictly an enzyme inhibitor, its presence enhances the natural inhibition process between thrombin and antithrombin III by forming a ternary complex. A specific pentasaccharide sequence containing a 3-*O*-sulphated D-glucosamine residue is essential for functional binding of antithrombin. Natural heparin is a mixture of glycosaminoglycans (Figure 8.21), with only a fraction of the molecules having the required binding sequence, and it has a relatively short duration of action. Partial hydrolysis of natural heparin by chemical or enzymic means has resulted in a range of low molecular weight heparins having similar activity but with a longer duration of action. **Certoparin, dalteparin, enoxaparin**, and **tinzaparin** are examples of these currently being used clinically.

Protamine, a basic protein from the testes of fish of the salmon family, e.g. *Salmo* and *Onchorhynchus* species (see insulin, page 417), is a heparin antagonist, which may be used to counteract haemorrhage caused by overdosage of heparin.

repeating disaccharide units, in which the amino functions and some of the hydroxyls are sulphated, producing a heterogeneous polymer. The carboxyls and sulphates together make heparin a strongly acidic water-soluble material.

AMINOSUGARS

Aminosugars are readily produced from ketosugars by transamination processes. Whilst many of the natural examples, e.g. glucosamine and galactosamine (Figure 8.10), demonstrate the results of this transamination, there are some further structures where the newly introduced amino group becomes part of a heterocyclic ring system. This arises by using the amino group as a nucleophile to generate an aminohemiacetal linkage, rather than a hydroxyl to produce a hemiacetal. This, of course, is the addition step in the formation of an imine (Schiff base) (see page 18). Should the anomeric hydroxyl then be removed in subsequent

modifications such as imine formation, the product will then be a polyhydroxy-piperidine or pyrrolidine. Any confusion with ornithine/lysine-derived alkaloids (see pages 292, 307) should be dispelled by the characteristic polyhydroxy substitution. The piperidine structures **deoxynojirimycin** and **deoxymannojirimycin** (Figure 8.22) from *Streptomyces subrutilis* are good examples.

The pathways to deoxynojirimycin and deoxymannojirimycin (Figure 8.22) start from the ketosugar fructose, which is aminated and then oxidized to **mannojirimycin**. This can then form a cyclic aminohemiacetal. Dehydration to the imine can follow, and reduction yields deoxymannojirimycin. **Nojirimycin** is an epimer of mannojirimycin, and analogous modifications then give deoxynojirimycin. Deoxynojirimycin is found in various strains of *Streptomyces* and *Bacillus*, as well as some plants, e.g. *Morus* spp. (Moraceae), and is attracting considerable attention in the search for anti-HIV agents. This and related structures are inhibitors of glycosidase

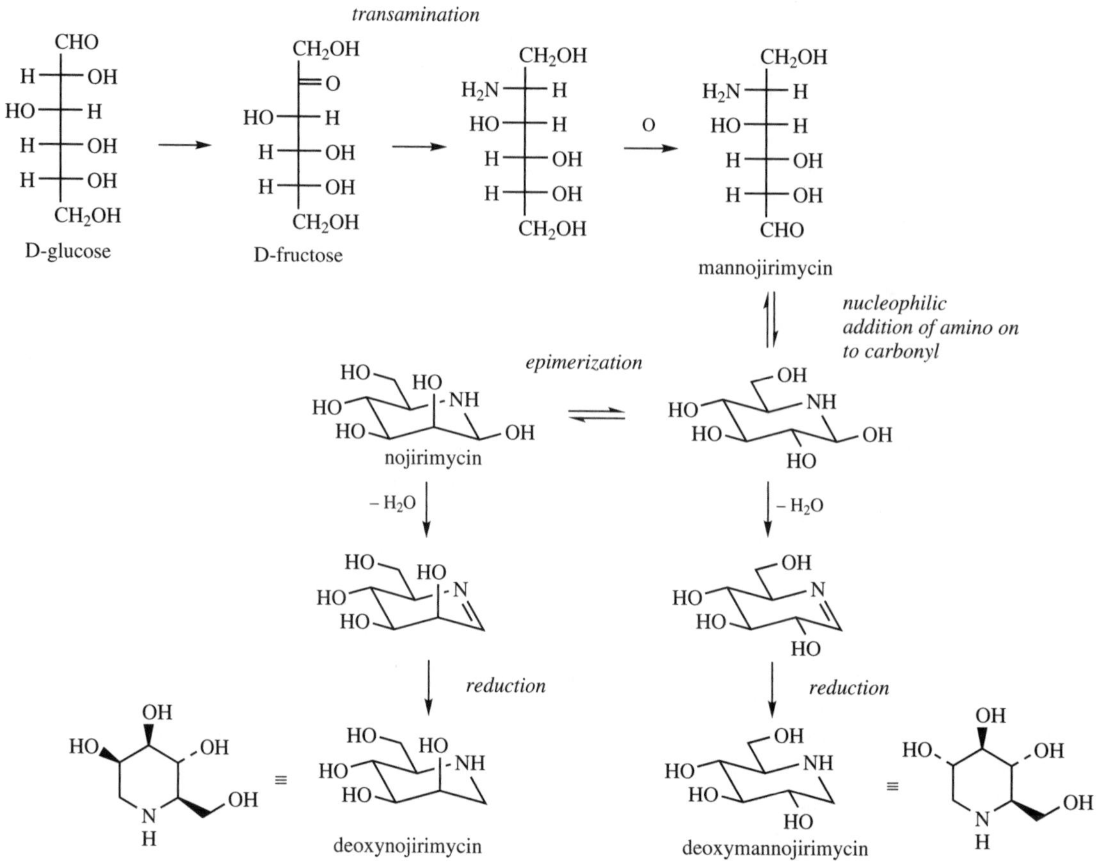

Figure 8.22

enzymes (compare indolizidine alkaloids such as castanospermine, page 310). By altering the constitution of glycoproteins on the surface of the virus, such compounds interfere with the binding of the HIV particle to components of the immune system.

AMINOGLYCOSIDES

The aminoglycosides* form an important group of antibiotic agents and are immediately recognizable as modified carbohydrate molecules. Typically, they have two or three uncommon sugars attached through glycoside linkages to an aminocyclitol, i.e. an amino-substituted cyclohexane system, which also has carbohydrate origins. The first of these agents to be discovered was **streptomycin** (see Figure 8.25) from *Streptomyces griseus*, whose structure contains the aminocyclitol **streptamine** (Figure 8.23), though both amino

groups are bound as guanidino substituents making **streptidine**. Other medicinally useful aminoglycoside antibiotics are based on the aminocyclitol **2-deoxystreptamine** (Figure 8.24), e.g. **gentamicin C$_1$** (see Figure 8.27) from *Micromonospora purpurea*. Streptamine and 2-deoxystreptamine are both derived from glucose 6-phosphate. The route to the streptamine system can be formulated to involve oxidation (in the acyclic form) of the 5-hydroxyl, allowing removal of a proton from C-6 and generation of an enolate anion (Figure 8.23). The cyclohexane ring is then formed by attack of this enolate anion on to the C-1 carbonyl. Reduction and hydrolysis of the phosphate produces *myo*-**inositol**. The amino groups as in **streptamine** are then introduced by oxidation/transamination reactions. **Streptidine** incorporates amidino groups from arginine, by nucleophilic attack of the aminocyclitol amino group on to the imino function

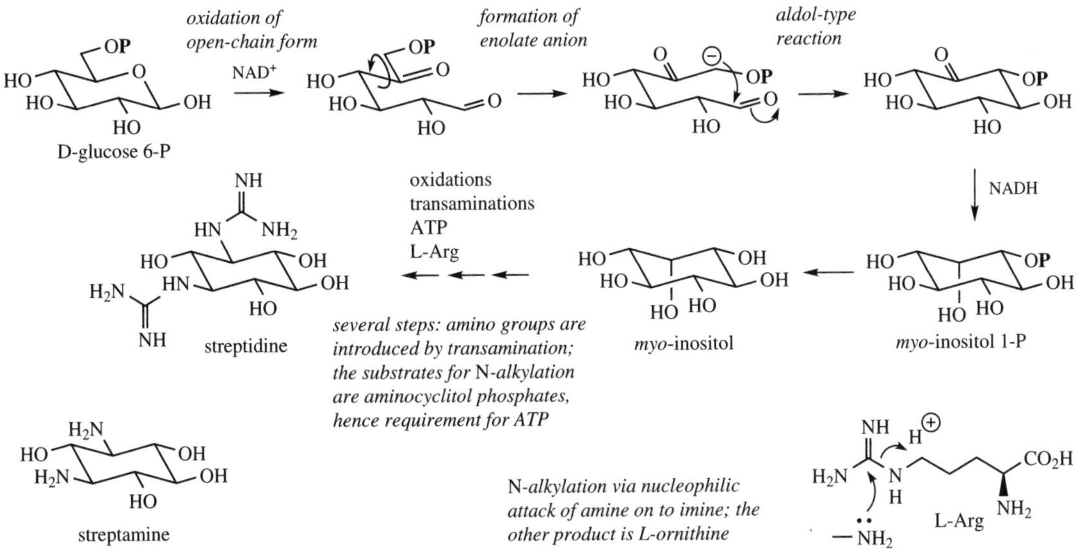

Figure 8.23

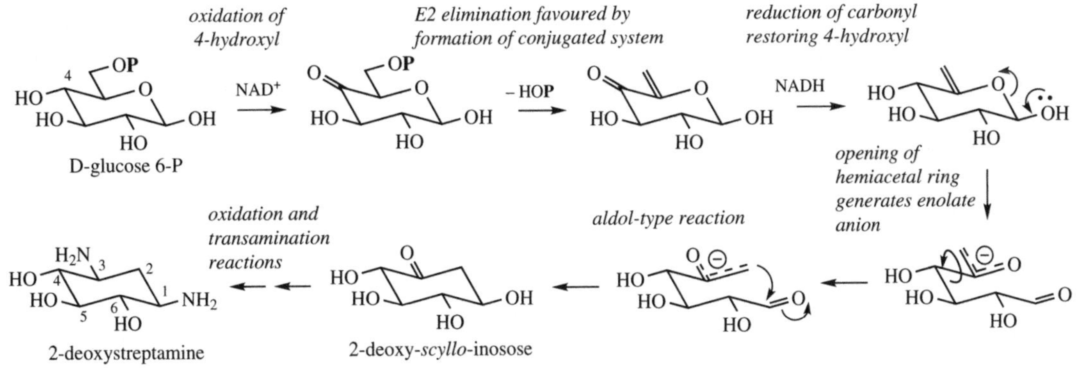

Figure 8.24

of arginine (Figure 8.23). However, streptamine itself is not a precursor, and the first guanidino side-chain is built up before the second amino group is introduced. The N-alkylation steps also involve aminocyclitol phosphate substrates. The biosynthesis of 2-deoxystreptamine shares similar features, but the sequence involves loss of the oxygen function from C-6 of glucose 6-phosphate in an elimination reaction (Figure 8.24). The elimination is facilitated by oxidation of the 4-hydroxyl, which thus allows a conjugated enone to develop in the elimination step, but the original hydroxyl is reformed by reduction after the elimination. The cyclohexane ring is then formed by attack of an enolate anion on to the C-1 carbonyl giving a tetrahydroxy-

cyclohexanone, and transamination reactions allow formation of **2-deoxystreptamine**. The pathway in Figure 8.24 is remarkably similar to that operating in the biosynthesis of dehydroquinic acid from the seven-carbon sugar DAHP in the early part of the shikimate pathway (see page 122).

The other component parts of streptomycin, namely L-streptose and 2-deoxy-2-methylamino-L-glucose (N-methyl-L-glucosamine) (Figure 8.25), are also derived from D-glucose 6-phosphate, though the detailed features of these pathways will not be considered further. Undoubtedly, these materials are linked to streptidine through stepwise glycosylation reactions via their nucleoside sugars (Figure 8.25).

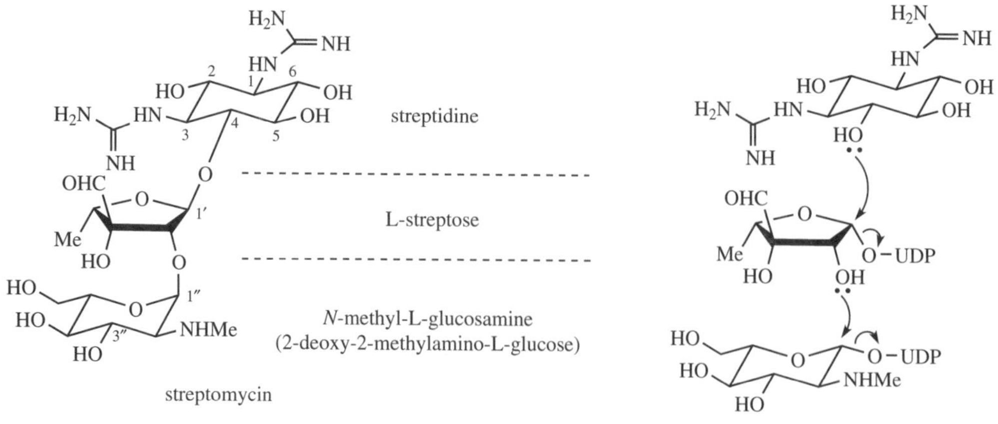

streptidine

L-streptose

N-methyl-L-glucosamine
(2-deoxy-2-methylamino-L-glucose)

streptomycin

Figure 8.25

Aminoglycoside Antibiotics

The aminoglycoside antibiotics have a wide spectrum of activity, including activity against some Gram-positive and many Gram-negative bacteria. They are not absorbed from the gut, and for systemic infections must be administered by injection. However, they can be administered orally to control intestinal flora. The widespread use of aminoglycoside antibiotics is limited by their nephrotoxicity, which results in impaired kidney function, and by their ototoxicity, which is a serious side-effect and can lead to irreversible loss of hearing. They are thus reserved for treatment of serious infections where less toxic antibiotics have proved ineffective. The aminoglycoside antibiotics interfere with protein biosynthesis by acting on the smaller 30S subunit of the bacterial ribosome. Streptomycin is known to interfere with the initiation complex, but most agents block the translocation step as the major mechanism of action. Some antibiotics can also induce a misreading of the genetic code to yield membrane proteins with an incorrect amino acid sequence leading to altered membrane permeability. This actually increases aminoglycoside uptake and leads to rapid cell death.

Bacterial resistance to the aminoglycoside antibiotics has proved to be a problem, and this has also contributed to their decreasing use. Several mechanisms of resistance have been identified. These include changes in the bacterial ribosome so that the affinity for the antibiotic is significantly decreased, reduction in the rate at which the antibiotic passes into the bacterial cell, and plasmid transfer of extrachromosomal R-factors. This latter mechanism is the most common and causes major clinical problems. Bacteria are capable of acquiring genetic material from other bacteria, and in the case of the aminoglycosides this has led to the organisms becoming capable of producing enzymes that inactivate the antibiotic. The modifications encountered are acetylation, adenylylation, and phosphorylation. (Note adenylic acid = adenosine 5′-phosphate). The enzymes are referred to as AAC (aminoglycoside acetyltransferase), ANT (aminoglycoside nucleotidyltransferase) (sometimes AAD (aminoglycoside adenylyltransferase)), and APH (aminoglycoside phosphotransferase). They differ with respect to the reaction catalysed, the position of derivatization (see numbering scheme in gentamicin, Figure 8.26), and the range of substrates attacked. Thus, some clinically significant inactivating enzymes are

(Continues)

(*Continued*)

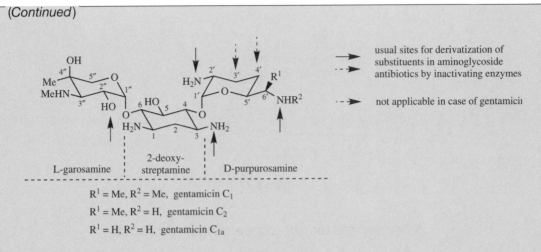

usual sites for derivatization of substituents in aminoglycoside antibiotics by inactivating enzymes

not applicable in case of gentamicin

$R^1 = Me, R^2 = Me,$ gentamicin C_1

$R^1 = Me, R^2 = H,$ gentamicin C_2

$R^1 = H, R^2 = H,$ gentamicin C_{1a}

Figure 8.26

- AAC(3) and AAC(6′), which acetylate the 3- and 6′-amino functions respectively in gentamicin, tobramycin, kanamycin, neomycin, amikacin, and netilmicin,
- ANT(2″), which adenylylates the 2″-hydroxy group in gentamicin, tobramycin, and kanamycin,
- APH(3′), which phosphorylates the 3′-hydroxyl in neomycin and kanamycin, and
- APH(3″), which phosphorylates the 3″-hydroxyl of streptomycin.

Other changes which may be imparted include acetylation of groups at position 2′, adenylylation of position 4′ substituents, and phosphorylation of the position 2″ substituent. Position 6 in the streptamine portion of streptomycin is also susceptible to adenylylation and phosphorylation.

Aminoglycoside antibiotics are produced in culture by strains of *Streptomyces* and *Micromonsopora*. Compounds obtained from *Streptomyces* have been given names ending in *-mycin*, whilst those from *Micromonospora* have names ending in *-micin*.

Streptamine-containing Antibiotics

Streptomycin (Figure 8.25) is produced by cultures of a strain of *Streptomyces griseus*, and is mainly active against Gram-negative organisms. Because of its toxic properties it is rarely used in modern medicine except against resistant strains of *Mycobacterium tuberculosis* in the treatment of tuberculosis.

Spectinomycin (Figure 8.27) is not strictly an aminoglycoside, but its structure does contain a modified streptamine portion linked by a glycoside bond to a deoxy sugar. It is sometimes written as a ketone at position 4, though this exists as a hydrate as shown in Figure 8.27. Spectinomycin is produced by cultures of *Streptomyces spectabilis*, and although it displays a broad antibacterial spectrum, it is only used against *Neisseria gonorrhoea* for the treatment of gonorrhoea where the organism has proved resistant to other antibiotics. It is known to inhibit protein biosynthesis on the 30S ribosomal subunit, but does not appear to cause any misreading of the genetic code.

(*Continues*)

(*Continued*)

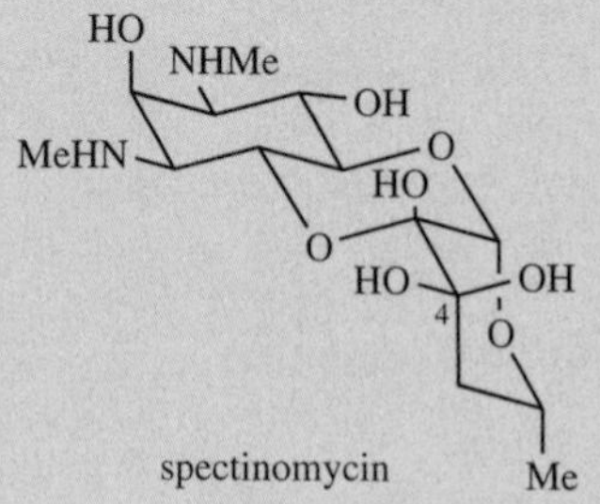

spectinomycin

Figure 8.27

2-Deoxystreptamine-containing Antibiotics

Gentamicin is a mixture of antibiotics obtained from *Micromonospora purpurea*. Fermentation yields a mixture of gentamicins A, B, and C, from which gentamicin C is separated for medicinal use. This is also a mixture, the main component being gentamicin C_1 (Figure 8.26) (50–60%), with smaller amounts of gentamicin C_{1a} and gentamicin C_2. These three components differ in respect to the side-chain in the purpurosamine sugar. Gentamicin is clinically the most important of the aminoglycoside antibiotics, and is widely used for the treatment of serious infections, often in combination with a penicillin when the infectious organism is unknown. It has a broad spectrum of activity, but is inactive against anaerobes. It is active against pathogenic enterobacteria such as *Enterobacter, Escherichia*, and *Klebsiella*, and also against *Pseudomonas aeruginosa*. Compared with other compounds in this group, its component structures contain fewer functionalities that may be attacked by inactivating enzymes, and this means gentamicin may be more effective than some other agents.

Sisomicin (Figure 8.28) is a dehydro analogue of gentamicin C_{1a}, and is produced by cultures of *Micromonospora inyoensis*. It is used medicinally in the form of the semi-synthetic *N*-ethyl derivative **netilmicin** (Figure 8.28), which has a similar activity to gentamicin, but causes less ototoxicity.

The **kanamycins** (Figure 8.29) are a mixture of aminoglycosides produced by *Streptomyces kanamyceticus*, but have been superseded by other drugs. **Amikacin** (Figure 8.29) is a semi-synthetic acyl derivative of kanamycin A, the introduction of the 4-amino-2-hydroxybutyryl group helping to protect the antibiotic against enzymic deactivation at several positions, whilst

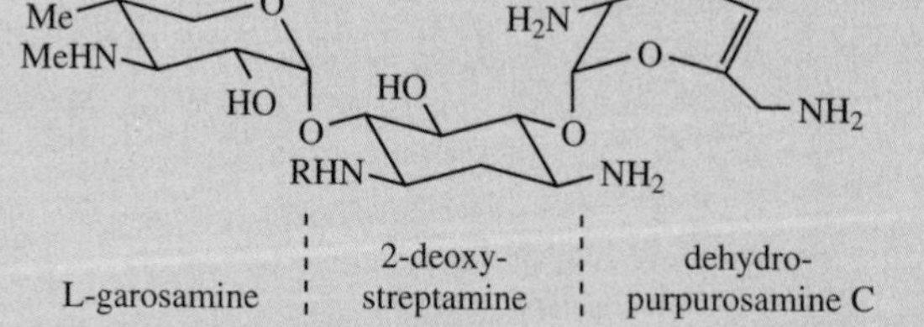

| L-garosamine | 2-deoxy-streptamine | dehydro-purpurosamine C |

R = H, sisomicin
R = Et, netilmicin

Figure 8.28

(*Continues*)

(Continued)

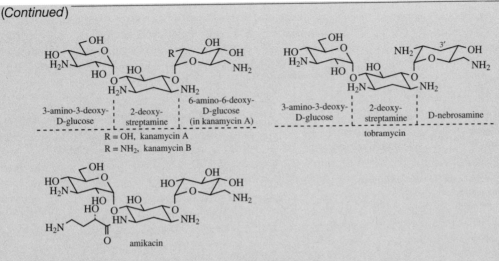

R = OH, kanamycin A
R = NH₂, kanamycin B

amikacin

Figure 8.29

still maintaining the activity of the parent molecule. It is stable to many of the aminoglycoside inactivating enzymes, and is valuable for the treatment of serious infections caused by Gram-negative bacteria which are resistant to gentamicin. **Tobramycin** (Figure 8.29) (also called nebramycin factor 6) is an analogue of kanamycin B isolated from *Streptomyces tenebrarius*, and is also less prone to deactivation in that it lacks the susceptible 3′-hydroxyl group. It is slightly more active towards *Pseudomonas aeruginosa* than gentamicin, but shows less activity against other Gram-negative bacteria.

    Neomycin is a mixture of neomycin B (**framycetin**) (Figure 8.30) and its epimer neomycin C, the latter component accounting for some 5–15% of the mixture. It is produced by cultures of *Streptomyces fradiae*, and, in contrast to the other clinically useful aminoglycosides described, contains three sugar residues linked to 2-deoxystreptamine. One of these is the common sugar D-ribose. Neomycin has good activity against Gram-positive and Gram-negative bacteria, but is very ototoxic. Its use is thus restricted to oral treatment of intestinal infections (it is poorly absorbed from the digestive tract) and topical applications in eyedrops, eardrops, and ointments.

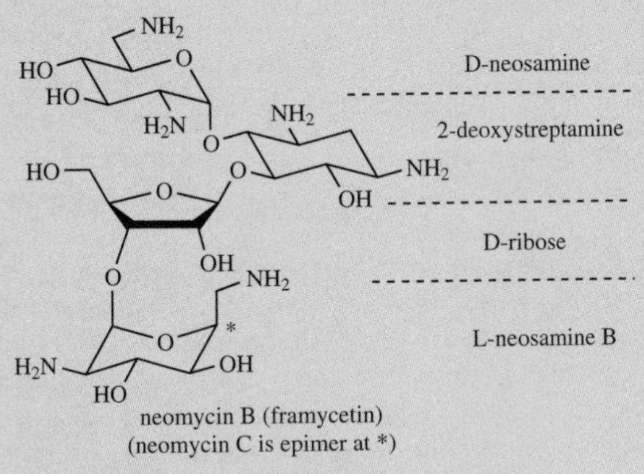

neomycin B (framycetin)
(neomycin C is epimer at *)

Figure 8.30

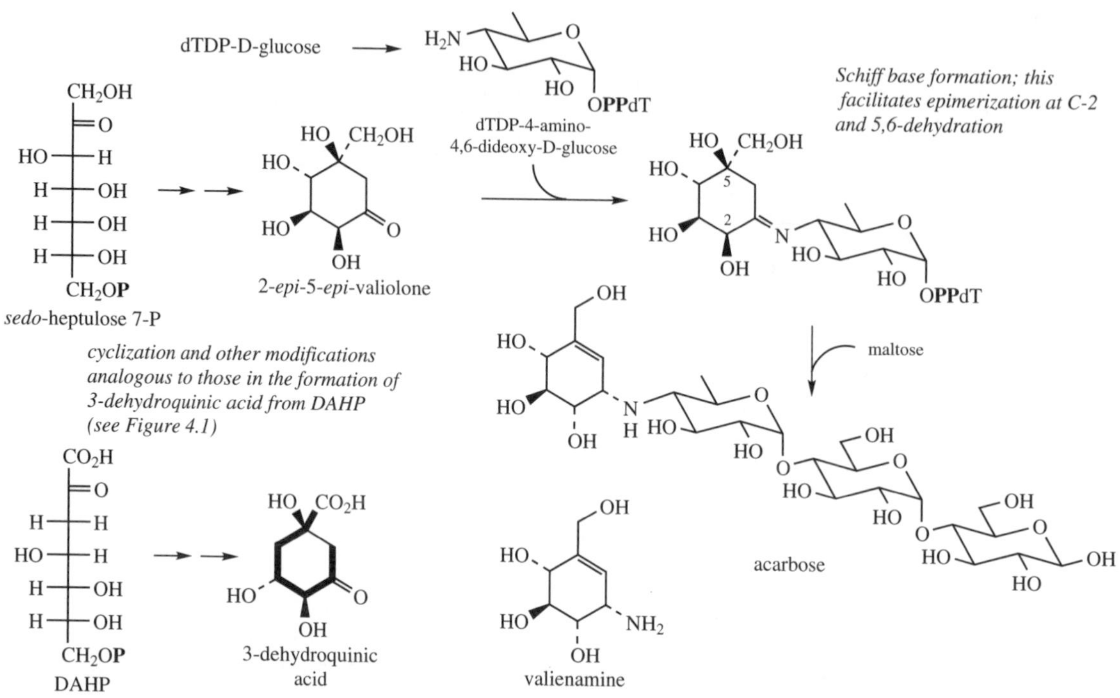

Figure 8.31

The aminocyclitol found in **acarbose*** (Figure 8.31) is based on **valienamine**, though this is not a precursor, and the nitrogen is introduced via an imine with the aminosugar, 4-amino-4,6-dideoxyglucose in the form of its deoxyTDP derivative. The cyclitol involved is 2-*epi*-5-*epi*-valiolone, and this appears to be produced from the seven-carbon sugar derivative *sedo*-heptulose 7-phosphate. The reaction sequence is exactly analogous to that seen in the transformation of DAHP into 3-dehydroquinic acid at the beginning of the shikimate pathway (page 122). The valienamine moiety requires subsequent epimerization and dehydration steps, and these are readily seen to be facilitated by the imine function. Unusually, the two further glucose units are not added sequentially, but via the preformed dimer maltose. Acarbose is produced by strains of *Actinoplanes* sp., and is of clinical importance in the treatment of diabetes.

The antibiotic **lincomycin*** (Figure 8.32) from *Streptomyces lincolnensis* bears a superficial similarity to the aminoglycosides, but has a rather more complex origin. The sugar fragment is termed methyl α-thiolincosaminide, contains a thiomethyl group, and is known to be derived

Acarbose

Acarbose is obtained commercially from fermentation cultures of selected strains of an undefined species of *Actinoplanes*. It is an inhibitor of α-glucosidase, the enzyme that hydrolyses starch and sucrose. It is employed in the treatment of diabetic patients, allowing better utilization of starch- or sucrose-containing diets, by delaying the digestion of such foods and thus slowing down the intestinal release of α-D-glucose. It has a small but significant effect in lowering blood glucose, and is used either on its own, or alongside oral hypoglycaemic agents, in cases where dietary control with or without drugs has proved inadequate. Flatulence is a common side-effect.

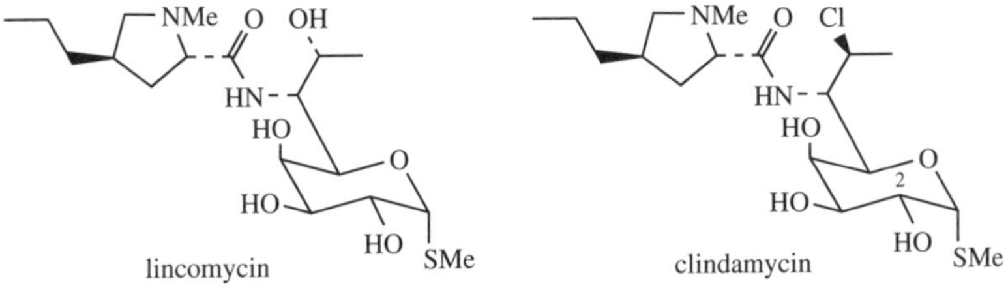

Figure 8.32

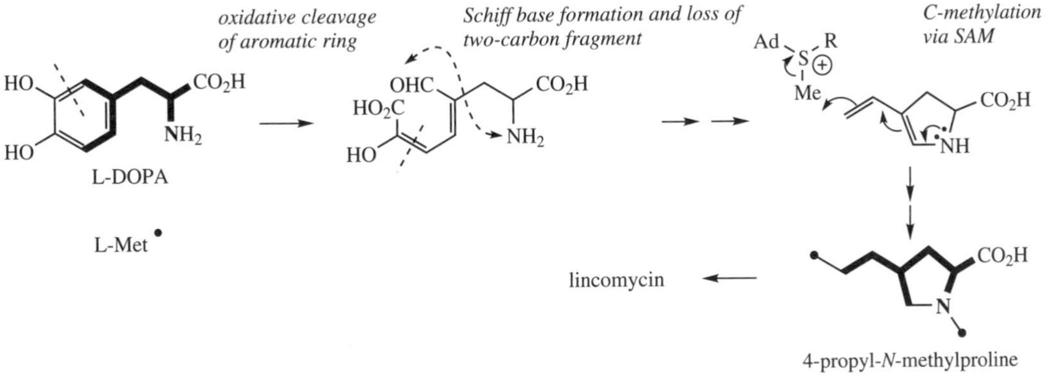

Figure 8.33

from two molecules of glucose, one of which provides a five-carbon unit, the other a three-carbon unit. The 4-propyl-*N*-methylproline fragment does not originate from proline, but is actually a metabolite from the aromatic amino acid L-DOPA (Figure 8.33). Oxidative cleavage of the aromatic ring (see page 27) provides all the carbons for the pyrrolidine ring, the carboxyl, and two carbons of the propyl side-chain. The terminal carbon of the propyl is supplied by L-methionine, as are the *N*-methyl, and the *S*-methyl in the sugar fragment. Two carbons from DOPA are lost during the biosynthesis.

Lincomycin and Clindamycin

Lincomycin (Figure 8.32) is obtained from cultures of *Streptomyces lincolnensis*. The semi-synthetic derivative **clindamycin** (Figure 8.32) obtained by chlorination of the lincomycin with resultant inversion of stereochemistry is more active and better absorbed from the gut, and has largely replaced lincomycin. Both antibiotics are active against most Gram-positive bacteria, including penicillin-resistant staphylococci. Their use is restricted by side-effects. These include diarrhoea and occasionally serious pseudomembraneous colitis, caused by overgrowth of resistant strains of *Clostridium difficile*, which can cause fatalities in elderly patients. However, this may be controlled by the additional administration of vancomycin (see page 426). Clindamycin finds particular application in the treatment of staphylococcal joint and bone infections such as osteomyelitis since it readily penetrates into bone. **Clindamycin 2-phosphate** is also of value, especially in the topical treatment of acne vulgaris and vaginal infections. Lincomycin and clindamycin inhibit protein biosynthesis by blocking the peptidyltransferase site on the 50S subunit of the bacterial ribosome. Microbial resistance may develop slowly, and in some cases has been traced to adenylylation of the antibiotic.

FURTHER READING

General

BeMiller JN (1992) Carbohydrates. *Kirk–Othmer Encyclopedia of Chemical Technology*, 4th edn, Vol 4. Wiley, New York, pp 911–948.

Weymouth-Wilson AC (1997) The role of carbohydrates in biologically active natural products. *Nat Prod Rep* **14**, 99–110.

Vitamin C

Emsley J (1995) A life on the high Cs. *Chem Brit* 946–948.

Gordon MH (1996) Dietary antioxidants in disease prevention. *Nat Prod Rep* **13**, 265–273.

Kuellmer V (1998) Vitamins (ascorbic acid). *Kirk–Othmer Encyclopedia of Chemical Technology*, 4th edn, Vol 25. Wiley, New York, pp 17–47.

Loewus FA (1999) Biosynthesis and metabolism of ascorbic acid in plants and of analogs of ascorbic acid in fungi. *Phytochemistry* **52**, 193–210.

Scott G (1995) Antioxidants – the modern elixir? *Chem Brit* 879–882.

Williams CM (1993) Diet and cancer prevention. *Chem Ind* 280–283.

Polysaccharides, Gums

Baird JK (1994) Gums. *Kirk–Othmer Encyclopedia of Chemical Technology*, 4th edn, Vol 12. Wiley, New York, pp 842–862.

Bugg TDH (1999) Bacterial peptidoglycan biosynthesis and its inhibition. *Comprehensive Natural Products Chemistry*, Vol 3. Elsevier, Amsterdam, pp 241–294.

Bugg TDH and Walsh CT (1992) Intracellular steps of bacterial cell wall peptidoglycan biosynthesis: enzymology, antibiotics, and antibiotic resistance. *Nat Prod Rep* **9**, 199–215.

Dea ICM (1989) Industrial polysaccharides. *Pure Appl Chem* **61**, 1315–1322.

Franz G (1989) Polysaccharides in pharmacy: current applications and future concepts. *Planta Med* **55**, 493–497.

Heparin

Bell WR (1992) Blood, coagulants and anticoagulants. *Kirk–Othmer Encyclopedia of Chemical Technology*, 4th edn, Vol 4. Wiley, New York, pp 333–360.

Petitou M and van Boeckel CAA (1992) Chemical synthesis of heparin fragments and analogues. *Prog Chem Org Nat Prod* **60**, 143–210.

Gunay NS and Linhardt RJ (1999) Heparinoids: structure, biological activities and therapeutic applications. *Planta Med* **65**, 301–306.

van Boeckel CAA and Petitou M (1993) The unique antithrombin III binding domain of heparin: a lead to new synthetic antithrombotics. *Angew Chem Int Edn Engl* **32**, 1671–1690.

Aminosugars

Hughes Ab and Rudge AJ (1994) Deoxynojirimycin: synthesis and biological activity. *Nat Prod Rep* **11**, 135–162.

Aminoglycoside Antibiotics

Lancini GC and Lorenzetti R (1993) *Biotechnology of Antibiotics and Other Microbial Metabolites*. Plenum, New York.

Lancini G, Parenti F and Gallo GG (1995) *Antibiotics – a Multidisciplinary Approach*. Plenum, New York.

McGregor D (1992) Antibiotics (aminoglycosides). *Kirk–Othmer Encyclopedia of Chemical Technology*, 4th edn, Vol 2. Wiley, New York, pp 904–926.

Lincomycin, Clindamycin

Bannister B (1992) Antibiotics (lincosaminides). *Kirk–Othmer Encyclopedia of Chemical Technology*, 4th edn, Vol 3. Wiley, New York, pp 159–168.

INDEX

Bold type has been used for major references and monograph material; asterisks indicate structural formulae are presented.